ulmination of mountain-building followed by erosion and moderate short-lived invasions of the sea. Early warm-
ng trends were reversed by the middle of the period to cooler and finally to glacial conditions. Subtropical
orests gave way to temperate forests and finally to extensive grasslands. Transition from primitive mammals to
aodern orders and eventually families. Evolution of humans during the last 10 million years. (See Figures 15-2
nd 15-3, pp 519 and 520.)

ast great spread of epicontinental seas and shoreline swamps. At the end of the period extensive mountain
uilding cooled the climate worldwide. Angiosperm dominance began. Extinction of archaic birds and many
eptiles by the end of the period. (See Figure 11-5, p. 373.)

limate was warm and stable with little latitudinal or seasonal variation. Modern genera of many gymnosperms
nd advanced angiosperms appeared. Reptilian diversity was high in all habitats. First birds appeared. (See Figure
1-4, p. 372.)

Continents were relatively high with few shallow seas. The climate was warm; deserts were extensive. Gymno-
perms dominated, angiosperms first appeared. Mammallike reptiles were replaced by precursors of dinosaurs
and the earliest true mammals appeared. (See Figure 11-3, p. 370.)

and was generally higher than at any previous time. The climate was cold at the beginning of the period but
warmed progressively. Glossopterid forests developed with the decline of the coal swamps. Mammallike reptiles
were diverse; widespread extinction of amphibians at the end of the period. (See Figure 11-2, p. 367.)

Generally warm and humid, but some glaciation in the Southern Hemisphere. Extensive coal-producing swamps
with large arthropod faunas. Many specialized amphibians and the first appearance of reptiles. (See Figure 11-1,
p. 365.)

Mountain-building produced locally arid conditions, but extensive lowland forests and swamps were the beginning
of the great coal deposits. Extensive radiation of amphibians; extinction of some fish lineages and expansion of
others.

The land was higher and climates cooler. Freshwater basins developed in addition to shallow seas. The first
forests appeared and the first winged insects. There was an explosive radiation of fishes, followed by the disappearance of many jawless forms. The earliest tetrapods appeared. (See Figure 8-1, p. 286.)

The land was slowly being uplifted, but shallow seas were extensive. The climate was warm and terrestrial plants
radiated. Eurypterid arthropods were at their maximum abundance in aquatic habitats and the first terrestrial
arthropods appeared. The first gnathostomes appeared among a diverse group of marine and freshwater jawless
fishes. (See Figure 4-4, p. 106.)

The maximum recorded extent of shallow seas was reached and the warming of the climate continued. Algae
became more complex, vascular plants may have been present, and there was a variety of large invertebrates.
Jawless fish fossils from this period are fragmentary but more widespread. (See Figure 4-3, p. 105.)

There were extensive shallow seas in equatorial regions. The climate was warm. Algae were abundant and there
are records of trilobites and brachiopods. The first remains of vertebrates are found at the end of this period.

Changes in the lithosphere produced major land masses and areas of shallow seas. Multicellular organisms
appeared and flourished—algae, fungi, and many invertebrates.

Formation of the earth and slow development of the lithosphere, hydrosphere, and atmosphere. Development
of life in the hydrosphere.

Vertebrate Life

Vertebrate Life

William N. McFarland
F. Harvey Pough
Tom J. Cade
John B. Heiser

CORNELL UNIVERSITY

Macmillan Publishing Co., Inc
New York

Collier Macmillan Publishers
London

Macmillan Publishing Co., Inc.
866 Third Avenue, New York, New York 10022

Collier Macmillan Canada, Ltd.

Library of Congress Cataloging in Publication Data

Main entry under title:

Vertebrate life.

Includes index.

1. Vertebrates.   2. Vertebrates, Fossil.   3. Vertebrates—Evolution.
I. McFarland, William Norman

QL605.V47   596   78-6990
ISBN 0-02-378870-4 (Hardbound)
ISBN 0-02-978880-3 (International Edition)

Printing: 1 2 3 4 5 6 7 8   Year: 9 0 1 2 3 4 5

Preface

A considerable time ago, perhaps longer than it should have been, several students in our course "The Vertebrates" suggested that we write this book. The course, which continues today, is structured around two themes—the evolution and the ecology of vertebrates. In spite of the several excellent textbooks of vertebrate morphology and evolution that are available, we could not find a text that underscored our view that a broad-based approach integrating traditionally separate specializations such as physiology and behavior or ecology and morphology is necessary to understand how animals function in their environment. This text, therefore, is an attempt to fill that gap. It is intended to provide students with a broad and detailed view of vertebrate biology. By better understanding the similarities of all vertebrates, one can also develop an appreciation of why vertebrates are so diverse. Our hope is that students who use this book will gain a keener perspective of themselves and, by doing so, develop a lasting reverence for living things—a commitment that is essential if vertebrate life, including human life, is to be sustained in our world.

The book's themes—evolution and ecology—are presented in phylogenetic order from fishes to mammals. In addition, other functional aspects of vertebrates are spread through several chapters. As a consequence, it is not possible to read in one chapter all we have said about kidney function and osmoregulation, or about social behavior and reproduction, or about body form and locomotion. Instead, aspects of these subjects are introduced in the context of the vertebrate taxa that best illustrate them. Major subjects such as these have been indexed for easier reference.

In addition to the phylogenetic sequence of chapters, five chapters are devoted to discussion of the geology and paleoecology of the time periods when major vertebrate groups arose. Because familiarity with the geological time record is so central to understanding the evolution of vertebrates, a time scale listing the various periods and eras is presented inside the front cover. A short Latin-Greek glossary is provided inside the back cover to assist students in deciphering the many compound words encountered in biology. Familiarity with only a few dozen Latin and Greek roots vastly simplifies the task of remembering and distinguishing the seemingly bewildering array of technical terms and animal names. In addition to the Latin-Greek glossary, a glossary of specialized English terms is included.

Many colleagues have provided suggestions, critical comments, and additional material in various stages of the development of the book. Dr. Frederick Test read the introductory chapters and the final chapter as well as the chapters on birds. Dr. Edwin Colbert reviewed the chapters covering geological events and paleoclimates, and Dr. Keith Thompson read these chapters as well as those concerned with fishes. Drs. George Bartholomew, Robert Carroll, Carl Gans, Rodolfo Ruibal, and Margaret Stewart reviewed the chapters on amphibians and reptiles. Dr. Dean Amadon read the chapters on birds, and Dr. Brian McNab those on mammals. Dr. John Repetski kindly provided the scanning electron micrographs of *Anatolepis* used in Figure 5–1. Our gratitude to Mary Beth Hedlund Marks is profound. She was involved in the book from its inception and typed large portions of the manuscript. More importantly she detected errors and inconsistencies and managed to bring a semblance of order to the diverse styles of the four authors.

Several students and former students, particularly Dr. Kentwood Wells, Willy Bemis, and Elaine Burke, read portions of the manuscript. Especially helpful was a review of the entire manuscript by Fredrica van Berkum, then a senior at Cornell. Rickie's perspective was valuable because she detected ambiguities that would bother a\student but escape the notice of a professional biologist. Dr. Alan Savitzky reviewed the glossary. Margaret Pough read much of the text, and Amanda Pough was a great help in compiling the index.

—William N. McFarland, F. Harvey Pough, Tom J. Cade , John B. Heiser

SOURCES AND ADDITIONAL READING

Assisting students to become acquainted with the literature of a field is an important function of an introductory textbook. To this end, we have cited a great variety of primary and secondary materials. The following books are important general references and any student of vertebrate biology should become familiar with them. We have cited these books in the text by author and date only—e.g. Romer 1966.

Colbert, E. H. 1969. *Evolution of the Vertebrates*, 2nd ed. John Wiley & Sons, New York.

Hildebrand, M. 1974. *Analysis of Vertebrate Structure*. John Wiley & Sons, New York.

Kluge, A. G. et al. 1977. *Chordate Structure and Function*, 2nd ed. Macmillan Publishing Co., Inc., New York.

Romer, A. S. 1966. *Vertebrate Paleontology*, 3rd ed. University of Chicago Press, Chicago.

Romer, A. S. 1968. *Notes and Comments on Vertebrate Paleontology*. University of Chicago Press, Chicago.

Romer, A. S. and T. S. Parsons. 1977. *The Vertebrate Body*. W. B. Saunders Co., Philadelphia.

Stahl, B. J. 1974. *Vertebrate History: Problems in Evolution*. McGraw-Hill, New York.

Young, J. Z. 1962. *The Life of Vertebrates*, 2nd ed. Oxford University Press, New York.

Several hundred additional references are provided at the ends of the chapters and in figure captions. To avoid repetition we have used the following system of coding for citations:

1. For references that appear only figure captions, the full citation is given in the caption.
2. Citations in figure captions that appear in the list of references at the end of the same chapter are indicated by a number—e.g. Noble [4].
3. Citations that appear in the list of references in a different chapter are indicated by the chapter number and reference number—e.g. Olson Ch. 5[7].

Contents

Vertebrate Life

1

The Basic Plan of Vertebrate Organization

Synopsis: The history of vertebrates covers a span of 500 million years or more. We think of the human as the most highly evolved vertebrate—specialized in many structures, hands, feet, vertebral column, cerebrum—but the structure and organization of the human body have been determined by a long and complex course of evolution. When we strip away all the special features of humans and compare the result with other vertebrates we can identify a "basic body plan," presumably the ancestral plan, which consists of a bilateral, tubular organization, possessing such characteristic features as notochord, pharyngeal slits, dorsal hollow nerve cord, vertebrae, and cranium, as well as other essential systems. One of the protochordates, amphioxus, and the ammocoete larva of lampreys offer suggestions of what the earliest vertebrates were like.

1.1 INTRODUCTION: OVERVIEW OF VERTEBRATE BIOLOGY AS A SUBJECT

The scientific study of vertebrates is a vast subject with a rich literature going back to the classical writings of Aristotle in the fourth century B.C. Among many other original contributions to our knowledge of vertebrates, Aristotle reported that whales are mammals and not fish, and he accurately described the peculiar reproductive system of the placental dogfish. Our subject also covers 500 million years or more of evolutionary history, as the earliest vertebrate fossils occur in the Cambrian Period. Biologists have described and studied tens of thousands of different vertebrate species, both living and fossil forms, each with a morphology and life of its own. Little wonder, then, that students making their first serious approach to vertebrate biology may hesitate, unsure how or where to begin.

1

A scholar can broach the subject in various ways, depending on a particular interest, but we have chosen to begin with some facts and concepts that are likely to be familiar to most biology students. From this general and familiar starting point, we can move into more specific, less well-known aspects of vertebrate life.

1.1.1 Some Familiar Facts About Vertebrates

Most students know that vertebrates belong to the Subphylum Vertebrata and that their name derives from the serially arranged vertebrae, which comprise a major portion of an axial endoskeleton that vertebrates share as a common diagnostic character (Figure 1-1). Anteriorly skeletal elements have been elaborated into a cranium or skull, which houses various sense organs and a complex brain. Another name sometimes used for the group is Craniata, because the cranium is also characteristic as, for that matter, is the tripartite brain inside. In fact, there is reason to believe that the distinctive vertebrate cranium and brain evolved before the vertebral column and are, therefore, more fundamentally characteristic of vertebrates than the backbone.

Most students no doubt also know that the vertebrates share some fundamental morphological features with certain marine invertebrates, and that by virtue of these common structures they are classified in the Phylum Chordata. These common chordate structures are the notochord, dorsal hollow nerve cord, and pharyngeal slits. Only the nerve cord remains as a definitive and functional entity in the adult stage of many vertebrates, but all three chordate features are clearly evident at some stage in the development of all vertebrates.

We now have the minimum information needed to define a **vertebrate**. A vertebrate is a special kind of chordate animal that has a cartilaginous or bony endoskeleton. The axial components of this endoskeleton consist of a cranium housing a brain, which is divided into three basic parts, and a vertebral column through which the nerve cord passes. No other group of animals possesses this constellation of fundamental and related characters which have existed in vertebrates since the late Cambrian and Ordovician.

1.1.2 The Diversity of Vertebrates

What about the different kinds of vertebrates? Everyone knows that humans are vertebrates, that dogs and cats and cows and chickens are all vertebrates, but few people realize that there are some 40,000 living species that share this distinction, not to mention the many extinct fossil forms. Most biology students know that there are different major groups of vertebrates—jawless fishes, cartilaginous fishes, bony fishes, amphibians, reptiles, birds, and mammals—each possessing certain distinctive features that set it apart from the others. In our system of zoological classification, these groups correspond to the Classes of Vertebrates (Table 1-1).

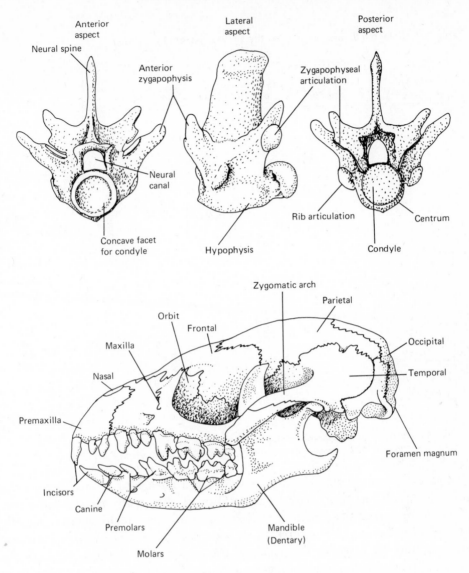

Figure 1-1 Examples of a typical reptilian vertebra and a mammalian cranium.

1.1.3 The Significance of Similarity and Differences

Each of the major groups or classes of vertebrates differs from the others in some fundamental way, but all share the common chordate-vertebrate characters, which in turn set them apart from all other animals. What is the meaning of the underlying similarity? Since Darwin's *Origin of Species* we have understood that the sharing of such fundamental similarities, or **homologs**, among widely different groups of species indicates that they all evolved from a common ancestor that also possessed the same features. In general the more homologs two species share, the more closely related they are (see Chapter 3).

Table 1-1. Selected Characteristics of the Living Vertebrate Classes

	Jaws	Endoskeleton	Locomotory Appendages	Respiratory Surface	Extra-Embryonic Membranes	Body Temperature Energetics	Integument	
AGNATHA	AGNATHA	CARTILAGE	FINS (Pisces)	GILLS		ECTOTHERMAL	GLANDULAR (Mucous secretions)	Naked
CHONDRICHTHYES	GNATHOSTOMATA				ANAMNIOTA (Yolk sac and chorion)			Placoid Scales
OSTEICHTHYES		BONE		Note 1				Dermal Scales
AMPHIBIA			LIMBS (Tetrapoda)	Note 2				Naked
REPTILIA				LUNGS	AMNIOTA (Yolk sac, chorion, allantois and amnion)		AGLANDULAR (Dry)	Epidermal Scales
AVES						ENDOTHERMAL		Feathers and Epidermal Scales
MAMMALIA							SECONDARILY GLANDULAR (Oily and watery secretions)	Hair

Notes: [1] Primarily gills; secondarily gut and integument specializations
[2] Primarily lungs; secondarily integument and neotenic retention of gills

What is the meaning of the diversity within a lineage of related groups and species? The differences relate to adaptation to different environmental conditions or opportunities. Each species has an ecological **niche** that is different from all others and that is expressed, at least in part, by altered body form and function, and the diversity of species in the higher taxa, at the level of genera, families, orders, and even classes, gives an indication of the genetic plasticity or responsiveness of that group to environmental differences.

Evolution and adaptation are the major themes of this book. Throughout the subsequent pages we shall be directing attention to the following sorts of questions. What were the historical, ancestral precursors of any structure, behavior, or function under consideration? How does the structure, function, or behavior promote survival and reproduction of the organism in its natural environment?

1.1.4   Teleology Versus Teleonomy

Adaptations can be tricky subjects to write about in a textbook, because one of the most obvious features of adaptation is functional design. An adaptation, presumably, serves some useful purpose in the life of the organism possessing it. Because human beings are purposive animals, perceive the means to ends, and anticipate results prior to their achievement, some philosophers and even scientists of an earlier era ascribed a guiding principle or divine purpose as the cause for useful adaptations and for organic evolution. This philosophy, which is not widely held by scientists today, is called *teleology*. Teleologic explanations for adaptations are based on the assumption that final causes exist and that design in the universe presupposes the existence of a designer.

When biologists discuss adaptations they are referring to alterations in structure or function that result from natural selection operating on the random genetic variability of organisms. These alterations confer improved fitness for survival and reproduction on the altered individuals. By this process, adaptive design results from mechanistic interactions between the inheritable variability of organisms and selective pressures from the environment. Design emerges without the existence of a prior purpose for it. This scientific explanation of adaptations has been termed *teleonomy*.

In this book where we use word-saving phrases such as "legs evolved for jumping," "wings adapted for flight," "feathers that function to conserve heat," and so forth, our meaning is teleonomic, not teleological. Since we humans are so accustomed to think in anthropomorphic and teleological terms, it will be necessary for most students to make a conscious effort to keep these distinctions in mind.

1.2 THE HUMAN AS A FAMILIAR EXAMPLE OF A VERTEBRATE

Most of us are fairly conversant with the human body, and we can therefore consider ourselves as a familiar example of a vertebrate and examine some of our structural and functional details in relation to other vertebrates as a starting point for the study of vertebrate biology. We all know that the human is a "highly evolved" and unique form of life, and the beginning student therefore may wonder whether there is any meaning to be derived from a view of the human as a vertebrate animal.

1.2.1 The Relevance of Vertebrate Biology as a Science

Even a hundred years after Darwin, man's exalted view of himself still persists. It has very deep roots in the Judeo-Christian religious view of western civilization and has been buttressed historically by the idea that man was created in the image of God—a little lower than the angels—but with very clear dominion over the beasts of the field. Such arrogance has been further strengthened by human achievements in modern technology—atomic bombs, space travel to the moon, green revolutions, and the like. Can we human beings find meaningful roots among the lower animals when our philosophy and technology have transported us so far beyond them?

It is a curious anthropological fact that even aboriginal people have much the same ethnocentric view of themselves. Most tribal names—"Navajo," for example—translate to mean something equivalent to "the people," "the chosen ones," or "those set apart." In fairness to these peoples, it must be said that in many cases they also feel a certain kinship with animals, usually in a religious or mystical sense.

When biologists look at humankind as an animal species they quickly see that every feature of human anatomy and physiology, and much of the behavior and social organization, have quite clear counterparts in other living vertebrates and direct antecedents in the history of vertebrate evolution tracing back hundreds of millions of years in some cases. "Know thyself," is an old Greek injunction that humanists are fond of invoking; however, to pursue it in the fullest biological sense requires not only a study of yourself as an individual organism and human beings as a distinctive species but also a study of all forms of life to which the human is related by direct, lineal ancestry. Most particularly, it requires knowledge of the human's closest relatives, the vertebrates.

Viewed in that perspective, there is no need for questions about the "relevance" of biology. Biology does not have to be made relevant by some gimmick, such as creating a new course, because by the very nature of its subject—life in all its myriad aspects—it is relevant to the human condition. Biology is the one study that makes a natural bridge between the "hard sciences" and the humanities.

Humans have always prided themselves in their uniqueness, in their *humanness*, in what some have even referred to as our godlike qualities. But how unique is humankind really?

1.2.2 The Uniqueness and Nonuniqueness of *Homo sapiens*

In more than a metaphorical sense the history of vertebrate evolution is reflected in the structure and organization of the human body. We can begin to see how this assertion is true by examining some of our unique and some not-so-unique features.

Table 1–2 lists some of the most distinctive anatomical and behavioral traits of humans. As far as our structural features are concerned, the most crucial anatomical development for the evolution of the human condition was the acquisition of a fully upright posture and a strictly bipedal mode of locomotion. As we shall see later, many other vertebrates have independently evolved bipedalism, but only the human has become a fully erect bipedal strider. The other main peculiarities of human anatomy, such as the manipulative hand, the S-shaped vertebral column, and the oversized brain, follow from that posture or are coadaptations with it. The major behavioral traits in human adaptive achievement have no doubt been tool making and language, which in turn have been dependent upon the evolution of the human hand and brain.

The scientific name that humans have given to themselves—*Homo sapiens*—

Table 1-2. Distinctive Traits of *Homo sapiens*

Anatomical	*Behavioral and Psychological*[1]
1. Normal posture upright	1. Curiosity, imitation, attention, memory, imagination all more highly developed than in other animals
2. Legs longer than arms	
3. Toes short, the first usually longest and not divergent	
4. Vertebral column with S curve	2. Ability to improve adaptive nature of behavior by rational thought
5. Hands prehensile and thumb strongly opposable	3. Uses and makes tools in great variety
6. Body mostly bare with only short, sparse, inconspicuous hair	4. Self-conscious—reflects on past, future, life, death, consequences of own behavior, etc.
7. Joint for neck in middle of base of skull	5. Makes mental abstractions and develops related symbolism—especially language
8. Brain uniquely large for body size with very large, complex cerebrum	
9. Face short, almost vertical under front of brain	6. Sense of beauty
10. Jaws short, with rounded dental arch	7. Religious emotions in broad sense—awe, superstition, animism, belief in supernatural spirit
11. Canines usually no larger than premolars, normally not separated by gaps in tooth row.	8. Moral and ethical values
12. First premolar like the second, tooth structure generally distinctive	9. Culture and social organization unique in complexity
	Major factors in human adaptive achievement have been toolmaking and language.
Most crucial anatomical development for the evolution of the human condition: acquisition of upright posture and strictly bipedal locomotion; the other main peculiarities of human anatomy follow from that or are coadaptations with it.	

[1] Based on Charles Darwin, *The Descent of Man,* 1871.

suggests one set of traits that we consider unique, our wisdom. Humans know more about themselves and their world than does any other animal. Consequently we can do an infinite number of things and think an infinite number of thoughts. We can predict or know the consequences of an action before it happens, and this prescience, together with our highly social nature, has led to the development of a moral conscience. We may be the only animal that has a conscience, although some canids and apes act as though they may have.

All our vaunted intelligence we owe to a large, complex brain, the intricate workings of which are just beginning to be understood. Especially we owe it to our forebrain, the cerebral hemispheres, which have become so enlarged with respect to the rest of the brain that they cover over most of it and have become highly folded, thereby greatly increasing the surface area available for associational neurons (Figure 1-2). The most important component of the cerebrum—the roof or **neopallium** (*neo* = new, *pallium* = cloak)—is a structure that man shares with all other mammals. It is also very weakly developed in some reptiles but not at all in birds or other vertebrates.

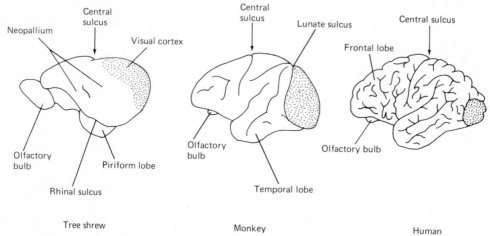

Figure 1-2 Cerebral hemispheres of tree shrew, monkey, and human drawn to same size and showing increase in degree of folding in cortex and change in relative size of the olfactory bulbs and visual cortex (stippled). Arrows indicate central sulcus.

Probably the neopallium first began to enlarge in mammal-like reptiles of the late Paleozoic, around 250 million years ago, so that the neuroanatomical basis for much of the human's distinctive intelligence has had a long history. Nor is this the end of the story, because the neopallium itself derives from a still older structure, the olfactory forebrain or **prosencephalon** (*pro* = before, *encephalon* = brain). Like the midbrain and the hindbrain, the prosencephalon has a history of existence back to the earliest known vertebrates, about 500 million years ago; and its exact origin as an element of the primitive tripartite brain of the first vertebrates or prevertebrates is still earlier.

We can consider the human hand in a similar way. It is usually taken for granted that the hand of no other species compares with ours for dexterity and manipulation. One has only to watch a good typist or pianist or jeweler at work to be impressed by the subtlety and infinite variety of movements of which the human hand is capable. Clearly humans can do more things with their hands than any other animal.

The human hand is a tool-using and tool-making organ *par excellence*, and the development of human culture and civilization from the earliest bone and flint tools to modern spaceships, superhighways, and multimillion dollar astrodomes has been based not only on our superior intellect but also on our ability to make things and manipulate things with our hands. Our technology is quite firmly based on the evolution of our hands.

Even with our complicated speech, the human hand is also used for all sorts of social communication, much more so than we are consciously aware. Emotive communication especially is likely to be manual. We salute the flag, we point accusing fingers, we shake our fists in anger, and we swear in a variety of finger signs. Good speakers continually add emotional content to their speeches with hand gestures.

Sign language no doubt preceded vocal language in human social relations and has probably played a larger role in the evolution of symbolic communication than most anthropologists had thought until recently. Comparative anthropologists and psychologists of the 1920s and 1930s, concerned with the evolution of human language and symbolic thought, were interested to know whether any subhuman primates could be taught to speak and learn to use language. Attempts to train chimpanzees to communicate by speech were failures because these apes lack the complex vocal cords, jaw structure, and tongues to form the necessary sounds of human speech; and no doubt the appropriate areas of the neocortex also are insufficiently developed for vocal language. Recently, however, chimpanzees have been trained to communicate with humans in sign language. One of the most interesting experiments involved a female chimp named Washoe who was trained to use the gestural sign language of the deaf, the American Sign Language. In 22 months of training, she learned to use 34 hand signs and to link as many as three of them together to form crude sentences. Chimpanzees are naturally gesticulatory in their social communication and no doubt have a highly organized cerebral area for integrating sensory and motor aspects of hand signals. Once investigators adopted the appropriate mode of communication, it became possible to demonstrate symbolic understanding and crude language in a nonhuman primate. Such studies have interesting possibilities for providing new insights into the evolution of human language and thought.

For the moment, however, we simply want to emphasize that the structure of the human hand has played an extremely important role in the evolution of two of our most distinctively human traits, our technology and our language, and through language our symbolic thinking processes.

The human hand is capable of performing four basic kinds of movement— divergence, convergence, prehensility, and opposability (Figure 1-3). The first two are general mammalian characteristics. The first mammals evolved **divergence**, the ability to spread the toes of their forepaws and hindpaws, in association with their load-bearing function for quadrupedal locomotion, and this ability of the forepaws has not been lost in those forms that later evolved bipedalism. **Convergence**, or cupping of the hand, is achieved by flexion at the joints of the metacarpals and phalanges. Two convergent paws can be used like one prehensile hand, and many mammals—squirrels and sea otters are familiar examples—manipulate their food in two convergent paws while eating. **Prehensility**, which refers to the ability to wrap the fingers around an object, is a characteristic that has been especially perfected by primates in association with **brachiation**, swinging by hands and arms through branches of trees,

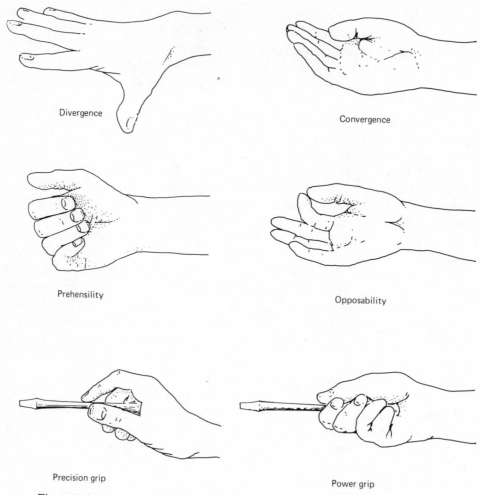

Divergence

Convergence

Prehensility

Opposability

Precision grip

Power grip

Figure 1-3 Diagram of basic hand movements (modified from J. Napier [3]).

although it has been independently evolved in other arboreal mammals like the opossums. **Opposability** is the ability to pass the thumb across the palm while rotating it around its long axis to allow the ventral ball to touch the tips of the other fingers. Although many primates can perform this movement, the underlying bone and muscular structures associated with it are best developed in the human (Figure 1-4).

During the course of evolution humans capitalized on the basic movements of prehensility and opposability, which they had inherited from their tree-dwelling ancestors, to develop a power grip based mainly on prehensility and a precision grip based on opposability, which allowed for the development of tool-using and tool-making. Yet the human hand remains capable of the postures and movements of the primitive mammalian paw, retaining many of the anatomical structures that go with the The later capabilities have merely been added to the older ones by sub modifications of bony structure and muscles.

The evolutionary sequence in development of the human hand and its elaborate movements can be seen by comparing the skeletons of the manus of various primates (Figure 1-5). The trend toward development of the human hand began some 70 to 75 million years ago among the early arboreal ancestors of the primates. All these specialized primate and human ways of working the hand are based on a skeletal pattern of the manus—the phalanges, metacarpals, and carpals—that trace back in time to the earliest fish-amphibian progenitors of the land vertebrates, as indeed does the entire appendicular skeleton, some 400 million years ago.

We could consider any other feature of the human body in the same fashion. Our uniqueness is relative and is based on some prior condition in our ancestors, which stretch back to the first fishlike vertebrates of the Cambrian seas more than 500 million years ago. In the same manner, also, we could examine any other species of vertebrate and show how it consists of a number of specialized traits added to or modified from an older, more generalized form. All living organisms are mixtures—mosaics—of advanced, special characters peculiar to a few forms and older, more ancestral ones, which are more widely distributed among groups of distantly related species.

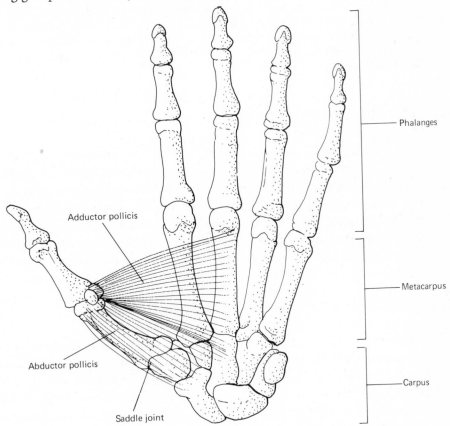

Figure 1-4 Human hand bones, showing the anatomical basis for opposability (modified from J. Napier [3]).

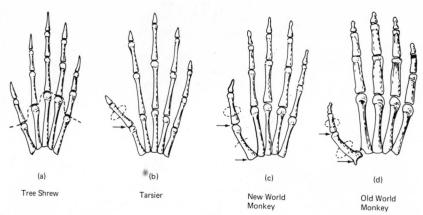

(a)   (b)   (c)   (d)

Tree Shrew   Tarsier   New World   Old World
                                  Monkey   Monkey

Figure 1-5 Evolutionary changes in primate hands associated with increasing manual dexterity (modified from J. Napier [3]). Drawn to the same size. The tree shrew (a) shows a start toward development of the unique thumb. The thumb is more distinct in the tarsier (b) and can rotate at the joint between its proximal phalanx and metacarpal. In the New World monkey (c) the angle between the thumb and adjacent finger is wider and movement can be effected at the joint at the proximal end of the metacarpal. The Old World monkey (d) has a saddle joint at the base of the metacarpal, allowing full rotation of the thumb, which is set at a wide angle. In humans the thumb becomes longer in proportion to the index finger and is set at a still wider angle. Development of the saddle joint allows the thumb to be rotated 45 degrees around its own longitudinal axis.

1.3   THE BASIC VERTEBRATE BODY PLAN

Suppose we try to strip away all the special features of the different kinds of vertebrates to answer the following questions. What is the least common denominator that applies to the organization of all vertebrates? What is the minimum functional organization that we can call vertebrate? Do any vertebrates actually conform to this generalized plan, which also represents a hypothetical ancestral form, or is it merely an abstraction based on an *a priori* assumption that evolution has proceeded from simple to complex organization?

1.3.1 Symmetry and Fundamental Organization

The symmetry of this hypothetical ancestral vertebrate would certainly have to be bilateral with definite head and tail ends. The basic internal organization would be some kind of modified tube-within-a-tube arrangement with the major internal organs lying within a body cavity or coelom (Figure 1–6). In these respects our basic vertebrate is not unlike many higher invertebrate animals, as these are all general body features that developed early in the course of metazoan evolution and have persisted in the organization of such diverse groups as mollusks, annelids, arthropods, and vertebrates.

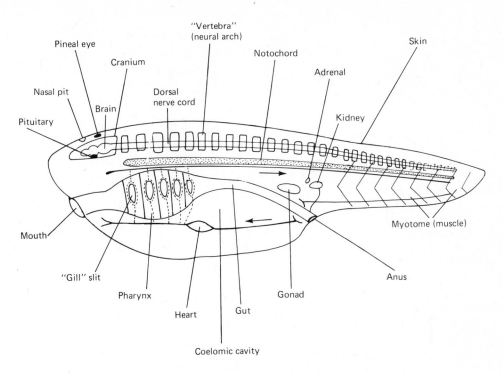

Figure 1-6 Hypothetical body plan of ancestral vertebrate.

To qualify as a bona fide vertebrate, the hypothetical animal would also have to possess, as functioning structures of the organism, all the diagnostic chordate-vertebrate characters that we briefly considered at the beginning of this chapter: notochord, dorsal hollow nerve cord, pharyngeal slits, vertebrae, cranium, and tripartite brain with associated sense organs. Bone is also a unique vertebrate tissue, which had evolved as early as the late Cambrian; however, since not all vertebrates have bony skeletons, we must omit it from the list of required features. Indeed, it is reasonable to assume that the first vertebrates lacked true bone since their soft-bodied chordate ancestors had none.

So far we have only blocked out the fundamental, identifying features of vertebrate morphology. Other essential organs and structures are required to make even the simplest vertebrate into a self-maintaining reproducing organism.

1.3.2 Required Functional Systems

Table 1–3 lists ten organ systems involved in the vital functions of all vertebrates. Some of the necessary organs and structures making up these systems have been mentioned in the previous section, but others have not. The following descriptions indicate the minimum requirements for a simple, ancestral form.

Table 1-3. Basic Vertebrate Systems

System	Basic Functions	Major Components
Integumentary	(1) Protect underlying tissues from injurious physical, chemical, thermal, radiant, and biological stimuli (2) Prevent excessive loss or absorption of water and the consequent effect on tissues (3) Aid excretion and absorption of specific metabolites and ions (4) Site of the cutaneous sense organs (i.e., those of touch, heat, cold, etc.)	Skin: Composed of epidermis above and the dermis below (Greek, *epi*, upon; *derma*, skin) and the derivatives of these two layers such as scales, feathers, and hair
Skeletal	(1) Provide a framework for all body systems and give the body adaptive form (2) Provide attachments for muscles, tendons, fascia (connective tissue coverings of muscles), etc. (3) Enclose and protect vital organs such as the heart, brain, lungs, and sense organs (4) Serve as a reserve storehouse for certain minerals	Bones (except in certain lower classes), cartilage, ligaments (connective tissue bands that bind bones to bones), etc. These tissues are divided into: (1) Axial skeleton of skull, vertebral column, and (when present) ribs, etc. (2) Appendicular skeleton of pectoral and pelvic girdles and limbs
Muscular	(1) Movement of body and body parts (hence, capture of food, escape from enemies, reproduction, etc.) (2) Maintenance of posture (3) Internal transport and expulsion (hence, movement of food through digestive tract, blood through vessels, germ cells through reproductive tract, bile from gall bladder, urine from kidneys, feces from alimentary canal, etc.) (4) Homeostatic (stable state or equilibrium) adjustments such as size of opening of the pupil of the eye, the pylorus, the anus, blood vessels; heat production in some vertebrates.	Smooth (nonstriated) muscles of involuntary control found primarily in wall of digestive tract, genital ducts, blood vessels, etc. Cardiac muscle of involuntary control restricted to the heart Striated muscles generally under voluntary control found attached to the skeleton so intimately that the name musculoskeletal system is often applied; tendons (the connective tissue bands that bind striated muscles to bone)

Table 1-3. Basic Vertebrate Systems (continued)

System	Basic Functions	Major Components
Digestive	(1) "Capture" and physical/chemical disintegration of food (2) Absorption, detoxification, alteration, storage, and controlled release of the products of digestion and metabolism	Alimentary canal: mouth and oral cavity with associated teeth, tongue, and jaws (when present) Pharynx (associated intimately with the respiratory system) Esophagus Stomach (when present) Intestine divided and specialized in various ways Accessory glands: salivary (when present) Liver: [responsible for most of the functions listed in (2)] Pancreas, etc.
Circulatory	(1) Transport of materials to cells (i.e., nutrients, water, oxygen, hormones, etc.) (2) Transport of materials from cells (i.e., and wastes such as carbon dioxide, lactic acid, urea, hormones from glands and food absorbed by intestinal cells) (3) Transport, formation, storage, and destruction of blood cells for O_2 transport, defensive, and immunogenic functions (4) Drain fluids from between cells and return it to the regular circulatory system from which it "leaked"	Heart Arteries (from the heart to the tissues) Arterioles (small arteries) Capillaries (extremely small vessels connecting arterioles and venules) Venules (small veins) Veins (from tissues to the heart) Spleen (and other sites in various vertebrate types, but always intimately associated with the digestive tract and/or skeletal system) Lymphatic system
Respiratory	(1) Exchange of gases (primarily intake of oxygen and discharge of carbon dioxide) between the organism and its environment (water or air) (2) Various accessory functions from production of sound to nest building (in certain fishes)	Lungs, gills, and/or skin, depending on which groups of vertebrates are under discussion; lungs and gills are derived from and intimately connected with the pharyngeal region of the digestive system

Table 1-3. Basic Vertebrate Systems (continued)

System	Basic Functions	Major Components
Excretory	(1) Chemical (and to a lesser extent physical) homeostasis or maintenance of a constant internal environment by: (a) Excreting toxic and metabolic waste products, especially those containing nitrogen (b) Maintaining proper water balance (c) Maintaining proper concentration of salts and other substances in the blood (d) Maintaining proper acid-base equilibrium in body fluids	Kidneys Excretory ducts (tubelike passages) variously aided by the gills, lungs, skin, and/or intestines The mode of development and use of common ducts makes this and the following system inseparable morphologically so that the two are often referred to as the "urogenital system."
Reproductive	(1) Perpetuation of the species by formation of zygotes which are the union of two specialized gametes to produce new individuals of the same biological variety	Primary sex characters in the form of male (testes) or female (ovaries) gonads Secondary sex characters concerned with transport of gametes from their site of formation to their site of union, etc. Accessory sex characters assuring union of gametes, such as glands and external genitalia (claspers, penis, vulva, etc.)
Endocrine	(1) Regulation and correlation/integration of body activities through chemical substances (*hormones*) carried by the blood. As opposed to the method of action of the following system, the endocrine system is slower acting—being limited by the rate of blood flow—but is capable of long, continuous action	A large number of cell types discharge secretions of regulatory effect on other cells after transport in the circulatory system In primitive vertebrates, the general rule is for these cells to be widely scattered in other tissues Later vertebrates have discrete aggregations of these cells to form endocrine glands

Table 1-3. Basic Vertebrate Systems (continued)

System	Basic Functions	Major Components
Nervous	(1) Regulation and correlation/integration of body activities through conduction of a wave of protoplasmic change in the individual nerve cells or neurons, which eventually cause a response in some other system (especially muscular contractions). As opposed to the method of action of the endocrine system, the nervous system is faster acting, and conduction may be faster than 90 m/sec (human sciatic nerve). An impulse is a single rapid phenomenon that must be reinitiated if its effect is to be maintained.	Central nervous system (CNS): brain, spinal cord Peripheral nervous system (PNS): Craniospinal nerves, which exit from the protective skeletal sheath of the cranium and vertebrae and may be either of a voluntary nature (to striated muscles) or involuntary (to smooth muscles); nerves of the latter type are often referred to collectively as the autonomic nervous system

The integumentary system, which consists of two basic components, the epidermis derived from the embryonic ectoderm and the deeper lying dermis derived from mesoderm, has become variously modified and elaborated in the course of vertebrate evolution; however, even in its least specialized form it provides an effective external, protective skin over the entire surface of the body. In amphioxus, a vertebratelike protochordate, the epidermis consists of a single layer of columnar cells that secrete a cuticle in the adult stage, but in all true vertebrates the epidermis consists of a more complex stratified epithelium.

The supportive system consists of an endoskeleton, which again is a morphological feature peculiar to chordates, and associated connective tissues. In its simplest form in vertebrates the endoskeleton consists only of axial elements, the notochord, cranium, and neural arches of the vertebrae, no appendicular skeleton being present in the most primitive swimming forms.

The muscular system of vertebrates consists of three distinctive types of tissues, smooth muscle, striated muscle, and cardiac muscle. Whether all three types should be included as necessary components of a simple, basic vertebrate body is debatable, but it is certain that some form of contractile tissue, probably not unlike striated muscle, was present in the trunk and tail region of the first vertebrates and produced movements involved in swimming.

The alimentary system of advanced vertebrates is extremely complex, but in its simplest, general form it consists of a tube specialized from anterior to posterior for ingestion, digestion, assimilation, and elimination. Ingestion can be accomplished with a fixed orifice for a mouth and some sort of filter-feeding mechanism involving excurrent pores (pharyngeal slits). Digestion requires nothing more complicated than a storage space into which digestive enzymes and acids can be secreted. Assimilation requires highly vascularized surfaces, the area of which can be increased by lengthening the gut or by development of such surface increasing structures as spiral valves, and so on, and elimination can be effected through an anal opening closable by muscle fibers.

Similarly, the respiratory system can involve complex structures like lungs and gills, which are designed primarily to increase the surface area for gaseous exchanges between the blood and environment (air or water), but in simple and small aquatic forms respiration can be accomplished by gas exchanges across the skin. Complex respiratory structures can, therefore, be eliminated from our construction of a simplified, ancestral vertebrate.

In all known vertebrates the circulatory system is a closed system with a muscular heart acting as a force pump to move the vascular fluid under pressure in a one-way direction through the vessels. Whether the earliest organisms to possess the full array of chordate-vertebrate diagnostic characters also possessed a circulatory system with a single, well-differentiated heart is uncertain, since we know that other chordates with closed systems move the blood by means of several discrete contracting areas in the system.

All vertebrates excrete the end products of nitrogen metabolism and other chemical wastes in urine, which is produced by paired kidneys. The functional unit of the kidney is the kidney tubule, the **nephron**, which has no homologous counterpart in any other group of animals, including the lower chordates. Just when the vertebrate nephron began to evolve is unknown, but the embryonic tubule structure of the lower vertebrates suggests that the first kidneys may have functioned by means of tubules with ciliated funnels opening into the coelomic cavity, from which wastes and excess water were removed. It seems likely that some kind of primitive kidney was present in the earliest vertebrates, but the evolutionary origin of this organ cannot be surmised from a study of any known protochordate or invertebrate.

The central nervous system no doubt consisted of a brain and dorsal hollow nerve cord. At just what stage of vertebrate evolution the brain developed the characteristic tripartite division into **prosencephalon, mesencephalon,** and **rhombencephalon** is problematic; this development occurred at about the same time as the evolution of the cranium and the vertebrae and the various cephalic sense organs. In a simple ancestral vertebrate we might expect that peripheral nerves would be involved mainly in spinal reflex arcs associated with swimming movements.

The endocrine system was probably already well-developed in the earliest vertebrates with homologs of such key hormone producing glands as the pituitary, thyroid, parathyroid, adrenal, and gonadal interstitium developed. Other characteristic vertebrate hormones no doubt evolved early in the history of the group.

Finally, the reproductive system would consist of paired gonads, which usually differentiate during embryonic development into either male testes or female ovaries. Bisexuality was no doubt fully developed in the prevertebrate ancestors that gave rise to the first true vertebrates.

These are not the only important features of the basic vertebrate body plan, but they serve to emphasize that vertebrates as a group are highly active, perceptive, strongly motivated animals. The generalized ancestral body plan possessed great potential for the evolution of forms adapted to all kinds of living conditions and has generated through time adaptive variations that have achieved a greater degree of freedom from environmental limitations than has been possible in any other phylum.

It is no accident that the generalized beast depicted in Figure 1–6 bears a strong resemblance to the lowly amphioxus and ammocoete larva of lampreys. These are living forms, one a protochordate, the other a larva of the most ancient group of vertebrates, that come closest to possessing the kind of simple, generalized body plan that could have been ancestral to all the vertebrates. Obviously, neither *is* that ancestor, since they are modern day forms, but their existence provides insights into the likely construction of the first vertebrates.

References

[1] Diamond, I. T. and W. C. Hall. 1969. Evolution of neocortex. *Science* 164:251–262.
[2] Gardner, R. A. and B. T. Gardner. 1969. Teaching sign language to a chimpanzee. *Science* 165:664–672.

[3]   Napier, J. 1962. Evolution of the human hand. *Scientific American,* 207(6) Reprint no. 140. This and the following article are clearly illustrated essays about two unique human characteristics, manipulation and locomotion.

[4]   Napier, J. 1967. The antiquity of human walking. *Scientific American* 216(4): 55-66, Reprint no. 1070.

[5]   Romer, A. S. and T. S. Parsons. 1978. *The Vertebrate Body.* Fifth edition. W. B. Saunders, Philadelphia. The latest edition of a standard and classic introduction to vertebrate anatomy.

[6]   Simpson, G. G. 1966. The biological nature of man. *Science* 152:474-478. A revealing essay by an outstanding American paleontologist on human traits and their origins.

[7]   Simpson, G. G. 1969. *Biology and Man.* Harcourt, Brace Jovanovich, New York.

[8]   Singer, C. J. 1959. *A History of Biology.* Third revised edition. Abelard-Schuman, London. A widely read general history that describes the developments during the eighteenth and nineteenth centuries important to our initial insights into vertebrate evolution.

2

Vertebrate Ancestors and Origins

Synopsis: The earliest known vertebrates are the ostracoderms, primitive jawless "fishes" that are related to lampreys and hagfishes and first appear in the late Cambrian and Ordovician. They were completely encased in heavy dermal bony armor. There are no fossils intermediate between them and any of the presumed invertebrate progenitors of the first vertebrates. It is only by comparison of living forms that some idea of the possible origin of vertebrates from invertebrates can be inferred. The notochord, pharyngeal slits, and dorsal hollow nerve cord are shared with certain "protovertebrate" animals, and it is most probable that the ancestral vertebrate arose from a chordate with the general features of amphioxus. Garstang's theory of vertebrate evolution from a tunicate larval stage by paedogenesis is commonly accepted. Still further back in evolutionary history it appears likely that the chordates are more closely allied to the Echinoderms and certain lophophorate groups than to any other invertebrate phyla. The first animals that can be called vertebrates probably evolved in early to middle Cambrian seas, but the fossil record suggests that the first great evolutionary radiation of vertebrate types occurred in fresh-water habitats of the Silurian and Devonian periods. Although this sequence of events is not an absolutely *proven* history of vertebrate origins, it is a reasonable one. It is a good example of how evolutionary theory is built up by inference from "circumstantial evidence."

2.1 INTRODUCTION: EARLIEST KNOWN VERTEBRATES

The earliest known vertebrates are fossil aquatic animals collectively called **ostracoderms** ("shell-skinned"). Their name refers to an outer covering of dermal bony armor, which is especially elaborate on the head and around the thoracic region. Ranging from a few to more than 50 cm in length, at first glance the ostracoderms look like nothing else on earth (see Figure 2–1), but on

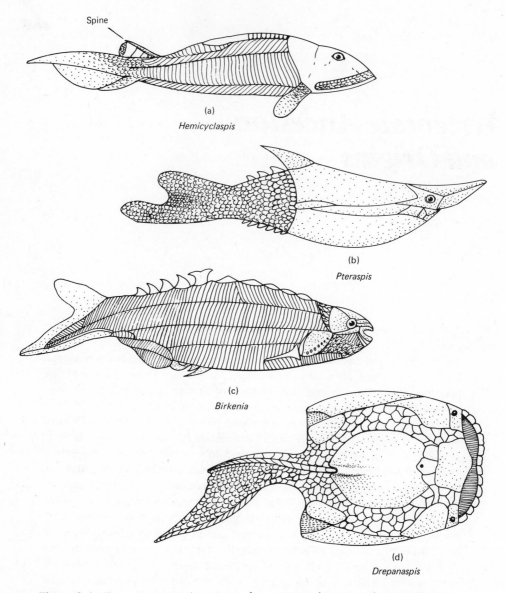

Spine

(a)
Hemicyclaspis

(b)
Pteraspis

(c)
Birkenia

(d)
Drepanaspis

Figure 2-1   Four representative ostracoderms, Monorhina (a and c); Diplorhina (b and d).

closer examination paleontologists have concluded that they share fundamental similarities with the living hagfishes and lampreys (cyclostomes).

Like cyclostomes, ostracoderms lacked jaws, although some of them may have had movable mouth parts based on peculiar structural arrangements not found in any other vertebrates. Typically their mouths are fixed circular or slitlike openings, which appear to have functioned as intakes that filtered small food particles from the water or from bottom detritus. They also lacked paired appendages homologous with the paired appendages of other vertebrates.

Their respiratory apparatus consisted of a variable number of separate pharyngeal gill pouches that opened independently along the side of the head or through a common passage. The notochord was presumably persistent throughout adult life and was the main axial support, as no vertebrae of ostracoderms have ever been found. Because ostracoderms and cyclostomes share these characteristics they are usually placed together in the Class Agnatha and are regarded as representatives of a primitive level of vertebrate organization. As we shall see, the ostracoderms were also highly specialized in the development of their bony armor, whereas the cyclostomes became specialized in entirely different ways for burrowing or parasitic modes of existence (sections 5.1 and 5.2).

2.1.1 Origin of the Ostracoderms

Unfortunately the fossil record of the ostracoderms is broken by a gap between the earliest known specimens and later examples, so that the evolution of the group is not well understood. The most complete and best known fossils occur in the late Silurian to middle Devonian Old Red Sandstone formation in southwestern England and Wales and in similar rock found in Scotland, Norway, and Spitzbergen, dating from 400 million years ago. (Refer to frontispiece and Figure 3–7 for geological periods and time.)

The Silurian and Devonian ostracoderms were contemporaries of a variety of archaic jawed fishes. This diversity indicates that the first vertebrates must have arisen at a considerably earlier time. The oldest fossils that are certainly vertebrates occur in the late Cambrian and middle Ordovician, between 50 and 125 million years prior to the time when ostracoderms became abundant, but there is no trace of vertebrate life in the interval between the Ordovician and late Silurian. The Ordovician specimens are of two sorts. A group of small denticles from the early Ordovician, found near Leningrad, have no distinctive traits, and at present it is only an assumption that some early ostracoderm produced the dentine and enamellike material of which these denticles are composed. The other set of specimens consists mainly of fragments of dermal bone found in late Cambrian and Ordovician formations exposed in South Dakota, Wyoming, Colorado and several other North American localities.

These latter specimens are particularly interesting because their microscopic structure is similar to that of the dermal bony plates of certain later ostracoderms, the Heterostraci (Figure 2–2) (see section 5.1). They indicate that this type of ostracoderm, characterized by laterally placed eyes, double nostrils, and other features suitably generalized for an ancestral stock of vertebrates, existed in the Cambrian and Ordovician prior to the Silurian radiation of jawless and jawed fishes. These bony fragments and the denticles from Russia also tell us that a variety of bony tissues had developed early in the course of vertebrate evolution, disproving an old idea that the cartilaginous fishes are primitive and ancestral to the bony fishes.

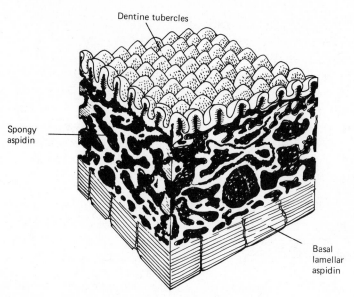

Figure 2-2 Three-dimensional block diagram of heterostracan dermal bone. (Based on B. Stahl [9] and on L. Halstead [2].)

2.1.2 Paucity of the Fossil Evidence

So much for the fossil record of the first vertebrates. It serves mainly to whet our appetites for more information. It reveals little about the course of evolution from the time of the earliest known vertebrates in the Cambrian to the rather sudden appearance of a great variety of agnathous and jawed forms in the late Silurian. Also, it provides not one clue about the early transformations that led to the evolution of vertebrate organization from an invertebrate progenitor. Nor does it shed any light on the evolution of the jawed vertebrates from their agnathous ancestors.

Many biologists conclude that the backboned animals were the last of the great groups to evolve. The absence of vertebrate fossils in rocks older than the late Cambrian, however, allows only speculation about the earlier history of the vertebrates. Most of the other animal phyla appear more than 50 million years earlier, in the oldest strata of Cambrian rock sufficiently undistorted by geophysical forces to yield good fossil materials. The best guess is that the earliest "vertebrates" were, in fact, rather rare, small, soft-bodied forms whose existence in those early seas was unlikely to be recorded by the chancy circumstances that lead to fossilization of soft parts. Also, the first dermal ossifications were perhaps so small or so delicate that none has survived in a recognizable state. Bone may have appeared and differentiated rather suddenly in the course of vertebrate evolution, after the basic vertebrate body plan had been established. In any case, from what paleontologists know about rates of evolution, it seems probable that the vertebrate pattern of organ systems had been evolving for tens of millions of years before the appearance of the first, late Cambrian fossils. Some day a lucky fossil find may clarify this speculation.

2.2 A SEARCH FOR THE ANCESTORS OF THE FIRST VERTEBRATES

Which group of animals was ancestral to the first true vertebrates? This question has been a major evolutionary puzzle ever since Darwin's *Origin of Species* set the theoretical framework for its formulation in the latter part of the nineteenth century. The fact that no fossils are intermediate between the ostracoderms, which are already vertebrates when first encountered, and any other group of animals presents great difficulties for a biologist interested in the origin of the vertebrates. Other kinds of indirect evidence must be used to construct possible evolutionary relationships between the vertebrates and other animals.

Fortunately evidence from comparative anatomy and embryology of living forms is helpful. This evidence is based on the principle of **homology** (see Chapter 3). Anatomically similar structures in different organisms are **homologs** if they are inherited from a common ancestor. Homology therefore represents the phenotypic evidence for genetic relatedness. There are two problems in putting the principle into practice. One is to distinguish resemblances that are homologous from those that are not; the other is to discover transformations of structures that look different but are in fact homologous. The details of embryonic development are particularly helpful in solving both of these problems, and comparative anatomists accept the ultimate phenotypic definition of homologous structures as being those that arise from the same germ layers in the same relative location in the embryo. Homologs are derived from the same group of early differentiating, organ-forming cells, which are often identifiable by the end of the early stages of development.

Following the principle of homology, it is most reasonable to expect to find the ancestral source of the vertebrates among the invertebrate chordates, which possess the notochord, dorsal hollow nerve cord, pharyngeal slits, and postanal tail so fundamentally associated with the vertebrate body plan. No other group of animals shows closer structural affinities to the vertebrates than these rather unlikely looking marine animals.

2.2.1 Amphioxus as a Model Prevertebrate

The cephalochordates, particularly the well-studied amphioxus or lancelet (*Branchiostoma lanceolatum*), are an interesting group of marine animals because they possess the basic chordate characters in diagrammatic form and because their filter-feeding is similar in some respects to that of the ammocoete larvae of the lampreys. These facts indicate a relationship with the vertebrates and suggest that the morphology of the cephalochordates should provide some hints about the early chordate ancestors of the vertebrates.

The Subphylum Cephalochordata of the Phylum Chordata consists of some 14 species, divided into two genera, *Branchiostoma* (eight species), and *Asymmetron* (six species). The latter differs from the former primarily in having gonads only on the right side, as their generic name suggests. All are small, fusiform, superficially fishlike, aquatic animals usually under 5 cm long. Lancelets are widely distributed in oceanic waters of the continental shelves, and, although they can swim freely, they are burrowing, sedentary animals as adults. Many of their special features are adaptations for this mode of existence. Amphioxus burrows head first into sandy ocean bottoms, turns around, and rests with just the anterior part of its body protruding above the bottom surface. There it employs a highly elaborate mechanism for filtering microscopic forms of life from the surrounding water. The main features of the lancelet's morphology relate to its methods of locomotion and feeding (Figure 2–3).

A significant feature of amphioxus is its fishlike locomotion, resulting from the contraction of serially arranged **myotomes,** blocks of striated muscle fibers positioned dorsolaterally along both sides of the body and separated by sheets of connective tissue, the **myocommas.** Within the posteriorly pointing, V-shaped myocommas, each muscle fiber runs parallel to the anteroposterior axis. The

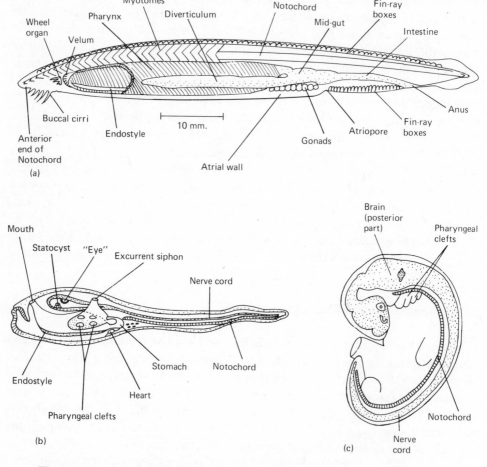

Figure 2-3 (a) Longitudinal, parasaggital section of amphioxus with posterior myotomes removed, compared with (b) a tunicate larva and (c) a mammalian embryo. (Based on J. Z. Young [10] and B. Stahl [9].)

contractions of the myotomes bend the body in a way that results in forward propulsion. (For a discussion of vertebrate swimming, see Chapter 7.) The notochord plays an essential role in the mechanics of these movements. It is an incompressible, elastic rod, running down the full length of the cephalochordate body and so positioned as to prevent the whole body from shortening, which would otherwise result from contraction of the myotomes. Because the notochord prevents this anteroposterior shortening, the myotomal contractions bend the body instead. The notochord of amphioxus differs from the vertebrate notochord in that it extends from the tip of the snout to the end of the tail, projecting well beyond the region of the myotomes at both ends. This condition apparently aids the burrowing habits of amphioxus (Figure 2-4).

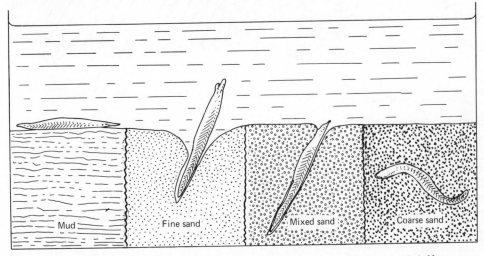

Figure 2-4 The behavior of amphioxus in relation to substrates of different particle size (after Webb and Hill, 1958; from J. Z. Young [10]).

Amphioxus is a filter-feeder extracting small food particles from a stream of water drawn through its mouth and pharynx by the movements of cilia (Figure 2-5). As in all animals that use cilia for filtering, a large surface is required to obtain sufficient food, and the pharyngeal apparatus of amphioxus occupies more than half of the animal's length, for it is the pharynx that performs the main operation of collecting the food. Its walls are perforated by up to 200 oblique, vertical slits, the number of which increases as the animal grows older. The slits are separated by bars with internal skeletal rods, and the entire sidewall of the perforated pharynx is covered over by the outer protecting walls of the **atrium**. The atrial cavity between the walls and the pharynx communicates to the outside of the animal through a posterior **atriopore**. A current of water passes in at the mouth, through the pharynx, and out through the atrium and atriopore. This current is generated by the beating of cilia on the sides and inner surfaces of the gill bars, driving water out through the atriopore and thereby drawing in a food-laden current of water at the mouth. Solid particles caught up in this current encounter a variety of organs that

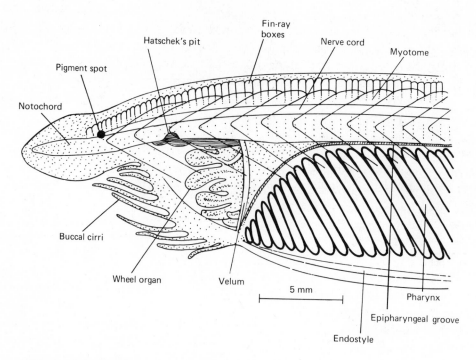

Figure 2-5 Anterior end of Amphioxus showing details of structures involved in filter-feeding. (Based on J. Z. Young [10].)

separate edible from inedible matter and conduct the entrapped food to the digestive tract. **Buccal cirri**, attached to the margin of the **oral hood** in front of the mouth, possess sensory cells and form a funnellike sieve that prevents the entry of large particles. The mouth itself is surrounded by a ring of sensory tentacles, the **velum**, which presumably functions as an additional screening mechanism against unwanted items. Within the cavity of the oral hood there apparently is an area of still water where some food particles may fall out of the incurrent stream. These are caught on a complex set of ciliated tracts known as the **wheel organ**. Mucus from **Hatschek's pit** is discharged onto the wheel organ, and the food particles caught in this material are swept by cilia towards the mouth and then enter with the main stream of water.

On the floor of the pharynx lies the **endostyle**, another mucus-secreting organ. The endostyle consists of columns of ciliated cells alternating with mucus-secreting cells and produces a series of sticky threads in which food particles become entangled. Various ciliary currents then draw the food-carrying mucus strands up the sides of the pharynx and into a median, dorsal **epipharyngeal groove**, in which cilia move the material backwards into the midgut. The midgut also has cilia that rotate the cord of food and mucus during digestion. Amphioxus employs both intracellular and extracellular digestion.

The mechanics of feeding outlined here for amphioxus have interesting and probably homologous counterparts in other groups of animals. The homology is clear in the case of another subphylum of chordates, the Urochordata or tunicates. They use exactly the same basic mechanisms for feeding, including an incurrent mouth, excurrent pharyngeal slits, cilia, and a mucus-secreting endostyle. Similar use of cilia and mucous cells on *external* structures occurs on the proboscis of the acorn worms (Hemichordata), the oral tentacles of lophophorate feeders (pterobranchs, brachiopods, bryozoans, and phoronid worms), and the ambulacral system of the crinoid echinoderms. Among vertebrates, the filtering mechanism of the ammocoete larvae in lampreys is the same, except that a muscular pump is used to move the water instead of cilia.

Amphioxus shows several other similarities to the fundamental plan of vertebrate organization. It has a closed circulatory system in which slow waves of contraction occur in various, separate parts, driving the blood forward in the ventral vessels and backward in the dorsal ones. Below the hindend of the pharynx there is a large sac, the *sinus venosus*, which collects blood from all parts of the body, much as the vertebrate heart does, although the sinus venosus is not the exclusive pumping organ. Amphioxus reveals no indication of the complex vertebrate brain, possessing little more than an expansion of the anterior end of the dorsal hollow nerve cord into a so-called **cerebral vesicle**. The nerve cord itself is similar to that of vertebrates and is connected to the periphery by sets of dorsal and ventral nerve roots on each side of the body segments. The roots do not join as in vertebrates, but the ventral roots lie opposite the myotomes to which they carry motor fibers that end on muscle fibers with motor end-plates; the dorsal roots run out between the myotomes, carrying all the afferent sensory fibers of the segment and motor fibers for the nonmyotomal muscles (Figure 2-6). These details conform to the fundamental pattern of the spinal nerve roots in all vertebrates (see section 5.3.2).

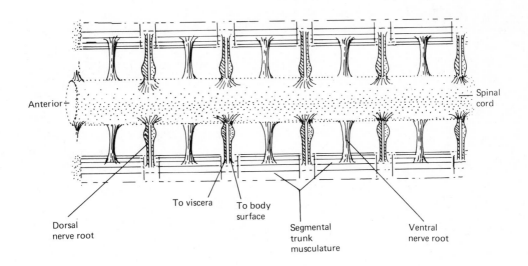

Figure 2-6 Relation of the dorsal and ventral nerve roots to the spinal cord in Amphioxus.

Despite the several probably homologous similarities in structure between the cephalochordates and the vertebrates, there are some basic differences indicating that amphioxus and its relatives could not be in the direct line of ancestry leading to the vertebrates. For one thing, the excretory system is completely unlike that of the vertebrates, consisting of flame cells called **solenocytes**, which are more like the excretory structures of platyhelminths, mollusks, and annelid worms (Figure 2–7). At the dorsal end of each primary gill bar there lies a sac that opens by a pore to the atrium and is studded with several hundred elongated flame cells. These solenocytes do not open internally into the coelom but lie in close contact with special blood vessels, which separate the flame cells from the coelomic epithelium. Embryonically the solenocytes appear to originate from ectoderm, whereas the vertebrate kidney cells come from mesoderm, indicating that these excretory organs could in no way be homologous.

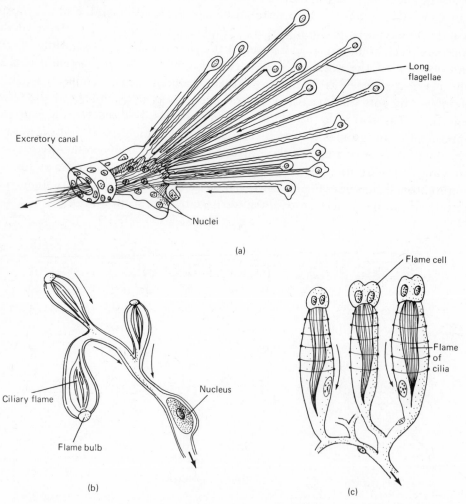

Figure 2-7 (a) Solenocyte of amphioxus compared with flame cells of (b) flat-worm and (c) annelid worm. Compare with vertebrate kidney nephron in Figure 7–19.

The lack of strong cephalization and the unique extension of the notochord to the anterior end of the rostrum also makes cephalochordates unlikely candidates as animals that could evolve the large, vertebrate-type brain. In addition, there are no sense organs in amphioxus that suggest themselves as possible homologs of the eyes, ears, nose, or other cephalic sense organs of vertebrates. We can conclude that the cephalochordates are the closest living relatives of the vertebrates and that they tell us something about the likely ancestral body plan of the vertebrates, but they are not themselves ancestral. Cephalochordates and vertebrates represent divergent paths of evolution from a common ancestry far back in time.

2.2.2 Other Distant Relatives

The prospects seem even less likely for a vertebrate ancestor among other known protochordate groups, if we consider only the adult body forms of these animals. The adult tunicates (Phylum Chordata, Subphylum Urochordata) are either sessile or free-living forms whose body plans are so divergent from those of vertebrates that it is impossible to imagine an evolutionary transformation from one to the other (Figure 2-8). The same conclusion applies to the acorn worms and pterobranchs (either Phylum Hemichordata or Phylum Chordata, Subphylum Hemichordata). Pharyngeal slits and a (probably nonhomologous) section of hollow nerve tube are the only chordatelike features they possess. In fact, their body plans are otherwise so highly divergent that they are now usually put in a separate phylum of their own (Figure 2-9). Their larvae are ciliated, bean-shaped creatures that look so much like the larvae of echinoderms that they have frequently been misidentified as such, a fact to which we shall return (Figure 2-10).

In contrast, the larvae of tunicates are tadpolelike animals with pharyngeal slits, notochord, and dorsal hollow nerve tube fully developed. In addition, they possess a muscular, postanal tail, which moves in a fishlike swimming pattern (Figure 2-11). They are sufficiently generalized in their chordate and other structural features to suggest that they represent only slightly modified living forms of the ancestral chordates that gave rise to the vertebrate line of evolution. But tunicate larvae, after a brief free-swimming existence, metamorphose into adult animals that bear no resemblance to vertebrates. By what evolutionary process could the larval traits of the tunicate tadpole be passed on to succeeding generations of adult animals? A resolution of this difficulty was first proposed by W. Garstang in 1928. He suggested that vertebrates originated from tunicatelike larvae that failed to metamorphose but nevertheless developed functional gonads: The genotype could be passed on without the necessity of a separate adult morph in the life cycle.

The development of gonads in an otherwise larval body form should be termed **paedogenesis** or **paedomorphosis**. Unfortunately it is also often incorrectly called **neoteny**, with the result that a good deal of confusion exists in

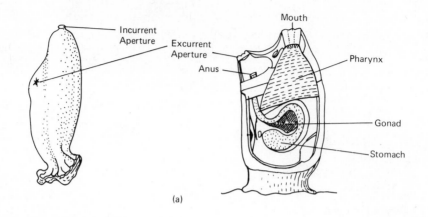

(a)

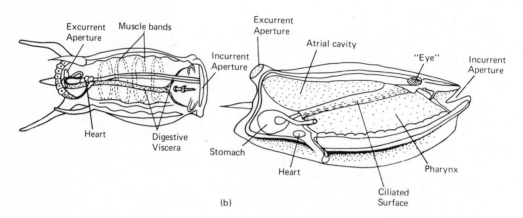

(b)

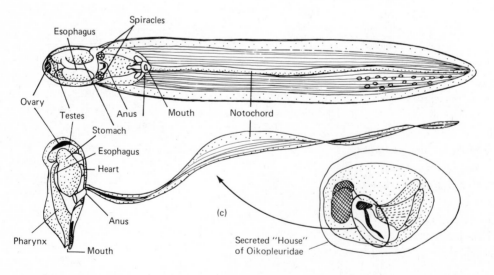

(c)

Figure 2-8 Adult examples of the three classes of tunicates: (a) Ascidiacea, sessile; (b) Thaliacea, free-floating; (c) Larvacea, swimming (based on various sources including A. Alldredge, Sci. Amer. 235(1):95–102, July 1976).

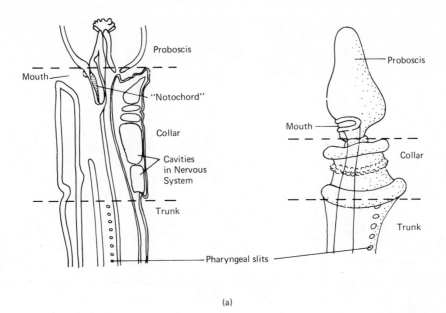

(a)

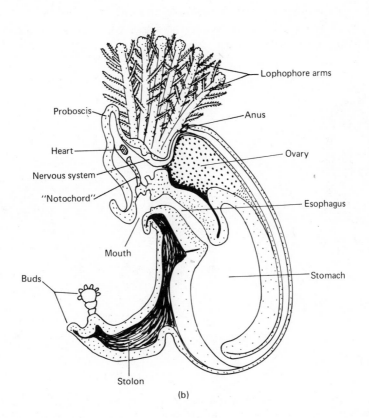

(b)

Figure 2-9 (a) Structure of acorn worm—anterior end. (b) Section of pterobranch.

the literature on two rather different evolutionary-developmental processes. Neoteny properly refers to the retention of a larval or embryonic trait in the adult body. It results when some part or parts of the growing animal fail to develop at the normal rate or become arrested in development before the typical, definitive adult condition is attained. Neotenic structures may result from developmental abnormalities that lessen the individual's chances for survival and so are selected against, or may confer advantages and become incorporated into the genotype by natural selection. Some familiar examples of neotenic traits that have evidently been selected for adult expression during the course of vertebrate evolution are the following: the lack of bone in adult Chondrichthyes (elasmobranchs, chimaeras), which is generally considered to be a secondary condition resulting from loss of the ancestral bony armor and retention of the embryonic cartilaginous skeleton in the adult; the external gills of certain adult aquatic salamanders in which the larval gills are only slightly modified; and the small face and expanded brain case of human beings, traits that are widespread in mammalian embryos but are retained in the postfetal stages of man, presumably as a necessary correlate of increased brain capacity. Although neoteny emphasizes the retention of embryonic or larval characteristics in one or a few structures of an otherwise typically adult body, paedogenesis stresses the precocious development of the gonads in an otherwise larval body, which becomes permanently arrested in larval condition and never undergoes metamorphosis.

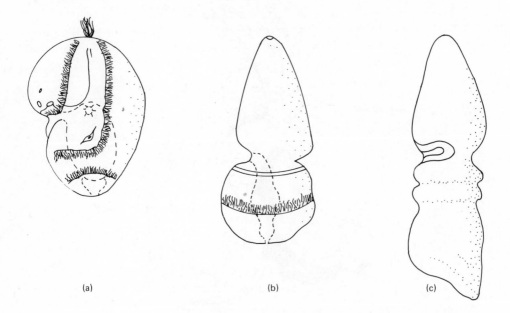

(a)   (b)   (c)

Figure 2-10 Larva of acorn worm and stages in transformation to adult form. (a) larva, (b) proboscis forming, (c) definitive anterior organization nearly complete (modified from P. A. Meglitsch, *Invertebrate Zoology*, Oxford University Press, 1967).

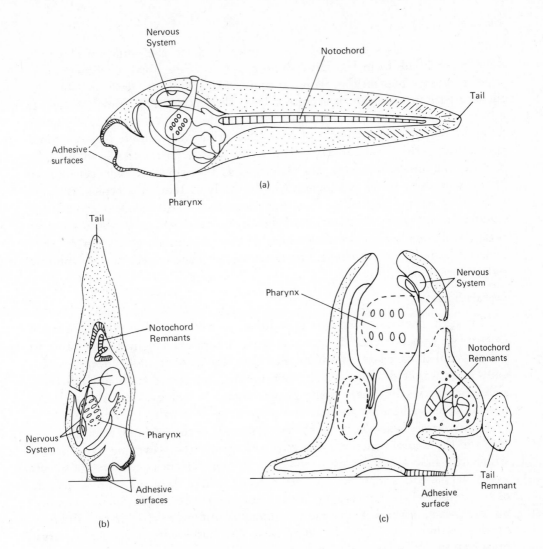

Figure 2-11 Tunicate larva (*Clavelina*) compared with adult; (a) free-swimming larva before metamorphosis; (b) attached larva in metamorphosis; (c) adult tunicate.

2.2.3 Evidence of Paedogenesis Among Chordates

Because the concept of paedogenesis is central to the most commonly accepted theory of vertebrate origins, it is especially important for students of vertebrate life to know the factual basis for the existence of such a phenomenon. It is not just a convenient idea that happened to pop into Garstang's mind; a number of living animals, both chordate and nonchordate, demonstrate varying manifestations of sexual maturity as larvae. Others have adult bodies that clearly indicate that neoteny has been involved in their evolutionary histories. We summarize here only the evidence from chordates.

Tunicates

The tunicate class Larvacea is so named because the adult form is only slightly modified from the larva (Figure 2–8c). These small, free-swimming, adult tunicates live in the plankton and build a housing, with an elaborate filtering structure, of mucus secretion from their skins. The animal, which has a broad propulsive tail, attaches to a point inside the filter-trap and, by undulating its tail, sets up a current of water that brings very minute flagellates into contact with the food-collecting apparatus, which passes the food particles back to the animal's mouth. Its pharynx has two slits, an endostyle, ciliary bands, and the general organization of its body is that of a typical tunicate tadpole. It seems probable that the larvaceans have evolved from their tunicate ancestors by acceleration of the rate of development of the alimentary organs and gonads with the elimination of drastic metamorphic change. They appear, therefore, to represent a parallel case with the vertebrates of evolution by paedogenesis.

Cephalochordates

During the famous *Challenger* expeditions of 1872 to 1876, which did much to give rise to the science of oceanography, naturalists caught a large (10 cm) pelagic cephalochordate to which they gave the generic name *Amphioxides*. Only later was it discovered that these ocean-going lancelets are not a distinct genus or species but only giant pelagic larvae of amphioxus. Although the normal cephalochordate larva is free-swimming, it usually restricts its movements to waters near shore, performing mainly cyclic 24-hour vertical movements in the littoral zone. Metamorphosis in these littoral-dwelling larvae occurs within the first 140 days after hatching. Under certain conditions, however, young larvae are swept out to deeper water where they take up a pelagic existence and where, for reasons that are not known, larval life can be extended up to 3 years, during which time the animal continues to grow in size and may develop as many as 25 to 30 pharyngeal slits. Otherwise it retains its larval organization. In some individuals the gonads start to develop *before* metamorphosis. Amphioxides comes close to demonstrating just the sort of larval history that is required for Garstang's theory.

Amphibia

The salamanders and other tailed amphibians (Order Urodela) present an array of adult forms, from well-developed four-legged, lung-breathing types with eyelids and a thickened skin with a *stratum corneum*, adapted for terrestrial life, to fully aquatic, gilled animals with reduced legs, no eyelids, thin skin, and median tail fins (Figure 2–12). Although there are exceptions that we will mention later (Chapter 10), many salamanders lay eggs that hatch into aquatic larvae which grow for a time as larval forms with gills and associated skeletal structures. These larvae also possess finned tails and reduced pectoral and pelvic appendages (particularly the hind limbs) and have a thin skin and no eyelids. At some point in the life cycle, they metamorphose relatively quickly into the

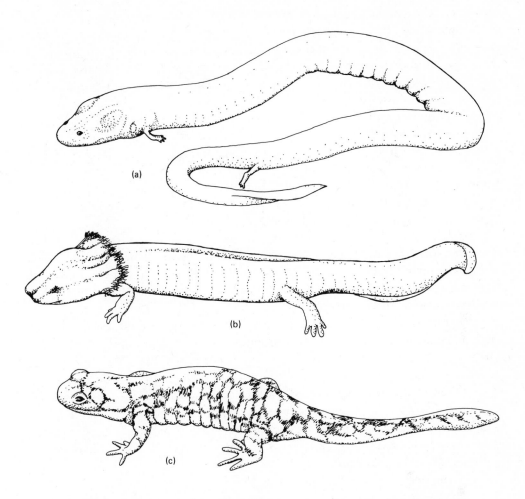

Figure 2-12 Body forms of aquatic and terrestrial salamanders; (a) *Amphiuma*, exclusively aquatic; (b) *Necturus*, aquatic; (c) *Ambystoma* adult, terrestrial.

adult form. The metamorphic transformation is most striking in those terrestrial salamanders that have aquatic larvae; less pronounced changes occur during the development of the permanently aquatic forms, and biologists recognized long ago that varying degrees of larval body organization are retained in the adult stages of these species. Some appear to be permanently arrested larvae or larvae that fail to undergo full metamorphosis, but the gonads reach sexual maturity.

Do the aquatic urodeles represent examples of neoteny or paedogenesis? The distinction between the two processes becomes blurred in these amphibians. If the ancestral populations that gave rise to these "permanent larvae" were animals that underwent a pronounced metamorphosis, then they should be considered paedomorphic. If the ancestors had a more direct mode of development, as in some modern amphibians, then it would be correct to consider these modern peculiar aquatic forms as neotenic. A study of the development of these amphibians shows that various structural changes take place at such different and disconnected periods of early life that none of them can be said to undergo metamorphosis, which generally indicates a large-scale reorganization of structure that occurs in a limited time. If the embryologist focuses his attention on the obvious changes that occur during typical amphibian metamorphosis—the development of limbs, of the maxillary bones, the loss of gills, and reduction of the branchial arches—it becomes obvious that natural selection has acted to retard developmental rates or developmental end points of various structures and organs to different degrees in these aquatic urodeles. The result has been parallel or convergent adaptation for permanent aquatic existence in several phylogenetic lines of tailed amphibians whose ancestors were adapted for terrestrial life as adults.

The most informative example of paedogenesis among vertebrates is the Mexican axolotl (an Aztec name), a salamander known to science as *Ambystoma mexicanum*, which occurs naturally only in Lake Xochimilco in the highlands of Mexico. It belongs to a widespread genus of North American salamanders that have a typical life cycle consisting of an aquatic larva that metamorphoses to a terrestrial adult. There is a tendency for the occasional appearance of paedomorphic individuals in various populations of ambystomid salamanders, but sexually mature larval individuals predominate in the Lake Xochimilco population, and only occasionally do transformed adults occur in nature. Environmental stresses such as cold temperature or insufficient iodine (a component of thyroxin hormone that induces amphibian metamorphosis) may be involved in the failure of some individuals to metamorphose. It appears more likely, however, that the genetic basis for metamorphosis in salamanders is multifactorial, variable, and subject to selective pressures. For some reason the environment of Lake Xolchimilco favors a permanent larval morph. A similar phenomenon occurs in the high altitude races of the closely related tiger salamander in Colorado. The genes for transformation have become suppressed in these populations, but they have not entirely disappeared, for there is the occasional appearance of transformed adults!

The genetic basis for a permanently paedogenic form of axolotls is revealed in the history of the laboratory strain so commonly used for experiments in embryology and regeneration. The laboratory animals now used worldwide are from one of the purest lines of experimental animals in existence, for they all descend from two original collections of axolotls shipped to Paris in 1864 and 1868. Individuals of this strain, unlike those of the wild population, *never* metamorphose unless forced to do so by experimental manipulation. Indeed these transformed individuals do not breed with much success in the laboratory. In this domesticated stock of axolotls we see a case of an inadvertent but strong selection favoring the nontransforming genome. Although the result of "human selection" rather than natural selection, the laboratory axolotl provides a clear example of the process postulated to account for the origin of the vertebrate line.

2.2.4 Changes Leading to the First Vertebrates

Once the occurrence of paedogenesis has been established, it is not difficult to imagine how slight morphological changes of an adaptive nature could lead from sexually competent tunicatelike larvae through various evolutionary stages similar to amphioxus and ammocoetes to the first full vertebrates with crania and backbones (Figure 2–13). Increase in body size, the development of various sense organs and an anterior brain, and extension of segmental muscles forward onto the trunk resulting in more powerful locomotion would have allowed our prevertebrate ancestors to reach a new level of organization that permitted successful life as active, swimming filter-feeders in the surface waters of the sea.

Only the origin of bone remains a major puzzle in our story. The chemistry and fine structure of calcified materials offer some suggestions about the way in which this distinctively vertebrate tissue evolved. Although bone cannot be homologized with any skeletal tissues found among invertebrate phyla, it does share some common features with the mineralized tissues of other animals. Like the exoskeletons of arthropods, the shells of mollusks, and the hard parts of other metazoa, bone results from the deposition of minerals into a matrix or ground substance produced by special cells. Among vertebrates and invertebrates alike, the cells involved in the formation of calcified tissues derive from both the embryonic ectoderm and mesoderm, and the matrices they produce are alike in that they usually contain polysaccharide and protein. The minerals deposited in the matrix generally consist of calcium carbonate or calcium phosphate, either in amorphous form or as crystals. Although bone is unique in its particular combination of characteristic cells, matrices, and minerals, all of the components of the vertebrate hard skeleton exist in other animals and presumably had existed long before the first forms with backbones evolved. With these facts in mind, one can conclude that bone did not originate in early vertebrates as a completely new tissue. Rather it was a new combination and elaboration of phenotypic features already well established in the genome of the chordate and deuterostome ancestors of the first vertebrates.

The main mineral constituent of bone is calcium phosphate, laid down initially in amorphous form (new bone) and subsequently organized into crystals of hydroxyapatite (mature bone). Calcium carbonate is the mineral usually found in the hard tissues of invertebrates, but calcium phosphate and apatite do occur in groups other than vertebrates. Certain "lamp shells" (Phylum Brachiopoda) have been producing shells of calcium phosphate (apatite) since Cambrian times, and many systematists consider the brachiopods to be related to the hemichordates and chordates. Of even greater interest are the curious microfossils known as conodonts, which are widespread and abundant in marine deposits from the late Cambrian to the late Triassic and disappear completely before the end of the Cretaceous. These are small, spiny fossils of apatite with a structure that indicates they were laid down in soft tissues by the apposition of concentric layers exactly as were primitive

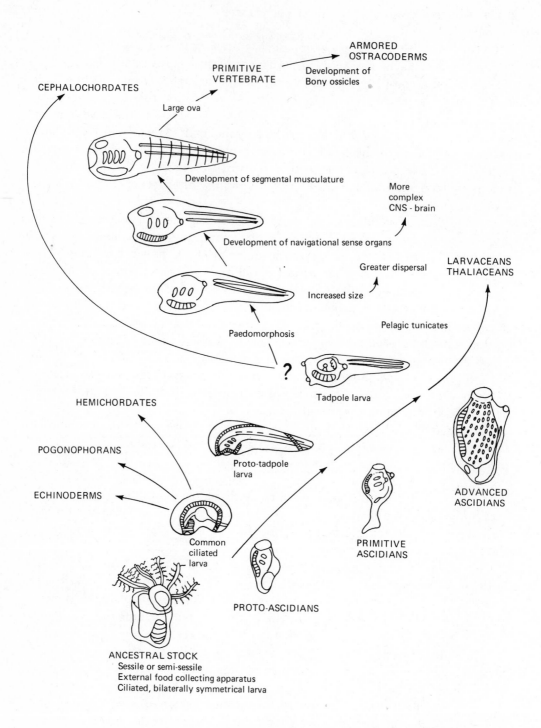

CEPHALOCHORDATES

PRIMITIVE
VERTEBRATE

ARMORED
OSTRACODERMS

Development of
Bony ossicles

Large ova

Development of segmental musculature

More
complex
CNS - brain

Development of navigational sense organs

Greater dispersal

LARVACEANS
THALIACEANS

Increased size

Pelagic tunicates

Paedomorphosis

?

Tadpole larva

HEMICHORDATES

POGONOPHORANS

Proto-tadpole
larva

ADVANCED
ASCIDIANS

ECHINODERMS

Common
ciliated
larva

PRIMITIVE
ASCIDIANS

PROTO-ASCIDIANS

ANCESTRAL STOCK
  Sessile or semi-sessile
  External food collecting apparatus
  Ciliated, bilaterally symmetrical larva

Figure 2-13 Scheme of evolutionary changes leading from an ancestral deuterostome lophophorate to vertebrates (based on Romer and Parsons, Ch. 1 [5] and other sources).

fish scales. Students of these fossils have remarked about the close similarity in chemical composition and microstructure of the conodonts to vertebrate bony scales, but until recently no other parts of the organisms to which they belong had been found. Consequently, conodonts were formerly variously described as skeletal parts of marine algae, as phosphatic annelid jaws, fish teeth, gill rakers, gastropod radulae, nematode copulatory spicules, and as arthropod spines (Figure 2–14).

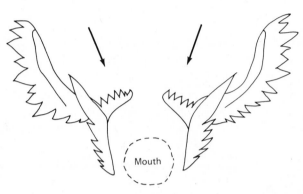

Figure 2-14 Conodont "denticles" arranged to suggest possible position with respect to the mouth of the organism and feeding currents (from M. Lindström [3]).

Another interpretation has been that conodonts are structures that were arranged on the tentacles of free-swimming or floating "lophophorates," of which the bryozoans, branchiopods, and pterobranchs are existing sedentary forms. They presumably functioned in some way in the lophophore method of feeding. Carboniferous fossils discovered recently in Montana, however, represent the first complete conodont animals. They bear a striking resemblance to cephalochordates (Figure 2–15) and show that conodont structures are associated with the central gut region, confirming that they function to support soft tissues. Other features of the anatomy of these animals indicate that they are some kind of a protochordate assemblage and, therefore, possibly in the direct ancestry of vertebrates. In any case, the geological distribution of conodonts shows that they were highly successful forms and are old enough to have preceded the late Cambrian and Ordovician vertebrates. Their mineralized skeletons, together with those of equally ancient brachiopods, demonstrate that the early lophophorate-deuterostome-chordate assemblage of marine animals already possessed the genetic potential for the development of bone-like structures in early Cambrian times.

One final point about the origin of bone needs brief mention. Although we usually think of bone as primarily skeletal or protective in function, it also serves as a mineral store of calcium and phosphate. Mineral regulation in the vertebrate organism involves both the deposition of calcium and phosphate in bony tissue and the decalcification of bone with the mobilization of calcium and phosphate back into the circulatory system for distribution and use in

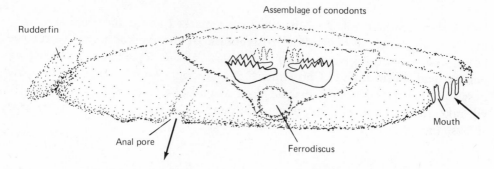

Figure 2-15 Outline drawing of a conodont animal showing some anatomical features (modified from W. Merton and H. W. Scott [5]).

other parts of the body. It is quite likely that the need for a mineral store particularly for phosphorus, a relatively rare element in natural environments, was one of the early selective forces involved in the evolution of primitive dermal ossifications of the first vertebrates. Once evolved, these ossicles enlarged and became transformed, perhaps under other selective influences, into the dermal teeth and armor plates of the ostracoderms. Later still, endochondral bone evolved, primarily in association with increased locomotor and supportive functions of the axial skeleton in some aquatic and all terrestrial vertebrates.

2.3   LOPHOPHORATES, DEUTEROSTOMES, AND CHORDATES

We have yet to consider the origin of the chordate ancestors of the vertebrates. Almost every invertebrate phylum has been suggested as the source of the chordate line. Fairly strong cases were once argued for annelids and arachnids, but a close study of the basic developmental patterns in the various phyla of coelomate animals has revealed a different and rather surprising conclusion. The Chordata and Hemichordata are actually more closely related to the Echinodermata, marine forms with radial symmetry as adults, and to certain obscure marine phyla such as the arrow worms (Chaetognatha) and the beard-worms (Pogonophora).

Phylogenists now realize that the early coelomate animals evolved two basically different developmental patterns for forming the early embryo and larval stages. The Mollusca, Arthropoda, and Annelida, with a similar early developmental pattern, are grouped together as relatives in the Annelid Superphylum. Because the other five phyla share a distinctive embryology, they are collectively placed in the Echinoderm Superphylum. Table 2-1 summarizes the main differences between these superphyla.

Certain phyla in which the mouth is of the annelid type do not fit easily into this phylogenetic scheme because they also possess some of the developmental features of the Echinoderm Superphylum. These are the so-called "lophophorates" (Phyla Phoronida, Ectoprocta, and Brachiopoda), sessile forms with a crown of ciliated tentacles, the **lophophore**, which is used to

Table 2-1. Basic Differences Between the Annelid and Echinoderm Superphyla

Feature Compared	Annelid Superphylum	Echinoderm Superphylum
Cleavage	Spiral and determinate	Radial and indeterminate
Blastopore	Forms the mouth	Forms the anus
Coelom formation	Schizocoelous (a rift or split within the mesoderm)	Enterocoelous (outpocketing from the dorso-lateral gut wall) (*except vertebrates*)
Type of larva	Trochophore (Figure 2–17a)	Tornaria (Figure 2–17b) or bipinnaria (*except chordates*)

capture food. They show several striking similarities with the pterobranch hemichordates, also sessile, lophophore-feeders as adults (Figure 2–16). Like the echinoderms these lophophorates have free-swimming, ciliated larvae, usually of the tornaria type. Quite probably the common ancestor of the echinoderms, chordates, and related phyla was a generalized sessile or semisessile lophophorate in the adult stage with a free-swimming ciliated larval stage for dispersal (Figures 2-13, 2-17). Table 2-2 summarizes the distribution of lophophorate, deuterostome, and chordate traits among these phyla. In the line leading to the chordates, parallel changes took place both in the motile larva, improving its locomotor and feeding mechanisms, and in the sessile adult stage, in which the feeding apparatus altered radically. The adult lophophorate method of food capture with external tentacles was gradually replaced by food entrapment inside the pharynx through the use of excurrent slits (trema) in the walls of the pharynx and mucus secretion by an endostyle. One of the existing genera of pterobranchs, *Cephalodiscus*, has a single pair of pharyngeal slits in addition to the lophophore, showing the potentiality for transitional stages between the two modes of feeding (Figure 2–9b).

The ancestral tornaria larva gradually changed into an elongate, prototadpole larva, which at first was still propelled by bands of cilia while developing, in parallel with the adult, slits in the pharynx (**pharyngotremy**) and the internal food-trapping mechanism of the chordates. Muscle fibers later evolved in association with the notochord and postanal tail, and according to Garstang the longitudinal ciliary bands moved to the middorsal line and became transformed into the hollow nerve cord (still ciliated in vertebrates), and the adoral cilia developed into the endostyle. These transformations produced the typical chordate tadpole larva (Figure 2–17). Freed from the severe limitations imposed by ciliary locomotion, such a larva could increase in size under the influence of natural selection, and with the potential for paedogenesis various evolutionary paths became possible. It is likely that paedogenesis occurred inde-

Table 2-2. Distribution of Selected Characters in Some Metazoans.

	Radial cleavage	Deutero-stomous	Entero-coelous	"lopho-phore"	"tornaria" larva	Pharyn-geal slits	Dorsal nerve cord	Noto-chord	Post-anal Tail	Myo-tomes	Cranium bone	Verte-brae
Bryozoans			?	■	?							
Brachiopods				■	■							
Phoronids			?	■	?							
Echinoderms	■	■	■	■	■							
Pogonophores	■	■			?							
Pterobranchs (Hemichordate)	?	?	?	■	■	■						
Enteropreusts (Hemichordate)	■	■	■		■	■						
Larval Tunicates	■	■				■	■	■	■	■		
Cephalochordates	■	■	■			■	■	■	■	■		
Agnathan Vertebrates	■	■	?			■	■	■	■	■	■	■
Gnathostomes	■	■				■	■	■	■	■	■	■

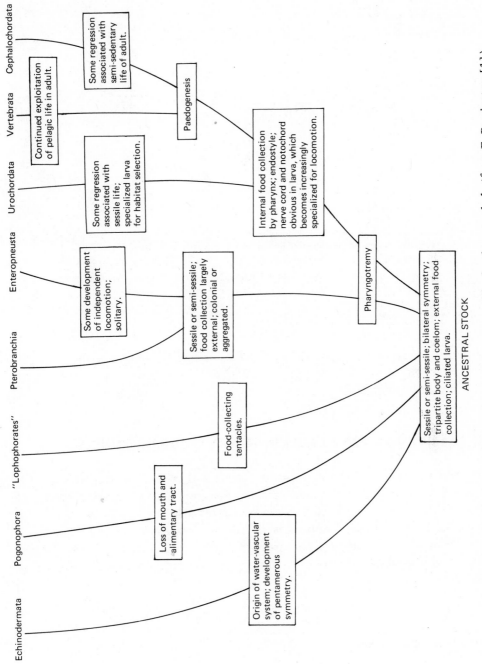

Figure 2-16 Suggested relationship of deuterostome and lophophorate phyla (from E. Barrington [1]).

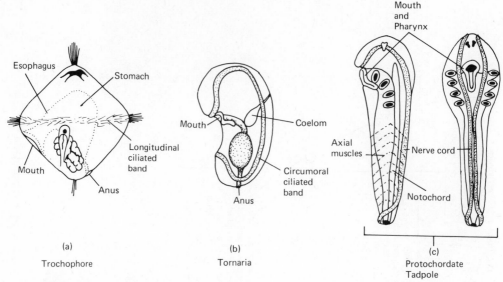

Figure 2-17 Comparison of trochophore and tornaria larvae of coelomates and Garstang's concept of how a protochordate tadpole might have evolved from a ciliated larva of the tornaria type (from J. Z. Young [10]). As a larva evolves larger size, a point is reached when locomotion by cilia becomes inadequate, because of surface to mass ratios. Locomotion by contracting muscles is not limited by surface to mass ratios, favoring the evolution of axial muscles along the body.

pendently in several diverging lines of these early tunicatelike larvae, with the vertebrates, the cephalochordates, and the larvaceans representing three cases of such parallel evolution.

2.4 ENVIRONMENT IN RELATION TO VERTEBRATE ORIGINS

By the late Silurian armored ostracoderms and primitive-jawed fishes were abundant in both fresh water and marine environments. For years students of evolution have argued whether the vertebrates had a fresh-water origin or marine origin. Under what conditions did the first vertebrates evolve? Although we will detail the geological and paleontological evidence in Chapter 5 and consider physiological aspects in Chapter 7, it is worthwhile to consider those arguments briefly now.

2.4.1 Geological Evidence

A marine origin has been postulated for all of the animal phyla. The Cambrian and Ordovician seas were rich environments for the life of 500 to 600 million years ago. Fossils of most major groups of animals appear to occur only in marine deposits of these periods. The first land plants appear in the Silurian, and the first land animals in the Devonian. Life therefore probably remained aquatic and marine for at least the first half of evolutionary history. Geologists have found few rocks earlier than those of the Silurian that were formed in fresh-water deposits. Consequently we know little about the fresh-water biota of the Ordovician period or of earlier times. Nonetheless, Alfred Sherwood Romer, one of the world's foremost paleontologists, and Homer Smith, a renowned American renal physiologist, were staunch advocates for the fresh-water origin of the first vertebrates and presented evidence to support this view.

Ostracoderm fragments found in the Middle Ordovician Harding Sandstone of Colorado occur in association with fossils of marine origin. Their location in the sediments and their worn condition, however, were interpreted by Romer as indicating that these fragments had washed downstream from rivers or lakes into estuaries and were fossilized in marine sediments. In addition Romer was impressed by another aspect of the fossil record: At the time he formulated his ideas, vertebrate remains were absent from the rich fossil-bearing marine rocks of the Cambrian and were rare in Ordovician rocks, and yet vertebrates are present in late Silurian and abundant in Devonian marine formations. If vertebrates arose in the sea, Romer wondered why they left no significant fossil record before the Silurian. If, on the other hand, vertebrates evolved in fresh water, Romer reasoned their absence from the exclusively marine Cambrian and Ordovician formations is explained. Their sudden appearance in a variety of forms in the late Silurian and Devonian corresponds with the earliest occurrence of abundant fresh-water sediments in the geological record. Romer further proposed that active swimming by means of a muscular postanal tail evolved in the early vertebrates in response to the downriver sweep of stream currents.

2.4.2 Physiological Evidence

Homer Smith's studies of vertebrate kidney structure and function led him to support Romer's idea of a fresh-water origin. The blood and body fluids of most marine organisms contain salts at concentrations similar to those of the sea water that bathes them. Any alteration of this concentration equilibrium will produce a flow of water from the more dilute to the more concentrated fluid—a process called **osmosis** (see Chapter 7). In bony fishes, body fluid concentrations are very dilute compared to sea water. In fresh-water habitats this lowering of body fluid concentration, although reducing the inward osmotic flow of water, nevertheless cannot be reduced sufficiently to stop hydration of the body completely. A fresh-water fish therefore takes up water and must have a mechanism for excreting the excess water and for retaining needed ions. The vertebrate glomerular kidney is well adapted for these functions (see Figure 7–19), and Smith proposed that dilute body fluids and the glomerular kidney evolved as adaptations to fresh-water conditions.

2.4.3 Arguments for a Marine Origin

In spite of Romer's and Smith's arguments, evidence for a marine origin of vertebrates is now overwhelming. All of the protochordate and deuterostome invertebrate phyla are exclusively or primitively marine forms. Because of chordate affinities to these phyla, there must have been a marine transitional form in the evolutionary line leading to the vertebrates. All known Cambrian and Ordovician vertebrates occur as marine fossils. The Harding Sandstone Formation, which Romer and others interpreted to be estuarine, is now known to extend over many thousands of square kilometers and is more reasonably interpreted as an offshore shelf deposit. Denticles from Lower Ordovician Russian rocks came from a type of sandstone that is formed only under true marine conditions, and the same is true for all of the more recently discovered fragments of early ostracoderms from North American locations. The chance of a fresh-water vertebrate being swept so far out from land seems very small.

The exclusively marine hagfishes (*Myxiniformes*) have body fluids that are similar in salt concentration with sea water, as do the tunicates and other deuterostomes. It is not unreasonable, therefore, to assume that the first vertebrates were also in osmotic equilibrium with the sea. Nevertheless, the hagfishes have a well-developed glomerular kidney and so do the marine cartilaginous fishes, which are also osmotically similar to sea water. A functional glomerulus is not necessarily associated with the need for powerful osmotic regulation. The function of the glomerulus cannot be considered independently from that of the convoluted tubule making up the rest of the kidney microstructure (see Chapter 7). The glomerulus functions to produce a fluid from which red blood cells and blood proteins have been excluded. The tubule cells then act on this fluid to produce the final urinary product. The tubule cells can return water, salts, glucose, and other "useful" materials back to the circulatory system, while allowing toxic substances, nitrogenous wastes, excess salts, and the like to remain in the filtrate for excretion. The kidney system therefore operates to separate certain ions and molecules, including water, from others. Its primary function is excretory and not osmoregulatory, and the kidney would be valuable to either a fresh-water or a marine organism. Professor James Robertson at the University of Glasgow has suggested that the selective pressure resulting in evolution of the glomerular kidney was the necessity of increasing the filtering rate to rid the body of rapidly accumulating wastes associated with high metabolism. This hypothesis certainly coincides with one basic attribute of vertebrate life: vertebrates are very active and motile. In this view, whether a vertebrate is adapted to the sea, to fresh water, or to life on land, the glomerular kidney is valuable and could as easily have first evolved in a marine vertebrate as in a fresh-water one. Once a powerful and efficient glomerular filtration system had been achieved, however, only slight modifications of the nephron would produce an efficient water-excreting device for osmotic regulation in rivers and lakes. Further details of kidney structure and function are presented in Chapters 7 and 20.

2.4.4 The Significance of Armor in Ostracoderms

Reasoning by analogy from the function of certain crustacean exoskeletons Homer Smith postulated that the dermal bony plates of ostracoderms functioned as a barrier against the intake of water, acting as an auxiliary mechanism in osmoregulation. The bony plates, however, are porous and probably were overlain by epidermis. Romer doubted the armor could have been an effective barrier to permeation by water.

The plates probably were protective armor to thwart the attacks of powerful predators. Several observations suggest this function. The body design of early ostracoderms indicates that they were sluggish creatures, incapable of performing sustained movements over distances. Most were tadpole-shaped and probably swam only for short distances before coming to rest on the bottom. They had no true paired appendages and therefore little maneuverability to avoid predators. Lacking jaws, they had little in the way of defense mechanisms. The only adaptive recourse against predation would be protective armor to discourage the onslaughts of grasping or biting feeders.

The ostracoderm fossils are frequently found in association with fossil eurypterids, giant scorpion-like arthropods. Eurypterids were dominant predators of the Cambrian, Ordovician, and Silurian periods (Figure 2–18). They first were marine, but they became abundant in fresh waters during the Silurian. They were close ecological associates of the ostracoderms, and it is likely they were the predators that exerted the selective force responsible for the evolution of dermal bony armor in these small, vulnerable fishes.

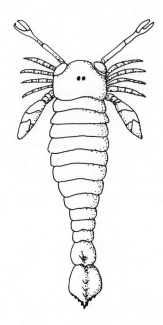

Figure 2–18 A Devonian Eurypterid.

2.4.5 Summary of Environmental Influences on Vertebrate Origin

It seems reasonable that animals essentially vertebrate in character existed in the early to middle Cambrian seas. They were no doubt small (probably under 10 cm in length), had all the primary chordate characters fully developed and, in addition, had a tripartite brain, possibly a cartilaginous cranium and neural arches surrounding the nerve cord, a muscular postanal tail, a primitive glomerular system (perhaps like the embryonic nephrons of some present day vertebrates), and body fluids similar to sea water. They may also have had small, dermal ossicles of calcium phosphate, which served as mineral stores. By late Cambrian times dermal bony armor had evolved in the ostracoderms.

Later vertebrates became adapted to estuarine conditions at the mouths of rivers, where the waters were less saline than in the sea (see Chapter 7). The osmotic stress imposed by the variable salt concentration in estuaries may have been a powerful selective force operating on these vertebrates to reduce their body fluid concentrations and to evolve a renal system effective in water removal (Chapter 7). This new environment may also have affected calcium and phosphate regulation, leading to a further function of the dermal bony tissues. We know, for example, that as a divalent ion calcium decreases cell membrane permeability and its presence in dilute sea water can greatly increase the osmoregulatory ability of marine organisms. In fact, some marine fishes penetrate fresh-water streams that have a high calcium content. Similarly, a marine fish that could carry a readily mobilizable store of calcium in its tissues might have been able to live in fresh water.

References

[1] Barrington, E. J. W. 1965. *The Biology of Hemichordata and Protochordata.* W. H. Freeman, San Francisco.

[2] Halstead, L. B. 1969. *The Pattern of Vertebrate Evolution.* Oliver and Boyd, Edinburgh. A stimulating, opinioned short book by a student of the earliest vertebrates.

[3] Lindstrom, M. 1964. *Conodonts.* Elsevier, Amsterdam.

[4] Lovtrup, S. 1977. *The Phylogeny of Vertebrata.* John Wiley, New York. A recent opposing view of the origin of vertebrates and chordates, that emphasizes biochemical as well as morphological characters, and relies primarily on cladistic analysis of the resulting data.

[5] Merton, W. and H. W. Scott. 1973. Conodont-bearing animals from the Bear Gulch limestone, Montana. Geological Society of America, Special Paper 141.

[6] Noble, G. K. 1931. *The Biology of the Amphibia.* McGraw-Hill, New York.

[7] Smith, Hobart. 1969. The Mexican axolotl: Some misconceptions and problems. *BioScience* 19:593–597.

[8] Smith, Homer. 1953. *From Fish to Philosopher.* Little, Brown, and Co., Boston. A delightful, easy to read, speculative book by one of the leading proponents of a fresh water origin of vertebrates.

[9] Stahl, B. J. 1974. *Vertebrate History: Problems in Evolution.* McGraw-Hill, New York. A thorough evolution of the vertebrates, highly recommended to students in search of an advanced and recent overview of the subject.

[10] Young, J. Z. 1962. *The Life of Vertebrates.* Second edition. Oxford University Press. A widely known work that provides numerous examples of the natural history of vertebrates.

3

Diversity and Classification of Vertebrates

Synopsis: Classification is important both in the theory and in the practice of biology. It is based on the binomial nomenclature and hierarchical categories (taxa) of Linnaeus, and its aim is to organize and to explain the vast diversity of species that have evolved on earth. Since Darwin, classification has been based on phylogenetic interpretations. Similarities and differences among species are used to determine degrees of relationship. Only those similarities that are homologous can be used to indicate relationship. Difficulties of interpretation arise in the case of close convergences and parallelism, as well as when homologs become transformed for different functions. Classification remains to some extent a compromise between a practical and orderly representation and a true reflection of phylogenetic relationships.

As bisexual organisms, vertebrates are naturally grouped into interbreeding populations that are reproductively and genetically isolated from each other. Each such reproductively isolated population is, by definition, a species. Species are the units of biological organization that specialize for particular modes of existence and that shift adaptations through successive generations in response to environmental change. For these reasons, the species remains a key unit of organization in studies of evolution, ecology, behavior, physiology, and population biology.

Variation and isolation, primarily geographic, are prerequisities for the formation of new species. The completeness of the process of speciation is tested in zones of secondary contact after isolation; speciation is the process by which most evolutionary change takes place.

3.1 INTRODUCTION: GOALS OF THIS CHAPTER

Today, few students receive instruction about classification or have an opportunity to learn why it is important in theory as well as in practice.

This chapter has three goals: To make some general points about classification and its uses in biology, to present some specific applications to the vertebrates as an aid to understanding and organizing the content of the following chapters, and to describe the nature of "species" in the Vertebrata.

3.2 LINNAEAN SYSTEM OF CLASSIFICATION

Our present-day system of classifying organisms traces back to methods that were worked out by the naturalists of the seventeenth and eighteenth centuries and came to be especially associated with the name of Carl von Linné, a Swedish naturalist, better known by his Latin pen name, Linnaeus. The Linnaean system is based on **binomial nomenclature** for species and on their arrangement into hierarchical categories or higher **taxa** for classification.

3.2.1 Binomial Nomenclature

The scientific naming of species became standardized through the publication of Linnaeus's monumental work, *Systema Naturae*, which went through many editions between 1735 and 1758. He attempted to give an identifying name to every known species of plant and animal. His method involves assigning a double name to each species: a generic name, which is a Latin noun, Latinized Greek, or a Latinized vernacular word, and a species name, usually a Latin adjective, or similar derivative. Some familiar examples follow: *Homo sapiens* for human beings (*homo* = man, *sapiens* = wise), *Passer domesticus* for the house sparrow (*passer* = sparrow, *domesticus* = belonging to the house), *Ursus horribilis* for the grizzly bear (*ursus* = bear, *horribilis* = dreadful), and *Tyrannosaurus rex* for an extinct carnivorous dinosaur (*tryannosaurus* = tryant lizard, *rex* = king).

Why use Latin words? Aside from the historical fact that Latin was the early universal language of European scholars and scientists, it has provided a uniform usage that has continued to be recognized worldwide by scientists regardless of their vernacular languages. The same species may have many different colloquial names in the same language. For example, *Felis concolor* ("the uniformly colored cat") is known in various parts of North America as the cougar, the puma, the mountain lion, the American panther, the painter, and the catamount. It has other common names in Spanish, French, German, and other modern languages, but all trained mammalogists regardless of nationality recognize the name *Felis concolor* and understand which species of cat is being discussed. On the other hand, the same common name is often applied to entirely different species. For example, the name "gopher" refers to a turtle in Florida; in the midwest it is applied to ground squirrels (genus *Spermophilus*), whereas in the far West it refers to an entirely different kind of fossorial rodent, the pocket gopher (genus *Thomomys*). The English name "robin" was originally applied to a small common European bird with a red breast, *Erithacus rubecula*.

Everywhere the English colonized around the world they found some local bird that reminded them of the familiar songster at home and to which they applied the name "robin." The American robin, *Turdus migratorius*, except for a reddish breast, is not at all like the English robin and, in fact, is more closely related to the European blackbird (*Turdus merula*). (The name "blackbird" is another that has been applied to a variety of different, often unrelated, species which happen to be black.)

In scientific nomenclature, no two species can have the same combination of generic and specific names. The use of these binomials under a set of international rules has led to a universally understood system for naming species. Certain conventions are adhered to in writing species names. The generic name is always capitalized, but in zoology the species name is not; the whole name is italicized in print or underlined to indicate italics. Note also that the generic name implies a grouping of species. All the species with the generic name *Passer* are rather similar sorts of sparrows.

3.2.2 Hierarchical Groups; the Higher Taxa

The European naturalists of the sixteenth and seventeenth centuries were bewildered by the increasing array of species that were being recorded through voyages of discovery to the New World and other previously unknown places. The museums began filling up with a variety of specimens that taxed human verbal description. The embarrassing difficulties presented by this accumulation of material are revealed by an English naturalist, Thomas Moufet, writing about 1590 in his book, *Theatre of Insects.* His description of grasshoppers and locusts runs in part as follows:

> Some are green, some black, some blue. Some fly with one pair of wings, others with more; those that have no wings they leap, those that cannot either fly or leap, they walk; some have longer shanks, some shorter. Some there are that sing, others are silent. And as there are many kinds of them in nature, so their names were almost infinite, which through the neglect of Naturalists are grown out of use. Now all Locusts are either winged or without wings. Of the winged some are more common and ordinary, some more rare; of the common sort, we have seen six kindes all green, and the lesser of many colours . . . .

This passage shows how naturalists before the time of Linnaeus were simply overwhelmed by the wealth of specimens collected from all parts of the world. Today there are more than 350,000 described species of plants and more than a million species of animals. Clearly some practical way had to be found to organize all the information about species and to arrange them in orderly fashion for convenient study and reference.

Linnaeus and other naturalists of his time developed what they called a "natural system" of classification, in which all similar species are grouped together in one **genus**, based on certain shared characters that define the genus.

The most commonly used characters were anatomical, because they are the ones most easily preserved in specimen form. Thus all doglike species—various wolves, coyotes, and jackals—were grouped together in the genus *Canis* because they all share certain common anatomical features, such as an erectile mane on the neck and a skull with a long, prominent saggital crest, on which massive temporal muscles originate for closure of the jaws. In the same way, all genera possessing certain common characters were grouped together in a higher, more inclusive category or taxon, the **Order**, and similar orders were grouped together in a **Class**, and classes in a **Kingdom**.

Although Linnaeus used only these three taxa higher than the genus in his system, the method of arrangement that he worked out forms the basis of the hierarchical classification still in use today. Modern classification employs seven basic taxonomic categories, listed below in decreasing order of inclusiveness:

> Kingdom
> > Phylum
> > > Class
> > > > Order
> > > > > Family
> > > > > > Genus
> > > > > > > Species

Depending on the complexity of a particular assemblage of related species and on the degree of knowledge about them, various intermediate taxa may be added at any particular hierarchical level in the system. These intermediate categories are usually designated by the use of prefixes such as sub, super, and infra. Thus in any particular scheme of classification one may encounter a Subphylum, a Superfamily, an Infraorder, and so forth.

The complete scheme that is now required to express all the relationships among mammals is as follows:

Kingdom
Phylum
Subphylum
Class
Subclass
Infraclass
Cohort
Superorder
Order
Suborder
Infraorder
Superfamily
Family
Subfamily
Tribe
Subtribe
Genus
Subgenus
Species

Sometimes "subspecies" are still given formal descriptions and names too, but this practice has given way to more quantitative methods for expressing the geographic variation within a species.

The names for some taxa have more or less standardized endings that indicate the hierarchical levels of their groups. Family names end in *idae*, subfamilies in *inae*, tribes in *ini*, and among the vertebrates, especially for birds, there is a trend toward the use of *iformes* as the ending for orders. There are no widely adopted endings for any of the other taxonomic categories.

Linnaeus's method of grouping species in classification proved to be theoretically sound as well as practical, because it was based on anatomical, and to some extent on physiological and behavioral, similarities and differences. Unknowingly he used taxonomic characters that we understand today are genetically determined biological traits and that, within limits, express the degree of genetic similarity or difference among groups of organisms. His own interpretation of why the system works was, however, completely wrong.

3.2.3 The Typological Theory of Classification

Linnaeus adhered to the prevailing view of his times about the "fixity of species." He expressed himself variously on this point: "There is no such thing as a new species." "There are just as many species as there were created in the beginning." In short he had no conception of evolutionary processes or of phylogeny based on genetic relatedness.

Instead, he explained the higher taxa as the material or earthly, and somewhat imperfect, expressions of universal or divine "ideas" or "archetypes." This theory of classification is based on the philosophy of Plato and Aristotle, particulary as interpreted by medieval churchmen such as Thomas Aquinas. According to this philosophy, the various kinds of substantial objects that we can perceive by our senses are imperfect material copies of perfect ideas that are the reality underlying the universe. Guided by these notions, the taxonomists studied the material specimens of their science in order to determine the underlying idea or archetype for each species, each genus, each order, and so on. Individual variations were regarded as sports or freaks, bad coinage to be cast aside in the search for the one "type specimen" that best conformed to the ideal of the species. Type specimens are still in use today, but only for the purpose of fixing a particular scientific name to a particular specimen, which serves as the "legal representative" of its species.

3.3 PHYLOGENETIC AND EVOLUTIONARY INTERPRETATION OF CLASSIFICATION BEGUN BY DARWIN

It was precisely the variation among individuals of a species that Charles Darwin fixed upon as one of the basic facts in support of his theory of organic evolution by natural selection, and a fundamental change in the principles of classifying organisms came with the acceptance of evolutionary theory in the late 1800s. Biologists came to realize that species are not separate, unchanging

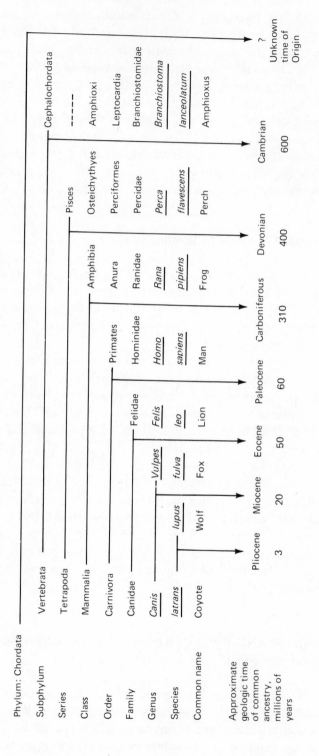

Figure 3–1 Schematic representation of hierarchical classification of some chordates. Phylogenetic and evolutionary implications of the taxonomic categories are indicated by lines that bracket related forms at each taxonomic level. Arrows point to approximate time in geological history when taxa at a given hierarchical level diverged from common ancestry.

creations, fixed by type since the beginning of time, but rather that they have evolved one from another through a long course of history by natural selection acting on the inherited variability of individuals within a species population. More importantly for their impact on classification, the relationships among species were no longer considered to be mere abstractions or reflections of metaphysical archetypes for genera, families, orders, and other higher taxa. Instead, scientists began to see that species are related in the fully material sense that they have descended generation by generation from common ancestors.

Thus, similar species in a genus share common, genetically determined patterns of form and function because they are all derived from a single, common ancestral species. In the same way, all the genera in a family share certain common, familial characters because they are derived from the same ancestral stock. The same principle applies to increasingly remote ancestry right through the higher taxa of classification. Conformance with a series of increasingly broader archetypes is no longer the principle for arranging species in classification. Today they are grouped into higher taxa according to their phylogeny, according to how closely they are related in their ancestry.

3.4 MODERN USES OF CLASSIFICATION

3.4.1 An Example of Using Classification to Express Relationships Among Vertebrates

Figure 3–1 shows schematically how the higher categories in classification reflect the degree of phylogenetic relationship existing among several common vertebrate species and one protochordate. A close study of this simplified version of classification will help to fix in mind the modern, evolutionary interpretation of hierarchical classification.

The coyote and the wolf are two very similar carnivorous species that differ mainly in size, the coyote being a smaller and more delicately built "dog" than the wolf. Actually, some of their most conspicuous differences have to do with nonmorphological traits such as vocalizations, social behavior, hunting tactics, and habitat requirements. Morphologically they are so similar that anatomists have difficulty telling their teeth, skulls, and other bones apart, except on the basis of size. Nonetheless, they are perfectly good species in that individuals of the two populations normally are completely reproductively isolated from each other in nature. The coyote and wolf are only two of several closely similar species—which include various jackals and "wolves"—all of which are classified in the genus *Canis* because their close morphological similarities indicate that they have all derived from a common ancestral species not very remote in time or number of generations. In fact, the fossil record shows that the earliest member of the genus *Canis* occurred in the Pliocene, so we can assume that the common ancestor of the coyote and wolf is about 3 million years old, possibly a little more.

The red fox is another doglike species that shares some common traits with coyotes and wolves (the same kinds and numbers of teeth, other canid cranial characters, digitigrade stance, and so on), and yet it shares more traits in common with other species of foxes than with any species in the genus *Canis*. It is therefore placed in the genus *Vulpes*, which includes all closely similar red fox species and which emphasizes the differences that *Vulpes fulva* has from *Canis* species. The genus *Vulpes* is, however, aligned with the genus *Canis* in the family Canidae in recognition of the several common traits shared by all coyotes, wolves, foxes, and similar doglike forms. The fossil record shows that species possessing the generic characters of *Vulpes* were in existence by late Miocene time, indicating that the common ancestry of *Canis* and *Vulpes* must be around 12 to 20 million years back.

In terms of degree of relationship, these taxonomic arrangements mean that all species included in the genus *Canis* are more closely related to each other than any is to other species of doglike predators, the same being true for all foxes in the genus *Vulpes*. At the same time, all species of foxes and "dogs" are more closely related to each other than to any other species—such as "cats"— and so they are grouped together in the family Canidae along with other genera that share the "canid" set of family characters.

Cats and dogs, although sharing the common characters of the order Carnivora, have fundamental differences in tooth structure, skull shape, and foot structure that were already clearly expressed in fossils of the Eocene, 50 million years ago. These differences are best represented in phylogenetic classification by placing "dogs" and "cats" in separate but related families, the Canidae and Felidae. Similarly, all carnivores and primates share a common mammalian ancestry, and yet they had already diverged from one another as separate phyletic lines in the Paleocene some 60 million years ago. Moving backward in time to still more remote ancestors, one finds the common origin of the phyletic lines leading to amphibians and ray-finned fishes in the Devonian, 400 million years ago. Though only distantly related now, tetrapods and bony fishes still show more similarities in structure (vertebrate characters) and are more closely related to each other than either group is to amphioxus, which is set apart from all vertebrates in the subphylum Cephalochordata. Still, all the species listed in Figure 3–1 are chordates and share a common ancestry, probably back in the Cambrian 600 million years ago. They are more closely related to each other than any is to species in other animal phyla.

3.4.3 Difficulties in Using Similarities and Differences to Determine Phylogenetic Relationships

In order to apply the phylogenetic principle to classification, the similarities that systematists use to group species into higher taxa must be those that are inherited from a common ancestry, that is, they must be **homologous** similarities. Likewise, the differences among species must be ones that reflect a dissimilar or unrelated inheritance. Unfortunately, many biological similarities are not homologous, and some homologs are not necessarily very similar in their definitive form or function.

One class of similarities that are not homologous consists of so-called *analogous* or functional similarities (= analogs), which serve the same general function but which have nevertheless been evolved from entirely unrelated ancestries. If such structures are also similar in appearance they are **homoplastic** as well as analogous. The examples in Figure 3–2 will help to distinguish among homologous, homoplastic, and analogous similarities. The legs and wings of insects bear homoplastic similarities to those of vertebrates. The forelimbs of a praying mantis and a kangaroo rat are analogous as well as homoplastic, as both animals use them to manipulate food. On the other hand, the forelimbs of the kangaroo rat and of the horse are homologous but not analogous, as the horse's forelimb serves primarily the function of cursorial locomotion.

The wings of a house fly are analogous and homoplastic to the wings of flying vertebrates, but the wings of a pterosaur, a bird, and a bat are both analogous and homologous. It is important to understand the distinctions here. All flapping wings have evolved under selective pressures to perform the same function: to maintain the animal in flight. But the wings of insects have evolved in an entirely different morphogenetic way than the wings of vertebrates and are derived from ancestors that possess only a very remote coelomate relationship with chordates. The wings of pterosaurs, birds, and bats are homologous tetrapod forelimbs, as clearly revealed by the details of their bony structure, by their similar modes of development, and by their fossil ancestors, not because they are wings with the same function for flapping flight.

Other morphological similarities are explained by the evolutionary processes of **convergence** and **parallelism**. It is a well known fact that distantly related species with similar modes of life may have superficially very similar morphological structures and body shapes, as the result of the same, highly selective forces acting on different genotypes (convergence). There are a number of clear examples among the classes and subclasses of vertebrates. Figure 3–3 shows the similarity in body shape of a pelagic fish (sword fish), an aquatic reptile (ichthyosaur), and an aquatic mammal (porpoise) The fusiform body shapes have been independently selected as that form which is least resistant to movement through water, and the similarity in structure and position of the median and lateral "fins" is most striking between the fish and reptile, indicating that a similar mode of locomotion by lateral undulations of the trunk and tail evolved in both. On the other hand, the porpoise (and other cetaceans) diverge from this pattern and evolved a swimming movement of trunk and tail in the vertical plane, up and down. Another striking example of convergence is seen in comparing the Australian marsupial mammals with placental orders (Figure 3–4).

In the preceding examples the ancestries of the animals compared are so distant that the convergent character of their resemblances is obvious. However, the principle of the independent evolution of similar form to fit a similar mode

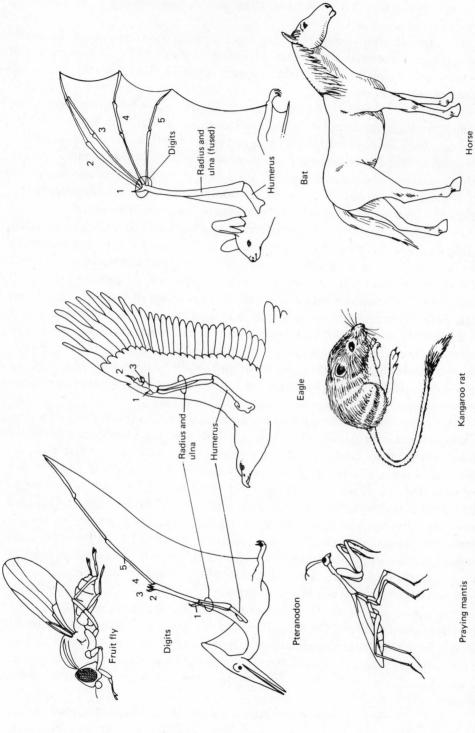

Figure 3-2 Examples of homologous, homoplastic, and analogous similarities (see text for details). (Modified from Simpson *et al.* [8].)

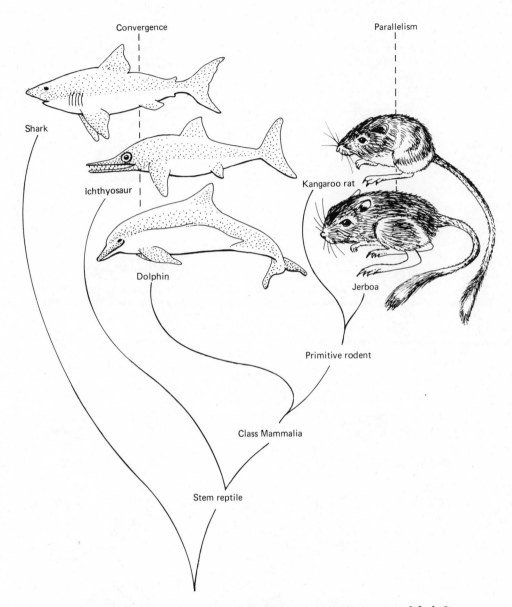

Figure 3-3 Convergence in body form of some vertebrates (modified from M. Hildebrand, *Analysis of Vertebrate Structure*, J. Wiley and Sons 1974).

of life becomes more difficult to determine when the ancestries are not very remote; or when such similarities have evolved in descendant lines from a common ancestor that did not possess the characteristics under consideration, but passed on a genetic potential for the same adaptation in divergent lines. The term parallelism is usually applied to the latter sort of case, but obviously there is no sharp distinction to be made between convergence and parallelism. Bipedal jumping (saltation) has been independently evolved as a mode of locomotion in several groups of rodents. It occurs in jumping mice (Zapodidae), kangaroo

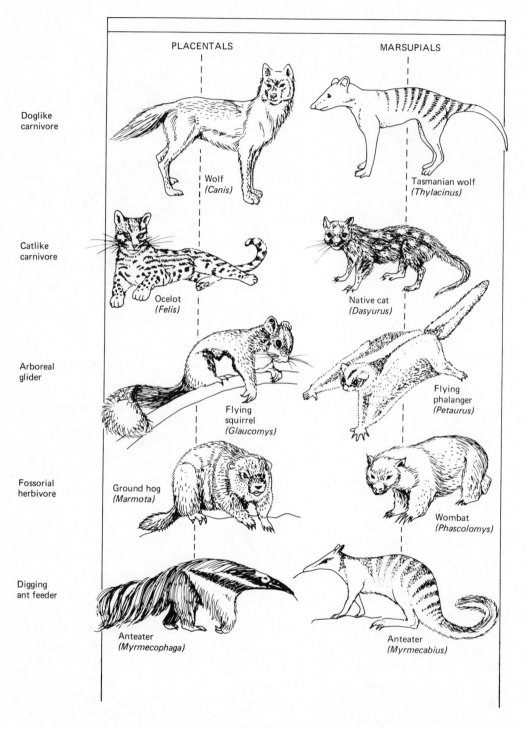

Figure 3–4   Convergences in body form and habits between Australian marsupials and placental mammals (Modified from Simpson *et al.* [8].)

rats and mice (Heteromyidae), jerboas (Dipodidae), gerbils (Gerbillinae), the Cape jumping hare (*Pedetes*), one genus of cricetid rodent on Madagascar, and in several Australian murid rodents. The similarities in structure involve enlargement of the hind limbs, reduction of the forelimbs, inflation of the auditory bullae (capsules housing the middle ear), and frequently also reduction in the number of hind toes and the development of a long tail with a conspicuous terminal tuft. Neck shortening and fusion of cervical vertebrae are other associated modifications (Figure 3–5). The fusion involves the axis (second to anterior vertebra) and elements posterior to it, with coalescence of the centra, neural arches and spines, and the tips of the transverse processes (refer to Figure 1–1). Bipedal saltation is a mode of locomotion that appears to be especially adaptive in open desert habitats, and some kind of ricochetal (jumping) rodent has evolved independently and in parallel in all the major desert areas of the world, except in South America.

Some structures that are true homologs have become so modified during the course of evolution that they are no longer very similar in form or function in the various descendant and related lines. This evolutionary process is called **transformation**, and transformed homologs can be extremely difficult to identify, unless there is a good fossil record of their ancestries or the details of their embryological development can be worked out. Two examples among the vertebrates serve to indicate the kinds of problems involved in detecting these transformations.

Hearing has become an important sense in tetrapod vertebrates, but a different problem in sound reception exists on land than in water. Sound waves in air are relatively weak and ordinarily can have little direct effect in producing vibrations in the fluid (endolymph) of the inner ear, as can happen under water in the case of sound reception by fish. Some sort of sound-amplifying system is necessary for vertebrates living on land. The early amphibians solved this problem by modifying the hyoid gill arch into a new bony structure called the **columella** or **stapes** for the mechanical transmission of sound waves from the external tympanic membrane to the oval window of the internal ear cavity, where the now amplified vibrations are transmitted to the endolymph (Figure 3–6). Earlier in evolution, the hyomandibular element had served mainly as a support for the jaws of fishes, but it could also transmit sound waves. This transformation is clearly evident from comparative embryological studies. Although many modern amphibians have departed from this generalized and primitive method of sound amplification and transmission from the external to the inner ear regions, reptiles and birds all retain its essential feature: a single bony element, the stapes.

Mammals are different. They have three middle-ear ossicles involved in sound transmission. In addition to the stapes, which is the innermost bone with its expanded foot against the oval window, as in other tetrapods, there is the **incus**, a middle element, and the **malleus** with its expanded surface against the eardrum (Figure 3–6). What are the evolutionary origins of the malleus and incus? They are derived from ancestral reptilian jaw bones. During the course

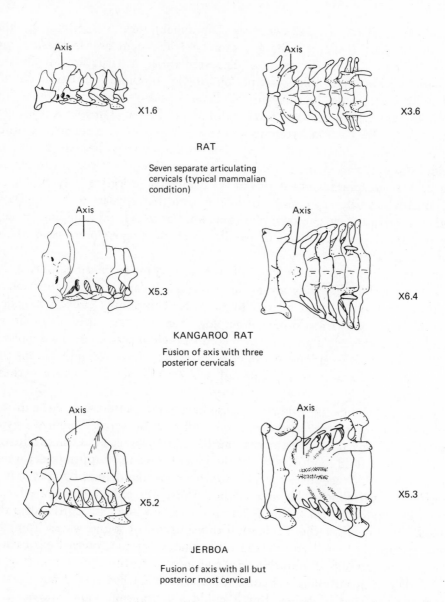

Figure 3-5 Parallel evolution of fused cervical vertebrae in bipedal, saltatorial rodents. Note by comparison with the cervicals of the quadrupedal rat that the jerboa has achieved a more complete fusion than the kangaroo rat. (Modified from R. T. Hatt, 1932, Bull. Amer. Mus. Nat. Hist. 58:599-738.)

of evolution from reptile to mammal, a new method of articulation between the upper and lower jaws gradually developed between the dentary bone of the lower jaw and the squamosal bone of the upper jaw. At the same time the old reptilian articular became transformed into the mammalian malleus, and the quadrate transformed into the incus. Recognition that the malleus is homologous with the articular, and the incus with the quadrate, first came from embryological studies, which showed that the early developmental stages of the

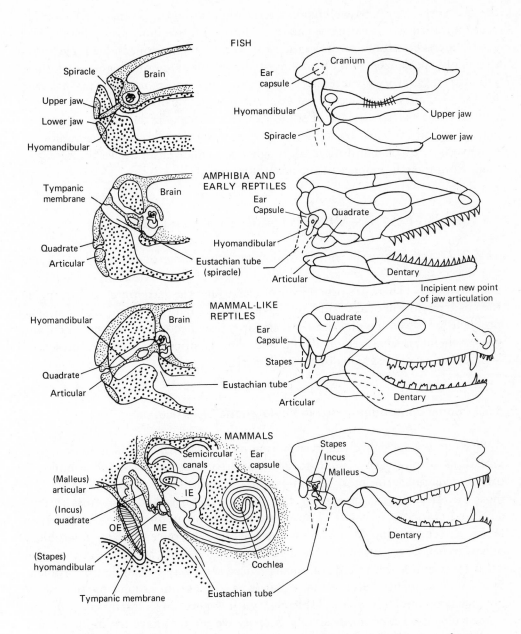

Figure 3-6 Evolution of the sound-transmitting structures in the ears of vertebrates. OE = outer ear; ME = middle ear; IE = inner ear (Modified from Simpson *et al.* [8].)

incus and quadrate are the same, as are those of the malleus and articular. Later paleontological discoveries of fossil mammallike reptiles reveal some of the intermediate evolutionary stages of these transformations.

A more tenuous case exists for homology between the gas bladder of teleost fishes and the lungs of tetrapods. The swim bladder of teleosts has become highly specialized as a hydrostatic organ for adjusting the net specific gravity of the fish with respect to water pressure at a given depth. In other fishes it has been modified as a hearing apparatus. More primitive bony fishes have paired gas bladders that function in a crude form of air respiration, and it is clear that the highly specialized swim bladders of the teleosts are homologous with these primitive respiratory structures. Comparative anatomists can also make a case for the similarity in structure between the primitive paired lungs of early tetrapods and the paired air sacs of some fish, but there is little paleontological or embryological evidence to support the contention that they are true homologs. It remains a reasonable notion, although a few anatomists derive the tetrapod lungs from the gill pouches of the agnathans.

3.5 CLASSIFICATION AND PHYLOGENY OF VERTEBRATES

Table 3–1 and Figure 3-7 on the following pages show one way of classifying the major taxa of vertebrates and give a view of their evolutionary sequence and phylogeny. There are other ways of classifying the vertebrates and arranging the taxa phylogenetically, and we have tried to indicate some of them. The problems involved in phylogenetic classification become apparent when any given scheme is analyzed in detail, and the classification we have presented can be used to discuss some of these difficulties.

3.5.1 Some Conventions Used in Phylogenetic Classification

First, it will be helpful in interpreting classification lists and graphic representations of phylogeny (phylogenetic trees) to know some of the commonly used conventions. In most classification lists, the most primitive, oldest, or most ancestral taxon appears at the top of the list, and the most recent, advanced taxon appears last. Thus, Agnatha appears in our list as the first class under Vertebrata, and Mammalia is the last one. At once there is the problem of how to handle groups that evolved parallel to each other and arose at the same time, for example, birds and mammals. There is no way to handle this problem in a simple sequential arrangement, but it can be gotten around by the use of phylogenetic trees (compare Table 3–1 with Figure 3–7). Mammals are conventionally listed after birds simply because we humans have the intuitive and, perhaps, conceited view that mammals are in some sense more highly evolved than birds.

Another convention often seen in classification lists is the use of dagger or asterisk symbols before the names of taxa that are represented only by fossil forms. †Ostracodermi, †Placodermi, and †Acanthodii are examples among the higher taxa of vertebrates.

A note of caution is also in order about the use of classical names that are no longer valid names for taxa. The International rules of Zoological Nomenclature have established a priority ruling that says that the first scientific name un-

Table 3-1. A Classification of the Major Taxa of the Phylum Chordata

System Followed in this Text	Alternative Systems
Phylum Chordata	Hemichordata previously included (now considered nonchordate)
Subphylum Urochordata (tunicates)	
Subphylum Cephalochordata (lancelets, amphioxus)	
Subphylum Vertebrata	
Class Agnatha	Superclass Pisces versus Tetrapoda; 'Class' Ostracodermi followed by Orders
Order †Osteostraci	
Order †Anaspida	
Order †Heterostraci	
Order †Coelolepida	
Order Cyclostomata (lampreys and hagfishes)	Class Cyclostomata for hagfishes and lampreys or complete separation of hagfishes from lampreys (see Table 5-1)
Suborder Myxinoidea (hagfishes)	
Suborder Petromyzontia (lampreys)	
Class Placodermi	Class Elasmobranchiomorpha containing two or three subclasses for placoderms, elasmobranchs and holocephalians
Order †Arthrodiriformes	
Order †Antiarchiformes	
Class Chondrichthyes	
Subclass Elasmobranchii	Elasmobranchii and Holocephali as separate classes.
Order †Cladoselachiformes (extinct Paleozoic sharks)	
Order †Xenacanthiformes (Paleozoic fresh-water sharks)	
Order Selachii (typical sharks)	Recent classifications divide these into three to five distinct groups of equal ranks (see Table 5-1)
Order Batoidea (skates and rays)	
Subclass Holocephali	
Order Chimaeriformes (chimaeras or ratfishes)	
Class †Acanthodii (extinct fishes of doubtful relationships)	
Class Osteichthyes (higher bony fishes)	
Subclass Actinopterygii (ray-finned fishes)	
Infraclass Chondrostei (primitve ray-finned fishes)	
Infraclass Holostei (dominant ray-finned fishes of Mesozoic)	

Table 3-1. (continued)

System Followed in this Text	Alternative Systems
Infraclass Teleostei (dominant ray-finned fishes of Cenozoic and Recent times)	
Subclass Sarcopterygii	('Crossopterygii,' Choanichthyes)
Order Crossopterygii (ancestors of land vertebrates)	(or Superorder, see Table 6-1)
Order Dipnoi (lungfishes)	(or Superorder, see Table 6-1)
Class Amphibia	
Subclass †Labyrinthodontia (extinct earliest land vertebrates)	
Subclass †Lepospondyli (extinct forms of late Paleozoic)	
Subclass Lissamphibia (modern amphibians)	
Order Anura (frogs and toads)	Salientia is an old name for this order
Order Urodela (salamanders and newts)	Caudata is an old name for this order.
Order Apoda or Gymnophiona (caecilians)	
Class Reptilia	
Subclass Anapsida	
Order †Captorhinomorpha (extinct stem reptiles)	
Order Testudinata (turtles)	
Subclass †Synapsida	
Order †Pelycosauria (primitive mammallike reptiles)	
Order †Therapsida (advanced mammallike forms)	
Subclass Lepidosauria	
Order †Eosuchia (early lepidosaurs)	
Order Squamata (lizards, snakes, amphisbaenians, and the tuatara)	
Subclass Archosauria	
Order †Thecodontia (extinct ancestors of dinosaurs, birds, etc.)	
Order †Pterosauria (extinct flying reptiles)	
Order †Saurischia (dinosaurs with pubis extending anteriorly)	

Table 3-1. (continued)

System Followed in this Text	*Alternative Systems*
Order †Ornithischia (dinosaurs with pubis rotated posteriorly)	
Order Crocodilia (crocodiles and alligators)	
Subclass †Euryapsida (contains several extinct orders of marine reptiles)	
Class Aves	
Subclass †Archaeornithes (extinct Jurassic fossil birds)	
Subclass Neornithes (all other birds including many modern orders)	
Class Mammalia	
Subclass Prototheria	
Infraclass †Eotheria (Triassic and Jurassic fossils)	
Infraclass Ornithodelphia (monotremes)	
Infraclass †Allotheria (multituberculates)	
Subclass Theria	
Infraclass †Trituberculata (Jurassic fossil mammals)	
Infraclass Metatheria (Marsupalia; pouched mammals)	
Infraclass Eurtheria (higher mammals with placenta; includes many modern orders)	

† Indicates taxa represented only by fossil forms.

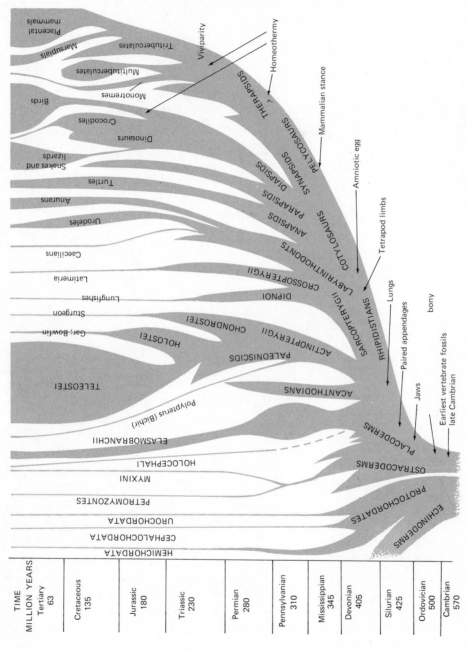

Figure 3-7 Phylogenetic tree of the vertebrates. Width of branches indicates the relative number of recognized genera for a given time level on the vertical axis (time in millions of years indicates beginning of geological periods).

ambiguously applied in print to a given genus and species is its proper and legal name. Unfortunately many species have been given more than one set of scientific names, and in some cases generic and species names long in use for common species have turned out not to be the ones that are acceptable under the International Rules. The lancelet (Cephalochordata) is a good example. The correct scientific name is *Branchiostoma lanceolatum*, the name given to it by Costa in 1834, but for years it was known as *Amphioxus lanceolatus*, the name applied to it by Yarrell in 1836. In the meanwhile, the generic name *Amphioxus* had become such a popular appelation for this species that it has now become vernacularized, and most biologists still refer to *Branchiostoma lanceolatum* as "amphioxus." A similar history of change in use of names can be told for the California newt, now known as *Taricha torosa* but still often referred to by the vernacularized form of its old generic name, "triturus." Students of vertebrates must be aware that such changes have occurred and learn to distinguish between proper scientific names and old Latinized names that have been retained in common or vernacular use.

3.5.2 Some Problems of Representing Relationships in Classification

Many existing classifications are still to some degree "practical" and are based on the degree of similarity existing among the included species whether the similarities reflect common ancestry or not. The Class Agnatha is an example. The known forms can be grouped into two subdivisions, the Monorhina and the Diplorhina, based on whether they have a single specialized nasal aperture or two. It is not known whether these groups represent true **clades** (phyletic lines), which diverged from a common ancestry early in agnathan evolution, or whether the specialized monorhine condition may have evolved independently. Such doubts extend especially to the existing hagfishes (Myxnoidea) and lampreys (Petromyzontia), which are so dissimilar as to suggest convergent evolution of a single nasal opening. The fossil record shows that a perfectly good lamprey existed as far back as the Carboniferous, which means that the common ancestor of hagfishes and lampreys is very old indeed and may have been coeval with some of the armored ostracoderms. The principal classification presented in Table 3–1 is used primarily for its standard reference value to the student at a time when the higher classification of vertebrates is undergoing active revision.

Classifiers try to compromise by striving for a practical classification that is based on monophyletic taxa. A monophyletic taxon is one that includes groups of species all of which are related to a common ancestral group at the next earlier evolutionary level. For example, a group of genera all of which evolved from a common ancestral genus makes up a monophyletic family. Even this rule cannot always be rigidly applied. Breaking up presumed polyphyletic taxa does not lead to a more practical or meaningful system if one does not know where to place the fragments in the system of classification.

The large, flightless birds called "ratites" are an example. These include such species as the ostrich, rheas, emus, cassowaries, kiwis, and the extinct moas of New Zealand and elephant birds of Madagascar. They are frequently grouped together as related orders in the Superorder Palaeognathae, based on certain similarities of palatal structure and other morphological details. Many ornithologists feel that these species are unrelated and only secondarily similar in connection with the increase in body size made possible by the loss of flight; in other words, their similarities are convergent rather than homologous. Placing these ratites into five unrelated orders does not really simplify avian classification, especially since it is not known to which groups of flying birds the flightless forms are most closely related. Consequently they are still grouped together in most bird classifications, and in fact there is some new evidence to indicate that they may be derived from a tinamou-like ancestor and are monophyletic after all.

Another kind of difficulty, not easily resolved by phylogenetic classification, results when similar levels or grades of morphological organization are reached independently by different clades. The so-called polyphyletic origin of the mammals from several lines of advanced mammallike therapsid reptiles is an example. This is a subject to which we shall return later (see Chapters 15 and 18), but for now it is sufficient to point out that all mammals have routinely been grouped together in the Class Mammalia, which implies a common ancestry from one line of therapsid reptiles; yet the fossil record, which is rather rich in forms transitional between reptiles and mammals, indicates that from four to nine different clades of therapsids gave rise to forms with typically mammalian dentition, jaw articulation, and other features that we associate with mammals. Should the Class Mammalia be broken up into several classes? Few taxonomists would opt for such a chaotic solution. Perhaps the more primitive subclasses should be included as taxa within the Therapsida and only the Subclass Theria included in the Mammalia. Another possible solution would be to include all the therapsid reptiles and all mammals in one taxonomic class; but it may be that the therapsids themselves were derived polyphyletically from different groups of synapsid reptiles. Similar difficulties exist in the attempt to construct a phylogenetic classification of the fishes and amphibians.

Another problem for phylogenetic classification arises from unequal rates of evolution in different clades. Consider the reptilian subclasses in relation to the Class Aves. The crocodilians and birds are both derived from diapsid, archosaurian reptiles (Subclass Archosauria), whereas snakes and lizards come from an entirely different line of diapsids (Subclass Lepidosauria), and the turtles derive from the ancestral Subclass Anapsida (see Figure 3–7 and Table 3–1). No logical classification can express the fact that crocodiles and alligators are more closely related to birds than they are to snakes, lizards, and turtles, even though the crocodilians are reptiles.

These are just some of the problems involved in making and using a system of classification. We will confront others as we consider the major groups of vertebrates in the chapters that follow.

3.6 SPECIES AND SPECIATION IN VERTEBRATES

What are species, and what is the biological significance of the fact that organisms can be grouped into discrete populations that we human beings recognize as species? These questions have puzzled naturalists since the time of Aristotle but have been the particular focus of a great deal of biological discussion since Darwin's publication of *The Origin of Species*. We cannot enter upon a detailed review of the historical changes in man's concept of the nature of species, beginning with Plato's doctrine of the *eidos* or archetype and culminating in a modern biological concept of species, or even to review the controversy that followed Darwin over whether species are real units of nature or mere abstractions of human thinking. Instead, we will begin with the proposition upon which most present day biologists, and particularly those who study vertebrates, agree: species have biological reality and the individual organisms included within the limits of a particular species represent something more than the best opinion of a competent taxonomist, to paraphrase Darwin's conclusion.

3.6.1 Species in Nature

Our modern concept of vertebrate species is critically dependent upon a study of animal populations in nature and not just on examining the morphological details of museum study specimens. When a biologist catalogs any local fauna, he finds that it consists of well-defined "kinds" of animals. For example, in the Cayuga Lake basin of central New York State there are 160 distinct kinds of breeding bird populations. Each of these kinds, or species, is separated from all the others by observable "gaps" or differences in morphology, physiology, behavior, and ecology. Intermediates or "hybrids" between any two of these separate populations seldom occur in nature. Among birds not more than about one in 50,000 individuals shows mixed characteristics of two species. Such hybrids are usually less adapted to the environment than normal individuals and seldom reproduce. The members of one species only breed with **conspecifics**—with other members of the same species. They are reproductively isolated from the members of all the other coinhabiting (**sympatric**) species with which they come in contact, and they are also ecologically somewhat different from the members of other sympatric species. Each species has an ecological niche of its own.

The five species of thrushes breeding in eastern North America provide an instructive example of the effectiveness of the reproductive isolation that characterizes species populations in nature (Figure 3–8). They are all rather similar in size and proportions, with drab greyish to brownish plumage; nevertheless they are separated by recognizable gaps, especially in their distinctively different songs and in behavior. Even though the five species are basically similar kinds of birds, no intermediates or hybrids are ever found in the regions where they occur together as breeding birds. They do not interbreed.

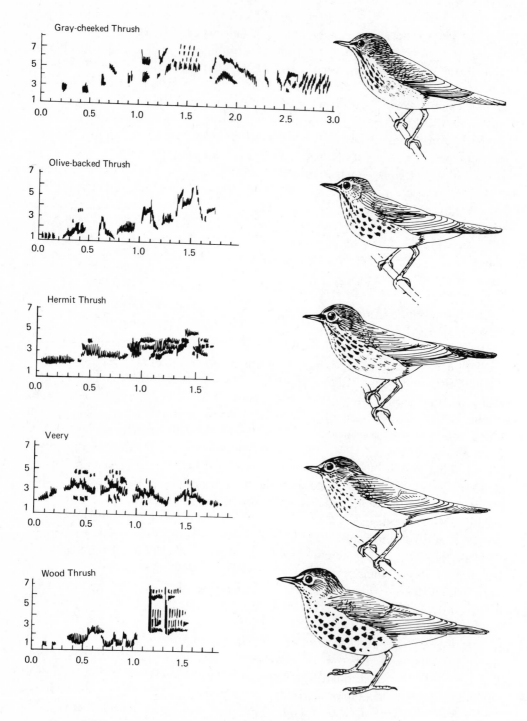

Figure 3-8 Differences among some eastern North American species of thrushes and their songs. (Based on original figures from W. C. Dilger, *Auk* 73:313–353, 1956 and R. C. Stein, *Auk* 73:503–512, 1956.)

In a similar way one can examine any pair of species in a local biota, and, no matter how similar they may be in appearance, close examination will show that they are separated by some kind of gap or difference that keeps them reproductively isolated from each other. This is even true of so-called **sibling species** that may be morphologically so similar that dead specimens cannot be separated one from the other; yet behavioral or physiological differences exist that can be recognized in the living organisms.

The discontinuity that exists between natural populations in local faunas greatly impressed the early naturalists such as Thomas Ray and Linnaeus. This phenomenon, which results from reproductive isolation, remains a cornerstone of the species concept in modern systematics.

3.6.2 Definition and Concept of "Biological Species"

For sexually reproducing organisms, the definition that has been particularly propounded by Ernst Mayr remains the most workable and general one: "Species are groups of actually (or potentially) interbreeding natural populations which are reproductively (genetically) isolated from other such groups." Such a definition works well for the vertebrates, virtually all of which are bisexually reproducing organisms, but students of asexual organisms, and some botanists, obviously find difficulties with it. We avoid further consideration of the exceptions to the definition, since it does apply well to the species included in the Vertebrata.

What is the biological and evolutionary significance of the fact that organisms occur in nature as discrete groups that are reproductively isolated from each other? Reproductive isolation means that each species consists of a closed genetic system of individuals that contribute to and share in a common "gene pool," and have a common epigenetic system of development. The genetic implications of reproductive isolation mean, furthermore, that species are the fundamental units of evolutionary change, because the genes of no other population can influence the evolutionary potential of a closed gene pool. Thus, closed gene pools are what natural selection works on. (Actually natural selection works directly on the phenotypic expression of the individual genotypes making up closed gene pools—the morphological, physiological, behavioral, and ecological manifestations of the genetic species.)

Species, or isolated populations of them, are the units of biological organizations that evolve specialized modes of existence, or ecological niches, and that shift adaptation through successive generations in response to environmental changes. It is for these reasons that the species has remained a central unit of organization in studies of evolution, ecology, behavior, physiology, and population biology, and the central task of the student of speciation is to determine the population structure of organisms and the causal factors involved in the origin of closed genetic systems (species).

Each species, then, is the expression of an independent genetic system having the properties of being reproductively isolated from and ecologically compatible with other sympatric species. Speciation is the process by which populations of organisms acquire these properties. This view of species stresses the fact that a species must not only maintain a closed genetic system but also must be able to survive in the face of competition from other species.

Some of the main contributors to our present biological concept of species have been: Ernst Mayr, a systematist; Sir Julian Huxley, a general biologist; George G. Simpson, a paleontologist; and Th. Dobzhansky, a geneticist. Readers interested in the details of their work and philosophy should consult the references at the end of this chapter.

3.6.3 Intraspecific Variation

No two individuals of a species are exactly alike phenotypically and—except for cases such as identical twins—no two share exactly the same complement of gene alleles. As we have already mentioned, this inherited variability among individuals is one of the cardinal facts that led Darwin to postulate natural selection as the mechanism of evolution; however, not all inherited variations have the same potential to give rise to the formation of new species. Thus, we need briefly to consider the kinds of intraspecific variation that exist and to identify those variations that are likely to lead to speciation under the right conditions.

One sort of variation within a species is ontogenetic variation. Morphology, physiology, and behavior of individual animals may vary strikingly at different developmental stages in the life cycle of a species. This is especially true of those vertebrates that have larval stages and undergo metamorphosis, although it is true in lesser degree of all vertebrate species. Anuran tadpoles are very different organisms from metamorphosed frogs and toads and occupy entirely different ecological niches or adaptive zones. This sort of variation has evolved, in part as one way in which the members of a species avoid intraspecific competition for resources—in this case competition between adult and juvenile individuals. Such variation, which is based on different ontogenetic expressions of the same genotype, is exactly the sort proposed in Chapter 2 as the basis for the origin of the vertebrates. Nevertheless, it appears that relatively few taxa have evolved in this manner, but evolutionary processes like paedogenesis may account for the major discontinuities that exist among animal phyla.

Seasonal variations in morphology or behavior are also frequent among vertebrate species. Many birds have distinct summer and winter plumages that result from complete or partial molt of their feathers (Figure 3-9). Most mammals undergo similar but less striking seasonal changes in pelage, although the color changes from dark summer fur to white winter fur are most obvious in some northern mammals such as the ermine, snowshoe hare, and arctic fox. These sorts of variation represent adaptation of the individual for different

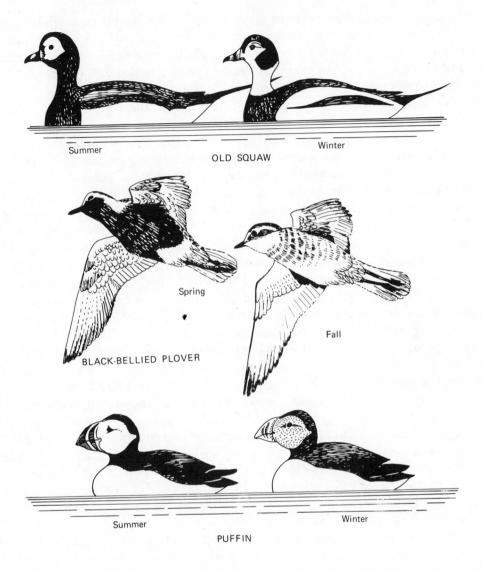

Figure 3-9   Seasonal differences in the plumage of some birds.

environmental problems occurring annually in the life cycle. As developmental expressions of one genotype, they are not the sort of variation likely to give rise to new species.

Still another mode of intraspecific variation is called **polymorphism.** Not infrequently the individuals of a species, regardless of sex or age, occur in two or more distinct forms or **morphs,** which are based on genetic differences and not on different ontogenetic expressions of the same genotype. The morphs that are most easily detected in vertebrates, and thus are the ones that have been most thoroughly studied, involve disjunctive differences in color or pattern of the epidermal structures. Many species of birds have "color phases," over one hundred of which were originally described as separate species. Some examples are the blue goose–snow goose complex, the Florida great white heron (a white phase of the great blue heron), and the white goshawk of Australia (a morph of the more common gray form) There are also many cases in other classes of vertebrates, for example, the striped and ringed variants of the king snake (*Lampropeltis getulus*) in southern California.

The adaptive value of polymorphic variants of the sort described above remains obscure in most cases. Correlations between morph frequencies in populations and climatic or other ecological factors have seldom been demonstrated, even though morph-ratio changes do occur along geographical transects in many species. Although the phenotypic expressions of polymorphism may not be directly adaptive in many cases at a given time and point in space, the adaptive and evolutionary significance of the underlying genetic polymorphism may relate to the more general problem of storing and protecting genetic variation. A species' gene pool must allow for adaptive shifts to occur quickly and with minimum loss of population in the face of climatic or other environmental changes that demand some adjustment if the species is to continue to survive. There is some indication that phenomena such as "lemming cycles"— cyclical changes in population from high to low densities—may be associated with temporal shifts back and forth between the frequencies of polymorphic variants, one of which is better adapted for individual survival at high density, the others for survival and reproduction at low density.

Sexual dimorphism is a special kind of polymorphism that is widespread among vertebrates. Males and females often differ not only in their primary and accessory sexual organs but also in so-called "secondary sexual characters" that may be only indirectly related to sexual processes as such. Because sexual dimorphism results from selection acting differentially on males and females within the same species or closed gene pool, a good deal of evolutionary and adaptational change can occur among the members of one interbreeding population of bisexual organisms, particularly in cases where the population exists in the absence of other competitive species.

Most sexual dimorphism has evolved in connection with various aspects of mating. Familiar examples are the male adornments so frequently evolved as structures that function in courtship and pairing (Figure 3–10), or the larger body size of males in species that have evolved a polygynous mating system (Figure 3–11). Darwin coined the term "sexual selection" to designate the evolutionary process by which sexual dimorphism comes about as the result of competition, usually among males, for mates. Like natural selection, sexual selection is only a special case of individual selection, and it may not always be possible to distinguish between the two in actual cases of the evolution of sexually dimorphic characters.

In quite a few vertebrate species, the individuals of a population may exploit the resources available to them in slightly different ways so that, as individuals come to occupy different subniches or adaptive zones, adaptive radiation occurs within the species population. Such a process can result in expansion of the total niche or adaptive zone of a species, especially when the subniches are associated with the sexes.

Figure 3-10 Male sexual adornments in vertebrates.

Sexual differences in the size and shape of the beaks and tongues of some birds are associated with different foraging habits of males and females (Figure 3-12). An extreme example is the extinct wattlebird of New Zealand. The male had a somewhat woodpecker like bill used to chip away at bark and rotten wood to get at wood-boring insects, whereas the female had a much longer, decurved bill adapted to probing into holes, which often were first exposed by the foraging behavior of the male. Among reptiles, the female Barbour's map turtle (*Graptemys barbouri*) has a broad head with greatly enlarged alveolar

(a) Sage Grouse

(b) Hooded Seal

Figure 3-11   Sexual size dimorphism in polygynous vertebrates.

surfaces, whereas the male has a small, narrow head with normally pro-
portioned alveolar surfaces. The females feed on mollusks, but the males
presumably do not.

The difference in body size between small male and large female birds of
prey—especially exaggerated in bird-killing hawks and falcons—is associated with
a difference in the sizes and kinds of prey killed by males and females (Figure
3-13). Many biologists now think that this kind of sexual dimorphism—like
the difference in primary feeding structures mentioned previously—represents

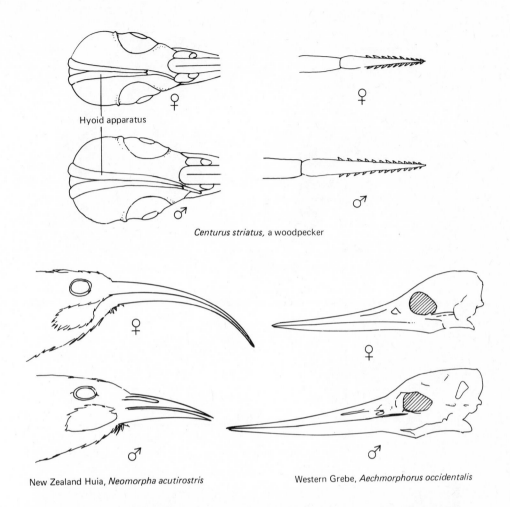

Hyoid apparatus

Centurus striatus, a woodpecker

New Zealand Huia, *Neomorpha acutirostris*   Western Grebe, *Aechmorphorus occidentalis*

Figure 3–12 Sexual differences in beaks and tongues of some birds. These differences correlate with average differences in kinds and sizes of prey consumed.

adaptive selection alleviating intraspecific competition between males and females of the same population. Such differences have been termed **ecological polymorphism**, but in the case of sexually dimorphic characters it is usually impossible to determine with high probability whether the differences between male and female have resulted from selective forces associated with reproductive biology (sexual selection) or with niche-relationships (natural selection). There is probably some interplay among several selective forces acting simultaneously to produce differences between the sexes.

Frequently superimposed on all the forms of variation we have so far considered is a spatial variable, **geographic variation**. As this mode of individual and populational variation is central to the mechanism of speciation, it requires special consideration.

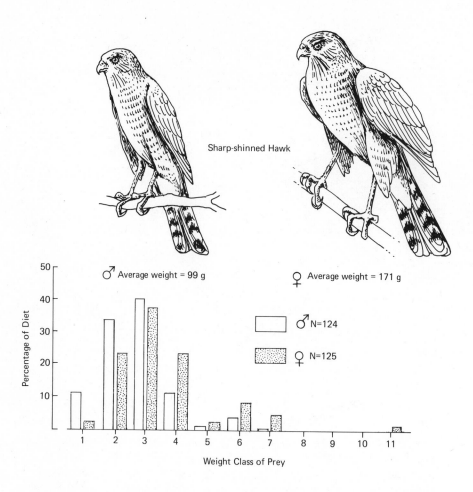

Figure 3-13 Sexual size dimorphism in the bird-eating sharp-shinned hawk in relation to prey captured. Weight class 1 = 3–8 g birds, class 2 = 8–16 g, 3 = 16–27 g, 4 = 27–43 g, 5 = 43–64 g, 6 = 64–91 g, 7 = 91–125 g, 8 = 125–166 g, 9 = 166–216 g, 10 = 216–275 g, and 11 = 275–343 g.

3.6.4 Geographic Variation

Except for island endemics and a few other kinds of locally restricted species that have small total populations, most vertebrate species occupy rather large and diverse total ranges on the order of tens to hundreds of thousands of square kilometers and consist of tens of thousands to hundreds of millions of individuals. A few highly successful vertebrate species, such as humans, rats, and herring, attain populations numbering into the billions or even trillions of individuals, and some are so widely distributed that their ranges encompass many millions of square kilometers. Within a single genus of birds, the falcons, one finds extremes ranging from *Falco punctatus*, whose total population of less than 20 individuals is restricted to the island of Mauritius in the Indian Ocean, to the peregrine (*Falco peregrinus*) which, although numbering only in the few thousands of individuals, occurs as a breeding bird on every continent except Antarctica and on many of the oceanic islands of the world.

Wide-ranging species usually show considerable geographic variation in their biological traits and are often referred to as **polytypic species** by taxonomists who like to divide them into "subspecies." A polytypic species is one that taxonomists divide into two or more subspecies, whereas a **monotypic species** is one that is not so divided; that is, its biological traits are so uniform over its entire range that recognizable subspecies cannot be delimited. The peregrine varies considerably in size and color over its wide range and has been divided into as many as 22 subspecies. The osprey, or fish hawk (*Pandion haliaetus*), which is nearly as wide-ranging as the peregrine, is considered to be a monotypic species. It is doubtful, however, that there is such a thing as a monotypic species in the sense that all populations have the same frequencies of genes, except in the case of those species with a very limited range and small population.

Geographic variation in the biological traits of a species may be continuous or discontinuous. Both types of variation can be found among the populations of one species, particularly when some of its populations are geographically isolated whereas others occupy contiguous ranges. The various populations of the song sparrow provide examples of both continuous and discontinuous variations (Figure 3–14). These geographic variations often change in some consistent or directional way—an increase in size from south to north, or from low to high altitude, or an increase in melanization of the epidermal covering from arid to humid environments. Such "directional" trends in geographic variation are called **clines**. Clinal variation in a trait results from two conflicting tendencies: natural selection, which tends to make every population uniquely adapted to its local environment, and gene flow, which tends to make all populations of a species identical. The pattern of geographic variation in one trait, such as size, is not necessarily paralleled by the variation in another, such as color. This lack of concordance in the geographic variation of biological traits within one species is the reason why the "subspecies" concept is so difficult to apply in a consistent and meaningful way and why it is no longer used much by systematists.

It is geographical variation that provides, under the right circumstances, the greatest potential for the formation of new species from an existing closed gene pool. What can cause geographical variations to become so extreme in a population that reproductive isolation, that is, speciation, occurs?

3.6.5 Factors Promoting Speciation

An old proverb says that one swallow does not make a spring. Neither, usually, does a single gene mutation make a new species. According to the micromutational theory of evolution, a population of animals generally must

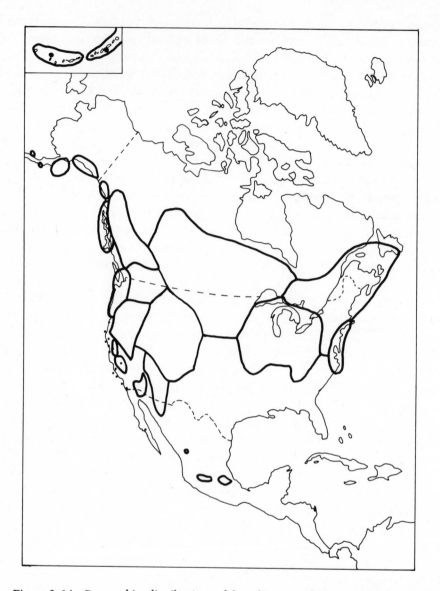

Figure 3-14 Geographic distribution of breeding populations (subspecies) of the song sparrow. Large, dark sparrows occur in the Pacific Northwest; small, pale ones in the southwestern deserts; and other geographic variants in different parts of the total range. Note isolated populations in Mexico. (From A. H. Miller, *Evol.* 10:262–277, 1956.)

accumulate small genetic differences at numerous loci on chromosomes before it can become reproductively isolated and ecologically different from the original population. For this reason, it is extremely difficult, if not impossible, for a freely interbreeding species population to split into daughter species. Some kind of external block to the free exchange of genes must be interposed between segments of the population. In other words, some kind of spatial or **geographic isolation** is generally required before **reproductive isolation** can evolve between segments of a species population. Conceivably there might be a temporal separation between two segments of a population that come to breed at different times of the year in the same region, but there are no unambiguous instances of such a situation having arisen among species of vertebrates.

The way geographic isolation promotes speciation is particularly well seen on clusters of oceanic islands well separated from continental land masses, as Darwin himself was the first to observe in his examination of the endemic forms of animal life in the Galapagos Islands, which lie about 1000 km off the coast of Equador (Figure 3–15). Darwin's finches (Subfamily Geospizinae) have become the classic example of evolution at the species level (Figure 3–16). These 13 or 14 species apparently have evolved from a single, ancestral population of seed-eating, ground-dwelling finches that arrived from South America and became established on one or more of the Galapagos Islands in relatively recent geological time.

What conditions made it possible for a single, small, founding population of finches to diversify into as many as 13 or 14 species on these islands? Although we know nothing about the actual mechanism of dispersal from the mainland, the founding population, a minimum of one male and one female, had to arrive

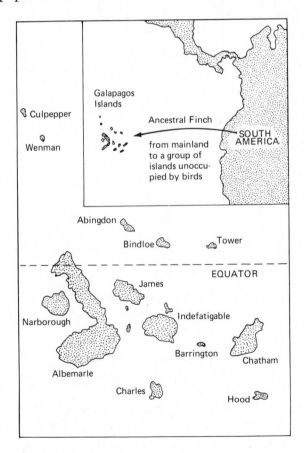

Figure 3–15 The Galapagos Island archipelago.

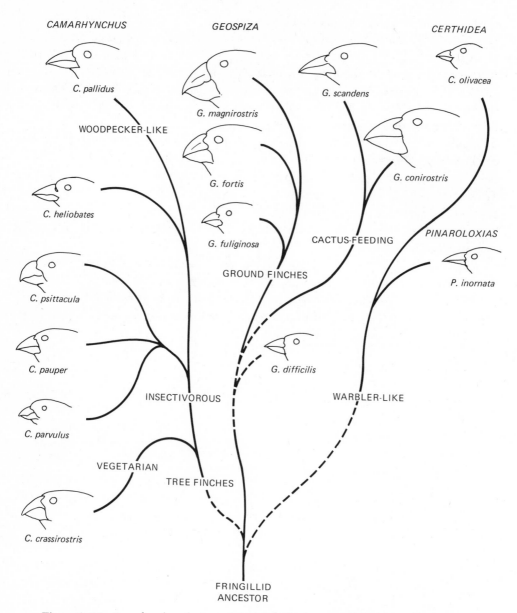

Figure 3-16 Postulated evolution of Darwin's finches. (Modified from Simpson, *et al.* [8].)

in the Galapagos. Presumably a small group of birds could have been blown to the islands by a storm from the mainland, or they might have rafted to the Galapagos on some floating debris and vegetation, which washed away from the mainland shore. Once in the Galapagos the birds had to have an ability to survive and reproduce under the new environmental conditions, and finally it would be important that there be no close, ecologically competing species already well established in the ecosystem. Given that the latter two conditions must have existed at the time the first finches arrived in the Galapagos, natural selection would steadily improve the overall adaptation of the finches to the new environment over several generations. In time, an entirely new kind of finch adapted to the local conditions would result.

From time to time individuals spread to other islands in the Galapagos from the original, founding species, and there, in these new environments, with ecological opportunities for which no competition from other species existed, the new founders were able to repeat the process of establishment and eventual adaptation to local conditions. In time many of these isolated populations accumulated sufficient genetic differences, so that, when secondary contact was established among them by reinvasion or multiple invasion from one island to another, they proved to be reproductively isolated from one another and could coexist as sympatric species, each with an ecological niche of its own. Today each island has as many as three to ten species, depending on the diversity of the vegetation and ecological opportunities.

For land-based animals such as finches, oceanic islands are obvious geographic features for the isolation and speciation of populations. The ocean is a potent barrier against easy dispersal from a continental source, but it is not an absolute barrier, so that given sufficient time a few successful colonists are bound to arrive and to become established. Habitats on mountain tops are isolated from each other in much the same way as islands are, and it is easy to see how a surrounding lowland forest or desert may serve as an effective barrier against frequent dispersal of animals that are adapted to live only in the highland situation.

The rapid evolution of numerous related species like Darwin's finches, each of which represents a different adaptive type, is a recurrent process in the evolution of living organisms. It has been documented in all classes of vertebrates from fishes to mammals. Usually the basis for the rapid evolution of such groups of species is the opening up of new environmental opportunities, each of which requires a different set of adaptations for successful exploitation. This phenomenon is called **adaptive radiation**.

Until recently it has not been so easy to explain how speciation has occurred on continents, particularly in regions where environmental conditions remain uniform over vast areas, such as in the tundra and boreal forests of the Northern Hemisphere, or the Amazonian forest of South America. It now appears that a mechanism similar to isolation on islands has been at work repeatedly and that much speciation of land vertebrates on continents can be accounted for by geographic isolation.

The Pleistocene period—approximately the last 2 million years of geological history—has been characterized by repeated climatic fluctuations, which have been particularly expressed as changes in mean annual temperature and rainfall (see Chapter 21). One result of these climatic alternations has been a repeating cycle of glacial and interglacial periods varying from about 40,000 to 70,000 years in duration. During maximum glaciations much of the Northern Hemisphere and some mountainous parts of the Southern Hemisphere have been covered by ice sheets, whereas during interglacials, such as the one we now enjoy, the ice retreats, and the boreal and arctic-alpine regions of the world support life. Less obvious but very important vegetational and biotic changes associated with the overall climate have also occurred in the middle and equatorial latitudes of the world.

Both geological and biogeographic evidence now support the conclusion that islandlike refugia of the tundra and taiga ecosystems persisted in northern latitudes even during maximum glaciations (Figure 3–17), while during interglacials these refugia have become rejoined as continuous biomes of great expanse. Thus, species populations that were widely distributed in these uniform biomes during interglacials became fragmented and isolated in refugia during

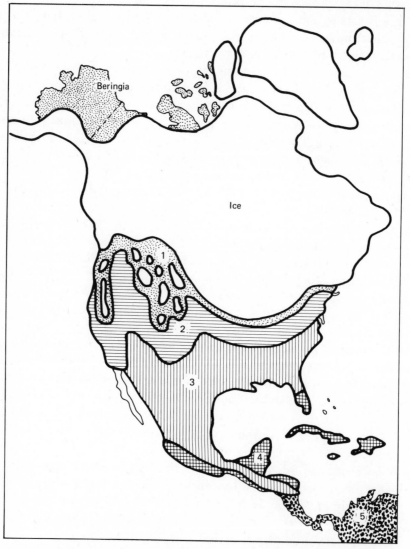

Figure 3–17 Northern biomes during Pleistocene glaciations. Biome types are: 1) tundra, 2) taiga, 3) temperate forests, 4) sub-tropical forests, and 5) tropical forests. (Modified from R. M. Mengel, *The Living Bird* 3:9–43, 1964.)

glaciations. This is precisely the sequence of events needed for speciation, and the present distributions of a number of bird and mammal species in North America indicate that they have evolved in this way (Figure 3–18).

Similar events have also been postulated for environments outside the regions of active glaciation, for example, in the forest biomes south of the glaciated region of North America. Evidently biomes south of the ice have not remained stable in position or species composition either but have fluctuated with the climatic cycle and have been fractured and reformed many times during the Pleistocene. As many as 10 to 17 of the 46 species of wood warblers (Parulidae) occurring in North America can be explained as the products of isolation in forest refugia created during the Pleistocene (Figure 3–19).

Evidently the changes in temperature and precipitation during the Pleistocene have been great enough to have effected changes in most biomes of the world, including those in the tropics (see Chapter 21). The Amazonian lowlands now comprise the largest continuous tropical rain forest in the world, and it is especially rich in vertebrate species, particularly birds. The Amazonian forest has apparently been broken up repeatedly into refugia that survived during periods of lower than average rainfall in the tropics, only to coalesce into a

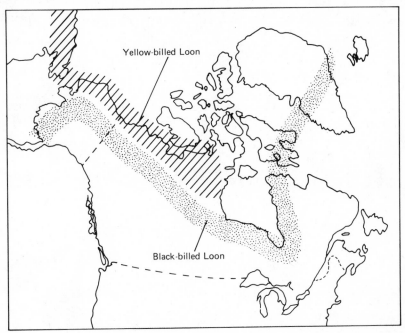

Figure 3-18 Distribution of the black-billed (common) loon and yellow-billed loon in North America. Northern breeding limits of the former species and the southern and eastern breeding limits of the latter species are indicated. (Modified from A. L. Rand, *Evolution* 2:314-321, 1948.) Presumably the yellow-billed form was restricted to "Beringia" during the maximum of the last glaciation, whereas the black-billed form was isolated in refugia to the south of the glaciers. They have subsequently expanded their ranges to meet but not overlap in north-western Canada.

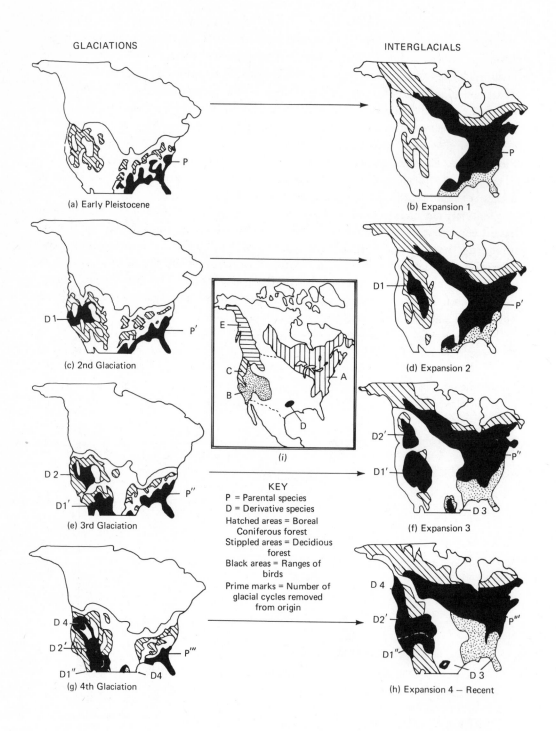

Figure 3-19 Model of speciation in wood warblers (Parulidae) compared with actual species groups in the genera *Dendroica* and *Vermivora*. (Modified from R. Mengel, *The Living Bird* 3:9–43, 1964.)

continuous biome again during so-called pluvial periods (Figure 21–2). Similar shifts in the distribution of savanna and forest biomes in Australia, also associated with long term climatic changes, account for much recent speciation among birds on that subcontinent (Figure 3–20).

In Africa, during cool, glacial periods the montane forests grew down to an elevation of 500 m and became continuous habitat, while the lowland forest and nonforest habitats became fragmented, forming numerous geographic isolates along the margin of the continent and in low river basins, a condition that existed about 20,000 years ago. During warm interglacials, just the reverse occurs. The montane forest retreated up to 1500 m, forming discontinuous "islands" of isolated highland habitat, while the lowland habitats expanded and coalesced over large regions. This latter condition exists today, and the pattern of modern day bird species distributions in Africa indicates that the rate of speciation has recently been high in the lowland habitats, which support many groups of contiguously allopatric species, whereas there is great uniformity in the species composition of the avifaunas in the isolated patches of montane forest, indicating recent fragmentation of a formerly continuous biome.

Most biogeographers and paleoclimatologists now agree that climate, especially temperature and precipitation, and the major biotic associations of plants and animals, have fluctuated repeatedly on a worldwide scale during the Pleistocene. The result has been the repeated fragmentation of biomes alternating with reformation of more continuous, continental distributions, plus some

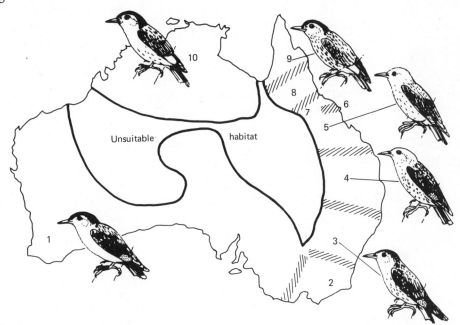

Figure 3-20 Distributions and hybrid zones (hatched) of tree runners (*Neositta* spp) in Australia. Numbers indicate distinct forms. (Modified from A. Keast in E. Mayr [4].)

change in species composition of the biomes. Whereas geographic isolation on continents, as a requisite for speciation, was once difficult to explain, the Pleistocene and earlier cyclic events now provide abundant possibilities. A few years ago ornithologists thought that most bird species dated from the Pliocene and were several million years old. Now most of them feel that speciation has been very rapid during the Pleistocene, especially among small birds with limited dispersal abilities—characteristics equally applicable to many other vertebrates.

There are many kinds of natural barriers that can result in geographic isolation, depending on the type of vertebrate under consideration, and particularly whether it is aquatic or terrestrial. Generally speaking, any area that does not provide for the requirements of a particular species and is more extensive than the usual distances moved by individuals can serve as a distributional barrier. In addition to insular situations, rivers are sometimes barriers for land animals, whereas aquatic vertebrates find no easy dispersal from the river system or lake basin in which they occur. Barriers and factors promoting isolation in oceans are more subtle, and they differ for littoral, pelagic, benthonic, and deep-sea fishes. Pelagic forms encounter few restrictions and tend to have unusually wide distributions.

Although we tend to think of geographic barriers as purely mechanical devices that physically block a species from dispersing, in fact the geographic features of a given environment—its mountains, rivers, deserts, and oceans—affect the populations of different species in different ways. Each species has physiological and behavioral characteristics that cause it to react in its own special way to particular environmental circumstances. One of the most critical of these intrinsic factors in relation to barriers and geographic isolation is the dispersal ability of individuals of the species. The ability of animals to move over long or only over short distances and the method of movement are important. The strength of the tendency to remain at its natal locality is also related to an individual animal's dispersion. Many birds, for example, are strongly **philopatric**, and even though they have wings that could carry them hundreds or thousands of miles across barriers to new lands, the intrinsic tendency to remain near home prevents them from moving far. **Habitat selection** is another important intrinsic factor leading to localization of populations and restriction of species to their particular ecological niches. Physiological tolerances to temperature, humidity, salinity, or other factors of the physical environment also play a role.

3.6.6   Tests for the Completeness of Speciation

The critical stage in the process of speciation when potential genetic, reproductive, and ecological isolating mechanisms are put to the test is when two populations that have been geographically isolated for a period of time come together in a **zone of secondary contact**. Several possible genetic, behavioral, or ecological interactions can occur between populations that establish contact after such a period of isolation, and the outcome reveals the extent to which speciation has resulted.

If the period of isolation has been relatively short and only a few genetic differences have developed between the two populations, individuals of the two groups will freely interbreed and exchange genes in the zone of contact; after several generations the two formerly isolated populations will share much the same frequencies of genes. No distinguishing genetic or biological differences result.

In other cases, a number of genetic differences accumulate and become expressed phenotypically, for example, as differences in color or pattern, but complete reproductive isolation and ecological separation do not occur between the two populations. Hybridization occurs in the zone of contact, but either the hybrids are at a selective disadvantage and disappear without reproducing or, in some instances, they are able to compete successfully with the nonhybrid genotypes in the area of contact and establish more or less permanent zones of hybridization (Figure 3-21). Nevertheless, the hybrids are unable to become established throughout the entire range of the two populations, evidently because they are at a selective disadvantage as they move away from the immediate zone of contact. Such situations, which are rare, suggest that incipient speciation has occurred between the former isolates.

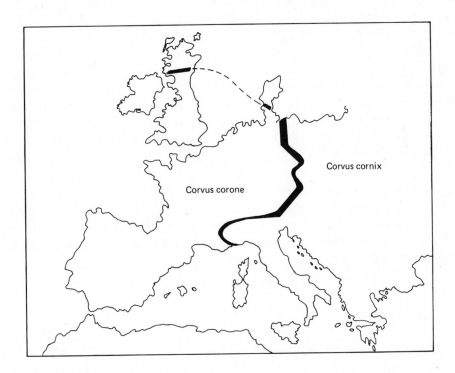

Figure 3-21 Permanent hybrid zones (black) between the carrion crow (*Corvus corone*) and hooded crow (*Corvus cornix*) in Europe. (Modified from E. Mayr, [4].)

There are more examples in which the isolates upon contact prove to be reproductively isolated from each other and therefore no longer exchange genes but have remained essentially ecologically the same, or at least so similar that they do not overlap in range but maintain mutually exclusive adjacent distributions; they are said to be in **contiguous allopatry.** Such populations are often called **allopatric species.** They meet the test of reproductive isolation for full species but not the ecological requirement of niche separation that would allow them to become sympatric species (see Figures 3–18 and 3–20).

Finally, the test of full species status is met when the isolates on contact prove to be reproductively isolated and also ecologically different, so that the two populations come to overlap broadly in their ranges and achieve **sympatry,** without excessive competition with each other for resources.

3.6.7 Species and the Evolution of Higher Taxa

The fundamental role that species play in the evolutionary processes that give rise to the major phylogenetic units (clades) that taxonomists include in the higher taxa is based on several characteristics. (1) Each species represents a different aggregation of genes that control a uniquely adapted epigenetic system. (2) Each species occupies a unique niche, which is the expression of a particular teleonomic solution for the exploitation of the environment. (3) Because it is to some degree polymorphic and polytypic, a species is able to adjust to changes and variations in its environment through time. (4) Through geographic isolation, a species has the potential to break up into separate populations that come to occupy new ecological niches. Occasionally such a shift to a new niche is so adaptive that it allows the species to occupy an entirely novel **adaptive zone,** as was the case with the first lobe-finned fishes that began to spend part of their life on land. Any population that makes such a shift is to some degree an evolutionary pioneer that may become the ancestral founder of a new higher category. One, or at most a few, species of lobe-finned fishes were the direct progenitors of all the tetrapods.

Thus, as Ernst Mayr has said, species are the real units of evolution, and speciation, which is the production of new gene complexes capable of ecological shifts, is the process by which evolutionary change progresses. Without speciation, there would be little diversification of living forms, little adaptive radiation, and hence little evolutionary progress.

References

[1]   Brown, J. L. 1975. *The Evolution of Behavior.* W. W. Norton, New York. A modern penetrating analysis of the evolution of behavior and its molding by environmental circumstances.

[2]   Dobzhansky, T. 1970. *Genetics of the evolutionary process.* Columbia University Press, New York. An eminent geneticist describes how evolution occurs in a classic

book on speciation, with heavy emphasis on the principles of population genetics.

[3]   Huxley, J. S. 1942. *Evolution, the Modern Synthesis.* Allen and Unwin, London. An old, but nevertheless informative work on the theory of evolution.

[4]   Mayr, E. 1963. *Animal Species and Evolution.* Harvard University Press, Cambridge, Mass. A truly classic and widely used reference on the evolutionary process. It is filled with detailed examples about vertebrates.

[5]   Selander, R. K. 1971. Systematics and speciation in birds. In *Avian Biology*, vol. 1, edited by D. S. Farner and J. R. King. Academic Press, New York.

[6]   Simpson, G. G. 1953. *The Major Features of Evolution.* Columbia University Press, New York. This and the following references are the efforts of one of America's leading vertebrate paleontologists. His works are readable, thought-provoking, and highly recommended to all students of vertebrate biology.

[7]   Simpson, G. G. 1961. *Principles of Animal Taxonomy.* Columbia University Press, New York.

[8]   Simpson, G. G., C. S. Pittendrigh, and L. H. Tiffany. 1957. *Life.* Harcourt, Brace Jovanovich, New York.

4

Geology and Ecology During Vertebrate Origins

Synopsis: Here we consider some of the evidence for Continental Drift—the theory that the various continental land masses have drifted across the face of the earth. Continental movement is particularly important to an appreciation of the complexity of vertebrate evolution. Understanding this complexity requires not only a knowledge of the phylogeny of vertebrates (which we consider in detail in succeeding chapters), but also *when, where* and under *what* conditions each particular group originated. During the early Paleozoic, in periods when the first vertebrates, the Ostracoderms, appear and become increasingly abundant in the fossil record, the ancient land masses were coalescing into a giant single continent called Pangaea. Geologists have been able to identify the locations on earth of the ancient marine sediments in which the first Ostracoderms are found. All of these sediments were located near the paleo-equator and deposited in what must have been tropical seas.

Parts of Pangaea, especially the warmer parts, were repeatedly inundated by shallow epicontinental seas. Productivity was probably high, for marine plants (mostly diatoms and simple algae) flourished. Simple fresh-water plants occur in Ordovician deposits, but complex fresh water forms did not become common until the Devonian. It is likely that the Ostracoderms and several types of invertebrates did not penetrate into fresh waters until plants were firmly established and productivity was high. Therefore, the extensive radiation of early fresh water fishes was largely a Devonian event. The climate throughout the early Paleozoic, which from the late Cambrian until the end of the Devonian covers slightly more than 150×10^6 years, was equitable. This extremely long period of fairly constant climatic conditions must have promoted high productivity and favored most forms of life. The vertebrates, then as now, were highly motile animals adept at exploiting a rich and diverse supply of food. The tenets of modern geology teach us that this climatic equitability, which was so important to the evolution and radiation of the early vertebrates, was largely the result of chance—the formation of Pangaea.

99

4.1 EARTH HISTORY, CHANGING HABITATS, AND VERTEBRATE EVOLUTION

To examine the evolution of vertebrates in any meaningful way requires that we also consider when and where specific events occurred. For example, did the first vertebrates originate in the sea or in fresh water? Did amphibians originate in the Silurian or later in the Devonian when the first fossils are found? Answers to these questions and thousands more like them provide the structure from which we outline the chronology and different modes of vertebrate evolution. There are seldom, of course, complete records of evolutionary events in the past, only probabilities that one interpretation is more likely than another. For instance, as we have seen (section 2.4) interpretations of the fossil record and other types of evidence about the place of origin of vertebrates have been very controversial.

Generally, time-dating the stratum in which a fossil occurs can be accomplished accurately. Where uncertainty exists about the time of origin of a particular class, it is more often caused by as yet undiscovered or unrecognized transitional fossils between two groups of vertebrates than by doubt about the age of the strata. If rapid evolution characterized the transition, fossil information may never be available. When ancestral fossils are missing, we must extrapolate backward to approximate the time of origin of a particular group. For the major classes of vertebrates the fossil record is sufficiently complete to provide a firm and consistent chronology of origins.

Determining *where* and under *what* ecological conditions the various vertebrates evolved is, however, another matter and a much more complex problem. How can one explain the enigmatic fact that marsupial mammals are mainly characteristic of Australia and South America? Why were horses and camels absent from North America when the Spaniards first arrived, and yet the fossil record indicates that they probably originated there? There still is a great deal to be learned about where different vertebrates arose—but the initial solutions to these questions are startling.

To evaluate where, and in what kinds of habitats, specific vertebrate groups have evolved we must consider concepts of modern geology, especially the geophysical formulations known as plate tectonics, which deal with the movement of the continents and have led to the current theory of continental drift. That the continents have moved through geological time and gradually assumed their present locations is no longer an heretical idea espoused by a limited number of scientists. It is rather an established fact—an axiom—that most accept. Continental drift can explain a large array of otherwise baffling features of distributions of both living and fossil vertebrates and, furthermore, it is supported by an ever increasing body of fossil and geophysical evidence. How and when the continents move, and how their movements relate to the origin of each class of vertebrates, are considered in this and several subsequent chapters that introduce each major group of vertebrates.

4.2 WHY THE CONTINENTS MOVE

The continents, although a part of the earth's crust, stand above the oceans because they literally float on other elements of earth's crust. Many geologists believe earth consists of: (1) a bilayered central core, (2) a very thick middle mantle, and (3) a thin outer shell called the crust or lithosphere (Figure 4-1). The crust and mantle are solid, although each possesses properties we associate with plastic substances, whereas the core is believed to possess an outer liquid layer and a dense solid center. Let us consider some of the evidence and possible mechanisms that are believed to cause continental drift.

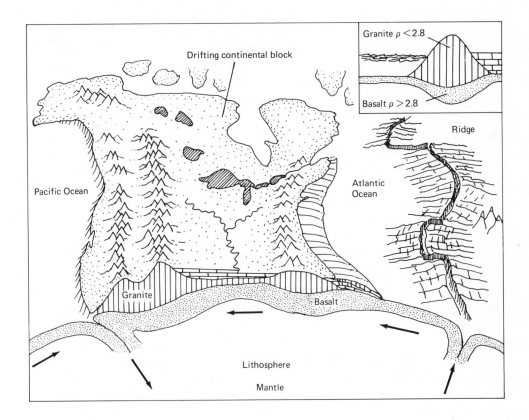

Figure 4-1 Generalized geological structure of the North American continent. The crystalline crustal layers (lithosphere, which includes granite and basalt), are thinner beneath the oceans than the continents. With increased continental thickness (altitude) there is also an increase in the depth of the lithosphere (see insert). Granitic and sedimentary rocks are less dense (2.6–2.7 g cm^{-3}) than the underlying basaltic rocks (2.8–3.0 g cm^{-3}). Thus, the continents float on a basaltic crust and the entire lithosphere floats atop the underlying, noncrystalline plastic mantle. Arrows depict the general movements of the crustal elements and their interactive circulation with the mantle that ultimately causes continental drift.

Why do the continents float above rather than sink beneath the oceans? The answer is a simple one—a basic principle of physics described by the word buoyancy. The average density of the rocks that characterize the continents is less than that of the underlying lithosphere or sea floor. The immense size of the continents, however, tends to obscure any intuitive idea of their buoyant nature. A crystal block of a given mass will sink into the mantle until it displaces its own weight, much like an ice cube in a glass of water. Given time, localized disruptions in the lithosphere, such as those caused by mountain building (orogeny), are eroded to lower lying areas. Erosion of the continental blocks and the subsequent deposition of this eroded material lead to the formation of the sedimentary rocks characterized as sandstones, limestones, shales, and so on. The weight of mountains therefore becomes redistributed as the weight of sediments. We may expect, then, continents to float higher or lower in the underlying denser lithosphere and to have changing shorelines and relations to the hydrosphere as a result of erosion and deposition. This continual adjustment is part of what geologists term isostatic movement.

Since the boundary between mantle and crust is plastic, the probability that the continents might move is very high. The evidence for such movement is convincing. Perhaps the most convincing evidence centers on studies of the sea floor. In the Atlantic, the Pacific, and the Indian Oceans a series of ridges protrude from the sea floor. The mid-Atlantic Ridge, which is actually a large underwater mountain range, is a classic example (Figure 4-1). Basaltic materials flow upward from the mantle to create the ridge. At varied distances from the ridge, subduction, or withdrawal of the sea floor into the earth's mantle, takes place. The subductive zones are represented in many instances by oceanic trenches, such as the very deep Mariana Trench near the Philippine Island arc. Generally, areas of subduction are associated with vulcanism and seismic activity. Subduction produces a horizontal movement of the sea floor from ridge to trench. Since the continents float on top of basaltic plates that are continuous with the sea floor (Figure 4-1), as the sea floor spreads the continents are dragged along.

The actual mechanism driving sea-floor spread, that is, the addition of material at the ridges and its subduction at the trenches, is a matter of considerable speculation. Current interpretations favor a mechanism involving convection currents in the earth's mantle. What is unresolved are the actual source of the energy necessary to drive such a system and the actual location of movements in the deeper layers of the earth. Whatever the mechanism that causes the sea floors to spread, it is subject to change. A change in the rate of sea-floor spreading is reflected in continental geology. For example, at increased rates of sea floor spreading ridge material accumulates. This material displaces sea water and raises the sea level relative to the continents to produce shallow, epicontinental seas. It also increases the rate at which the continents drift. The rate of sea-floor spread, therefore, very probably has profoundly influenced the course of vertebrate evolution.

4.3   EVIDENCE FOR CONTINENTAL DRIFT

To estimate the contours and position of Paleozoic continents is most difficult. Most paleogeologists accept the idea that the six major land masses recognized today were once grouped together as a single supercontinent. This early continent was christened **Pangaea** by the German meteorologist and geologist Alfred Wegener, who as early as 1915 championed its existence. Wegener suggested that this land mass broke in late Paleozoic or early Mesozoic time into discrete sections, the modern continents, which slowly drifted to their present positions. Wegener, therefore, is considered the father of today's theory of continental drift. There is now little argument that Pangaea did exist. In fact, it seems certain that it was formed by the coalescence of yet earlier separate continents and was more than a Paleozoic phenomenon, lasting from the late Devonian to Jurassic time. Several types of data attest to the existence of Pangaea during the Paleozoic, but two lines of evidence—sedimentary geology and paleomagnetism—are particularly convincing.

Areas of extensive sedimentation, termed **geosynclines**, tend to border or form low-lying margins of the continents. Geosynclinal sediments are extensive in Paleozoic strata. Today these formations are represented, for example, by the massive sandstone, shale, and limestone deposits of the eastern United States. These deposits, of course, have been greatly modified by later deformation and are exposed in the Appalachian Mountain chain. Indeed, North America is circumscribed by a series of such Paleozoic geosynclinal deposits: The Cordilleran on the west coast, the Quachita to the south, the Appalachian to the east, and the Franklin and other synclinal complexes to the north. The other major continents today have analogous Paleozoic geosynclinal deposits; in some cases they circumscribe the land mass, as in North America; in others they are present only along one maritime border.

If similar depositional conditions have prevailed in the past, then an appropriate rearrangement of our modern continents into a single land mass should incorporate these ancient Paleozoic geosynclinal sediments in such a way that they lie on the margins of Pangaea. The fitting together of this geological jigsaw puzzle has only recently been accomplished with precision (Figure 4-2). The fit seems most reasonable. In fact, the geosynclincal sediments suggest that Pangaea actually resulted from the coalescence of five separate ancient continents that probably existed from Cambrian through Silurian time (Figures 4-3 and 4-4). Four of these primitive continental blocks were represented by most of North America, by Europe, by Siberia, and by Indochina. The fifth unit consisted of a single mass including South America, Africa, southern India, Madagascar, Antarctica, and Australia—New Guinea—a massive continent referred to as **Gondwanaland**.

The second type of evidence comes from the study of paleomagnetism, that is, the orientation of ferromagnetic minerals in ancient rocks. Inferences about past continental positions rely on the fact that the spatial orientation of magnetic minerals in rocks is fixed in reference to the earth's magnetic field at

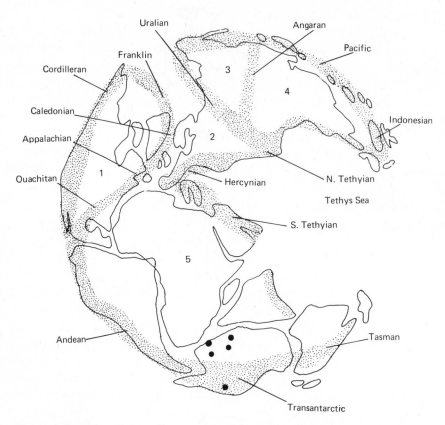

Figure 4-2 Major geosynclines associated with early Paleozoic continents. The continents are arranged according to their probable position in Carboniferous time. Geosynclines today are formed largely along continental margins. The positions of early Paleozoic geosynclines (stippled areas) suggest that five continental blocks existed (numbers), although their respective longitudes are uncertain. When the various continental blocks are arranged as indicated, paleomagnetic data from each continent place the probable Carboniferous position of the South Pole in Antarctica (black circles). There is some uncertainty with respect to India. Recent data place it close to Antarctica and together with a large part of south Asia that includes the Himalayas. This and the following continental position diagrams are based on those of C. K. Seyfert and L. A. Sirkin in *Earth History and Plate Tectonics*, 1973, a convenient single source to which the student is referred for further information.

the time when a particular rock was formed from molten material. Determinations of the declination and inclination (position in three-dimensional space) of magnetic minerals *in situ* can reveal the past orientation of a given formation relative to the north and south magnetic poles. This fixes the rock's paleolatitude. As a result, using these magnetic orientations as a compass, it is possible to approximate the position of ancient continents with respect to the magnetic poles. For example, the South Pole of Pangaea during the Carboniferous was centered in Antarctica (Figure 4-2), as it is today. Paleomagnetic evidence alone is attractive, but other data, especially fossil distributions, also support the historical reality of Pangaea.

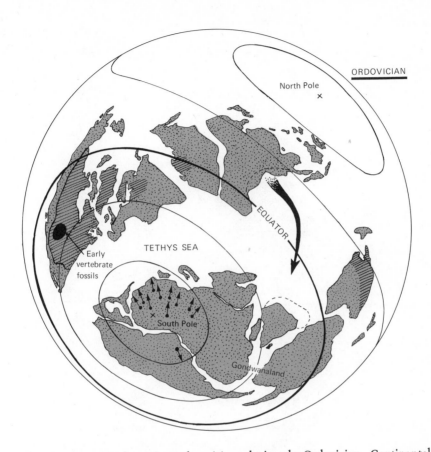

Figure 4-3 Presumed continental positions during the Ordovician. Continental land areas are indicated by stippling; epicontinental seas by cross hatching. Palaeomagnetic fixation places the South Pole in North Africa during Ordovician time (x). Indications of extensive continental glaciation in North African rocks of Ordovician age (scouring, glacial till, and so on; small arrows indicate direction of glacial movement) confirm the palaeomagnetic data. India, very possibly, included most of the Himalayan region and some part of Southern China (areas within dashed lines). For purposes of easy identification the ancient land masses are drawn to include well known present geographic features. In reality, only gross resemblances existed. Thus, the North American continental plate probably existed, but Greenland was a part of this unit; Mexico and the southeastern United States, however, may have been a part of Gondwanaland or nonexistent. Gondwanaland represented a continental conglomerate that existed from at least the Cambrian until the Jurassic. The Taconic Revolution at the close of the Ordovician produced folding and uplift of the Cambrian and earlier Ordovician sediments of the Caledonian geosyncline (see Figure 4-2). This orogeny most likely resulted from plate compressional forces produced as North America coalesced with Europe. In North America shallow seas transgressed the continent on three occasions. Gondwanaland in contrast shows little evidence of shallow epicontinental seas. Early definitive vertebrate fossils are associated with middle Ordovician marine strata of Colorado, Wyoming, the Dakotas and Montana (blackened area). Their early distribution suggests that vertebrates may have arisen in shallow epicontinental seas.

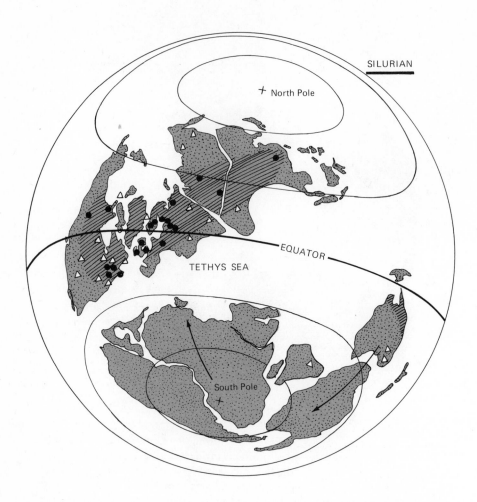

Figure 4-4 Presumed continental positions during the Silurian. Gondwanaland, as based on palaeomagnetic fixation of the South Pole (x) and using Africa as a reference, drifted from south to north and Antarctica and Australia from north to south during Ordovician-Silurian time (compare Figures 4-3 and 4-4; and large arrows Figure 4-4). In the equatorial region and the northern hemisphere, the coalescence of North America-Europe with Siberia and China was virtually complete, although compressional forces from the plate collisions continued to produce orogeny into the Carboniferous. By the close of the Silurian two large land masses existed, Gondwanaland to the south and Laurasia to the north. The name Laurasia is more often used to describe this continental region during the late Mesozoic and early Cenozoic eras. During Silurian time shallow epicontinental seas were prevalent in Laurasia (three marine transgressions are recorded in North America, cross hatching), but as in the Ordovician they were not extensive in Gondwanaland. Reefs dominated by solitary corals (Δ) are widespread. During the Silurian both monorhinid and diplorhinid ostracoderms flourished and were widespread throughout most equatorial seas of Laurasia (•). By the late Silurian they were established in fresh water as well as in the seas.

4.4 CONTINENTAL POSITIONS IN THE EARLY PALEOZOIC

In Carboniferous times vertebrates were well established in the seas, fresh waters, and land areas of Pangaea. If we want to relate continental drift to vertebrate origins, therefore, we must explore earlier geologic periods. Where were the continents located in the Cambrian and, especially, in the Ordovician to Silurian when vertebrates become common in the fossil record? Both the geosynclinal evidence and the paleomagnetic data suggest that Pangaea did not coalesce into a single unit until the Devonian. From paleomagnetic data, however, the positions of the continents can be approximated, and these are indicated for Ordovician and Silurian times in Figures 4–3 and 4–4. Examination of the continental positions shown in these figures emphasizes several features of importance about vertebrate origins:

1. North Africa and northern South America migrated from a south polar position toward the equator. During the Ordovician, extensive glaciation occurred in central Africa and South America, as indicated by scouring of the rocks from glacial movement. This fits precisely with the position of the South Pole during the Ordovician. At the same time, Antarctica and Australia migrated from the paleoequator toward the South Pole. Most of Gondwanaland, therefore, must have drifted across the South Pole (Figures 4–3 and 4–4).

2. North America, Europe, Australia, and to a lesser extent Siberia and Indochina, occupied latitudes close to the paleoequator. As a result, the paleoclimates of North America, Europe, and north Australia must have enjoyed much warmer conditions during most of the late Cambrian, Ordovician, and Silurian; whereas Africa, South America, and Antarctica were probably cool to frigid. Siberia and ancestral China occupied more temperate latitudes in the Northern Hemisphere (Figures 4–3 and 4–4). The presence of extensive limestone, dolomite (another calcareous, biotically produced rock type), and coral formations in North America, Europe, and north Australia is indicative of a tropical location during their deposition. Deposits of these types today are associated mainly with shallow warm seas within 30° of the equator. The presence of limestones, corals, and other tropical marine index fossils in North America, Europe, and Australia further confirms the paleomagnetic fixation of these continents close to the equator.

These findings are of major importance in the examination of vertebrate origins, for they suggest why the earliest vertebrate fossils are associated with late Cambrian to Silurian sediments in North America and Eurasia and a single occurrence in Australia. Fossils of early ostracoderms are now considered to occur mainly in sandstones and limestones deposited in shallow seas. We presume, because of their paleoequatorial location, that these first vertebrates were characteristic of *warm* shallow seas. Their absence in Ordovician–Silurian sediments in Siberia and China must reflect the fact that these seas were too cool (Figure 4–4). Similar reasoning suggests that the shallow seas for all of Gondwanaland, except parts of Australia, were also too cool for the first vertebrates. There are few early Paleozoic marine sediments associated with Gondwanaland except in eastern Australia and western parts of South America (Figure 4–4). Presumably, a higher altitudinal relief of the land prevented the transgression of the sea over Gondwanaland.

Interestingly, only one species of ostracoderm is known from Ordovician–Silurian marine deposits of eastern Australia. Why this is so remains obscure, but possibly it reflects the great open-water distance and isolation of Australia from North America and Eurasia (Figure 4–4). Probably the numerous benthic and shallow-dwelling ostracoderms of Laurasia were not well adapted to crossing the deep Tethys Sea that separated these equatorial land masses (Figure 4–4). Here we encounter one of the first examples of biogeographic barriers to the dispersal of vertebrates. The conclusion we reach is that vertebrates must have evolved first somewhere in North America or western Europe under warm (or tropical) marine conditions that uniquely favored their existence and subsequent radiation.

4.5   THE EARLY HABITAT OF VERTEBRATES

Exactly how well can we describe the habitat of these earliest vertebrates? The subject is controversial, but most paleontologists accept the view that the late Cambrian, Ordovician and many of the Silurian ostracoderms were marine forms. Their association with brachiopods, crinoids, and corals—all marine invertebrates—attests to a marine origin. All are characteristic of shallow warm seas. During the Ordovician, North America was largely covered by a shallow continental sea (Figure 4–3), apparently providing the necessary conditions for evolution of a chordate into a vertebrate. We can say little about the microhabitats concerned, for the precise origin of these ostracoderms is obscure. But the general habitat, at least, of the first fossilized forms was probably benthic. Further, geochemists and geophsicists, such as E. J. Conway and W. W. Rubey, have developed rather strong arguments to suggest that the Paleozoic sea did not differ greatly in ionic composition and salinity from what it is today (Tables 7–1 and 7–2). The early ostracoderms therefore faced physiological problems, with respect to the physical and chemical properties of sea water, that are similar to those faced by modern tunicates, pterobranchs, and echinoderms. The aquatic medium in which they lived must have been highly saline, shallow, warm, and most likely highly transparent. These conditions are matched today by many shallow tropical seas.

Heterotrophic animals like vertebrates, which require preformed organic foods, ultimately depend on plants as a primary source of energy. In addition, many of the important characteristics of the habitats in which vertebrates live are determined by plants. In considering vertebrate origin and evolution, a corresponding knowledge of plant evolution is necessary since it reveals something about the biotic and climatic nature of the habitat. What kinds of plants existed in the early Paleozoic seas and in fresh waters when ostracoderms radiated? Simple, single-celled plants originated in Precambrian seas—blue-green algae are examples. By Ordovician time, however, more complex multicellular chlorophytes (green algae), phaeophytes (brown algae), and rhodophytes (red algae) had evolved. Phytoplankton such as diatoms and dinoflagellates, which are a major source of food in aquatic habitats today, were also abundant in Ordovician seas. Most fossilized green and red algae were lime secretors. Similar forms exist in the tropical seas today. These calcareous forms continued to proliferate in Silurian times, but compressed fossils of non-lime-secreting green, red, and brown algae suggest that noncalcareous multicellular algae were also an important element in these ancient shallow seas.

Similarly complex fresh-water plants of Ordovician and Silurian age are poorly known. It is probable that the invasion of fresh waters by early marine animals was initiated in Silurian times. Quite likely it accompanied the adaptation of marine plants to fresh water or, at least, followed their invasion closely. Not until the early Devonian did multicellular algae become abundant in fresh-water deposits. Many of these are much like the modern genus *Chara* and, indeed, are considered by many botanists to be in the same genus.

In general, then, an abundance of simple plants, including diatoms, blue-green algae, green algae, and other algae, was present from the Ordovician through the Devonian when vertebrates evolved in the sea and radiated in marine and fresh waters. One could reasonably guess that shallow Silurian and Devonian seas appeared as similar areas do today, with a profusion of algae. Perhaps fresh-water ponds and streams in Devonian time may have become choked with *Chara* and similar plants as they so often are today. The point of major importance is that these early habitats contained an abundant flora. Primary production must have been high. An abundant renewable source of energy was present. The biological challenge for vertebrates—as it has always been—was to exploit that resource.

4.6 EARLY PALEOZOIC CLIMATES

The Paleozoic from Cambrian through Silurian time is characterized by very thick limestone deposits along the edges and in the interior of the continents. To account for this it must be assumed that overall land profiles were low—much lower than today. Based on the present northerly position of warm-water fossils, earlier paleoclimatologists concluded that the entire earth must have been warmer, the climate perhaps more uniform. Such broad generalizations we now know are inaccurate, for they assumed fixation of continental geography. In actual fact, the early Cambrian may have been cooler than our present climate since, throughout the world, very extensive glaciation occurred at the close of the Proterozoic era. Continental glaciation, often considered a sign of extensive cooling, is not seen in the geological record again until the Ordovician when the African sector of Gondwanaland moved across the South Pole (Figure 4-4). Such contrasts between the fossil climates and present climates tempted biologists to assume that the entire earth has been either uniformly warm or frigid during a particular geologic period. But throughout the last 500 million years or more, the earth has probably displayed a general polar type of climate (as it does today) wherein the north and south polar regions are cooler than equatorial locations. No doubt paleoclimates showed seasonal changes from yearly procession of the earth about the sun. But slow movements of the continents and especially their latitudinal drift (north or south of the equator) must have dramatically affected long term continental climates. From the Ordovician throughout the remainder of the Paleozoic and on into the Triassic, the movements of the continents positioned much of North America, and to a lesser extent western Europe, close to the paleoequator (Figures 4-3, 4-4, 8-1, 11-1, 11-2, 11-3). As a result, over this long span of time these areas were exposed to a warm and more equitable year round climate, as tropical latitudes are today. Only land masses to the north or south of the tropics would have been subjected to strong seasonal changes in climate, as they also are today.

Continental movements have also modified climates by creating and obliterating entire oceans. For example, the mild climate of Great Britain in the twentieth century results largely from the northward transport of warm equatorial waters by the Gulf Stream. Without the Gulf Stream the British Isles would be a colder and much less habitable place.

Climates are also modified by topographic features, especially the presence of mountain ranges. Dramatic examples are the effect of the Andes of South America in producing the tropical forests and dry deserts of that continent, or the effects of the Himalayas on the seasonal monsoons of Southeast Asia. During the Cambrian and especially the Ordovician mountain-building occurred, perhaps as a result of initial contacts of North America, Europe, Siberia, and ancestral Indochina. But the orogenies were less extensive than those that occurred in later epochs. It is likely, therefore, that through most of the Ordovician and Silurian a very uniform tropical and subtropical climate prevailed in the vicinity of the equator where the first vertebrates evolved.

References

[1]   Hallam, A. (editor). 1973. *Atlas of Palaeobiogeography*. Elsevier, New York. A collection of papers that cover a variety of problems relating to the distribution of fossil organisms on the earth. Especially good for Paleozoic and Mesozoic fossils.

[2]   Matthews, S. W. 1973. This changing earth. *National Geographic Magazine* 5:143. A well-illustrated simple account of continental drift.

[3]   Nicolls, G. D. The geochemical history of the oceans. In *Chemical Oceanography*, vol. 2, Chapter 20, edited by J. P. Riley and G. Skirrow. Academic Press, New York.

[4]   Robertson, J. D. 1957. The habitat of early vertebrates. *Biological Reviews* 32:156–187. An earlier stimulating analysis about the likely conditions in which the first vertebrates might have arisen.

[5]   Seyfert, C. K. and L. A. Sirkin. 1973. *Earth History and Plate Tectonics, an Introduction to Historical Geology.* Harper and Row, New York. Beautifully illustrated and extensive in coverage, this geology textbook has been used as a basis of much of our analysis of where the various continents were located during the evoltuion of the vertebrates.

[6]   Takeuchi, H., S. Uyeda, and H. Kanamori. 1970. *Debate about the Earth.* Freeman and Cooper, San Francisco. A short, simple treatise on the complex data and arguments that underpin the theory of continental drift.

[7]   Valentine, J. W. 1973. *Evolutionary Paleoecology of the Marine Biosphere.* Prentice Hall, Englewood Cliffs, New Jersey. A general treatise that considers problems associated with reconstructing ancient habitats. Especially significant are discussions of how skeletal materials can be used to assess the ecology of extinct organisms.

[8]   Wilson, J. T. (editor). 1972. *Continents Adrift.* W. H. Freeman, San Francisco. A series of articles from *Scientific American* that reviews the developments of the theory of continental movements and plate tectonics.

5

Earliest Vertebrates

Synopsis: Despite earlier hypotheses to the contrary, current evidence indicates the initial evolution of the vertebrates occurred in the marine environment during the Cambrian. The evolution of bone, muscular pump filter feeding, and mechanisms to promote increased motility led to two distinct groups of early vertebrates (Table 5-1). The earliest were the Diplorhinae, characterized by the heterostracans; later the better known Monorhinae, typified by the osteostracans, appeared. Adaptive radiation into many bizarre forms of these fishes culminated in the Devonian. Only two types of survivors from these early agnathan radiations exist today: the hagfishes and the lampreys. Nevertheless, the biology of living agnathans is illustrative of the extreme specialization of which the primitive vertebrate body plan is capable. A new and major innovation in vertebrate evolution was the sudden appearance in the Silurian of fishes with jaws and paired appendages with internal skeletal supports. The first of these fishes, the acanthodians, were soon joined by a diverse array of distinctively different jawed fishes, the placoderms. The remains of these first gnathostomes do not give us definitive clues to the origin of jaws and paired appendages. However, studies of vertebrate embryology yield clear evidence of the development of jaws and their supports from modified anterior skeletal elements of the gills. No similarly definitive evidence for the origin of paired fins has been found. From these early jawed fishes arose an important group of living fishes, the chondrichthyans, which include sharks, skates, rays and chimaeras. The evolutionary history of the chondrichthyes shows periods of relatively rapid changes in feeding and locomotor mechanisms followed by adaptive radiations, subsequent further morphological changes and reradiations culminating in modern elasmobranchs and holocephalians. This pattern serves well as a model of vertebrate history The specializations and unique capabilities of these fishes offer us an opportunity to compare and contrast the evolutionary results of an alternative to the main line of vertebrate evolution.

Table 5.1. A classification of Agnathan, Placoderm and Cartilaginous fishes.

Phylum Chordata
  Subphylum Vertebrata (Craniata)
    Superclass and Class Agnatha
      Subclass Monorhina (Cephalaspidomorphi)
        Order †Osteostraci
          †Anaspida
          Petromyzontiformes (Lampreys)
      Subclass Diplorhina (Pteraspidomorphi)
        Order †Heterostraci
          †Coelolepida
          Myxiniformes (Hagfishes; relationships problematic)
    Superclass Gnathostomata
      Class †Placodermi
        Order †Arthrodiriformes
          †Antiarchiformes
          †Ptyctodontiformes
      Class Chondrichthyes
        Subclass Elasmobranchii
          Order †Cladoselachiformes
            †Xenacanthiformes
            †Hybodontiformes
            Lamniformes (Sharks)
            Squaliformes (Sharks)
            Squatiniformes (Angel sharks)
            Rajiformes (Rays)
        Subclass Holocephali
          Order †Iniopterygiformes
            Chimaeriformes (Chimaeras)
      Class †Acanthodia (uncertain position)

† = Extinct

5.1 THE FIRST EVIDENCE OF VERTEBRATES

As a major taxon, vertebrates are encountered first as fragmented fossils in the late Cambrian and Ordovician. When sectioned and examined microscopically, these unimpressive fragments (Figure 5-1) show an internal structure of considerable complexity. They represent osseous tissue as advanced as that found in vertebrate remains of a later date. This implies that these vertebrates were far advanced and had undergone a considerable but undetermined period of evolution *before* the first fossil evidence was laid down.

Are these 500 million year old fragments more fortuitous finds than imagined? Perhaps conditions correct for fossilization of this type of tissue simply did not exist during the evolution of vertebrates. A moment's reflection allows us to eliminate this hypothesis. As early as the Precambrian, delicate organisms, such as jellyfish medusae, were fossilizing in gently sedimenting environments leaving such excellent impressions that internal anatomy can be deciphered. Contemporaneous with fragments of the first vertebrates are brachiopods of enormous variety. Some of these invertebrates form their shell of hydroxyapatite, the same calcium phosphate salt found in bone. They have fossilized perfectly. Perhaps the lack of early vertebrate fossils is due to the particular habitat in which they lived which was inimical to fossilization.

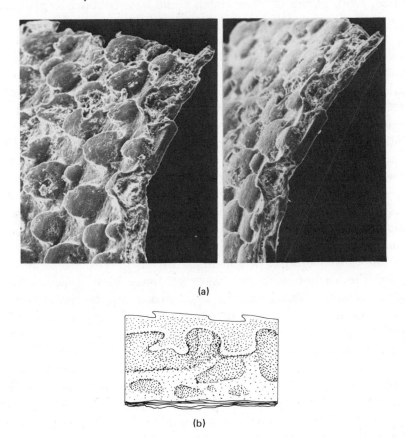

(a)

(b)

Figure 5-1 The earliest known remains of vertebrate origin. Scales of the heterostracan *Anatolepis* from the late Cambrian have definitely been identified from Oklahoma, Washington and Wyoming. By the early Ordovician, *Anatolepis* was widely distributed in North America and to Spitzbergen, and related genera occurred in Australia. Two views of the same specimen show (left) the tonguelike surface ornaments, which probably pointed posteriorly and (right) an edge showing a break through the plate revealing the solid surface layer, the cavity-rich middle layer, and the thick internal lamellar layer (compare with Figure 2–2). Scanning electron micrograph at approximately 220x courtesy of J. E. Repetski, U. S. National Museum.

An entire school of students of vertebrate evolution, led by the late A. S. Romer, has claimed that vertebrates evolved in fresh-water streams and rivers (see Section 2.4). These are areas of erosion and not deposition and hence would not have left a good record. In addition Romer's hypothesis explains the fragmentary nature of the early finds: dead vertebrates were rolled, tumbled, and broken apart while being washed down to their final resting place. Although this hypothesis is attractive, no definitive ostracoderms have been found in fresh-water deposits before the Silurian.

There are difficulties with Romer's hypothesis, however. The known fossil fragments are neither excessively worn nor sorted into large and small size groups, as would be expected if they had been transported downstream. The fragmented, worn nature of these fossils could be attributed to predators and scavengers and the motion induced by waves and tides.

X-ray diffraction of the fragments and their surrounding matrices indicate from the crystalline structure that these specimens were fossilized where they are found today. Spectrographic analysis allows precise determination of the chemical elements incorporated into the fossilized material. Boron is an important element in determining the salinity under which a fossil was produced because its concentration correlates with the salinity of the environment during sedimentation. Additional ecological reconstruction comes from the older technique of stratigraphic analysis, which compares existing fossils in one formation with other and perhaps better understood assemblages elsewhere.

X-ray diffraction, boron analysis, and stratigraphic techniques have led to the conclusion that the earliest fishes died and were deposited in a marine environment, where some critical factor prevented the fauna from becoming very rich. Boron concentrations indicate that this critical factor was variable salinity. The most logical environment to fit all the evidence is an estuarine zone. Here the environment may have at times been hypersaline due to evaporation or more dilute than the adjacent sea due to fresh-water runoff. Even today the stress of coping with such salinity changes is met by relatively few species. Apparently, some early vertebrates were capable of tolerating the stress of variable salinity as are many modern fishes (see Section 7.2).

We have yet to answer the question, "Why are the vertebrates so late in appearing in the fossil record?" The most likely hypothesis (but the least susceptible to verification) is that the earliest part of vertebrate evolution occurred in an environment that prevented the fossilization of soft tissues. It is likely that our earliest ancestors lacked hard skeletal material—certainly they were without bone.

5.1.1   What Were the Earliest Vertebrates Like?

Two early sites have yielded most of the oldest vertebrate fossils, but without microscopic examination the fossils would hardly have been recognized. One site is a small island near the east coast of the Baltic Sea along with adjacent shores as far as Leningrad. The second site is an enormous deposit which includes the Harding Formation extending from Arkansas to Montana. The more widespread Silurian fossil vertebrates are similar to those of the Ordovician but are occasionally articulated and recognizable as fish. The richest finds are in the North American Silurian. The construction of the Interstate Highway System has provided paleontologists with extensive fossil series, but they contain little new information about Ordovician vertebrates. Therefore we shall have to be content with Silurian vertebrates as a starting point for analysis of vertebrate life.

In addition to bone these Silurian organisms possessed another innovation of major importance: the switch from a ciliary mode to a muscular pumping mode of filter-feeding. Filter-feeding has a venerable history among invertebrates (including our deuterostome relatives such as lophophorates, echinoderms, hemichordates, urochordates, and the very vertebratelike cephalochordates; see Section 2.2.1). All these invertebrate forms have one characteristic in common: currents of water are wafted past food-snaring structures by the activity of great numbers of ciliated cells. Because close contact between cilia and water is required, cilia are not effective for moving large volumes of water suddenly, especially from a distance.

Living jawless vertebrates, however, suck water into the pharynx as a result of muscular activity that contracts the oral chamber. The structure of the pharyngeal region of Silurian vertebrates suggests they did the same. This novel method of moving water for feeding as well as respiration is not subject to the same restrictions as ciliary pumping. Water can be moved in rapid powerful pulses and, therefore, considerable suction can be exerted to engulf larger particles even if they are at some distance from the oral opening. Presumably this feeding mode permitted highly active, larger organisms to evolve. Most of the early vertebrate forms were only 20 to 30 cm long, but this represents a considerable size increase compared to the average invertebrate.

Although all early vertebrates have long been collectively called ostracoderms as we have done in Chapter 2, there is increasing evidence that they were not all equally related and at least two distinct groups are recognized (Table 5–1).

Subclass Diplorhina

The earliest jawless filter-feeders are found in the fossil record until near the very end of the Devonian. Because of their curious shelled appearance these particular ostracoderms were named the Heterostraci ("those with a different shell"; Figure 5–2). They were encased anteriorly by bony articulating pieces with distinctive shapes differing between species. In some early groups this shell extended all the way to the anus. Posterior to the anus was a short, probably mobile tail covered by smaller, protruding barblike plates. The head section of the shell had an ornamented dorsal plate, one or more lateral plates, and several large ventral elements. Impressions on the inside of the dorsal plate suggest that the brain had two separate olfactory bulbs. Because it is

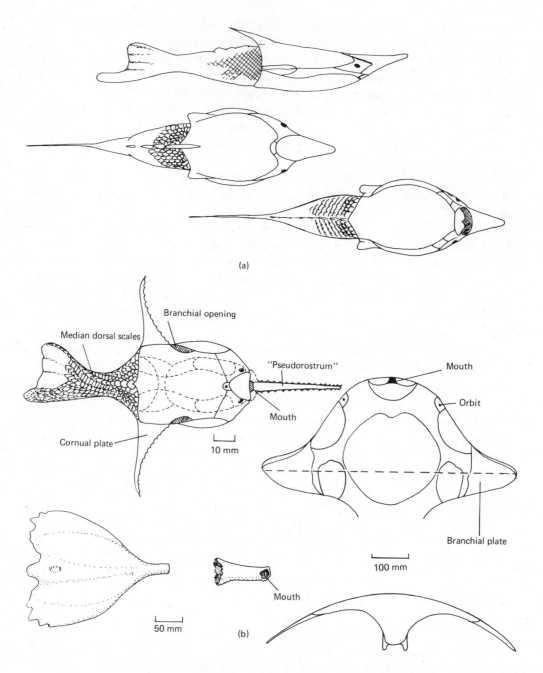

Figure 5-2   (a) Heterostracan *Pteraspis rostrata,* a typical, relatively unspecialized species.   Reconstruction of an 18–cm specimen from the lower Devonian. Lateral view above, dorsal surface left, ventral surface right. (b) Specialized heterostracans (in dorsal view). Clockwise from the upper left: a sawfishlike form *Doryaspis,* lower Devonian; form with enlarged cornu ('horns') and ventral 'runners' (cross-section of head region below), *Pycnosteus,* middle Devonian; an eyeless, tube-snouted form *Eglonaspis,* lower and middle Devonian. (Modified after Moy-Thomas and Miles [5].)

assumed that these were connected with two separate nasal openings, these earliest vertebrates are placed in the subclass Diplorhina ("two nares").

Olfaction was not the only sense the heterostracans used to guide activities in the shallow sea in which they lived. Openings for two eyes are placed laterally, one on each *side* of the head shield. In the middle of the dorsal plate was a small opening for a third, median eye or **pineal organ**. The mouth is near the end of the body (terminal) but opens ventrally in many forms. From this downward pointing opening, the large branchial cavity with its series of loosely articulated and therefore moveable ventral plates expand. It seems very likely that these plates were encased in a fleshy membrane and were protrusible. These plates may have folded tightly against the head to improve hydrodynamics when moving from place to place but protruded to aid in scraping up unconsolidated detritus when feeding. The tail is disproportionately large below its median axis. This lower lobe contains the axial support element, a tail fin construction called **hypocercal**. The body is generally round in cross section and in early heterostracans shows little sign of stabilizing projections. Possibly these fishes were erratic swimmers and resembled tadpoles by swimming with something less than precisely controlled locomotion! While feeding they may have oriented head down and plowed their protruding mouth through the bottom sediments.

As might be guessed, evolutionary trends in the heterostracans led to the improvement of locomotor capabilities. During their later history, the cross section of benthic species flattened ventrally but remained arched or rounded dorsally (Figure 5–2). The head shield developed solid lateral stabilizing projections called **cornu** (horns) just in front of the single pair of gill openings. The head shield shortened and except for a dorsal ridge of plates the bony covering became restricted to the anterior end of the animals. Although specialized edges around the mouth to allow for biting and grasping did not evolve, some of the oral plates developed enlarged toothlike projections that may have been used for scraping.

As they evolved a more hydrodynamically efficient form, the heterostracans also gave rise to species of bizarre appearance. Some (Figure 5–2) developed enormous cornu that may have acted as water-planing surfaces (hydrofoils) to produce lift for the heavy head when swimming. In addition some had two narrow sledlike "runners" on the ventral surface, which at rest presumably held the head above the substrate. Several forms developed dorsally directed mouths and a much reduced skeleton. Perhaps these species fed at the water's surface by filtering large quantities of plankton-rich water. Forms with long toothed rostra projecting from the edge of the mouth in a manner not unlike that of the living sawfish are also known (Figure 5–2). The rostrum's function is difficult to understand because the mouth was superior to the "saw," just opposite to morphologically similar forms among today's more advanced vertebrates. Perhaps it was used to stir up organisms from the bottom.

About the middle of the Silurian, when the diverse array of heterostracans entered the fossil record, another distinct but less perfectly known group of jawless vertebrates appeared. Rather than large plates, these small fish (Figure 5–3) boasted numerous tiny **denticles,** small toothlike structures not unlike that of living sharks. Actually, isolated denticles from the Ordovician may belong to these fishes. No articulated specimens are known before the mid-Silurian, and by the mid—Devonian they are extinct. They were fusiform or torpedo shaped with ridges that correspond to the dorsal and anal fins in more advanced fishes. In addition, broad flanges project from their sides where anterior paired appendages occur in later vertebrates. Like the heterostracans with which they are often associated, they had lateral eyes, a pineal opening, a jaw-less mouth, and a hypocercal tail. Several names have been given to this group of fishes, all referring to characteristics of the scales: Coelolepida ("hollow scale"), Thelodonti ("nipple tooth"), and so on.

The size range of the coelolepids was rather narrow; most seem to have been between 10 and 20 cm long. Their body shape indicates their mode of life. Anteriorly they are dorsoventrally depressed and posteriorly laterally compressed. Living fishes with this shape have a bottom-feeding mode. Coelolepids, therefore, may have fed like their relatives the heterostracans by skimming organic deposits off of the bottom into their small mouth with the aid of muscular suction. Apparently, coelolepids remained in shallow coastal estauries, whereas the heterostracans radiated into fresh water, living in flowing waters and resting on the bottom perhaps in the shelter of boulders. Where they occurred together, the small, lightly armored coelolepids were probably behaviorally very different from the larger, armored, and heavy-bodied heterostracans.

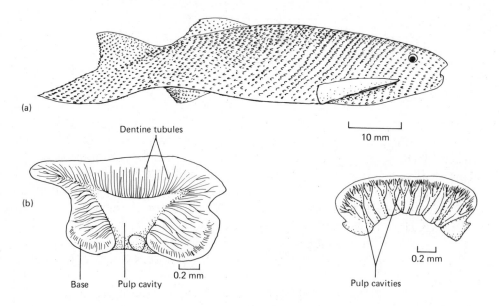

Figure 5–3 (a) A typical coelolepid (thelodont), *Phlebolepis*, upper Silurian. (b) Cross sections of two types of coelolepid scales. (Modified after Moy-Thomas and Miles [5].)

5.1.2 The Appearance of the Monorhina and Vertebrate Radiation

A second major group of vertebrates appeared with the coelolepid diplorhina, survived through the late Devonian, and made a brief impressive stand. Three distinct groups are collected in the subclass Monorhina. They are related by common possession of a single, large, slitlike nasal opening in the center of the head between and in advance of the eyes and just anterior to the smaller pineal foramen. The first group is known as the Osteostraci ("bony shells"), the second as the Anaspida ("shieldless ones"), and the third—if indeed it can properly be related to the other two—the still extant Cyclostomata ("round mouths"), including the Carboniferous *Mayomyzon*.

Like the heterostracans, the osteostracans were heavily armored (Figure 5-4). However, the osteostracan shield was a single, solid element devoid of sutures on its dorsal surface and with a field of small irregular plates on its flattened ventral surface. The heterostracan plates show evidence of periodic growth around their margins. The solid construction of the shield in osteostracans, however, and the lack of incremental growth marks indicate that their head shield did not grow throughout life. Interestingly, for each species so far found, the head shields are all of the same size. The conclusion drawn by some workers is that osteostracans had a naked larval life not unlike that of a lamprey ammocoete (see Section 5.2), and then metamorphosed into a stage where a head shield and other bony armor were deposited without further growth. It is also possible that the different species may be different stages in the life cycle of fewer species. Between the stages it would have been necessary for the dermal armor to be resorbed and the form changed before a new shield was laid down. Also like the early heterostracans, early osteostracans had extensive shields and no paired lateral stabilizers. Later types had short shields, moveable paddlelike extensions of the body in the position of pectoral appendages, and hornlike extensions of the head shield just anterior to these paddles (Figure 5-4).

Unlike heterostracans, osteostracans had a **heterocercal** tail, in which the lobe above the midline of the body was much larger and stiffer than the more flexible lower lobe. Their heterocercal tail may have resulted in a locomotor system that provided considerable lift, which increased their overall mobility (see Chapter 6.5.1 for a discussion of heterocercal tails). The body was armored posterior to the head shield (usually) by two rows of narrow plates, a similar series of belly plates, and a dorsal row of thickened and enlarged plates forming a dorsal "fin." Apparently osteostracans, like many diplorhinids, fed by expanding the pharynx and thus sucking material from the bottom.

The osteostracans are generally referred to as models of early jawless vertebrates even though they appear later in the fossil record than heterostracans. Two features have encouraged their study and made osteostracans better known than other agnathans of the Paleozoic. The first is their single-piece head shield, which resists disintegration better than a series of articulated plates. Secondly, within this shield was a brain case of cartilage with channels for the passage of nerves and blood vessels whose inner surface developed a thin layer of bone.

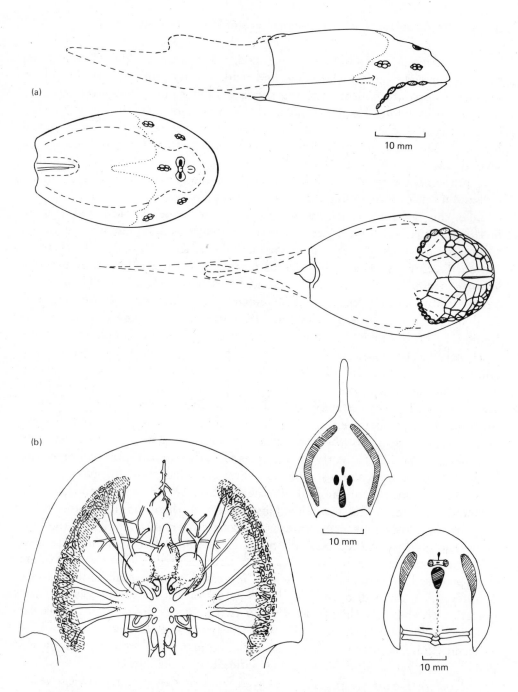

Figure 5-4 (a) Primitive osteostracan, *Tremataspis*, upper Silurian, in lateral (above), dorsal (left) and ventral (right) reconstructions. (b) More specialized osteostracans are: reconstruction of the brain and cranial nerves of *Kiaeraspis*, lower Devonian, showing the detailed information obtainable from impressions left on the inner surface of the head shield; the long rostrum *Boreaspis* of the lower Devonian; and *Tyriaspis* of the upper Silurian. (Modified in part after Moy-Thomas and Miles [5].)

These internal characters have been preserved in sufficient fossils to allow reconstruction of the soft anatomy of the head. Professor Eric Stensiö and his collaborators have patiently polished and ground away layer after layer of the precious fossils, taking photographs of each successive layer until they had serial photographs through the complete head shield. From these they could trace in three dimensions the canals and cavities that in life, had been paved with bone (Figure 5–4). Astonishingly, the internal anatomy of the brain and nervous system of osteostracans 425 million years ago was very similar to that found in the modern lamprey.

Along the dorsolateral edges of the head shield, and sometimes in the center behind the pineal opening, are placed peculiar fields of thin, irregular, and separate small plates (Figures 2–1 and 5–4). These fields form depressions connected to the cranial cavity by four or five huge canals that run through the shield and into the inner ear. Whether these were forerunners of the lateral line, electroreceptors, or even electrogenic organs as the polygonal nature of the small plates has suggested to some, is unknown. Obviously these structures were of some importance, since all osteostracans had them and the canals often had a volume equal to the rest of the nervous system combined. The osteostracans became abundant and diverse during the Devonian even though competing with older groups and surviving in the presence of more advanced jawed vertebrates. In part, their success may have been related to these mysterious, unique adjuncts to their nervous system.

In late Silurian and with less frequency in late Devonian sediments fossils of the second group of monorhinids, the anaspids, have been found. All about 15 cm long, these entirely fresh-water, minnowlike fishes (Figure 5–5) were counterparts to the coeloepids. Like their osteostracan relatives, the anaspids had a single median nasal opening anterior to the pineal foramen. Narrow scale rows (when they were present) invested the body in a manner similar to those along the posterior part of the osteostracans. The head, however, was covered in most species by a complex of small plates or was naked. The anaspids also differed from the osteostracans in having a hypocercal tail. Initially they were considered mid- or surface-water plankton feeders, but their small mouths make this unlikely. Now they are considered to have been bottom detrital feeders who fed in a head-down position reminiscent of that proposed for the heterostracans. Their stabilizing dorsal, anal, and lateral fins, the spines and scutes associated with these "fins," and the compressed shape of their fusiform bodies undoubtedly provided considerable agility and locomotor prowess not known in the heterostracans or osteostracans. Quite likely these fishes were found in areas not accessible to the other fresh-water agnathans. Escape from predators may have been eaiser for anaspids than for the heavirer armored heterostracans and osteostracans. All these forms occurred in fresh water and must have encountered eurypterids, the giant predatory aquatic scorpions of the Paleozoic. It is not hard to imagine a school of minnowlike anaspids fleeing, and catfish-like heterostracans and osteostracans burrowing or hiding in crevices under on the arrival of one of these pincered beasts.

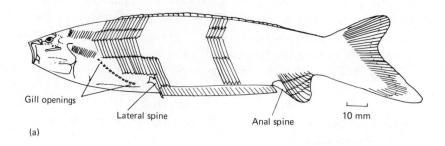

Gill openings

Lateral spine

Anal spine

10 mm

(a)

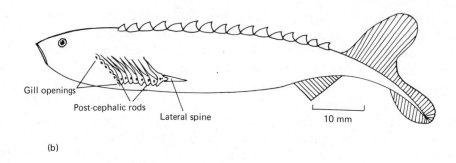

Gill openings

Post-cephalic rods

Lateral spine

10 mm

(b)

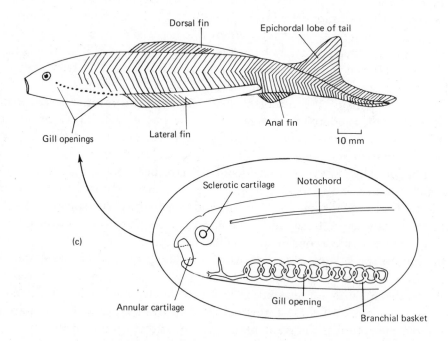

Dorsal fin

Epichordal lobe of tail

Gill openings

Lateral fin

Anal fin

10 mm

(c)

Sclerotic cartilage

Notochord

Annular cartilage

Gill opening

Branchial basket

Figure 5-5 Reconstructions of typical Upper Silurian anaspids. (a) *Pharyngolepis*, (b) *Lasanius*, (c) *Jamoytius* showing (inset) internal structures known from the head region. (Modified after Moy-Thomas and Miles [5].)

During the late Silurian and Devonian all four types of jawless vertebrates coexisted (Figure 5–6). In estuaries they probably exploited similar resources. Fresh waters harbored three of these four extinct agnathan forms at the same time. The species of each type number in the tens to hundreds, each a bit different from its closest relatives in food-getting apparatus, defensive mechanisms, locomotor organs, and a host of nonfossilizable soft parts, physiological characters, and behaviors.

Whatever the pre-heterostracan vertebrate was, the evolution of muscular filter feeding coupled with the energy (feeding) rewards of increased mobility and the protection that dermal bone afforded seem to have triggered a proliferation of variations on the vertebrate theme that spread into the waters of the world. Wherever organic production gave rise to small particulate matter capable of being sucked up and digested, vertebrates invaded. Each source of nutrient must have presented its special problems of procurement and sorting. The open, sunlit waters of bays demanded different adaptations than the detritus beds of turbid estuaries or the organically coated rocks of streams. The basic agnathan body plan was exceptionally adaptable to new modes of existence. The basic primitive forms and the evolutionary trends just described hardly do justice to

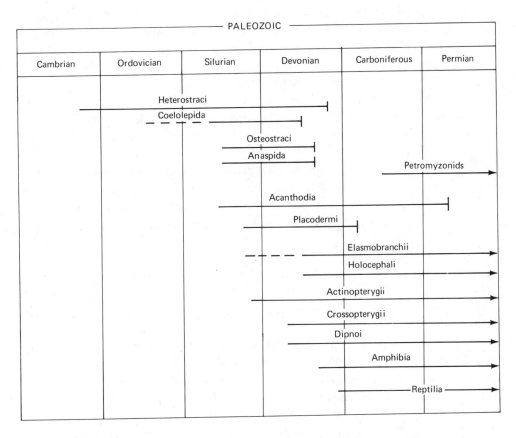

Figure 5-6   Occurrence in the fossil record of the Paleozoic vertebrates.

the emerging picture of these forms. Eyeless, tube-snouted heterostracans, and nearly naked, sucker-mouthed anaspids who left their sucking marks on other organisms shared a bizarre world of jawless forms (Figures 5–2 and 5–5).

Here we observe for the first time a phenomenon repeated over and over again in the record of vertebrate life. By chance, evolution "invents" a basic modification on the vertebrate framework, and a flood of forms using this new modification in conjunction with specialized adaptations explodes onto the scene. From a central generalized adaptation, vertebrate life radiates in scores of different directions to exploit the resources that this new "invention" has made available. These adaptive radiations (Chapter 3.6.5) are documented more easily in recent geological time and we shall examine modern examples in greater detail.

5.2   LIVING AGNATHANS

Two living groups of fishes are jawless and therefore included with the extinct forms in the class Agnatha. Both taxa are distinctive. Few systematists agree on their relationship to each other or to the Silurian and Devonian agnathans. A single fossil (Section 5.1.2) similar to one of the living groups but showing no linkage to older fossil Agnatha was found in Carboniferous sediments. Fossils from any other period are unknown.

Living agnathans are the most primitive and also the most specialized of extant fishes. Their primitive characters harken back to the origin of vertebrates, for they lack jaws and have no paired appendages to aid them in their locomotion. Some, however, are unlike any other fish, living or fossil, in being specialized as obligate ectoparasites of other vertebrates. Because they each possess a round jawless mouth these fishes may be combined in the order Cyclostomata: lampreys in the family Petromyzontidae and hagfish or slime-hags in the family Myxinidae. The two families show great differences in morphology. Some of the differences can be attributed to their long phylogenetic separation, but many reflect their different habits and habitats.

5.2.1   Hagfishes

The hagfishes (Figure 5–7) are entirely marine. Approximately 15 recognized species are placed in six genera with a more or less worldwide distribution. Living primarily on the continental shelves of the world, they also occur at great depths. The largest concentrations seem to occur on the continental shelf off river mouths where sediments are thick, soft, and unconsolidated and waters are cool. Hagfishes are never caught much above the bottom. They live in colonies, each individual in a mud burrow with a volcanolike mound at the entrance. Since burrowing polychaete worms form the most common item in the diet of many species, they probably live a molelike existence, finding their

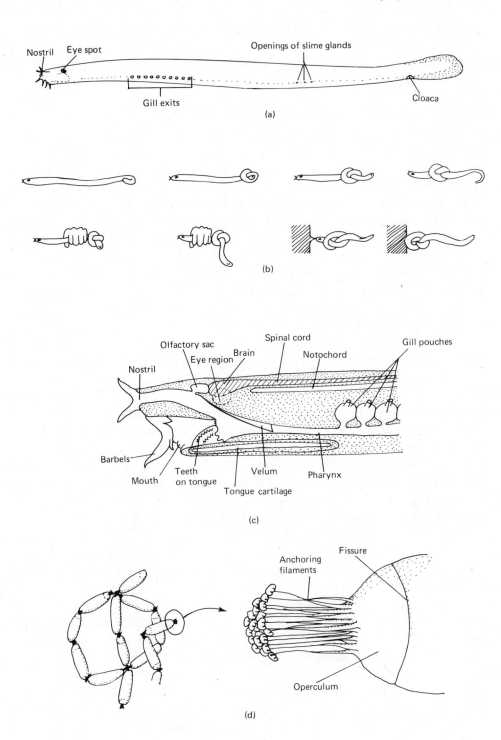

Figure 5-7   Hagfish. (a) Lateral view; (b) knot tying activity; (c) saggital section of the head region; (d) eggs attached to each other (and to the substrate) by filaments with anchorlike terminals. (Jensen, D. Sci. Amer. **214**(2):82–90, Feb. 1966).

prey beneath the mud and ooze. They must be active when out of their burrows, for they are quickly attracted to bait and moribund fishes caught in gill nets. Small morphological differences between populations indicate that hagfish are not wide ranging, but rather tend to live and breed with members of a single colony.

Elongate, scaleless, and pinkish to purple in color, hagfishes have a single terminal nasal opening that connects via a broad tube directly with the pharynx. The eyes are degenerate and covered with a thick skin. The mouth is surrounded by six cartilage-supported tentacles that can be radially erected and swept to and fro by movements of the head when the hagfish is searching for food in the open. Within the mouth two multitoothed, horny plates border the sides of the protusible "tongue." These plates spread apart when protruded and fold together, the "teeth" interdigitating in a pincerlike action when retracted. When feeding on something the size and shape of another fish, hagfishes concentrate their pinching efforts on surface irregularities, such as the gills or the anus, where they can more easily "grasp" the flesh. Once attached they tie a knot in their tail and pass it forward along their body until they can brace themselves against their prey and tear off the flesh in their pinching grasp (Figure 5-7). It is little wonder that hagfishes take only dead or dying vertebrate prey or that they often begin by eating only enough flesh to enter the coelomic cavity where dining on soft parts is possible! Once a food parcel reaches the gut it is enfolded in a mucoid bag secreted by the gut wall. This peritrophic membrane is permeable to digestive enzymes and digestate and is excreted as a neat wrapper around the feces. No functional significance is known for this curious practice.

Different genera and even different species in the same genus of hagfishes have variable numbers of external gill openings. From 1 to 15 openings occur on each side, but they do not correspond to the number of internal gills because the external openings occur as far posterior as midbody and the pouch-like gill chambers are just posterior to the head. The long tubes leading from the gills variously fuse to reduce the number of external openings to which they lead (Figure 5-7). The posterior position of the gill openings is probably an adaptation to burrowing.

The internal anatomy of hagfishes is also peculiar. They have only one semicircular canal on each side of the head, and their kidneys are of a very primitive type. Along the sides of the coelom and opening through the body wall to the outside are large mucus glands (Figure 5-7) that secrete enormous quantities of mucus and tightly coiled proteinaceous threads. The latter straighten on contact with sea water to entrap mucus close to the hagfish's body. This thoroughly obnoxious defense mechanism is so well developed that it is claimed a single hagfish can, in a few seconds, easily fill a two gallon bucket with jelling slime. Not only is this slime disgusting to a would-be molester, but it seems to be less than appreciated by the hagfish. When danger seems past a hagfish agains draws a knot and squeezes out of the mass of mucus, then sneezes sharply to blow its nasal passage free!

In addition to the heart near the gills and in contrast to all other vertebrates, hagfishes have accessory hearts in the caudal region and capacious blood sinuses. Despite this they have very low blood pressure. The blood vascular system seems devoid of the immune reactions characteristic of other vertebrates, and has the same or slightly higher osmotic concentration as the environment, although the specific ion concentrations are not necessarily those of the external medium (see Chapter 7). Through examination of the gonads it has been claimed that at least some species are hermaphroditic, but nothing is known of mating. The eggs (Figure 5-7) are oval and over 1 cm in length. They are encased in a tough clear "shell" that has clusters of hooklets at either end. Secured to the bottom by these hooklets, the yolky eggs develop to yield small completely formed hagfish, thus bypassing any larval stage. Adult hagfish are generally under 1 m in length.

5.2.2 Lampreys

Although superficially similar in size and shape to hagfishes, the nine genera and over 30 species of silver-gray lampreys (Figure 5-8) are in most other respects radically different from hagfish. All lampreys are **anadromous**; that is, they must ascend rivers and streams to breed. Although originally the adults were probably marine, landlocked species are known that spend their adult lives in lakes or streams. Some species are known only from fresh water and have abbreviated adult lives that function as little more than reproductive stages since they neither feed nor migrate downstream to some larger body of water.

Except for the tropics, the lampreys are worldwide in distribution. Several species ring the northern hemisphere, being absent only from the Arctic Sea and Hudson Bay area. Fourteen species are recorded from North America. Two other interrelated groups inhabit the temperate portion of the western Andes and the New Zealand–Tasmania–Southern Australian Region. Anadromous species attain the greatest size, though 1 m in length is about the upper limit. The smallest species are less than one quarter of this size.

Very little is known of the life habits of the adults since they are generally only observed during reproductive activities or when captured along with their host. In the latter case, they are firmly attached to the body of another vertebrate by suction and have rasped a shallow seeping wound through the integument of the host. Held in captivity, lampreys swim sporadically with exaggerated, rather awkward lateral undulations. The single nasal opening is high on the head and continues as a *cul de sac* beneath the brain in close proximity to the pituitary gland. The eyes are large and well developed as is the pineal body, which lies under a poorly pigmented spot just posterior to the nasal opening.

Despite these anatomically well-developed senses no clear picture has emerged of how the lamprey locates or initially attaches to its prey. The round mouth and tiny esophagus are sheltered by and located at the bottom of a large fleshy

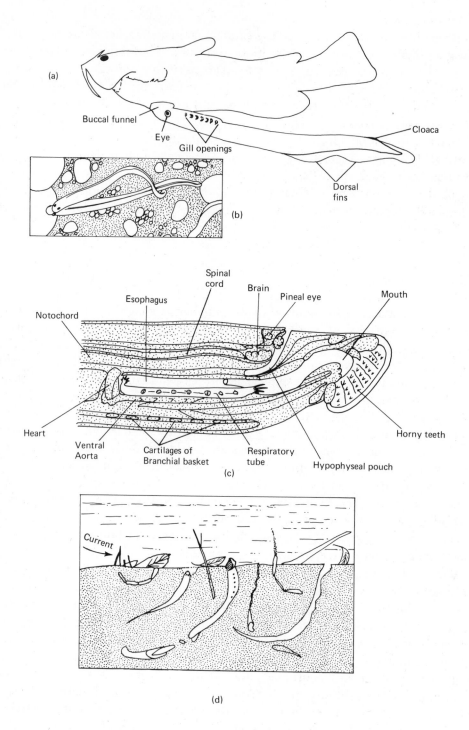

(a)

Buccal funnel

Eye

Gill openings

Cloaca

Dorsal fins

(b)

Spinal cord

Brain

Pineal eye

Mouth

Esophagus

Notochord

Heart

Ventral Aorta

Cartilages of Branchial basket

Respiratory tube

Hypophyseal pouch

Horny teeth

(c)

Current

(d)

Figure 5-8 Lampreys. (a) Adult attached to prey (host) catfish. (b) Spawning behavior of a pair on their nest. (c) Saggital section of the head region. (d) Larval lampreys (ammocoetes) feeding from the openings of their stream bed burrows.

funnel whose inner surface is studded in a species-specific pattern with horny conical teeth. The protrusible "tongue" is studded with similar teeth and together these structures allow for tight attachment and rapid abrading of the host's integument. An oral gland secretes an anticoagulant. When attached, feeding is probably continuous.

Lampreys do not generally feed on their host until its death but detach, leaving a weakened animal with an open wound exposed to the septic environment. At sea, lampreys have been found feeding on several species of whales and porpoises. Swimmers in the Great Lakes, after having been in the water long enough for their skin temperature to drop, have reported initial attempts by lampreys to attach to their bodies. No one seems to have carried the investigation on long enough to determine if the lamprey can successfully feed on humans! The bulk of an adult lamprey's diet consists of body fluids of fishes. The digestive tract is reduced, as is appropriate for such a rich and easily digested diet as blood and other body fluids.

Seven pairs of gills open to the outside just behind the head. Internally two semicircular canals occur on each side: a condition shared with the extinct osteostracan agnathans. Well-developed mesonephric kidneys regulate ions and maintain a particular osmolarity in the body fluids, allowing the lamprey to exist in a variety of salinities (see Chapter 7). Females produce hundreds to thousands of eggs about a millimeter in diameter and devoid of any specialized shell such as that found in the hagfish.

Spawning of lampreys is a complicated process. After temperature-triggered migration to the upper reaches of streams where the current flow is moderate and the stream bed composed of cobbles and gravel, the fasting lampreys construct nests to receive the spawn. For an organism without lateral appendages and lacking jaws, building a nest is not an easy proposition! First males, later joined by females, select a site, attach themselves by their mouths to the largest rocks in the area, and thrash about violently. Smaller rocks are thus dislodged and carried a short distance by the current. If the current does not distribute the gravel to suit the lampreys, they will carry individual pebbles held in their mouths by suction. The nest is complete when a pit is rimmed upstream by large stones and downstream by a mound of smaller stones and sand that acts as an eddy producer. Water in the nest proper is well mixed and oxygenated but turbulent and not flowing strongly in a single direction. The weary nest builders spend the last of their energy, the female attached to one of the upstream rocks and the male wrapped around her, laying eggs and fertilizing them —a process that may take 2 days (Figure 5–8b). At last the female exhausts herself and dies, to be followed in a day or two by the male. The corpses often remain to decompose on streamside trees and bushes when the high waters of the spring spawning season recede.

The larvae hatch in about 2 weeks. They are so radically different from their parents that they were originally described as a distinct genus, *Ammocoetes*. This name has been retained as a vernacular name for the larval form. A week to 10 days after hatching, the tiny 6- to 10–mm ammocoetes leave the nest.

They are pink, wormlike organisms with a large fleshy oral hood and nonfunctional eyes hidden deep beneath the skin. Currents carry the ammocoetes downstream to backwaters and quiet banks where they burrow into the soft mud and take up a sedentary filter-feeding life for 3 to 7 years. The protruding oral hood funnels the muscularly pumped water through the pharynx where food particles are entrapped in mucus and subsequently swallowed. If conditions permit, the ammocoete may spend its entire larval life in the same burrow without any major morphological or behavioral change until it is 10 cm or more in length and several years old. Metamorphosis into the eyed, silver-gray, and usually parasitic stage begins in midsummer, but downstream migration may not occur until the following spring. Adult life is usually no more than 2 years, and many lampreys return to spawn after 1 year in a lake or at sea. Some nonparasitic lamprey species transform and leave their burrows to spawn immediately and die.

During this century man and the lamprey have increasingly come to odds. Although the sea lamprey, *Petromyzon marinus*, seems to have been indigenous to Lake Ontario, it was unknown from the other Great Lakes of North America before 1921. The St. Lawrence River proved no barrier to colonization by lampreys, and the rivers and streams that fed into Lake Ontario provided acceptable conditions for landlocked populations to develop. For some reason until the 1920's lampreys presented little problem, and their populations remained at a low level. It is known that by slowly creeping upward using their sucking mouth, lampreys can negotiate a vertical falls of at least 2 m in height (as long as the flow of water is not strong). The 50-m height of Niagara Falls was, however, too much for the most amorous lamprey. The Welland Canal connecting Lake Erie and Lake Ontario was opened in 1829, but it was not until 100 years later that lampreys were known to be spawning successfully in Lake Erie's drainage basin.

Since the 1920s there has been a rapid expansion of lampreys across the entire Great Lakes basin. The surprising fact is not that they were able to invade the upper Great Lakes but that it took them so long to initiate the invasion. Lake Erie is the most eutrophic and warmest of all the lakes and has the least appropriate feeder streams. Many of these streams run through flat agricultural lands that have been under intensive cultivation since the 1800s. The streams are highly silted and frequently have had their courses changed by the activities of man. Because of the terrain, flow is slow and few rocky or gravel bottoms occur. Perhaps the lampreys simply could not find appropriate spawning sites in Lake Erie to develop a strong population.

Once in the upper end of Lake Erie, they quickly gained access to the other lakes. There they found suitable conditions and were able to expand unchecked until sporting and commercial interests became alarmed at the reduction of economically important fish species, such as the lake trout. Chemical lampricides, electrical barriers at the mouths of spawning streams, and mechanical weirs at similar sites have all been employed to bring the Great Lakes' lamprey populations down to their present level. Man's inadvertent mismanagement (or initial lack of management) of the lamprey has been to our own disadvantage. The virtual disappearance of the lake trout is just one of the many unfortunate experiences that animals have suffered at our hands.

5.3 THE FIRST JAWED VERTEBRATES

In the mid-Silurian, some 425 million years ago and coincidentally with the appearance of the monorhine agnathans, jaws, perhaps the most significant "invention" in vertebrate evolution, were preserved in a record of stone. It may seem strange that a major new morphological adaptation like jaws should arise before the extensive radiation of agnathans occurred *instead* of arising from some later product of that radiation. This pattern of evolution, however, is seen over and over again in an examination of vertebrate life. Two basic phenomena account for adaptive advances occuring at stages that precede the adaptive radiation of seeminly more primitive forms.

The first is the unplanned, designless mechanics of natural selection and evolution. Patterns or direction in vertebrate evolution are concepts derived from our viewpoint, looking back along the path of the past as exposed by a study of the fossil record. But during the life of organisms, adapting or failing to adapt to their immediate environment is the only "plan." For evolving organisms there is no "design," only trial and error. A fossilized sequence that displays orderly progression in our thinking, therefore, may bear little resemblance to actual events in the past.

Second, studies of a wide variety of animals indicate that adaptation often occurs in the presence of competition. Competition must act as a powerful force in order to give selective advantage to the acquisition of new habits and the occupation of new habitats. It is very likely that the advent of jawed vertebrates actually competitively stimulated the diversification of each of the agnathan groups into new environments where their set of morphological attributes proved superior. Some ecological niches were so suited to the agnathan body plan that no evolutionary advance yet realized has been able to replace it. The living lampreys and hagfishes provide undeniable evidence. The great majority of agnathans, however, succumbed to what is generally thought to have been superior competition from jawed vertebrates. Actually, no one has yet been able to determine exactly what elements of superiority eventually overcame the agnathans.

5.3.1 Acanthodians and Placoderms

Frustratingly, jawed fishes appear in the fossil record fully developed, without intermediates. The Silurian fossils consist of detached spines, scales, teeth, and the first examples of jaws. Jaws gave strength and form to the rim of the mouth that was not possible in agnathan mouths. In addition, from the first evidence jaws are solidly braced against the cranium (and thus via the rest of the skeleton connected to the entire body), yet they are mobile and are fitted with a specialized musculature. Some of the advantages of jaws, especially when studded with teeth as they seem to have been, are obvious. Other advantages that are not so obvious may be equally important. They will be discussed in the next section along with the methods used in determining the origin of structures, such as jaws, that appeared without a previous fossily history.

Some of these new fishes are called acanthodians because of the stout spines (Greek: *acantha*) anterior to their well-developed dorsal, anal, and often numerous paired fins (Figure 5–9). Although the earliest fossils are fragmentary and it is not until the Devonian, Carboniferous, and up to their extinction in the Permian that complete fossils are found, the acanthodians were so conservative in form that reconstructions of early forms are probably quite accurate. Usually not more than 20 cm in length (some 2-m species are known), these marine and fresh-water fishes were mostly clad in small, square-crowned scales that grew in size as the animal grew. The head was large and blunt and housed large eyes. The mouth was also large, and the jaws were studded with teeth. The teeth were fused directly to the jaw cartilages or found in replacement series attached to connective tissue bands bound to the jaw cartilages, not unlike those of present day sharks. The head was covered with small scales similar to those of the body. In some, the scales were enlarged to form a cover for the gills. In others, the head scales were lost completely except along the courses of sensory canals that crossed the head. The brain was encased in a well-developed cartilaginous box, which also housed three semicircular canals. There was a vertebral column, evidenced by remains of neural and hemal arches but no centra. The heterocercal tail, the fusiform body, and the arrangement of the fins indicate that acanthodians were good swimmers and probably did not associate with the substrate as hypothesized for most of the agnathans.

This same basic form lasted throughout acanthodian history, but several trends in their evolution are evident. Early forms had robust pectoral spines ventrolaterally behind the gill opening and another large posterior pair of spines in the position of pelvic fins. Between these were numerous other pairs of spines of variable size. Although numerous, these spines were superficial and not embedded deeply in the body. Often the pectoral spines were associated with ventral dermal plates in a supportive girdle. In later species this dermal skeleton along with some areas of scales and intermediate spines were lost, and the remaining spines became deeper set in the body. In addition, there was improvement in the respiratory apparatus and a loss of teeth. Jaw articulation and gill changes accompanying the loss of teeth indicate that edentulous types were surface to midwater plankton feeders who swam in eellike fashion with their enormous mouths open to strain the water for smaller organisms.

If acanthodians partly avoided competition with agnathans by living high in the water column, the second group of jawed fishes to appear certainly did not. Superficially the earliest placoderms, the arthrodires (Table 5–1), resembled heterostracans and osteostracans in appearance and habitat (Figure 5–10). As

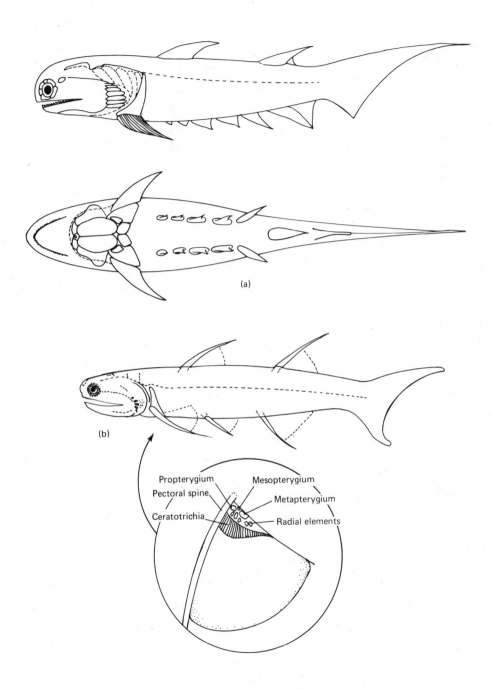

Figure 5-9   Reconstructions of acanthodians. (a) Primitive type: *Climatius,* lower Devonian, with multiple superficially attached spines (lateral view above, ventral view below). (b) More advanced type: *Ischnacanthus,* lower Devonian, with fewer, more deeply embedded spines. (Inset) Detail of pectoral fin and spine of *Acanthodes,* lower Carboniferous.

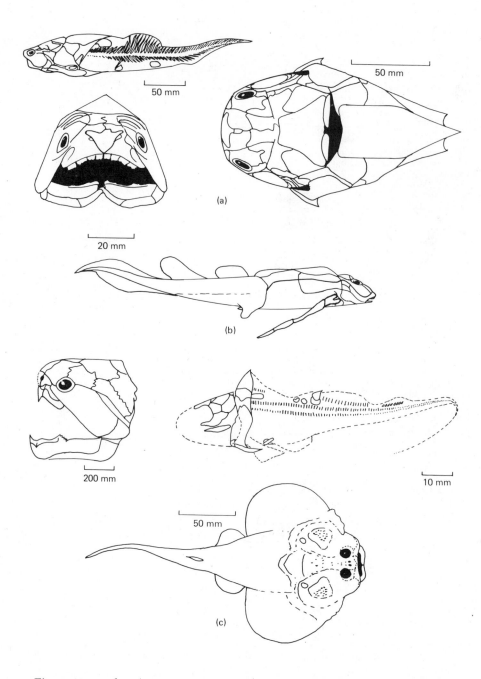

Figure 5-10   Placoderms.   (a) Lateral, frontal and dorsal views of a generalized placoderm, *Coccosteus*, middle Devonian. (b) The peculiar *Bothriolepis*, with a jointed exoskeleton that supported pectoral appendages; (c) Three widely varying types of placoderms (from top to bottom): left, the giant predator *Dunkleosteus*, upper Devonian; right, the chimaeralike *Rhamphodopsis*; and below, the raylike *Gemuendina*. (Modified after Moy-Thomas and Miles [5].)

the name placoderm ("plate skinned") implies, they were covered, at least early in their evolutionary history, with a thick, often ornamented bony shield over the anterior one half to one third of their bodies. Unlike the bottom-dwelling agnathans, from their first record in the late Silurian the placoderms had pectoral appendages functionally assisted by spinelike plates on the head shield. The pectoral fin divergently evolved in various placoderms. By their extinction in the early Carboniferous, some had muscular, mobile structures that must have contributed to their active predatory existence. Other placoderms, the antiarchs, developed the pectoral appendages into stiff props by encasing all soft tissue in jointed, bony tubes reminiscent of arthropod appendages. We can picture them sifting and sorting through the organic detritus on which their small weak jaws allowed them to feed perched on these stiltlike pectoral appendages and rolling their small dorsally situated eyes around on the alert for predators (Figure 5–10).

In early placoderms the head shield of numerous large plates was separated from the multipieced trunk shield by a narrow gap. A mobile connection between the anterior vertebrae and the skull allowed the head to lift when certain of the head-to-trunk muscles contracted. The craniovertebral joint permitted the mouth to be opened wider than would lowering only the mandible. During their evolution, certain of the increasingly agile and obviously predatory placoderms, among the arthrodires, developed a curious adaptation furthering this gape-increasing ability. The space between the head and the thoracic shields widened and a pair of joints evolved, one above each pectoral fin on a line which passed through the older craniovertebral joint. This allowed for much greater flexibility between the shields and contributed not only to an enormous head-up gape but probably also to respiratory efficiency and some steering control during locomotion. The name given to this order of placoderms pays tribute to this strange adaptation: *arthros* (a joint) plus *dira* (the neck).

Placoderms were mostly creatures of the Devonian. Only scant and usually fragmentary remains have been unearthed in deposits outside that period. Nevertheless, the placoderms radiated into a large number of lineages and adaptive types. Because of the convergence in different lineages (due to adaptations to similar habitats) and the adaptive radiation of related forms (due to the differences in the environments in which they lived), the systematics and relationships of placoderms are poorly understood. *Dunkleosteus* was a 10 meter long voracious predatory arthrodire. *Bothriolepis*, an antiarch, supported itself on stilt-like pectorals (Figure 5–10). Another group of placoderms had the palatoquadrate firmly attached to the cranium and complex, solid tooth plates for crushing shellfish. In some placoderms, sexually dimorphic pelvic appendages indicate that internal fertilization occurred, undoubtedly coupled with appropriately complex courtship. Other groups show a tendency for dorsoventral flattening, eyes on top of the head, and subterminal mouths, all indicating benthic specialization. Some benthic forms such as *Gemuendina* bear a striking resemblance to modern skates, although they were completely armored with a mosaic of small plates and could not have used their broad pectoral fins in an undulating skatelike manner for locomotion.

A curious characteristic of all placoderms, and one which may have been involved in their eventual extinction, concerned their dentition and feeding apparatus in general. No placoderm ever developed teeth of a modern type as did the acanthodians. Slightly modified dermal bones lined the jaw cartilages of placoderms and, though they had long knifelike cutting edges and strong picklike points for slicing and piercing prey, they were subject to wear and breakage without apparent means of replacement. The jaws in placoderms, although an outstanding milestone in vertebrate evolution, were often immovably bound to the cranium or tightly articulated to the rest of the head shield. This prevented their participation in any sucking action, a very successful feeding process as evidenced by its success in the agnathans and again in the vast majority of living jawed fishes. According to two different views placoderms either became extinct without giving rise to any surviving forms or gave rise to two groups still very much with us—elasmobranchs and holocephalians. Before discussing these possible relatives, we shall consider what advantages jaws have and how vertebrates came to acquire them.

5.3.2 The Origin of Jaws and the Cranium

Professor A. S. Romer stated in *The Vertebrate Body* that, "Perhaps the greatest of all advances in vertebrate history was the development of jaws and the consequent revolution in the mode of life of early fishes." Why did Romer elevate the evolution of jaws to such an eminent position? Functionally jaws allow behaviors that otherwise would be difficult, if not impossible, to perform. The presence of strengthening bars (jaws) around the mouth, manipulated to open and to close by muscles, allows an organism to grasp objects "firmly." The importance of this innovation cannot be overemphasized. Armed with teeth the jaws' grip becomes surer; teeth sharpened with cutting edges allow food particles to be reduced to edible size; large, flattened, and opposed teeth provide a mill in which to grind harder, less assimilable foods. When the vertebrates evolved jaws, therefore, new food sources must have become available; carnivorism and herbivority became exploitable ecological and behavioral options. The development of jaws must have placed the placoderms and acanthodians in a commanding position, for many soon increased in size and, during the Devonian, completely replaced many more primitive vertebrates.

Jaws also provide for defense. One need examine few aggressive stances of vertebrates to realize the central importance of jaws in protection. The snarl of mammals, the pecking of birds, the gaping threat of alligators, all call attention to the major generalized weapon of vertebrates—the jaws.

Another advantage of jaws must have been the ability to manipulate objects. The ostracoderms and earliest jawed vertebrates did not possess moveable fins. Only the mouth plates of some heterostracans were movable and these most likely were utilized only in feeding. A grasping, moveable jaw provides for a new mode of living—holes can be dug in search of prey or for nests; mates or other objects can be grasped. It is little wonder that Romer placed so much importance on jaws.

It has long been held that the vertebrate jaw was derived through modifications of an anterior gill arch. An impressive array of facts supports this theory. The facts are derived from descriptive studies and experiments on embryogenesis of the head region of vertebrates and on studies of the anatomy and function of the cranial and spinal nerves of chordates. Let us begin by asking what we need to know to unravel the morphological complexities that underlie the origin of jaws. The most revealing facts are (1) embryonic development of the cranium and jaws, (2) development of the cephalic and branchial muscles, and (3) functional morphology of the spinal and cranial nerves.

Embryonic Development of the Cranium and the Jaws

The cranium. Portions of the skull of all vertebrates first appear as cartilage. In most gnathostomes the cartilages are replaced by bone. These cartilages are specific in location and usually appear as paired bars (Figure 5–11). The trabecular cartilages form anterior to and the parachordals adjacent to the tip of the notochord. They form ventral to the orbit and the otic region of the head, respectively, and fuse along the midline to form a single basal plate. The nasal sac and auditory capsule cartilages soon form and fuse with the basal plate to create a ventrolateral "cranial plate" that extends from the anterior end of the notochord to the tip of the snout (Figure 5–11). The otic region of the plate expands upward and arches over to enclose the main portion of the forming "brain." The occipital segments, posterior centers of chondrification, similar in most respects to vertebrae, fuse to the posterior margin of the parachordals and otic capsule. They also arch up and over to form the occipital arch. The fenestra or opening created is the **foramen magnum**, and marks the posterior margin of the cranium and serves as the cranial exit for the central nervous system. Details in the extent of chondrification vary, but the basic pattern is similar in all vertebrates.

The cranium has evolved through coalescence of an anterior segmented pattern, presumably present in the "protovertebrate." Indeed, the anterior region of amphioxus remains metameric (Figure 2–3). In the middle 1800s several famous biologists—A. von Geothe, Lorenz Oken, and Sir Richard Owen—postulated that the skull represented fused vertebrae. Thomas Huxley, crushed this "vertebral" theory of the skull. Nevertheless, the segmental nature of the head region was subsequently established on firm anatomical grounds, especially by the elegant descriptive studies of the elasmobranch chondrocranium by Carl Gegenbauer in the last half of the nineteenth century. The occipital segment(s) in many vertebrates form in a segmented manner and then fuse (Figure 5–11). In fact, the segments look like the rudiments of vertebrae and are derived from similar axial skeletal embryonic tissue, the sclerotome (Figure 5–12). Possibly the parachordals (even the otic capsules?) are derived similarly, but

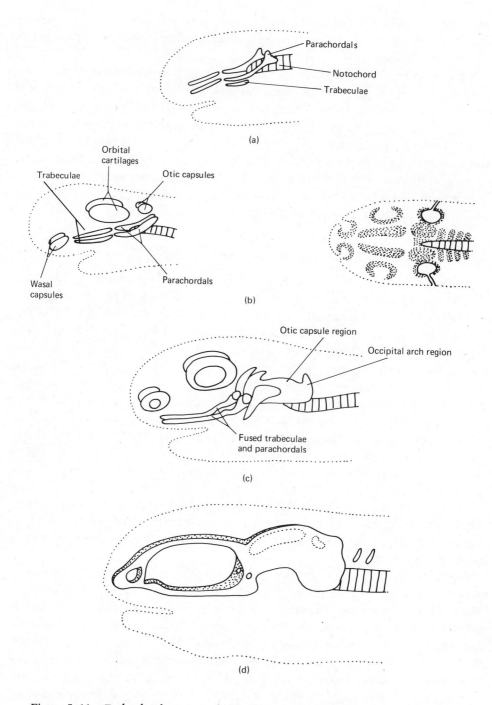

Figure 5-11 Early development of the cranium. (a) through (d) are progressively more advanced stages of development. In (b) lateral view left, dorsal view right. The trabecular and parachordal cartilages are usually formed first, followed by the sensory capsules and the occipital arch. These centers of chondrification fuse to form the braincase. (Modified after Ballard, W. W. Comparative Anatomy and Embryology. 1964. Ronald Press, N.Y.)

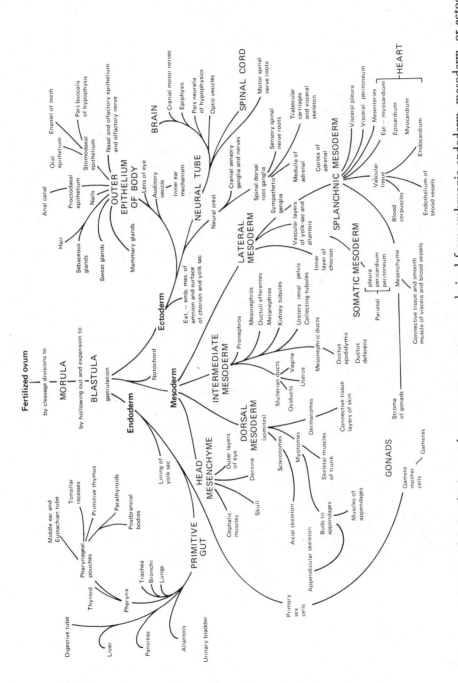

Figure 5-12 Schematic diagram showing basic vertebrate organ structures as derived from embryonic endoderm, mesoderm, or ectoderm. The diagram indicates origin of epithelial part of organ only. Most organs have supporting structures of mesodermal origin. (Modified after Patten, B.M. 1964. Foundations of Embryology. McGraw-Hill, New York).

this is unclear from their embryogenesis. What is clear is that the posterior chondrocranium, the box that encloses most of the brain, develops in close association with the anterior end of the notochord, a region that remains segmented in amphioxus.

The Jaws. The visceral skeleton, composed of the jaws, their supporting elements, and the gill arches, is distinctly segmental (Figure 5-13). In fact the jaw elements (mandibular arch = mandibular + palatoquadrate) and the suspensorium (hyoid arch = hyomandibular) and a varible number of smaller elements seem to represent anterior gill arches of early agnathans (Figure 5-13). Gill slits pierce the body wall from inside the pharynx to the outer body surface between each gill arch. Gill arches are formed in cartilage in several pieces which, in general, form an upper (pharyngo-, epi-) and a lower (cerato-, hypo-, basi-) branchial series. Each visceral arch originates in the ventral portion of a body segment. In a typical fishlike vertebrate, the upper jaw (palatoquadrate) is a transformed epibranchial element, the lower jaw (mandible) a ceratobranchial element. It is unclear whether the hypo- and basi-branchials become a part of the lower jaw or were possibly lost in the gill arch transformation. The next posterior gill arch was transformed into a support from which to suspend the hinge of the jaws (Figure 5-14).

Prima facie evidence has suggested to some anatomists that a first "primitive" gill arch (= premandibular arch) contributed to the formation of the trabecular cartilages (Figure 5-13). Other anatomists view these cartilages as isolated elements of the mandibular gill arch, arguing that a premandibular arch did not exist. The trabecular cartilages along with all elements of the visceral skeleton are derived from a unique vertebrate embryonic tissue, the **neural crest**. No other cartilage or skeletal element of the vertebrate body has its origin in neural crest cells. The parachordals and occipital segments, which form the rest of the

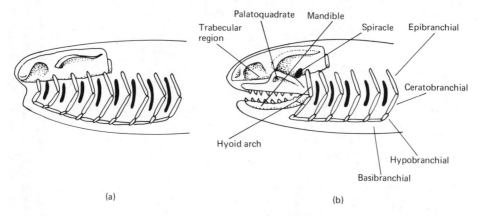

(a)   (b)

Figure 5-13 Evolution of the vertebrate jaws from anterior gill arches. (a) agnathous condition; (b) gnathostomous condition. The fate of the first gill arch (or arches) is unclear. The next gill arch likely formed the trabecular cartilage while a more posterior arch gave rise to the jaws. Note that an additional arch (the hyoid) acts as a support for the jaw hinge. The gill slit between the jaws and the hyoid arch was reduced to a spiracle, as seen in modern sharks.

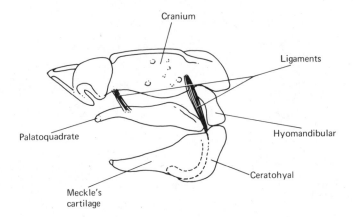

Figure 5-14   Relationships of the hyoid arch in supporting the jaw hinge as seen in the shark *Scyllium*. The hyomandibular, which acts as a strut to the jaw hinge, allows the jaw to be thrust forward for grasping and to enlarge the gape.

"cranial plate," are derived not from neural crest tissue but from sclerotome, a dorsal mesodermal tissue (Figure 5-12).

Neural crest proliferates as a unique **mesenchymous** tissue issuing from the dorso-medial junction of the neural tube of vertebrates (Figure 5-15). From there its cells migrate to various portions of the body and participate in formation of distinctive organs. The visceral skeleton and the dorsal ganglia of the spinal and cranial nerves are examples, and each is important to considerations of the origins of the cranium and the jaws. Removal of the neural crest prior to invasion of the head region results in an embryo that lacks cranial nerves, a visceral skeleton, and all of the cranium *anterior* to the notochord. Significantly, the cephalochordate amphioxus which lacks neural crest, also lacks dorsal ganglia and visceral cartilage (Figure 2-3).

Why were these visceral, neural crest elements incorporated into the chondrocranium? The trabeculae must have provided a platform on which the nasal capsule and the eyes could be stabilized; the parachordals likely served a similar function for the inner ears (otic capsules). Until the pre-notochordal region of the "protovertebrates" was stiffened by the trabecular cartilages, there must have been little stabilization and protection of the sensory systems of nose and eyes. In addition, the trabecular cartilages, when fused in the "cranial plate," must have created a stiff platform from which a jaw could be supported. In most modern fishes, amphibians, and many reptiles the upper jaw has some mobility (in holocephalians, birds, and mammals the upper jaw fuses to the cranium) but it is usually tied to the skull at one or more points by ligaments. The hyomandibular flexibly supports the jaws at the hinge and is itself attached to the otic region (Figures 5-11, 5-13, and 5-14).

The need for a stable jaw support can be easily demonstrated. Try to sharpen a pencil in an unattached pencil sharpener! To convert force effectively into

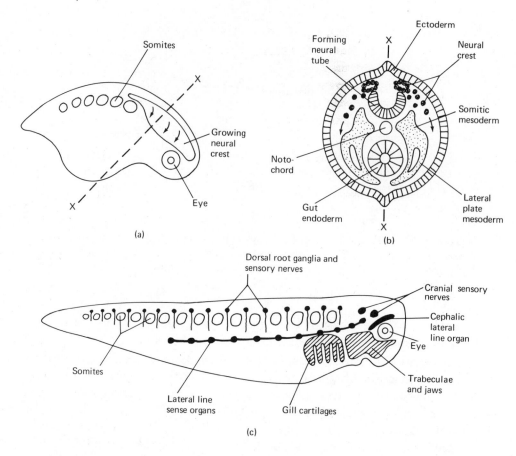

Figure 5-15 The growth of neural crest from the neural tube to form several unique vertebrate structures. (a) lateral view of early embryo; (b) cross section of same embryo to show site of neural crest origin; (c) lateral view of more advanced embryo showing neural crest derivatives (all black and hatched structures).

work a stable platform is needed—for jaws this platform was the "cranial plate." Ultimately the entire cranial skeleton, derived from branchial and somitic elements, came to serve several functions: (1) stabilization of the major sense organs, (2) protection of the sense organs and evolving brain, and (3) the solid platform to support a working, moveable jaw.

Development of the Cephalic and Branchial Muscles

Vertebrate muscles are derived from embryonic mesodermal tissue, the mesoblasts, but can be divided into two types—somatic and visceral muscle. The same is true for the muscles of amphioxus. The mesoblast (Figure 5-12) separates into dorsal somites and a ventral lateral plate of mesoderm. In amphioxus these mesodermal derivatives are metameric throughout the body; that is, they occur in repeated similar segments. In vertebrates, however, only the dorsal somites retain metamerism. The lateral plate mesoderm is unsegmented. The reasons for the loss of segmentation in the visceral region of vertebrates are obscured by millions of years of change, but we suspect that the loss relates to the increasing mobility and changing diets of the early vertebrates. Modifications in the digestive system must have been required to coincide with the new habits. In the body of some cyclostomes and the head region of most vertebrates, however, traces of the original visceral metameric pattern are seen during development.

The metameric, dorsal somites give rise to the major trunk muscles of the body. In the head region of all vertebrates they are reduced. Generally only the extrinsic ocular muscles that move the eye and the muscles that move the tongue are metameric and form from somites. Because these muscles are pre-otic in position they provide additional evidence that the head region once was segmented. The hypoglossal muscles migrate from their dorsal origins ventrally into the pharynx and then forward to produce the tongue. The three or four postotic somites are lost or greatly reduced in most vertebrates.

The visceral muscles participate mostly in digestion, circulation, and secretion. In the head region, however, they form a branchiomeric series that activates the jaws and gills. In addition, the muscles of facial expression in mammals are visceral. Most of the muscles of the head, therefore, are derived from the lateral plate mesoderm, not from somites (Figure 5–12).

Spinal nerves. From the spinal cord in each body segment of all vertebrates issue two nerves that supply motor (efferent) and sensory (afferent) fibers to the somatic and visceral components of that segment (Figure 5–16). Somatic

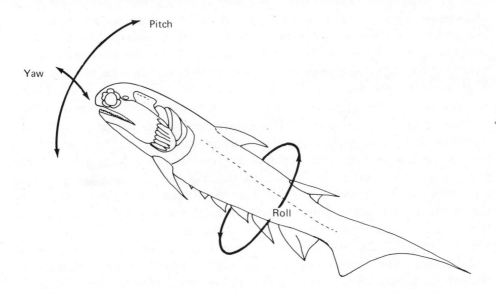

Figure 5–16   An acanthodian, *Climatius,* seen obliquely from in front and below to show the orientation of pitch, yaw, and roll and the fins whose surfaces are advantageously placed to counteract these movements.

components include the skeletal muscles, visceral components the gut and circulatory musculature. Although the morphology of these nerves varies a great deal between vertebrates, a ventral and a dorsal root tract are always recognizable. In the primitive condition the ventral root supplied only motor nerves to the myotomes or trunk muscles of each body segment (somatic components). The dorsal root, however, was a mixed nerve that contained both visceral motor nerves and the sensory nerves.

Ventral root nerves are conservative in that the metameric relationship between the muscle forming myotome (somite) and the nerve supply to that somite have been retained throughout vertebrate evolution. If a particular somite is lost (e.g., some head somites) so is the ventral nerve; if a myotome is modified and moved to serve other functions (e.g., the appendicular or limb muscles of tetrapods) the nerve preserves the metamerism because it exits from the segment of the cord embryonically associated with that myotome. The dorsal root nerves appear to be more plastic; inter-segmental capture is common. The dorsal root nerve contains the visceral motor fibers of the continuous gut and the somatic sensory nerves of the skin, two parts of the body which lack a segmentation.

Cranial nerves

Armed with this concept of what constitutes a primitive spinal nerve, we are ready to ask if the cranial nerves represent modified spinal nerves. If the cranial nerves represent modified spinal nerves, then we have additional evidence that the vertebrate head and jaws were derived from a fusion and modification of anterior segments. Ten to twelve pairs of cranial nerves are recognized in vertebrates (Table 5–2). Cranial nerves were first described by anatomists studying mammals, especially human cadavers. Their numbers refer to the anterior to posterior sequence of exit of the nerves from the mammalian brain. Their names broadly describe what the nerves do (e.g., oculomotor = eye mover) or their appearance (e.g., trigeminal = a nerve with three major branches).

Three of the cranial nerves, the Olfactory (I), the Optic (II), and the Auditory (VIII), are special sensory nerves that are not related to spinal nerves, but are each associated with separate portions of the tripartite brain. The cranial nerves most obviously derived from ancient, anterior spinal ventral roots are nerves III, IV, VI and XII (Table 5–2). Like the ventral roots of the spinal nerves which send motor fibers to the segmental muscles of the trunk and tail, these cranial nerves send motor fibers to the somatic muscles of the head (i.e., eye and tongue). What was the fate of the dorsal roots of the ancient anterior spinal nerves? The Glossopharyngeal nerve (IX) apparently has been modified the least from its primitive state. Its function is clear and analogous to that of a primitive dorsal root of a spinal nerve. It supplies both somatic and visceral sensory fibers, as well as visceral motor fibers (and is therefore a mixed nerve) to the first gill arch of fishlike, jawed vertebrates. Because the Glossopharyngeal nerve is located just posterior to the hyoid arch of fishes we assume that it

Table 5–2. The Cranial Nerves and Their Primary Function

Number	Name	Somatic motor	Somatic sensory	Visceral motor	Visceral sensory
0	Terminalis		+		
I	Olfactory		+		
II	Optic		+		
III	Oculomotor	+			
IV	Trochlear	+			
V	Trigeminal		+	+	
VI	Abducens	+			
VII	Facial		+	+	+
VIII	Acoustic		+		
IX	Glossopharyngeal		+	+	+
X	Vagus		+	+	+
XI	Accessory			+	
XII	Hypoglossal	+			

Functions boxed as *mixed nerves*.

was associated with the third gill arch of gnathostome ancestors (Figure 5–13). The Facial nerve (VII), which is a mixed nerve, innervates the hyoid apparatus or suspensorium derived from the second gill arch (Figures 5–13 and 5–14). The Trigeminal (V), again a mixed nerve, innervates the jaws which are believed to have been derived from the first gill arch (Figure 5–13). Thus cranial nerves V, VII and IX, on the basis of their mixed sensory and motor components, most likely represent dorsal roots of ancient anterior spinal nerves whose body segments have fused. They can be aligned with specific gill arches.

Attempts to align the ventral root components of these presumed old spinal nerves (cranial nerves III, IV and VI) with specific gill arches as can be done with the dorsal roots, however, would be highly speculative. Similar arguments can be used to explain the fate of the primitive spinal nerves associated with more posterior gill arches. For example, the Vagus nerve (X) probably represents a coalescence of several dorsal roots, and the Hypoglossal the remnant of one or more ventral roots. The specific innervations and function for each of these cranial nerves testifies to their origins from ancient spinal nerves that were associated with the head region.

It is possible that one or more anterior body segments which included the original mouth, have been lost. A small "cranial nerve" that parallels the olfactory nerve was described in 1894 by F. Pinkus in dipnoans (lungfish). It has been found in many, but not all, vertebrates. It is called the Nervus Terminalis (Table 5-2) and has been assigned the numerical position 0. Its function is unclear, although it seems to be sensory. Perhaps the Terminalis represents the dorsal root vestige of a terminal body segment long lost or so modified that it is unrecognizable.

Our considerations of the varied evidence for the origin of jaws have provided us with a glimpse of how anatomical structures (in this case gills) can be modified to serve a different function (a jaw). A basic attribute of this evolutionary process is the fact that seemingly new structures are derived from older structures, not produced *de novo*. As we shall see later, these anatomical transformations have been never ending, for even the jaws of fishes have been partly transformed to serve other functions in higher vertebrates.

5.3.3   The Origin of Fins

Jaws are an advantage only when applied to an object. In an aquatic environment, suction can draw many objects to the mouth, but over modest distances and for all but the smallest objects the body must be guided to the graspable object. This sounds simple, but in practice the guidance of a body in three-dimensional space is quite complicated. Yaw (swinging to the right or left) combines with pitch (tilting up or down) to make accurate contact with a target difficult. Because jaws are bisymmetrical, grasping is effective primarily in one direction. Roll (rotation around the body axis) is therefore involved in effective jawed grasping. Especially if the target moves, perhaps evasively, a change of course and quick adjustments of roll, pitch, and yaw are necessary. It is little wonder that the development of strong mobile fins was coincident with the evolution of jaws. Even agnathous fishes developed analogs of all the gnathostome appendages, although they were never as successfully refined.

Fins provide broad surfaces that resist the flow of water and act as hydrofoils applying pressure to the surrounding water. Since water is practically incompressible, force applied by a fin in one direction against the water is opposed by an equal force in the opposite direction (Figure 5-16). Thus fins which project from the body surface can resist roll if pressed on the water in the direction of the roll. Fins projecting horizontally near the anterior end of the body similarly counteract pitch. Yaw is controlled by vertical fins along the middorsal and midventral lines. Especially effective when placed near the posterior part of the body, vertical fins tend to pivot to the right or left, veering fish back on course. Fins have a great many other functions as well. They increase the surface area of propulsive tails for greater thrust. Presented at angles to a flow of water, they produce considerable lift (see Section 5-5). Hardened portions of fins are used in defense. Combined with glandular secretions, they produce poisonous injection systems. Colorfully marked fins are used to send visual signals. Designed so that they may collapse or unfold by muscular action, fins are used, abruptly and often startlingly, to increase the apparent size of their possessor. Fins obviously provide a high selective advantage.

Fin structure of early fishes was quite variable. Agnathans had spines or enlarged scales with finlike functions derived from dermal armor. In addition to cephalic cornu, osteostracans had paddlelike pectorally located structures supposedly without internal supports. Some anaspids had long, flat finlike sheets of tissue running nearly the full length of the flanks. Acanthodians had a variable number of spines (sometimes with attached membranes extending in two lengthwise rows along the ventro-lateral aspects of the trunk). The internal skeleton of acanthodian fins (Figure 5-9c) was composed of at least two rows of tightly packed rod shaped cartilages; one row distal to the other. These basal elements were associated with horny, threadlike rays which did not protrude far into the fin. Placoderm pectoral appendages, in contrast, had in their later evolutionary stages a narrow base with few elements, but extending into the fin were numerous rod shaped elements and horny fin rays. It is clear that fins, although always in approximately the same positions, were radically different in internal and external construction and number.

There is no fossil evidence of the origin of fins, especially the paired pectoral and pelvic fins that were of significance in the later stages of vertebrate evolution. Vertebrate biologists are left to formulate and to test hypotheses about fin origins. Unfortunately, very little evidence has come to light about the origin of paired fins compared to the anatomical and embryological evidence on the origin of jaws. Although some early workers fancied a resemblance between the pectoral fins and the gill arches, the detailed embryological studies discussed in the last section make any hypothesis of a branchial origin doubtful. Fins are derived from mesoderm and have somatic, not visceral, musculature and innervation.

Long held in high regard was the fin-fold theory of the origin of paired appendages. The fin-fold theory used as prototypes certain anaspids with a pair of long-based lateral flaps extending from gills to anus. Early acanthodians had pairs of spines in the same region. In fact this was merely a morphotypic series of organisms between which no relationship could be shown. By comparing this hypothetical series to the fin folds of amphioxus, discrete fins were pictured as originating from continuous fin folds. These fin folds were thought to be paired laterally but single dorsally and posteriorly. Broken into short segments and reduced in number, fin folds provided for the origin of the fins seen in today's fishes.

Because no fossil evidence of any phylogenetic sequence depicting these events has been uncovered, the fin-fold theory has become less appealing in recent years. In line with our current understanding of evolutionary principles, it seems best to consider fins so beneficial to any form which by chance mutation developed lateral projections that multiple evolutions, involving perhaps more than one mechanism, have occurred. That so many vertebrates tended to have the same basic fins can be attributed to the fact that, as vertebrates, early fishes were more closely related to each other than to other organisms and hence shared a larger number of common genes and evolutionary potentials. In addition, a multiple evolution with similar results is to be expected when only a limited number of fin positions and shapes provide the hydrodynamic advantages we attribute to fins.

5.4   CHONDRICHTHYES—THE CARTILAGINOUS FISHES

Few groups of fishlike vertebrates have had so much information published, filmed, or reported about them as the sharks and their relatives. No other group remains as poorly understood by the scientific community or so misunderstood by the public. From their first appearance in the fossil record of the early Devonian to the present, they have a history of waxing and waning success. In some of their systems morphological refinement and sophistication of anatomy has evolved to levels that are surpassed by few other living vertebrates; yet they retain such simple constructions of other portions of their anatomy that they are used universally as examples of the primitive vertebrate body form. Identified by a cartilaginous skeleton, living forms are divided into those with a single gill opening on each side of the head, shielded by a fleshy operculum, and those with multiple gill openings on each side, protected by individual modest flaps of skin. Because of the undivided nature of the head, the first group are called Holocephali. Their common names of ratfish and chimaera come from their bizarre form: long whiplike tail, fishlike body, and a head with big eyes and buckteeth that resembles a caricature of a rabbit. The second group is the Elasmobranchii, meaning "plate-gilled." Elasmobranchs are divisible into two groups: those with five to seven (almost always five) gill openings on each side of the head at about the midline (Plerotremata or Selachii); and those that are flattened and adapted for a benthic existence with five regular gill slits on the ventral surface (Hypotremata or Batoidea). Elasmobranchs are commonly known as sharks, skates, and rays.

5.4.1   Evolutionary Adaptations in Elasmobranchs

In spite of the ever increasing and rather good fossil record, understanding of the phylogeny of cartilaginous fishes remains somewhat unclear. This haziness results partly because early elasmobranchs, like modern species, were diverse in habits and habitats. Because of this diversity, no phyletic relationships (that is, lineages) can be based on continuous transitions in morphology in any sequence of fossils. When elasmobranchs first appeared in the Devonian they shared common morphological characteristics and we assume they were monophyletic, that is, they stemmed from a common ancestor. Their initial radiation, however, was rapid and emphasized changes in teeth, jaws, and fins and, therefore, in feeding and locomotor mechanisms. Apparently within different lineages the feeding and locomotor apparatus evolved at different rates. In some, advanced dentition was coincident with primitive fin structures, in others the opposite. As a result, when examined in detail and compared, fossils display a confusing mosaic of both primitive and advanced characters.

Perhaps the different habitats in which early elasmobranchs lived selected for a general improvement in feeding and locomotion. Improvements in feeding structures involved a suite of characters (teeth, jaw suspension, muscles, and so on) the components of which evolved at different rates. Yet, through time, different lineages of elasmobranchs tended to accumulate similar but not identical modifications in their feeding and locomotor structures because of similar selective pressures. This pattern of evolution of similar adaptations in related lineages is an example of **parallel evolution.**

Because of parallel evolution, certain modifications can be expected to appear independently, perhaps repeatedly, in the course of time. Parallel evolution of characters is very difficult to distinguish from that of characters inherited intact from a common ancestor. Adaptations found in several lineages are known as **general** or **broad adaptations.** Examples are paired appendages, and perhaps even jaws and muscular pump filter feeding. Broad adaptations have penetrating effects on the organism's integration of behavior, physiology and morphology. Broad adaptation define an organizational level in a horizontal (nonphylogenetic) classification. The species showing characters of a particular organizational level belong to the same **grade.** A grade may contain different phyletic lines. Each phyletic line is called a **clade.** With the elasmobranch fishes we can define three grades in their evolution, but at present any attempt to determine the clades involved in this evolutionary story is premature (Figure 5–17).

5.4.2 The Earliest Elasmobranch Radiation

The earliest recognized grade of elasmobranch evolution is a radiation identified by the form of the teeth common to the majority of the species in the grouping. The teeth (Figure 5–18) are basically three-cusped with little root development. Although there is evidence of bone around their bases, the teeth are primarily dentine structures capped with enamel. The central cusp is the largest in *Cladoselache,* the best known genus (Figure 5–18a) but smallest in *Xenacanthus,* a more specialized form (Figure 5–18b).

Remains of *Cladoselache* are amazingly well preserved. Bashford Dean, a student of archaic fishes, described specimens from the Devonian inland sea that left Illinois, Indiana, and Ohio with rich fossil beds. He found *Cladoselache* to have distinct gill filaments, kidney tissue showing tubules, and skeletal muscle fibers whose cross striations were still visible after 300 million years. *Cladoselache* was sharklike in appearance, about 2 m long when fully grown,

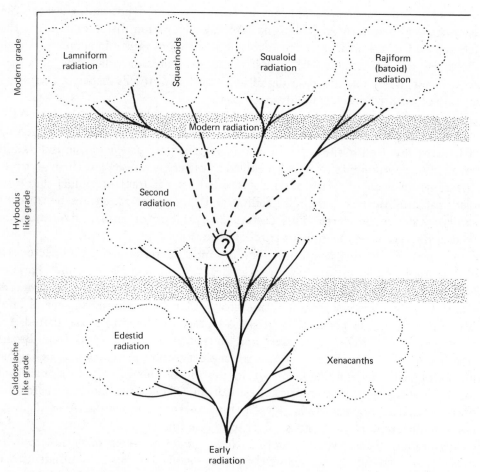

Figure 5-17 A diagrammatic representation of a possible elasmobranch phylogeny. The earliest radiation, with its xenacanth and edestoid offshoots as well as the second radiation, are discussed in the text. The four branches of modern level sharks indicated are not described separately but may be characterized as generally large, voracious tropical sharks such as the tiger shark *Galeocerdo* (lamniform radiation); the often smaller, schooling temperate forms exemplified by *Squalus* (squaloid radiation); a single genus of bottom dwelling dorsoventrally flattened angel sharks, *Squatina* (squatinoid radiation); and the large number of skates and ray (rajiform radiation).

with large fins and mouth and five separate external gill openings. The mouth opened terminally without an overhanging rostrum or snout, and the cartilaginous chondrocranium had several large areas for the attachment of the jaw. The upper cartilage of the jaw, the palatoquadrate, had several tight ligamentous attachments to the chondrocranium. One was at the symphysis between the right and left halves of the jaw, a second via a dorsal process extending up from the palatoquadrate behind the eye, and a third via a process to the ear region and with some support from the second visceral arch, the hyoid-arch. The name **amphistylic** (*ampho* = both, *styl* = pillar or support) is applied

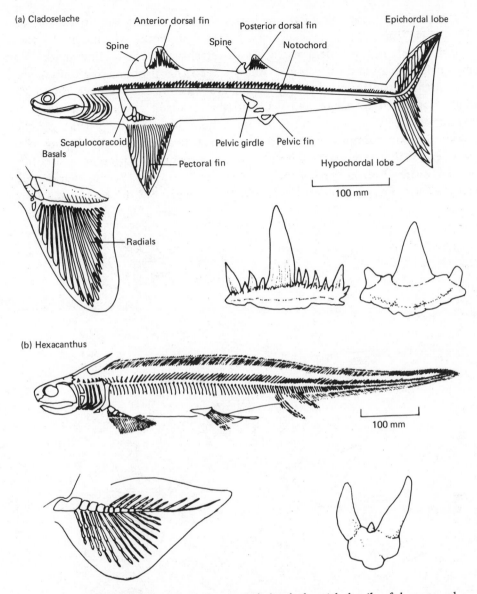

Figure 5-18 Early elasmobranchs. (a) *Cladoselache* with details of the pectoral structure and teeth of the '*Cladolus*' type. (b) *Xenacanthus*, a freshwater elasmobranch with details of its archipterygial pectoral fin structure and peculiar teeth. (Modified after Moy-Thomas and Miles [5].)

to this upper jaw suspension. The lower jaw was attached to the upper jaw by a curious double joint. The posterior end of both lower jaw cartilages had a ball and a socket that articulated with a similar ball and socket from the upper jaw cartilage. The gape was solidly bordered and large, since the jaws extended from the tip of the snout to well behind the rest of the skull. The three-pronged teeth were probably especially efficient for feeding on fishes of a size that could be swallowed whole or severed by the knifelike cusps.

Enamel and dentine, although the hardest tissues known, are brittle and subject to wear on any friction surface. Erosion of teeth, which renders them less functional, is a problem faced by all vertebrates. The earliest sharks (and some acanthodians and early bony fishes) solved this problem in a unique way. Each tooth on the functional edge of the jaw was but one member of a "tooth family," attached to a ligamentous band that coursed down the inside of the jaw cartilage deep below the fleshy lining of the mouth (Figure 5-19a). Aligned in a file directly behind the functional tooth were a series of developing teeth. In modern sharks essentially the same dental apparatus is present. Tests done on living sharks indicate that tooth replacement is rapid. Young sharks under ideal conditions replace each lower jaw tooth every 8.2 days and each upper

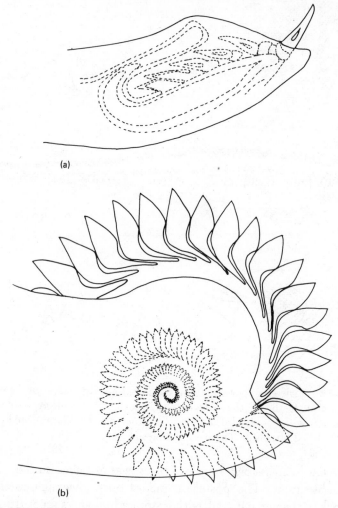

(a)

(b)

Figure 5-19   (a) Diagramatic cross section of the jaw of a living shark showing the single functional tooth backed by a band of replacement teeth in various stages of development. (b) Lateral view of the symphysial (middle of the lower jaw) tooth whorl of the edestoid Cladodont *Helicoprion*, showing the chamber into which the life-long production of teeth spiraled.

jaw tooth every 7.8 days! If *Cladoselache* replaced it's teeth, as seems likely, a significant advance in feeding structures is indicated for elasmobranchs compared to their placoderm competitors.

The body of *Cladoselache* was supported only by a notochord, but cartilaginous neural arches gave added protection to the spinal cord. Five gill arches followed immediately behind the two anterior visceral arches (mandibular and hyoid). The fins of *Cladoselache* consisted of two dorsal fins, paired pectoral and pelvic fins, and a well-developed forked tail. The first and sometimes second dorsal fins were preceded by stout superficial spines, triangular in cross section. Blunt and posteriorly hooked, they do not seem likely contributors to defense or effective as cutwaters. Some specimens lacked spines whereas seemingly identical specimens had spines. The suggestion that the dorsal spines may have been sexually dimorphic characteristics is intriguing. The dorsal fins were broad based triangles whose internal structure probably consisted of a triangular basal cartilage and a parallel series of long radial cartilages that extended to the margin of the fin. The pectoral fins were larger, but similar in construction. A scythe-shaped pectoral girdle on each side did not meet its mate ventrally but provided a broad attachment for the basal fin elements.

Each type of cladodont shark seems to have had a different sort of pectoral arrangement (Figure 5–18), but all possessed basal elements that anchored the fin in place. Extending laterally from these basal elements were long unsegmented branching cartilages, the radial cartilages that extended to the tips of the fins. From their structure these pectorals appear to have been hydrofoils with little capacity for alteration of their angle of attack. The pelvic fins were much smaller but otherwise designed like the pectorals. Some species show evidence of **claspers**: male copulatory organs. No anal fin is known; it is also lacking in many modern sharks.

The caudal fin of *Cladoselache* is distinctive (Figure 5–18). Externally symmetrical, its internal structure was nonsymmetrical and contained a band of subchordal elements resembling the hemal arches that protect the caudal blood vessels in modern sharks. Long unsegmented radial cartilages extended into the hypochordal (lower) lobe of the fin. At the base of the caudal were paired lateral keels that are identifying characteristics of modern rapid pelagic swimmers (see Section 5–5).

The integument was composed of scales that resembled teeth. Cusps of dentine were covered with enamel and contained a cellular core or pulp cavity. Unlike a tooth, each scale had several pulp cavities to match its several cusps. These scales were limited to the free margins of the fins and the circumference of the eye, where they formed an orbital ring. Their occurrence within the mouth behind the teeth and their structural similarity leave little doubt that the teeth of early sharks and of vertebrates in general are derived from specialized elements of the integument.

From their morphology and the stratigraphy of their fossil localities, we can piece together the life of cladodonts. Probably pelagic (open water) predators, most early sharks and *Cladoselache* in particular swam after their acanthodian, placoderm, and newly evolved bony fish prey in a sinuous manner, engulfing them whole or slashing them with their daggerlike teeth. The lateral keels on the base of the tail indicate they were fast swimmers. The lack of body denticles and calcification in many structures indicate a tendency to reduce weight and therefore increase buoyancy.

Reproduction in some forms involved internal fertilization, implying that a highly developed behavioral system existed to insure successful mating. If the possession of claspers is a primitive character shared with the placoderms, as some authors contend, the lack of claspers in *Cladoselache* may mean that the males of this "well-known" fossil species have been placed in separate genera.

One of the score or so of genera assigned to this early grade of elasmobranch evolution is *Xenacanthus*, whose dentition was an obvious modification of the cladodont type. *Xenacanthus* also had a brain case, jaws, and jaw suspension identical with those of *Cladoselache*. But there the resemblance ends. The xenacanths were fresh-water bottom dwellers of rather slow habit, as judged from their long compressed body, pointed (diphycercal) caudal fin, and limb-like paired fins (Figure 5–18b). The pectorals seem to have been derived by freeing the elongate subcutaneous basal anchor typical of other early sharks and the development of radials on the medial side. The pelvic fins lacked these medial radials, but the males had well developed medial claspers. A curious two-unit anal fin and a long stout spine extending from the center of the head complete the weird picture of xencanths left for us in the fossil record. The idea that they were bottom dwellers derives from their similarity to living lungfish (which have a benthic habit) and the fact that their cartilaginous skeletal elements were heavily calcified decreasing their buoyancy. The xenacanths appeared in the Devonian and survived until the Triassic when they died out without leaving descendents.

Another group of chondrichthyan fishes shared morphological characteristics of pectoral fins and very stiff, equilobate, but strongly forked tails with other fusiform, fast swimming marine sharks. Nevertheless these sharks, the **edestoids**, were morphologically considerably removed from the main lines (clades) of elasmobranch evolution. The basis for this separation from other early sharks is the peculiar dentition of the group. Most of the tooth families of edestoids were greatly reduced, but the symphysial tooth row of the mandible was tremendously enlarged and interlocked at each tooth base. Apparently, several members of this tooth family were functional at the same time. Blunt for crushing in some forms and compressed to create a series of knifelike blades in others, the mandibular tooth row bit against small flat teeth associated with a poorly developed palatoquadrate. In other types these small upper teeth were directly attached to the ventral surface of the chondrocranium, the palatoquadrate having been reduced to a simple articulating element for the mandible. Most edestoids replaced their teeth rapidly, the oldest worn teeth being shed from the tip of the mandible. In contrast, *Helicoprion* retained all of its teeth in a specialized chamber into which the life-long production of dentition spiraled (Figure 5–19b). Perhaps teeth no longer efficient in size or shape provided a solid foundation for the functional teeth.

5.4.3 The Second Elasmobranch Radiation

The next grade of elasmobranch evolution involved reorganization in both feeding and locomotor systems. These modifications began in Carboniferous sharks and lasted until the late Cretaceous sharks. To clearly identify the adaptations that characterize this second grade, we shall describe a late and well-known genus of the Triassic and Cretaceous, *Hybodus*. Paradoxically many of the earlier sharks thought to be ancestral to *Hybodus* are less well known than species of the first elasmobranch radiation, because less skeletal material remains. This means that these later sharks were either less calcified and the skeleton rotted along with the soft tissues or that sharks such as *Cladoselache*, lived (or at least died) in environments more favorable to fossilization. *Hybodus* however has left complete 2-m skeletons (Figure 5–20a).

The dentition of this radiation of sharks seems pivotal to their success. *Hybodus* teeth are readily derived from *Cladoselache* type teeth but they show a diversity within a single individual not previously known in vertebrates. The anterior teeth were sharp-cusped and excellent for piercing, holding, and slashing softer foods. The posterior teeth were stout, blunt versions of the anterior teeth. Instead of becoming functional one at a time, they appeared above the fleshy lining of the mouth in batteries consisting of several teeth from each individual tooth family. A modern form with a similar dentition (Figure 5–20b) indicates how the business end of a *Hybodus*-like shark must have looked. This simple alteration in dentition, without change in jaw suspension and only minor changes in the length of the jaws and proportions of the chondrocranium, gave these sharks a decided edge over other sharklike forms. The living Port Jackson shark (*Heterodontus* sp.) which has *Hybodus*-like dentition, feeds on small fish, crabs, shrimp, sea urchins clams, mussels, and oysters. The sharp teeth near the symphsis seize and dispatch soft-bodied food, but shelled foods are thoroughly crushed by the batteries of pavementlike posterior teeth before being swallowed. Because they lived in rich shallow marine environments, the variety of foods available to such sharks must have been abundant and varied.

The coming together of land masses to form Pangea brought together faunas that had been living and evolving in isolation on the shelves of distant continents for millions of years (see Chapter 8). If this meant competition and extinction for many forms, as the fossil record indicates, a stenophagous (*stenos* = narrow, *phagos* = to eat) vertebrate, adapted to feed on a limited variety of organisms, would be at a disadvantage compared to a euryphagous (*eury* = broad) form. Therefore, *Hybodus* and its relatives were preadapted trophically to survive the radical changes in invertebrate faunas which characterized the Permo-Triassic periods.

Also characteristic of the hybodonts were their fins. The pectoral girdle remained divided into separate right and left halves, but the articulation between girdle and the fin consisted of three narrowed platelike basals instead of the long series seen in earlier sharks (Figure 5–19). This tribasal arrangement was also found in the pelvic fins, and both pairs of fins were more detached from the body than in earlier sharks, supported on a mobile stalk composed of the three basals. Mobility of the distal portion of the paired fins was also increased. The cartilaginous radials did not extend to the fin margin and were segmented along their shortened length (Figure 5–20a).

Proteinaceous, flexible structures called **ceratotrichia** extended from the outer radials to the margin of the fin. Intrinsic fin muscles permitted flexion

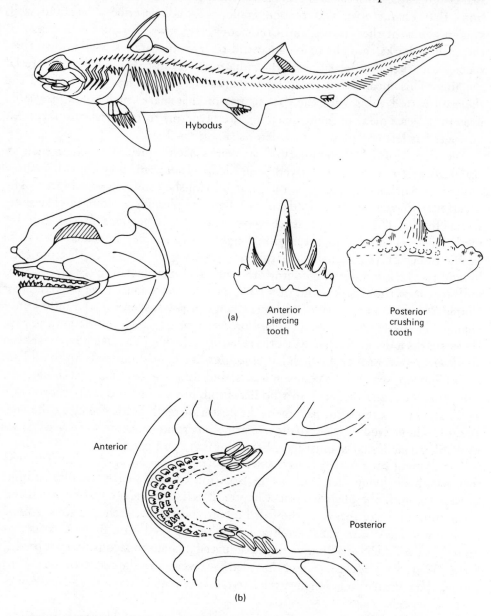

Hybodus

(a)

Anterior
piercing
tooth

Posterior
crushing
tooth

Anterior

Posterior

(b)

Figure 5–20   (a) The second radiation of elasmobranch evolution represented by *Hybodus*. Detail of head skeleton and teeth   (Modified after Moy-Thomas and Miles [5]). (b) Upper jaw and pharynx of the living hornshark, *Heterodontus*, whose dentition is similar to that of many hybodonts.

between the basals (arching the fin from anterior to posterior) and between the basals and proximal radials, as well as between the proximal and distal radials (arching the fin along its long axis). This greater mobility allowed the paired fins, especially the large pectorals, to be used hydrodynamically in ways which seem impossible with the fin construction characteristic of *Cladoselache.* By assuming different shapes, the pectorals could produce lift anteriorly, aid in turning, or function as simple hydrofoils. Along with changes in the paired fins, the caudal fin assumed new functions and an anal fin appeared (Figure 5–20a). The caudal shape was altered by reduction of the hypochordal lobe, division of its radials and addition of flexible ceratotrichia. The epichordal lobe containing the unconstricted notochord and neural and hemal arches was thus larger and stiffer than the lower lobe. This construction is known as **heterocercal** (*hetero* = different, *kerkos* = tail). Undulated from side to side, the fin twisted due to water pressure so that the flexible lower lobe trailed behind the stiff upper one. This distribution of force produces forward thrust that combined with the variable planning surfaces produced by flexible pectorals, could counter the sharks' tendency to sink or could lift it from a benthic resting position to capture food, escape enemies, or to meet a mate. Whatever mechanism the earlier sharks used to remain afloat, it is clear that the dynamic design of the fins in this second radiation of sharks allowed them more behavioral adaptability.

Other morphological changes include the appearance of a complete set of hemal arches to protect the arterial and venous trunks running below the notochord, well developed ribs, and narrow, more pointed dorsal fin spines closely associated with the leading edge of the dorsal fins. These spines were deeply inserted in the muscle mass, ornamented with ridges and grooves, and studded with numerous barbs on the posterior surface indicating defensive functions. Claspers are common to all species, leaving little doubt about the development of courtship and internal fertilization. In addition, male *Hybodus* had one or two pairs of hooked spines above the eye which probably functioned as cephalic claspers during copulation.

In the terminal mouth, amphistylic jaw suspension, unconstricted notochord, and multicusped teeth, *Hybodus* and its relatives resembled their presumed *Cladoselache*-like ancestors. But a direct line cannot be drawn between the two in time or in morphology. Some forms considered related to *Hybodus* had *Cladoselache*-like dentition but tribasal pectorals. One genus with *Hybodus*-like dentition lacked the tribasal pectoral but developed a very tetrapodlike support for its highly mobile pectorals. Because the caudal was reduced, they probably moved around on the sea floor using limblike pectorals. Known only from a 5-cm, immature individual, another form had a paddlelike rostrum one third its body length. Other types were 2.5-m giants with blunt snouts and enormous jaws. Despite their variety, which permitted them to flourish throughout the Mesozoic, this second radiation of elasmobranchs became increasingly rare and disappeared from earth at the close of the Mesozoic.

5.4.4   The Modern Radiation—Sharks and Rays

As early as the Triassic and perhaps even in the late Carboniferous, indications of a new grade of elasmobranchs appears. By the Jurassic the fossil evidence clearly indicates that sharks of an entirely modern appearance had evolved, and the Cretaceous contains many genera that are still living. Paleontologists are not in agreement as to the origin of modern sharks. They may have evolved from *Hybodus*-like sharks or from *Cladoselache*-like ancestors and acquired characteristics in parallel with *Hybodus* and its relatives. The most obvious difference between earlier radiations and modern sharks is the almost ubiquitous rostrum or snout that overhangs the ventrally positioned mouth. Modern sharks differ from the earlier grades of elasmobranchs in their jaw suspension and dentition, both of which have effected trophic changes of great importance. In addition, postcranial skeletal modifications have increased efficiency in both feeding and locomotion.

Modern elasmobranchs have an enlarged dorsal element of the hyoid arch, the hyomandibular cartilage, which braces the posterior portion of the palatoquadrate and firmly, but movably, attaches to the otic region of the cranium (Figure 5–21). The only other direct connection to the chondrocranium is via paired palatoquadrate projections that fit on either side of the brain case just behind the eyes and are attached to it by elastic ligaments. Jaw suspension of this type is known as **hyostylic** (see amphistylic suspension, Section 5.4.2). Hyostyly allows for multiple jaw positions each appropriate to different feeding opportunities.

The two pectoral girdles are fused together ventrally into a single U-shaped scapulocoracoid cartilage. Muscles run from the ventral coracoid portion to the symphysis of the lower jaw and function in opening the mouth. Simple opening of the mouth is used to engulf small prey or floating food particles. Since the mandibular teeth are well adapted to pierce and impale, this dropping of the lower jaw can be used to scoop up food from the sea floor. The advantages of the jaws of modern elasmobranchs are displayed when the upper jaw is protruded. Muscles swing the hyomandibula laterally and anteriorly to increase the distance between the right and left jaw articulations and thereby increase the volume of the orobranchial chamber. When performed rapidly, this sucks water and food forcefully into the mouth.

With hyomandibular extension, the palatoquadrate is protruded to the limits of the elastic ligaments on its orbital processes. This allows delicate plucking of benthic foodstuffs. Protrusion also increases exposure of the upper dental arcade, drops the mouth away from the head, and allows a modern shark to bite into an organism much larger than itself despite its large, sensitive rostrum. The dentition of the palatoquadrate is specialized; the teeth are usually stouter than in the mandible and often recurved and strongly serrated. When feeding on large prey a modern shark opens its mouth, sinks its lower and upper teeth deeply into the prey, and proceeds to protrude its upper jaw ever more deeply into the slash initiated by the teeth. As the jaws reach their maximum

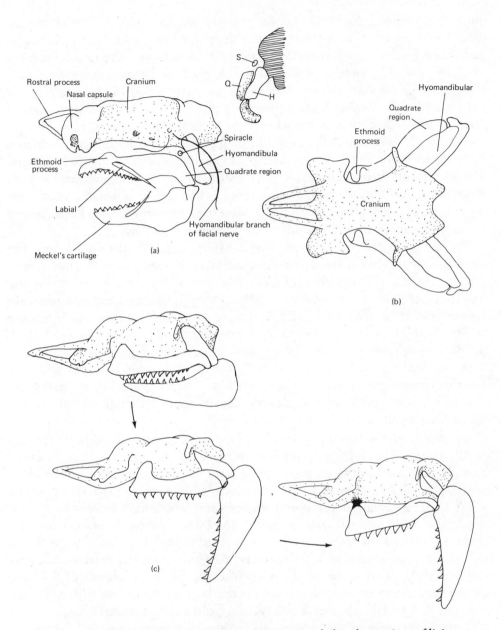

Figure 5-21 Anatomical relationships of the jaws and chondrocranium of living hyostylic sharks based on *Scyllium* and *Carcharhinus*. (a) Lateral, and cross-sectional views of the head skeleton of *Scyllium* with the jaws closed. (b) Dorsal view of *Carcharhinus* with the jaws closed. (c) Lateral views of *Carcharhinus* during jaw opening culminating with the upper jaw maximally protruded. The hyomandibula rotates during jaw opening and upper jaw protrusion from a position parallel to the long axis of the cranium to a position nearly perpendicular to that axis. S = spiracle; Q = quadrate region of the palatoquadrate; H = hyomandibula. (Modified after (a) E. S. Goodrich, *Studies on the Structure and Development of Vertebrates,* 1930, Macmillan, London: (b) and (c) S. A. Moss, J. Zool. Lond. 167:1972.

initial penetration, the shark throws its body into exaggerated lateral undulations, resulting in a violent side-to-side shaking of the head. The head movements bring the serrated upper teeth into action as saws to sever a large piece of flesh from the victim. Mobility in the head skeleton, known as **cranial kinesis**, in combination with movement of the jaws allows the consumption of large food items. Cranial kinesis permits inclusion of large items in the diet of vertebrates, such as elasmobranchs, without excluding smaller, more diverse foodstuffs. It has allowed the evolution of gigantism, one advantage of which is avoiding predation. The approximately 225 species of sharks (Figure 5–22) and 350 species of skates and rays known today are all large organisms, even for vertebrates. An average sized shark species is about 2 m in length, an average ray half that length. Nevertheless a few interesting miniature forms have evolved and inhabit mostly deeper seas off of the continental shelves.

The green dogfish, *Etmopterus virens*, attains a length of only 25 cm. The black and grey bodies of these schooling sharks are punctuated with an elaborate pattern of green-glowing photophores. The stomach contents of green dogfish, like those of much larger relatives, indicate they feed on organisms larger than themselves. More than half of their food is squid and octopus, the eyes, and beaks of which are sometimes so large that one wonders how they could have passed through the jaws and throat! Dr. Stewart Springer, long a student of living elasmobranchs, concludes that "green dogfish hunt in packs and may literally swarm over a squid or octopus much larger than they are biting off chunks . . . and perhaps maintaining the integrity of their school visually through their distinctive lighting system."

Another miniature shark, *Isistius brasiliensis*, is even more bold in its feeding on large prey. This brilliantly luminescent shark lives in tropical deep waters in what is known as the **deep scattering layer** (DSL), a concentrated band of vertically migrating animals of great variety. Many surface predators descend to the DSL in feeding forays; porpoises and tuna, which may weigh 400 kg, are among the voracious vertebrates who visit the DSL. These giant, fast swimming predators often suffer wounds of curious origin, the shape of a silver dollar and about 1 cm in depth. Only recently have these wounds been matched in size and shape to the jaws of *Isistius*! Whether the photophores of a 40-cm *Isistius* play a part in some deception of the large predators on which it feeds or are used as social cues awaits observation from a deep submersible.

On the other end of the modern elasmobranch size spectrum are the largest living fishes: the whale shark and the basking shark. The basking shark, *Cetorhinus maximus*, 10 m long or more, lives in subpolar and temperate seas feeding exclusively on zooplankton, such as millimeter-long copepods. A feeding basking shark swims with its mouth wide open. Over a thousand long, erectile, and whiplike denticles on the inner surface of the gill arches strain from the surface waters the hundreds of pounds of food needed each day. During the winter months when plankton stocks are at a low, basking sharks disappear and are thought to rest on the bottom after shedding their gill rakers for the season. In contrast, whale sharks, *Rhincodon typus* which grow

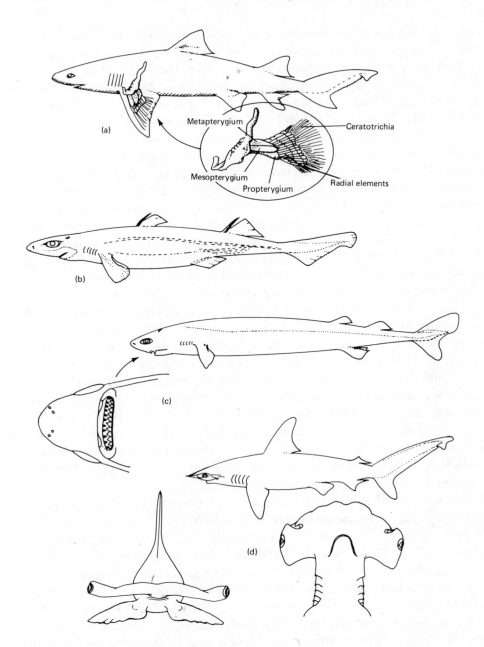

Figure 5-22 Some interesting examples of modern sharks. (a) *Negaprion brevirostris*, the lemon shark, is widely used in elasmobranch research but can attain a size sufficient to threaten humans because it inhabits warm waters of the Atlantic frequented by bathers and divers. The internal anatomy of the pectoral girdle and fin are shown superimposed in their correct relative positions. (b) *Etmopterus vierens*, the green dogfish, is a miniature shark only 25 cm in length, yet it feeds on much larger prey items. (c) *Isistius brasiliensis*, the "cookie cutter" shark, is another miniature species whose curious mouth (inset) is highly adapted to take chunks from fish and cetaceans much larger than itself. (d) The hammerhead shark (*Sphyrna* sp.) in lateral, ventral, and frontal views.

to 20 m are more tropical and able to feed on plankton year around. Special branchial adaptations have converted their gills to function as enormous sieves as well as respiratory surfaces. These solitary, white spotted and striped Goliaths feed head up and tail down at the surface. They lift above the water in the middle of a shoal of small fish until all the water in their orobranchial chamber has drained out the gill slits. With gills closed and mouth open the whale shark sinks tail down until surface water floods over the rim of its terminal mouth, bringing with it sustenance for its 10 ton body. Recently humpback whales (*Megaptera novaengliae*) have been observed feeding in a similar manner.

Although it seems paradoxical that the largest vertebrates feed on tiny motes floating in the sea and the smallest species tackle organisms many times their own size, there are good ecological reasons behind these phenomena. Sunlight is the ultimate source of energy for life on earth. No vertebrate, however, has the ability to capture this energy directly for nutritional use, and so we must all obtain our food from indirect sources. The enumeration of the indirect sources through which solar energy passes is called a **food chain** or, because few vertebrates feed on a single food source, more properly, a **food web**. At each point of transfer in a web a great deal of energy is "lost." On the average only about 10 percent of the energy in each step is passed to the next. Clearly a vertebrate feeding on predatory tuna, as tiny *Isistius* does, is far removed from the primary source of energy. Because their individual and population requirements are comparatively low, the feeding methods employed by *Isistius* are effective. If a whale shark fed primarily on tuna, it would simply not be able to find enough food to maintain its enormous bulk. The solution for whale sharks and all extremely large vertebrates—most dinosaurs, moas, elephants, blue whales—is or has been to feed near the sunny side of the food web by eating plants or organisms that eat plants (Figure 5–23).

One extinct shark of potentially enormous size seems not to fit this pattern of nature. An awesome attraction in the halls of the American Museum of Natural History is the reconstruction of the jaws of *Carcharodon megalodon* (Figure 5–24). This "great" white shark, whose fossil teeth (as much as 15 cm in length) are found in deposits of early Cretaceous to Pleistocene age all over the world, had individual teeth practically identical except for size to those of the living white shark *Carcharodon carcharias*. The preparators for the museum used the largest teeth they could find to reconstruct the jaw. In an uncharacteristic lapse of attention to detail, most of the teeth used came from the middle of the upper jaw. This particular position produces the largest teeth in today's white shark. The reconstructed jaw of *C. megalodon* suggests it was nearly one third larger than it probably was in life. Instead of being 20 m long and weighing 50 tons, Dr. John Randall of the Bishop Museum in Honolulu has calculated that *C. megalodon* was "only" about 14 m long! The largest *C. carcharias* on record was an Australian specimen slightly over 11 m in length and weighing at least 2 tons. The white sharks so feared by inshore swimmers are perhaps not entirely representative of the species; larger ones probably exist. Captures of white sharks below 1000 m depth suggest that enormous specimens

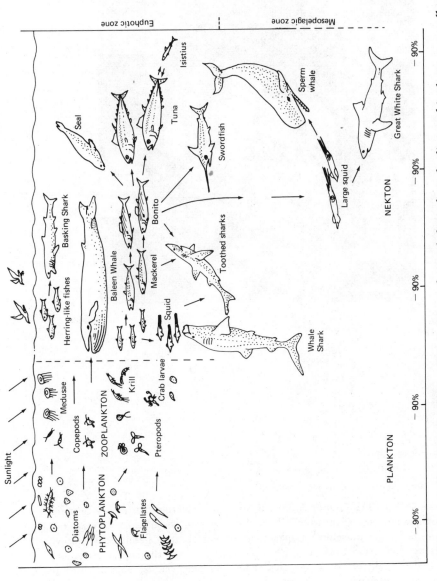

Figure 5-23 Food web in the open sea indicating the trophic position of several elasmobranchs discussed in the text as well as several other vertebrates to be discussed in the chapters that follow. Note the relatively low (near the sun) position of the largest elasmobranchs, the whale shark and basking shark, and the extreme ecological distance from the sun's energy of such specialists as *Isistius*, which is one of the smallest of sharks.

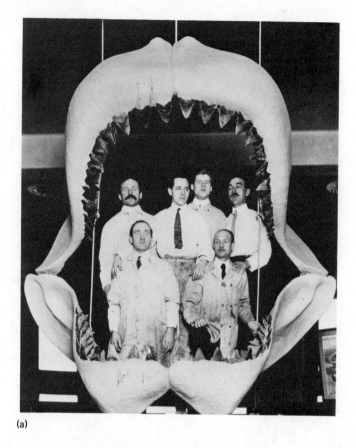

(a)

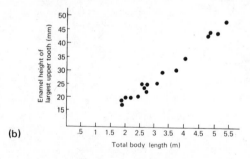

(b)

Figure 5-24 (a) The famous American Museum of Natural History reconstruction of jaws of the giant fossil shark *Carcharodon megalodon*. Overzealous preparators in the museum's Department of Vertebrate Paleontology (shown framed by their handiwork in a 1909 photograph) used the largest teeth they could find to make the most spectacular jaw. In living sharks, teeth diminish in size as they approach the corners of the jaw but the preparators used teeth primarily from the anterior positions of the upper jaw, much exaggerating the final reconstruction. Estimates of the length of *Carcharodon megalodon* have ranged from 18.2 to 30.6 m (60 to 100 ft). (American Museum of Natural History.) (b) A regression of tooth size on actual body length for the living *Carcharodon carcharias* indicates by extrapolation that *Carcharodon megalodon* was "only" 13 m (43 ft) in length! (Randall, J.E., *Science*, 181:169–170. 13 July 1973.)

may live at these depths, perhaps feeding on giant squid in competition with sperm whales! Whatever the diet, white sharks are rare and exceptionally mobile. These two factors allow them to rely on prey animals that are also far from the solar end of the food web.

Given the enormous range in size of modern elasmobranchs, a great diversity in skeletal development is expected. However, all modern sharks have common skeletal characteristics that earlier shark radiations lacked. The continuous notochord of earlier sharks was replaced in the modern radiation by cartilaginous centra that calcify in several distinctive ways. Between centra, spherical remnants of the notochord fit into depressions on the opposing faces of adjacent vertebrae. Thus, the axial skeleton is an exquisite laterally flexible structure with rigid central elements swiveling on a ball-bearing joint of calcified cartilage and notochordal remnants. The axial skeleton is incompressible to longitudinal forces, compression being evenly transmitted from centrum to centrum by the notochordal remnants. In addition to the neural and hemal arches, extra elements not found in the axial skeleton of other vertebrates (the intercalary plates), protect the spinal cord above and the major arteries and veins below the centra.

Perhaps in conjunction with the locomotor advances inherited from the hybodonts and these new structural adaptations, shark squamation also changed. The scales of modern sharks are single-cusped and have a single pulp cavity. The size, shape, and arrangement of these **placoid** scales influences the flow of water adjacent to the body surface and therefore the efficiency of swimming. It seems most likely that this simplification and standardization of denticles relates to increased locomotor efficiency. Some pelagic sharks have secondarily evolved pectoral fins whose radial cartilages extend to the tips of their fins in a manner similar to pelagic Devonian sharks. That these are secondary derivatives is indicated by the persistence of long ceratotrichia over the surface of the radials.

Skates and rays are derivatives of the modern shark radiation adapted for benthic habitats. They too have radial cartilages that extend to the tips of their pectoral fins. However, in skates and rays, these fins are greatly enlarged and the anterior most basal elements fused with the chondrocranium in front of the eye or with each other in front of the rest of the head. These specializations are related to the very different mode of locomotion employed by skates and rays.

A characteristic that differentiates all chondrichthyans from other living jawed fishes is that they never evolved a gas-filled bladder to make them buoyant or to use as an accessory respiratory device. Although the well-oxygenated waters of the marine habitat in which most elasmobranchs live eliminates selective advantage for the latter, buoyancy is still vital. Elasmobranchs (and holocephalians) utilize their liver to counteract the weight of their dermal denticles, teeth, and calcified cartilages. Dr. H. D. Baldridge of the Mote Marine Laboratory has found the average tissue densities of Florida sharks with their livers removed to be 1.062 to 1.089 grams per milliliter (g/ml). Because sea water has a density of about 1.030 g/ml a shark not swimming to stay afloat would sink. The liver tissue of sharks is well known for its oil content, the primary oil (a carotene derivative) is known as squalene and is contained in large vacuoles within the liver cells. Because of its high fatty content, shark liver tissue has a density of about 0.95 g/ml. The liver may contribute as much as 25 percent of the body weight. By adjusting the oil content and size of their livers, sharks can adjust their buoyancy as well as store energy. A 4-m tiger shark (*Galeocerdo cuvieri*) weighing 460 kg on land may weigh as little as 3.5 kg in the sea! Not surprisingly, benthic sharks have livers with fewer and smaller oil vacuoles in their liver cells.

Long before it was realized that sharks may weigh very little in their natural habitat due to the squalene stored in their livers, a very different mechanism was credited with allowing sharks to float. As introduced in the description of the *Hybodus*-like elasmobranchs, the peculiar hydrodynamics of a laterally oscillated heterocercal tail was thought to produce an upward lift as well as a forward thrust (see also Chapter 7, Section 7.1). If anterior surfaces such as the pectoral fins and perhaps the snout are properly oriented, they too will produce lift as the shark moves through the water. Thus, although gravity pulls a shark downward, two areas of lift—one anterior and one posterior— were considered to act on a swimming shark permitting it to float. Two objections may be raised to this long used explanation. First, through the work of Baldridge and others we now know that a motionless pelagic shark may be very nearly neutrally buoyant. Any lift, either anterior or posterior, during forward locomotion would therefore cause a shark to rise in the water column—something we know does not necessarily occur. Secondly, the amount of lift resulting from the heterocercal tail and the pectoral fins is proportional to the forward velocity; the gravitational force it is supposed to counteract is constant. Hence a shark would be neutrally buoyant and able to remain at the same height in the water column only over a narrow range of swimming speeds. If the shark swam too slowly, it would sink; if it swam too rapidly, it would rise. Such limitations and difficulties are not apparent from observations of living sharks.

Dr. K. S. Thomson of the Peabody Museum of Natural History at Yale University has carefully analyzed motion pictures of the swimming action of several shark species. He finds that the heterocercal tail is capable of delivering thrust that can be oriented in a wide range of angles, not simply forward and up as previously supposed. This new analysis stems primarily from the fact that the heterocercal caudal attaches to the undulating axial skeleton along a considerable length of the vertebral column. The more ventral portions of the caudal fin attach further anteriorly than their dorsal counterparts and thus react to the rearward passage of a lateral undulation at a different (earlier) time. In addition the portions of the heterocercal caudal that lack the rigidity afforded by the presence of the axial skeleton do have strong muscular and connective tissue elements that actively alter the shape of the various lobes to produce differing hydrodynamic results. Thomson believes that in general the thrust produced by the heterocercal tail projects forward and through the plane of the pectoral fins' insertions. The orientation of the pectorals efficiently controls the shark's position in the water column. Indeed, Thomson thinks that the heterocercal tail allows sharks to develop extremely powerful dives and climbs in the water, permitting them to make oblique attacks and shear off flesh from large prey. It will be interesting to see how researchers apply this functional analysis to the many forms of fossil fishes with heterocercal tails.

Although not unique to elasmobranchs, the sensory systems of modern sharks, skates, and rays are certainly refined, diverse, and undoubtedly significant to the success of living forms. Initially sharks may detect prey via their **lateralis system**, an interconnected series of superficial tubes, pores, and sensitive cells distributed over the head and along the sides which respond to vibrations transmitted through the water. It has been shown that the **ampullae of Lorenzini**, mucus filled tubes with sensory cells and afferent neurons at their base, are exquisitely sensitive to electrical potentials and can even detect prey from their weak electrical fields (see Chapter 7, Section 7.2.2). Sharks have been described as "swimming noses," so acute is their sense of smell. They can detect many substances, including blood, diluted to 1 part in over a million of sea water. Hammerhead sharks, *Sphyrna* sp. (Figure 5–22d) may have enhanced the directionality of their olfactory apparatus by placing the nostrils far apart on the odd lateral expansions of their heads.

Finally, vision is very important at close range to the feeding behavior of sharks. Especially well developed are mechanisms for vision at low light intensities at which humans would find vision impossible. This is primarily due to a rod-rich retina assisted by numerous platelike crystals of guanine in the choroid layer located just behind the retina. Called the **tapetum lucidum**, the crystals function like microscopic mirrors to reflect light back through the retina and increase the chance that light rays will be absorbed. This mechanism, although of great benefit at night or in the depths, has obvious disadvantages in the bright sea surface of midday. To regulate the amount of bright light, melanin pigment expands over the reflective surface to occlude the tapetum lucidum and absorb all light not stimulating the retina on first penetration. In the lemon shark, *Negaprion brevirostris* (Figure 5–22a), the tapetum may be completely occluded in as little as 1 hour of exposure to bright light. Regaining full reflectivity is an even faster process when a shark is placed in darkness. With so many sophisticated sensory systems, it is perhaps not surprising to find that the shark brain is proportionately heavier than the brains of other fishes and approaches the brain-to-body weight ratios of some tetrapods.

Like the sensory apparatus, the reproductive biology of modern grade elasmobranchs is not unique, but a great deal of their success as a group results from their "advanced" breeding mechanism. Internal fertilization is universal in modern elasmobranchs. The pelvic claspers of modern species have a consolidated skeletal structure that may increase their copulatory efficiency. During copulation (Figure 5–25a) a single clasper is bent at 90° to the long axis of the body. Since the right clasper flexes to the left and vice versa, the dorsal groove present on each clasper comes to lie directly under the cloacal papilla.

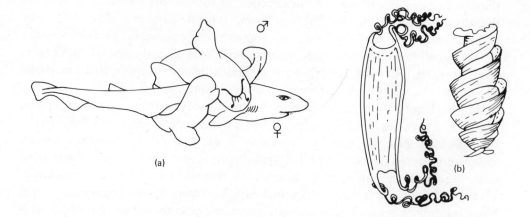

Figure 5-25   (a) Copulation in the smooth dogfish *Scyliorhinus*. Only two other species of shark and fewer species of ray have been observed *in copulo* but all assume analogous postures so that one of the male's claspers can be inserted into the female's cloaca. (b) The egg cases of two oviparous species of shark, *Scyliorhinus* (left) and *Heterodontus* (right). (Not to same scale.)

The single flexed clasper is inserted into the female cloaca and may be locked there by an assortment of barbs, hooks, and spines near the clasper's tip. Male sharks of small species secure themselves *in copulo* by wrapping around the female's body. Male skates accomplish much the same thing by biting down on the female's pectoral fin.

Exactly what position the larger sharks assume is conjectural. A pair of 3-m sharks held in captivity have been seen swimming side by side with their heads held slightly apart and the posterior half of their bodies in close contact. Whether this was copulation or courtship is questionable. The same male was later seen to swim close to the female and suddenly sink to the bottom in a curled position for periods as long as 4 minutes. Prolonged curled positions are known to occur during copulation of smaller sharks. Whatever the position taken by the pair, when the single clasper is crossed to the contralateral side and securely inserted into the female cloaca, sperm from the genital tract are ejaculated into the clasper groove. Simultaneously a muscular subcutaneous sac extending anterior beneath the skin of the pelvic fins contracts. This **siphon sac** has a secretory lining and is filled with sea water by pumping activity of the fins before copulation. The siphon sac's seminal fluid washes sperm down the groove into the female's cloaca, from which point the sperm ascend the female reproductive tract.

The female has specialized structures at the anterior end of the oviducts, the **nidimental glands**, which secrete a proteinaceous shell around the fertilized egg. All common shark eggs are large (the size of a chicken yolk or larger) and contain a very substantial store of nutritious yolk. Consequently **oviparous** (*ovum* = egg, *pario* = to bring forth) elasmobranchs have very large egg cases with openings for sea water exchange and some bizarre protruberences to tangle or wedge themselves (as far as known, without the mother's help) into protected portions of the substrata (Figure 5–25). During the prolonged 6–to 10-month developmental period (it can hardly be called incubation), the zygote obtains nutrition exclusively from the yolk. Inorganic molecules including dissolved oxygen are taken from a flow of water induced through swimming-like movements made by the developing fish. Upon hatching, the young are generally miniature replicas of the adults and newly hatched sharks and skates seem to lead a life much as they do when mature.

A most significant step in the evolution of elasmobranch reproduction was prolonged retention of the fertilized eggs in the reproductive tract. Today many **ovoviviparous** (*vivus* = alive) species retain the developing young within the oviducts until they are able to lead an independent life. Few modifications in morphology are necessary: reduction in the nidimental gland's shell production, and increased vascularization of the oviducts and yolk sacs are the only notable differences between oviparous and ovoviviparous forms. All nutrition comes from the yolk, only inorganic ions and dissolved gases are exchanged between the maternal circulation and that of the developing young. The shell is very thin and may be an elongate tube, called a candle, that contains several embryos within the female's oviduct. The eggs often hatch within the oviducts, and the young may spend as long in their mother after hatching but before birth as they did within the shell. As many as 100, but more often about a dozen, young are born at a time.

A natural next step from the ovoviviparous condition is **viviparity** in which the embryos are not only protected but also where their nutritional supply is continuous and not limited to the yolk. Elasmobranchs have independently evolved viviparity several times, judging from the diversity of mechanisms used to get nourishment to the young. Some elasmobranchs develop long spaghetti-like extensions of the "uterine" (oviduct) walls that penetrate the mouth and gill openings of the internally hatched young and secrete a milky nutritive substance. Other species simply continue to ovulate, and the young that hatch in the uteri feed on these eggs! The most common and most complex adaptation to true viviparity is the yolk sac placenta. The yolk sac of each embryo becomes intimately associated with the uterine wall of the mother, and nourishment is obtained from the maternal blood stream via the highly vascular yolk sac of the embryo. A curious feature of viviparous sharks is that the left ovary is rudimentary and the right ovary supplies both oviducts. Perhaps this is possible because of the efficiency of viviparity in producing young.

For the most part we have been examining the characteristics of **pleurotremate** elasmobranchs—the sharks. These forms number about 300 living species. Surprising to many is the fact that the **hypotremate** elasmobranchs—the skates and rays—are more diverse than are the sharks! Approximately 400 living species of skates and rays are currently recognized. These fishes are closely interrelated and have a long history of phylogenetic isolation from all living sharks. The suite of adaptations characteristic of skates and rays relates to their early assumption of a benthic, durophagus (*duro* = hard, *phagus* = to eat) habit.

The teeth are almost universally hard, flat crowned plates which form a pavementlike dentition. The mouth may be highly and rapidly protrusible to provide a powerful suction used to dislodge shelled invertebrates from the substrate.

Locomotion in these dorsoventrally depressed fishes is by undulation of the massively enlarged pectoral fins. Generalized members of the group produce deep waves of depression and elevation, which pass back along the length of the expanded pectorals (Figure 7–6c) and push on the water, often resulting in startling accelerations. This great flexibility of the fins is accomplished by loss of extensive dermal armor. The placoid scales so characteristic of a shark's integument are absent from large areas of the body, especially the pectoral fins. The few remaining denticles are often greatly enlarged to form sharp, stout bucklers along the dorsal midline. Perhaps to compensate for this relative nudity, or perhaps as a predatory ruse, many of these skates and rays cover themselves with a thin layer of sand. They spend hours partially buried and nearly invisible except for the dorsally prominent eyes and spiracles through which they survey their surroundings and take in fresh respiratory water. More advanced forms, such as sting rays, have a very few greatly elongate and venomous modified placoid scales at the base of the tail—an apparently sufficient defense for their otherwise entirely naked bodies. The most highly specialized rays are derived from these entirely naked types but spend little of their time resting on the bottom. Using powerfully extended pectorals, these devilfish or manta rays (up to 6 m in width) "fly" through the open sea with motions that look quite different from those of their benthic relatives. Closer examination of their locomotion reveals that it is proportionately rather than qualitatively distinct from that of other hypotremates. It is interesting to note that skates and rays are primarily benthic invertebrate feeders (occasionally managing to capture small fishes), but the largest rays, like the largest sharks, are plankton strainers.

5.4.5   A Second Clade of Chondrichthyans—Holocephali

Though the bulk of living chondrichthyans are contained in the Elasmobranchii, a most interesting if small portion are grouped as ratfish or chimaeras (subclass Holocephali) Concerning the roots of Holocephalian phylogeny, there is little agreement. Two distinct fossil groups have been proposed as chimaeran ancestors. A group of peculiar placoderms, the ptyctodontids, are known primarily from the mid-Devonian. Rarely exceeding 20 cm in length, they show a reduction in the extent and number of head and thoracic shield plates. A short anteriorly situated palatoquadrate tightly bound to the cranium carried a single pair of large upper tooth plates that were opposed by a smaller mandibular pair. The gills were covered by a single operculum and postcranial characters, such as fin spines, paired appendages, claspers and caudal development, are strongly reminescent of living holocephalians. The belief that they were ancestors of living chimaeras may have had a great deal to do with the

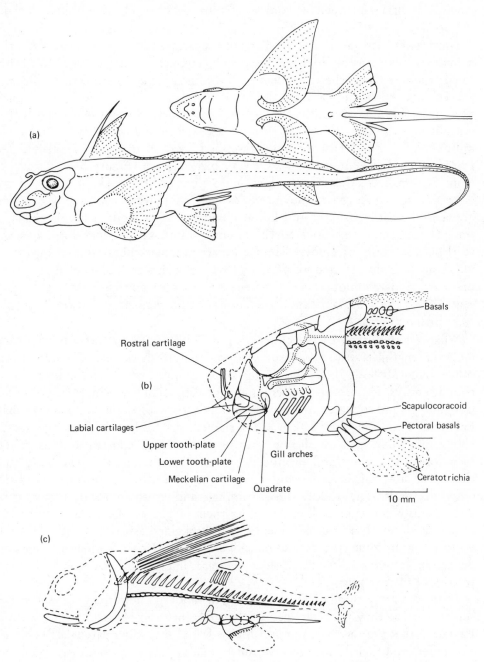

Figure 5-26 (a) The common chimaera *Hydrolagus* sp., a living holocephalian, in lateral and ventral views. (Modified after Bigelow, H. B. and W. C. Schroeder; (1953. Fishes of the Western North Atlantic. Part two.) (b) *Ctenurella* an upper Devonian ptyctodont placoderm believed by some paleontologists to be a direct chimaera ancestor. (Modified after Moy-Thomas and Miles [5]). (c) *Phomeryele,* a Carboniferous iniopterygian "shark" proposed by another group of paleontologists as an ancestor to the modern ratfishes. (Modified after Greenwood et. al. [3]).

obvious similarity between ptyctodont reconstructions and living holocephalians (Figure 5–10c and 5–26a and b).

An unfortunate gap in the fossil record separates these placoderms from the chimaeras. The first undoubted holocephalians are of Jurassic age, and other fossils proposed as earlier members of the chimaera lineage do little to link the ptycodonts and modern forms. Ptycodonts are more like *modern* holocephalians than the more primitive holocephalians which are in some cases rather sharklike. Since the 1950's Dr. Rainer Zangerl and coworkers at the Field Museum of Natural History have described a group of Pennsylvanian fishes from the central United States. Of very bizarre form (Figure 5–26c) and obviously specialized, these Inioterygia (*inion* = back of the neck, *pteron* = wings) have characteristics that convinced Zangerl they were evidence for a link between the earliest sharks and holocephalians. As in modern holocephalians, the palatoquadrate is fused to the cranium (autostylic suspension), but the teeth, unlike those of modern chimaeras, are in replacement families like those of elasmobranchs. If iniopterygians prove to connect holocephalians with primitive grade elasmobranchs, then the ptyctodonts and chimaeras will represent one of the most outstanding examples of convergent evolution in groups from different eras.

Whatever the origin of the chimaeras, the 25 living forms (none much over a meter in length) have a soft anatomy more similar to sharks and rays than to any other living fishes. They have long been grouped with elasmobranchs as chondrichthys because of their common specialization involving a loss of bone. Generally found in water of over 80 m depth, the Holocephali move into shallow water to deposit their 10-cm horny shelled eggs from which hatch minature chimaeras. They feed on shrimp, gastropod mollusks, and sea urchins. Their locomotion is curious and produced by lateral undulations of the body that throw the long tail into sinusoidal waves and by fluttering movements of the large pectorals. The solidly fused nipping and crushing tooth plates grow continuously throughout life, adjusting their height to the wear they suffer. Of special interest are the armaments: a poison gland associated with the stout dorsal spine in some species, and macelike cephalic claspers of males. A curious detour in vertebrate life, the Holocephali are, like the other chrondrichthyes, worthy of further study.

From the array of adaptations present in modern chondrichthyans, it is not surprising that they have survived unchanged since the Mesozoic. The surprising fact is that they are not more diverse or found in a wider variety of habitats. To understand their limitations we must examine their competition—the bony fishes (Chapter 6).

References

[1]   Various authors 1977. Recent advances in the biology of sharks. *American Zoologist* 17(2). "State of the art" symposium papers review a wide variety of topics with extensive references to the important literature.

[2]   Brodal, A. and R. Fänge (editors). 1963. *The Biology of Myxine.* Universitetsforlaget. Oslo, Norway. A compilation of data and references on what many zoologists consider the most primitive living vertebrates.

[3]   Greenwood, P. H., R. S. Miles, and C. Patterson (editors). 1973. *Interrelationships of Fishes.* Academic Press, New York. A technical compendium of views of world-recognized authorities on the evolution of fishes.

[4]   Hardisty, M. W. and I. C. Potter (editors). 1971. *The Biology of Lampreys.* Academic Press, New York. Two volumes survey much of what is known about these fascinating vertebrates.

[5]   Hodgson, E. S. and R. F. Mathewson (editors). 1978. *Sensory Biology of Sharks, Skates, and Rays.* Tech. Information Div., Naval Res. Lab., Wash., D. C.

[6]   Moy-Thomas, J. A. and R. S. Miles. 1971. *Palaeozoic Fishes.* W. B. Saunders, Philadelphia. The classic, concise, and well illustrated review of knowledge of early vertebrate evolution.

[7]   Repetski, J. E. 1978. A fish from the Upper Cambrian of North America. *Science* 200:529–531. Contains references to several reports on the earliest vertebrate remains.

[8]   Romer, A. S. 1966. *Vertebrate Paleontology.* Third edition. University of Chicago Press, Chicago. The classic introductory text on the subject and basic reference for this and many of the following chapters.

[9]   Romer, A. S. 1968. *Notes and Comments on Vertebrate Paleontology.* University of Chicago Press, Chicago. A series of ideas, arguments and insights the author thought inappropriate in his regular text but which make interesting reading.

6

The Bony Fishes — Osteichthyes

Synopsis: At the time of their first appearance in the fossil record, osteichthyans are separable into distinct lineages, each with unique adaptations and evolutionary potentials. The Dipnoi (lungfishes), the Crossopterygii (lobe-finned fishes) and the Actinopterygii (ray-finned fishes) show morphological indications of common ancestry, but they are not equally interrelated (Table 6–1). Ray-finned fishes were distinct as early as the Devonian. The lungfishes and lobe-fins may be grouped together as the Sarcopterygii (fleshy-finned fishes). Although living sarcopterygians are few in number, they offer exciting glimpses of adaptations evolved in Paleozoic environments. The evolution of actinopterygians is the great success story of vertebrate history. They inhabit over seventy per cent of the earth's surface and are the most numerous and speciose of vertebrates. Several levels of development in food gathering and locomotory structures, each followed by adaptive radiations, characterize actinopterygian evolution. These radiations are represented today by two relict groups, the chondrosteans (the sturgeon and related forms) and the holosteans (the gars and *Amia*), and by the very successful teleosteans which number close to 30,000 living species. Two groups of teleosts, ostario-physans in fresh water and acanthopterygians, characteristically in sea water, constitute a large proportion of these species. Teleostean adaptations of morphology, behavior and life history are so numerous and diverse that their dominant position in the aquatic and marine ecosystems of the world is understandable. The special adaptations of teleosts to life on coral reefs and in the deep sea serve as illustrations of the plasticity which is the hallmark of the teleostean form of vertebrate life.

Table 6-1. A Classification of Osteichthyes, the Bony Fishes

Class Osteichthyes (Bony fishes)

    Subclass Sarcopterygii (Fleshy finned fishes)

      Superorder Dipnoi (Lungfishes)

      Superorder Crossopterygii (Lobe-finned fishes)

        Order †Rhipidistia

            Actinistia (Coelacanths)

    Subclass Actinopterygii (Ray-finned fishes)

    Infraclass Chondrostei

      Order †Paleonisciformes

          Polypteriformes (Bichirs)

          Acipenseriformes (Sturgeons and paddlefishes)

    Infraclass Holostei

      Order Semionotiformes (Gars)

          Amiiformes (Bowfins)

    Infraclass Teleostei

      Superorder Clupeomorpha (Herrings)

          Elopomorpha (Tarpons and eels)

          Osteoglossomorpha (Bony tongues)

          Protacanthopterygii (Trouts and lantern fishes)

          Ostariophysi (Catfish and minnows)

          Paracanthopterygii (Cods and anglerfishes)

          Acanthopterygii (Spiny rayed fishes)

† = Extinct

6.1   MAJOR GROUPS OF BONY FISHES

Remains of the bony fishes first appear in the Lower to Middle Devonian; nevertheless, they represent a radiation of forms already in full bloom. By early Devonian three major and distinctive **osteichthyian** (*osteo* = bone, *ichthys* = fish) types were well developed. These osteichthyans possessed a combination of locomotor and trophic characters that gave them sufficient advantage to become the dominant fishes during the Devonian—the "Age of Fishes." Some produced the major adaptations that led to land vertebrates. Most, however, retained fishlike characteristics and these gave rise to modern bony fishes—fishes so numerous that they constitute the largest group of living vertebrates.

When and where did bony fishes originate? Who were their progenitors? How are the distinguishable groups interrelated? The answers are not clear. The distinctive and well-diversified forms that appeared must have had a considerable time to evolve. Fragmentary bony fish remains are known from the Upper Silurian, and the class must have had its origin sometime during that period. In certain details of the head structure, osteichthyans resemble acanthodians, but the latter possess an array of unique specializations that preclude them as ancestors of the Osteichthyes. The similarities may point to a common ancestor for acanthodians and osteichthyans in the early Silurian.

6.1.1 The Earliest Osteichthyes

Articulated and often abundant remains of osteichthyans are known from the Middle Devonian. The three basic types were the lungfishes or Dipnoi (*di* = double, *pnoe* = breathing; Figure 6-1a and b), the Crossopterygii (*cross* = a fringe or tassel, *pterygium* = fin; Figure 6-1c and d), and the Actinopterygii (*actino* = stout ray, *pterygium* = fin; Figure 6-1e and f). All were rather similar and not easily confused with any other Devonian vertebrates. Characteristic of the Osteichthyes were similar patterns of lateral line canal distribution; similar opercular and pectoral girdle dermal bone elements; fin webs supported by bony dermal rays; and, perhaps most important to their future evolution, all osteichthyans appear to have an air-filled diverticulum of the esophagus functioning as an accessory respiratory organ *and* bouyancy device. In addition, many of the forms had two dermal bones (a premaxillary and a maxillary) forming the upper biting edge of the mouth, with teeth typically fused to them, and a neurocranium composed of anterior and posterior ossified sections separated by a fissure that was cartilaginous or completely free, allowing movement between the two halves of the skull.

Note that the presence of bone, although common to all three groups of Osteichthyes, is not a unifying characteristic since agnathans, placoderms, and acanthodians also possessed true bone. The name Osteichthyes was coined before the occurrence of bone in other primitive vertebrates was realized. Likewise, the various named subgroups of the actinopterygians, having first been described from living and often degenerate forms, imply an increase in the ossification of fishes as an evolutionary trend (for example, chondrosteans, the "cartilaginous bony fishes," gave rise to holosteans, "entirely bony fishes," which culminated in teleosteans, "final bony fishes"). The fossil record, however, indicates the opposite is true.

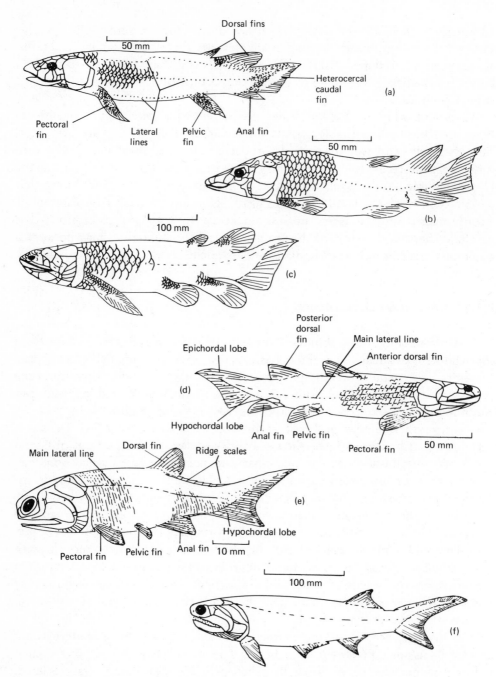

Figure 6–1 Primitive osteichthyans. (a and b) Dipnoans, (c and d) Crossopterygians, (e and f) Actinopterygians. (a) Relatively unspecialized dipnoan *Dipterus*, middle Devonian. (b) Long-snouted dipnoan *Griphognathus*, Upper Devonian. (c) Laterally compressed porolepiform *Holoptychius* Upper Devonian to Lower Carboniferous. (d) Cylindrical osteolepiform *Osteolepis*, Middle Devonian. (e) Typical early actinopterygian *Moythomasia*, Upper Devonian. (f) Fine-scaled actinopterygian *Cheirolepis*, Middle to Upper Devonian. (Modified from Moy-Thomas and Miles Ch. 5 [5]).

The three groups of early osteichthyans are not equally interrelated. Crossopterygians and dipnoans have similar body shapes and sizes (20 to 70 cm) and a similar arrangement and structure of fins. Each has two dorsal fins, an epichordal lobe on the heterocercal caudal fin, and paired fins with a notably fleshy, scaled central axis. The paired fins' rays extend in a feather or compound leaflike manner in contrast to the fanlike form of actinopterygian paired fins. In addition, crossopterygian and dipnoan affinities are strengthened by similarities in the exact position of the cheek lateral line, the shape and construction of the scales, the mode of ossification of the vertebral centra (when present), and so on.

For some time both groups were thought to possess a third pair of nasal openings, called the **choanae**, connecting each olfactory capsule with the palate. Due to these shared characters the crossopterygii and the dipnoi were combined in a single group, known as the Choanichthyes. It has been shown subsequently that dipnoi do not have a third nasal opening, nor do all crossopterygians. A more appropriate name including both crossopterygians and dipnoans is now employed: Sarcopterygii (*sarcos* = fleshy, *pterygium* = fin). Though abundant and prolific in the Devonian, the sarcopterygians dwindled in the late Paleozoic and Mesozoic and today only four genera remain: among the dipnoans *Protopterus* in Africa, *Lepidosiren* in South America, the endangered *Neoceratodus* in Australia, and the rare deep-water crossopterygian *Latimeria* of the Comoro Archipelago off East Africa. Nevertheless, these fishes are of importance to us, for terrestrial vertebrates originated within their ranks.

6.1.2 Sarcopterygii

Dipnoans

Even the earliest fossil dipnoans, which were marine, indicate a separation from the other osteichthyans. The dipnoi are immediately distinguishable from other osteichthyans (except coelacanths, see page 185) by the lack of marginal tooth-bearing dermal bones, the premaxillaries and maxillaries. The palatoquadrate is fused to the undivided cranium. In these earliest dipnoans, the teeth were scattered over the entire palate and, in addition, concentrated into fused tooth-ridges along the lateral palatal margins. Powerful adductor muscles of the lower jaw spread upward over the cranium. Throughout their evolution, this specialized **durophagous** (that is, feeding on hard foods) crushing apparatus has persisted. During the Devonian the lungfish body underwent rapid evolution, terminating in a form quite dissimilar to the other Osteichthyes. The anterior dorsal fin was reduced and eventually lost. The remaining median fins elongated and fused around the posterior third of the body. The caudal fin, originally heterocercal, became symmetrical through enlargement of the epichordal lobe. The early complicated mosaic of small dermal bones of the skull (often covered by a continuous sheet of cosmine, an enamellike substance) evolved into a pattern of fewer large elements. The extensive sheet of cosmine that covered the head and body was lost. Having undergone rapid changes during the Devonian, the dipnoi have remained conservative and changed very little since that time. Living dipnoans, therefore, probably have lives similar to those of their ancestors.

The monotypic Australian lungfish, *Neoceratodus forsteri* (Figure 6–2a), is morphologically most similar to the majority of Paleozoic and Mesozoic dipnoi. Unlike some fossil lungfishes, but characteristic of all extant forms, *Neoceratodus* is restricted to fresh waters. Were it not for recent transplantings into other fresh waters of Australia, *Neoceratodus* would be limited to a dwindling population in southeastern Queensland. The Australian lungfish attains a length of 1.5 m and a reported weight of 45 kg (100 lb). Although a powerful fighter when netted, it is a slow moving, docile invertebrate feeder. It swims by lateral body undulations or slowly 'walks' across the bottom on its pectoral and pelvic appendages. The chemical senses seem important to lungfishes: their mouths are reported to contain numerous taste buds. The nasal passages are located near the upper lip with the incurrent opening on the snout just outside the

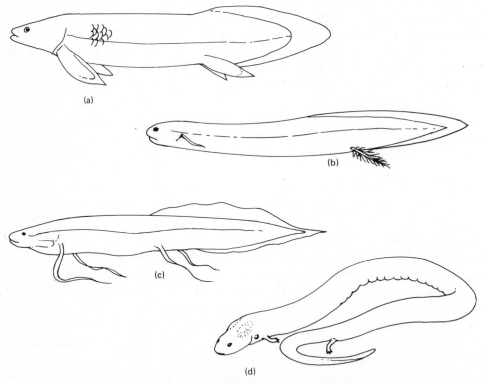

Figure 6-2 Living Dipnoans. (a) The Australian lungfish, *Neoceratodus forsteri*. Note its considerable similarity to the lungfishes of the Paleozoic, Figure 6–1a and b. (b) South American lungfish, *Lepidosiren paradoxa*, male. Note the specialized pelvic fins of the male during the breeding season. (c) African lungfish of the genus *Protopterus*. (d) The amphibian *Amphiuma* is shown for comparison.

mouth and the excurrent openings within the oral cavity. Thus each gill ventilation draws water across the nasal epithelium. Under normal circumstances, *Neoceratodus* respires almost exclusively via its gills and uses its single lung only when stressed. Little is known of the behavior of *Neoceratodus* although pairs supposedly spawn by laying and fertilizing individual eggs on plants and debris and leaving them to develop without parental care.

More is known about the natural history of the single South American lungfish, *Lepidosiren paradoxa*, and the four or so closely related species of African lungfishes, *Protopterus*. These two genera are distinguished taxonomically by a differing number of weakly developed gills. Because of this reduction in the gills, South American and African lungfishes will drown if prevented from using their paired lungs. Nevertheless, the gills remain important in eliminating carbon dioxide. These seemingly scaleless, 1- to 2-m, heavy snakelike forms have unique filamentous but highly mobile paired appendages. For a time they were not considered fishes at all but considered to be specialized urodele amphibians, some of which they superficially resemble (Figure 6–2b and c). These modern lungfishes are mostly cartilaginous and therefore would not be expected to fossilize well. Nevertheless fossils have been found of very similar species.

One habit, **aestivation**, considerably increases the chance of fossilization of the African species. This habit is similar to hibernation but is induced by the extensive drying of the habitat rather than by cold. Both living genera frequent areas that flood during the wet season and "bake" during the dry season. The vast majority of other fishes migrate in and out of these areas to take advantage of the high productivity in the warm and shallow but often oxygen-poor flood waters. The lungfishes, too, enjoy these flood periods, feeding heavily and growing rapidly, but unlike other fishes, which leave the area during periods of drought, the African lungfishes remain and aestivate. When the swamp waters begin to recede, these lungfishes dig meter-deep vertical burrows in the mud that end in an enlarged chamber. As drying proceeds, the lungfish becomes progressively more lethargic and breathes atmospheric air from the burrow opening. The mucus produced at the body surface helps prevent the muddy water of the burrow from clogging the gills, which remain essential to ridding the body of carbon dioxide. Eventually even the water of the burrow dries up, and the lungfish enters the final stages of aestivation. The lungfish retires to the deep chamber, curls into a ball with its tail over its eyes, and secretes a special heavy mucoid envelope around its body. Only an opening at its mouth remains to permit breathing.

Particular kinds of mud and the positioning of the tail over the eyes seem of great importance to successful aestivation. Depriving a drying lungfish of either usually results in its death. Both may relate to sustaining respiratory gas exchange. The late renal physiologist, Professor Homer Smith, who travelled to Africa in the 1930's to study the lungfish, pointed out that certain muds, when dried, crack excessively and tear the mucoid cocoon. This event leads to rapid desiccation of the lungfish's tissues and its death. Although the rate of energy consumption during aestivation is very low, nevertheless it continues, and muscle proteins are utilized as an energy source. The waste products from protein utilization and the problems and advantages to the lungfish are discussed in Chapter 7. Lungfishes normally spend less than 6 months aestivating, but they have been revived after 4 years of enforced aestivation! When the rains return, the withered and shrunken lungfish becomes active and, in less than 1 month, regains its previous size by rehydrating and feeding voraciously on mollusks, large crustaceans, and fishes.

Apparently aestivation is not a recent adaptation of dipnoi. Fossil burrows containing lungfish tooth plates have been found in Devonian, Carboniferous and Permian deposits of North America and Europe. These early lungfishes were as devoid of ossification as the living forms, making fossilization rare. Without the unwitting assistance of lungfish, which initiated fossilization by burying themselves, their fossils would probably not exist.

Both the South American and African lungfishes have more complicated reproductive behaviors than the Australian species. The males of *Protopterus* dig oval pits or holes at the base of tall swamp grasses. They entice several females to lay their salmon-colored, large eggs on the nest's mud floor. As many as 5000 eggs may be laid in a single nest, and after fertilization the male vigorously guards them against all intruders, including humans. During the 50 or so days before the young leave the nest the male performs bouts of surface thrashing which serve to aerate the water, a practice known for many species of osteichthyans. The male *Lepidosiren* digs an elaborate tube-like burrow extending horizontally over a meter just beneath the floor of the flooded swamp. The eggs are laid in a leaf-lined chamber at the far end of this burrow. As yet unconfirmed is the report that male *Lepidosiren* supply oxygen to the cloistered offspring in the brood chamber by use of the pelvic fins. These structures become highly modified during the breeding season into vascular tufts (Figure 6–2b). The male leaves the burrow periodically to gulp atmospheric air and, allegedly, oxygen diffuses from the freshly aerated blood in the modified pelvic fins to the vicinity of the eggs and young.

Crossopterygians

Coincident with the appearance of the dipnoi in the early Devonian, the earliest known crossopterygian fossils are found and were already well differentiated. They shared many similarities with the other Devonian osteichthyans, but their most characteristic feature is the division of the cranium into anterior (nasal) and posterior (otic) portions, moveably articulated behind the orbits. This joint allowed dorsoventral flexion in the middle of the head, the function of which is not understood, except that it possibly allowed changes in the orientation of the open mouth. Two crossopterygian clades are well known: the larger, predatory rhipidistians (many from 1 m to over 4 m in length) and the usually smaller coelacanths (= actinistians), the largest about 1.5 m.

The rhipidistians are divisible into two major types. The heavy bodied, somewhat laterally compressed porolepiformes (Figure 6–1c) had lobate pectoral fins placed high on the sides, both incurrent and excurrent external nostrils and choanae (internal nasal openings) within the oral cavity, no vertebral centra, and a heterocercal tail. The osteolepiformes (Figure 6–1d) were slender-bodied, broad-headed forms typically with thick scales, a single external nostril, with olfactory currents passing directly from the nasal capsule via the choanae into the oral cavity, and variable caudal structure. Importantly, many had ring-like vertebral centra with various accessory ossifications associated with the spinal nerves. Though all rhipidistians were basically freeswimming predators of shallow waters, many osteolepids were adapted for life at the water's edge. Most paleontologists agree the tetrapods arose from these osteolepids. (see Section 9.2.1).

Coelacanths, as yet unknown before the mid to late Devonian, most likely were derived from a rhipidistian. Their hallmarks were the unlobed, anteriorly situated first dorsal fin and the unique symmetrical three-lobed tail, whose central lobe was fleshy and ended in a fringe of rays (Figure 6–3a). Coelacanths also differ from the rhipidistians in the pattern of head bones (they lacked, among other elements, a maxillary), in details of the fin structure, in the absence of internal choanae, and the presence of curious rostral organs. Although the differences of coelacanths from their rhipidistian ancestors suggests rapid evolution during the Devonian, the coelacanths then entered into an unparalled history of evolutionary stability. Even the conservative

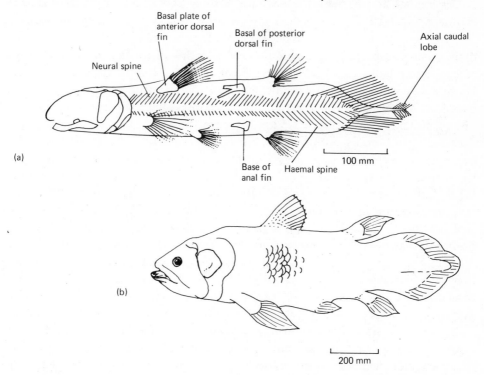

Figure 6–3 Representative coelacanths. (a) Restoration of *Coelacanthus*, upper Carboniferous to late Triassic. (b) *Latimeria chalumnae*, the living coelacanth.

dipnoans changed more. The earliest Devonian coelacanths differ from the most recent fossils (of Cretaceous age) only in the degree of skull ossification. The earliest coelacanths lived in shallow fresh waters, but during the Mesozoic along with a horde of other osteichthyans (see Section 6.2) they inhabited the sea. Most osteichthyans rapidly radiated into a variety of niches, but the coelacanths retained their peculiar form. They remained abundant until the Cretaceous, when they disappear from the fossil record.

Just before Christmas of 1938, a trawler captain out of East London, South Africa, bent over an unfamiliar catch and nearly lost his hand to its ferocious snap. Imagine the astonishment of the scientific community when Dr. J. L. B. Smith of Rhodes University announced that the captain's catch was a living species of coelacanth! Unfortunately, hours on the deck in the heat of the southern hemisphere summer, taxi drivers reluctant to accommodate odoriferous finned passengers, and holiday postal delays prevented Smith from recovering more than a taxidermist's mount of the skin. The characteristics of this large fish (Figure 6–3b) were so similar to those of Mesozoic fossil coelacanths that its systematic position was unquestionable. Dr. Smith named this "living fossil" *Latimeria chalumnae* in honor of Ms Courtenay-Latimer, curator of the East London museum who recognized it as unusual and brought the specimen to his attention. The specific name refers to the site of capture: about 4 km offshore from the mouth of the Chalumna River.

Despite an active public appeal no further specimens of *Latimeria* were captured until 1952. A second damaged individual was taken in the Comoro Archipelago far to the north between Madagascar and Zanzibar. Since then more than 80 specimens ranging in size from 3/4 m to slightly over 2 meters and weighing from 13 to 80 kg have been caught, all in the Comoros. Despite attempts by well-equipped expeditions, all known specimens have been caught by native fishermen. The coelacanths are hooked near the bottom usually in 260 to 300 m of water about 1.5 km offshore. Strong and aggressive, *Latimeria* lives for several hours out of water, but none have survived to be maintained in captivity, perhaps because of a 15°C difference between the bottom and surface water temperatures. *Latimeria* is steely blue-grey with irregular white spots and highly reflective golden eyes. The reflective nature of the eyes results from a tapetum lucidum that enhances visual ability under conditions of low light levels (see section 5.4.4). Along the side of its well-developed olfactory capsules, *Latimeria* has two richly innervated sacs, filled with gelatinous material, that open to the surface through a series of six pores. These rostral organs are perhaps electroreceptors as suggested by Dr. D. E. McAllister of the National Museums of Canada (see section 7.2.2). Whatever sensory modalities are utilized, *Latimeria* is a predator, for stomachs have contained fish and cephalopods.

In spite of its rarity a surprising wealth of data has accumulated on *Latimeria*'s anatomy and physiology. A major asset derived from *Latimeria*'s discovery has been the confirmation of the earlier reconstructions based on coelacanth fossils. As a result we know that these reconstructions were surprisingly accurate. A case in point is that of coelacanth reproduction.

In 1927 D. M. S. Watson described two small skeletons of a Jurassic coelacanth, *Undina*, from inside the body cavity of a larger individual and suggested that coelacanths gave birth to their young. Because copulatory structures have never been found on any coelacanth fossil, some dismissed Watson's specimen as a case of cannibalism. Female *Latimeria* containing up to 19 9-cm shell-less eggs have now been captured. Drs. R. W. Griffith and K. S. Thomson surmised Watson was correct because of the small number of eggs and their lack of a shell to provide osmotic protection. Recently C. L. Smith and others at the American Museum of Natural History were dissecting a 1.6 m specimen and discovered five advanced, 30-cm young in the single oviduct! Internal fertilization must occur. How clasperless male *Latimeria* achieve copulation is unknown.

Of all the crossopterygians *Latimeria* is the sole survivor. Although conservative for osteichthyans, nevertheless the crossopterygians gave rise to the vertebrates that were to master the land. They did not contribute to mastery of the seas; this is the role of their relatives the actinopterygians.

6.2 THE EVOLUTION OF THE ACTINOPTERYGII

The early actinopterygians are referred to as paleoniscoids (in reference to a primitive clade now extinct) or as chondrosteans. Although fragments of late Silurian vertebrates exist, possibly referable to the actinopterygii, remains complete enough to reconstruct entire fishes have not been found earlier than the mid to late Devonian. Compared with contemporary sarcopterygians, they were rather small fishes (5 to 25 cm) with a single dorsal fin, a strongly heterocercal but more or less forked caudal fin, paired fins with a long thin (not fleshy) base, large eyes, and reduced snout (Figure 6-1e and f). The interlocking scales, although thick and heavy like those of the sarcopterygians, were quite distinct in structure and growth pattern. The internal structure supporting the fins was a parallel array of closely packed radials. The bony rays supporting the fin membrane were greater in number than the supporting radials. Two morphological aspects of the paleoniscoids deserve special attention: adaptations for locomotion and adaptations for feeding. It is in these two areas that active evolution occurred, ultimately producing the fantastic array of teleosts that far outnumber all other forms of vertebrates alive today.

The same basic functional analysis of locomotion described for agnathans and elasmobranchiomorphs can be applied to paleoniscoids (Figure 6–1e and f). The heavy armor apparently was not a significant buoyancy factor in the paleoniscoids. Undoubtedly this was due to the flotation provided by the air bladder.

The jaws of these small predators functioned like snap traps. Supported by the hyomandibular, which extended obliquely down and back from the otic region of the neurocranium, the lower jaw was snapped closed by the adductor mandibulae muscle to drive small conical teeth into prey (Figure 6–4a). The adductor mandibulae originated in an enclosed cavity between the maxilla and the palatoquadrate and extended ventrally to insert on the lower jaw just

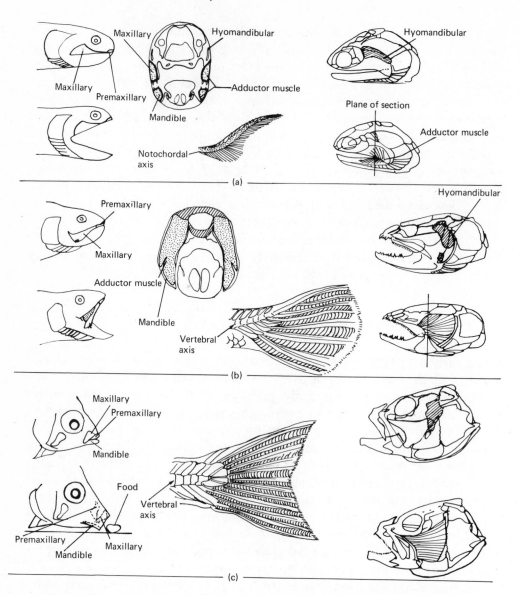

Figure 6-4 Morphological adaptations characteristic of each of the three grades of actinopterygian evolution. (a) Paleoniscoid jaw size as indicated by gape (left), hyomandibular support (upper right), musculature (in cross section of the head, and cut away lateral view, lower right) and caudal construction. Head morphology is that of *Pteronisculus* a paleoniscoid. Caudal morphology is of the living sturgeon, *Acipenser*. (b) Holostean jaw size and partially circular gape (left), nearly vertical hyomandibular orientation (upper right), enlarged jaw musculature (cross section of head, and cut away view, lower right), and caudal construction. Head morphology is that of *Amia*, the living bowfin. Caudal morphology is that of *Lepisosteus*, the living gars. (c) Teleostean jaw protrusion to form a tube with a circular opening (left) as illustrated in a perciform fish, the vertical (sometimes even anteriorly directed) hyomandibular (upper right), the complex jaw musculature as it appears on the cheek surfaces, and the homocercal caudal construction. Caudal morphology is that of the salmon, *Salmo*.

anterior and medial to its articulation with the quadrate. The solid nature of the cheeks prevented expansion of the orobranchial chamber beyond that required for respiration.

The paleoniscoids were very successful for roughly 200 million years. During this time several divergent types evolved that would be difficult to interpret if similar types were not present among living teleosts (Figure 6-5).

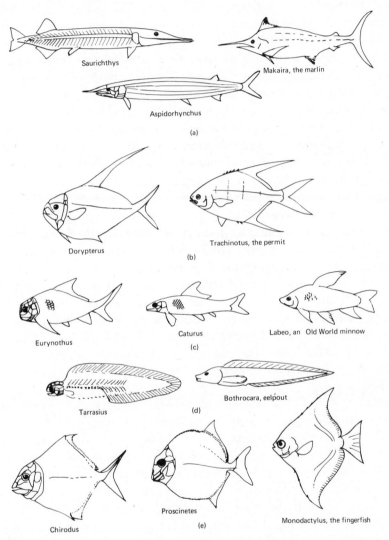

Figure 6-5 Convergence in adaptations of body form in actinopterygians from the paleoniscoid, holostean, and teleostean grades of evolution. Paleoniscoids of Carboniferous to early Triassic age are shown in the left column, Holosteans of late Triassic to Cretaceous age in the middle column, and living teleosts are shown on the right. No attempt to show the fishes to scale or to strictly match habitats (when they are known) has been made, but the detail of morphological convergence is readily apparent. (a) Piscivorous fishes with long bill-like rostra and/or jaws. (b) Fork-tailed strong swimmers with trailing fins. (c) Broad finned bottom-feeding fishes. (d) Eellike fishes with confluent dorsal, caudal, and anal fins. (e) Laterally compressed round fishes designed for maneuverability.

Near the end of the Paleozoic paleoniscoids showed signs of evolutionary change. The upper and lower lobes of the caudal fin were often nearly symmetrical in size and all fin membranes were supported by fewer bony rays—about one for each internal supporting radial. The increased caudal symmetry produced less posterior lift, and the fins became more flexible. The dermal armor also showed reduction in some forms through loss of layers of heavy osseous material in the scales and in others in the extent of squamation. The changes in fins and armor were complementary: more mobile fins mean more versatile locomotion and increased ability to avoid predators. The ability to escape predators permitted a reduction in heavy armor. This reduction of weight further stimulated the evolution of increased locomotor ability. This positive feedback type of evolution was undoubtedly initiated and constantly perpetuated by perfection of the air bladder as a delicately controlled hydrostatic device.

Not only does better locomotion assist in predator avoidance, but it also opens up a wider range of available food items; that is, it enhances predatory capability. The food gathering apparatus of several clades underwent radical changes to produce in the upper Permian a new grade of actinopterygians—the holosteans. The holostean jaw mechanism was characterized by a vertical orientation of the hyomandibular, a shortening of the maxilla, and a freeing of the posterior end of the maxilla from the other bones of the cheek (Figure 6-4b). Because the cheek no longer was a solid shield, the vertical hyomandibular could swing *laterally*, thus increasing the volume of the orobranchial chamber in a rapid motion to produce a powerful suction useful in capturing prey. The power of the sharply toothed jaw was greatly increased since the adductor mandibulae muscle was not limited in size. No longer enclosed, the jaw muscle mass expanded dorsally through the space opened by the freeing of the maxilla. In addition, an extra lever arm—the coronoid process—developed at the site of insertion of the adductor mandibulae, adding torque to the closure of the mandible.

The bones of the gill cover (operculum) were connected to the mandible in such a way that the expansion of the orobranchial chamber aided in opening the mouth. The anterior, articulated end of the maxilla developed a ball-and-socketlike joint with the rostral portion of the neurocranium. Because of its ligamentous connection to the mandible, the free posterior end of the maxilla was rotated forward as the mouth opened (Figure 6–4b). This directed the maxilla's marginal teeth forward, aiding in grasping prey. The folds of skin covering the maxilla changed the shape of the gape from that of a "smile" to a circular opening. This enhanced the directionality of suction and also eliminated a possible "side door" escape route for small prey.

Thus the first holosteans had considerable trophic and locomotor advantages. A strong selection for these particular types of adaptations occurred, for different clades of paleoniscoids independently gave rise to forms with the jaw and fin structures recognized as "holostean." The holostei therefore represent a polyphyletic assemblage of fishes that evolved in both fresh and marine waters.

Radiation of the paleoniscoids was primarily a phenomenon of the last half of the Paleozoic, and their numbers became greatly reduced by the end of the Triassic. The holosteans first appeared in the Permian (for example, *Acentrophorus*, Figure 6-6a) and became the dominant fishes of the Mesozoic. During the Jurassic and perhaps in the late Triassic several clades of holosteans, like the chondrosteans before them, gave rise to fishes with superior trophic and locomotor adaptations. These fishes constitute yet another polyphyletic taxon, known as the Teleostei. Although the teleosts probably evolved in the sea. they soon radiated into fresh water. During the Jurassic, the teleosts became abundant and by late Cretaceous time replaced most of the holosteans. By the end of the Mesozoic, most of the 200 to 300 families of modern teleosts had evolved. The first adaptations that led to the teleosts seem to have been locomotor, especially a change in the caudal fin (Figure 6-4c). Although heterocercal in appearance during larval development, in adult teleosts fin support is derived from a few enlarged hemal spines attached to the tip of the abruptly upturned vertebral column.

The caudal fin external to the muscle mass of the body, therefore, is symmetrical in shape and of great flexibility. This type of caudal fin structure is known as **homocercal**. No longer is an unsymmetrical force produced during

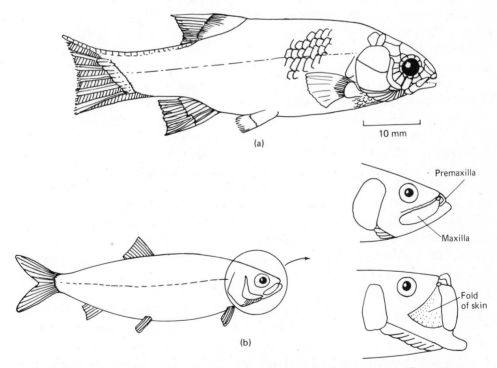

Figure 6-6   (a) An early holostean *Acentrophorus* of the Permian illustrating the generalized type from which the holostean radiation began. (b) *Leptolepis* an early Jurassic teleost with enlarge mobile maxillae, which form a nearly circular mouth when the jaws are fully opened. Membranes of skin close the gaps behind the protruded bony elements. Modern herrings have a similar jaw structure.

lateral undulation of the tail. This homocercal tail in conjunction with a well controlled air bladder allows a teleost to swim horizontally without continuously employing its paired fins as rigid control surfaces. Drag is thus reduced. Relieved of this function, the paired fins of teleosts became more flexible, mobile, and diverse in shape, size, and position than in any previous fishes. There is hardly a function for which teleost fins have not become adapted—from food getting to love making, from sound production to walking, and even flying. As in the holosteans, improvements in locomotion brought about reduction in armor. Modern teleosts are almost universally thin-scaled (by Paleozoic and Mesozoic standards) or are naked. The few heavily armored exceptions generally show a secondary reduction in locomotor abilities.

An additional locomotor specialization, the **Mauthnerian system**, present in many fishes and in tailed phases of amphibians is especially well developed in teleosts. It is absent in sharks and rays. Centrally located in the medulla oblongata are two giant nerve cells, one on either side of the midline. Each cell body is accompanied by two enlarged dendrites that synapse with cranial nerve VIII, the acoustic nerve. A single, heavily myelinated giant axon issues from each cell and crosses to the opposite side of the medulla and then it descends the full length of the spinal cord. Each giant axon synapses with motor neurons in every segment on the same side of the body (Figure 6–7a). The giant nerve cells which are important in the study of the structure and function of vertebrate neurons, are known as Mauthner cells after the German anatomist who described them.

Stimulation of one of the Mauthner cells results in a very rapid, unilateral, forceful contraction of the trunk and tail myomeres. This reaction is similar and perhaps identical to the body snap or startle response that many teleosts exhibit when frightened by a sudden noise, mechanical shock, or change in illumination (Figure 6–7b). The simultaneous, strong contraction of the muscles on one side of the body propels the fish forward and slightly toward the opposite side and produces a very rapid acceleration. Following this snap start, normal, bilateral undulations are initiated. The adaptive value of such a startle response is obvious; not only is inertia overcome in preparation for rapid escape but the initial dart quickly alters the position of the fish in the water column. A predator incautious enough to produce stimulae activating the Mauthnerian system may loose potential prey due to the startle reaction and subsequent escape swimming.

Several neurophysiological properties of the Mauthnerian system enhance its effectiveness in confounding predators. Giant, heavily myelinated axons typically propagate nerve impulses at higher speeds than smaller nerve axons. So rapid is the conduction velocity of impulses to the muscles via this system that each segment contracts essentially simultaneously in producing the powerful bending of the body. Each Mauthner cell continuously and randomly varies in its sensitivity to stimulae, but when one cell is stimulated it completely inhibits the firing of the contralateral cell. Thus only *one* set of lateral muscles contracts during a "startle response." Which side contracts seems to be a

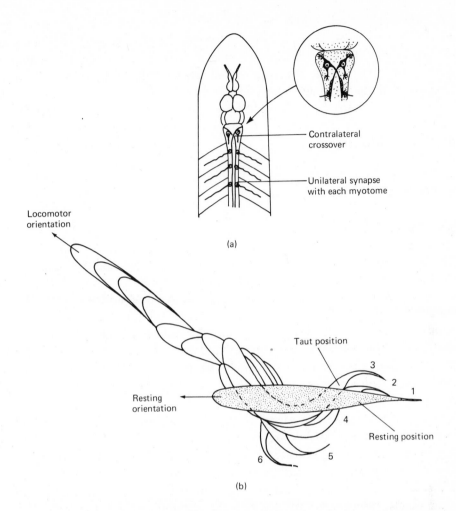

Locomotor orientation

Contralateral crossover

Unilateral synapse with each myotome

(a)

Taut position

Resting orientation

3

2

1

4

Resting position

6

5

(b)

Figure 6-7   (a) Anatomical relationships of the Mauthner neurons and associated structures in a teleost. (b) Body snap or startle response of a teleost (trout) as seen from above in tracings from a motion picture.

random phenomenon independent of the directionality of the stimulus. The position of a fish at the completion of a startle response is therefore not easily predictable. The adaptive value of the Mauthnerian system in escaping avian, mammalian, and probably piscine predators is underscored by the wide variety of teleosts with well-developed giant neurons. It is possible, but as yet unproven, that lurking predators also utilize this system. Only benthic forms, which depend on camouflage more than escape, and eellike forms, where a unilateral contraction of the entire body would produce little displacement, seem to lack Mauthner cells. Interestingly, large species of tunas, which possess specializations for fast, efficient swimming (Chapter 7), lack giant axons, whereas smaller species have a Mauthnerian system.

Teleosts, like most grades of fishes that preceded them, evolved not only special adaptations for locomotion but also improvements in their feeding apparatus. Trophic adaptations in the earliest teleosts showed only slight improvements over those in holosteans. Perhaps the only general early advance was a loosening of the premaxillae so that they moved during jaw opening to accentuate the round mouth shape. One early clade of teleosts showed an enlargement of the free-swinging posterior end of the maxilla to form a nearly circular mouth when the jaws were fully opened (Figure 6–6b). Later in the radiation of the teleosts distinctive changes in the jaw apparatus permitted a wide variety of feeding modes.

6.3   LIVING ACTINOPTERYGII

6.3.1   Chondrosteans

The chondrostean (paleoniscoid) grade was nearly replaced during the early Mesozoic by holosteans, but a few skeletally degenerate or specialized forms have survived. About 23 species of sturgeon, Ascipenseridae (Figure 6–8a), represent large (1 to 6 m) benthic fishes devoid of internal ossification and lacking much of the dermal skeleton of more primitive chondrosteans. They are found only in the Northern Hemisphere and are either anadromous (ascending into fresh waters to breed) or entirely fresh water in habit. Sturgeons are commercially important both for their rich flesh and as a source of the best caviar.

The two surviving species of paddlefish, Polyodontidae (Figure 6–8b), show affinities to the sturgeons but are apparently completely devoid of ossification. Their most outstanding feature is a greatly elongate and flattened rostrum, which extends nearly one third of their 2-m length The rostrum is rather soft and richly innervated with sensory organs that are believed to detect minute electric fields (Section 6.4.2) Despite the common notion that this 'paddle' or 'shovel' is used to stir food from muddy river bottoms, paddle fish are planktivores. They feed by actively swimming with their prodigious mouths agape and strain crustacea and small fishes from the water column. The two species of paddle fish present an intriguing zoogeographic distribution: one is found in the Yangtze River valley of China, the other in the Mississippi River valley of the United States!

One group of chondrosteans survives that, in many ways, is more like early actinopterygians, that is, the paleoniscoids. However, they are sufficiently specialized that their relationships to other fishes have puzzled systematic ichthyologists. They are the 11 species of African bichirs and reed fish, Polypteridae (Figure 6–8c). Like the sturgeons and paddlefish, these small (less than 1 m) fishes have heterocercal tails. Unlike those forms, they show little degeneration of skeletal ossification. In addition to a full complement of dermal and endochondral bones, they are well armored with thick, interlocking, multilayered scales. The peculiar series of flaglike dorsal finlets, as well as the fleshy base of the pectoral fins, must be considered specializations for a way of life about which we know very little.

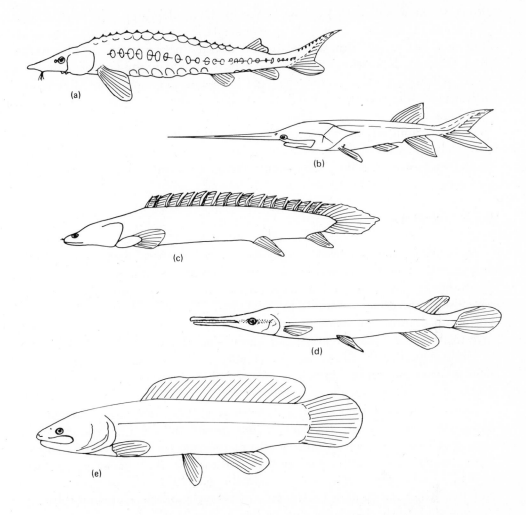

Figure 6-8 Living fishes of the chondrostean (paleoniscoid) grade of evolution (a through c) and those of the holostean grade (d and e). (a) *Acipenser*, one of the genera of sturgeons. (b) *Polyodon spathula*, one of two living species of paddlefish. (c) *Polypterus*, a genus of bichir. (d) *Lepisosteus*, the genus of gars. (e) *Amia calva*, the bowfin.

6.3.2 Holosteans

Only two genera of holostean grade fishes are extant. Both are limited to North America and represent widely divergent holostean types, perhaps due to having had different chondrostean ancestors. The seven species of gars, Lepisosteidae (Figure 6–8d), are medium to large sized (1- to 4-m) predators of warm-temperate fresh and brackish (estuarine) waters. The elongate body, jaws, and teeth are not typical of holosteans, but their interlocking multilayered scales are similar to Mesozoic holosteans. Gars feed on other fishes taken unaware when the seemingly lethargic and excellently camouflaged gar dashes alongside them and, with a sideways flip of the head, grasps them with needle-like teeth. Sympatric with gars is the single species of bowfin, *Amia calva* (Figure 6–8e). In its elongate body and dorsal fin, the bowfin is also unlike most Mesozoic holosteans. The head skeleton, especially the jaw mechanism, is more generalized than that of the gar. *Amia* is a predator on almost any organism smaller than its own 0.5- to 1-m length; they are even reputed to take ducklings from the surface. Bowfins move by body undulations and undulations of the long dorsal fin (as in Figure 7–6b). Scales of the bowfin are comparatively thin and made up of a single layer of bone as in teleost fishes. The caudal fin, however, is internally simple and asymmetrical, that is, heterocercal.

6.3.3 Teleosteans

Most living fishes are teleosts and share many advanced characters of caudal and cranial structure. Nevertheless, it has become increasingly clear in recent years that at least four (and perhaps more) different advanced holostean types gave rise to the present assemblage of teleosts. The descendents of these clades are grouped into four taxa of varying size and diversity. Active systematic investigation continues to adjust our knowledge of each of these four groups.

The group called the Clupeomorpha (Figure 6–9a) is specialized for feeding on minute plankton sucked into a modified holostean type mouth like that in Figure 6–6b. They are silvery, schooling fishes commonly called herrings and anchovies and inhabit marine and, to a lesser extent, fresh waters throughout the world. Clupeomorphs total about 300 species and are of great commercial importance. Another very similar group of teleosts, the Elopomorpha (Figure 6–9b) distinguishable on the basis of technical analysis of the skeleton, appeared by early Cretaceous times. They originally were like the living tarpon and bonefish. One outstanding difference between modern clupeomorphs and elopomorphs is the leptocephalus larvae, unique to the latter. These larvae, adapted to long life near the surface of the open ocean, permit wide planktonic dispersal of the species even though the adults may be restricted to shallow inshore habitats. As strange taxonomic bedfellows as they seem, the several families of eels, Anguilliformes (Figure 6–9c), are included in the elopomorphs. This relationship is not based on any fossil intermediates (none have as yet been discovered) but because of similar larvae.

Most of the 32 families and about 350 species of elopomorphs are marine and eellike. Some species are tolerant of fresh waters, however, and spend a portion of their life there. The most astonishing of these euryhaline types is the common American eel *Anguilla rostrata* (Figure 6–9c), which has one of the most spectacular life histories of any fish. After growing and reaching the age of sexual maturity (perhaps 10 to 12 years) in rivers, lakes, and even ponds often hundreds of miles from the sea, these catadromous (downstream migrating) eels leave their accustomed habitats and enter the sea. In the North Atlantic eels apparently continue their migration until reaching the Sargasso Sea

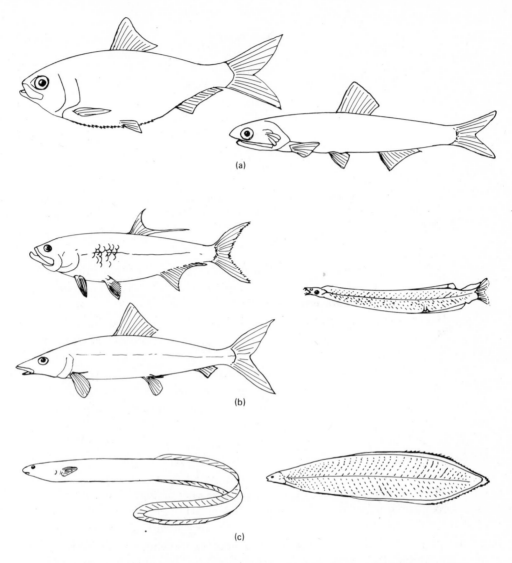

Figure 6-9   Living teleosts of isolated phylogenetic position. (a) Clupeomorpha represented by a herring (left) and an anchovy (right). (b) Elopomorpha represented by a tarpon (above) and a bonefish (below) and a typical fork-tailed leptocephalus larva (right). (c) Anguilliform elopomorphs represented by the common eel, *Anguilla rostrata* (left) and its 'tailless' leptocephalus larva.

of the infamous Bermuda Triangle. Here, presumably at great depth, they spawn and die, the eggs and newly hatched leptocephalus larvae float to the surface and drift in the currents. Larval life may extend over a year or more until the larvae reach continental margins, where they transform into miniature eels and ascend rivers to feed and mature.

A third group of teleosts which appeared in Cretaceous seas, the Osteoglossomorpha, are now restricted to tropical fresh water. Different morphological types are represented by *Osteoglossum* (Figure 6–10a), a meter-long predator

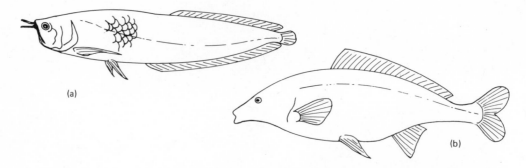

Figure 6-10   Living teleosts of the restricted Osteoglossomorpha. (a) *Osteo-glossum*, the arawana from South America. (b) *Mormyrus*, an elephant-nose from Africa.

from the Amazon, *Arapaima*, an even larger Amazonian predator, and *Mormyrus* (Figure 6-10b), a much smaller African bottom feeder that uses weak electric discharges to communicate with conspecifics (see section 7.2.2). As dissimilar as they may seem, the approximately 150 species (nearly all mormyrids) of osteoglossomorph fishes are united by unique, shared osteological characters of the head.

The three groups of teleosts mentioned do not account for a large percentage of the 20,000 or so recognized living teleost species. A fourth group evolved independently from advanced holosteans before the upper Cretaceous and gave rise to the vast majority of living teleosts (Figure 6-11). The primitive, basal stock for this group is today represented by the salmoniform fishes (Figure 6-12a) including the anadromous (upstream migrating) salmon, which usually spend a major portion of their lives at sea, and the trouts which live in fresh waters. The migrations of salmon are nearly as spectacular as those of eels, and their exceptionally accurate homing ability is well known (see Section 7-2). Also included in this group of economically important temperate-zone fishes are the pickerels and pikes, and the Southern Hemisphere galaxiids that live in similar habitats. Related to the salmoniform fishes but showing more advanced characteristics, such as those of jaw construction, are the mesopelagic lantern fishes, the myctophiforms (Figure 6-12b). Tiny, light-producing organs called photophores are arranged on their bodies in species- and even sex-specific patterns. They are probably used to maintain contact with conspecifics in the dim light of the deep sea (see section 6-5).

Of uncertain relationship to the preceding forms are the **Ostariophysi**, the predominant fishes of the world's fresh waters. The name Ostariophysi refers to a series of small bones (*ostar* = a little bone) which connect the air bladder (*physa* = a bladder) with the inner ear (Figure 6-13a). Using the air bladder as an amplifier and the chain of bones as conductors, this system, called the **Weberian apparatus**, enhances the hearing sensitivity of these fishes. Although all Ostariophysi possess Weberian ossicles, they are a very diverse taxon (Figure 6-13) and include the characins of South America and Africa, the

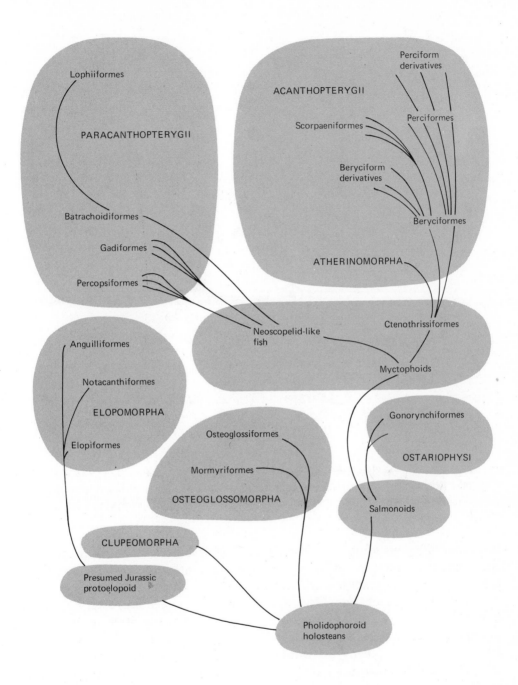

Figure 6-11 A phylogeny of living teleosts shows the three relatively isolated and limited groups Clupeomorpha, Elopomorpha, and Osteoglossomorpha and the dominant clade of living teleosts whose basal stock is represented by the living salmoniform fishes and the more advanced myctophoid fishes, often jointly grouped as the protacanthopterygii.

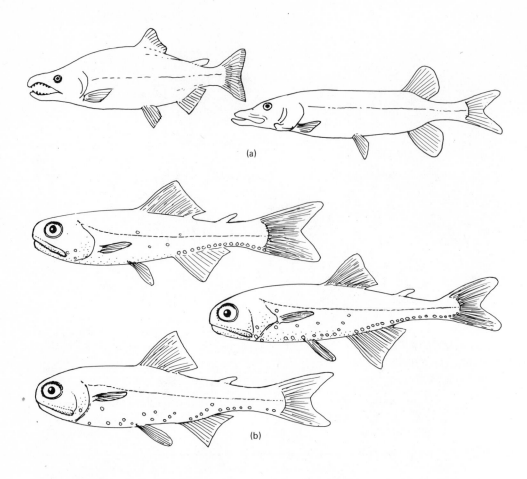

Figure 6-12   (a) Salmoniform fishes represented by the salmon (left) and the pike (right). (b) Myctophiform fishes, represented by lanternfishes, illustrate species specificity by the number and arrangement of light producing photophores concentrated on their ventral surface.

carps and minnows found on all inhabitable continents except South America and Australia, and the catfishes found on all continents with flowing fresh waters and in many shallow marine areas.

About four out of every five species in fresh water are ostariophysans. Nevertheless, it is difficult to attribute the success of nearly 5000 species solely to improved hearing. As a group they display a diverse variety of other adaptive traits. For example, many ostariophysans have developed protrusible jaws and thus are adept at obtaining food in a variety of ways. In addition, pharyngeal teeth act as "second jaws" to process food for better digestion. Many forms have developed fin spines or special armor for protection, and the skin often contains glands that produce substances used in olfactory communication. Although they show diverse reproductive habits, most lay sticky eggs or otherwise guard them to prevent their being swept downstream.

Another group of fishes derived from the salmoniforms (or perhaps from

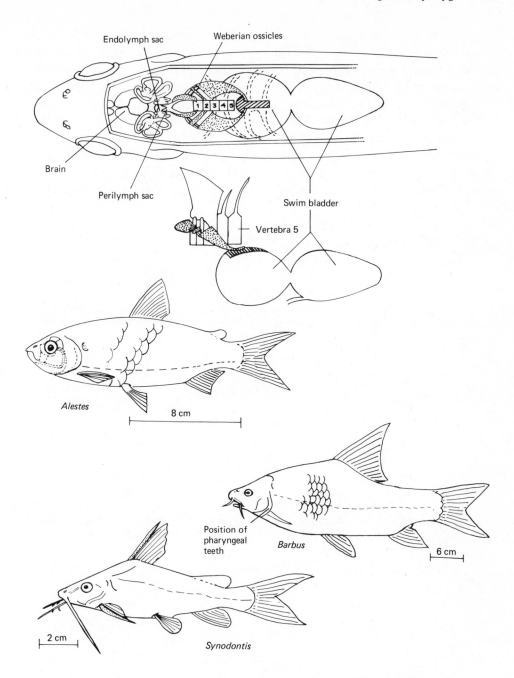

Figure 6-13 Ostariophysans, the predominant fishes of the world's fresh waters. Lateral (below) and dorsal (above) view of the Weberian apparatus showing the connections of the ossicles. The three main groups of ostariophysans are represented by a characin, *Alestes*; a minnow, *Barbus*; and a catfish, *Synodontis*. The pharyngeal teeth, especially well developed in the otherwise toothless minnows, are situated at the junction of the pharynx and esophagus posterior and ventral to the gill arches. (Modified after R. McN. Alexander [1975] *The Chordates.* Cambridge Univ. Press, London.)

primitive members of a more advanced group, as indicated in Figure 6–11) are the Atherinomorphs (Figure 6–14a). Atherinomorphs have very protrusible jaws and adaptations of form and behavior that suit them to shallow estuarine and marine habitats, although some well-known forms occur in fresh water. This group includes the silversides, grunions, flyingfishes, halfbeaks and egg-laying and live-bearing cyprinodonts—the latter including the guppies, mollies, and swordtails commonly maintained in home aquaria.

A tendency toward high mobility of the jaws and the development of protective, light weight, pungent spines in the median fins occurred in several groups of teleosts. Several fishes, including the cods and the anglerfishes, have

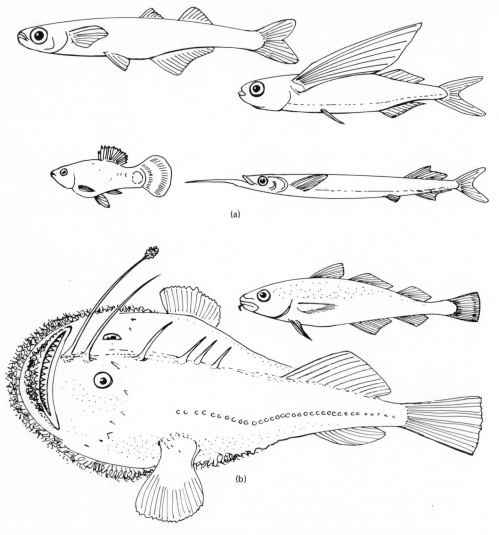

Figure 6–14   (a) Atherinomorph fishes represented by (clockwise from upper left) a silverside, a flyingfish, a halfbeak and a live-bearing killifish. Not to scale. (b) The so called "Paracanthopterygians" represented by the cod (above) and the goosefish angler (below). Not to scale.

been lumped together in what is likely an unnatural assemblage of species based on convergence rather than common ancestry (Figure 6–14b). One living assemblage that shows exceptional diversity, however, acquired these adaptations from a single ancestor. This monophyletic group of fishes also has developed exceptional precision in the control of locomotion. These fishes, the Acanthopterygii or true spiny-rayed fishes, dominate the surface and shallow marine waters of the world and may exceed 8000 species.

A distinguishing feature of acanthopterygians is the refinement of their feeding and locomotor structures. Acanthopterygians exhibit an unparalleled diversity of feeding modes, all of which relate to their precise control over jaw actions and body positions. This high degree of precision certainly was achieved through strong evolutionary pressures on the interactions between feeding and locomotion. These interactions are illustrated especially well by coral reef fishes, most of which are acanthopterygians. In his detailed day and night underwater observations of coral reef fishes, Dr. E. S. Hobson noted that the most primitive spiny rayed fishes are all predators (for example, squirrelfishes, cardinalfishes; refer to Figure 6–15). They disperse over the reef at night to feed, but during the daylight hours congregate in caves and holes in the reef—they are **nocturnal**. Other generalized predators such as the groupers are also solitary but feed heavily at dawn and dusk (that is, are **crepuscular**). They stalk prey and use the cover of the substrate to conceal their approach. These reef acanthopterygians rely on direct approach and a large mouth to seize prey that are fully exposed to attack. Hobson suggests that, as an early evolutionary response to predation, many reef invertebrates performed their activities at night and remained concealed during the day. In response to this nocturnality of prey, early acanthopterygians or their ancestors subsequently evolved the capacity to feed at night. To detect nocturnal prey they have developed very large, sensitive eyes functional at low light intensities.

A major advance in feeding types amongst reef acanthopterygians involved the evolution of fishes specialized to take food items hidden in the complex reef surface, generally by suction or a forcepslike action. These feeding tasks demanded sensory specializations, the most important of which was high visual acuity. Significantly, high visual acuity can only be achieved in the bright light of day. In addition, delicate positioning of the body and head was required to "aim" the jaws. The necessity of holding position against waves and tidal currents while undertaking precision feeding movements led to improved locomotor control. These interacting adaptations produced fishes capable of maneuvering around and through the cavity-rich reef in search of food. So accurate is their locomotor control, visual surveillance, and memory for hiding places and escape routes that these fishes can avoid predators and therefore can expose themselves and feed in broad daylight (that is, are **diurnal**). The refined feeding adaptations of diurnal reef fishes allow them to extract small invertebrates from their daytime hiding places or perform other specialized types of feeding, such as snipping coral polyps or nibbling sponges. In other species, mouths and dentition, as well as digestive systems, are specialized for

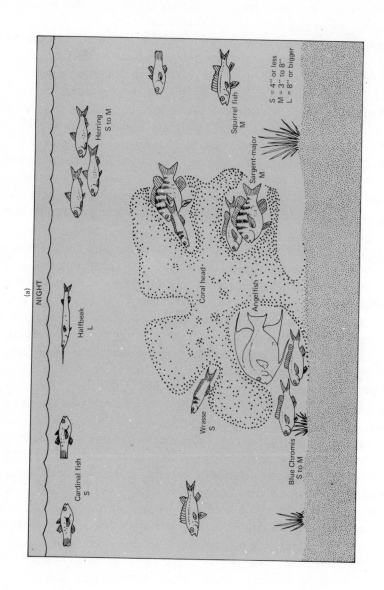

NIGHT

(a)

Cardinal fish
S

Halfbeak
L

Herring
S to M

Wrasse
S

Coral head

Angelfish

Blue Chromis
S to M

Sargent-major
M

Squirrel fish
M

S = 4'' or less
M = 3'' to 8''
L = 8'' or bigger

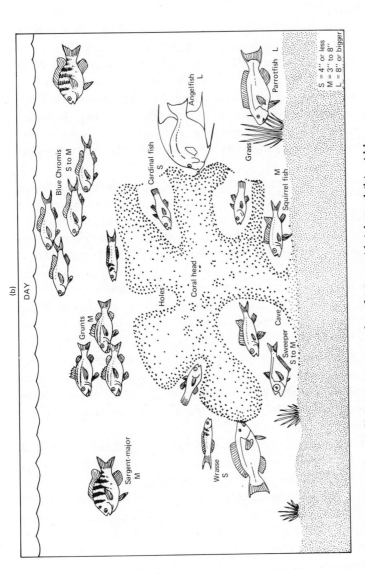

Figure 6-15 Fish population and activity differences on a coral reef (a) at midnight and (b) midday.

(a) At night plankton feeding herrings and piscivorous halfbeaks feed over the reef, whereas invertebrate feeding squirrelfish and grunts move out over the reef and adjacent sandy areas. Sweepers and cardinalfish feed on plankton. The brightly colored myriad of fishes to be seen during the day are modestly hidden in cracks, crevices, and holes in the reef.

(b) By day, few of the nocturnally active fishes can be found, most having taken refuge deep within the reef. The water column, reef surface, and adjacent sand flats are all alive with great numbers of often spectacularly colored fishes, each with distinctive feeding methods and/or localities. Note especially the plankton feeding damselfish (sargent majors and blue chromis), the algae scrapping grunts, parrotfish, the parasite removing wrasse, and the sponge eating angelfish. Cardinal fish, squirrel fish and sweepers utilize the reef for shelter during the day; at night they migrate some distance to feed.

herbivory; in yet others for removing small organic particles from a variety of different sites, including sand, the water column, and even the bodies, mouth, and gills of larger fishes (Figure 6–15b).

Released from heavy predation during the daytime, many of these advanced reef fishes have assumed gaudy signal colorations that communicate information significant to conspecifics and other competitors. These brilliantly colored fishes, seen feeding in incredible diversity, are one of the most impressive sights a person can experience. At dusk as these colorful diurnal fishes seek nightime refuge in the reef, the nocturnal fishes leave their hiding places to replace them in the water column. Thus, the entire fish community can be thought of as divided into species that are either nocturnal or diurnal. The precision in the time at which each species leaves and enters the protective cover of the reef day after day indicates an important ecological function (Figure 6–16). It is this partitioning of space, time, and trophic resources available on a reef that has permitted and maintained the great diversity of vertebrate life that is present on coral reefs.

The feeding and locomotor adaptations of coral reef fish reflect those found in all acanthopterygians. Reproductive modes in acanthopterygians are also very diverse. Indeed, teleosts show a wider variety of reproductive modes than any other vertebrate taxon. Many marine fishes are oviparous and release large numbers of small, buoyant, transparent eggs that are fertilized externally to develop and hatch in the open sea. The larvae settle out of the plankton when they wash into appropriate habitats. Other species lay adhesive eggs on rocks, leaves, or in gravel or sand where they may even be guarded by parents with the greatest of vigor. Sometimes the spawning fish leave the water to oviposit in the sand or on leaves. In the water nests are often prepared to receive the eggs. The nests represent simple structures from the clearing and cleaning of shallow depressions in sand or gravel to elaborate constructions of woven plant material held together by parental secretions. Floating nests of mucus-covered bubbles are constructed by many fishes. Others use natural holes or cavities or dig elaborate burrows in which to nest.

Parental guarding of the eggs may be supplemented or entirely substituted by unwitting assistance from other organisms near, on, or in which the eggs are laid. Among these "midwives" are stinging anemones, live mussels, crabs, sponges, and tunicates. Perhaps the ultimate in protection of the eggs is portage by one of the parents. Species are known that carry eggs on their fins, under their lips, in their mouth or gill cavities, or on specialized protuberances, skin patches, or even in pouches. Many fishes have developed internal fertilization, which permits development of the young within the mother's body. Both ovoviviparous or truly viviparous species are known, but are not as common in teleosts as in elasmobranchs and mammals (see Chapter 5).

Despite the astounding diversity of reproductive modes known in teleosts, the vast majority are oviparous. In fact, with the exception of elasmobranchs, in anamniotes oviparity is the prevalent reproductive mechanism. This widespread occurrence of oviparity relates to the qualities of the aquatic environ-

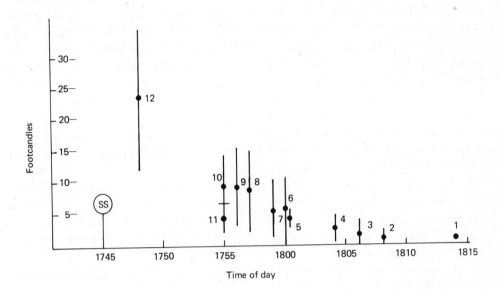

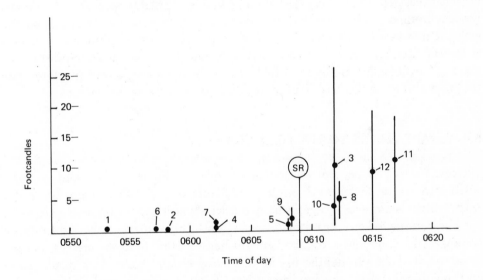

Figure 6-16 The precision of the time of disappearance of diurnal fishes at dusk and their reappearance at dawn is illustrated by these data obtained by S. B. and A. J. Domm at Hook Island on the Great Barrier Reef. The numbered point indicates the mean for all observations for that species and the vertical lines the standard deviation in light intensity for that species. The variation in time of disappearance and reappearance is not shown since most species of fishes appeared or disappeared within a regular, short interval of time every day. Note that the definite sequence of disappearances at dusk (SS = sunset) is more or less reversed in the order of appearances at dawn (SR = sunrise). Modified from Domm and Domm [1973] Pac. Sci. 17:128–135.

ment. For example, eggs suffer no desiccation as they might in air, gas exchange is facilitated, and dispersal may be automatic as eggs and larvae are wafted by currents. Food for the newly hatched fish is often available throughout the water column. Because of the general advantages of oviparity, both ovoviviparity and viviparity must be considered as specific adaptations to special conditions. Although they have repeatedly evolved in fishes, these latter modes have not become genetically favored throughout the anamniotes as a result of any overriding reproductive superiority.

Within oviparous teleosts, marine and fresh water species show contrasting adaptations. Marine teleosts tend to broadcast large numbers of small eggs that hatch rapidly, and the larvae immediately begin to feed on the plankton. This strategy is well adapted to the large volume, minute planktonic organisms, and long-term stability of the sea. In contrast, fresh water teleosts produce fewer, larger, nonplanktonic eggs that develop over a more extended period of time. This results in hatchlings that are more advanced and adultlike in body form and behavior. This reproductive strategy in fresh water relates to the flowing and ephemeral characteristic of upland waters, which could easily flush a less well-developed fish from its preferred habitat.

In summary, the feeding, locomotion, and reproductive mechanisms used by Osteichthyes represent exceptionally and highly refined general adjustments to the environment. In teleosts these adaptive refinements have resulted in the greatest number of existing species within any taxon of vertebrates that exists. A special case are the deep sea fishes, in which the general adaptations of teleosts have become highly modified. These deep sea specializations illuminate the adaptability of the vertebrate body plan to extreme environments.

6.4   ADAPTATIONS TO THE DEEP SEA

Oceans cover 71 percent of the earth's surface (Figure 6–17a) and are classified into several different depth zones (Figure 6–17b). Two major life zones exist: the pelagic, where organisms live a free-floating or swimming existence, and the benthic, where organisms associate with the bottom. Solar light rapidly attenuates in water and is totally extinguished by 1000 m in the clearest oceans or at much shallower depths in the less clear coastal seas. As a result, large portions of the oceans are aphotic (about 75 percent by volume). Much of the deep sea therefore is perpetually dark, interrupted only by the flashes and glow of bioluminescent organisms. Because of its remoteness and consequent mystery, the deep sea has long intrigued biologists. Zones from the epipelagic to abyssal depths have a long history that precedes the origin of life. Time, therefore, has been sufficient for the evolution of a distinctive and bizarre array of vertebrates, the deep sea fishes. Examination of these unusual vertebrates dramatizes how adaptations fit organisms to the properties of their environment.

Deep sea fishes decrease in abundance, in size, and in species diversity with increased depth. This also occurs in deep sea invertebrates. These trends are

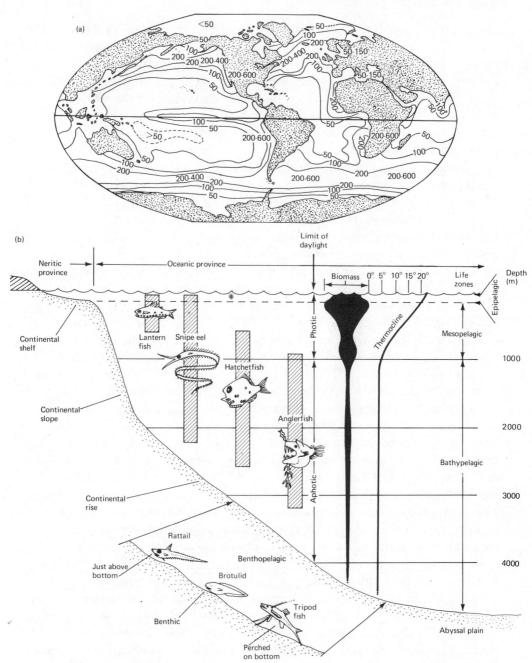

Figure 6–17 Life zones of the ocean depths. (a) Annual productivity at the ocean surface. Numbers are grams of carbon produced per square meter per year. Where highly productive waters overlie deep waters, rich assemblages of uniquely adapted sea fishes occur (Modified from Friedrich H. 1973. Marine Biology. Univ. Washington Press). (b) A schematic cross section of the life zones within the deep sea. Various pertinent physical and biotic parameters are super-imposed on the arbitrary "life zones" as are the vertical ranges of several characteristic fish species, some of which migrate on a daily basis (Modified from Marshall [5].)

not surprising, for all animals ultimately depend on plant photosynthesis, which is limited to the epipelagic regions (Figure 6–17b). Epipelagic animals exploit a ready supply of primary producers. Below the epipelagic, animals cannot directly exploit the primary producers but must depend on descent of food from above—a "rain" of detritus from the surface into the deep sea. The decrease of fishes with depth is inevitable, for the food upon which the deepest fishes depend must diminish if it is consumed during descent. Sampling confirms this decrease in food with depth. For example, in the Pacific Ocean surface plankton can reach biomass levels of 500 mg/m^3 wet weight. At 1000 m, where aphotic conditions commence (see Figure 6–17b), plankton decrease to 25 mg/m^3, at 3000 to 4000 m to 5 mg/m^3, and at 10,000 m to only 0.5 mg/m^3.

Fish diversity parallels this decrease: about 750 species of deep sea fishes are estimated to occupy the mesopelagic zones and only 150 species the bathypelagic regions. Regions of the deep sea that lie under areas of high surface productivity contain more and larger species of deep sea fishes than regions that underlie less productive surface waters. High productivity occurs in areas of upwelling, where deeper currents rise to the surface to recycle nutrients previously removed by the sinking of detritus. Today upwelling occurs in the eastern Pacific ocean off Central and South America and the eastern Atlantic ocean off west Africa and along the equatorial counter current (Figure 6–17a). In these places deep sea fishes tend to be most diverse and most abundant.

In tropical waters photosynthesis continues throughout the year. Away from the tropics it is more cyclic, following seasonal changes in light, temperature, and sometimes currents. Diversity and abundance of deep sea fishes decreases away from the tropics. Over 300 species of meso- and bathypelagic fishes have been caught in the vicinity of tropical Bermuda. In the whole Antarctic region, only 50 species are so far described, and yet Antarctic waters are fabled for their high surface productivities of phytoplankton and krill (a shrimplike crustacean). The high productivity of Antarctic waters, however, is restricted to a few months of each year. Apparently, the sinking of detritus through the rest of the year is insufficient to nourish a diverse assemblage of deep sea fishes.

In discussing deep sea fishes, let us emphasize that availability of energy (food) is the most formidable environmental stress they encounter. Energy affects their distributions and abundance and, indeed, their diversity, for many of their specific adaptations have been selected by food scarcity. Another major variable is hydrostatic pressure; lack of light and low temperatures are constants. One final important factor in this energy-short environment is gravity. Pelagic fishes must continuously counter it to avoid sinking. For benthic fishes gravity presents a problem only when venturing from the sea floor.

With increased depth each species of deep sea fish is further removed from a primary source of food. In general, the mesopelagic fishes and invertebrates, which inhabit depths from 100 to 1000 m, migrate vertically. At dusk they ascend toward the surface only to descend again near dawn, apparently following light intensity levels (Figure 6–17b and 6–18). The biological significance of these daily migrations is not completely clear, but several ecological benefits

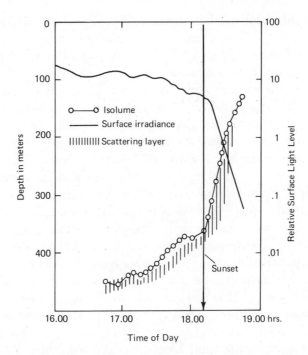

Figure 6-18 Upward migration of mesopelagic fishes as indicated by changes in depth of the deep scattering layer. Sonar signals readily reflect off the swimbladders of mesopelagic fishes; therefore, aggregations of them produce an acoustic scattering layer. As sunset occurs, the intensity of light at the surface (irradiance read on right axis) declines as does the light penetrating the deep sea. Plotting the depth of a single light intensity (an isolume read on the left axis) against time illustrates the progressively shallower depth at which a given light intensity occurs during dusk. If the light intensity chosen falls between that of starlight and full moonlight as measured at the sea surface, it corresponds closely with the upper surface of the deep scattering layer, which rises rapidly with the upward migration of mesopelagic fishes. (Modified after E. M. Kampa and B. P. Boden [1957]. Seep-Sea Res. 4:73–92.)

probably accrue to the fishes. By rising toward the surface at dusk, mesopelagic fishes enter a region of higher productivity where food is more concentrated. They also increase their chance of being eaten by surface predators. Many epipelagic carnivorous fishes are visual predators, however, and are more effective in capturing prey during daylight (tunas, dolphinfish, billfishes) when the vertically migrating fishes are at depth. Nocturnal ascent, however, is not without cost. In ascending over several hundred meters, especially in tropical and temperate seas, mesopelagic fishes are exposed to temperature increases that can exceed 10°C. The energy costs of maintenance and activity at these higher temperatures can double, even triple (see Chapter 7). In contrast, daytime descent into cooler waters lowers metabolism and therefore conserves energy. It also reduces the chance of predation, for fewer predators exist at depth.

It is far less certain that bathypelagic fishes, which inhabit aphotic regions, undertake daily vertical migrations. There is little to save in metabolic economy from vertical migration within this zone because temperatures generally are uniform (about 5°C). Furthermore, the cost and time of migration over the several thousand meters from the bathypelagic to the surface probably would outweigh the energy gained from invading the rich surface regions. Invasion of the less distant mesopelagic zone by bathypelagic fishes would provide little increase in food. As a result, bathypelagic fishes are specialized to live a less active life than their meso- and epipelagic counterparts.

Pelagic deep sea fish display a series of morphological adaptations related to their particular life styles. Figure 6–17 shows several typical mesopelagic, bathypelagic, and also benthic fishes. Eye size and function correlate with depth, as vividly discussed by N. B. Marshall, an Englishman with a life-long familiarity with the deep sea. Mesopelagic fishes possess large, often tubular eyes, bathypelagic fishes have smaller eyes, but benthic fishes vary in eye size. The large pupil of mesopelagic fishes maximizes the amount of light entering the eye, analogous to enlarging the diaphragm of a camera. In addition, the retina of deep sea fishes contains a high concentration of visual pigment, the photosensitive chemical that absorbs light in the process of vision. The visual pigments of deep sea fishes are most efficient in absorbing blue light—the color of light most readily transmitted through clear water. They are therefore maximally sensitive to the dim, blue solar light available. The smaller eye of bathypelagic and many benthic fishes suggest that vision retains some selective value, even below the depth to which solar light penetrates.

Many deep sea fishes and invertebrates are emblazoned with startling species-specific bioluminescent designs. The photophore organs emit an "eerie" blue light, a property in keeping with the wavelength of light best transmitted through water. The retention of eyes relates to the presence of bioluminescence in many deep sea organisms. Bioluminescence does not provide a general background light against which targets are viewed, as is the case under solar illumination. Rather it is presented as a point source(s) against a black background. To detect a point source, a large eye adapted to gather all light possible is apparently not necessary. Sensitivity is retained, however, in possession of a visual pigment that best absorbs the blue light of bioluminescence. Reduction of eye size provides a considerable economy for a bathypelagic fish, for the eye has a very high metabolic rate relative to other body tissues. In the bathypelagic region selection for smaller eyes that still fit visual needs can be related to the scarcity of food. Selective pressures for small eye size and even eye loss must be high, as exemplified by many blind cave fishes that live in total darkness and possess rudimentary eyes. Retention of the eye in most bathypelagic fishes therefore must depend on the presence of bioluminescent organisms.

Not only does the size of deep sea fishes tend to decrease with increasing depth, but there also is a reduction in bone and even skeletal musculature (Figure 6–19). These trends relate to the increased scarcity of food and to the problem of sinking. Reduced size increases the relative surface area of a fish per unit of volume (or weight). This would in turn increase its friction and slow the rate of sinking. However, increased surface friction is much more important

Figure 6-19 Reduction in the amount of bone in oceanic fishes that live at increased depths as seen in x-rays. Upper—surface limited jackfish(Carangidae); middle—mesopelagic, vertically migrating lanternfish (Myctophidae); bottom—mesopelagic deep living 'gulper-eel' (Gonostomatidae). Note also the change in relative size of the eye.

to very small animals like plankton than to fishes, because of a fish's ratio of surface area to volume is always rather low. Cytoplasm has a specific gravity slightly greater than that of sea water. Because bone is very dense (specific gravity about 2.0), its reduction in deep sea fishes decreases their total density and rate of sinking. Epipelagic fishes show little bone reduction and mesopelagic fishes less bone reduction than bathypelagic fishes. Yet they all must counter gravity. The differences find a partial answer in the varied modes of life of fishes in these regions.

Many epi- and mesopelagic fishes are active swimmers. To act efficiently, muscle requires a stiffening framework, the vertebral column. Surface fishes like tunas that swim continuously in search of food have strong ossified skeletons and large red muscles especially adapted for cruising (Section 7.1.4). Mesopelagic fishes, which swim mostly during vertical migration, have a more delicate skeleton and less axial red muscle. In many bathypelagic fishes, ossification of the axial skeleton and the mass of muscles whether red or white are greatly reduced and locomotion must be correspondingly limited as well.

A tendency toward strict carnivorism also occurs with depth. Jaws and especially teeth in deep sea fishes are usually enlarged to insure ensnarement of scarce prey (Figure 6-19). The ceratioid anglerfishes dangle a bioluminescent bait in front of their large mouths to lure prey. In many deep sea fishes the gape expands and the stomach stretches enormously to accommodate prey larger than the predator. Thus deep sea fishes, like most teleosts, show major specializations in locomotor and feeding structures (Section 6.3). Unlike surface teleosts, however, the scarcity of food has selected for structures that minimize the costs of "hunting" and maximize the capture of prey. Coincident with the reduction in bone and muscle is an increase in total body water content. Thus, the sinking rate of these fishes is reduced because their density is more nearly that of the supporting water.

Deep sea fishes provide a clear example of how the swim bladder functions to neutralize buoyancy. Located extraperitoneally, dorsal to the abdominal cavity but ventral to the vertebral column, is a gas-filled sac that arises as an evagination from the embryonic gut (Figure 6–20). This structure functions like a float to counteract gravity and keep a fish from sinking. To maintain neutral buoyancy a constant swim bladder volume is required. In marine teleosts the swim bladder occupies about 5 percent of the body volume, in fresh water teleosts 7 percent— volume differences just sufficient to produce neutral buoyancy in each medium. Thus, a teleost neither rises nor sinks and needs to expend little energy to maintain position when neutrally buoyant.

Not all fishes possess swim bladders (for example, many benthic fishes, tunas that migrate vertically, and many bathypelagic fishes). In those that possess them, special structures are required to regulate swim bladder volume. Consider a neutrally buoyant herring at 10 m depth. The swim bladder occupies 5 percent of body volume and the enclosed gas will be at 2 atmospheres (atm) pressure (for each 10 m of depth hydrostatic pressure increases by 1 atm). If the herring rises to the surface, the swim bladder will double in volume because of the reduction in hydrostatic pressure to 1 atm. Consequently, the herring will become positively buoyant. If the herring descends the swim bladder will compress, density will increase, and the fish will tend to sink. In both instances the fish will have to swim to maintain position.

If the swim bladder is to act as a float and to allow unimpeded vertical movements, a mechanism is needed to maintain its volume constant. This is accomplished by moving gas into it upon descent and removing gas upon ascent. The swim bladder wall is very impermeable to gas diffusion, so gas leaks out slowly. More primitive teleosts retain a connection, the pneumatic duct, between gut and swim bladder. These fishes, referred to as **physostomes**, often gulp air at the surface to fill the bladder and can burp gas upon ascent. Although the swim bladder functions as a float, many physostomes use the bladder as an accessory respiratory structure, that is, as a lung (Chapter 7.6). The pneumatic duct serves the same purpose as the wind pipe or trachea of terrestrial vertebrates. In adult advanced teleosts, the pneumatic duct is absent, a condition termed **physoclistic**; gas cannot be released through the gut as in physostomes. Regulation of constant swim bladder volume upon descent requires secretion of gas into the bladder in both physostomes and physoclists.

Many physostomes and all physoclists possess a structure, the gas gland, located in the anterior ventral floor of the swim bladder. Underlying the gas

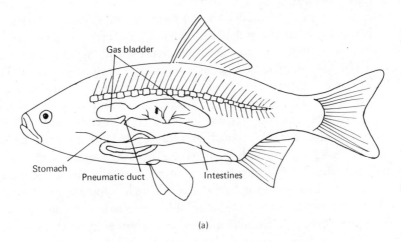

(a)

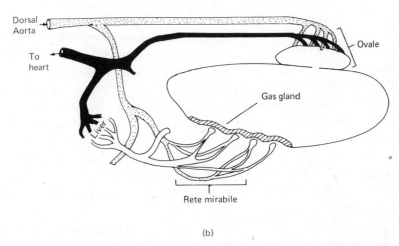

(b)

Figure 6-20   The swim bladder of typical osteichthyans.

(a) The swim bladder is positioned dorsal to the coelomic cavity just beneath the vertebral column. It may be partially constricted into multiple lobes, such as the two illustrated in this minnow. As in minnows, many relatively primitive teleosts have a duct connecting the swim bladder with the esophagus (**physostomous** fishes). More advanced fishes lose this connection.

(b) The vascular connections of a physoclistous swim bladder (one that lacks a duct to the esophagus) are similar to those in physostomes but represent the only avenue for filling and draining the bladder. This is accomplished by a combination of a rete mirabile countercurrent exchanger for gases (primarily oxygen) and a gas gland on the surface of the bladder that produces lactic acid, forcing oxygen out of solution. The efficiency of exchange of gas from the venous side of the rete to the arterial side, and therefore the final maximum pressure attainable in the swim bladder, is proportional to the length of the vessels in the rete. No elaborate mechanisms are necessary to remove gas from the bladder, since gas at a high partial pressure will diffuse into the blood readily across the ovale gland. Control of the ovale gland is by simple vasomotor action, shutting off or increasing blood flow to the wall of the bladder.

gland is a highly vascularized **rete mirabile** ("marvelous net"). This structure acts to move gas from the blood into the swim bladder, especially the oxygen carried by hemoglobin in the red blood cells. The secretion of oxygen occurs in many deep sea fishes at considerable depths and against enormous gas pressures within the bladder. Consider a mesopelagic lantern fish at 1000 m depth. Hydrostatic pressure and the gas pressure within the swim bladder are at 110 atm, yet the oxygen pressure in the arterial blood and the surrounding water does not exceed 0.21 atm, essentially the same as in surface waters. To secrete oxygen as a gas into the swim bladder requires that the partial pressure of oxygen in the blood be multiplied in excess of 500 times by the gas gland.

Although complex when considered in detail, basically the gas gland releases lactic acid into the rete mirable, and the resulting localized high acidity of the blood in the rete mirabile causes hemoglobin to release oxygen into solution. Oxygen pressure therefore builds up in the venous side of the rete mirabile. Because of the anatomical relations of the rete mirabile, oxygen diffuses from the venous side to the incoming arterial blood of the rete (Figure 6–20b). In this way the oxygen dissociated from the hemoglobin contained in red blood cells accumulates and is retained within the rete mirabile until its solution pressure exceeds the oxygen pressure within the swim bladder. When this occurs, oxygen diffuses into the bladder, adjusting its volume. The longer the capillaries of the rete mirabile, the greater the multiplication of gas partial pressure achievable (Figure 6–20). In the 1930s to 1950s, several physiologists demonstrated that the percentages of oxygen present in the swim bladder increased with depth and in deep sea fishes oxygen even reached 100 percent. These findings, which were initially baffling because the concentration of dissolved oxygen varies little with depth, are clear now that the secretion mechanism is understood.

To release gas from the swim bladder to compensate for the expansion of gas during ascent, physoclists simply open a valve, called the ovale, located dorsally in the posterior region of the bladder. It consists of a fold in the bladder wall that contains circular muscles that can be contracted to cover a local vascularized and more permeable part of the swim bladder. Outward diffusion of gas is reduced when the muscles are contracted, but upon opening the ovale the higher internal pressure of gases in the bladder (especially oxygen) causes diffusion into the blood.

Many deep sea fishes use a substitute means of maintaining buoyancy. For example many lanternfishes have oily deposits in the air bladder. Because fats have low density and are only partly compressible, they act as floats that function at different depths in spite of the changes in hydrostatic pressure. Thus total swim bladder expansion or compression is less upon ascent or descent because only a small volume of the bladder contains gas, and the amount of secretion required for a given vertical descent is less. Nevertheless, to achieve the high pressures required, a long rete mirabile is needed, and not surprisingly, the gas gland in deep sea fishes is very large.

As indicated, most adaptations of deep sea fishes relate to scarcity of food— they are small fishes (average length less than 5 cm) and not very abundant.

In so vast a habitat, for example, the numbers of anglerfishes (ceratioids) when most concentrated do not exceed 150,000 individuals per cubic mile (which is 4.17 km^3 or about 1 fish in 28,000 m^3). This estimate represents individuals of all ceratioid species, including males and females. The density of females in the most common species does not exceed five individuals per cubic mile and, more typically, less than one female per cubic mile. Imagine trying to find another human under similar circumstances, even without scaling for the enormous differences in body size! Yet to reproduce, each fish must locate a mate and recognize it as its own species.

The retention of eyes in bathypelagic fishes, even though the eyes are small, (Figure 6–19) relates in part to species recognition. Discrete bioluminescent patterns exist in the males and females and between species of many bathypelagic fishes. These differences must be useful in sex and species recognition. Anglerfish females possess a bioluminescent lure that is species specific in appearance (Figure 6–21). Although used to lure prey in feeding, the bait is probably also used to attract males. Detection of bioluminescent emissions over distances greater than a few hundred meters is not possible, however, because of the rapid extinction of light by water. Other senses must be used. Apparently scent trails, common in many terrestrial animals, are used by many deep sea fishes. The females are believed to secrete a characteristic odor, a sex pheromone, to attract males. Males usually have enlarged olfactory organs (Figure 6–21). Sensing the pheromone during random searching movements, males need then only swim upstream to an intimate encounter.

The life history of a typical ceratioid anglerfish dramatizes how selection adapts a vertebrate to its habitat. Ichthyologists now recognize about 200 ceratioid species. Although captured in all oceans, ceratioids are most common between 45°N and 35°S latitude. The adults, which represent the most bizarre and dimorphic vertebrates known, typically spend their lives in aphotic regions below 1000 m. Fertilized eggs, however, do not develop in the bathypelagic region, but rise to the surface where they hatch into larvae. Larvae remain in surface waters, mostly in the upper 30 m, grow, and later descend to the aphotic region. Descent accompanies metamorphic changes that differentiate females and males. The incredible differences in appearance between males and females during the various stages of the life cycle are depicted in Figure 6–21.

The sexes are similar when larvae, although the bioluminescent lure (a dimorphic characteristic of adults) is present in the female. Larvae of each sex have well-developed teeth with which to ensnare copepods and a balloonlike swelling of the skin, believed to reduce their density and slow their rate of sinking. Their bulbous shape must hamper their locomotor abilities; they drift passively with the surface currents. Body pigmentation is brown to black in adults but is absent in larvae. Larval transparency must act to reduce visual detection and capture by surface predators. The production of pelagic eggs and larvae adapted to an epipelagic existence, of course, is ecologically sound, for it places the young where food is abundant and therefore growth enhanced. The general importance of this reproductive pattern to survival is emphasized

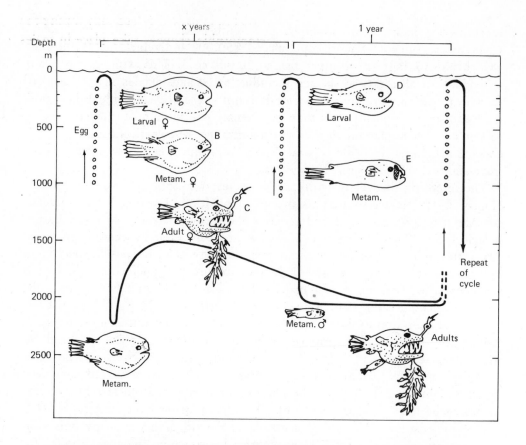

Figure 6-21   Diagrammatic representation of the life history of an anglerfish as a function of sex, age and depth distribution. (a) Larva of female, 25 mm; (b) larva of female, 43 mm; (c) adult female, 70 mm; (d) larva of male, 19 mm; (e) metamorphosed male, 23 mm. See text for details of life history. (Modified after E. Bertelsen [1951] The Ceratioid Fishes. Dana Report 39:1–281.)

by the discovery of pelagic eggs and larvae in a large variety of non-ceratioid deep sea fishes. This strategy probably increases the availability of food for the larvae by at least 50-fold.

During metamorphosis, young female ceratioids descend to levels below the adults (perhaps to escape competition with older fish), where they feed and grow slowly, reaching maturity after several years (Figure 6-21). Many species become little more than "a large mouth accompanied by a stomach," a common description that certainly fits their appearance. Dr. C. T. Regan, curator of fishes at the British Museum during the early 1900s, and the Danish ichthyologist E. Bertelsen, who specialized in the study of ceratioids collected during the famous Dana Deep Sea Expeditions, considered adult female anglers as floating fish traps. By extension and retraction, the bioluminescent lure is believed to mimic the movements of zooplankton and lure fishes and larger crustaceans into the region of the mouth. With a sudden opening of the gape prey is sucked in, ensnared in the teeth, and then engulfed.

Females feed throughout their lives. Males, in contrast, feed only during the larval stage. Upon descent, several metamorphic changes take place in males to prepare them for a very different future, for their function is reproduction, literally by life-long matrimony. The body may elongate and axial red muscles develop. No longer is food ever found in their guts; apparently, the enlarged liver provides sufficient energy for an extended period of several months of swimming. With metamorphosis males show enlargement of the olfactory organs, in accord with the presumed release of a sex phermone by the females, and the eyes continue to grow.

All of these changes suggest that adolescent males lead an active, short existence concentrated around finding a female. Nevertheless the search must be precarious, for one-fifth of a cubic mile, and usually much larger volumes, contains only a single female. Males must not only search this vast, dark region but also run the gauntlet of other deep sea predators. In the young adults there is an unbalanced sex ratio of unknown origin so that often more than 30 males exist for every female; this further assures that ultimate encounter will occur. All of the strategies used, however, whether morphological (feeding), behavioral (senses), or ecological (life history) seem directed to successful reproduction.

The story of the anglerfishes does not end here, for their most bizarre adaptation results in parasitism of the females by the males. Given the extreme dimorphisms, all of which increase the probability that a male will find a female, it still remains unlikely that either the female or the male will be in proper breeding condition when encounter occurs. Most ceratioids spawn in early summer, but the metamorphosed males descend several months later, out of phase with this spawning. Also, the testes of free-living metamorphosed males caught in the aphotic zone (presumably seeking females!) have all been immature. Males, therefore, upon contacting a female, bite into her flesh and attach themselves firmly. When collected, adult females often have an attached male, which invariably is much smaller than the female. Histological examinations have revealed that the tissues of each sex **anastamose**, that is, grow together and intermingle.

Preparation for this "encounter" begins very early for the teeth of larval males degenerate and strong incisorlike bones develop at the tips of the upper and lower jaws during metamorphosis (Figure 6–21). Males remain attached to the females for life, which covers several breeding seasons. In this parasitic state they grow some and their testes mature, nutrients being derived directly from the female. Monogamy prevails in this unusual parasitic marriage, for females invariably have but one attached male. As humans we must consider this lifestyle bizarre, and, indeed, it is extraordinary for it is known in no other group of vertebrates. But it also has been successful—some 200 species exist. The lesson these unique fishes provide is that vertebrate life is a very plastic venture capable through evolution of adapting to extreme conditions. In the ceratioids two features stand out: efficient energy utilization and reproduction. In the last analysis adaptations in all vertebrates relate to these two goals, but they are not often painted in such bold relief.

References

[1] Alexander, R. McN. 1967. *Functional Design in Fishes*. Hutchinson University Library, London. A short, quantitative and analytical morphology of fishes.

[2] Herald, E. S. 1961. *Living Fishes of the World*. Doubleday, Garden City, New York. A photographically illustrated survey of the most interesting and best known of living fishes.

[3] Herring, P. J. and M. R. Clarke. 1971. *Deep Oceans*. Praeger, New York. A well-illustrated introduction to the oceanography of the open sea with interesting sections on life in the depths.

[4] Hobson, E. S. 1975. Feeding patterns among tropical reef fishes. *American Scientist* 63(4):382–392.

[5] Marshall, N. B. 1971. *Explorations in the Life of Fishes*. Harvard University Press, Cambridge, Mass. Selected topics in fish biology analysed in a very comparative manner.

[6] Nelson, J. S. 1976. *Fishes of the World*. John Wiley and Sons, New York. A technical, family-by-family characterization of living and (in a more abbreviated treatment) fossil fishes.

[7] Norman, J. R. and P. H. Greenwood. 1975. *A History of Fishes*. Third edition. Ernest Benn, London. A very readable general introduction to fish biology.

[8] Schaffer, B. and D. E. Rosen. 1961. Major adaptive levels in the evolution of the actinopterygian feeding mechanism. *American Zoologist* 1(2):187–204. A technical analysis of jaw and jaw muscle function at different stages in ray-finned fish evolution.

[9] Smith, J. L. B. 1956. *Old Four-legs*. Longmans, London. A reminiscence of the discovery of *Latimeria* by its original describer.

[10] Thomson, K. S. 1969. The biology of the lobe-finned fishes. *Biological Review* 44:91–154. Reconstructions of the morphology, physiology and behavior of these nearly extinct fishes.

7

Life in Water: Its Influence on Basic Vertebrate Functions

Synopsis: A characteristic feature of vertebrates is their exceptional motility, which is reflected behaviorally in their diverse, usually agile, sometimes rapid and often graceful locomotor abilities. Because of this ability to move about, vertebrates occupy a dominant ecological position in the various ecosystems in which they live. High motility has several requirements; an efficient mode of locomotion, sensory systems for guidance, and an effective means of energy utilization to fuel the demands of motion. In the last analysis movement is work—and work takes energy in proportion to the rate of activity. Being an active, motile vertebrate, therefore, bears its price: a need for energy.

In most fishes, locomotion is accomplished by undulatory contractions of the body muscles a motility mechanism that has carried over into terrestrial vertebrates. Guidance is usually visual, but special senses involve the lateral line (for detection of low frequency pressure waves) and electroreception, a sense unique to fishes. High rates of energy utilization which are required to sustain the motion of fishes and their generally high activity levels, are provided for by an efficient respiratory system, the gills, which facilitate high rates of oxygen uptake for the oxidative combustion of food.

In general, active motile animals, such as the vertebrates, range widely and are less dependent on environmental stability than less active species. They must therefore be able to tolerate stressful changes in such environmental factors as temperature and availability of salt and water. Independence of environment requires internal regulation of body functions, a capacity defined in physiological terms as *homeostasis*.

Fishes show a considerable independence of the environment, a capability rooted in the development of homeostasis. All fishes except the hagfishes, show a high degree of osmotic and ion regulation, a condition achieved mostly by the kidneys and extra-renal structures such as those located in the gills of teleosts. Both the osmotic concentration and the internal level of the major ions in the blood (especially sodium, chloride, calcium, and magnesium) are maintained well below their concentrations in sea water.

In this chapter we describe several basic adaptations that have allowed fishes to become highly active and successful aquatic vertebrates. Many of the locomotor mechanisms, sensory systems, methods of energy utilization, and mechanisms for homeostatic regulation that are present in fishes set the stage and were preadaptive for life on land. Although our emphasis is on adaptation in fishes, much of what is discussed is common to all vertebrates. Where appropriate we have included considerations of analogous adaptations in terrestrial vertebrates.

7.1   LOCOMOTION IN WATER

Most vertebrates are capable of rapid motion for short periods. Many can sustain slower speeds over long periods. Vertebrates utilize diverse locomotor patterns; they swim, crawl, walk, run, jump, brachiate, fly, and even burrow. Only insects possess a similar diversity in locomotor patterns. As different as these diverse vertebrate locomotor modes appear, each is derived from the fundamental swimming pattern present in most aquatic chordates. Swimming results from sequential contractions of the muscles along one side of the body and simultaneous relaxation of those of the opposite side. Thus, a portion of the body momentarily bends and the bend is propagated down the body. When moving, therefore, successive points of the body of a fish alternately oscillate from side to side. These lateral undulations are most visible in elongate fishes, such as lampreys and eels, where the entire body bends into a sinelike wave (Figure 7-1). Identical undulations are observed in salamanders, snakes and many lizards. The basic movement in aquatic locomotion, therefore, has been carried on in much of vertebrate life.

If a variety of fishes are observed, different propulsive movements can be described. Flexible fishes oscillate the entire body, whereas stiffer fishes move mostly the caudal peduncle and tail, or, in fishes such as the trunk and box fishes, only the tail is moved (Figure 7-1). In his classic studies of fish locomotion, Dr. Charles Breder, at the American Museum of Natural History, classified the undulatory motions of fishes into three basic types: **anguilliform** —typical of highly flexible fishes capable of bending into more than half a wavelength; **carangiform**—undulations limited mostly to the caudal region, the body bending into less than half a wavelength; and **ostraciiform**—body inflexible, undulation of the caudal fin (Figures 7-1 and 7-6).

By recording body motions on film scientists have described the details of undulating locomotion in different vertebrates and determined how the oscillations are transformed into propulsive force. Especially important contributions to the study of swimming have been made by Sir James Gray, Professor R. Bainbridge, and Sir James Lighthill at Cambridge University. The number of factors involved and how they interact to produce motion is extremely complex. Indeed, complete hydrodynamic explanations have yet to be achieved. Many of the general adaptations in body form and structure, fins and muscle arrangements, however, function to increase the efficiency of the different modes of swimming. They are described here in qualitative terms.

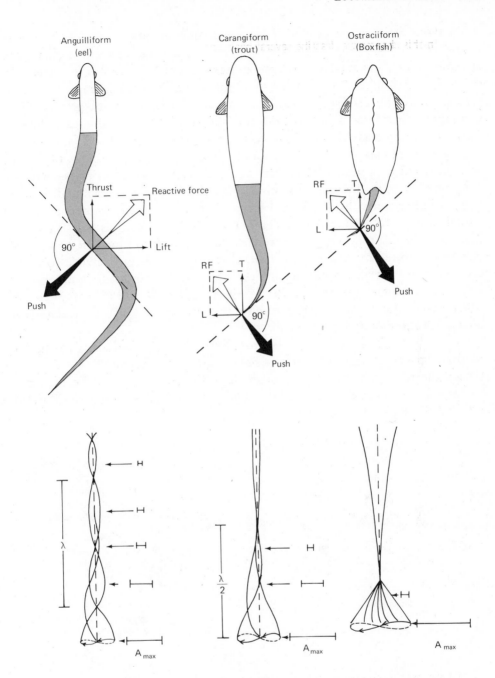

Figure 7-1 Basic movements of swimming fishes. Upper: outlines of some major swimming types showing regions of body that undulate (shaded). Lower: diagrammatic waveforms created by undulations of points along the body and tail. A_{max} represents the maximum lateral displacement of any point. Note that A_{max} increases posteriorly; λ is the wavelength of the undulatory wave. Ostraciiform swimming, as initially defined by C. M. Breder, refers to a limited number of very specialized fishes, like box fishes and trunk fishes. A variety of fishes swim by propulsion from various fins rather than the body (see Figure 7-6).

To swim a fish must produce *lift* to overcome gravity, and generate *thrust* to overcome *drag*. Drag (resistance to motion of an object through a fluid such as water or air), and gravity of course, must be countered by all vertebrates. To produce lift a foil, such as the wing of a bird or the fin of a fish, must be moved through water or air (Figure 7–2). Because it can act to counter gravity, lift is commonly considered as a vertical force (Figure 7–2b). Actually lift is defined as the force directed at right angles to the path of motion, and may or may not have a vertical component. For example, when a fish undulates its tail fin to swim forward it produces horizontal lift, a force whose direction is perpendicular and lateral to the path of motion (Figure 7–2a). When a fish undulates its body, the horizontal lift, or lateral force, generated caudally imparts a reactive lateral motion to the head (Figures 7–1 and 7–2a), known as *yaw*. This displacement causes the fish to deviate from the path of motion. In many fishes yaw is countered by a large rigid head that offers surface resistance to lateral displacement. Earlier we learned that the pectoral fins can act to counter *pitch*, and that the dorsal and anal fins can serve to minimize *roll* (Figure 7–2a), the other two major types of displacements that can cause a fish to deviate from course (Chapter 5).

Thrust, in contrast to lift, is the propulsive force directed along the path of motion. In vectoral analysis it acts at right angles to lift (Figure 7–2). To create thrust, the foil, as it is moved through the water (or air), must be canted from the path of motion. The angle created between the foil and the path of motion is called the *angle of attack*.

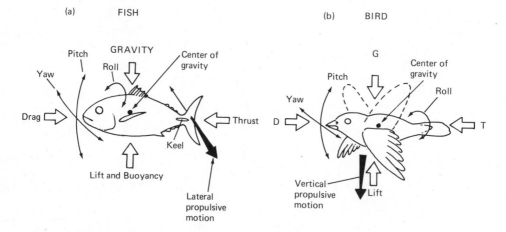

Figure 7–2   A comparison of forces associated with locomotion for a swimming and a flying vertebrate. (a) Motion of the caudal fin in the tuna produces a lateral movement far from the center of gravity. The reactive force to this movement causes the head to yaw in the same direction as the tail. Vertical lift and bouyancy of the fish are not a direct result of tail movements (see text). (b) In a bird the major propulsive stroke of the wings is downward. Since both the propulsive stroke and the reactive force (lift) act near the center of gravity the bird does not pitch. An up and down motion of the entire body occurs, however, because vertical lift is produced only during the downward stroke.

Most of us are familiar with how lift, gravity and drag interact. We all delight in feeling the wind tug on our hand when held out the window of a moving car. Angled up into the wind (producing an angle of attack) our hand rises effortlessly, angled down it declines. If the hand's angle to the wind is too great, however, unless countered with muscle force the hand is forced backward. To develop forward motion of the hand against the wind an additional element is required: an expenditure of muscular energy upon the angled hand to produce forward thrust. How do fishes overcome gravity, produce thrust, and reduce drag?

7.1.1 Overcoming Gravity—The Generation of Vertical Lift

Many fishes are neutrally buoyant and maintain position without body undulations. Only the pectoral fins backpaddle to counter a forward thrust produced in reaction to the water ejected from the gills during each respiratory cycle. Here neutral buoyancy is the result of the presence of an internal float—a gas-filled swim bladder lying below the spinal column (Section 6.4). Because the average tissue density of typical fishes exceeds that of water they sink, unless their total density is reduced with a float, or vertical lift is created by propulsive movements. Fishes that hover in the water, therefore, tend to have well developed swim bladders.

Surprisingly, many pelagic fishes like tunas, mackerels, swordfishes (all known collectively as scombroid fishes), and sharks do not possess swim bladders. They are all negatively buoyant. Tunas and many sharks must swim constantly to produce sufficient vertical lift to overcome gravity. In some cases lift is accomplished by extending the large pectoral fins like wings, or the diving planes of a submarine, and setting them at a positive angle of attack to the water that flows over them (Section 7.1.2). The amount of lift can be changed by adjusting the angle of these pectoral "wings" to the water's flow. The vertical lift created by the pectoral fins, however, requires forward velocity and this is produced by undulation of the caudal fin to produce thrust.

7.1.2 Overcoming Drag—The Generation of Thrust

Exactly how are the lateral undulations translated into thrust? Fishes swim forward by pushing backwards on the water with their body and fins. For every *active* force there is an equal and opposite *reactive* force (Newton's third law of motion). Undulations pass backwards, but the bending portions of the fish's body also push laterally. The *reactive* force, therefore, must be directed forward, but at an angle. In slender flexible fishes, like eels, the contractural waves propagate near the head and progress toward the tail. The amplitude of each undulation increases from head to tail. Wave velocity along the body is uniform and must exceed the swimming velocity. Lateral body undulations, therefore, characteristically have constant wavelength (λ) but increasing amplitude ($\dot{A}_{max}$) along the body (Figure 7–1). As demonstrated by Professor Bainbridge, to increase swimming speed most fishes merely increase the frequency of their body undulations—the tail beat frequency. For many fishes swimming speed is a linear function of the tail beat frequency (Figure 7–3). Increasing the frequency of body undulations increases swimming speed because it results in the application of more power (force per unit time) to the water by fish. Yet, different fishes achieve very different speeds through body undulations; some are slow (eels), others are very fast (tunas).

Anguilliform swimmers cannot depend on speed to obtain food and to escape enemies. The eel's long body limits speed because it induces too much frictional drag. The advantages of extreme flexibility, therefore, must compensate for their lack of speed. Fishes that swim rapidly are less flexible and utilize carangiform motion or modifications of it, where the body undulations are primarily observable in the caudal region (Figure 7–1). Muscle force from anterior segments is transferred through ligaments to the caudal peduncle and the tail. The morphological adaptations that accompany this swimming mode reach their zenith in fishes like tunas, where the caudal peduncle is slender and the tail greatly expanded vertically.

In both anguilliform and carangiform swimming the angle of attack of the undulating body or tail fin is critical to the efficient generation of thrust. Hydrodynamic studies reveal that the amount of lift produced by a foil moving through the water at constant velocity is proportional to the angle of attack. Importantly, the amount of thrust also increases with the angle of attack (Figure 7–4). As a result, the greater the angle of attack of a fin to the water, the more effective the generation of thrust. But there are limits to the magnitude of the angle of attack. Drag is exponentially proportional to the angle of attack, and at some critical angle (Figure 7–4) the amount of lift generated

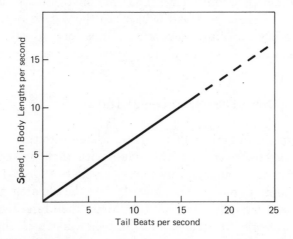

Figure 7–3   Relationship of tail beat frequency to specific speed of perchlike fish that swim with carangiform motion (after R. Bainbridge [1960] *J. Exp. Biol.* 37:129–153.)

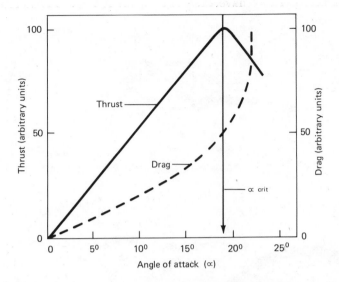

Figure 7-4 The effect of angle of attack on thrust and drag; α_{crit} is the angle at which a fin or wing (foil) stalls. The critical angle will vary with the shape of the foil and its speed through the fluid medium (water or air). Note that drag increases exponentially.

suddenly decreases and drag greatly increases. Thus thrust created by a foil's angle of attack is limited. This critical angle at which thrust reaches its limit is equivalent to the aerodynamic stall angle where the wing of an airplane ceases to produce vertical lift. To be effective therefore a fin must be set at a low angle of attack to the water, usually somewhere between 10° and 20° to the path of motion.

An important difference between carangiform and anguilliform motion is the evolution of a hinged coupling between the caudal fin and the caudal peduncle in carangiform swimmers. In eels the angle of attack at a given point on the body changes continuously throughout each power stroke (Figure 7-5b). Thrust generation in eels is maximal for a particular body segment as it passes across the midline of the path of motion, and falls to zero at the points of maximum displacement (Figure 7-5b). In carangiform swimmers the angle of attack changes in a similar manner as it does in anguilliforms, but the angle of attack is constant over a larger part of each power stroke (Figure 7-5a and b). Because of the caudal hinge and the resistance of the tail, the caudal fin lags behind the angle created by the caudal peduncle. As a result, the tail fin is maintained at a high and efficient angle of attack through a large portion of the power stroke, and this results in more thrust. The critical interrelationships involved are emphasized by the fact that most carangiform fishes are able to adjust the degree of caudal fin lag by tightening and loosening the pull on the tendons connecting the tail to the trunk region.

Thrust is proportional not only to the magnitude of the angle of attack of the tail fin, but also to the velocity at which the fin (or body segment) is moved

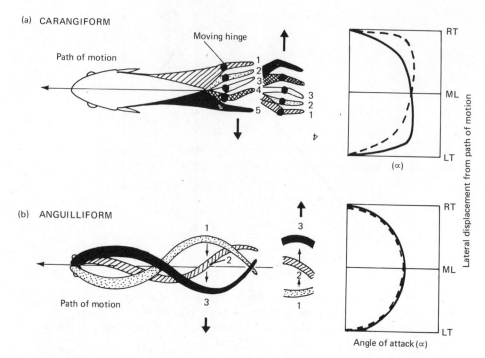

Figure 7-5 Attack angle of the propulsive surface throughout a complete undulatory cycle in two distinctive swimming modes. (a) Carangiform swimmers have a hinged couple between the caudal peduncle and tail fin that results in a more nearly constant angle of attack through an undulatory cycle. (b) Anguilliform swimmers lack a similar mechanism. The angle of attack varies continuously throughout each undulatory cycle.

laterally through the water. In carangiform fishes the velocity of lateral displacement, which declines to zero at the very end of each power stroke (that is, at maximum lateral displacement), is fairly constant over a large portion of each power stroke. Here again thrust generation is carefully controlled to reach a high efficiency.

In contrast to flexures of the body or beats of the tail, a diverse grouping of fishes seldom flex the body, but undulate the median fins (a swimming mode referred to as **balistiform** swimming). Usually several complete waves of undulation are observed along the fin (Figure 7–6a and b), and very fine adjustments in the direction of motion can be produced. The exquisite undulating fin motions of these fishes are a modification of the body undulations used to propel the great majority of fishes.

A large number of fishes (for example, chimaeras, surf perches, surgeon fishes, wrasses and parrot fishes) generally do not oscillate the body or median fins, but utilize rowing movements of the pectoral fins to produce movement (Figure 7–6d). Fishes that utilize such movements of the pectoral fins are said to use **labriform** swimming. Actually many fishes show combinations of the basic swimming patterns. Intermediates between anguilliform, carangiform, ostraciform, balistiform, and labriform movements exist, and most fishes can in-

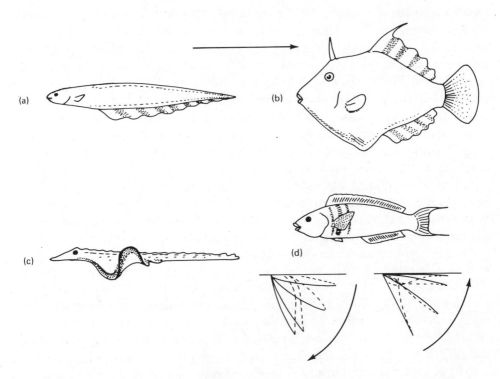

Figure 7-6 Propulsion by undulation of fins in various fishes. (a) Knifefish, and (b) filefish use various modifications of the balistiform mode of swimming. The skate (c) uses similar undulations of the pectoral fins (rajiform mode). Note that undulations usually show more than one complete waveform. Arrow indicates direction of wave during forward motion. (d) Wrasses primarily use their pectoral fins in the labriform mode of swimming. Bottom detail shows direction and attitude of pectoral fin during one complete cycle of its movement during forward locomotion, as seen from above. The leading edge of the pectoral fin is a dark line; the trailing edge is a dotted line.

dependently move the paired fins to adjust body position. The muscles and nerves that move these fins are derivatives of the segmental muscles and spinal motor nerves that produce body undulations. In fact, the nerve-muscle complexes that move the limbs of tetrapods have a similar origin. As different as swimming, crawling, running and flying appear in vertebrates, the machinery that drives the locomotion has a common origin—the segmental neuro-muscular complex.

7.1.3 Improving Thrust—The Reduction of Drag

Drag, the resistance to motion through a fluid, is of two forms: **viscous drag** from frictional forces between a fish's body and the water, and **inertial drag** from pressure differences along the fishes body created by the fish's displacement of water. At slow speeds inertial drag is low, but it increases rapidly with speed. Viscous drag is relatively constant over a range of speeds. Hydrodynamic studies demonstrate that viscous drag is affected by surface smoothness and inertial drag by body shape. Streamlined shapes produce minimum drag when their maximum width is about 1/4 of their length and occurs about 1/3 of the length from the leading tip (Figure 7–7). It is not surprising that the shape of many rapidly swimming vertebrates closely approximates these dimensions. Too thin a body has high viscous drag and produces turbulence just behind the leading edge. Too thick a body induces high inertial drag. Usually rapidly swimming fishes have very small scales or are scaleless, resulting in a smooth body contour and reduced drag. (For a variety of reasons many slow swimming fishes also are scaleless, emphasizing that universal generalizations are difficult to make.) In addition, the presence of mucus, which is so typical of the body surfaces of fishes, is believed by some to contribute to the reduction of viscous drag.

Drag is minimized in carangiform swimmers because the body is relatively stiff and retains an efficient shape during lateral undulations of the tail. The tail, however, in its thrusting movements against the water, like the body undulations of anguilliform swimmers, creates turbulence (one form of inertial drag) observed as vortices of swirling water in a fish's wake. The total drag created by the caudal fin also depends on its shape. When the **aspect ratio** of the fin (that is, the dorsal to ventral length divided by the anterior to posterior width) is large, the amount of lift produced relative to drag is high. The falcate (sickle) shaped fin of scombroids (mackerels, tunas) and of certain sharks, many of which are pelagic and swim continuously, results in a high aspect ratio and thus in efficient forward motion. Even the cross section of these caudal fins assumes a streamlined shape, which further reduces drag. A structural requirement necessary to maintain a fin's high aspect ratio is fin stiffness, and the fins of many carangiform swimmers are quite inflexible. Many forms with these adaptations swim continuously; rapid acceleration from a standstill is infrequent. High aspect ratio fins do not present adequate leverage with which to push on the water with sufficient force to initiate rapid accelerations.

The mode of swimming in which only the caudal peduncle and fin undulate is usually called "modified carangiform" motion. In scombroids and many pelagic sharks that swim with modified carangiform motion the caudal peduncle is constricted vertically but is relatively wide. Often the peduncle in carangids is studded laterally with bony plates called scutes. These adaptations present a knife-like profile to the water as the peduncle undulates from side to side and contribute to the reduction of lateral drag. The importance of these seemingly minor morphological changes to rapid swimming fishes is underscored by the fact that the tail stalk of whales and porpoises is also narrow and has a blunt, double-knife-like profile. But it is in the lateral dimension that the tail stalk is narrow, unlike the caudal peduncle of scombroid fishes. In these cetaceans the tail and stalk undulate dorsally and ventrally rather than laterally. These strikingly similar adaptations of modified carangiform swimmers—whether shark, scombroid, or cetacean—produce an efficient conversion of muscle contractions into forward motion.

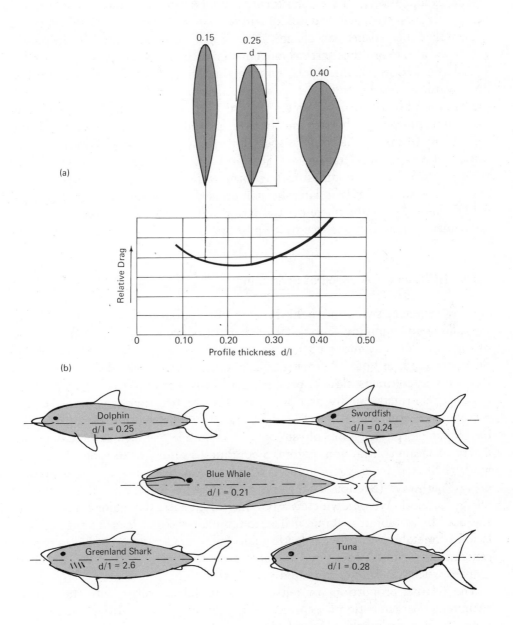

Figure 7-7 The effect of body shape on drag. (a) Streamlined profiles with width (*d*) equal to approximately one fourth of length (*l*) minimize drag. The examples are for solid smooth test objects with thickest section about two fifths of the distance from the tip. (b) Width to length ratios (*d/l*) for several swimming vertebrates. Like the test objects these vertebrates tend to be circular in cross section. Note that the ratio is near 0.25 and that the general body shape approximates a fusiform shape. (Modified after H. Hertel [1966] *Structure, Form, Movement.* Krauskopf-Verlag, Mainz, West Germany.)

The caudal fins of the carangiform swimmers such as trout, minnows, and perch are not stiffened and seldom have high aspect ratios. Bainbridge has shown that these fishes can change their caudal fin area to modify propulsive thrust and also to produce vertical movements of the posterior part of the body. This latter action is achieved by propagating an undulatory wave upward or downward across the trailing edge of the flexible caudal fin, as occurs in the median fins of balistiform swimmers (Figure 7–6). In carangiform fishes propulsion often proceeds in bursts, usually from a standstill with very rapid acceleration Many of these fishes possess special neural systems (the Mauthner cells) to initiate this rapid acceleration (see Section 6.2). The caudal peduncle of these fishes, unlike that of modified carangiform swimmers, presents a large surface area to the water for it is laterally compressed and deep (see Figure 6.7 and 6.12). Under these circumstances the latter half of the body as well as the tail contribute a great deal of force to propulsion.

7.1.4 Efficiency and Speed of Swimming

Two dominant swimming behaviors are observed in fishes: sustained "cruising" and rapid swimming "bursts." Each behavior depends upon a different type of muscle. For continuous power output when "cruising," so called red muscle fibers are used, in "burst" activity, distinctive white muscle fibers are utilized. Red muscle contracts slowly, producing a low power output. White muscle, in contrast, contracts rapidly and produces three to five times the power of red muscle. However, white muscle fatigues rapidly. The two muscle types differ in cellular morphology and physiology. Red muscle has high aerobic demands and to maintain contraction requires a continuous supply of oxygen. Its rate of work is limited by oxygen supply. White muscle relies in part on anaerobic energy pathways (Section 7.5.1) and is limited by the local cellular store of energy, often as glycogen which is a polymer of glucose, the major carbohydrate fuel used by vertebrates. Rapid utilization of energy stores leads to the accumulation of metabolites, often as lactic acid. Lactic acid build-up impedes further generation of the ATP required for contraction. Similar fatigue also occurs in humans when we exercise too rapidly, as in sprinting.

The relative proportions of red to white muscle correlate with the type of swimming characteristic of a species. For example, in a modified carangiform swimmer like the mackerel, a pelagic scombroid that swims continuously, red muscle constitutes 27 percent of the total muscle mass; in carangiform swimmers like the rainbow trout 24 percent, in the dogfish shark 18 percent, in the benthic halibut 7.5 percent, and less than 1 percent in deep sea benthic fishes. Thus, the more sluggish the swimming habits the less red muscle a fish has.

How efficient is swimming? A variety of models predict that 60 to 80 percent of the muscle force delivered to the caudal fin results in motion. This direct power conversion, however, is not the overall cost of motion. Dr. Paul Webb at the Pacific Biological Station, Canada, estimates for trout and salmon that the energetic conversion of food (glucose) into forward motion averages about 25 percent, a value similar to the fuel efficiencies achieved by well tuned cars.

How fast do fishes swim? It is worthwhile to consider both maximum speeds during "bursts" and sustainable speeds during "cruising." Burst speeds reach 33 meters/sec (ca. 75 mph), although top speeds of 18 to 22 m/sec (40 to 50 mph) are more usual. As might be expected, these very high speeds are for the specialized predatory scombroid fishes, especially swordfish and marlin. An alternate way of reporting speed is the specific swimming speed, i.e., the absolute speed divided by the fish's body length. James Gray found that top specific speeds for fishes swimming in the carangiform mode are about 10 body lengths/sec (for minnows, perches and trouts). In contrast, anguilliform fishes achieve top speeds of only 2–3 body lengths/sec but the scombroids attain speeds above 10 and some even 20 body lengths/sec. Size affects specific speed: in general, as size increases so does absolute velocity, but specific speed declines (Figure 7–8). Nevertheless, the classifications proposed by Breder to describe swimming styles correlate well with top speeds—anguilligorms are slowest, carangiforms intermediate, and modified-carangiforms (scombroids) fastest.

Slower, sustained "cruising" speeds vary in the same way for these three groups. Gray and Bainbridge found for minnows and salmonid fishes that specific cruising speeds are about 3 body lengths/sec. Eels apparently cruise at specific speeds of less than 1 body length/sec. Webb states that scombroids cruise between 5 and 10 body lengths/sec with average sprint speeds from 10 to 16 lengths/sec and suggests that scombroids are designed for high speed, and sustained activity, a conclusion congruent with their continuous swimming mode, their streamlined high aspect ratio fins, morphology, and muscle physiology. Cetaceans show similar trends: cruising is accomplished at specific speeds between 2 and 5 body length/sec whereas top speeds fall between 3 and 7 lengths/sec. Thus, their highly streamlined bodies and semi-lunate tails, like those of scombroids, produce high sustained speeds.

These various modes of swimming are compromises to particular aquatic habitats. Anguilliform locomotion is inefficient for rapid swimming, but its extreme flexibility allows exploitation of the interstices of complexly structured habitats—reefs, rocks and rubble, tree roots, burrows, etc. That this way of life is of considerable evolutionary importance is indicated by the large numbers of unrelated species which have converged on a flexible eel-like body form. In contrast, modified carangiforms, which are highly efficient in sustained motion, are mostly limited to vast expanses of open waters where continuous roaming in search of food is a daily necessity of life. Their extreme specialization for this mode of life has precluded them from exploiting other habitats. Most fishes occupy neither the open sea nor the close confines of interstices. These fishes are carangiform swimmers. Being both flexible and streamlined they are intermediate in specialization and therefore occupy the widest range of aquatic habitats. Their behaviors are usually performed with slow refined movements,

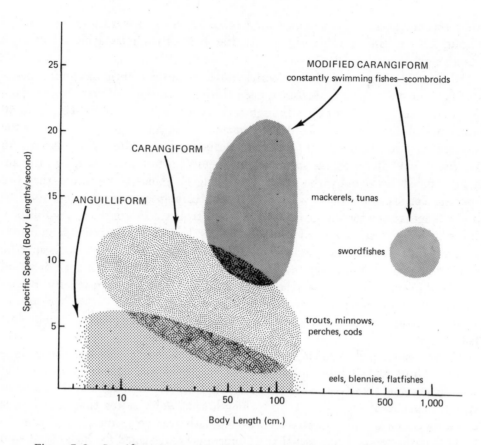

Figure 7-8   Specific swimming speeds for burst locomotions of fishes that swim with anguilliform, carangiform or modified-carangiform modes of locomotion. Actual data points for fishes swimming by each mode of locomotion fall within the indicated areas.

as exemplified by coral reef fishes (Section 6.3.3). Their burst activities produce rapid motion useful in many of their predator/prey interactions.

Most fishes and many other vertebrates propel themselves by undulating the body. Whether the motion is anguilliform (serpentine), carangiform, or even balistiform rippling of individual fins, undulation is an efficient propulsive system, more so than flying, walking or running (section 7.5.2). This results from the fact that every time a body segment contracts it produces a motion that generates thrust. In a mammal and in most birds, muscle effort is exerted to reposition the limbs before each power stroke. Therefore a portion of the energy expended in walking or flying produces little or no thrust, an inefficiency not found in vertebrates that undulate.

7.2 SPECIAL SENSES OF FISHES

Water represents a different physical phase than air and therefore possesses distinctive properties that strongly influence the behaviors of fishes and other aquatic vertebrates. We have already discussed how water, some 820 times as dense as air at sea level, affects locomotion. Also enlightening is a consideration of how water compares to air in the propagation of sensory stimuli that vertebrates normally use—light, chemical substances, sound, and displacement.

Light is attenuated more when passing through water than air. Indeed, objects become invisible in the very clearest water at a distance of a few hundred meters. In clean air, distance vision is virtually unlimited. Only the presence of aerosols, such as pollutants and water vapor, reduce distance vision in the atmosphere. In a similar way particles (silt, plankton) and many substances dissolved in water attenuate light to reduce visual range.

Nevertheless, fishes generally possess well-developed eyes. A major difference between aerial and aquatic vision relates to focusing light rays on the retina to produce a sharp image. The cornea in th eye of both terrestrial and aquatic vertebrates has an **index of refraction** (velocity of light in vacuum divided by the velocity of light in the medium) of about 1.37, which is close to the index of refraction of water (1.33). Terrestrial vertebrates rely on the refractive difference between air and cornea to focus incoming light rays on the retina. The lens of the terrestrial eye, which often is flattened in cross section and pliable, acts mainly as a fine focus adjustment. In water the cornea is eliminated as a focusing device for both fishes and terrestrial vertebrates, as opening our eyes underwater illustrates: objects appear fuzzy. Fishes possess a spherical, less pliable lens with higher refractive power to focus images on the retina. Aquatic mammals, such as cetaceans, also possess a large spherical lens to provide for adequate accommodation (focusing) of objects. Otherwise a fish's eye is similar to the eyes of terrestrial vertebrates—water primarily limits the distance over which objects can be seen.

Fishes possess taste bud organs within the mouth, and similar receptors of "general chemical sense" over much of the body surface for detecting nonvolatile and usually dissolved substances. In addition, well-developed olfactory organs to detect volatile substances are present on the snout. All function like those of terrestrial vertebrates—even the thresholds in highly sensitive species are similar. Sharks and salmons, for example, are capable of detecting odoriferous compounds at levels of less than 1 part per billion (ppb). Of course chemicals must reach the receptors by diffusion and by transport in currents of water and air. Because diffusion is slower in water than air, odors can be considered less effective as stimuli over long distances to fishes than to terrestrial vertebrates. Taste as a distance sense, however, would be more effective in fishes and less so in terrestrial vertebrates because it depends on substances in solution. There is an element of general truth to these intuitive ideas on how chemical stimuli reach the chemoreceptors, but such generalities are not universal. As expected there are feeding and searching responses in many fishes based on taste detection of dissolved substances that indicate nearby or upstream prey. Unexpectedly, adult migrating salmon are directed to their stream of origin from many thousands of miles by a chemical signature from the "home" stream permanently imprinted when they were juveniles. Plugging the nasal olfactory organs destroys the ability to "home," proving that odors can function as distance stimuli.

Mechanical receptors provide the basis for detection of displacement—touch, sound, pressure, and motion. Like all vertebrates, fishes possess an internal ear (labyrinth organ), including the semicircular canals, which functions to inform the animal of changes in speed and direction of motion. They also possess, at the base of the canals, gravity detectors that inform them of positional orientation—up from down. Most vertebrates also possess in this labyrinth complex an auditory region sensitive to sound pressure waves. Surprisingly, these diverse functions of the labyrinth are dependent on basically similar types of sense cells, the hair cells (Figure 7-9). In fishes and aquatic amphibians the hair cells are not limited to the labyrinth, but are also dispersed over the surface of the head and body. In fishes they are often organized into a series of canals on the head and one or more canals that pass along the sides of the body onto the tail. This unique surface receptor system of fishes and aquatic amphibians is referred to as the lateral line system.

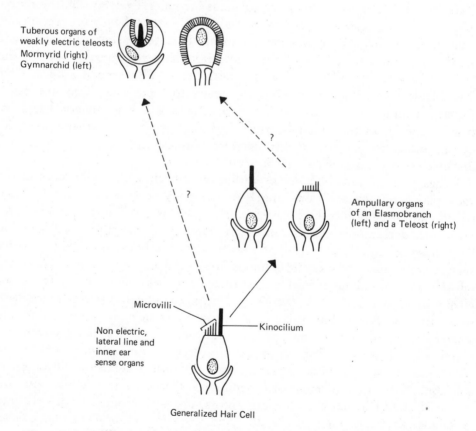

Figure 7-9 Different categories of hair cells present in vertebrate sensory structures. The sensory receptive apical portion of each type of cell is derived from a cilium and/or associated microvilli. The structures at the base of the sensory cells are neurons. The exact evolutionary sequence for the derivation of the sensory cells of weakly electric fishes is unknown (after Szabo, *in* Fessard [3]).

7.2.1 Water Displacement Detection—The Lateral Line

Hair cells are distributed on the body surface in two configurations: within tubular canals, or exposed in epidermal depressions, or both (Figure 7–10). Tubular canals protect the clumps of hair cells and their nerves from damage. Within each lateral line scale and in the depressions, groups of hair cells are arranged into units called neuromasts. As depicted in Figure 7–9 and 7–11, the apex of each hair cell has a kinocilium placed asymmetrically in a cluster of microvilli. Hair cells are arranged in pairs with the kinocilia positioned on opposite sides of adjacent cells (Figure 7–11). A neuromast contains many of these hair cell pairs; each cell is innervated by a separate neuron that transmits impulses to the brain. Each neuromast unit is serviced by two afferent lateral line nerves; one transmits impulses from hair cells whose kinocilia are of one orientation, the other impulses from cells whose kinocilia positions are reversed by 180°.

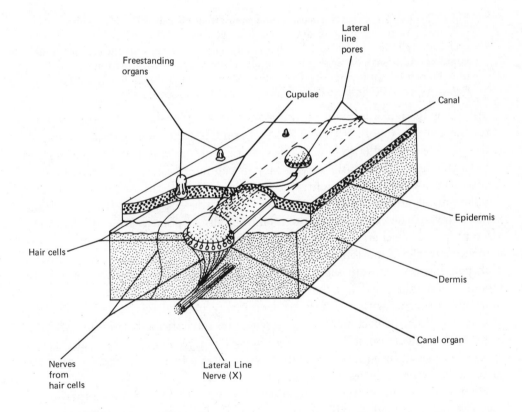

Figure 7-10　Representations of the two basic configurations of lateral line organs in fishes. Diagram is of freestanding and canal organs of a scaleless fish. Note that a gelatinous covering, the cupula, encapsulates the apical projections of each hair cell.

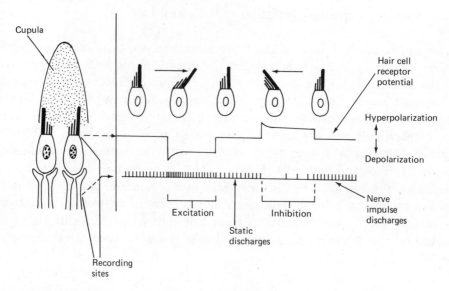

Figure 7-11   Hair cell deformations and their effect on hair cell transmembrane potential (receptor potential) and afferent nerve cell discharge rates. Note that each hair cell when static produces a steady, characteristic frequency of nerve impulses. Adjacent hair cells respond oppositely to the same deformation because their kinocilia bend in opposite directions. (Modified after Flock, in *Lateral Line Detectors*, ed. P. Kahn [1967] Indians U. Press, Bloomington, Ind.)

All kinocilia and microvilli are imbedded in an oval-shaped gelatinous structure, the cupula (Figure 7-10 and 7-11). Lateral displacement causes the cupula and embedded kinocilia to bend. The resultant deformation induces changes in the electro-chemical potential across the hair cell membrane. When a kinocilium is displaced away from the central axis of its hair cell, the result is membrane depolarization and increased excitation of its afferent nerve. If a kinocilium is bent toward the central axis, the membrane hyperpolarizes and inhibition of afferent nerve discharge results. Each hair cell pair, therefore, encodes unambiguously the direction of cupula displacement. The excitatory output of each pair has a maximum sensitivity to displacement along the line joining the kinocilia and falling off in other directions (Figure 7-11). In the canal organs, units are arranged with the axis of asymmetry parallel to the canal. The net effect of cupula displacement is to increase the static firing rate in one afferent nerve and to decrease it in the other nerve (Figure 7-11). These changes in lateral line nerve firing rates are utilized to inform a fish of the direction of displacement forces on different surfaces of its body.

What forces might produce cupula displacement? It is clear from experiments that currents directed on different portions of the body can be detected. Cutting the afferent nerves abolishes any response. Thresholds are extremely low. Water currents of 0.025 mm/sec are detected by the exposed neuromasts in the aquatic clawed frog, *Xenopus laevis*, with a maximum response at about 2 to 3 mm/sec (threshold of 5.6×10^{-5} mph). Similar responses occur in fishes. The lateral line organs also respond to low frequency sound, but considerable controversy exists amongst physiologists as to whether sound represents a natural lateral line stimulus. Many fishes also hear low frequencies with the ear, even after the lateral line organs have been destroyed. Sound induces traveling pressure waves in the water and also causes local water displacement as the pressure wave passes. It has been difficult to be sure whether neuromast output results from the accompanying water motions on the body surface or the sound's compressional wave. To solve this dilemma, careful behavioral experiments are required to tell us what stimulus modes are significant under natural conditions.

Several surface-feeding fishes and *Xenopus* provide a vivid example of how the lateral line organs can function under natural conditions. These species find insects on the water surface from the surface waves created by the prey's motions. In the killifish, *Aplocheilus lineatus*, three paired groups of naked neuromasts are located on top of the flattened head (Figure 7–12). E. Schwartz, in a series of clever experiments, has demonstrated that each neuromast group provides information about surface waves coming from a different direction. Thus, the nasally-located units are sensitive to waves arriving

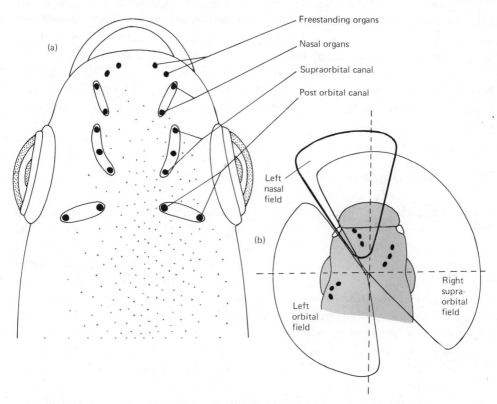

Figure 7-12 Distribution of the lateral line canal organs on (a) the dorsal surface of the head of *Fundulus notatus* and (b) the perceptual fields of the head canal organs in *Aplocheilus lineatus*. Note that fields overlap on each side as well as on the same side of the body. The wedge-shaped areas indicate the relative directional sensitivity for each group of canal organs. (Modified after Schwartz, *in* Fessard [3].)

toward the front of the head, the supraorbital groups detect laterally arriving waves, and the postorbital groups respond to waves arriving from a caudal direction. All groups, however, show stimulus field overlaps. Bilateral interactions between neuromast groups are indicated by the fact that extirpation of an organ from one side of the head disturbs the directional response.

Behavioral experiments on another surface-feeding fish, *Pantodon buchholzi*, have shown that directional sensitivity is very acute (about ±5°) over distances of three to four body lengths and sensitive to surface disturbances with amplitudes of only 4 μm, typical of waves created by insects. Presumably the large numbers of neuromasts on the head of some fishes assist in orienting their posture and motions with respect to water disturbances in the vicinity of the head. Several researchers have suggested that this might be important for schooling fishes in sensing vortex trails (Section 7.1.2) of adjacent schoolmates. Nevertheless many of the species of fish that form extremely dense schools lack lateral line organs along the flanks (herrings, atherinids, mullets, and so on) and retain only the cephalic canal organs. This reduction in the number of sensory elements would reduce the constant water-noise from turbulence that must be present within fish schools. The well-developed cephalic lateral line organs concentrate sensitivity to water motion in the head region where it is needed to sense the degree of turbulence into which the fish is swimming. In this way a fish might sense and choose the least turbulent direction. Over periods of time the reduction in drag through avoidance of turbulence could achieve a considerable metabolic swimming economy, as well as an even spacing between school members.

7.2.2 Electric Discharge and Electroreception

The beginning of human experience with electricity, other than with lightning, probably resulted from contact with the torpedo ray in the Mediterranean region and catfish in the Nile. Both of these fishes and the electric eel of South America can discharge sufficient electricity to stun other animals. The source of the electric current is modified muscle tissue. These cells, called **electrocytes**, have lost the capacity to contract but are specialized for generating an ion current flow. Generally, electrocytes are flattened cells with a smooth, innervated surface on one side and a highly infolded surface on the other side. As in normal skeletal muscle cells, each electrocyte is innervated by a spinal motor nerve which, when excited, causes the electrocyte to depolarize and to initiate a brief discharge called a spike potential. The spike potential is caused by cations that flow inward across the smooth cell surface and outward across the roughened surface (Figure 7–13). The spike achieves a peak potential of about 100 mV.

Because electrocytes are arranged in stacks or in series, like the batteries in a flashlight, the discharge potentials across each stack can add to produce higher voltages. Synchrony of discharge is required, however, and this is produced by

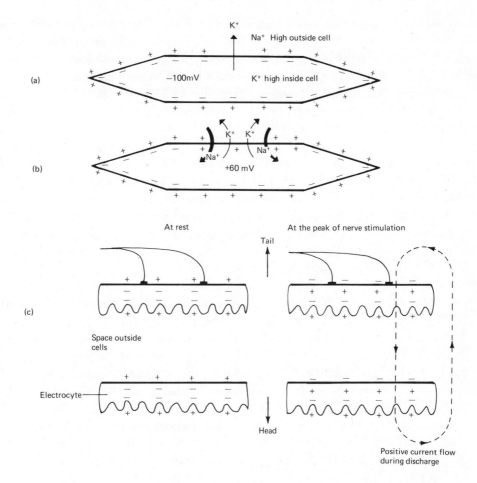

Figure 7-13   The use of transmembrane potentials by electric fishes to produce a discharge.

(a) When at rest the membranes of neurons or muscle cells (as shown) are electrically charged, with the intracellular fluids about −100 mV relative to the extracellular fluids. K^+ is maintained at a high and Na^+ at a low internal concentration by the action of a Na^+–K^+ cell membrane pump. At rest, permeability of the membrane to K^+ exceeds the Na^+ membrane permeability. As a result K^+ diffuses outward faster than Na^+ diffuses inward (arrow) and sets up the 100 mV resting potential.

(b) When appropriately stimulated the permeability of the membrane to Na^+ increases. As a result a large, rapid, but fleeting local influx of Na^+ is generated. This large influx inverts the membrane potential and excites adjacent membranes to depolarize. As a result, the disturbance propagates over the cell surface.

(c) In an electrocyte, which is a modified noncontractible muscle cell, one cell surface is rough and the opposite surface smooth. Nerve innervation is associated with the smooth surface and when excited into activity, only the smooth surface depolarizes. The resulting Na^+ flux across the smooth surface into the cell and the K^+ leakage across the rough surface and out of the cell yield a net positive current in the direction indicated. By arranging electrocytes in series, electric fishes can generate very high voltages. In the electric eel, for example, as many as 6000 cells are stacked and lead to potentials in excess of 600 V.

simultaneity of nerve impulses arriving at each electrocyte. In fishes like the African electric catfish and the electric eel, considerable potentials can be generated (in excess of 300 and 600 V, respectively). Current flow, of course, is limited in a single stack, but is increased by arranging stacks in parallel, as occurs in the electric eel. It was realized during Darwin's time that this unusual anatomical structure was present in several other species of fishes that did not produce electric shocks. Because of the anatomical similarities, these organs were considered to generate electric currents but not until the 1950s was their weak electric nature demonstrated. Electric fishes are now classified into two groups, the strongly electric and the weakly electric fishes (Figure 7-14).

Most electric fish species are limited to tropical fresh waters of Africa and South America. Electric marine forms are very limited: the torpedo ray, the ray genus *Narcine*, and some skates among elasmobranchs and only the stargazers amongst the teleosts are known to be electric. In the strongly electric fishes, the discharge is of sufficient intensity to be of use in defensive and predatory behavior. In all other electric fishes, the discharge voltages are too weak to be of direct defensive or offensive value. It is now realized that weak electric fishes use their discharges for electrolocation and social communications.

When a fish discharges its electric organ an electric field is established in the immediate vicinity. Because of the high energy costs of maintaining a continuous dc discharge electric fishes produce a discontinuous pulsating discharge and field. Some species produce a life-long continuous and constant-frequency discharge, others emit a variable-frequency discharge (see Figure 7-14). Frequencies are surprisingly high. Most weakly electric fishes pulse at rates between 50 and 300 Hz, but in the sternarchid knifefishes the frequency achieves constant rates of 1700 Hz, the most rapid continuous firing rate known for any muscle or nerve.

In the marine fishes the use of electric discharge for electrolocation is questionable. In skates, *Narcine*, and stargazers, which produce modest voltages, a specific behavioral function has yet to be demonstrated. The conductivity of sea water is so high, however, that the electric field produced even by a strong discharge is essentially short-circuited and, therefore, limited in range. In fresh water, where electric conductivity is much lower, the electric field from even weak discharges may extend outward from the fish for several meters and will be distorted by the presence of either conductive or highly resistant objects (Figure 7-15). Distortions will result in an increase or decrease of the electric potential distributed across specific regions of the body surface. Rocks, for example, are highly resistive but fishes, invertebrates, and plants highly conductive. An electric fish might thus detect their presence and position by sensing where on its body maximum distortion of its electric field occurs. If the fish could sense the discharges of another electric fish, the possibility for electrocommunication exists. To accomplish either electrolocation or electrocommunication, it is necessary to sense the electric field.

The skin of weakly electric fish contains two types of special sensory receptors: **ampullary** organs and **tuberous** organs. Each type represents modi-

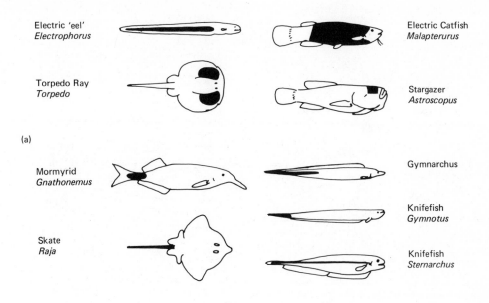

(a)

(b)

	Elasmo-branchii		Teleosts						
			Osteo-glosso-morpha		Ostariophysi				Acan-thop-ter-ygii
					Cypriniformes (Gymnotids)			Siluri-formes	
	Rajids	Torpedinids	Mormyrids	*Gymnarchus*	Electric eel	*Gymnotus*	Sternarchids	Electric catfish	Stargazers
Live in fresh water			+	+	+	+	+	+	
Strongly electric organ		+			+			+	+
Weakly electric organ	+	+	+	+	+	+	+		
Intermittently active	+	+			+		+	+	+
Continuously active			+	+		+	+		
Constant frequency				+			+		
Variable frequency			+		+				
Can cease firing			+	+		+			

Figure 7–14 Electric fishes: morphology and position of electric organs and convergent aspects of physiology and function. (a) Four strongly electric fishes (above) and five weakly electric fishes (below). Shaded areas indicate position of electrogenic organs. (b) The phylogenetic convergences in the evolution of electrogenic organ function are indicated by the similarities that cross broad phylogenetic categories. (Modified after Bennett [1].)

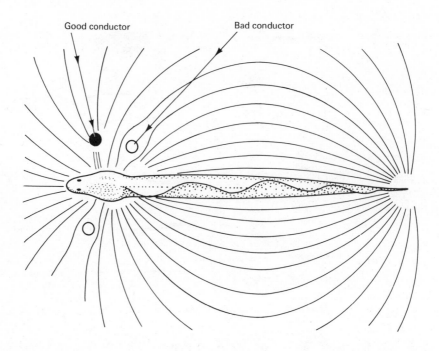

Good conductor   Bad conductor

Figure 7-15   Distortion of the electric field (lines) surrounding an electric fish (*Gymnarchus*) by conductive and nonconductive objects. Conductive objects concentrate the field on the skin of the fish where the increase in potential is detected by the electroreceptors; nonconductive objects spread the field and diffuse potential differences along the body surface.

fication of the lateral line organs and specifically the hair cells (Figure 7-9). Both receptors are capable of responding to minute changes in the electric field existing across the skin. They act like voltmeters, that is, the entire receptor resistance to current flow is very high except over the surface of the modified "hair cell." In this way potentials can be compared between discrete locations across the body surface. Voltage sensitivities are remarkable. For example, ampullary organs have excitatory thresholds of less than 1 μV/cm, a detection level otherwise achieved only by the best man-made voltmeters.

Electroreceptors like lateral line receptors have double innervation, an afferent channel sending impulses to the brain and an efferent channel which causes inhibition of the receptor. Theodore Bullock and his collaborators at Scripps Institution of Oceanography have shown that during each electric organ discharge (EOD), resulting from a central nervous system command, an inhibitory command is sent to the electroreceptors. Thus, during each period of electric discharge the fish is insensitive to its EOD. Between pulses, electroreceptors function to sense distortion in the electric field or the presence of a foreign electric field. It is now clear that the electric organs and receptors of weakly electric fish provide a "sixth sense," one that can only seem strange to humans with no sensory counterpart. The usefulness of this electric sensory system to the fishes is apparent, since the African and South American forms all tend to inhabit periodically turbid waters where vision is limited to short distances. Perhaps more important, most electric fishes tend to be nocturnal.

Although weakly electric fishes are restricted almost entirely to tropical waters of Africa and South America, the ability to detect electric fields is found in many fishes and, especially, the elasmobranchs. On the heads of sharks and the heads and pectoral fins of rays are unusual anatomical structures known as the **ampullae of Lorenzini** (Figure 7–16). The ampullae have long baffled biologists. We now know, especially from the elegant behavioral experiments of R. W. Murray, that the ampullae of Lorenzini are sensitive electroreceptors. It was first suggested that the ampullae served as mechanoreceptors, because deformation was shown to modify their discharges. In 1938 however, A. Sand demonstrated that the discharge rate of the organ was extremely sensitive to temperature. This was confirmed by others and the ampullae for some time served as a model for "temperature receptors." After World War II, Murray again suggested that the ampullae somehow function in mechanoreception, but revoked this interpretation in 1960 when he obtained the first evidence that the ampullae were very electrosensitive. He suggested tht they might function to detect nearby electric disturbances in the vicinity of the shark or ray.

Electric sensitivity had been known in fishes since 1917 when G. A. Parker, a famous American physiologist, and A. P. van Heusen demonstrated that blinded catfish were sensitive to metallic rods but not glass rods. They concluded that the catfishes were responding to local currents spontaneously flowing between the metal rod and aquarium water. Electroreception functions in nature as demonstrated by the "tremorous" fact that in Japan catfish react to an earthquake several hours before it takes place. Catfish become hypersensitive to stimuli when experimentally imposed voltage gradients reach 1 to 2 mV/cm. Similar voltages are generated in the earth's crust prior to and during earthquakes. Biologically the ability to detect electric disturbances of this magnitude does not function to predict earth tremors, but to inform a fish of the presence of other organisms, especially prey.

During locomotion, respiratory movements, and cardiac contractions (that is, during muscle activity) all animals generate local pulsating electric fields. Indeed, the medical use of EKGs in analyzing human heart function and EEGs in brain studies depends on these pulsating potentials. In addition to movement-related potentials, a standing dc potential also issues from an aquatic organism resulting from the organism's chemical disequilibrium with its surroundings and the differential electrical resistance that occurs across its body; dc potentials of 500 μV/cm have been measured close to the head of small marine teleosts. These fall off to about 2 μV/cm at 5 cm and 0.2 μV/cm at 10 cm. Because these voltage gradients are within the demonstrated sensitivity range of the ampullae of Lorenzini, electric detection of prey, predator, or mate exists.

In 1971, A. J. Kalmijn began a series of experiments demonstrating unambiguously that sharks detect their prey's electric gradients. To eliminate the

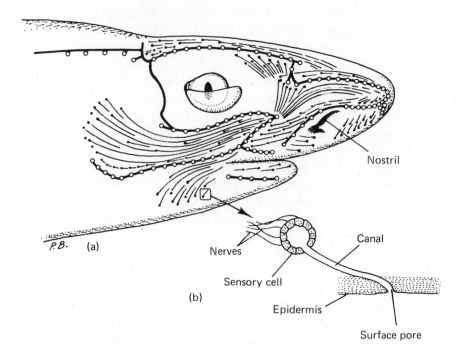

Figure 7-16   Ampullae of Lorenzini.

(a) The distribution and canal path of the ampullae of Lorenzini on the surface of the head of a spiny dogfish *Squalus acanthias*. Open circles represent the external pore, lines the canals, and the black dot the region of the sensory cells and nerves. Lateral line canals are also illustrated.

(b) Diagrammatic representation of a single ampullary organ. The canal is filled with a highly conductive "jelly" but the canal wall is an exceptionally strong insulator. Because the canal courses for some distance parallel to the epidermis, a sensory cell can "compare" the potential difference between the tissue in which it lies (which reflects that of the adjacent epidermis and environment) with that of the distant pore via the electrical connection of the canal.

possibility of visual detection Kalmijn used flatfish as prey because they bury themselves in the sand. When the odor from a piece of fish was introduced, the elasmobranchs would commence to search the tank bottom. When passing within 1 to 15 cm of a buried flatfish the shark would suddenly dive at the bottom and unearth its prey. To eliminate the possibility that the flatfish was detected by mechanical (water movement) or olfactory cues, rather than electrical potentials, a series of simple and elegant experiments were devised (Figure 7-17). It seemed clear to Kalmijn that neither visual, mechanical, or olfactory cues were operative, and that some property associated with living prey was sensed. When he placed a thin piece of electrically insulating plastic over the living fish, the shark could no longer find the fish. Furthermore, by

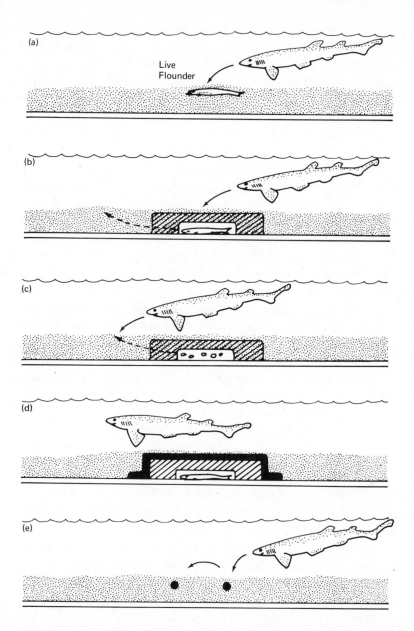

Figure 7-17 Kalmijn's experiments illustrating the electrolocation capacity of elasmobranchs. (Modified after Kalmijn, *in* Fessard [3].) (a) The shark *Sclyorhinus canicula* readily finds a living prey fish, the flatfish *Pleuronectes platessa*, even when it is buried beneath the sand. (b) An agar chamber transparent to electric potentials but diverting of olfactory cues (dotted arrows), does not alter the shark's reaction. (c) Pieces of dead fish, with little electric activity but odoriferous, placed in the same apparatus cause an incorrect response. (d) An electrically insulating film added to the apparatus in experiment (b), which also blocks olfactory cues, produces no response. (e) Electrodes simulating the bioelectric field in the absence of olfactory cues produces repeated and accurate responses.

burying electrodes close together in the sand and simulating flatfish potentials, sharks could be induced to attack the electrodes and to dig in their vicinity. This provided direct and convincing evidence of electric sensing of prey. Sharks and rays apparently respond only to dc potentials and slow pulsations up to about 8 Hz, frequencies associated with many bioelectric potentials.

By sensing the earth's electric fields, electric reception may function in another amazing way—navigation. Two ways are suggested as to how this might occur. The electromagnetic fields at the earth's surface produce tiny voltage gradients. A shark swimming across these fields, for example, can experience an actual voltage gradient as high as 0.4 $\mu V/cm$, well above the voltage sensitivity of its ampullae of Lorenzini. In addition, ocean currents themselves generate electric gradients as their dissolved ions move through the earth's magnetic field. Under appropriate conditions voltage gradients as high as 0.5 $\mu V/cm$ are achieved. In this case a fish drifting with the current could theoretically determine direction. Kalmijn has calculated that, by orienting in different directions, a fish might analyze its open ocean location. This remains to be proved, but experiments have shown that the migrating American eel, *Anguilla rostrata*, can detect voltages associated with the ocean currents when the voltage gradient is aligned perpendicular to but not parallel to their body axis. This ability could allow eels to orient themselves constantly along a migratory heading. Here is another example of how specific adaptations contribute to vertebrate life, in this case enhancing to a high degree of sophistication one of their basic attributes—mobility.

7.3   REGULATIONS OF IONS AND BODY FLUIDS: ION–OSMOREGULATION

Many geochemists believe that the ion balance and concentrations of Paleozoic seas were similar to those measured in the oceans today. The first vertebrates, the ostracoderms, must have had ion levels like their marine invertebrate ancestors which, presumably, are still reflected in most living marine invertebrates. Ion concentrations in most living vertebrates differ greatly from those of marine invertebrates, a result of their increased ability to regulate the internal ionic milieu.

Sea water is a mix of salts that contains mostly NaCl (Table 7–1 and 7–2). The salt concentrations in the body fluids of many marine invertebrates are similar to sea water, as are those of hagfishes, the most primitive living vertebrates. However, divalent ions (Ca^{2+}, Mg^{2+}, SO_4^{2-}, and so on) are less concentrated in hagfishes than in sea water. Na^+ and Cl^- concentrations are elevated in compensation, a condition also found in many marine crustaceans (Table 7–1). In contrast, in all other vertebrates, blood salt levels are greatly reduced; no single value can represent all vertebrates, but for most concentrations are less than 45 percent that of sea water. This reduction is characteristic of

Table 7–1. Representative concentrations of the major ions sodium and chloride, and osmolality of the blood in vertebrates and marine invertebrates. (Ions are expressed in millimoles per liter of water; all values are reported to the nearest 5 units.)

Type of animal	mOsm	Na$^+$	Cl$^-$	Other major osmotic factor	Source
Sea water	~1000	475	550	——	1
Marine invertebrates					
Coelenterates, mollusks, etc.	~1000	470	545	——	1
Crustacea	~1000	460	500	——	1
Marine vertebrates					
Hagfishes	~1000	535	540	——	2
Lamprey	~ 300	120	95	——	2
Teleosts	> 350	180	150		3
Coelacanth	1180	180	200	urea 375	4
Elasmobranch (bull shark)	1050	290	290	urea 360	5
Holocephalian	~1000	340	345	urea 280	9
Fresh-water vertebrates					
Polypterids	200	100	90	——	3
Chondrosteans	250	130	105	——	3
Holosteans	280	150	130	——	3
Dipnoans	240	110	90	——	3
Teleosts	< 300	140	120	——	3
Elasmobranch (bull shark)	680	245	220	urea 170	5
Elasmobranch (fresh-water rays)	310	150	150	——	6
Amphibians*	~250	~100	~ 80	——	7
Terrestrial vertebrates					
Reptiles	350	160	130	——	8
Birds	320	150	120	——	8
Mammals	300	145	105	——	8

*Ion levels and osmolality highly variable, but tend toward 200 mOsm in fresh water.

SOURCES

1. Potts, W. T. W. and Parry, G. J. 1964. *Osmotic and Ionic Regulation in Animals.* Macmillan, New York.
2. Robertson, J. D. 1954. *J. Exp. Biol.* 31:424–442.
3. Urist, M. R., *et al.* 1972. *Comp. Biochem. Physiol.* 42:393–408.
4. Pickford, G. E., and Grant, F. G. 1964. *Science* 155:568–570; Griffith, R. W. et al. 1975. *J. Exp. Zool.* 192:165–171.
5. Thorson, T. B., *et al.* 1973. *Physiol. Zool.* 46:29–42.
6. Thorson, T. B., *et al.* 1967. *Science* 158:375–377.
7. Bentley, P. J. 1971. *Endocrines and Osmoregulation.* Zoophysiology and Ecology Series, vol. 1. Springer-verlag, New York.
8. Prosser, C. L. 1973. *Comparative Animal Phsyiology.* Third edition. W. B. Saunders. Philadelphia.
9. Read, L. J. 1971. *Comp. Biochem. Physiol.* 39A:185–192.

Table 7-2. The intracellular concentration of major inorganic ions in marine invertebrates and vertebrates. (Concentration values are in millimoles per liter; compare with Table 7-1.)

	Na^+	Cl^-	K^+	Ca^{2+}	Mg^{2+}
Sea water	475	550	10	10	53
Marine invertebrates	(54–325)	(54–380)	(48–175)	(3–89)	(8–96)
Vertebrates					
Hagfish	122	107	117	2	13
All others	(8–45)	(11–30)	(83–185)	(2–9)	(7–11)

Note that the monovalent ions Na^+ and Cl^- are reduced relative to seawater and K^+ increased in all animals. For divalent cations, especially Mg^{2+}, a reduction is found in all vertebrates but not in all marine invertebrates.

vertebrates and is shared only with invertebrates that have penetrated estuaries, fresh waters, or the terrestrial environment.

The presence of solutes in sea water or blood plasma lowers the kinetic activity of water. Therefore, water flows from a dilute solution (high kinetic activity of water) to a more concentrated solution (low kinetic activity), a phenomenon termed **osmosis**. Dissolved particles not only alter the kinetic activity of water but lower the freezing point in proportion to their concentrations; 1 mole of solute particles depresses the freezing point by 1.86°C. It is possible to measure the freezing point of a solution and thereby establish its osmotic concentration. To express osmotic concentrations, units of **osmolality** are preferred. By definition 1 mole of solute particles in 1 kg of water has an osmolality of 1 osmole (Osm); 1 Osm is equivalent to a freezing point change of −1.86°C, 0.1 Osm of −0.186°C, and so on. Sea water varies somewhat in concentration geographically, but is generally near 1 Osm. For comparisons of different animals, osmolality is expressed in milliosmoles (mOsm) and thus sea water is equal to 1000 mOsm.

The milliosmolalities of various animals and of sea water are reported in Table 7-1. In most marine invertebrates and the hagfish, the body fluids are in osmotic equilibrium with sea water; that is they are **isosmotic** to sea water, there is no net flow of water into or out of their tissues. In marine teleosts and lampreys, body fluid concentrations are between 350 and 450 mOsm. Therefore water flows outward from their blood to the sea. In the coelacanth and chondrichthyans the blood osmolality is slightly higher than sea water. As a result, water flows from the sea into their blood. These different osmotic circumstances are specified by the terms **hyposmotic** (marine teleosts and lampreys) and **hyperosmotic** (coelacanth and chondrichthyans). This terminology compares the object of interest (in this case the fish) to that of its surroundings. One must be certain which object is the reference when using this terminology. For example, in fresh water teleosts the urine is hyposmotic (contains less solute) to blood, but hyperosmotic (contains more solute) to fresh water. But for unambiguous comparisons there is no substitute for the data—the osmolality of each fluid.

Fresh water fishes are hyperosmotic to the medium; nevertheless their blood osmolality is lower than that of their marine counterparts. This results primarily through reduction in NaCl (Table 7–1). Teleosts provide a clear example of how the problems posed by ion regulation are overcome. The physiological integrity of teleosts is constantly threatened by either (1) osmotic loss of water and salt gain when in sea water or (2) osmotic gain of water and loss of salts when in fresh water.

7.3.1 Fresh-water teleosts

Several mechanisms are involved in osmotic regulation in fresh-water fishes (Figure 7–18). First, the body surface has low permeability to water and to ions. Certainly fish scales contribute to this low permeability, but many teleosts are unscaled (catfish for example). The presence of mucus slime so typical of fishes and important in locomotion reduces body surface permeability. Scraping mucus from the body surface increases permeability and induces severe osmotic stress. This is one reason why fisheries biologists are very careful in handling fishes. However, fishes cannot lower the water and ion permeability of their entire surface and, therefore, totally prevent osmotic flow resulting from ion regulation. For respiratory gas exchange a large surface area is required. Membranes that are highly permeable to gases are also highly permeable to water and to a lesser extent permeable to ions. Because the major respiratory surfaces in fishes, the gills, are in direct contact with water and gill surface area exceeds the area of the body, most ion and water movements take place across them.

Second, to compensate for hydration, the kidney in a fresh-water teleost produces a copious volume of urine. To reduce salt loss, the urine is diluted by actively reabsorbing salts. Indeed, urine processing in a fresh-water teleost provides a simple model of vertebrate kidney function (Figure 7–19). Each kidney consists of millions of individual tubular units, the **nephrons**, each of which produces urine. A primary function of the nephron is to cleanse the blood of waste metabolites and foreign substances. To accomplish this the blood is first filtered through the **glomerulus**, a structure unique to vertebrates. Each glomerulus is a complex unit composed of a leaky arterial capillary tuft encapsulated within a sievelike filter (Figure 7–19). Due to arterial blood pressure, fluid is forced into the lumen of the nephron to form an **ultrafiltrate**. This ultrafiltrate is composed of the blood minus the blood cells and larger molecules such as proteins. The ultrafiltrate is then processed to return small but essential metabolites (glucose, amino acids, and so on) and water to the general circulation. In addition, and when necessary, the walls of the nephron

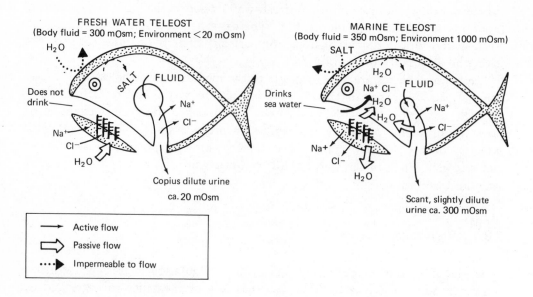

Figure 7-18   General scheme of the osmotic and ionic gradients encountered by fresh-water and marine teleosts. See text for further discussion.

actively excrete toxins. This takes place in the **proximal convoluted tubules** **(PCT)** and **distal convoluted tubules (DCT)** of the nepron. Finally, wastes are excreted as urine to the exterior.

Another important function of the vertebrate kidney is osmotic and ionic regulation. In fresh-water teleosts, very large glomeruli allow production of a high volume of ultrafiltrate to permit later formation of copious urine. But a glomerular ultrafiltrate is always isomotic to blood and contains essential blood salts. To conserve these ions, they are actively reabsorbed across the PCT and DCT. The DCT in fresh-water teleosts is impermeable to water and, therefore, as the ions are reabsorbed the urine is diluted, becoming hyposmotic to blood. In this way osmotic hydration across the gills is countered and salts conserved. Nevertheless, some salt is lost in the urine in addition to that lost by diffusion across the gills (Figure 7–18). In time, therefore, a fresh-water teleost would encounter mineral deficiency. Usually this is compensated by the uptake of salts from food. Fresh-water teleosts, however, utilize an additional method to gain essential salts. Special cells located in the gills absorb Na^+ and Cl^- from fresh water. This extrarenal absorption of ions is an up-hill process, that is, ions are absorbed against a concentration gradient, and this requires energy.

7.3.2 Marine teleosts

In marine fishes, where dehydration and salt gain are serious problems (Figure 7–18), the integument is highly impermeable so that most osmotic and ion movements occur across the gills. The kidney glomeruli are often reduced in size and possess less filtration area than in fresh-water teleosts (Figure 7–19). In marine teleosts, therefore, the glomerular filtration rate is lower, less urine is formed and the volume of water lost in urine is reduced. In contrast to fresh-water teleosts, a water impermeable DCT is absent. As a result, urine leaving the nephron is scant in volume and more concentrated than in fresh-water teleosts, although it is always slightly hyposmotic to blood and some water is lost in eliminating wastes. To compensate for osmotic dehydration, marine teleosts do something unusual—they drink sea water. From this briny beverage Na^+ and Cl^- are actively absorbed across the lining of the gut, but most other ions remain in the intestine. Because the gut is permeable to it, water flows osmotically into the blood with the actively absorbed Na^+ and Cl^-. Estimates of sea water consumption vary for different species, but many fishes drink in excess of 25 percent of their body weight per day and actually absorb as much as 80 percent of the ingested water into their extracellular spaces! Because they remain at constant weight, the large volume of the water imbibed by marine teleosts must equal the water they lose osmotically. This loss, there-fore, amounts to as much as 20 percent of the body weight per day. The mechanisms by which water is absorbed across the gut also increases the total influx of Na^+ and Cl^-. To compensate for salt loading discrete cells sometimes called **chloride cells** are located in the gills; these pump Na^+ and Cl^- outward against a large concentration gradient, a process that requires energy.

To summarize, in fresh-water teleosts ion-osmoregulation involves: (1) lowered integument permeability to water and salts, (2) formation of a copious urine, and (3) extrarenal uptake of salts. In marine teleosts regulation is achieved by: (1) low permeability of integument to water and salts, (2) a scant urine formation only slightly hyposmotic to blood, (3) compensating osmotic dehydration by drinking sea water, and (4) excreting salt by extrarenal means.

7.3.3 Other fishes

The chemical work of ion regulation is minimized in marine hagfishes by regulating only divalent ions, and osmotic work is avoided by being isosmotic to sea water or very nearly so. In Chondrichthys and coelacanths a different mechanism counters the dehydration potential created by their regulation of internal ions at lower than sea-water concentrations. By retaining and tolerating urea (360 mM/liter) these fishes raise their total blood osmolality to a level slightly hyperosmotic to sea water (Table 7–1). Rather than constantly losing water these fishes gain water by osmotic diffusion across the gills. As a result, chondrichthyans need not drink sea water to compensate for dehydration. This influx of water permits larger kidney glomeruli, as in fresh-water teleosts, to produce high glomerular filtration rates and, therefore, rapid cleansing of the blood. Urea, however, is very soluble and readily diffuses through most bio-logical membranes. To conserve urea chondrichthyan gills are impermeable to it, and the kidney tubules actively reabsorb it. Although hyperosmotic to sea

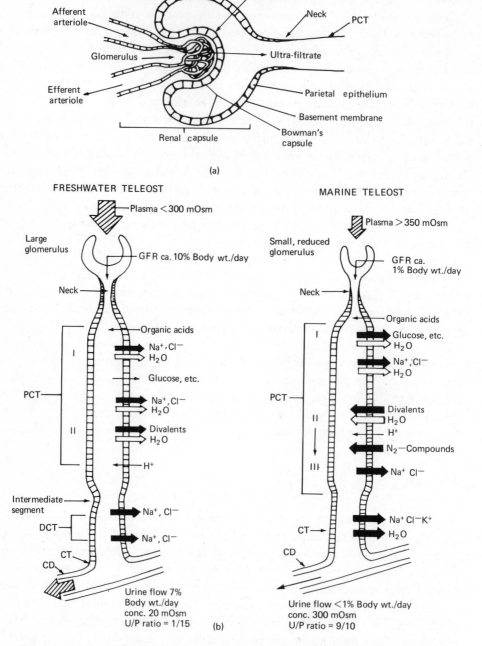

(a)

(b)

Figure 7-19 (a) Schematic diagram of the renal corpuscle of the vertebrate kidney. (b) A schematic comparison of nephron structure and functions in a fresh-water and marine teleost fish. GFR is the glomerular filtration rate at which the ultrafiltrate is formed, expressed in percentage of body weight per day. PCT is the proximal convoluted tubule (in fishes sometimes referred to as the proximal tubule). Two segments (I and II) of the PCT are recognized in both fresh-water and marine teleosts. Segment III of the marine teleost's PCT is

water, chondrichthyans nevertheless regulate Na^+ and Cl^- at levels approximating one half that of sea water concentrations. They consequently experience, as do marine teleosts, an ion influx across the gills. Chondrichthyans and coelacanths generally do not possess "extrarenal" salt excreting cells in the gills. Rather, they achieve ion balance via the rectal gland, an organ that empties into the cloaca. This gland secretes a fluid isosmotic to their blood but one which is hyperionic in Na^+ and Cl^-. The rectal gland therefore is the functional analog of the chloride cells of marine teleosts.

Like many other fishes, some elasmobranchs tolerate a wide range of salinities; sawfish, some stingrays, and bull sharks are examples. In sea water bull sharks retain high levels of urea, but when in fresh water their blood urea levels decline (Table 7–1). Stingrays in the family Potamotrygonidae, which are found in the Amazon basin, spend their entire lives in fresh water. Urea is present but only at very low blood concentration. Interestingly blood Na^+ and Cl^- concentrations are 35 to 40 percent below those maintained by sharks that are found in fresh water and only slightly above levels typical of fresh-water teleosts (Table 7–1). The potamotrygonid rays have existed in the Amazon basin for a long period, perhaps since the Tertiary, and their blood chemistry has evolved a reduced ion-osmotic gradient adapted to fresh water. When exposed to increased salinity, potomotrygonid stingrays do not increase blood urea, as do euryhaline elasmobranchs, even though the enzymes required to produce urea are present. Apparently, their long evolution in fresh water has led to a reduction in the amount of urea produced, an increase in the permeability of the gills to urea and reduced the ability of the kidney tubules to reabsorb it. The presence of high concentrations of urea in the blood of fishes appears to relate to sea water not fresh water habits.

The polypterids and dipnoans (Chapter 6) have resided in fresh water longer than any other living fishes and have even lower blood salt concentrations than potamotrygonids (Tables 7–1 and 7–3). Thus, there appears to be a relationship between historical residence in fresh water and the osmolality of the blood of vertebrates—the longer the residence the lower the osmolality. Lowered osmolality has the advantage of reducing the ion and osmotic gradients between fish and fresh water and, as a result, the amount of osmotic and ionic work that a fish must perform to maintain homeostasis. The opposite should hold for fishes in sea water, that is, the higher the body fluid concentration the less the osmotic gradient and the less the energy expended to correct for osmotic fluxes. Why then are not all fresh water fishes similar to *Polypterus* (lowest blood os-

---

sometimes equated with the DCT (distal convoluted tubule) of fresh-water teleosts. Darkened arrows represent active movements of substances, open arrows passive movements, and hatched arrows indicate by their size the relative magnitude of fluid flow. Note that Na^+ and Cl^- are reabsorbed in the PCT segment I and in the CT (collecting tubule) in both fresh-water and marine teleosts; water flows osmotically across the PCT in both fresh-water and marine teleosts but only across the CT of marine teleosts. Water permeability of the CT (and also the DCT) is therefore low in fresh-water teleosts. (See text).

molality = 200 mOsm) and all marine fishes isosmotic or slightly hyperosmotic to sea water?

Peter L. Lutz suggests the answer lies in the interplay between cell activities and ion concentrations. Cellular enzymes require many specific ions as cofactors, and their concentrations can strongly alter biochemical rates and, therefore, cell activity. Indeed intracellular ion concentrations and ratios are rigidly maintained by energy-demanding ion pumps located in the cell membrane. These ion pumps, in addition to affecting enzymatic reactions, operate to maintain cell volume and transmembrane potentials, properties also vital to cell function. In colonizing a hyposmotic environment, an organism confronts osmotic hydration as well as a loss of body ions (Figure 7–18). As seen in fresh-water teleosts, a variety of mechanisms can be used to counter hydration, but each demands extra energy. Simple reduction in the concentration of body fluids lessens the threat of hydration. However, for a metazoan to reduce its concentration the cells must suffer the consequences, for they remain isosmotic to the body fluids. Lowering ion concentrations, therefore, dictates concomitant adjustments in those cell processes that depend on specific ion levels to maintain function. Apparently these intracellular adjustments to different ion concentrations require considerable evolutionary time. Cell chemistry appears a very conservative trait and one, therefore, that has constrained vertebrate evolution. Careful analysis of the evolution of fishes, as Lutz points out, attests to this fact.

The order of first appearance of different groups of fresh-water vertebrates in the fossil record clearly correlates with the extent of reduction in blood osmolality and major salts (Table 7–3). Each of these groups has retained fresh-water representatives throughout all subsequent periods of geologic time. The majority of marine fishes have been derived from fresh-water ancestors at various points in geologic time, the myxinoids or hagfishes probably being the sole exception. Assuming their ancestors showed some degree of adaptation to fresh water by internal reduction of ions, each reinvasion of the sea must have required reversal in the osmotic adjustments of body fluids and cell ion sensitivities to sea water.

Again the order of first appearance of different fish groups in the marine fossil record correlates with an increase in the major blood ions (Table 7–3). However to combat the severe osmotic dehydration that would exist as a result of their low salt concentrations relative to sea water, two of the most ancient marine groups retained urea as an osmotically active substance. But the oldest group, the hagfishes, in contrast, have blood salt concentrations that we believe represent the condition typical of the first vertebrates. Their high internal ion levels represent some of the evidence supporting the idea that hagfishes have always been marine. Teleosts, the most recent invaders of the sea (which may well have been derived from marine holosteans), have ion levels slightly above their fresh-water predecessors but, nevertheless, have not utilized other substances to eliminate the osmotic flow of water from their bodies. Rather, they expend additional energy in ion-osmoregulation. Although the

Table 7-3. The correlation between the first appearance of different fish groups and the blood osmolality and salt (Na^+ and Cl^-) concentrations of living representatives. (Osmolalities and salt concentrations are referenced in Table 7-1.)

Habitat type and vertebrate	First appearence in habitat type	Blood concentration in living species mOsm	Na + Cl (mM/liter)
Fresh water			
Paleoniscoids	Early Devonian	200	190
Dipnoans	Early Devonian	240	200
Chondrosteans	Pennsylvanian	250	235
Holosteans	Late Permian–Early Triassic	280	280
Teleosts	Late Cretaceous	<300	260
Potomotrygonids	Pliocene (?)	310	300
Bull sharks, stingrays, etc.	Recent	680	465
Marine			
Myxinoids	Ordovician–Silurian	Isosmotic with sea water	~1000
Elasmobranchs	Late Devonian	Hyperosmotic to sea water	580
Coelacanths	Early Triassic	Hyperosmotic to sea water	380
Teleosts	Late Cretaceous	>350	330

reasons are unknown, maintaining lower ion concentrations in body fluids and cells must be beneficial. In fact even the coelacanths and chondrichthyans have allowed their body fluid *ion* concentrations to rise only slightly toward sea water concentrations, in spite of their long residence in the sea. Again, this is suggestive of some benefit derived from reduced ion concentrations.

7.3.4 The Basis of Salinity Tolerance

Most fishes are **stenohaline** (*steno* = narrow, *haline* = salt); that is, they inhabit either fresh water or sea water and survive only modest changes in salinity. A considerable number of fishes, nevertheless, are **euryhaline** (*eury* = wide); that is, they inhabit both fresh and sea water and tolerate large changes in salinity. In many salmon a breeding migration from sea water into fresh water precedes spawning. After an initial growth phase in fresh water the young return to a marine environment and grow to adulthood. Fishes that migrate from sea water to fresh water for reproduction are termed **anadromous**. Other fishes, like the European and American eels in the genus *Anguilla*, are **catadromous**; they grow to maturity in fresh water and migrate to the ocean to breed.

Anadromous and catadromous fishes, although euryhaline, often are intolerant of sudden changes in salinity. The ion and osmoregulatory mechanisms that underlie their euryhalinity take time to adjust. For example, immature eels living in fresh water die when placed in sea water. Only sexually mature eels can withstand a rapid salinity change from fresh to salt water. Many euryhaline fishes live where salinity fluctuates with the daily tides. For example, flounders can withstand instantaneous change from sea water to fresh water, primarily by special adaptations that regulate ion fluxes across their gills.

Over the past decade Drs. R. Motais and J. Maetz and their collaborators, working in France, have applied radioisotope techniques to measure and compare the influx and efflux of ions in fishes when in sea water and fresh water. These studies have established that both ion influx and efflux are very large in stenohaline marine fishes. In the sea bass, *Serranus scripta,* the net influx of Na^+ from sea water to blood is equal to nearly 10 percent of the total body Na^+ each hour! In a fresh-water teleost the Na^+ efflux from blood to water, however, is less than 1 percent of the total body Na^+ each hour. The net permeability of a fresh-water teleost to ions is therefore much less than in a marine teleost.

In euryhaline teleosts, such as the flounder, *Platichthys flesus,* the net influx of Na^+ when in sea water is similar to that of *Serranus* and other marine teleosts. If marine teleosts are to maintain ion balance, the very large Na^+ influx must be countered by an equal excretion of Na^+. This ion balance is accomplished in euryhaline and stenohaline fishes by the activities of the chloride cells in the gills (Figure 7–18). Stenohaline marine fishes cannot tolerate fresh water because of their inability to reduce the high salt permeability of their gills—they continuously leak salt and die. What mechanism allows euryhaline fishes to withstand sudden salinity changes? They differ in being able to reduce salt permeability across the gills rapidly when exposed to fresh water and vice versa when in sea water. In contrast, fresh-water teleosts have low ion permeability across the gills. When exposed to a salt load they are incapable of actively excreting ions, as do marine and euryhaline teleosts, for they lack chloride cell function in the gills.

7.4 NITROGEN EXCRETION IN VERTEBRATES

By retention of urea, which is usually considered a waste product, marine chondrichthyans and coelacanths achieve approximate osmotic equivalence to sea water. Because urea is derived predominantly from proteins, which are common in the diet of vertebrates, we should consider how different vertebrates process nitrogenous wastes. This subject is a fascinating example of the interactions between the biochemical activities of vertebrates and their overall ecology.

Vertebrates are mobile, active animals. Because activity requires energy, the more vertebrates move the more food they must consume. In utilizing carbohydrates and fats (composed of carbon, hydrogen, and oxygen) vertebrates produce CO_2 and H_2O, which are easily voided. Proteins and nucleic acids are another matter, for they contain nitrogen (N). In oxidizing protein for its energy the nitrogen is enzymatically reduced to ammonia (NH_3) through a process called deamination. Ammonia is highly diffusible and very soluble in water but also extremely toxic to cellular activities. Rapid NH_3 excretion is, therefore, crucial. NH_3 requires a lot of water to excrete to keep it at a low nontoxic level. Differences in NH_3 processing are partly a matter of the availability of water and partly the result of vertebrate phylogeny. Although vertebrates excrete several minor nitrogenous wastes (creatinine, nucleic acid residues, and so on) most nitrogen is eliminated as NH_3, as urea, or as uric acid (Figure 7-20).

Because NH_3 is highly soluble and diffusible in water, aquatic invertebrates excrete NH_3 directly, as do vertebrates with gills, "leaky" skins, or other permeable membranes that contact water. Excretion of nitrogenous wastes predominantly as NH_3 is termed **ammonotelism**. Some animals accumulate and excrete waste nitrogen as urea; this is termed **ureotelism**. Sharks are one example of vertebrates that synthesize urea from NH_3 in a cellular enzymatic process, sometimes called the **urea cycle**. Urea synthesis requires an additional expenditure of energy, compared to ammonia production. The value of ureotelism, therefore, lies not in energetic economy but rather in the benefits derived from urea itself.

Urea has two main advantages. First, recall that its retention by some marine vertebrates counters osmotic dehydration. In other ureotelic vertebrates, mammals being the most conspicuous example (Figure 7-20), urea is not retained in the blood for osmotic purposes but is excreted. Because water is often scarce for mammals, they cannot rely on excreting NH_3 because it requires copious urination to flush NH_3 from the body. The less toxic urea can accumulate and be concentrated in urine, thus conserving water. A second function of urea synthesis, therefore, is NH_3 detoxification in environmental circumstances that prevent its rapid elimination. In contrast to the chondrichthyans, where urea synthesis occurs in most cells of the body, other ureoteles localize these enzymes in the liver. Thus, NH_3 arising in all tissues must be transported to the liver for conversion to urea. To avoid a local cellular toxicity, NH_3 is combined with glutamate to form glutamine, a nontoxic amino acid.

Many amphibians switch from ammonotelism to ureotelism when they emerge onto land. In frogs the urea synthesizing enzymes, inoperative in tadpoles, are activated during metamorphosis. A surprising response to osmotic dehydration takes place in the crab eating frog, *Rana cancrivora*. This frog inhabits intertidal mudflats in Southeast Asia and each day is exposed to 80 percent sea water. During sea-water exposure, the frog allows its blood ion concentrations to rise and thus reduces the ionic gradient. In addition, proteins are deaminated and the NH_3 rapidly detoxified by conversion to urea. This urea is released into the blood, whose osmolality it significantly increases. Urea levels rise from 20 to 30 mM/liter and the frogs become hyperosmotic to the surrounding water. In this sense *Rana cancrivora* functions like an elasmobranch and absorbs water osmotically. Frog skin, unlike that of elasmobranchs,

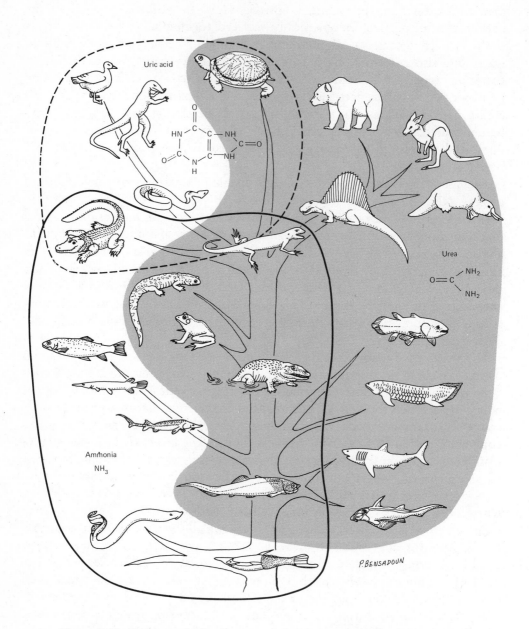

Figure 7-20 Phylogenetic distribution of the three major nitrogenous wastes in vertebrates. The types of wastes excreted by extinct vertebrates are unknown; examples merely provide visual continuity to the phylogeny. (Modified after Schmidt-Nielson [12].)

is permeable to urea and, as a result, urea is rapidly lost. To compensate for this loss, the urea synthesizing enzymes are very active and concentrated. Like most tadpoles, those of *Rana cancrivora* lack urea synthesizing enzymes. Like marine teleosts they possess extrarenal salt excreting cells in the gills and maintain their blood hyposmotic to sea water.

A complete set of urea synthesizing enzymes has not been demonstrated in cyclostomes, but is present in teleosts, albeit at very low activities. Ureotelism probably evolved independently several times. In one case it may have evolved as a strategy to avoid osmotic dehydration in saline water. Perhaps ureotelism developed in early fresh-water fishes as they reinvaded the sea, as exemplified by chondrichthyans and coelacanths. In other instances ureotelism may have evolved in response to dessication. In invasion of the land ureotelism was preadaptive, for it allowed retention of waste nitrogen in a detoxified state until sufficient water was available for excretion.

A fascinating example of this preadaptation was discovered in the African lungfish, *Protopterus*, by the late Homer Smith. Lungfishes are ammonotelic when swimming about in rivers and ponds. During periods of drought when the rivers and ponds dry out, lungfishes aestivate (see Section 6.1.2). While aestivating they slowly oxidize their carbohydrate, fat, and protein stores to sustain life. The NH_3 is detoxified and urea accumulates, increasing body fluid osmolality. This in turn reduces the vapor pressure across the lung surface and therefore the tendency for water to be lost through evaporation. When rains return, the lungfishes rapidly take up water and excrete the accumulated urea. An analogous process takes place in many toads that inhabit arid environments (see Chapter 10). Ureotelism similarly may have been advantageous to the first Devonian tetrapods.

Nitrogen excretion in all birds and many reptiles follows a different chemical pathway. Their mode of nitrogen excretion, via uric acid, is termed uricotelism. A sequence of enzymes with some similarities to those of the urea cycle converts NH_3 to uric acid, which is nontoxic but, unlike urea, is very insoluble. The excretion of nitrogen wastes as uric acid requires less water than does urea. Thus the urine of birds contains little water and is pasty. As pointed out by the English embryologist Joseph Needham in the 1930s, the evolution of uricotelism was as much a response to embryonic needs as to adult needs. The first completely terrestrial vertebrates, the reptiles, evolved the cleidoic egg—a major vertebrate "invention." When eggs are laid in dryer sites, water conservation is essential to embryonic survival. Water conservation was accomplished by a decreased permeability of the egg shell to water and also by producing metabolic water from the large rich yolk. However, the yolk contains not only considerable fat but also protein, a potential source of NH_3. Uricotelism in cleidoic eggs serves two purposes, detoxification of NH_3 and economical waste storage through precipitation of uric acid. Urea could serve these functions too, but it would require maintaining a large volume of unusable water in the egg. Excreting uric acid avoids this problem, for it requires little water and prevents the rise in embryonic osmolality that a ureotelic egg would face. Retention of uricotelism in adults is also a water-conserving measure.

All birds are uricotelic, but among reptiles nitrogen excretion pathways vary. In alligators, ammonotelism persists along with uricotelism. *Sphenodon*, the tuatara (Section 14.1.2), is both ureotelic and uricotelic. In turtles all three pathways exist and switches occur from predominantly ammonotelism in aquatic habitats, to ureotelism in semiaquatic habitats, and to uricotelism in terrestrial conditions. It appears that those reptilian stocks that gave rise to modern snakes, lizards, alligators, and birds lost many of the enzymes required to complete urea synthesis.

In summary, it can be said that the earliest vertebrates were probably ammonotelic. Ureotelism developed in response to the osmotic dehydration resulting when fishes with low blood osmolalities reinvaded saline environments, and also in response to desiccation in marginal fresh-water habitats. Ureotelism was partly preadaptive to aerial desiccation in amphibious fishes and semiterrestrial amphibians. Uricotelism is a response to totally terrestrial habits and especially to increasing aridity and lack of water. Oddly, mammals are ureotelic, except in monotreme egg embryos, which are uricotelic. In viviparous mammals, the cleidoic egg is absent, and viviparity allows direct transfer of embryonic wastes from fetus to mother. Why all adult mammals abandoned uricotelism (or never evolved it) as an adaptation to terrestrial habits remains puzzling. It can be said that excreting urea rather than uric acid affords one economy, for urea contains less carbon (a potential source of energy) per nitrogen atom than uric acid.

A final note emphasizes how flexible pathways of nitrogen excretion are and how intertwined with water economy they remain. Terrestrial amphibians have long been considered ureotelic, aquatic forms ammonotelic. In 1970 J. P. Loveridge discovered that, during the dry season, a period when any self-respecting frog would retire to a burrow and aestivate, the South African frog, *Chiromantis xerampelina*, remained active. Generally retreat from "aridity" by an amphibian is considered a necessity for their skin is quite permeable to water. *Chiromantis* skin, like the skin of a reptile, was much less permeable to water. Even more surprising, its excretions contain uric acid—biochemically *Chiromantis* is like a reptile! It has subsequently been shown that the South American frog, *Phyllomedusa sauvagii*, responds in precisely the same way to aridity. There is a lesson in these unusual findings: evolutionary convergence works on all levels of biological organization—anatomical, behavioral, physiological, biochemical—but its direction is determined by interactions with environment.

7.5   ENERGY UTILIZATION—A CHALLENGE TO ALL VERTEBRATES

In order to understand some of the physiological and biochemical adaptations of vertebrates we must examine energy utilization, a common problem confronting both fishes and terrestrial vertebrates. The complexity of vertebrate adaptations relates to their characteristic mobility which, in turn, has allowed them to occupy a commanding position in most ecosystems that they inhabit. This prominence has not evolved without a price, however, for activity requires energy, and energy often is limited. Vertebrates usually burn more energy and therefore require more than most other metazoan animals. As a result, they expend much of their active time in the search for food. As a consequence the impact of vertebrates on the environment, that is their demand on resources, is great.

7.5.1   Aerobic Metabolism

Vertebrates are obligatory aerobes, for oxygen is required to complete the breakdown of their food. The basic foods (carbohydrates, fats, and proteins) are oxidized to CO_2 and H_2O. To accomplish this task O_2 must be supplied to the mitochondria of each cell where the final stages of oxidation and ATP formation takes place.

The advantages of aerobic metabolism lie in its complete oxidation of foods and, therefore, the higher yield of ATP produced per gram of food utilized. This efficient use of food seems an obvious strategy to an active animal. Nevertheless, activity performed at too high a level cannot be sustained for long and fatigue sets in. Many invertebrates are capable of supplying their energy needs via anaerobic metabolism. In this process oxygen is not utilized, although foods are partly oxidized (in this instance by removal of electrons to an acceptor molecule). Because foods are incompletely oxidized, the amount of ATP generated per gram of food is reduced and the food residues are excreted. As a result anaerobic animals can be considered wasteful; certainly their levels of activity fall far below that of aerobic animals like vertebrates. Anaerobic metabolism does allow animals to live in oxygen-depleted environments, a feat that aerobes cannot accomplish.

In the initial steps of the cellular breakdown of foods, however, all vertebrates accomplish the "oxidations" anaerobically. Thus, glucose is split into pyruvate through a sequence of biochemical reactions known as glycolysis, and the energy released by this degradation is converted in part into ATP. Generally the pyruvate is decarboxylated and further "oxidized" aerobically to CO_2 and H_2O to yield additional ATP. In active skeletal (striated) muscle, the rate of glycolysis may be sufficiently high that pyruvate accumulates faster than oxygen can be supplied for its complete oxidation. As a result the tissue becomes anoxic, that is, devoid of oxygen. Accumulation of the end product of glycolysis, pyruvate, further inhibits breakdown of glucose and thus ATP generation.

By temporarily diverting the accumulating pyruvate to lactic acid, however, glycolysis can proceed. As lactic acid accumulates in the muscle cells it causes fatigue. A vertebrate can swim, fly, or run at its maximum rate for only so long until it drops from fatigue. This is an important ecological point: there is a time limit as well as mechanical limits placed on vertebrate motion or mobility. As a result, most behaviors take place at levels of energy expenditure that exceed only slightly resting metabolic levels. The long migrations that many vertebrates undertake, in which they sustain motion over days, even weeks, must be performed at levels of energy utilization where the limits of their aerobic metabolism are not taxed. Vivid exceptions are the brief pursuits of prey by predator, the courtship "fights" of male versus male, and the nuptial gyrations between mates.

Because vertebrates are active aerobes, they possess special adaptations in the oxygen-consuming tissues and effective respiratory structures to exchange oxygen and CO_2 with the external medium. Vertebrate white muscles, especially well developed in fishes function in bursts of locomotion where oxygen depletion is likely (Section 7.1). Often high energy phosphates are stored in white muscle cells as creatine phosphate, which can be converted to ATP rapidly, permitting muscle contraction under circumstances of limited oxygen. Because these white muscles are poorly vascularized, they operate anaerobically. Stored glycogen is decomposed, generating ATP, and lactate accumulates. During and subsequent to burst activity, this lactate is oxidized to CO_2 and H_2O generating additional ATP to replace the depleted store of high energy phosphates. This is only partially accomplished in the muscle cells, for much of the lactate diffuses into the blood and is transported to the liver for oxidation. Burst activities are thus followed by a period of higher than normal aerobic metabolism, referred to as payment of the incurred **oxygen debt**. Oxygen debts are also incurred in vertebrates by any behavior that restricts normal access to oxygen. Diving below the surface of water by lung breathing vertebrates such as whales, seals, and birds is a prime example.

In the circulatory system the respiratory pigment, hemoglobin, found in the red blood cells of vertebrates, facilitates oxygen transport within the body. In fact, correlation exists between the amount of hemoglobin present in each milliliter of blood and the general activity level of a vertebrate (Table 7–4). In addition the extent of development of the circulatory and respiratory apparatus of vertebrates correlates with their levels of activity. Simply to operate and maintain these complex circulatory and respiratory systems, which are adapted for high levels of activity, requires energy Thus, not only is the cost of vertebrate activity high, but the maintenance cost of the machinery that permits this activity is also high.

7.5.2 Metabolic Levels in Vertebrates

The amount of energy that vertebrates burn up is reflected by the amount of oxygen they consume, for all vertebrates are obligatory aerobes. The general reaction for carbohydrate oxidation reveals that for every oxygen molecule consumed a CO_2 molecule is produced

$$C_6H_{12}O_6 + 6O_2 \leftrightarrows 6CO_2 + 6H_2O$$

When burning protein and fat as an energy source, the ratio of CO_2 produced to O_2 consumed is 0.8 and 0.7 respectively. This ratio is called the **respiratory**

Table 7-4 The relation between general level of activity, rate of oxygen consumption, respiratory structures and blood characteristics in fishes of three activity levels.

Activity level	Species of fish	Oxygen consumption $(ml\ O_2\ g^{-1}\ h^{-1})$	No. secondary gill lamellae mm^{-1} of primary gill lamella.	Gill area $(mm^2\ g^{-1}$ of body wt.$)$	Oxygen capacity $(ml\ O_2\ 100\ ml^{-1}$ of blood$)$	Hemoglobin $(g\ 100\ ml^{-1}$ of blood$)$
High	Mackerel* (*Scomber*)	0.73	31	1160	15.8	37.1
Intermediate	Porgy (*Stenotomus*)	0.17	26	506	7.3	32.6
Sluggish	Toadfish** (*Opsanus*)	0.11	11	197	6.2	19.5

*Modified carangiform swimmers; swim continuously.
**Benthic fish.

quotient (RQ). Vertebrates feed on a mixture of carbohydrate, fat, and protein and, therefore, the RQ is less than 1.0 but greater than 0.7. There are exceptions to this, however, and some vertebrates burn only one type of food (for example, birds utilizing fat reserves during migration, vampire bats feeding on blood protein, and so on).

In general, the minimum rate at which aerobic animals consume oxygen is limited by the level necessary to support life and a maximum level sustainable only for short periods (burst activity; see above and Section 7.1). The upper limit of aerobic metabolism is determined by the respiratory and circulatory systems, that is, the rate at which oxygen and nutrients can be supplied to cells. The lower aerobic limit, often referred to as the **standard metabolic rate** (SMR), is set by the minimum energy required to maintain life in an organized state. If the SMR is to be useful in comparative studies, it is necessary to define carefully the conditions under which it is measured. As a rule, vertebrates seldom operate at their SMR. For example, if the resting oxygen consumption of a fish, a lizard, or a bird is measured following a meal, the metabolism will be 5 to 30 percent higher than their SMR's. This increase results from the costs of digestion and is particularly high for protein digestion. Other factors such as visual or mechanical disturbance also must be eliminated, for they cause significant increases in oxygen uptake by inducing stress. Low oxygen or high CO_2 levels in the environment have pronounced effects on energy metabolism. Perhaps the most pervasive environmental factor, however, is temperature. Because an increase in temperature usually leads to an increase in the rates of chemical reactions, including aerobic metabolism, it can have an overriding influence on organisms. Ambient temperature (T_a) can vary from summer to winter, from day to night, and even between microhabitats at the same time.

1. SMR and Body Size

SMR is reported in terms of oxygen consumed per unit of time. For most vertebrates the units are volume of oxygen at standard pressure and temperature (STP), abbreviated as $\dot{V}_{0_2}$. Obviously one would expect a large vertebrate to consume more oxygen than a small vertebrate, and they do (Figure 7–21). To allow for comparison the SMR is adjusted for body size to yield a weight-specific SMR ($\dot{V}_{0_2}$/body mass). It is immediately apparent that within any vertebrate species the weight-specific SMR decreases as body size increases (Figure 7–14). The slope of the regression of SMR on body mass is strikingly similar between different vertebrates, and also holds for a wide variety of invertebrates. Most physiologists assume that a basic mechanism underlies this phenomenon. This relationship has intrigued biologists for over 100 years, but a clear, unambiguous explanation has proved elusive.

Several biologists have pointed out that the increase in weight-specific SMR with decreased size might result from the relative increase in body surface area associated with decreasing size. For example, the surface area of a sphere increases only as the $\frac{2}{3}$ power (0.67) of the diameter. If the surface area of a vertebrate varies in a manner similar to a sphere, then small vertebrates must have a relatively higher surface area than larger vertebrates. About 100 years ago the German physiologist Max Rubner showed that the heat loss across each unit of body surface of small and large dogs was the same (about 100 kcal/m² under his experimental conditions), even though the weight-specific SMR was higher in smaller dogs. To explain this "paradox" he reasoned that small dogs with relatively high surface area must lose body heat per unit weight more rapidly than larger dogs. Small dogs compensate for their increased heat lost by

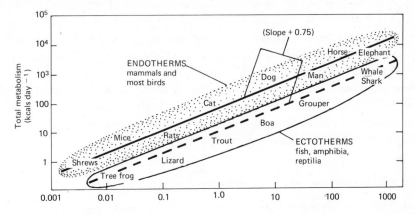

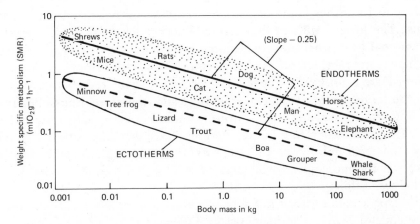

Figure 7-21 Comparison of the relationship of body size on metabolic rate in ectotherms and endotherms. Upper graph: the relationship of the log of total metabolism (in kcal) when plotted against the log of body weight yields a straight regression line with a slope of approximately +0.75. Lower graph: the total metabolism is converted to oxygen consumption per unit of body mass per hour to yield a weight specific metabolism (SMR). The relationship of the log of SMR against the log of body weight yields a slope of −0.25. Small vertebrates therefore consume more energy per unit of mass than larger ones. Both the total metabolism and the SMR of endothermic vertebrates are approximately six times as high as for ectotherms. Names and stippled areas are used to emphasize these differences (see text).

an increase in weight-specific SMR to maintain body temperatures. Rubner felt the regression of SMR on mass obtained from these data was sufficiently close to a slope of 0.67 that the phenomenon was explained by body geometry.

For many different vertebrates this slope has now been established to have a value near 0.75, that is, the SMR varies as the $\frac{3}{4}$ power and not the $\frac{2}{3}$ power of body mass. Of course one can argue that vertebrates are not spheres and their surface area to mass relationships may lie on a different slope. Thomas Mac-Mahon of Harvard claims that the actual surface area to mass relationship of vertebrates does lie on a different slope, but the value is closer to 0.62 rather than to either 0.67 or 0.75. Obviously the relationship between a vertebrate's body surface area and its mass does not totally explain SMR as related to body size. To be sure for birds and mammals, which maintain a high constant body temperature by metabolically generating heat, Rubner's hypothesis may partly explain the relationship between SMR and size. But why then is the relationship between weight-specific metabolism and size the same in reptiles, amphibians, and fishes that do not maintain constant body temperature? Obviously this phenomenon remains unexplained, but one fact is clear: metabolic rate is related to body size in a similar way in all vertebrates.

2. Metabolic Differences Between Ectotherms and Endotherms

In spite of the similar slopes (Figure 7–21), a single regression line does not fit all the metabolic data for all vertebrates. At least two lines are required: one for fishes, amphibians, and reptiles and another for birds and mammals. The lower SMR's of fishes, amphibians and reptiles result from their lack of internal heat generation to maintain a high constant body temperature (T_b). These vertebrates are referred to as **ectothermic** (heat from outside); birds and mammals are **endothermic** (heat from within). Usually ectotherms are at or close to the temperature of their surroundings and, therefore, can be described as **poikilothermic** (variable temperature). Endotherms usually maintain a constant T_b and are therefore **homeothermic** (constant temperature). As with so many biological definitions there are exceptions to this classification (see discussion on thermoregulation in sections 14.2 and 20.3). Nevertheless, the terminology is useful.

The SMR of a poikilotherm averages one sixth that of a homeotherm of the same size (Figure 7–21). The high cost of T_b maintenance demands that birds and mammals consume more food than poikilotherms of similar size. But vertebrates are motile animals and, whether poikilothermic or homeothermic, all require more energy when active.

7.5.3 The Cost of Motility

Whether differences exist between the *active* metabolic rates of poikilothermic and homeothermic vertebrates is a difficult question to answer. Over the last three decades, however, the studies of F. E. J. Fry and his colleagues at the University of Toronto, Knut Schmidt-Nielsen and Vance Tucker at Duke University, and Richard Taylor at Harvard University have provided a much clearer idea of the costs of vertebrate activity. Most vertebrates move either by swimming, flying, running, or walking. The oxygen consumption of vertebrates has been measured during locomotion over a range of different but sustained velocities. Predictably, the active rate of metabolism tends to increase with increased velocity. The energetic "cost" of moving a unit mass a unit distance is closely tied to the mode of locomotion. Specifically, for travel over the same distance—independent of the speed—swimming costs a fish less than flying costs a bird, which in turn is less than the cost of running for a mammal! Why?

Because SMR is a function of body size it might be expected that sustained active metabolism would also be related to body size. In any vertebrate species the weight-specific **active metabolic rates** (AR) of smaller individuals are indeed greater than those of larger individuals (Figure 7–22). For each size of vertebrate, however, the specific AR is linearly related to the velocity of motion and, therefore, it is possible to calculate the cost of movement independent of speed. Only body size must be taken into account. For example the units in Figure 7–22 are in ml O_2 per gram per hour (ml O_2 g^{-1} hr^{-1}) and in kilometers per hour (km hr^{-1}). The slope of each line is ml O_2 g^{-1} hr^{-1}/km hr^{-1}, which reduces to ml O_2 g^{-1}/km—the metabolic "cost" of moving 1 g over 1 km. Plotting this "cost" against body mass for each mode of locomotion, three

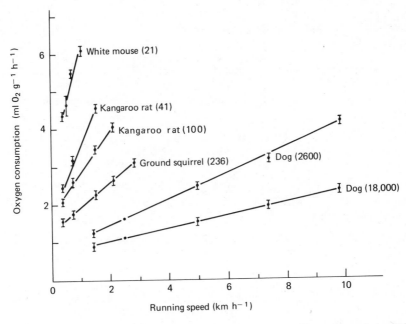

Figure 7-22 A comparison of the cost of running at different speeds in different sized dogs and rodents. Weights are in grams. Note that the regressions are linear, but the slopes decrease with increasing size. (Modified after Schmidt-Nielson [12].)

separate lines result (Figure 7–23). These results, lucidly pointed out by Schmidt–Nielsen, revealed that swimming costs less than flying, and flying less than walking or running. Each vertebrate, of course, has evolved a body form and physiology adapted to its particular mode of locomotion. The importance of such adaptations is illustrated by the excessively high cost of swimming for man, a nonswimmer by nature (Figure 7–23).

The ecological implications of these fascinating findings are just beginning to find their way into environmental studies. The cost of walking or running is ten times more expensive than swimming and, therefore, the impact of a tetrapod on the environment is potentially much greater than that of a fish. Within normal limits the speed of movement has very little influence on the total cost of locomotion. Thus when a mammal runs or a fish swims 1 km the ultimate cost is constant, no matter what the speed; only the time that it takes to traverse the distance varies. In the proverbial race of the tortoise and hare the energetic "cost" of crossing the finish line for each is similar, only the speeds differ! Why does the tortoise always win?

The cost of a particular locomotor mode seems to be about the same whether a vertebrate is a poikilotherm or homeotherm. Although this may appear paradoxical, the force required to move a unit mass a unit distance is invariable. The force a fish or whale needs to move 1 kg of weight a given distance requires

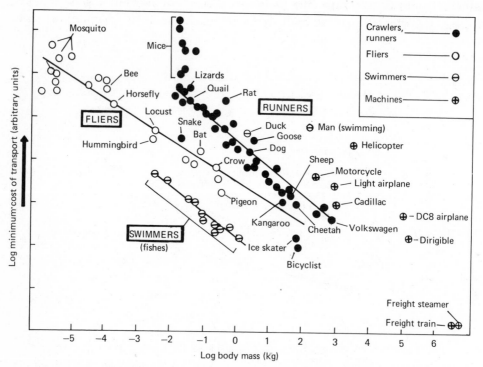

Figure 7-23 The minimum cost of transport for swimmers, fliers, and runners. Each locomotory mode tends to fall along a separate regression line. Note that a duck swims with motions similar to those used in running and a snake crawls with undulations similar to those used by swimmers. (Modified after Tucker [14].)

the expenditure of an equal minimum amount of force. To be sure precise studies do reveal slight differences in the costs of movement—some fishes are more efficient than others and some birds are especially effective flyers. Certainly adaptations, such as streamlining, have resulted in improvements in the conversion of energy into motion. But the essential point of the Schmidt-Nielsen, Tucker, and Taylor studies is not the slight differences, but the overall similarity within swimmers, within flyers, and within runners, regardless of their taxonomic position or particular physiology.

The problem reduces therefore to understanding why swimming is less expensive than flying or running. As yet the solutions are not clear. The relative "costs" of swimming to flying to running seem to contradict intuitions about locomotion. Water is dense and offers very high resistance to motion and therefore one would expect swimming to be costly.

Recall how a fish moves by sinuous undulations (Figure 7-1). This motion in part explains the low cost of swimming. Every time a trunk muscle contracts thrust is produced. Little energy is expended in returning the body to an appropriate position for the next propulsive contraction. In a tetrapod, forward motion is created only when the limb is thrust backwards. To return the limb to an appropriate forward position for the next propulsive motion requires that different muscles expend additional energy on the limb without producing thurst. Biomechanical analysis of tetrapod locomotion does show that some of this additional work is stored in the elastic tissues of the limbs and body and later released to produce additional thrust. The resultant cost of motion with limbs on land is, nevertheless, very high. These considerations in part explain the energetic differences between the three major modes of locomotion. A recent experiment on garter snakes—land undulators—shows that their "cost" falls between that of swimmers and runners (Figure 7-23). This result fits predictions because snakes undulate like swimmers but, in addition, must overcome greater friction than if suspended in water.

7.5.4 Responses to Temperature

Air temperatures on earth range from winter lows of about $-20°C$ to summer highs of about $+40°C$. In addition, diurnal changes in deserts and at high altitudes often exceed 30 C°. Daily temperature fluctuations in aquatic habitats are less pronounced because of the high heat capacity of water (1 cal ml^{-1} °C^{-1}) compared to that of air (about 0.0003 cal ml^{-1} °C^{-1}). The vertebrates, more than any other animal group—arthropods come close—actively occupy habitats from cold polar latitudes to very hot deserts. To understand this adaptability let us consider how temperature effects a poikilotherm like a fish. We shall consider how homeotherms respond to different temperatures in Section 18.3, 20.2 and 20.3.

Organisms have been described as a "bag of chemicals catalyzed by enzymes." This view, although narrow, emphasizes that organisms are subject to the

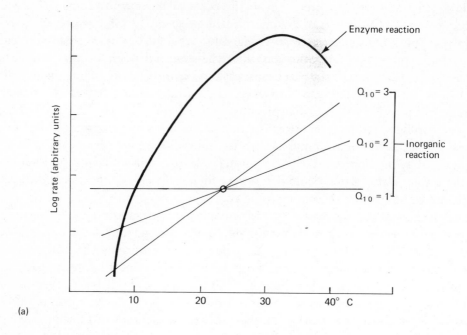

(a)

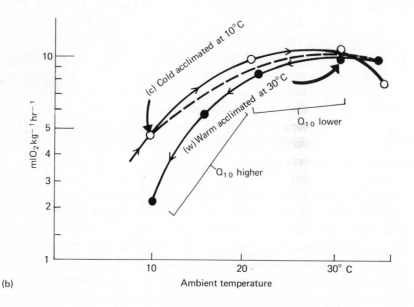

(b)

Figure 7-24 Rate-temperature responses in nonliving and living systems. (a) Typical Q_{10} responses of inorganic reactions and an enzyme catalyzed organic reaction. (b) Acute (solid lines) and chronic (dashed line) oxygen consumption responses of cold and warm acclimated goldfish (see text for details; modified after M. Kanungo and C. Prosser [1959] J. Cell. and Comp. Physio. 54:259–264).

"laws" of physics and chemistry. Because temperature influences the rates at which virtually all chemical reactions proceed, ambient temperatures vitally affect the life processes of organisms. In general, for every rise of 10°C most chemical reactions double or triple in rate. We describe this change in rate by saying the reaction has a Q_{10} of 2 or 3, respectively. A reaction that does not change rate with temperature has a Q_{10} equal to 1 (Figure 7–24a) If the rates of life processes relate to their underlying chemical reactions, we can expect these rates to increase with rises and decrease with declines in temperature. For example, if the SMR of a fish is 2 ml O_2 min^{-1} at 10°C and the Q_{10} response 2, then the fish will consume 4 ml of O_2 at 20°C and 8 ml at 30°C. This represents a substantial increase in the cost of maintenance. Most fishes cannot withstand sudden temperature changes that exceed 10 to 15°C, but many can withstand changes of 20°C or more if the changes occur over several hours. The goldfish is an excellent example; its resting metabolism at different temperatures is shown in Figure 7–24b. In this figure curves *W* and *C* are the "acute" (rapid) response curves for two groups of fish: one maintained for 1 week at 30°C (warm acclimated = *W*), and a second maintained for 1 week at 10°C (cold acclimated = *C*). Examination of these curves reveals several features of importance.

1. In both groups the reduction of oxygen consumption is most pronounced at lower temperatures. The Q_{10} for SMR of goldfish is not constant, as it is for many chemical reactions; it is in fact highest at the lower temperatures (compare Figures 7–24a and b). This type of response is typical of most poikilotherms and a wide variety of enzymatic reactions. Because oxygen consumption of a resting fish represents the energetic costs of all its maintenance processes, the data indicate that virtually all life processes must slow as the ambient temperature (T_a) falls. Sudden temperature declines therefore may induce a state of "torpor" in fishes, amphibians, and reptiles as a result of the slowing of the chemical reactions of life. Metabolism can show a temperature optimum where its rate peaks, a response that is analogous to the temperature optima of most enzymes (Figure 7–24a and b).

2. Although both curves are similar in shape, in general the oxygen consumption rates, when measured at the same temperature, are greater for the cold-acclimated fish than for the warm-acclimated fish. Obviously the cold-acclimated fish are able to compensate partly for the depressant effects of lowered temperature. When warm-acclimated fish are cooled from a temperature of 30° to 10°C, the oxygen consumption rate declines nearly five-fold, but in fish acclimated to cold, metabolic adjustment is significant. Their metabolism is only depressed two-fold (Figure 7–24). These types of adjustments to temperature are referred to as temperature compensations. If a series of goldfish are acclimated to different temperatures

and then each fish's oxygen consumptions measured at its temperatures of acclimation, an acclimation response curve can be established. This curve generally has a flatter overall shape than the "acute" curves (Figure 7–24b). Goldfish are related to a primarily tropical group of teleosts and can withstand very high temperatures. Teleost fishes with a primarily temperate evolutionary history favor cooler temperatures, as is reflected in the position of their oxygen consumption/temperature curves (Figure 7–25).

Fishes that encounter large seasonal changes in ambient temperature usually temperature compensate, and thus partly escape what has been called the "tyranny of temperature." Temperature compensation occurs in other rate processes, as well as in oxygen metabolism. Examples are heart rate, respiratory rate, nerve conduction velocities and discharge frequencies, and so on. As a result, when a fish is suddenly exposed to cold the heart rate slows, but with chronic exposure the heart slowly adjusts its rate upward. These "chronic" responses compensate for the "acute" effects of temperature on rate processes. In ecological terms, temperature compensations allow poikilothermic vertebrates to function effectively over a wider range of temperatures than would otherwise be possible.

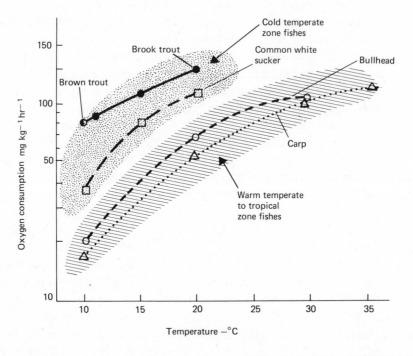

Figure 7-25 A comparison between the standard metabolic rate (SMR) of fishes that have evolved in different latitudinal (climatic) regions. Note that fishes typical of cold waters have higher SMRs than fishes typical of warmer waters. (Modified after F. W. H. Beamish and P. S. Mookherjii [1964] Canad. J. Zool. 42:177–188.)

Temperature compensation also occurs at the cellular level. In fishes an enzymatic process is often catalyzed by several similar enzymes (**isozymes**) each of which functions best over different ranges of temperature. Chronic exposure to a particular temperature causes certain genes to be activated or inactivated, and this results in the presence of more or less of each isozyme. Via these adjustments, a particular cell reaction can proceed at adequate rates over a wider temperature range in a fish that possesses inducible isozymes compared to a fish that produces only a single enzyme. The lactate isozymes (LDH's) of trout, which catalyze the conversions of pyruvate to lactate during anaerobic muscle contractions, are a classic example of an inducible system. Compensation also effects changes in hormone levels and in ion-osmoregulation. Professors P. W. Hotchachka and G. N. Somero have summarized much of this work on temperature in their text, listed at the end of this chapter.

The active metabolic rate also shows temperature compensation. This has been demonstrated clearly in fishes by Dr. F. E. J. Fry and his students. Working at the Nanaimo fisheries laboratory on Vancouver Island, Dr. J. Brett acclimated salmon to different temperatures. He then measured their SMR and their active metabolic rates (at maximum sustainable speeds) at each acclimation temperature. Subtracting the SMR from the active rate establishes the amount of energy required for swimming. This difference, defined as the **scope for activity** by Dr. Fry, varies at each acclimation temperature (Figure 7-26).

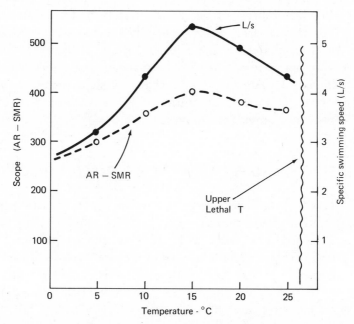

Figure 7-26   The relationship of swimming speed and the scope for activity to various acclimation temperatures in salmon. AR − SMR: active metabolic rate minus standard metabolic rate is defined as scope for activity. L/s: the number of total body lengths that a fish swims per second is defined as specific swimming speed (data from J. Brett [1971] Am. Zool. 11 (1):99–113.)

Salmon have highest scope when acclimated at 15°C. Coincident with this peak in energy expenditure is a peak in swimming speed. Thus, salmon display thermal optima; that is, there is a specific acclimation temperature at which they swim fastest. Fishes typical of warmer waters, like goldfish, tend to show higher thermal optima. These physiological examples of thermal optima are often reflected in behavior. Given the opportunity to select from different temperatures, fishes tend to seek out those temperature ranges at which they perform best—the range of their thermal optima.

The net effect of temperature compensation is to extend the temperature range over which poikilotherms remain active, an ecological benefit that permits them to exploit their environment more effectively. Temperature compensation takes time. Fishes adjust to temperature shifts over periods of days and weeks, not minutes or hours. Nevertheless, these slow biochemical and physiological adjustments represent regulation in energy utilization by the organism. Compensation is more effective in countering slow seasonal changes in temperature and is most prevalent in aquatic vertebrates, where water buffers sudden changes in temperature. In contrast, terrestrial vertebrates encounter sudden fluctuations, as well as seasonal changes, in temperature. The mechanisms used by reptiles to cope with the fluctuating temperatures of the terrestrial environment will be discussed in sections 14.2 and 14.3.

Although water acts as a thermal buffer and allows temperature compensation to be an effective adaptive strategy, it poses serious respiratory problems. Oxygen has a low solubility in water, and a variety of fresh-water and marine habitats may frequently lack oxygen entirely (anoxic conditions). Such circumstances are seldom if ever found in the terrestrial habitats of vertebrates. Thus aquatic vertebrates like fishes can be placed under severe stress in certain marginal habitats. Adaptations of the respiratory structures to cope with low oxygen levels were perhaps more significant than any others in the preadaptations of Devonian vertebrates for life on land.

7.6   GILLS, LUNGS, AND PREADAPTATIONS TO LIFE ON LAND

Most aquatic vertebrates possess gills, evaginations from the body surface over which respiratory gases are exchanged. Gills are subject to abrasive damage when they are exposed externally, as in the mud puppy (*Necturus*). In fishes gills are protected within covered pharyngeal pockets (cyclostomes—Figure 5–7; Osteichthys—Figure 7–27). In Osteichthys water is generally pumped across the gills in one direction by the action of muscles that expand the buccal cavity and the opercular plates that cover the gills (Figure 7–27a). The buccal flaps just inside the mouth and the flaps at the posterior and ventral margin of the opercula act as one-way valves to direct the movement of water from pharynx across the gills to the exterior. The gill respiratory surfaces are delicate, vascularized projections from the lateral side of each gill arch (Figure 7–27a). Two columns, made up of numerous gill filaments, extend from each gill arch.

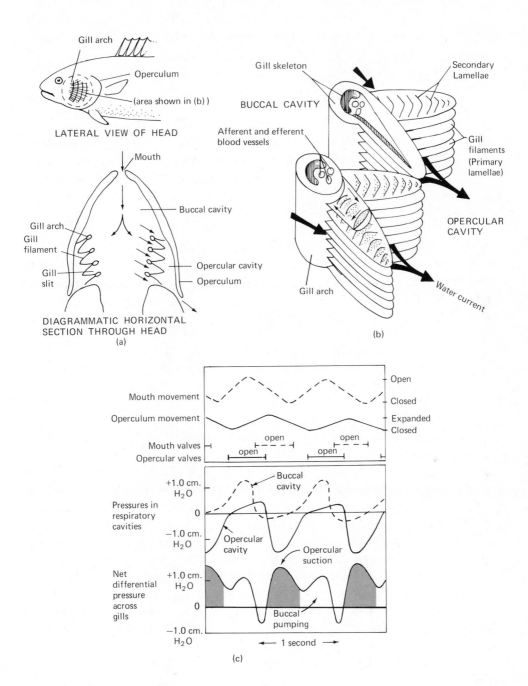

Figure 7-27 Anatomy and functional morphology of teleost gills. (a) Position of gills in head and general flow of water. (b) Water flow patterns across the gills. (c) Water pressure changes across the gills during various phases of ventilation (modified after Hughes [9].)

The tips of these filaments from adjacent arches meet, and thus water as it exits the buccal cavity must intimately contact the vascularized surfaces. Actual gas exchange takes place across numerous microscopic projections, the secondary lamellae.

The pumping action of mouth and opercula is pulsatile, but the combined action of the buccal and opercular motions tends to sustain a positive pressure from pharynx to opercular cavity so that the respiratory current is only slightly interrupted during each pumping cycle (Figure 7–27b). In many pelagic fishes (tunas, sharks, and so on) the ability to pump water across the gills has been lost or greatly reduced. In these fishes a respiratory current is created by swimming with the mouth slightly opened, a method known as "ram jet ventilation." It has the advantage that water flowing across the gills does so without interruption, but the fishes must perpetually swim.

The vascular arrangement in the teleost gill is an engineering marvel, for it maximizes oxygen exchange. Each gill filament is supplied with two arteries, an afferent vessel running from the gill arch to filament tip and an efferent vessel returning blood to the arch. Each secondary lamella is a blood space covered only by a thin epithelium (through which gas exchange occurs) connecting the afferent to the efferent vessel (Figure 7–28a). Blood flow through the lamellae is counter to the flow of water across the gill. This structural arrangement, known as a **countercurrent exchanger**, assures that the blood in the gills takes up as much oxygen as possible. The exchange itself results from the inward diffusion of oxygen down its concentration gradient. As depicted in Figure 7–28b and c, if the blood and water flow parallel, the blood exiting the gills would contain oxygen at the same partial pressure as in the vented water. Because the two flows are opposed, however, the blood achieves a partial pressure nearly equal to that of the incoming water.

A correlation exists between the level of activity of fishes and the extent of development of the respiratory structures and the hemoglobin concentration (Table 7–4). Active pelagic fishes, like tunas, which sustain activity over extended periods of time have relatively larger gill exchange areas and a higher oxygen carrying capacity per milliliter of blood than sluggish benthic fishes, such as toadfishes and flatfishes. In spite of this, seemingly sluggish fish are capable of considerable activity for short periods (minutes), although fatigue quickly sets in when the activity is prolonged.

Most vertebrates seldom encounter low oxygen concentrations. This is especially true for terrestrial vertebrates, for air contains 20 to 21 percent oxygen. In aquatic habitats oxygen concentrations are more variable. The amount of oxygen that can dissolve in water is much less than present in a similar volume of air (1 liter of air at standard temperature and pressure contains 210 ml of oxygen; 1 liter of fresh water at STP contains 10.29 ml of oxygen). Increasing temperature reduces the solubility of oxygen in water and, as a result, oxygen content in warm water can be seriously reduced. This is particularly true in tropical fresh waters where high temperatures and the metabolism of microorganisms and phytoplankton often produce anoxic conditions.

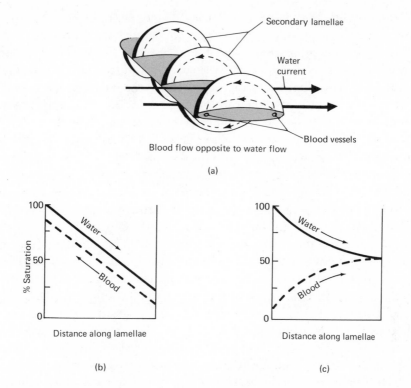

Figure 7-28   Countercurrent exchange in the gills of teleost fishes. (a) The direction of water flow across the gill opposes the flow of blood through the secondary lamellae. (b) This results in a higher oxygen loading tension in the blood and lower oxygen tension in the water leaving the gills. (c) If water and blood flowed in the same direction over and within the secondary lamellae, a lower oxygen tension would occur in the blood leaving the gills (modified after Hughes [9]).

Fishes that live in these conditions cannot obtain enough oxygen via gills and have alternate respiratory structures. Accessory respiratory structures have evolved independently in several lines of teleosts enabling them to breath air. The electric eel, in addition to gills, has an extensive series of highly vascularized papillae in the pharyngeal region. Electric eels rise to the surface to gulp oxygen-rich air, which diffuses across the papillae into the blood. Other fishes swallow air and extract the oxygen through vascularized regions of the gut. The anabantid fishes (bettas, gouramies) have elaborately vascularized chambers in the rear of the head, called labyrinths, which function in a similar manner. Many fishes with accessory respiratory structures use them facultatively, that is, uptake switches from the gills to the accessory organs when oxygen in the water becomes low. Others, like the electric eel and the anabantids, are obligatory air breathers, for the gills alone are inadequate to exchange sufficient oxygen. Denying these fishes access to air causes them to suffocate.

Lungfish avoid desiccation by burrowing in mud during periods of drought, and breathe air with the aid of their lungs (Chapter 6). In these fishes the lungs are derived from an evagination of the gut, and similar structures are typical of osteichthyans including, perhaps, most of the early Devonian acanthodians. These evaginations may have served as floats to produce a more nearly neutral density. Indeed, this lunglike structure became the air bladder of teleosts, contributing significantly to their refined locomotion. The early Devonian fresh-water fishes very likely encountered environmental stresses like those faced by tropical fresh-water fishes today. Periodically drought restricted ponds and rivers and the slow moving, often stagnant waters must have become oxygen deficient. Thus, the fishes were exposed to episodes of anoxia and hypercapnia (high CO_2). It is easy to envision these early fishes rising to the surface to gulp air, or wriggling overland from drought restricted ponds to the next pond (as anabantids are known to do today). Obviously a lung or other accessory respiratory structure would be adaptive. The air filled evaginations of the gut in early fishes undoubtedly served this function as well as participating in hydrostatic adaptations and possibly in special sensory adaptations.

For atmospheric breathing, gills are inadequate. To function efficiently the secondary lamellae of gills are very close together, structurally thin-walled and weak. Exposed to air they stick together and thus reduce the gas exchange surface areas. This is further complicated by the copious mucous secretions in the gills, which normally reduce frictional resistance to water flow but, in air, only add to filament adhesiveness. To breathe air a more rigid or enclosed exchange structure must be used. Obviously the higher the aerobic respiratory needs of the animal, the greater the surface gas exchange area must be. As one examines air-breathing vertebrates in a series from fishes, amphibians, and reptiles to birds and mammals the structure of the respiratory surface (generally a lung) becomes more and more complex. In fact, in many fishes and amphibians (and even some turtles) aerobic metabolic requirements are so low that air-breathing is supported across the skin, and neither gills nor "lungs" are utilized.

When oxygen and/or CO_2 is exchanged across a respiratory structure, the movement occurs because of diffusion down a concentration gradient. The amount of diffusion follows Fick's equation

$$dQ = - DA \, (du/dx) \, dt$$

the amount of a substance (Q) moving a given short distance (dx) in a given time interval (dt) is proportional to the concentration difference (du) across (dx), to the diffusing area (A) and the diffusion constant (D). D is a property of the diffusing substance and the material through which it passes and is inversely related to the molecular weight of the diffusing substance. Because a molecule of water is lighter than a molecule of oxygen, given equal pressure gradients water diffuses more rapidly than oxygen. In air the vapor pressure gradient of water across the respiratory structure from animal to air can be very large. Therefore respiratory gas exchange can be accompanied by a considerable loss of body water. Of course this is less severe in a humid environment like a damp forest than a dry environment like a desert because the water's vapor pressure gradient is less.

The structural characteristics of respiratory membranes require that they remain moist to function. Respiratory gas exchange in air therefore almost always leads to water loss by evaporation. All terrestrial animals, vertebrates or invertebrates, must cope with this problem. All have evolved anatomical, physiological and/or behavioral strategies to cope with the inescapable laws of diffusion.

Enclosing the accessory air-breathing structures has been a major solution in both vertebrates (lungs) and invertebrates (tracheal structures of insects for one). Enclosing the respiratory exchange surface provides a space wherein the air in the lung can achieve a very high vapor pressure. The problem becomes therefore one of minimizing water loss to the outside, not one of drying the delicately thin respiratory membranes and thus upsetting gas exchange. An alternate to enclosing accessory structures is skin breathing, as found in many amphibians. It is restricted to very humid environments or at least habitats adjacent to water or moist soil where the animal can periodically remoisten the skin.

For several chapters we have considered the evolution and adaptations of fishes for life in water—a life style of which they are masters. Their enormous diversity attests to the fact, however, that no single species can master all of the waters on earth. Like terrestrial habitats, aquatic habitats vary greatly from location to location, and the phylogeny of fishes has been tailored by these varied demands. In general, habitats in water tend toward greater stability than do terrestrial ones. This stability, derived from the properties of water, minimizes temperature variation and prevents dehydration, and so reduces changes in metabolic rates. Thus, buffering by the water medium provides a portion of the regulation required to maintain the constant "internal milieu" so vital to an active animal.

If homeostasis is to be maintained in a less stable environment, as in transgressions upon the land, it will cost more in regulation energy. Not only will the costs be high but the complexities of the morphological and physiological means required to maintain homeostasis in an unbuffered environment are enormous. Unquestionably, eons of evolutionary time were required for transitions through amphibious stages. In fact the fossil record testifies to these difficulties for the slow accumulation of distinctly terrestrial traits seen in reptiles required nearly 100 million years, from the mid-Devonian to the late Carboniferous. By the end of the Devonian all of the major groups of fishes had evolved, life was teeming in fresh waters, and competition must have been intense. Interrelated adaptations of these fishes to marginal fresh-water habitats were significant preadaptations to the rigors of terrestrial existence, especially those that allowed for locomotion on land and for aerial respiration. Even if the first invasions of the land were crude attempts, they must have reduced competition for the transgressor and have been favored. But competition between species was not the only factor that stimulated vertebrate evolution relentlessly toward the land. The unique history of the continents during and following the Devonian also forcefully molded the pattern of vertebrate evolution.

References

[1] Bennett, M. V. L. 1971. Electric organs. In *Fish Physiology*, edited by W. S. Hoar and D. J. Randall, vol. 5, chapter 10. Academic Press, New York. This review extensively describes the properties of the electrocytes, their distribution in fishes and their neural control. Slanted toward physiology, there is adequate treatment of morphology and function to make the review useful to all biologists.

[2] Bennett, M. V. L. 1971. Electroreceptors. In *Fish Physiology*, edited by W. S. Hoar and D. J. Randall, vol. 5, chapter 11. Academic Press, New York. A detailed description of electroreceptors in elasmobranchs and teleosts, the review covers physiology, morphology, distribution, function and evolution.

[3] Fessard, A. (editor). 1974. Electroreceptors and other specialized receptors in lower vertebrates. In *Handbook of Sensory Physiology*, vol. 3, part 3. Springer-Verlag, New York. Included are chapters on lateral line receptors and electroreceptors. A comprehensive review, the various chapters are pitched to the intermediate to advanced biology student.

[4] Flock, A. 1971. The lateral line organ mechanoreceptors. In *Fish Physiology*, edited by W. S. Hoar and D. J. Randall, vol. 5, chapter 8. Academic Press, New York. A clear, concise, general account of the morphological and physiological properties of lateral line hair cells and their integration and interactions with the brain.

[5] Gordon, M. S., G. A. Bartholomew, A. D. Grinnell, C. B. Jorgensen and F. N. White. 1977. *Animal Physiology: Principles and Adaptations*. Third edition. Macmillan, New York. A general treatise of comparative physiology intended for undergraduate biology majors. Illustrations and examples concentrate on vertebrates.

[6] Gray, J. 1968. *Animal Locomotion*. Weidenfeld and Nicolson, London. A classic monograph that describes the modes by which animals from bacteria to vertebrates move. Clearly illustrated, it also provides a mathematical and physical approach to the analysis of animal motion.

[7] Hill, R. W. 1976. *Comparative Physiology of Animals, an Environmental Approach*. Harper and Row, New York.

[8] Hochachka, P. W. and G. N. Somero. 1973. *Strategies of Biochemical Adaptation*. W. B. Saunders Co., Philadelphia. An excellent review of biochemical adaptations in animals, with emphasis on vertebrates. Particular attention is given to environmental factors. Although helpful, a course in biochemistry is not required to appreciate the many examples used to illustrate how biochemical systems, as well as morphology, are adaptive.

[9] Hughes, G. M. 1963. *Comparative Physiology of Vertebrate Respiration*. Harvard University Press, Cambridge, Mass. A short, excellent treatise on respiratory systems in vertebrates. Emphasis is on function and directed to introductory biology students.

[10] Lockwood, A. P. M. 1964. *Animal Body Fluids and their Regulation*. Harvard University Press, Cambridge, Mass. A short monograph that is excellent for beginning biology students.

[11] Prosser, C. L. 1973. *Comparative Animal Physiology*. Third edition. W. B. Saunders Co., Philadelphia. A very detailed account of the principles and facts of comparative physiology. The work is very useful as a reference to the literature.

[12] Schmidt-Neilsen, K. 1975. *Animal Physiology. Adaptation and Environment*. Cambridge University Press, London and New York. A textbook designed for beginning students in comparative physiology. The narrative style is charming and especially suitable for explanation of concepts. Particularly good is the coverage of energy utilization and related processes in vertebrates.

[13] Tavolga, W. N. 1971. Sound production and detection. In *Fish Physiology*, edited by W. S. Hoar and D. J. Randall, vol. 5, chapter 6. Academic Press, New York.

[14] Tucker, V. A. 1975. The energetic cost of moving about. *American Scientist*, 63(4): 413–419. A concise review of vertebrate locomotion which includes a consideration of muscular work.

[15] Webb, P. W. 1975. Hydrodynamics and energetics of fish propulsion. *Bull. Fisheries Res. Board of Canada*. No. 190. A comprehensive review of hydrodynamic considerations of fish locomotion. The approach is amply illustrated, but intended for advanced students.

[16] Walters, V. 1962. Body form and swimming performance in scombroid fishes. *American Zoologist*, 2(2):143–149.

[17] Hoar, W. S. and D. J. Randall (eds). 1978. *Locomotion*, Vol. VII, *Fish Physiology*. Academic Press, N. Y. A very recently received multiple authored volume with an up to date account of locomotion in fishes. Intended primarily for advanced graduate level students.

8

The Geology and Ecology
of Tetrapod Origin

Synopsis: The amphibians originated during the Devonian, when Pangaea coalesced into a single land mass and equitable climates prevailed. Slight continental uplift reduced the transgression of shallow epicontinental seas during this period, but swamplike fresh-water habitats were common. The earliest terrestrial plants required moist soils for growth and reproduction, and therefore were limited to the margins of waters. Early arthropods (crustaceans, insects, and arachnids) became abundant in the late Devonian and provided a new food resource on land. Whether in response to this new terrestrial resource or to the abundant predatory fresh water fishes, the Devonian crossopterygian fishes gave rise to the first amphibians, the ichthyostegids. Anatomically and physiologically this feat was complex and required millions of years. The long equitability in climate, due in part to the location of much of Pangaea in tropical latitudes, favored this transition from water to land.

8.1 CONTINENTAL GEOGRAPHY IN THE LATE PALEOZOIC

Coalescence of the ancient continents was nearly complete in Devonian time (Figure 8-1). Some geologists believe that only a narrow sea, an extension of the ancestral Tethys, separated Gondwanaland from ancestral North America, Europe, and Asia. With the advent of the Carboniferous, the Tethys Sea closed completely; Pangaea became a single continent that persisted through the Permian and the Triassic periods. During this long period of time, however, crustal movements did not cease. Coalescence of these ancient lands produced orogeny (mountain building), especially along the geosynclines sandwiched between the colliding plates (Figure 4-2). Extensive mountain chains, such as the Appalachians and Urals, arose. The shallow epicontinental seas so typical of the

285

Figure 8-1. Continental positions during the Devonian.

The arm of the Tethys Sea between North American and Gondwanaland did not close completely until the end of the Devonian when Pangaea was formed. Epicontinental seas were more restricted than in the early Paleozoic. Only one major transgression, which covered much of southern Europe, middle North America, and Asia, occurred in the middle Devonian. The Acadian revolution in the eastern United States and Canada resulted in uplift and deep folding of the sediments in what is now the northern Appalachian region. This Acadian orogeny was reflected in the mid-Devonian rocks of Great Britian with less intensity. The Acadian, along with the Caledonian orogeny at the end of the Silurian (Figure 4-4) caused sufficient uplifting to produce an extensive series of continental and marine deposits that covered most of middle and northeastern America, Greenland, the northern British Isles, and Scandinavia. This complex is often collectively referred to as "the Old Red Sandstone Continent." Included in the sandstones are many of the early fish fossils and the first amphibians. The Caledonian and Acadian orogenies probably resulted from compressional forces generated by the coalescence of North America and Europe in the Silurian.

Ordovician and Silurian periods were far less extensive (compare Figure 4–3 and 4-4 with Figure 8–1). Devonian, Carboniferous, and Permian geological formations in North America and Europe clearly show an increase in the thickness and extent of terrestrial deposits. The massive deposits of evaporites, of red beds, and of coal are indicative of terrestrial conditions. During the late Permian, however, the tectonic forces that created Pangaea waned. As a result, the folding and thrusting of strata into mountain belts ceased and erosion reduced the mountains in size. The resultant low profile of Pangaea was accompanied by the return of epicontinental seas, as shown by widespread deposits of limestone during portions of the Permian.

8.2 DEVONIAN CLIMATES

As we have seen, in Devonian times early bony fishes dominated both marine and fresh-water habitats. Their extensive radiation must have resulted in part from intensive interspecific competition for food and for suitable habitats. To eat and avoid being eaten, and to reproduce represented daily tactics and lifetime strategies as important to survival then as they are today. In the late Devonian the class Amphibia originated from one group of fishes—the lobefins (crossopterygians). Amphibious habits must have provided a selective advantage. Whether the advantage was associated with a reduction in predation or a new food source is a question we shall discuss. But leaving the water, even for short periods, exposed vertebrates to new and formidable difficulties. What climates prevailed during this period of the earth's history? What terrestrial organisms existed when protoamphibians ventured onto land?

In Chapter 4 we suggested that the climate of equatorial latitudes in the early Paleozoic was tropical and reasonably uniform. During 250 million years, from the early Cambrian to the end of the Carboniferous, only a single major (worldwide) climatic interruption took place. This interruption, indicated by extensive glaciation in the Ordovician, must have lowered worldwide temperatures by several degrees. Its effect on the emerging and then marine ostracoderms is

---

Two floristic regions characterize the Devonian: zone 1—a tropical region of enormous extent, which in early Devonian time contained mosses and simple herbaceous vascular plants and in mid to late Devonian time a flora dominated by arborescent club mosses (lycopods); zone 2—a late Devonian region lying north of the Angaran geosyncline (Figure 4–2) that was dominated by heterosporous plants as well as lycopods. Unlike the plants of zone 1, the Angaran flora showed seasonal growth annuli. The Australian flora (zone 3) maintained close affinities to the plants of zone 1. The distribution of Devonian ostracoderms (●) lies within floral zone 1. Also in this region are the earliest vertebrates of probable freshwater origin: upper Silurian cephalaspidomorphs. During the Devonian jawless vertebrates radiated widely in the fresh waters of zone 1. The first amphibians, the ichthyostegids, are recorded from Greenland (⊕) and more advanced forms in younger formations of West Virginia and Nova Scotia.

hard to assess because the tropical seas in which ostracoderms originated would have buffered the effect of climatic change. In the Silurian and Devonian, climate was again warmer in the tropics and undisturbed by glaciation.

A glance at Figure 8–1 shows that North America, Europe, and western Asia, as well as much of Indochina, and Australia, lay in the tropics and subtropics, that is, between 30°S and 30°N latitudes. Warm conditions prevailed. Productivity of plants must have been high, providing an abundant primary food source for the invertebrates and vertebrates of the time. The close proximity of the warm equatorial Tethys Sea to this land mass provided abundant rainfall on the continents through the evaporation of sea surface water into the atmosphere. Lakes, rivers, and streams in the Silurian and early to middle Devonian were extensive in North America and Europe. Conditions were ripe for the radiation of Devonian fresh-water fishes.

8.3 TERRESTRIAL HABITATS OF THE DEVONIAN

The picture that one obtains of early Devonian habitats suggests that in shallow seas plant and animal life was abundant and diverse; plant colonization and radiation in fresh water was extensive, and animal radiation was well underway. On land, however, the emergence of plants and animals had only begun. Terrestrial fossil beds from the early Devonian are strikingly devoid of plant and animal remains. Most botanists suspect that green algae, which are aquatic, were ancestral to our present day terrestrial plants. The first land plants were primitive mosses and simple vascular plants. Like mosses today they needed moist conditions. Fossils of the first vascular plants, which were short, leafless, and possessed only simple branching stems, are found in late Silurian strata in Bohemia. As Figure 8–1 indicates, in the Silurian this area was close to the equator. Its climate must have been tropical and moist. Thus, predictions from drift theory are consistent with the probable ecological requirements of the first land plants.

By the early Devonian, mosses and vascular plants had become common. Indeed, by mid-Devonian the flora consisted of large horsetails and a variety of scale trees (lycopods). This Devonian flora suggests that terrestrial habitats were moist. An important innovation in the flora, the appearance of heterosporous plants (progymnosperms), occurred at the beginning of the late Devonian. The first seed-bearing plants (gymnosperms), derived from the progymnosperms, appeared in the Devonian, but they were not abundant. They are known only from fossilized seeds in Devonian deposits.

Once plants had colonized the land, animals soon followed to exploit this new source of food. Scorpions appear in the late Silurian, perhaps the first land animals, and spider-like arachnids occur in early Devonian sediments. As with the first land plants, fossils of these animals are found in Germany. Insects, represented by springtails (collembolids) and bristletails (embiopterids), occur in late Devonian sediments, but they are not abundant in fossil deposits until the Carboniferous.

Because the Devonian land profile was low, these early terrestrial communities were swamplike. Reproduction and distribution of plants were dependent on vegetative processes and spore transport which, for the most part, require moist soils. Only with the evolution of the seed-bearing plants, where the embryo is protected from desiccation, could plant dispersal and the colonization of arid areas have effectively taken place. Because the seed bearing gymnosperms were rare, it seems clear that in the Devonian terrestrial plants were mainly limited to the margins of water. Early terrestrial organisms must have strayed little from the major watercourses that crisscrossed the land.

A unique feature of the middle to late Devonian flora should be stressed. Most of the plants were uniformly distributed along the tropical equator between 30°S and 30°N latitude from the western United States, across Europe, western Russia, and into China (Figure 8-1). That the climate was predominantly tropical (uniform) is shown by the absence of seasonal growth rings in the stems and trunks of these fossil plants. In Siberia, however, a distinctive flora of upper Devonian age does show seasonal growth rings. Drift theory locates Siberia between 30 and 60°N latitude in the Devonian. This Siberian flora probably represents one of the first extensive terrestrial plant communities adapted to cooler conditions.

8.4 IMPORTANCE OF DEVONIAN CONTINENTAL LOCATIONS TO THE ORIGIN OF AMPHIBIANS

In the tropical fresh-water habitats of the late Devonian, the tetrapods, in the form of a fishlike amphibian, first encountered a terrestrial habitat—a habitat with uniquely different characteristics from water. Obviously, adjustments to such new and different conditions could not be bridged in a few generations. Millions of years were to pass before a truly terrestrial vertebrate was to walk on earth. But a slow, ponderous start was made in the Devonian. There must have been advantages in slight degrees of amphibious behavior. What were they? What changes in form, in function, in behaviors were required to make this transition from water to land? There were several, and we shall consider them in subsequent discussions of amphibians and reptiles. But reflection on the origins of amphibians in relation to continental drift points out why an understanding of vertebrate evolution requires diverse types of information.

We have emphasized the view that the earliest vertebrates are associated with uniform tropical conditions. Why did they not also occur in more temperate regions, even boreal regions? Modern fishes and amphibians certainly do. To emerge from water onto land and function as a terrestrial vertebrate requires many changes in anatomy and physiology. The most likely site for amphibious behavior to succeed must have been where the number of radical changes encountered was minimal. Emergence in tropical rather than cooler climates would certainly minimize the temperature change between water and air. In tropical swamps and forests, there is a surprising uniformity of temperature throughout the habitat and at different times of day; humidity is also high. In Chapter 7 we learned that a sudden change in temperature can upset the metabolism of ectotherms, like fishes and amphibians, and can dramatically modify their ability to be active. It is small wonder, then, that the first amphibians are found in strata from Greenland, which was then on the equator, and not strata from Japan or from South Africa (Figure 8–1). Combining information about animal functions and requirements with the concept of continental drift, which tells us a great deal about probable past climates, clarifies what otherwise would remain a puzzle.

References

[1]   References 1, 2, 5, and 9 in Chapter 9 should be consulted for more details on Late Paleozoic continental positions and climates.

[2]   Banks, H. P. 1970. *Evolution and Plants of the Past.* Wadsworth, Belmont, California. A well-illustrated account of the origin and evolution of plants, especially useful in considering the changes that occurred in terrestrial floras.

9

Origin and Radiation of Amphibians

Synopsis: The fossil record of tetrapod origins is very incomplete. Rhipidistian fishes are plausible ancestors of tetrapods, but there are no fossils that link the rhipidistians known from the mid-Devonian with the amphibians known in the late Devonian. The elaboration of limbs, limb girdles, ribs, and vertebrae that characterizes tetrapods must have begun among still aquatic protoamphibians that lived in the same habitats as more conservative rhipidistian lineages. Amphibians radiated rapidly after their appearance, and the earliest fossils known, the ichthyostegids of the late Devonian, show specializations indicating that they are an offshoot from the stock that gave rise to other Paleozoic amphibians.

Amphibians continued to radiate extensively through the Carboniferous, Permian, and Triassic. By the end of the Triassic all of the ancient groups had become extinct, and the transitional forms that gave rise to modern amphibians are unknown. Three major groups of Paleozoic amphibians can be defined. The subclass Lepospondyli included two orders of small, primarily aquatic animals. The subclass Labyrinthodontia also contained two orders, the Anthracosauria and the Temnospondyli. Anthracosaurs were primarily terrestrial, although one suborder, the Embolomeri, consisted of forms secondarily specialized for an aquatic life. The terrestrial anthracosaurs gave rise to reptiles in the late Carboniferous. None of the anthracosaurs persisted through the Permian, possibly because their reptilian descendants progressively excluded them from terrestrial habitats.

The second order of labyrinthodonts, the temnospondyls, were primarily aquatic with one suborder, the Rhachitomi, that was largely terrestrial. The aquatic stereospondyl temnospondyls of the late Permian and Triassic were enormous animals. Some approached the size, body form, and probably the feeding habits of modern crocodilians.

A third group of Paleozoic amphibians, the Microsauria, is often included among the lepospondyls but may represent a distinct lineage. Microsaurs were small animals, similar in size to modern salamanders, and they lacked the heavy scales that characterized labyrinthodonts. In their habits and behavior they were probably more similar to modern amphibians than they were to their enormous contemporaries.

Table 9-1 Taxonomic arrangement of paleozoic amphibians

Class Amphibia

Subclass † Labyrinthodontia

Order † Ichthyostegalia (*Ichthyostega* and related forms)

Order † Temnospondyli

Suborder † Rhachitomi (semiaquatic and terrestrial forms including *Eryops* and *Cacops*)

Suborder † Stereospondyli (specialized aquatic forms including *Mastodonsaurus* and *Capitosaurus*)

Order † Anthracosauria

Suborder † Embolomeri (secondarily aquatic forms characterized by dispondylus vertebrae)

Suborder † Gephyrostegida (terrestrial anthracosaurs including *Gephyrostegus* and related forms)

Suborder † Seymouriamorpha (includes both terrestrial forms like *Seymouria* and aquatic forms like *Kotlassia* and *Discosauriscus*)

"Batrachosaurs" (an artificial group of anthracosaurs specialized for terrestrial life but not closely related to the lineage that gave rise to reptiles. *Solenodonsaurus* and *Limnoscelis* are included. Some paleontologists consider *Diadectes* and its relatives to be anthracosaur amphibians of this sort)

Subclass † Lepospondyli

Order † Aistopoda (legless aquatic or semiaquatic forms including *Ophiderpeton* and *Phlegethontia*)

Order † Nectridia (aquatic forms including *Diploceraspis* and *Urocordylus*)

Status uncertain

† Microsauria (a diverse group of small aquatic and terrestrial forms)

† = extinct

9.1   INTRODUCTION

The history of amphibians falls into three parts separated by gaps in the fossil record. The origin of labyrinthodont amphibians (Subclass Labyrinthodontia) from rhipidistian crossopterygians is clearly indicated although no fossils of the transitional forms are known. From this origin the separation of labyrinthodonts into at least two major lines of evolution, the Orders Anthracosauria and Temnospondyli, can be assumed. The origin of reptiles from the anthracosaurs completes the first part of amphibian history. The second part consists of the lepospondyl amphibians (Subclass Lepospondyli) that arose in the mid-Carboniferous, radiated in the late Carboniferous, and had disappeared by the end of the Permian. The microsaurs are yet another group of Carboniferous amphibians. They have traditionally been grouped with the lepospondyls, but recent work has cast doubt on that assumption. The third part of amphibian history is the modern amphibians, the Subclass Lissamphibia, which appear as well-diversified fossils in the Jurassic (indicating an origin sometime earlier) and can be traced to living forms. The origins of the lepospondyls and lissamphibians remain buried in yet-undiscovered fossil deposits, but the morphological similarity of labyrinthodonts, lepospondyls, and lissamphibians justifies their recognition as three subclasses of the Class Amphibia (Figure 9–1).

9.2   ORIGIN OF AMPHIBIANS

The fossil record provides evidence of the origins of terrestrial vertebrates. Important paleontological discoveries of amphibian progenitors and early amphibians have been made in sediments of the middle and late Devonian age in Scotland and Greenland. These areas, now far north of the geographic ranges of most modern amphibians, were warm tropical regions located near the equator during the Devonian (see Chapter 8).

9.2.1   Tetrapod Ancestors

Three major groups of fishes have been considered possible ancestors of terrestrial vertebrates—the lungfishes, the bichirs (represented by the living genus *Polypterus*), and the crossopterygians. Each has had its supporters, and superficially any one seems a possible amphibian forerunner. All had lungs, which we have seen were a very early development in vertebrates. In each the fins were modified in the direction of tetrapodlike limbs which—to judge from the living forms—were used in crawling across the bottom of ponds and lakes.

Closer examination, especially of the fin skeleton and the skull, eliminates the lungfishes and bichirs from consideration as amphibian ancestors. Lungfish are highly specialized, and Devonian fossils are scarcely less specialized than the living species. In particular, the dermal bones of the lungfish skull have a dif-

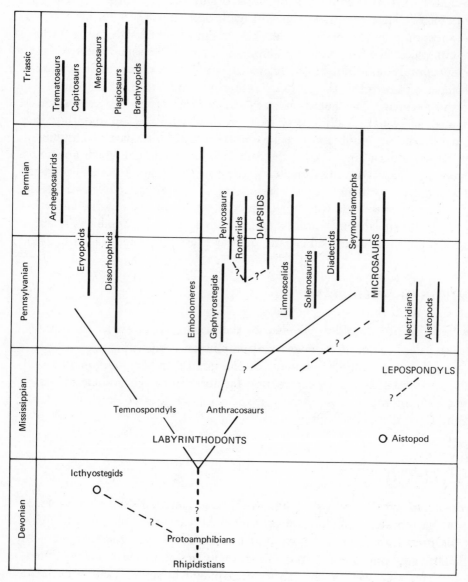

Figure 9-1 Phylogenetic relationships within the Class Amphibia. The fossil record is very incomplete and relationships between the groups are tentative. A single aistopod, indicated by a circle, is known from the Mississippian.

ferent pattern from those of early amphibians. Furthermore, lungfish lack the premaxillary and maxillary bones which, in tetrapods, are the important tooth-bearing bones of the upper jaw. In lungfish, the teeth are borne on large plates on the palate. In addition, the skeleton of a lungfish's fin consists of a central axis with symmetrical rays forming a leaf-shaped structure or **archipterygium** (Figure 9-2a). There are no obvious homologies between this limb structure and the tetrapod limb seen in the earliest known amphibians.

The bichirs, too, can be eliminated from consideration as tetrapod ancestors by an examination of the skull and limb bones. In this case the broad-based pectoral fin with a single row of bones is clearly unlike the starting point of the evolution of a tetrapod limb (Figure 9-2b). The skull bones, too, cannot be homologized with the pattern of bones in the skulls of the earliest amphibians.

In contrast, comparison of the morphology of crossopterygians, specifically the rhipidistian crossopterygians, with early tetrapods reveals extensive similarities, indicating that amphibians probably arose from this group sometime in the mid-Devonian. Early crossopterygians had an archipterygium like that of lungfish, but in the later forms the central axis was shortened and branches were confined to the anterior margin. Homologies between the limb bones of rhipidistian crossopterygians and early amphibians are clearly apparent (Figure 9-2c, d). The skulls of rhipidistians and amphibians also are similar. The major differences are found in the overall proportions of the skull and the bones of the nasal region (Figure 9-3). In crossopterygians the nasal region is short and covered by a mosaic of small bones, while amphibians have a much longer snout and a pair of large nasal bones. The parietal bones make a useful landmark, although they initially caused paleontologists considerable confusion. The parietal foramen in rhipidistians lies so far anteriorly that the bones surrounding it were first identified as the frontals. Although the similarities of rhipidistians and amphibians were compelling evidence of their relationship, paleontologists were hard-pressed to explain the apparent change in position of the parietal foramen from the frontals to the parietals. T. S. Westoll offered a solution— the bones surrounding the parietal foramen in crossopterygians really are the parietals as in tetrapods, but the skull proportions have changed so greatly in the transition that the bones which lay between the orbits of crossopterygians now form the rear of the skull of tetrapods.

A third similarity between rhipidistians and early amphibians lies in the structure of the teeth. Both rhipidistians and tetrapods appear to have been predators; they were armed with sharp, sturdy teeth. In cross section the teeth of both groups show the **labyrinthodont** pattern of complex infoldings of the walls of the pulp cavity (Figure 9-4). The function of the labyrinthine structure is not clear; possibly such a tooth is particularly strong. In any case, the presence of the same tooth structure in the two groups is considered evidence of their close phylogenetic relationship. Similarly, the position of the internal and external nares is more likely to indicate homology than convergence (see section 9.3.2).

Examination of the earliest amphibians yet discovered, the ichthyostegids from Upper Devonian deposits in Greenland, has provided additional support

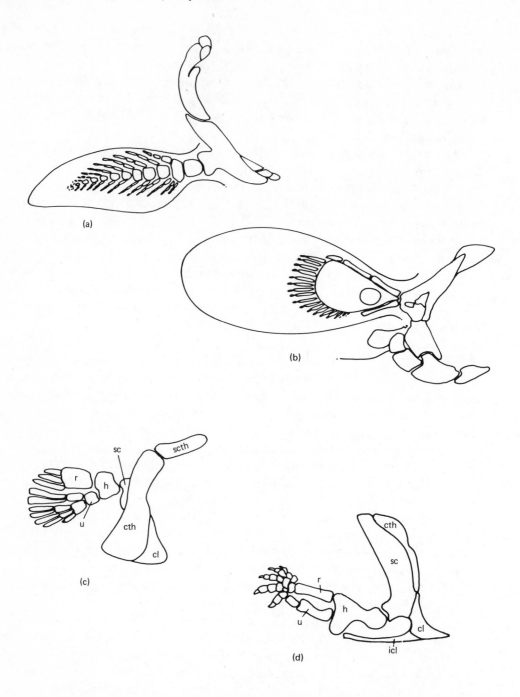

Figure 9-2 Forelimbs and shoulder girdles of a primitive amphibian and several suggested ancestors. (a) lungfish; (b) bichir; (c) crossopterygian; (d) *Ichthyostega*. Only the crossopterygian limb contains elements that can be homologized with the humerus, radius, and ulna of the amphibian. Code: cl = clavicle, cth = cleithrum, icl = interclavicle, h = humerus, r = radius, sc = scapula, scth = supracleithrum, u = ulna. (Modified from Romer and Parsons, Ch. 1[5].)

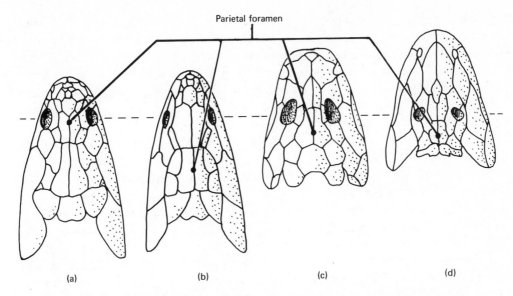

Figure 9-3 Skulls of crossopterygians (a and b) and labyrinthodonts (c and d). In both rhipidistians and labyrinthodonts the facial region of the skull lengthened. The changes can be seen by comparing the proportions of the skull ahead of and behind the eyes. The elements behind the parietal foramen became progressively shorter. (a) *Osteolepis* (Middle Devonian); (b) *Eusthenopteron* (Upper Devonian); (c) *Ichthyostega* (Upper Devonian); (d) *Eryops* (Lower Permian). (Modified from various sources.)

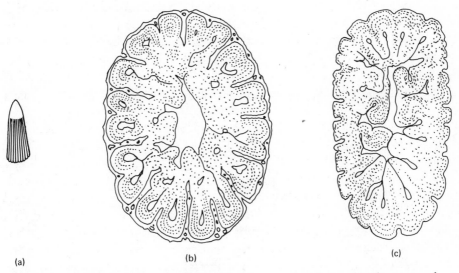

Figure 9-4 One of the morphological features shared by rhipidistians and Paleozoic amphibians was the labyrinthodont tooth. A complex folding of the walls of the pulp cavity produced the labyrinthine structure that identifies this sort of tooth. (a) External view showing striations on outside of tooth. (b) Detailed cross section of a rhipidistian tooth (*Polyplocodus*). (c) Detailed cross section of labyrinthodont amphibian tooth (*Benthosuchus*). (Modified from Byskrow, A.P. [1935] *Acta Zoologica* 16:65–141.)

for the origin of amphibians from crossopterygians. In addition to the homologies already discussed which are visible in advanced Paleozoic amphibians, ichthyostegids have fishlike features that were lost in later amphibians. While the opercular bone of fishes has disappeared in *Ichthyostega*, a remnant of the preopercular persists. The sensory canals of the lateral line system are embedded in the dermal bone of the skull, and show the pattern seen in rhipidistians. Unlike any other known amphibians, *Ichthyostega* had a caudal fin that was supported by fin rays dorsal to the neural spines of the caudal vertebrae. Finally, the discovery of *Ichthyostega* solved a problem that had been a stumbling block for paleontologists who favored rhipidistians as amphibian ancestors. In rhipidistians the braincase was formed in two units which could move in relation to each other, and there was a depression in the floor of the rear part of the braincase that probably accommodated the anterior part of the notochord. Neither feature was known in amphibians until *Ichthyostega* was discovered. In *Ichthyostega* the braincase was a solid structure, as it is in all other amphibians, but the suture between the two parts is visible. Furthermore, the notochordal canal persisted in the posterior part of the braincase.

9.2.2   Possible Polyphyletic Origin of Amphibians

Although the origin of tetrapods has been convincingly traced to the rhipidistian crossopterygians and earlier theories of the evolution of some or all tetrapods from lungfishes have been abandoned, there remain many questions about the exact group (or groups) of rhipidistians that gave rise to tetrapods. Much of the confusion stems from a gap in the fossil record; we have a variety of mid-Devonian rhipidistian fossils that show all the characters needed to be amphibian ancestors, and then the record jumps some 30 million years to the late Devonian and we find amphibians present and already beginning to diversify. The gap comes at a critical time, and until fossil deposits from that period are found and examined by paleontologists, we can only speculate about the probable significance of anatomical differences between late rhipidistians and early amphibians.

Very early in their evolution, amphibians split into several lineages. In early Carboniferous sediments there are amphibians with two very different vertebral structures, the labyrinthodonts and the lepospondyls (Figure 9–1). Proponents of a polyphyletic origin of amphibians have suggested that this very early differentiation, as well as differences among the three groups of living amphibians, may indicate that amphibians evolved independently from more than one group of rhipidistian fish. Erik Jarvik, in particular, strongly supports a diphyletic origin of Recent amphibians. He separates rhipidistians into two orders, the Porolepiformes and the Osteolepiformes, and traces the modern salamanders through the lepospondyls to the Porolepiformes and frogs via labyrinthodont amphibians to the Osteolepiformes. His views are not widely accepted; paleontologists take issue not with his anatomical descriptions but his interpretation of them. The differences between the Osteolepiformes and Porolepiformes are found in the palatal region and branchial skeleton. Osteolepiformes have broader heads than Porolepiformes, and many paleontologists argue that the differences on which Jarvik bases his theories are, in fact, merely reflections of the relative width of the head in the two rhipidistian groups and do not warrant the importance Jarvik has assigned to them.

Opponents of the theory of polyphyletic origin of tetrapods point to the details of the skeleton which show remarkable similarity in all tetrapods. They argue that it is highly unlikely that independent origins would all lead to a limb skeleton consisting of a single bone in the upper limb, paired bones in the lower limb, and five digits. Even more unlikely is the independent acquisition of a pelvic girdle consisting of three paired bones instead of two or four. Speculation based solely on inferences from anatomy cannot provide a definitive answer to the question of monophyletic or polyphyletic origin of tetrapods. That must await discovery of more fossil material from the transitional period.

9.3   EVOLUTION OF TERRESTRIAL VERTEBRATES

Tantalizingly incomplete as the skeletal evidence is, it is massive compared to the information about the ecology of the rhipidistians, protoamphibians (the group or groups of rhipidistians that actually evolved into amphibians), and early amphibians. This is the sort of information we need to assess the selective forces that shaped the evolution of tetrapods. What sorts of lives did the rhipidistians lead? What was their habitat like? What were their major competitors and predators? What were the protoamphibians able to do that rhipidistians could not? There must have been advantages to the morphological differences that slowly accumulated in the evolutionary lineages that led eventually to amphibians, but we cannot evaluate them in the absence of knowledge of the natural history of the animals. Again the gap between the mid-Devonian rhipidistians and the late Devonian ichthyostegids is frustrating. If we knew the sequence in which the differences between rhipidistians and amphibians appeared, we could infer much about the lives of the protoamphibians and the selective value of the differences. Unfortunately, *Ichthyostega* and its relatives are thoroughgoing amphibians when they appear in the fossil record. They show similarities with rhipidistians but there is no record of the sequence of their evolution from the protoamphibians which intervened between the fish grade and the amphibian grade of organization.

9.3.1   How Does a Land Animal Evolve in Water?

Certain inferences can be drawn from the fossil material available. We start from the basis that any animal must function in its habitat. If it does not, it does not leave descendants and its phylogenetic lineage becomes extinct. Evolutionary change occurs because the naturally occurring variation in organisms is subject to selection. Any evolutionary trend is perceived because progressive changes conferred an advantage on the individuals possessing them; it is merely a fortunate coincidence that some features may be preadaptive for a different (in this case a terrestrial) way of life. Our challenge is to interpret the fossil record in the light of these principles and infer what we can about the changing lives of the protoamphibians.

9.3.2 Ecology of Rhipidistians

Rhipidistians were large fish, as much as a meter long, with heavy cylindrical bodies and large teeth (Figure 6-1c and d). They probably either stalked their prey or lay in ambush and made a sudden rush. One can picture a rhipidistian prowling through the dense growth of weeds on the bottom of a Devonian pond. The flexible lobe fins would support them as they waited motionless for prey, and the evolutionary change from the elongate archipterygium of the early crossopterygians to the stouter limb of later ones appears to reflect selection for strengthening the limb for this sort of movement and to bear the weight of the fish in very shallow water.

In most fishes the nasal passages have two pairs of external openings on the head. Water enters the anterior opening, flows across the sensory epithelium where odors are detected, and exits from the posterior opening. The force that produces the water current is derived from the forward motion of the fish. In sedentary fishes forward motion does not produce a strong current, and such species have nasal openings close to the lips where water movement associated with respiration facilitates flow through the nasal passages. In a predatory fish such as a rhipidistian, hunting in dense vegetation, a chemosensory system would be very important, but the slow movement of the fish would not generate sufficient current through the nasal passages. In the evolution of rhipidistians a third pair of openings developed in the roof of the mouth at the front of the palate. These openings are the internal nares, or **choanae**. Choanae would be ideal for a predator like a rhipidistian. By keeping its mouth closed it could draw large volumes of water across the nasal epithelium as its opercula pumped water across the gills. Furthermore, it could dispense with the rhythmic opening and closing of the mouth that is a conspicuous part of the respiration of most fishes. Anything that made it less conspicuous should be advantageous to a fish like a rhipidistian that caught its prey by stealth.

Although there is no direct fossil evidence of lungs in rhipidistians, we can assume they were present because every related group (bichirs, dipnoans, coelacanths, and amphibians) is known to have lungs or the remnants of lungs. In the warm swampy areas rhipidistians inhabited, oxygen levels in the water were probably low much of the time. Rhipidistians presumably breathed atmospheric oxygen either by swimming to the surface and gulping air, or in shallow water, by propping themselves on their pectoral fins to lift their heads to the surface.

9.3.3   Evolution of Amphibian Characters in an Aquatic Habitat

The features we have discussed so far are very general features that apply to most rhipidistians and in some cases to all sarcopterygians. What were the selective forces that could have started some crossopterygians along an evolutionary lineage that culminated in the amphibian structural grade?

One of the most striking differences between rhipidistians and ichthyostegids is the extension of the facial region of the snout in amphibians (Figure 9-3). Examination of a variety of rhipidistians suggests that this trend already existed among the fishes. In living aquatic vertebrates, a long snout is associated with utilization of a sudden sidewards movement of the head to capture food. Crocodilians are the best examples of this technique, although it is observed in other vertebrate classes as well. A longer snout could quite reasonably be advantageous for fishes that depend upon waiting in ambush for prey, because it would increase the volume in front of the fish in which prey could be captured. A fish's morphology, however, prevents it from taking full advantage of a long snout compared to a tetrapod with a snout of the same length. Because the pectoral girdle of a fish is attached to the rear of the skull, a fish does not have a functional neck. A fish cannot turn its head; it must bend or rotate its whole body as does a gar. Thus its capacity for a sudden sidewards snap is less than that of a tetrapod. A neck is produced when the pectoral girdle loses its attachment to the rear of the skull, and the greater the separation of the pectoral girdle and skull, the more flexible the neck becomes.

It seems likely that the development of a tetrapod pectoral girdle, separated from the skull, was an early stage in the evolution of protoamphibians. The ability to strike sidewards to capture prey through a larger arc than could a rhipidistian with a fishlike pectoral girdle may have been an early adaptive advantage leading to exploitation of a new ecological niche by protoamphibians.

Freeing the pectoral girdle from its attachment to the skull would have little effect on its locomotor function. When a protoamphibian was immersed in water the limbs bore little weight and would function equally well whether the girdles were fused to the skull or not. In shallow water the forelimbs bore some of the weight of the body and, through selection for this ability, the muscles holding the pectoral girdle to the trunk probably increased in mass, eventually reaching the ichthyostegid condition.

9.3.4   What Were the Advantages of Terrestrial Activity?

The reasoning outlined above explains an evolutionary process by which selection for features that were advantageous to animals in an aquatic habitat could have produced an organism with enough preadaptations for terrestrial life to have emerged onto land, but they do not explain why it might have taken that step. All of the morphological features that were preadaptive for life on land were originally adaptive in the aquatic habitat in which they evolved. Why did some vertebrates leave the aquatic habitat for which they were so well suited and begin to exploit a radically different habitat on land?

This question has fascinated biologists for a century or more, and there is no shortage of theories. Their ideas, although frequently presented as "the" explanation for the evolution of tetrapods, are seldom mutually exclusive, and can readily be combined.

The classic theory was put forth most forcefully by A. S. Romer. He proposed that the Devonian was a time of seasonal droughts. Shallow ponds that formed during the rainy seasons often evaporated during the dry season, stranding their inhabitants in rapidly shrinking bodies of stagnant water. Fishes trapped in such situations are doomed unless the rainy season begins before the pond is completely dry. Romer suggested that certain rhipidistians had limbs sufficiently well developed to allow them to crawl from these drying ponds and move overland to other, larger bodies of water in which their chances of surviving the dry season were better. Romer believed that millions of years of selection of the rhipidistians best able to escape death by finding their way to permanent water would produce an evolutionary lineage showing increasing agility on land.

The classical theory has been criticized on several grounds. It is questionable whether intermittent catastrophic selection of the type Romer suggested could have the effect he envisioned. Further, a fish that succeeds in moving from a drying pond to one that still holds water has enabled itself to go on leading the life of a fish—that seems a backward way to evolve a terrestrial animal.

The evidence that the Devonian was a time of seasonal drought is debatable. It rests on the widespread occurrence of Devonian "red beds" in which the red color is produced by the mineral hematite (ferric oxide). Hematite is thought to form in warm climates under conditions of alternating moisture and dryness— just the sort of habitat Romer suggested existed in the Devonian. Modern red beds, however, are forming in tropical regions in which moisture is continuous. Furthermore, hematite, once formed, is stable. It can be eroded and redeposited elsewhere without losing its red color. Thus, Devonian red beds may represent erosion and redeposition in Devonian times of hematite formed in some earlier geologic period. Fossil evidence and inferences based on knowledge of continental drift indicate that equatorial areas were moist in the Devonian.

A number of alternative theories have been proposed that have the advantage of stressing positive selective values associated with increasing terrestrial activity by protoamphibians. One emphasizes the contrast between terrestrial and aquatic habitats in the Devonian. The water was swarming with a variety of fishes that had radiated to fill a multiplicity of ecological niches. Active, powerful predators and competitors abounded. In contrast, the land was free of vertebrates. Any protoamphibian that could occupy terrestrial situations had a predator- and competitor-free environment at its disposal. The exploitation of this habitat can be seen as proceeding by gradual steps. Juvenile rhipidistians might have congregated in shallow water, as do juveniles of living fishes, to escape the attention of larger predatory fishes that are restricted to deeper water. At the edge of a pond many morphological and physiological characteristics of terrestrial vertebrates would have been useful to a still-aquatic protoamphibian. Warm water holds little oxygen, and shallow pond edges are likely to be especially warm during the day. Thus, lungs are important to a vertebrate in that habitat, whether it be fish or amphibian. Similarly, legs would have borne the weight of the animal in the absence of water deep enough to float it. In shallow water an air-breathing, upstanding fish would have lifted its head above the water, and the changes in the shape of the lens of the eye associated with the differences in the refractive indices of water and air could have started to occur. These behavioral and morphological features can be found among a number of living fishes, such as the mudskipper, climbing perch, and walking catfish that make extensive excursions out of the water, even climbing trees and capturing food on land.

Starting from fishes that snapped up terrestrial invertebrates that fell into the water, one can envision a gradual progression of increasingly agile forms capable of exploiting the terrestrial habitat for food as well as shelter from aquatic predators. Terrestrial agility might have developed to the stage at which small rhipidistians moved overland from the pond of their origin to other ponds. Many vertebrates include a dispersal stage in their life history, usually in the juvenile period. In this stage individuals spread from the place of their origin to colonize suitable habitats, sometimes long distances from their starting point. This type of behavior is so widespread and adaptive in living vertebrates that we may assume that it occurred in some form in the rhipidistians as well.

It is possible that the main selection for terrestrial life occurred in the juvenile stages of rhipidistians, and that the adults were far more aquatic. Adult rhipidistians were large, clumsy-looking animals. It is hard to imagine an adult rhipidistian a meter long being sufficiently agile to capture a terrestrial invertebrate such as a scorpion scuttling about in its own terrestrial habitat. It is much easier to visualize a 15-cm long juvenile rhipidistian chasing the scorpion under a log and grabbing it. Adult protoamphibians were probably aquatic fish eaters occupying much the same habitat as a rhipidistian fish. They might also have been scavengers, eating dead animals that had floated into water too shallow for fishes to reach.

A different perspective has been suggested by herpetologists, who reason from the biology of living amphibians rather than of fishes. Although the conventional view of amphibians emphasizes their dependence upon water, especially for reproduction, there are a large number of forms that have evolved considerable independence of bodies of water, including the ability to lay eggs on land (Chapter 10). Even some fishes lay eggs that can develop in air, and it is possible that rhipidistians did so as well. This reasoning has led to several suggestions that the evolution of tetrapods is a case in which the egg came before the chicken. That is, the original selective advantages involved terrestrial reproduction, and exploitation of the terrestrial habitat for feeding was secondary.

9.4   STRUCTURAL ADAPTATIONS FOR TERRESTRIAL LIFE

Most of the structural features distinguishing aquatic from terrestrial animals are a consequence of the differences between water and air. The high frictional resistance of water makes a streamlined body form important for all but the most sedentary aquatic animals, but only the speediest terrestrial animals need to cope with significant air resistance. Water buoys an animal up, but air gives no such support. Resistance to gravity is of minor importance in the skeleton of a fish—the demands of locomotion are paramount. In contrast, resistance to the downward pull of gravity is a major factor shaping the skeleton of a terrestrial vertebrate. The sensory world of a terrestrial animal also differs from that of an aquatic animal. Light, sound, odor—all have different qualities in air.

9.4.1   The Structure of Ichthyostega

Although *Ichthyostega* is the earliest tetrapod of which we have a record of the axial skeleton, it is far beyond the transitional stage (Figure 9-5). By the Upper Devonian, ichthyostegids were not only fully at the amphibian structural grade, but they had started to diversify—three genera of ichthyostegids have been distinguished in the Greenland deposits.

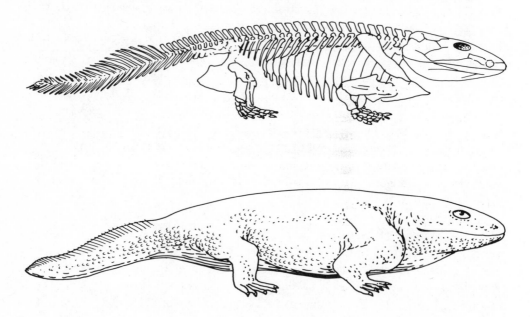

Figure 9-5   Skeleton and restoration of the external appearance of *Ichthyostega*. A number of fishlike characters are evident in *Ichthyostega*, most notably epidermal scales and a tail fin supported by fin rays. Other features are fully amphibian, especially the well-developed ribs, girdles, and limbs. (Modified from E. Jarvik [1955]. The Scientific Monthly 80:141–154.)

When a terrestrial vertebrate stands, its body hangs from the vertebral column which, like the arch of a suspension bridge, supports the weight of the trunk and transmits it to the ground by two sets of vertical supports, the girdles and legs. The vertebral column must be rigid and the girdles and legs sturdy and firmly connected with the vertebral column. Without all of these features, a terrestrial animal cannot stand. In *Ichthyostega* the vertebral structure was still the rhipidistian type, but additional ossification and more articulation between vertebrae lent strength. In later amphibians the vertebral column was further strengthened by additional ossification and expansion of the articular surfaces between adjacent vertebrae (Figure 9-6). Classically paleontologists have considered that rhipidistians and ichthyostegids share a common vertebral structure with primitive amphibians. Recently several paleontologists have suggested that the elements of rhipidistian and ichthyostegid vertebrae are not, after all, homologous with those of tetrapods and that the tetrapod condition arose subsequent to the evolution of ichthyostegids.

In the pectoral girdle of *Ichthyostega* the posttemporal and supracleithral bones, which in fishes attach the cleithrum and clavicle to the skull, have disappeared completely (Figure 9-2). In the ventral midline a new bone, the interclavicle, braced the ventral part of the pectoral girdle, and the scapulocoracoid provided a strong dorsal attachment for the muscles that bound the pectoral girdle to the trunk. In the pelvic region, the two bony ventral plates of fishes had been replaced by three paired bones that form the pelvic girdles of all tetrapods—the pubis, ischium, and ilium. The pubis and ischium of the left and right sides met in the ventral midline in a firm symphysis while the ilium on each side articulated with the sacral rib and vertebra, forming a rigid connection from the spine through the hind legs to the ground.

In a large, heavy-bodied animal like *Ichthyostega* a rib cage is necessary to support the body when the animal lies on the ground. Without that rigid, bony support, the weight of the body would squash the internal organs and collapse the lungs. *Ichthyostega* had a well-developed rib cage that had probably developed at the protoamphibian stage where it would support the body as the animal ventured into shallow water. *Ichthyostega* may have breathed by expanding and contracting the rib cage to change the pressure in the lungs. Although it may have been able to supplement its respiration by pumping air in and out of the buccal cavity as many modern amphibians do, the head seems too small in proportion to the rest of the body for buccal pumping to supply more than a fraction of the oxygen needed.

Two external features were retained from the rhipidistian ancestors—a tail fin supported by fin rays, and a scaly body covering. Reasoning by analogy from living amphibians, biologists long assumed that most Paleozoic amphibians had naked skins, or at most a heavy keratinized layer on the outer surface of the skin. That idea is probably wrong. Although fossilization of soft tissues such as skin is rare, a number of examples have been found among Paleozoic amphibians and these indicate that many, possibly most, had a scaly covering.

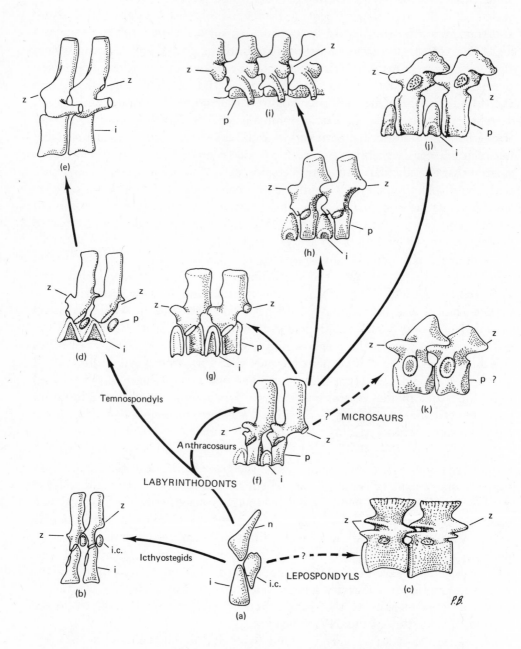

Figure 9.6 Evolution of amphibian vertebral structure. In rhipidistians such as *Osteolepis* (a) the vertebrae contained two central elements in addition to the neural arch. Paired intercalary cartilages extended downward from the dorsal side of the notochord and the intercentrum extended upward from the ventral side. Two major evolutionary changes appeared in amphibians that were associated with the changed function of the vertebral column in a terrestrial animal: 1. The neural arches increased in size and developed articulating surfaces (zygapophyses) that transmitted gravitational forces from vertebra to vertebra.

9.4.2 The Phylogenetic Position of Ichthyostegids

Although ichthyostegids show numerous primitive features that place them closer to the fish–amphibian transition than any other known fossils, they have several specialized features that have led paleontologists to conclude that they were not on the main line of amphibian evolution, but represent an offshoot from the first amphibian stocks (Figure 9-1). For example, the intertemporal bone has been lost in ichthyostegids although it is present in many later labyrinthodonts. Thus, we are presently forced to base our inferences about the first amphibians and their origins on animals which were not only long separated in time from the transition, but which had branched off on their own.

---

2. One of the central elements enlarged. The increased robustness and complexity of articulation of the neural arch had developed in *Ichthyostega* (b), and the intercentrum was larger than the intercalary cartilages. Lepospondyls like *Urocordylus* (c) had complex zygapophyses and a single central element that cannot be homologized with either the intercentrum or intercalary cartilage of rhipidistians. The pleurocentrum of the temnospondyl and anthracosaur labyrinthodonts is usually considered homologous with the rhipidistian intercalary cartilage. In the temnospondyl lineage (primarily aquatic animals) the pleurocentrum became increasingly smaller through the rhacitomous condition seen in *Eryops* (d) to the stereospondylous stage in which the pleurocentrum had entirely disappeared (e, *Mastodonsaurus*). In the anthracosaur lineage of largely terrestrial labyrinthodonts there was an early reduction in the intercentrum (f, *Proterogyrinus*). This trend was reversed in one offshoot from the anthracosaurs, the embolomeres which retained a dispondylous condition with pleurocentrum plus intercentrum (g, *Eogyrinus*). The embolomeres were an aquatic lineage of anthracosaurs and the persistently dispondylous condition may have been an adaptation that facilitated swimming by lateral undulation of the body. Terrestrial anthracosaurs showed a continued reduction of the intercentrum, and in advanced forms like *Gephyrostegus* (h) the ossified portion of the intercentrum was very small. In life there was probably a cartilaginous extension of the intercentrum (indicated by a dotted line). This cartilaginous material would not have been preserved during fossilization, but it seems likely that there was some material present in life to fill the voids that would otherwise have existed between the large pleurocentra. In primitive reptiles, which were derived from advanced anthracosaur labyrinthodonts, the pleurocentrum was the sole central element (i, an unnamed form, probably an eosuchian). A similar reduction of the intercentrum occurred in the reptile-like amphibians, the batrachosaurs, in a parallel adaptation to terrestrial activity (j, *Seymouria*). Microsaurs do not fit readily into this scheme. The single central element of microsaurs (k, *Pantylus*) may be homologous with the pleurocentrum of labyrinthodonts. Alternatively, microsaurs have traditionally been grouped with the lepospondyls. Code: i = intercentrum, i.c. = intercalary cartilage, n = neural arch, p = pleurocentrum, z = zygapophysis. (Sources: (a)–(d), (f)–(h), (j) and (k) are modified from Panchen [3]; (b) is Panchen's revision of Jarvik's reconstruction of the arrangement of central elements in *Ichthyostega*. (e), from Romer [1967]; (j), from R. Carroll and D. Baird [1972] Bulletin of the Museum of Comparative Zoology 143:321–364.)

9.5   RADIATION OF PALEOZOIC AMPHIBIANS

For 60 million years, from the mid-Devonian to the mid-Carboniferous, amphibians were the only terrestrial vertebrates. For an additional 70 million years, well into the Permian, only primitive reptiles competed with them for terrestrial niches. Thus, for 130 million years—six times as long as it has taken humans to evolve from protohominid to the present stage—amphibians diversified into an unimaginable variety of niches. It is no wonder that the fossil record is both prolific and confusing. Parallel and convergent evolution were widespread, and it is hard to separate phylogenetic relationship from convergent evolution to a similar way of life. In the following discussion, emphasis will be placed on morphological adaptations associated with different ecologies rather than on details of phylogeny.

9.5.1 Phylogeny of Amphibians

A degree of order is provided by examination of the fine details of the structure of the vertebral column and skull. Recent discoveries and reanalysis of old material have led to substantial changes in proposed phylogenetic relationships. Further work will undoubtedly lead to further changes, and the scheme suggested in Figure 9–6 is no more than a working hypothesis.

Many rhipidistians, presumably including the ancestors of tetrapods, had vertebrae with two centers of ossification. One center was dorsal to the notochord and extended upward. Ichthyostegalians and some primitive labyrinthodonts appear to have had a similar vertebral structure, although some paleontologists doubt that the dorsal elements in fishes and amphibians are truly homologous. In advanced amphibians vertebral structure was modified to provide more rigid support. These modifications involved both the neural arches and the centra. In all lines including the ichthyostegalians the arches enlarged and the facets that articulate with adjacent neural arches, the prezygapophysis and the postzygapophysis, increased their area of contact. These changes allowed the spinal column to transmit forces from vertebra to vertebra and thus distribute the weight of the internal organs over the entire length of the vertebral column. The changes in the centra also reflect selection for increased rigidity; a single centrum is more rigid than the intercentrum-plus-pleurocentrum arrangement of the primitive rhachitomous condition with two centra, and increased ossification provides additional support. From an intercentrum-plus-pleurocentrum starting point, an evolutionary trend toward a single central element could lead either to an intercentrum alone or a pleurocentrum alone, and in fact both trends can be identified. This dichotomy provides the basis for the division of labyrinthodonts into the Orders Anthracosauria and Temnospondyli. In the temnospondyl line the rhachitomous condition was retained, initially with little modification. The intercentrum remained the dominant element, but both intercentrum and pleurocentrum

became increasingly ossified (Figure 9–6). In the **stereospondylous** condition
that arose from the rhachitomes, the pleurocentra had vanished and the intercentrum became the entire centrum. The opposite occurred in the other labyrinthodont line, the anthracosaurs. The intercentrum became progressively
more reduced in size until in primitive reptiles it was represented by only a
small nubbin of bone. In advanced reptiles, birds, and mammals the pleurocentrum makes up the entire centrum. Many paleontologists believe that the
single centrum of advanced microsaurs is homologous with the pleurocentrum
of reptiles, birds, and mammals. There is at present no evidence of the origin
of microsaurs. The **embolomeres** are an aberrant group of anthracosaurs in
which the pleurocentrum and intercentrum both form complete rings around
the notochord. The embolomeres appear to have been a secondarily aquatic
group of anthracosaurs, and the flexibility of a vertebral column with this sort
of dispondylous condition may be an adaptation for anguilliform locomotion.

Unfortunately, it is not possible to fit the other subclasses of Amphibia into
this scheme with any degree of confidence. The lepospondyls show a completely different vertebral arrangement, in which only one central element
ossifies, but it is impossible to homologize it with an intercentrum or pleurocentrum. In living amphibians the adult centra are also simple cylinders that
cannot be clearly homologized with either the intercentrum or pleurocentrum
of early vertebrates.

9.5.2   Order Temnospondyli

The temnospondyls were the most successful and abundant labyrinthodonts.
Two suborders are recognized, the Rhachitomi and the Stereospondyli, nomi-
nally on the basis of vertebral structure, but in practice a variety of characters
are used. Many stereospondyls do not, in fact, have fully stereospondylous
vertebrae. There were many side branches with parallel and convergent evolution and classification at the family level is in a state of flux.

The main line of temnospondyl evolution consisted of stocky, short-legged,
heavy-bodied, large-headed semiaquatic predators. In rhachitomes the body was
relatively rounded but in stereospondyls there was a progressive tendency
toward flattening the body. A few groups showed lengthened legs, but in most
of the later temnospondyls the legs were reduced, indicating increasing restriction to an aquatic habitat (Figure 9–7).

D. M. S. Watson has traced a combination of evolutionary changes from early
temnospondyls such as *Eryops* to the very specialized stereospondylous aquatic
forms. (1) The skull became progressively flatter. (2) The plane of the articulation of the lower jaw with the skull moved into the vertical plane of the occipital condyles (Figure 9–8). (3) A retroarticular process developed on the lower
jaw behind the articulation with the quadrate. (4) In the palate interpterygoid
vacuities enlarged, and ossification of the nasal capsules, orbital, and otic regions
was reduced. Watson interpreted these changes as structural modifications that

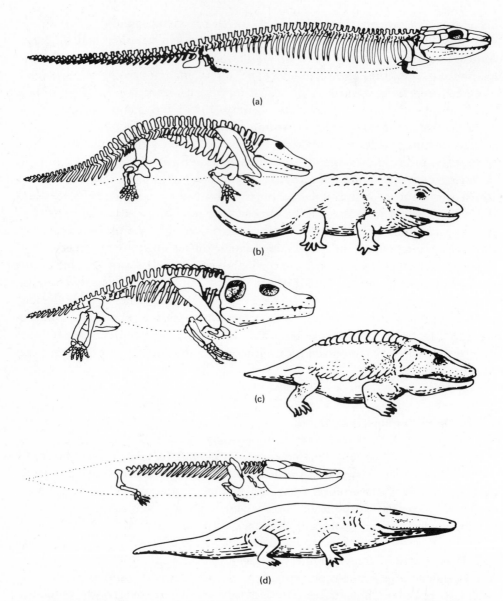

Figure 9-7 Temnospondyl labyrinthodonts. (a) *Pholidogaster*, a primitive temnospondyl from the early Carboniferous. (b) *Eryops*, a Permian rhachitome with sturdy legs and a rounded body. (c) The Permian rhachitome *Cacops* was among the most terrestrial of the temnospondyls. The skeleton was sturdy and well ossified. Dermal armor protected the back, and the head was large in proportion to the body. (d) *Cyclotosaurus*, a stereospondyl from the upper Triassic, was an aquatic form. It had small legs and a dorsoventrally flattened body. It was probably incapable of lifting its body from the substrate without the support of water. (Modified from the following sources: (a) from Romer, Ch. 5[7] (b) from Gregory in Romer, and from E. A. Colbert [1969] *Evolution of the Vertebrates*. Wiley and Sons, New York. (c) from Williston in Romer, and from Raymond in Stahl, Ch. 2[9]; (d) from C. L. Fenton and M. A. Fenton [1958] *The Fossil Book*. Doubleday, New York.)

accompanied the flattening of the body which, in turn, was associated with an increasingly aquatic life.

A vertebrate normally opens its mouth by dropping the lower jaw. In an animal like *Eryops* that stood sturdily on its legs, this mechanism is satisfactory. The mouth can be opened until the bottom of the jaw touches the ground. If a wider gap is needed, the head can be lifted slightly by bending the neck upward. An animal like *Cyclotosaurus* that normally rests with its ventral surface on the ground has a problem. Its lower jaw is already in contact with the ground, and the only way it can open its mouth is by bending the neck to raise the skull. Doing that not only puts a strain on the cervical vertebrae, but it also pushes the lower jaw forward, scraping against the ground. The closer the rotational axes of the jaw articulation and occipital condyles are, the less will be the strain on the cervical vertebrae, and the less the movement of the lower jaw along the ground. The neck muscles are at a considerable mechanical disadvantage in trying to lift the skull in an animal like *Eryops* because they run nearly horizontally from the rear of the skull to the vertebral column and there is little vertical component to the force they produce on contraction. In the later temnospondyls, a retroarticular process developed on the lower jaw that allowed muscles to pass from the process nearly vertically to insert on the dermal skull roof (Figure 9-8d-f). These muscles had a large vertical component to their force and, when the lower jaw rested on the ground, their contraction lifted the skull (Figure 9-8f). All flat-headed labyrinthodonts have a retroarticular process, supporting Watson's view that it is a necessary corollary to the flattening of the skull. The same interpretation probably applies to the development of interpterygoid vacuities and reduction of ossification. As the skull became flatter, space for the large eyes and their associated muscles, as well as for the jaw muscles, was reduced. Additional space was provided by the development of vacuities in the palate. This loss of bone, and the reduction of ossification in some areas, also reduced the weight of the skull, making it easier to raise in order to open the mouth.

The most generalized temnospondyls appear to have been semiaquatic predators that probably fed on other small amphibians. *Eryops* of the late Paleozoic was about 1.6 meters long. Its head comprised one-fifth of its total length. Its contemporary *Cacops* (Figure 9-7c) was only about 40 cm long, but its head was nearly one-third of the total length of the body; thus *Cacops* could ingest considerably larger prey in proportion to its own body size than could *Eryops*. *Cacops'* skeleton indicates that it was quite terrestrial; the limb girdles are sturdy and the legs and toes are long in proportion to the body. Rhachitomes like *Cacops* appear to represent the height of temnospondyl terrestrial development. The capitosaurs, stereospondylous Triassic labyrinthodonts, were large, flat-headed amphibians. In some forms the skull alone was more than a meter long. These impressive labyrinthodonts must have represented the top of the food chain, feeding on their smaller relatives as well as any reptiles incautious enough to approach them.

A group of small labyrinthodont amphibians had long puzzled paleontologists.

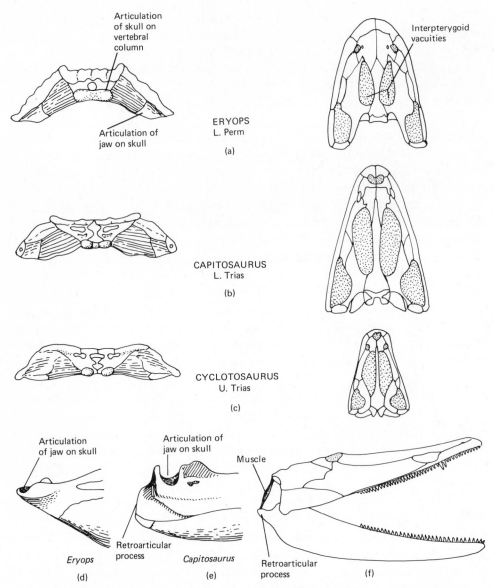

Figure 9-8 Progressive changes in the skulls of temnospondyl labyrinthodonts (a–c). Rear views of skulls on left and palatal views on right. As temnospondyls became increasingly specialized for aquatic life their skulls became flatter and the palatal vacuities proportionally larger. The flattening of the skull brought the articulation of the lower jaw with the skull into the same horizontal plane as the articulation of the skull with the vertebral column. The coplanarity of these articulations simplified the jaw opening mechanism. (d) In *Eryops* the articulation with the skull was at the posterior tip of the lower jaw. (e) In *Capitosaurus* and other specialized forms a retroarticular process had developed, extending posterior to the jaw articulation. (f) This process provided an attachment site for muscles that could exert a vertical force to lift the skull. (Modified from D.M.S. Watson [1951] *Paleontology and Modern Biology*. Yale University Press, New Haven.)

These appeared to be similar to rhachitomes in many respects, but the bones were poorly ossified and the skulls were short and broad. So difficult to classify were these animals that a separate order, Phyllospondyli, was created for them. Detailed study indicated the presence of gills and branchial arches in some specimens and the group is frequently referred to as "branchiosaurs" (Figure 9-9a, b). When large series had been assembled, it became obvious that the larger a branchiosaur was, the more it looked like the labyrinthodonts found in the

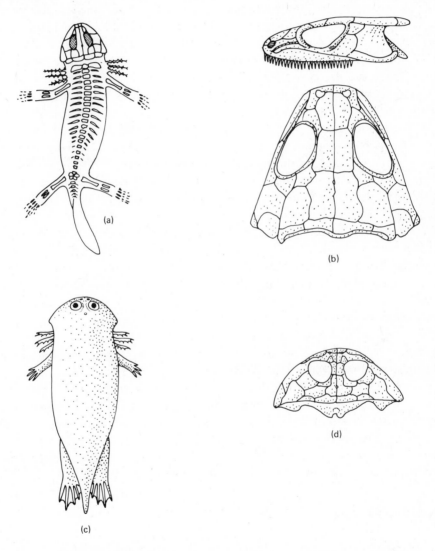

(a)

(b)

(c)

(d)

Figure 9-9   A "branchiosaur" (a, b) shows a number of characters that mark it as a larval labyrinthodont. The most prominent of these are external gills, the broad, short skull, and incomplete ossification of the skeleton. Plagiosaurs (*Gerrothorax*, c, d) show a similar combination of characters, suggesting that they were derived by paedeogenesis from labyrinthodonts. ((a) modified from Bullman and Whittard, (b) modified from Watson, (c) modified from Nilsson, (d) modified from Romer. All in Piveteau [4].)

same deposits, and from that realization it was but a step to recognize branchiosaurs as larval labyrinthodonts. Thus, the "Order Phyllospondyli" is a figment of paleontologists' imaginations, and can be erased from phylogenetic diagrams.

Some of the most interesting ecological specializations among temnospondyls are found in forms that returned secondarily to a completely aquatic way of life. The Plagiosauridae (several genera which are probably not as closely related as their inclusion in one family suggests) and the Brachyopidae are bizarre, flat-headed, short-snouted amphibians. *Gerrothorax*, a late Triassic plagiosaurid, was about one meter long (Figure 9–9c). The body was very flat and armored dorsally and ventrally. The dorsal position of the eyes suggests that *Gerrothorax* lay in ambush on the bottom of the pond until a fish or another amphibian swam close enough to be seized in a sudden rush. The broad head and wide mouth would enable *Gerrothorax* to attack large prey. The broad, short skull (Figure 9–9d) and retention of external gills in at least some forms indicate that the brachyopids and plagiosaurids evolved by paedogenesis from the branchiosaur larvae of temnospondyls. Because brachyopids and plagiosaurids were separated in both distance and time, it is likely that they represent at least two separate derivations and quite possibly genera within the two families were independently derived.

The trematosaurids are among the most remarkable temnospondyls. As early as the Permian some rhachitomes reversed the trend to a broad, flat skull and evolved the elongate snout characteristic of specialized fish eaters. Forms like *Archegosaurus* (Figure 9–10a) may later have given rise to even more specialized fish eaters including *Aphaneramma* (Figure 9–10b) or the later forms may represent a convergent evolutionary lineage. The trematosaurids are found in lower Triassic marine beds. Salt water is an unusual habitat for amphibians. The scaly covering which trematosaurs probably retained from rhipidistian ancestors could have helped to protect the adults from the osmotic stress of either fresh or salt water, but what of the presumed larval stages? Were trematosaurs viviparous like a few living amphibians, or could their larvae tolerate at least moderate salinities as do tadpoles of a few frogs and toads? Furthermore, if trematosaurs were able to invade the sea, why weren't other amphibians? Romer has suggested that the movement of actinopterygian fishes into the sea was followed by these highly specialized piscivorous amphibians.

9.5.3 Anthracosaurs

The anthracosaur branch of the labyrinthodonts contains fewer species and less morphological and ecological diversity than the temnospondyl branch. Ultimately the anthracosaurs gave rise, via the reptiles, to both birds and mammals, thereby earning a type of immortality, but the anthracosaurs themselves did not persist past the Permian.

There are a number of minor features that distinguish anthracosaur labyrinthodonts from temnospondyls, but the major character is vertebral structure.

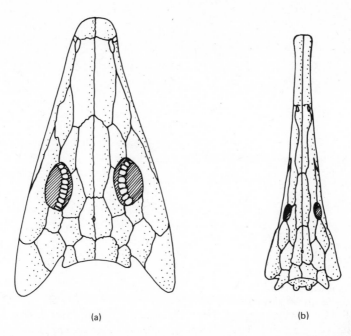

(a)   (b)

Figure 9-10   Specialized, long-snouted fish-eating rhachitomes like *Archego-saurus* (a) evolved in the early Permian. These may have given rise to the early Triassic trematosaurs like *Aphaneramma* (b). Trematosaurs, which were marine, had very elongate snouts, similar to those of living crocodilians. (Part (a) modified from Hofker, part (b) modified from Piveteau and Dechaseauz, both in Piveteau [4].)

The anthracosaur line is characterized by increasing size of the pleurocentrum. Anthracosaurs appear to have branched in three directions: (1) In the embolomeres the intercentrum enlarged and eventually formed a second ring, so that each vertebra included two centra. (2) In seymouriamorphs the intercentrum was reduced in size, and (3) a similar reduction occurred independently and even more rapidly in the line that evolved into reptiles. The skull retained its primitive deep, relatively narrow proportions in anthracosaurs. In many anthracosaurs the body was elongate and the legs relatively small. Others, like *Gephyrostegus* (Figure 9-11), had sturdy legs and were probably terrestrial. Toward the end of the Carboniferous several groups of terrestrial animals with still greater adaptations for terrestrial life evolved from the anthracosaurs. The precise relationship of these groups of animals is in doubt. It is clear that many of the similarities between the groups represent parallel and convergent evolution. These groups evolved from the anthracosaurs include the true reptiles, of which the romeriid captorhinomorphs of the Pennsylvanian are the earliest known examples. In addition to the true reptiles there are several groups of late Carboniferous and early Permian animals that were very reptilelike in appearance, but which were probably amphibians. (The distinction between reptiles and amphibians is based on the mode of reproduction, and we have no direct evidence of how these animals reproduced. The evolution of reptiles is

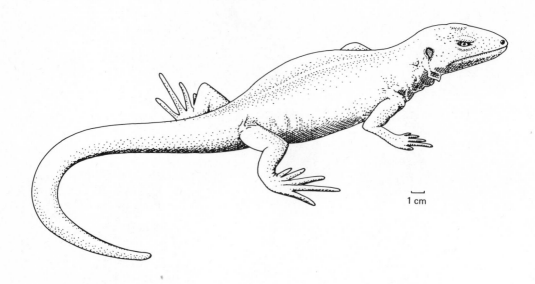

1 cm

Figure 9-11   *Gephyrostegus*, a late Carboniferous anthracosaur, had sturdy legs supporting a deep body. It appears to have been terrestrial, and may be close to the lineage from which reptiles arose. (Modified from R. L. Carroll [1972] *Encyclopedia of Paleoherpetology*. Part 5B, pp 1–19.)

discussed in Chapter 11.) These reptilelike animals can be lumped in a group called "batrachosaurs." It must be understood that this is a catchall term that is used for convenience. It is not presented as a formal taxonomic designation, and it does not imply a close relationship among the evolutionary lineages included. Rather it describes a structural grade of tetrapod, well adapted for terrestrial life as an adult in most cases, but probably retaining aquatic reproduction and larvae. The principal batrachosaurs are the limnoscelids, solenodonsaurids, seymouriamorphs, and possibly the diadectids (Figure 9–12). Limnoscelids and solenodonsaurids were about a meter in length. They appear to have been terrestrial or semiaquatic carnivores. The seymouriamorphs include *Seymouria*, a terrestrial predator similar in size to solenodonsaurids, and other forms that were more amphibianlike. *Discosauriscus*, a seymouriamorph from the early Permian of Czechoslovakia, had the flattened skull and reduced ossification characteristic of aquatic amphibians. "Branchiosaur" larvae associated with *Discosauriscus* suggest that this seymouriamorph, at least, retained aquatic reproduction.

Some paleontologists consider *Diadectes*, a large Permian form, an advanced anthracosaur amphibian and others believe that it is a reptile. *Diadectes* was unusual in having laterally expanded cheek teeth and chisel-shaped front teeth which suggest that it was an herbivore. If so, it was possibly the first terrestrial herbivorous tetrapod. Its contemporaries had sharp-pointed teeth and were apparently carnivorous.

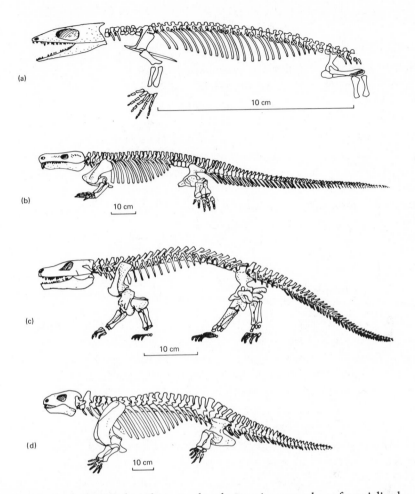

Figure 9-12   In the late Carboniferous and early Permian a number of specialized anthracosaur amphibians shared the terrestrial habitat with the early reptiles. None of these lineages persisted past the mid-Permian. Although the body forms of these anthracosaurs resemble those of some reptiles the similarities appear to be the results of parallel and convergent evolution. These anthracosaurs do not appear to be closely related to reptiles, and they are probably not very closely related to each other. For convenience these anthracosaurs can be called "batrachosaurs", but that term does not imply phylogenetic similarity. (a) *Solendonsaurus*, a late Carboniferous form. (b) *Limnoscelis*, from the late Carboniferous and early Permian. (c) *Seymouria*, a mid-Permian representative of the anthracosaur suborder Seymouriamorpha. Although *Seymouria* was terrestrial, other seymouriamorphs appear to have been aquatic. (d) *Diadectes* was an herbivorous diadectid. This Permo-Carboniferous group is considered reptilian by some paleontologists and amphibian by others. *Diadectes* is included here to illustrate the differences in body form among terrestrial tetrapods that had appeared by the late Carboniferous. The pattern of relatively active carnivores and ponderous, sedentary herbivores will be repeated in reptiles and in mammals. (Modified from various sources in R. L. Carroll [1969] pp 1–44 in *Biology of the Reptilia*, vol. 1, edited by C. Gans, A. d'A. Bellairs, and T. S. Parsons. Academic Press, New York.)

9.5.4 Microsaurs

The microsaurs are often included among the lepospondyls, but a recent reanalysis of phylogenetic relationships among primitive tetrapods by A. L. Panchen suggests that they should be considered a separate group. The name Microsauria (= small reptile) is inappropriate, but this confusion with reptiles is understandable. Some microsaurs had moderately elongate bodies with small but functional limbs and deep, sturdy skulls (Figure 9–13). Like the reptiles they were probably small, active terrestrial predators that specialized in feeding on invertebrates, and some were not greatly unlike modern lizards and sala-manders in body proportions. In addition to these generalized predators, the microsaurs included a specialized elongate form, *Lysorophus*, with very small limbs and about 70 trunk vertebrae.

9.5.5 Lepospondyls

The lepospondyls are an assemblage of amphibians that appeared in the fossil record in the mid-Carboniferous without any indication of their evolutionary origin, flourished modestly in the late Carboniferous, and had disappeared by the end of the Permian without leaving any descendants that can be unequivo-cally traced. In an age of giant amphibians, lepospondyls were small—most were less than 30 cm long and the largest only about one meter. Two orders are grouped in this subclass without very convincing evidence of a common evolutionary origin. The feature they share is a lepospondylous vertebra in which the centrum ossifies around the notochord and is fused to the neural and hemal arches. Both orders contain highly specialized forms, and in each there is a lineage of elongate amphibians with the legs reduced or entirely absent.

The Aistopoda is the smallest of the orders, containing only two or three genera of legless amphibians found in Mississippian and Pennsylvanian deposits. Aistopods range from a few centimeters to about one meter in length. Early re-constructions showed them with external gills, but recent study has indicated that the structures thought to represent gills were actually scales. Aistopods are usually likened to snakes, but in appearance and ecology they probably were more like modern legless lizards—elongate animals that foraged in leaf litter and under rocks and logs. The skulls were moderately sturdy and pointed ante-riorly (Figure 9–14a, b), but were not nearly as modified for digging in the soil as is the case in living fossorial amphibians and reptiles.

The Nectridia are a second order of lepospondyls. Two lines of specializa-tion can be discerned among nectridians. Both types were elongate, but the elongation occurred mainly in the tail (unlike aistopods in which both body and tail lengthened). In the Urocordylidae the skull was sharply pointed, and the legs so small it seems certain that the animals were fully aquatic (Figure 9–14c). The other lineage, represented by the family Keraterpetontidae, showed

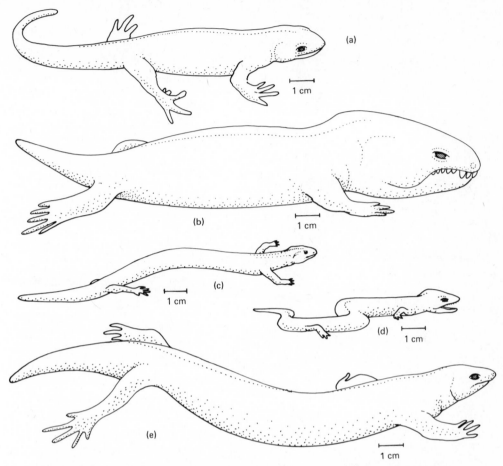

Figure 9-13 Microsaurs were a diverse group of amphibians of unknown phylogenetic affinity. Many were terrestrial whereas others were aquatic or burrowing. (a) *Tuditanus*, (b) *Pantylus*, (c) *Goniorhynchus*, (d) *Cardiocephalus*, (e) *Pelodosotis*. (Modified from R. L. Carroll and P. Gaskill [1978] Memoirs of the American Philosophical Society, 126:1–211.)

extreme dorsoventral flattening of the skull to form crescentic structures often called horns that, at their greatest extent, were more than five times the width of the anterior part of the skull (Figure 9–14d). The ontogenetic development of the skull can be traced from the normal-looking oval skull of small individuals to fully developed horns in adults. The mouth was small, and the eyes far forward and directed upward, indicating that the horned nectridians lived on the bottoms of bodies of water. These peculiar animals have aroused considerable speculation, but little in the way of satisfactory suggestions about the function of the horns. Were they covered in life with a flap of skin that extended back to the shoulder? The fossil evidence does not show such a skin flap, but one would not necessarily expect it to be preserved. If there was a flap, was it a fin to help the animal swim, or a highly vascularized tissue used to extract oxygen from water?

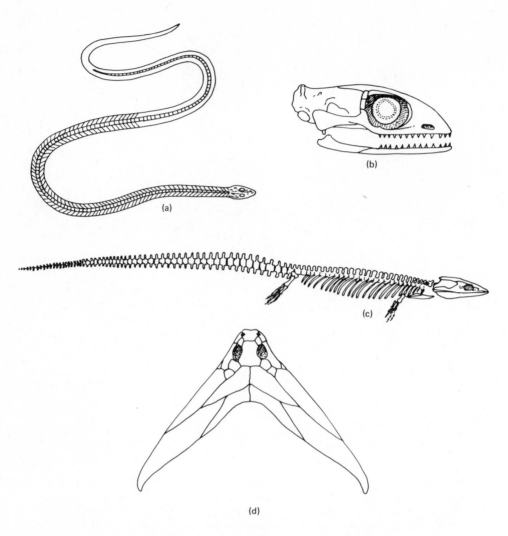

(b)

(a)

(c)

(d)

Figure 9-14 Lepospondyls showed a number of body forms. The aistopods were legless. Only three genera are known; all were small. *Ophiderpeton* (a) was about 75 cm long. The skulls of some aistopods such as *Phlegethontia* (b) were well ossified in the facial region, perhaps as an adaptation to probing in the ground litter. The jaw suspension may have had considerable flexibility, allowing aistopods to swallow large prey. The nectridians were more varied than the aistopods. One type, illustrated by *Urocordylus* (c) had elongated bodies and small legs, laterally flattened tails and sharply pointed snouts. The "horned" nectridians were apparently more sedentary (d) *Diploceraspis* . They show a combination of morphological features similar to those seen in the specialized aquatic stereospondyls. The body and head were dorsoventrally compressed, the legs small, and ossification of the skeleton was reduced. During the growth of an individual the tabular bones extended backward to form greatly elongated "horns". These may have counterbalanced the weight of the skull when the mouth was opened. (Part (a) modified from Fritsch, part (b) modified from Gregory and Turnbull, part (c) modified from Steen, all in Pivateau [4]. Part (d) modified from J. R. Beerbower [1963] Bull. Mus. Comp. Zool. 130:31–108.)

Was it merely a device to conceal the animal as it rested on the muddy bottom of a pond by distorting its outline? On the other hand, if there was no skin flap, what did the horns do? Do the small serrations on the outer tips of the horns in *Diploceraspis* suggest that the horns were weapons used in defense, and if so, why were the horns of other nectridians not serrated? One interesting suggestion is that the horns were a counterbalance that helped these flat-headed animals to lift the skull in order to open the mouth. By this interpretation the horns, like the retroarticular process of temnospondyls, are a solution to the mechanical problem posed by the lack of mechanical advantage of the neck muscles.

9.6 THE DECLINE OF PALEOZOIC AMPHIBIANS

With the close of the Permian, the diversity of amphibians, which had begun to decline in the late Carboniferous, plummeted drastically. It is tempting to suggest that competition with reptiles, which were radiating into ecological niches occupied by amphibians, was the cause of the reduction in species diversity among amphibians. To a certain extent this is probably a reasonable explanation, but it cannot be the only answer, for some amphibians disappeared without being replaced by reptiles. Inferences of competitive replacement of one animal by another are easy to make, but hard to prove. We have too little information about the ecology of Paleozoic amphibians and reptiles to provide a factual basis for speculations about their interactions.

Geological evidence has been interpreted as indicating that the Permian was a time of increasing aridity. Possibly the reduction of lowland swampy areas and the appearance of seasonal fluctuations in rainfall spelled the end of some of the specialized aquatic and semiaquatic amphibians.

References

[1]   Carroll, Robert L. 1977. Patterns of amphibian evolution: an extended example of the incompleteness of the fossil record. In *Patterns of Evolution*, edited by A. Hallam. Elsevier, Amsterdam.

[2]   Panchen, A. L. 1975. A new genus and species of anthracosaur amphibian from the lower Carboniferous of Scotland and the status of *Pholidogaster pisciformis* Huxley. *Proc. Roy. Soc. London B* 269:581–637. Concludes that *Pholidogaster* is a temnospondyl, not an anthracosaur, and therefore previous schemes of labyrinthodont vertebral evolution are in error.

[3]   Panchen, A. L. 1977. The origin and early evolution of tetrapod vertebrae. In *Problems in Vertebrate Evolution,* edited by S. Mahala Andrews, R. S. Miles, and A. D. Walker. Academic Press, London. Reviews theories of the evolution of vertebral form from rhipidistians through the Paleozoic tetrapods.

[4]   Piveteau, J. 1952–1968. *Traité de Paléontologie.* 7 vols. Mason et Cie, Paris. Four volumes of this extensive well illustrated work are important sources in the study of vertebrate evolution.

[5]   Romer, A. S. 1972. Skin breathing—primary or secondary? *Respiration Physiology* 14:183–192. Discusses the implications of the scaly covering of Paleozoic amphibians.

[6]   Schaeffer, B. 1965. The rhipidistian–amphibian transition. *American Zoologist* 5:267–276. This and the following reference provide two contrasting views of the selective forces responsible for the evolution of tetrapods.

[7]   Szarski, H. 1962. The origin of the amphibia. *The Quarterly Review of Biology* 37(3):189–241.

[8]   Thomson, K. S. and K. H. Bossy. 1970. Adaptive trends and relationships in early amphibia. *Forma et Functio* 3:7–31.

10

Modern Amphibians

Synopsis: Three orders of living amphibians are placed in the Subclass Lissamphibia. There is little agreement about the evolutionary origin of the three orders, and they may well have had independent origins. The three orders of living amphibians share a number of morphological characters, and the similarities appear to justify grouping them as one subclass despite the lack of firm information about their phylogenetic relationships.

Beyond their underlying similarities, lissamphibians have evolved very diverse morphologies and reproductive methods. Locomotor adaptations distinguish the three living orders. Salamanders (order Urodela) usually have sturdy legs that are used with a lateral undulation of the body in walking. Aquatic salamanders use lateral undulations of the body and tail to swim, and some specialized aquatic species are elongate and have very small legs. Frogs and toads (Order Anura) are characterized by specializations of the pelvis and hindlimbs that permit both hind legs simultaneously to deliver a powerful thrust. This movement is used for both jumping and swimming. Many anurans walk quadrapedally when they move slowly and are agile climbers. The apodans (Order Apoda) are so little known that they lack a common name. They are legless tropical amphibians that construct tunnels through moist soil.

Fertilization is internal in all but the most primitive salamanders, but most frogs rely upon external fertilization of the eggs. Apodans have internal fertilization. Many species of amphibians have aquatic larvae, but direct development that omits the larval stage is common. Direct development is often combined with some parental care of the eggs. A few urodeles and anurans retain the fertilized eggs in the oviducts and give birth to living young. Viviparity is common among apodans and the embryos obtain nourishment from the mother's oviduct as well as from the egg yolks.

Amphibian skin contains an abundance of glands, including glands that produce toxic substances, and many amphibians rely on these substances to protect them from predators. Some of the most toxic species are brightly colored and appear to advertise their noxious qualities to deter predators. Several examples of mimicry have been demonstrated in experimental studies of amphibians.

The living amphibians differ from Paleozoic amphibians in a number of features, and the lives they lead are probably equally different from those of labyrinthodonts. The small body size of most lissamphibians and the permeability of their skins to water and gases are two characters that profoundly affect their ecology and behavior. The rate at which water evaporates from an amphibian's skin limits its activity in time and space. Most amphibians require moist microhabitats in which they can regain the water they lose by evaporation during activity. Because they are small, such microhabitats are readily available to them under rocks and logs or in the leaf litter on a forest floor. At the other end of the scale, the versatility of amphibian water relations is illustrated by animals like spadefoot toads and tiger salamanders that occupy deserts by using the permeability of their skins to extract water from soil. Clearly living amphibians are specialized vertebrates. Consequently they are not appropriate models for generalizations about the evolution of tetrapods in the Paleozoic.

Table 10-1 Taxonomic arrangement of living amphibians. The major families mentioned in the text are included.

Class Amphibia
  Subclass Lissamphibia
    Order Urodela (tailed amphibians: salamanders)
      Family Cryptobranchidae (hellbenders)
           Ambystomatidae (mole salamanders)
           Salamandridae (newts)
           Proteidae (mudpuppies and olms)
           Plethodontidae (lungless salamanders)
    Order Anura (tailless amphibians: frogs and toads)
      Family Pipidae (specialized aquatic frogs in South America and Africa)
           Ranidae (true frogs)
           Bufonidae (true toads)
           Pelobatidae (spadefoot toads)
           Hylidae (treefrogs)
           Rhacophoridae (Asian treefrogs)
           Leptodactylidae (terrestrial, aquatic, or arboreal frogs)
      Families Atelopidae and Dendrobatidae (arrow-poison frogs)
    Order Apoda (legless amphibians: caecilians)

10.1 INTRODUCTION

Living amphibians are placed in the Subclass Lissamphibia, which includes three diverse orders, each specialized for a particular mode of locomotion. These locomotor specializations have presented the opportunity for a wide range of ecological specializations. Underlying this diversity, however, are a number of common features. These shared characters may indicate a common evolutionary origin, and some are critical in shaping the lives of modern amphibians. Particularly important are features of the skin, which lacks external scales and contains an abundance of glands. These characteristics of the skin are reflected in many facets of the ecology and evolution of living amphibians.

All living adult amphibians are carnivorous, and there is relatively little morphological specialization associated with different dietary habits within each order. Amphibians eat almost anything they are able to catch, kill, and swallow. In aquatic forms the tongue is broad, flat, and relatively immobile, but some terrestrial amphibians can protrude the tongue from the mouth to capture prey. The size of the head is an important determinant of the maximum size of prey that can be taken, and sympatric species of salamanders frequently have markedly different head sizes, suggesting that this is a feature that reduces competition. Frogs in the tropical American genus *Ceratophrys*, which feed largely on other frogs, have such large heads that they are practically walking mouths (Figure 10–5d). In this respect they are reminiscent of some Paleozoic amphibians like *Cacops*.

10.1.1 Urodeles

The salamanders (order Urodela or Caudata) have the most generalized body form and locomotion of the living amphibians. Salamanders are elongate, and all but a very few species of completely aquatic salamanders have four functional limbs. Their walking gait is probably similar to that employed by the earliest tetrapods. It utilizes the lateral bending characteristic of fish locomotion in concert with leg movement (Figure 10–1). The two legs of each pair are always in opposed positions; the trunk curves convexly away from the side on which the foreleg and hind leg are most widely separated, giving the maximum stride length. Urodeles are almost entirely limited to the northern hemisphere; their southernmost occurrence is in northern South America. North and Central America have the greatest diversity of salamanders—there are more species of salamanders in Tennessee than in all of Europe and Asia combined. Two trends have occurred repeatedly in salamander evolution—paedomorphosis and the reduction of lungs. Paedomorphosis is very widespread, and several families are constituted solely of such paedomorphic forms. These can be recognized by the retention of larval characteristics, including larval tooth and bone patterns, the absence of eyelids, retention of a functional lateral line system, and (in some cases) retention of external gills.

The largest living salamanders are the Japanese and Chinese giant salamanders, which reach lengths of a meter or more. The related North American hellbenders grow to 60 cm (Figure 10–2c). All members of the Family Cryptobranchidae and are paedomorphic and permanently aquatic. As their name indicates, (*crypto* = hidden, *branchus* = gill) they do not retain external gills, although they do have other larval characteristics. Another large aquatic salamander, the mudpuppy (*Necturus*), is paedomorphic and does retain external

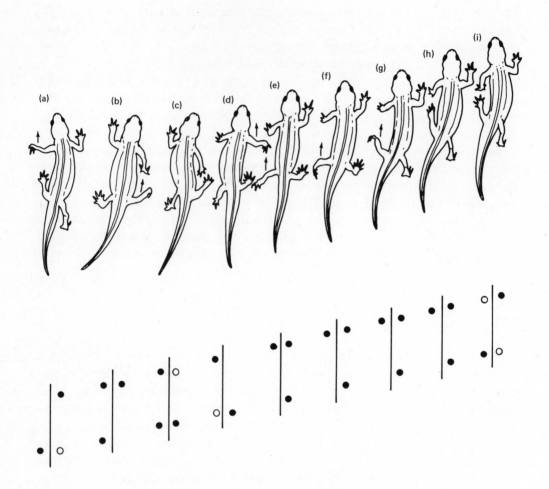

Figure 10-1. Salamanders combine lateral undulation with leg movement. As the body bends to the right (a–c) the salamander advances its left front and right rear legs. As it bends to the left it advances its right front and left rear feet (d–i). The body is supported by at least two diagonally opposite legs at all times. In the diagrams below the drawings, the position of the feet is shown for the corresponding drawing: Solid circle = foot firmly planted, open circle = foot being lifted, blank = foot in air. (Modified from B. Schaeffer [1941] Bulletin of the American Museum of Natural History 78:395–473.)

gills (Figure 10–2a). Mudpuppies occur in lakes and streams in eastern North America.

Several evolutionary lineages of salamanders have become adapted to life in caves (**troglodyty**). The constant temperature and moisture of caves make them good salamander habitats, and food is supplied by cave-dwelling invertebrates. The brook salamanders (*Eurycea*, Family Plethodontidae) include a number of species that form a continuum from those with fully metamorphosed adults inhabiting the twilight zone near cave mouths to fully paedomorphic forms in the depths of caves or sinkholes. Some North American cave-dwelling plethodontids are more specialized. The grotto salamander, *Typhlotriton spelaeus*, is the only specialized troglodyte that metamorphoses. Its larvae live in surface streams and are practically indistinguishable from other salamanders, but at metamorphosis the adults move into caves. The eyes become covered by skin, and dark pigmentation fades leaving a greyish white animal. The Texas blind salamander, *Typhlomolge*, is a highly specialized troglodyte—blind, white, with external gills, extremely long legs, and a flattened snout used to probe underneath pebbles for food (Figure 10–2e). The unrelated European olm (Family Proteidae) is another troglodyte that has converged on the same body form (Figure 10–2f).

Although paedogenesis is the rule in families like the Cryptobranchidae and Proteidae and characterizes most troglodytes, it also appears as a variant in the developmental pattern of salamanders that usually metamorphose. A particularly intriguing example of the balance between aquatic and terrestrial life is seen in the red-spotted newt (*Notophthalmus viridescens*) (Figure 10–9) which is widespread in eastern North America. Its life history is more complex than that of most salamanders. The adults are aquatic and lay eggs that hatch into aquatic larvae which spend several months in their home pond. Larvae may metamorphose into a bright orange terrestrial salamander called a red eft. The efts spend 3 to 7 years in terrestrial habitats, and then undergo a "second metamorphosis" to aquatic adults. The red color changes to olive-green dorsally with a yellow ventral surface, the skin becomes smoother, and a fleshy tail fin develops.

Some populations of newts do not adhere to this developmental sequence. Efts are generally abundant in montane regions but rare or entirely absent in lowland populations. In these lowland populations, larvae develop directly into adults—in some cases by complete metamorphosis, in other cases into paedomorphic adults that retain gills. Larvae that remain aquatic and develop into adults mature earlier than those that transform into efts. The rate of growth is directly related to the amount of food consumed. Terrestrial efts do most of their feeding during rainy periods in the summer; dry periods interrupt their feeding and growth. In contrast, the aquatic forms can continue to feed almost all year. As a result it takes an aquatic individual an average of only 2 years to mature, whereas a terrestrial eft takes at least twice as long.

A large number of factors interact to determine whether it is advantageous in a particular population to have a terrestrial stage. In eastern North America terrestrial habitats in mountains are generally moist, and there are many microhabitats for salamanders; however ponds suitable for aquatic newts may be rare and ephemeral. In lowlands, terrestrial habitats are drier and less salubrious for salamanders, but ponds and quiet river backwaters are abundant. By maintaining a terrestrial stage, a population of newts exploits two habitats. The terrestrial efts, which represent as many as 7 years' reproductive effort, are a reservoir of individuals that can establish a population in a new pond if the old

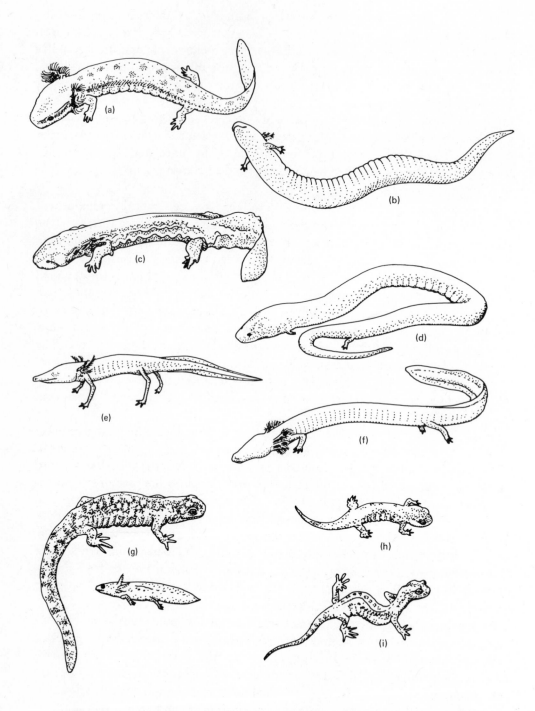

Figure 10-2. Urodele body forms reflect differences in life histories and habitats. Many aquatic salamanders like the mudpuppy (*Necturus*, a) and siren (*Siren*, b) retain external gills throughout their lives. Other aquatic salamanders lose their gills at metamorphosis but retain other larval characteristics; for example, the hellbender (*Cryptobranchus*, c) and the "congo eel" (*Amphiuma*, d). Specialized

one disappears. The disadvantages of maintaining the terrestrial stage may center around the increased average time to maturity; a lineage that omits the terrestrial stage brings all of its offspring quickly to reproductive age and can increase in numbers rapidly. Local environmental conditions determine which alternative is most advantageous in a given population.

Lungs seem an unlikely organ for a terrestrial vertebrate to abandon, but among salamanders the evolutionary loss of lungs has been a very successful tactic. An entire family, the Plethodontidae, is characterized by the absence of lungs. Plethodontids have more species and a greater geographic range than any other group of salamanders. The selective pressures that led to loss of lungs are unknown, but life in mountain streams has been suggested as a possibility. The cold water of the streams is well oxygenated, so salamanders would be able to absorb the oxygen they needed through the skin. Loss of lungs would reduce the buoyancy of the animals, and thus make it easier for them to creep about on the stream bottom without being washed away by the current. Furthermore, plethodontid salamanders have evolved specializations of the hyobranchial apparatus that allow them to protrude the tongue a considerable distance from the mouth to capture prey. This ability has not evolved in lunged salamanders, probably because the hyobranchial apparatus in these forms is an essential part of the system that ventilates the lungs. By using the tongue a plethodontid salamander can exploit a wider variety of prey than a lunged salamander can, and this widening of the food niche appears to be an important part of the success of plethodontid salamanders.

There are apparently some negative aspects to lunglessness as well. Because they lack lungs plethodontid salamanders depend upon cutaneous and buccopharyngeal gas exchange (pumping air in and out of the mouth and pharynx). Because they cannot obtain more oxygen by ventilating the lungs faster, any drying of the skin surface affects not just their water balance but also their ability to carry on gas exchange. As a result, lungless salamanders appear to occupy moister microhabitats than salamanders with lungs, and respond differently to rising environmental temperatures.

---

cave dwellers like the Texas blind salamander (*Typhlomolge*, e) and the European olm (*Proteus*, f) are white, lack eyes, and have long shovel-shaped snouts that fit into small spaces beneath rocks. Some terrestrial salamanders like the tiger salamander (*Ambystoma*, g) and the European salamander (*Salamandra*, h) have aquatic larvae. These larvae lose their gills at metamorphosis. The most fully terrestrial salamanders, the lungless plethodontids, include species like the slimy salamander (*Plethodon*, i) in which the young hatch from the eggs as miniatures of the adults and there is no aquatic stage. (Modified from the following: a, c, d, g, i from G. K. Noble, [1931] *The Biology of the Amphibia*, McGraw-Hill Book Co., Inc., New York. Republished in 1955 by Dover Publications, Inc., New York. The larva in g is from R. C. Stebbins [1951] *Amphibians of Western North America*, University of California Press, Berkeley. Parts b, h, from J. Z. Young [1962] *The Life of Vertebrates*, 2nd ed. Oxford University Press, New York; Part e from a photograph in D. M. Cochran [1]; Part f from H. Gadow [1909] *Amphibia and Reptiles*, Macmillan and Co., Ltd, London.)

10.1.2 Anurans

In contrast to the unspecialized locomotion of salamanders, the anurans (the Order Anura or Salientia: frogs, toads, treefrogs, spadefoot toads, and so on) have a highly specialized locomotion. The hind legs are elongate and, with powerful muscles, form a lever system that can catapult an anuran into the air (Figure 10–3). Numerous morphological changes are associated with this type of locomotion: a powerful pelvis strongly fastened to the vertebral column is clearly necessary, as is a stiffening of the vertebral column. The ilium is elongate and reaches far anteriorly, and the posterior vertebrae are fused into a solid rod, the urostyle. The pelvis and urostyle render the posterior half of the trunk rigid. There are only five to nine presacral vertebrae, and these are strongly braced by the zygapophyses to restrict lateral flexion. The strong forelimbs and flexible pectoral girdle absorb the impact of landing. The eyes are large and placed well forward giving binocular vision.

Aquatic species of anurans engulf food floating in the water, but most semi-aquatic and terrestrial species have highly specialized sticky tongues that can

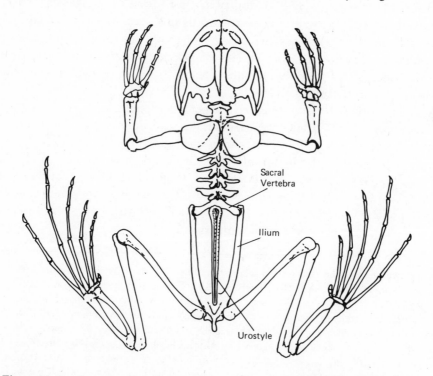

Sacral
Vertebra

Ilium

Urostyle

Figure 10–3. An anuran skeleton shows numerous adaptations for saltatory locomotion. The skull is broad and lightly built with large orbits. The number of vertebrae is greatly reduced and the vertebrae themselves are broad and allow little flexion. Behind the sacral vertebra, the remaining vertebrae are fused into a rigid urostyle. The ilia are elongated and form part of the lever system that also includes the long hind legs, feet, and toes. (Modified from Marshall in J. Z. Young [1962] *The Life of Vertebrates*, 2nd ed. Oxford University Press, New York.)

be flipped from the mouth to trap prey and carry it back into the mouth (Figure 10–4). Anurans have invaded a number of habitats without greatly modifying their skeletal morphology. A treefrog and a toad look very different, but most of the difference lies in the soft tissues. A generalized, semiaquatic anuran is represented, for example, by a member of the Family Ranidae which are the common frogs of the Northern Hemisphere and Africa (Figure 10–5a). It is moderately streamlined with a pointed head, slim body, and webbed hind feet. Similar morphology characterizes many terrestrial anurans. Other terrestrial forms like the toads (*Bufo*, Figure 10–5b) and spadefoot toads (*Scaphiopus*, Figure 10–5c) have heavier bodies and proportionately shorter legs. They move by walking or by short hops instead of the long leaps of frogs. Burrowing frogs have short legs and stout bodies (Figure 10–5e) Arboreal forms, the treefrogs in the Families Hylidae, Rhacophoridae, and Leptodactylidae, have independently converged on a long-legged, slim-waisted, blunt-headed morphology (Figure 10–5f). In many treefrogs the toe tips are expanded into discs that adhere to smooth surfaces. Treefrogs have considerably more flexibility in the trunk than other frogs and can twist the body to look to the side. Fully aquatic frogs, the Family Pipidae which includes the Surinam toad (*Pipa*) of South American and the clawed frogs (*Xenopus*; Figures 10–5g and 10–10f) of Africa, are at the opposite end of the morphologic scale from treefrogs. Pipids are dorsoventrally flattened and have thick waists, heavy legs, and large webbed hind feet. The forefeet are held beside the head as the frog swims. The toe tips have sensory structures, and the back of the hand is used to help stuff food into the mouth.

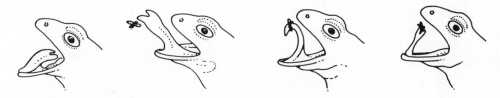

Figure 10–4 The tongues of most anurans are attached at the front of the mouth. They can be flipped forward to capture prey that adheres to the sticky tip. (Modified from H. Gadow [1909] *Amphibia and Reptiles.* Macmillan & Co., Ltd., London.)

10.1.3 Apodans

The third order of living amphibians is the least known and does not even have an English common name. These are the apodans (Order Apoda, Caecilia, or Gymnophiona), legless burrowing or aquatic amphibians that occur in tropical regions around the world (Figure 10–6). In addition to the loss of limbs, apodans have very reduced eyes, and some species have small scales buried in the dermis. Apodans have a pair of peculiar, perhaps unique, structures called tentacles that lie in grooves above the maxilla on each side of the head. They can be protruded by engorging them with blood. A sensory function has been suggested for the tentacles, but not demonstrated.

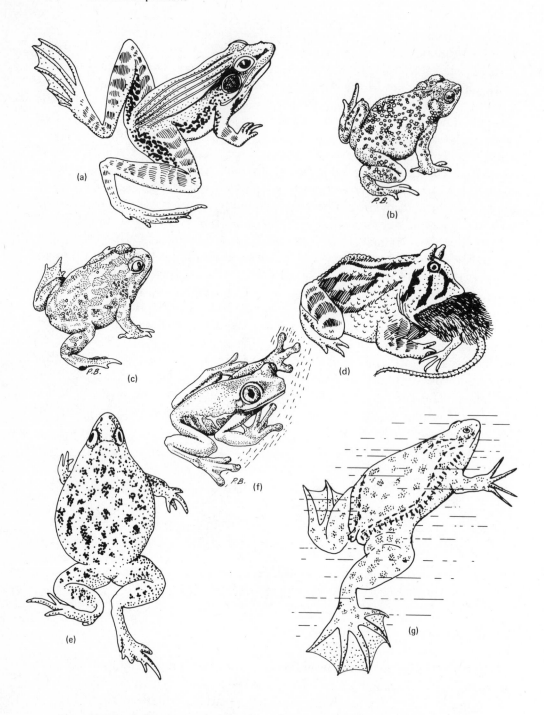

Figure 10-5 Anuran body forms reflect specializations for different habitats and different methods of locomotion. Semiaquatic frogs (*Ptychadena*, a, an African ranid) are moderately streamlined with webbed feet. Terrestrial anurans such as toads (*Bufo*, b) have blunt heads, heavy bodies, relatively short legs and little webbing between the toes. Spadefoot toads (*Scaphiopus*, c) take their name from a keratinized structure on the hind foot they use for burrowing. Some

10.2 EVOLUTIONARY ORIGIN OF LISSAMPHIBIANS

The origin of modern amphibians, the Subclass Lissamphibia, has long been an enigma because there are no fossils that demonstrate a transition between some Paleozoic group and any of the three modern orders. The oldest fossils that may represent modern amphibians are isolated vertebrae of Permian age that appear to include both urodele and anuran types. The earliest relatively complete lissamphibian fossil may be *Triadobatrachus*, a possible anuran from the lower Triassic (Figure 10–7). The skull is extremely froglike—broad, with the orbits enlarged into the cheek and temporal regions. The axial skeleton is less froglike than the skull, but the vertebral column is short and the ilium extends forward. The posterior vertebrae do not form a urostyle, however, and the anterior end of the ilium is not fused to a sacral rib. The absence of a long tail suggests that *Triadobatrachus* relied upon limb movements for swimming and was probably able to jump. Fossils later than *Triadobatrachus* come from the Jurassic and Cretaceous and are as specialized as modern forms. In most cases they can be tentatively assigned to modern families.

10.2.1 Monophyletic or Polyphyletic Origin of Lissamphibia?

In the absence of transitional material, paleontologists have had to depend upon presumed similarities between modern forms and Paleozoic fossils to decipher evolutionary relationships. For a long time a diphyletic origin was assumed. Similarities in the vertebral column seemed to trace anurans to labyrinthodonts, and urodeles to lepospondyls. Apodans were lumped with urodeles because they seemed more like salamanders than frogs.

Recently, Thomas Parsons and Ernest Williams suggested that the readily apparent differences between the three orders of living amphibians have blinded us to underlying similarities. The differences, they point out, are directly related to the different modes of locomotion of anurans, urodeles, and apodans and thus represent specializations of the three lineages. The similarities are not directly related to the ecological differences between the groups but, presumably, reflect the characteristics of the lineage from which the Lissamphibia

anurans like the horned frogs of South America (*Ceratophrys,* d) and Budgett's frog (Figure 10–13c) have extremely large heads and mouths. They feed on small vertebrates, including mammals. Many burrowing anurans have pointed heads, stout bodies, and short legs (*Hemisus,* e). Arboreal frogs have large heads and eyes, slim waists, and long legs (*Agalychnis,* f and *Centrolenella,* Figure 10–10a). Specialized aquatic frogs like the African clawed frog (*Xenopus,* g) and the Surinam toad (*Pipa,* Figure 10–10f) are dorsoventrally flattened and have thick waists, heavy legs with large webbed feet, and well developed lateral line systems. (Modified from the following sources: a, e from M. M. Stewart [1967] *Amphibians of Malawi,* State University of New York Press, Albany; b, c from R. C. Stebbins [1954] *Amphibians and Reptiles of Western North America,* McGraw-Hill, New York; d from various sources; f from a photograph by F. H. Pough, g from Noble [4].)

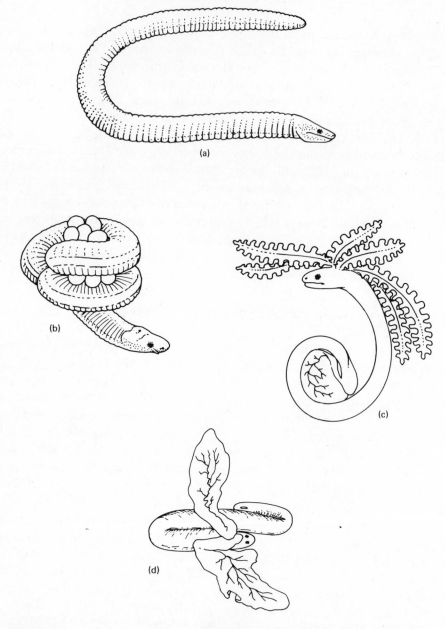

Figure 10-6. Apodans are limbless amphibians with the eyes nearly concealed under the skin (a). Females of some species brood the eggs (b) whereas others give birth to living young. The embryos of terrestrial species have long filamentous gills (c). In aquatic species the embryos have saclike gills (d). (Part a modified from J. Villa [1972] *Amfibios de Nicaragua.* Instituto Geografico Nacional y Banco Central de Nicaragua, Managua; b modified from H. Gadow, [1909] *Amphibia and Reptiles,* Macmillan and Co., Ltd, London; c and d modified from E. H. Taylor. [1968] *The Caecilians of the World.* University of Kansas Press, Lawrence.)

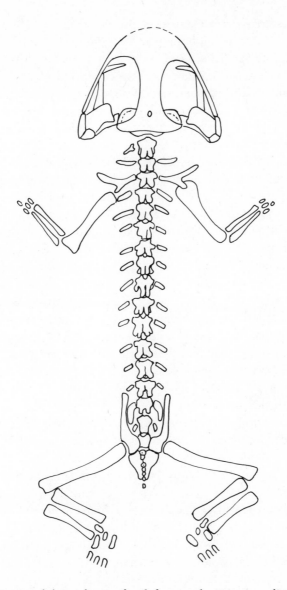

Figure 10-7 *Triadobatrachus*, a fossil from early Triassic sediments in Madagascar, shows some anuranlike characters, notably the broad, lightly built skull and a reduction in the number of vertebrae. There is only slight elongation of the ilia, however, and no fusion of postsacral vertebrae into a urostyle or elongation of the hind legs. (Compare Figure 10-3.) (Modified from R. Estes and O. A. Reig in Vial [6].)

arose. They listed 16 points in which modern amphibians resemble each other and differ from most or all other vertebrates. The most significant of the similarities involved the structure of the teeth, skull, ear, and skin.

All modern amphibians have **pedicellate teeth** in which the upper and lower parts are separated by a narrow zone of uncalcified tissue. In frogs and salamanders there is reduction in the dermal bones of the skull associated with the posterior and posteroventral expansion of the orbit. In the middle ear there is a second sound-transmitting bone, the operculum, fused to the stapes, and in the inner ear there is a special sensory area, the **papilla amphibiorum**, in the wall of the sacculus. The absence of a scaly skin covering makes the skin permeable to both gases and water. Parsons and Williams argue that it is extremely unlikely that all of these features could have evolved independently, and their occurrence in all modern amphibian orders makes a monophyletic origin virtually certain.

Not all workers agree with this conclusion. David Wake studied the vertebral column and concluded that there was no evidence there for a common origin of the living orders of amphibians. Robert Carroll has pointed out a number of ways in which apodans differ from anurans and salamanders in the features cited by Parsons and Williams. Carroll has suggested that the distinctive features of apodan skulls are primitive and that they indicate derivation of the group from microsaurs such as the Lower Permian genus *Goniorhynchus*.

If apodans are derived from microsaurs as Carroll speculates, where did anurans and urodels come from? There is no direct fossil evidence for a link with any group of Paleozoic amphibians. C. B. Cox has argued that dependence upon the skin for gas exchange is a key feature of living amphibians, although unfortunately not one that fossilizes. He thought it unlikely that temnospondyls, with their well-developed rib cages and presumed dependence upon thoracic ventilation by pressure changes, would have evolved the buccal pump and cutaneous diffusion respiratory processes of living amphibians. He suggested the gymnarthrids, a family of microsaurs, as lissamphibian ancestors because the skull, limbs, and axial skeleton do not have any specialized features that rule them out as protolissamphians.

Proponents of a labyrinthodont origin for lissamphibians have found a candidate in *Doleserpeton*, a fossil rhachitome from Lower Permian sediments in Oklahoma. This is the only Paleozoic amphibian fossil with pedicellate teeth. The general body form is salamanderlike but its large orbits, otic notch, and palatal vacuities make the skull froglike. At present *Doleserpeton* seems a potential candidate for a position somewhere in the family tree of either frogs or salamanders or perhaps a common ancestor of both groups.

10.2.2 Early Radiation of Modern Amphibia

The absence of fossils of early lissamphibians or of transitional forms is as puzzling as it is annoying. Why were protolissamphibians not fossilized with their contemporaries in the Paleozoic amphibian fauna? It is likely that protolissamphibians were small and fragile, but we have fossils of small and fragile labyrinthodont larvae and lepospondyls. Ivan Ivanovich Schmaulhausen, a leading Russian paleontologist, suggested that modern amphibians may have evolved in different habitats from those occupied by Paleozoic amphibians and that these habitats were not conducive to fossilization of small animals. Schmaulhausen believed that modern amphibians might have evolved in extremely shallow water and upland mountain streams, places that were too shallow for large amphibians to pursue them and too cold for the early reptiles. He pointed to the streamlined body form of modern anuran tadpoles and urodele larvae as evidence that they evolved in running water and secondarily invaded still-water habitats.

The anuran body form probably evolved from a more urodelelike starting point. Both jumping and swimming have been suggested as the mode of loco-motion that made the change advantageous. Urodeles and apodans swim as fish do by passing a sine wave down the body. Anurans have inflexible bodies and swim with simultaneous thrusts of the hind legs. Some paleontologists have proposed that the anuran body form evolved because of the advantages of that mode of swimming. An alternative hypothesis that seems more probable traces the anuran body form to selection for an amphibian that could rest near the edge of a body of water and escape aquatic or terrestrial predators with a rapid leap followed by rapid locomotion on either land or water.

The earliest fossils of all the orders of modern amphibians are very like living forms. The oldest frog, *Vieraella*, from the Lower Jurassic of Argentina shows affinities to two modern families, the Ascaphidae and Discoglossidae. There are some rather specialized salamanders known from the Upper Jurassic and an apodan from the Paleocene. Clearly the modern orders of amphibians have had separate evolutionary histories for a long time. The continued presence of such common characteristics as a permeable skin in the three orders after 250 million years suggests that the shared characteristics are critical in shaping the evolutionary success of modern amphibians. In other characters, such as re-production, locomotion, and defense they show tremendous adaptive radiation.

10.3   DIVERSITY OF LIFE HISTORIES OF LISSAMPHIABIANS

The living amphibians have long been used, consciously or not, as models of Paleozoic amphibians and, by extension, as an intermediate evolutionary stage between fully aquatic fishes and fully terrestrial reptiles. It should be obvious from the preceding discussion that this assumption is a dangerous one. Modern amphibians have had a very long evolutionary history independent of reptiles and of each other. Whether the Lissamphibia sprang from labyrinthodonts or microsaurs, their lineage had separated from that of reptiles by the end of the Devonian, 350 million years ago. It seems likely that at least two and probably all three of the modern orders had separated from each other by the Permian, more than 250 million years ago. For all the intervening period, amphibians have been evolving independently from reptiles, subjected to different selective pressures. Mammals, in fact, have more recent genetic continuity with reptiles than do living amphibians. Thus, it is only in the most general terms that modern amphibians can be considered to illustrate an amphibian structural grade transitional between those of fishes and reptiles. Nowhere in amphibian biology are the results of divergent, convergent, and parallel evolution seen as clearly as in the tremendous diversity of reproductive modes and life histories.

10.3.1 Apodans

The reproductive adaptation of apodans are as specialized as their body form and ecology. Internal fertilization is accomplished by a male intromittent organ that is protruded from the cloaca. Some species of apodans lay eggs, and the female may coil around the eggs, remaining with them until they hatch (Figure 10-6). Viviparity is widespread in the order, however, and in three of the four families there are species in which the eggs are retained in the oviducts and the female gives birth to living young. Recent studies by Marvalee Wake have provided fascinating details about this process. At birth young apodans are 30 to 60 percent of their mother's body length. A female *Typhlonectes* 500 mm long may give birth to nine babies, each one 200 mm long. Wake found that the initial growth of the fetuses is supported by yolk contained in the egg at the time of fertilization, but this yolk is exhausted long before embryonic development is complete. In *Typhlonectes* the fetuses have absorbed all of the yolk in the eggs by the time they are 30 mm long. Thus, the energy they need to grow to 200 mm (a 6.6-fold increase in length) must be supplied by the mother. The energetic demands of producing nine babies, each one increasing its length 6.6 times and reaching 40 percent of the mother's length at birth, must be considerable.

It appears that the juveniles obtain this energy by scraping material from the walls of the oviducts with specialized embryonic teeth. Wake observed that the epithelium of the oviduct proliferates and forms thick beds surrounded by ramifications of connective tissue and capillaries. As the fetuses exhaust their yolk supply, these beds begin to secrete a thick, white, creamy substance that has been called "uterine milk." When their yolk supply has been exhausted, the fetuses emerge from their egg membranes, uncurl, and align themselves lengthwise in the oviducts. The fetuses apparently bite the walls of the oviduct, stimulating secretion and stripping some epithelial cells and muscle fibers which they swallow with the uterine milk. Wake found that small fetuses are regularly spaced along the oviducts. Large fetuses have their heads spaced at intervals, although the body of one fetus may overlap the head of the next. She suggested that this spacing gives all the fetuses access to the secretory beds. The spacing and extension of the fetuses parallel to the body axis of the female may also be important in allowing her to carry several large fetuses without creating a bulge in her body that would interfere with burrowing.

Gas exchange appears to be achieved by apposition of fetal gills to the walls of the oviducts. All the terrestrial species have fetuses with a pair of triply branched filamentous gills (Figure 10-6). In preserved specimens the fetuses frequently have one gill extending above the head and the other stretched along the body. In the aquatic genus *Typhlonectes*, the gills are saclike but are usually positioned in the same way. Both the gills and the oviducal wall are highly vascularized, and it seems likely that exchange of gases, and possibly of small molecules such as metabolic substrates and waste products, takes place across the adjacent gill and oviduct. The gills are absorbed before birth, and cutaneous exchange may be important for fetuses late in development.

Differences in the details of fetal dentition in different species of apodans suggest to Wake that this specialized form of fetal nourishment may have evolved independently in different phylogenetic lines. Analogous methods of supplying energy to fetuses are also known in some elasmobranch fishes and in two species of salamanders.

10.3.2 Urodeles

Most groups of urodeles utilize internal fertilization; two primitive families, the Hynobiidae and Cryptobranchidae, retain external fertilization. Internal fertilization in salamanders is accomplished not by an intromittent organ but by the transfer of a packet of sperm (the **spermatophore**) from the male to the female (Figure 10-8). In the most primitive courtship patterns the female is passive and the male deposits the spermatophore on her body and uses his hind legs to maneuver it into her cloaca. An evolutionary sequence can be traced from this stage through a type of courtship in which the male deposits the spermatophore on the pond bottom and lowers the female onto it to advanced forms of courtship in which the female participates actively in following the male and picking up a spermatophore.

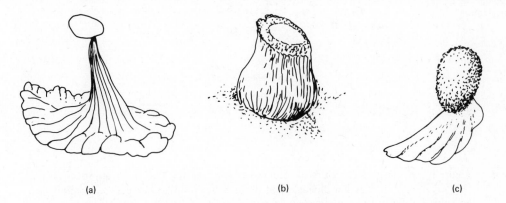

(a)   (b)   (c)

Figure 10-8. The form of the spermatophore differs in different species of salamanders but all consist of a sperm cap on a gelatinous base. The base is a cast of the male's cloacal cavity and in some species reproduces the ridges and furrows in accurate detail. Spermatophores from (a) red-spotted newt, (b) dusky salamander, (c) two-lined salamander. (Modified from Noble [4].)

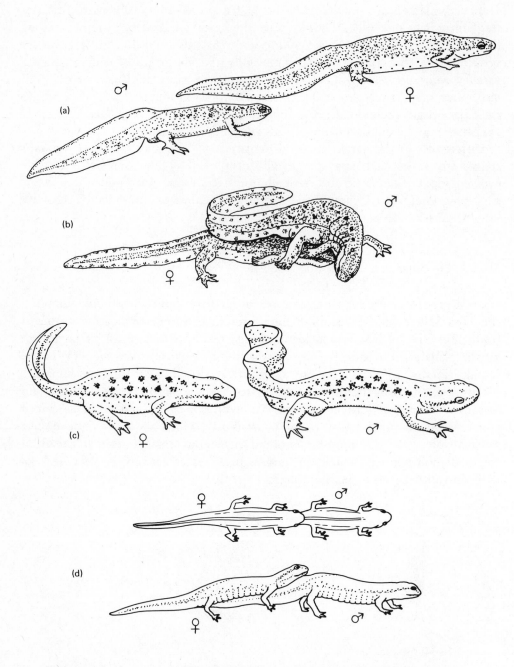

Figure 10-9. (a–c) Courtship of the red-spotted newt. The male approaches the female and nudges her cloaca, then mounts her holding her pectoral region with his forelimbs. The male moves forward on the female, shifting his grip to his hind legs and bending back to rub the female's snout with his chin. Finally the male moves forward off the female and walks in front of her waving his tail as if wafting the secretion of hedonic glands toward her. The female follows close behind the male and nudges his cloaca with her snout. This stimulates the male to deposit a spermatophore which the female picks up with her cloaca.

The evolutionary sequence appears to represent both differing ecological conditions, such as the differences between still and flowing water, and increased opportunity for premating isolating mechanisms to operate as the female's role becomes more active. Internal fertilization is believed to have evolved as an adaptation to life in running water where sperm released in the water would be swept away by the current. Once that pattern had evolved, increasing involvement of the female may have provided more stimulus-response steps that had to be completed properly in order for mating to occur. The more complicated the mating ritual, the less chance there is of two different species interbreeding. Movement from mountain streams to the quiet water of lowlands might have permitted more participation by the female because it was no longer necessary for the male to hold her continuously to prevent their being swept apart by the current.

Because courtship patterns are important species-isolating mechanisms, they show great interspecific variation. The description that follows of the courtship of the red-spotted newt is as close to a "typical" pattern as any one species can be. In this species, courtship and egg laying are both aquatic. Courtship starts with the male nudging the cloacal region of the female with his snout (Figure 10-9a). If the female is receptive, the male climbs onto her back, grasping her pectoral region with his forelegs, and then working his way forward until he is clasping her shoulders with his hind legs. By bending his body almost double, he rubs the female's head with his chin (Figure 10-9b). **Hedonic glands,** located on the surface of the male's body, secrete a substance that is believed to stimulate the female during this stage. The male moves off the female's back and leads her forward. The female's attention seems to be fixed on the cloacal region of the male, and in many species the male's cloacal region takes on a brilliant color in the breeding season. The male may wave his tail at the female as if he were wafting the secretions of hedonic glands toward her (Figure 10-9c). The female nudges the male's cloaca, and this seems to be the stimulus for the male to deposit a spermatophore on the pond bottom. The male continues to walk forward, leading the female directly over the spermatophore. As her cloaca passes over it, she picks the cap off with her cloacal lips. The gelatinous covering dissolves in the cloaca, releasing the sperm.

A complex behavioral pattern of this sort includes visual, tactile, and chemical cues, and all of them can be species-specific. The effectiveness of courtship patterns as isolation mechanisms is seen in the widespread occurrence of simultaneous breeding of different species of salamanders in the same pools. In northern New York, for example, two species of mole salamander breed

(d) Courtship is similar in species of salamanders that mate on land, but the female may maintain more intimate contact with the male. In many species of plethodontid salamanders the female straddles the male's tail with her chin resting on the base of his tail. (a, c drawn from photographs by Kentwood D. Wells; b modified from S. C. Bishop [1941] *The Salamanders of New York,* New York State Museum Bulletin, No. 324, Albany; d modified from J. A. Organ and D. J. Organ [1968] Copeia 1968:217–223.)

simultaneously in early spring. The Jefferson salamander, *Ambystoma jeffersonianum*, has a courtship pattern much like that described for *Notophthalmus* in which a single male clasps a female. The spotted salamander, *A. maculatum*, has a different pattern entirely. Several males court a female simultaneously. Courtship by a single male may not stimulate a female sufficiently to cause her to breed. The males cluster around the female, nudging her with their snouts, swimming off to deposit a spermatophore a short distance away, then returning again to nudge the female. When she is sufficiently aroused, the female moves to the small forest of spermatophores that have been deposited by the males and picks up one or more. The courtship behavior of these two species is so different that it is probably sufficient to isolate them even without the visual and chemical differences between them.

In most cases, salamanders that breed in water lay their eggs in water. The eggs may be laid singly, or in a mass of transparent gelatinous material. The eggs hatch into gilled aquatic larvae which, except in paedomorphic forms, transform into terrestrial adults. Some families, including the lungless salamanders (Plethodontidae), have a number of very terrestrial species that have dispensed in part or entirely with an aquatic larval stage. Plethodontid courtship on land is basically the same as in water, although the female may maintain closer contact with the male during the stage just preceding deposition of a spermatophore, straddling his tail as she follows him in a posture called the "tailwalk" (Figure 10-9d). The terrestrial environment does not offer the male the opportunity of wafting a species-identifying scent toward the female at this stage, and the tailwalk may be a modification of the pattern that has the same function. The dorsal surface of the tail has a concentration of glands (see section 10.5.2) and there may well be chemoreceptors on the ventral surface of the female's body.

Some plethodontid salamanders lay eggs in water, but many lay their eggs in moist microhabitats on land. The dusky salamander, *Desmognathus fuscus*, lays its eggs beneath a rock or log near water, and the female remains with them until after they have hatched. The larvae have small gills at hatching, and may either take up an aquatic existence or move directly to terrestrial life. The redbacked salamander, *Plethodon cinereus*, lays its eggs in a hollow space in a rotten log or beneath a rock. The embryos have gills, but these are reabsorbed before hatching and the hatchlings are miniatures of the adults.

There are a few salamanders that give birth to living young. The European salamander (*Salamandra salamandra*) produces 20 or more small larvae, each about one twentieth the length of an adult. The larvae are released in water and have an aquatic stage that lasts about 3 months. The closely related alpine salamander (*S. atra*) gives birth to one or two fully developed young about one third the adult body length. A female alpine salamander produces as many eggs as a European salamander, but only one egg in each oviduct develops. The remaining eggs break up into a mass that provides food for the developing embryo.

10.3.3 Anurans

Anurans are the most familiar amphibians, largely because of the vocalizations associated with their reproductive behavior. It is not even necessary to get outside a city to hear them. In springtime a weed-choked drainage ditch beside a highway or a trash-filled marsh at the edge of a shopping center parking lot is likely to attract a few toads or treefrogs that have not quite yet succumbed to human usurpation of their habitat. Anuran calls are diverse; they vary from species to species, and most species have two or three different sorts of calls used in different situations. The most familiar calls are the ones usually referred to as "mating calls" although a less specific term such as "**advertisement calls**" is preferable. These calls range from the high-pitched *peep* of a spring peeper to the nasal *waaah* of a spadefoot toad or the bass *jug-o-rum* of a bullfrog. The characteristics of a call identify the species and sex of the calling individual. Many species of anurans are territorial, and in these it is possible that territorial defenders identify each other individually by voice.

A species-specific call is a conservative evolutionary character, and among related taxa there is often considerable similarity in the calls. Superimposed on the basic similarity are the effects of morphological factors, such as body size, as well as ecological factors that stem from characteristics of the habitat. Toads (*Bufo*) have a trilled mating call, but the pitch of the call varies with the body size. In the oak toad (*B. quercicus*), which has a body length of only 2 or 3 cm, the dominant frequency of the call is 5200 Hz. In the larger southwestern toad (*B. microscaphus*), which is 8 cm long, the dominant frequency is lower, 1500 Hz, and the giant toad (*B. marinus*), with a body length of nearly 20 cm, has the lowest pitched call of all, 600 Hz.

Biological interactions can modify the basic characteristics of anuran calls. The eastern narrow-mouthed toad (*Gastrophryne carolinensis*) occurs in the southeastern United States, and the Great Plains narrow-mouthed toad (*G. olivacea*) in Texas and Oklahoma. At the junction of their ranges the two species are found in the same breeding sites. As one would expect from congeneric species, the calls are very similar—a peep followed by a short buzz. In the area of sympatry, the difference between the dominant frequencies is 1300 Hz, which is considerably greater than the difference in frequency outside the area of overlap. This has been interpreted as an example of character displacement. In the area of overlap there is the possibility for the toads to confuse the calls of the two species, and thus there has been selection for those individuals in each species with calls that were distinctly different from the other species. The advantage of not being mistaken for the other species might lie in avoidance of interspecific hybrid matings, which would lead to offspring less fit than those from conspecific mating and thus represent a waste of gametes. Alternatively (or additionally), the call displacement may represent a partitioning of the sound environment to allow each species to broadcast on a clear channel, free from interference from the other species.

In tropical habitats, where ten or more species of anurans may call simultaneously from the same breeding area, division of the auditory spectrum between species is very marked. In general, each species has a dominant frequency that is distinct from the dominant frequency of all other species in the area. When there is an overlap of frequencies, the species involved often have two-part calls, each part with a different dominant frequency. Each of the two parts may overlap another species, but each species' combination of frequencies is unique. Neurophysiological recordings from the brains of anurans have shown that there are individual cells in the auditory region that are closely tuned to the dominant frequency of the call of that particular species. Those cells respond strongly to sounds of a species' own frequency and give very little response to sounds of slightly different frequencies.

Three other categories of vocalization are widespread in anurans. Male release calls are uttered by males when another male attempts to mate with them. They are low-pitched vibrations and signal that the male has chosen the wrong sex and is wasting his time. Grabbing a male anuran around the body with the thumb and forefinger will usually stimulate it to give the release call. Screams are emitted by some species of frogs when they are seized by a predator. The scream is astonishingly loud and mammallike and may be sufficiently startling to the predator to make it release its hold momentarily and give the frog a chance to escape. Male encounter calls are also found in territorial species of anurans. A territorial male frog uses the advertisement call to establish its territory. If another frog enters the territory, it is warned with the encounter call, which is quite distinct from the advertisement call. In the spring peeper the advertisement call is a high-pitched peep, the encounter call a low-pitched trill. In the green frog the advertisement call is a hollow *bonk*, the encounter call a rumbling growl. If a vocal warning is insufficient to drive away an intruder, the territorial male may attack.

Female frogs are responsive to the advertisement call of their own species for a very brief period when their eggs are ready to be laid. The hormones associated with ovulation are thought to sensitize specific cells in the auditory pathway. In nonterritorial species, male anurans are eager to mate with whatever comes within reach, and readily grab other male's spent females, or passing fish, turtles, and floating sticks. In such a situation a female has relatively little choice of mate, but she probably is attracted to the male with the loudest call. Selection for individuals with loud voices may explain the origin of the choruses of hundreds of frogs, sometimes including a dozen species, that can make a din heard for miles. Surprisingly, the chorus itself does not seem to help anurans find their way to a breeding site. In territorial anurans females may exercise more selection in the choice of a mate, and in these species the suitability of a male's territory as an egg-laying site may be paramount in determining how many clutches of eggs he will be able to fertilize.

Mixed choruses of anurans are very common in the mating season; a dozen or more species may breed simultaneously in one pond. A female's response to her own species' mating call is one isolating mechanism, and it is supplemented by a variety of behavioral and probably chemical differences between species. Calling site is one important difference. Among the Australian leptodactylids of the genus *Crinia*, for example, *C. sloanei* calls while floating in the water, *C. signifera* calls from beneath overhanging vegetation near shore, and *C. parinsignifera* calls from grass clumps out of water. Presumably a female frog seeks a male in the usual calling site for her species, and this behavior would reduce the chances of encountering a male of the wrong species.

Fertilization is external in nearly all anurans; the male uses his forelegs to clasp the female in the pectoral region (**axillary amplexus**) or pelvic region (**inguinal amplexus**). Amplexus may be maintained for several hours or even a day or more before the female lays eggs. During this period the male and female are in close contact. It seems likely that tactile and chemical cues play a role in species identification at this stage.

Anurans show even greater diversity in their modes of reproduction than urodeles. Similar reproductive habits have clearly evolved independently in different groups. The underlying selective mechanism appears to have been the advantage gained by reducing the amount of energy invested per offspring that survives to maturity. In different groups this has been achieved by some sort of parental care of the eggs and in some cases the tadpoles, by increasing the size of the egg and thus the size of the hatchling, by preparing a favorable nest site, or by some combination of these mechanisms. The discussion that follows is organized only in terms of increasingly extensive parental care of the young. It is not meant to imply an evolutionary sequence; the more elaborate forms of parental care did not necessarily evolve via the less complex forms.

The primitive anuran reproductive pattern is thought to consist of laying large numbers of small eggs. This pattern is retained in two of the most widespread genera of anurans, *Rana* (the true frogs) and *Bufo* (toads). For example, toads may lay as many as 10,000 eggs in one clutch, and the female puts about half the energy in her body into the eggs. She leaves the breeding pond immediately thereafter and accumulation of sufficient fat reserves to produce next year's eggs occupies her for the rest of the season. In temperate regions this reproductive pattern is almost universal among anurans, but in tropical areas as many as 80 percent of the species of anurans have other patterns.

One method of increasing the proportion of eggs that hatch successfully is to give them protection from predators, and a number of methods of accomplishing this have evolved. Many treefrogs lay their eggs not in water but in the branches of trees overhanging water (Figure 10-10a). The eggs undergo their embryonic development out of the reach of aquatic egg predators, and when the tadpoles hatch they drop into the water and take up an aquatic existence. Other frogs achieve the same end by constructing foam nests that float on the water surface. The female emits a copious mucous secretion that she beats into a foam with her hind legs, and the eggs are laid in the foam mass (Figure 10-10b). When the tadpoles hatch, they drop through the foam into the water.

Although these methods reduce egg mortality, the tadpoles are subjected to predation and competition. Some anurans avoid both problems by finding or

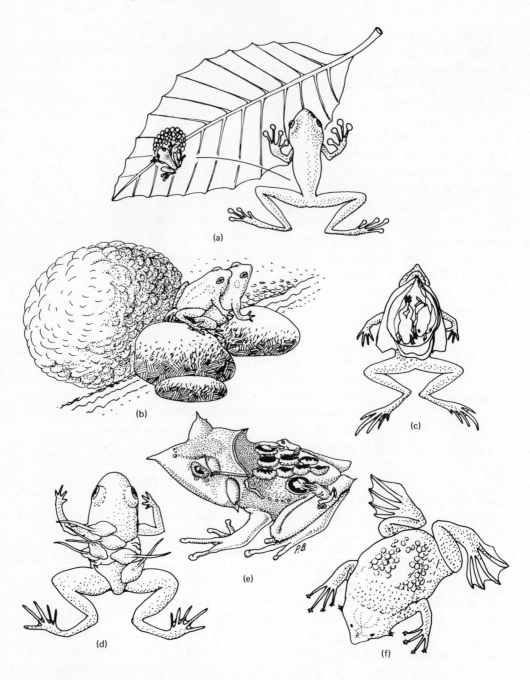

Figure 10–10. Not all amphibians lay eggs in water. Some species lay eggs on leaves or branches overhanging the water (a), and in some of those species one of the parents remains with the eggs until they hatch. Eggs may be laid in a nest of foam either on land or floating on water (b). Some anurans carry their eggs with them. The male Darwin's frog carries the eggs in his vocal sacs where they pass through metamorphosis. The drawing shows a dissection of a frog with two mature larvae in the sacs (c). Some species of small frogs in both Africa and South America lay eggs on land that hatch into tadpoles. The tadpoles adhere

constructing breeding sites free from competition and predation. Many tree-frogs, for example, lay their eggs in the water that accumulates in bromeliads—epiphytic tropical plants that grow in trees and are morphologically specialized to collect rainwater. A large epiphyte may hold several liters of water, and the frogs pass through egg and larval stages in that protected microhabitat. Many tropical frogs lay eggs on land near water. The eggs or tadpoles may be released from the nest sites when pond levels rise after a rainstorm. Other frogs construct pools in the mud banks beside streams. These volcano-shaped structures are filled with water by rain or seepage and provide a favorable environment for the eggs and tadpoles. The ultimate in reduction of larval mortality is achieved in some frogs that have eliminated the tadpole stage entirely. These lay large eggs on land that develop directly to little frogs. This pattern is characteristic of the tropical genera *Eleutherodactylus* and *Arthroleptis*.

A degree of parental care is widespread in anurans. Even among the forms with a primitive breeding pattern, the males of some species maintain territories and attack almost any small animal that intrudes. In the green frog (*Rana clamitans*), for example, males defend areas 2 or 3 m in diameter, and any eggs laid in that area are almost certain to be fertilized by the territorial male. Thus, the presence of a male guarding his territory may offer a degree of protection to his eggs even though he is probably oblivious of their presence. In many species of frogs, the adults guard the eggs specifically. In some cases it is the male, in others the female, and in most cases it is not clearly known which sex is involved because external sex identification is difficult in many anurans. Some of the treefrogs that lay their eggs over water remain with them. Some species sit beside the eggs, others rest on top of them (Figure 10–10a). Removing the guarding frog frequently results in the eggs desiccating and dying before hatching. Many of the terrestrial frogs whose eggs develop directly into adults remain with the eggs and will attack an animal that approaches the nest. Some of the arrow poison frogs of the American tropics deposit their eggs on the ground, and one of the parents remains with the eggs until they hatch into tadpoles. The tadpoles fasten themselves to the adult's back with the mouth, which is modified to a suckerlike structure, and the adult transports the tadpoles to water (Figure 10–10d).

---

to the back of one of the parents with mouthparts modified to function as suckers and are carried to water (d). In many species of frogs the eggs are carried on the female's back from laying to hatching. In some species of marsupial frogs, the eggs are exposed on the back (e) whereas in others they are covered by skin. In the Surinam toad each egg sinks into the skin which forms a capsule around it (f). (Modified from the following sources: a and b original; c from Noble [4]; d from H. Gadow [1909] *Amphibia and Reptiles*. Macmillan and Co., Ltd., London; e from W. E. Duellman [1970] *The Hylid Frogs of Middle America*. Monograph of the Museum of Natural History, The University of Kansas, No. 1, Lawrence; f from M. Lamotte and J. Lescure [1977] La Terre et la Vie 31:225–311.)

Other anurans, instead of remaining with the eggs, carry the eggs with them. The male of the European midwife toad (*Alytes obstetricans*) gathers the egg strings about his hind legs as the female lays them. He carries them with him until they are ready to hatch, at which time he releases the tadpoles into water. The male of the terrestrial Darwin's frog (*Rhinoderma darwinii*) of Argentina snaps up the two eggs the female lays and carries them in his vocal pouches which extend back to the pelvic region (Figure 10–10c). The embryos pass through metamorphosis in the vocal sacs and emerge as fully developed froglets. Males are not alone in caring for eggs. The females of a group of treefrogs carry the eggs on their back, either in an open oval depression, a closed pouch (from which comes the common name of the group, the marsupial frogs), or in individual pockets. The eggs develop into miniature frogs before they leave their mother's back (Figure 10–10e). A similar adaptation is seen in the completely aquatic Surinam toad, *Pipa*. In the breeding season the skin of the female's back thickens and softens. In egg laying the male and female in amplexus swim in vertical loops in the water. On the upward part of the loop the female is above the male and releases a few eggs which fall onto his ventral surface. He fertilizes them and, on the downward loop, presses them against the female's back. They sink into the soft skin and a cover forms over each egg, enclosing it in a small capsule (Figure 10–10f). The eggs develop through metamorphosis in the capsules. Tadpoles of the Australian frog *Rheobatrachus silus* are carried in the female's stomach and expelled through the mouth after metamorphosis. This is the only known case of gastric incubation among vertebrates.

There are even live-bearing anurans. An African bufonid (*Nectophrynoides*) retains the eggs in the oviducts and gives birth to little frogs, and the golden coqui, a Puerto Rican leptodactylid, has a similar mode of reproduction.

In view of the wide range of reproductive modes observed in anurans, it is surprising to find that species that inhabit arid regions, where the lack of water can prevent any breeding in some years, are not live-bearers nor do they lay eggs with direct development. Instead they adhere to the pattern of aquatic eggs that hatch into tadpoles. The modifications of the process that make them successful in desert regions involve the timing of reproduction and the rate of embryonic and larval development.

Desert regions are characterized by the scantiness and unpredictability of their rainfall. Generally there is a "rainy season" when the probability of rain is higher than at other times of year, but there is no way to know within a month when rain will fall on a particular area of desert or even if any rain will fall in a given year. Because of this unpredictability, desert anurans must be ready to breed as rapidly as possible after a rain; they depend upon the temporary pools formed by rainwater to lay their eggs. The tadpoles must metamorphose and leave the pools before they dry out.

In southern Arizona, the summer rainy season extends from July through September. As July approaches, spadefoot toads have fully developed eggs and sperm. They move from the bottoms of their burrows (see section 10.4.2) to within an inch or so of the surface in the evening and may rest with their eyes protruding above the ground. If it does not rain—and most days it does not— the spadefoot moves down into its burrow in the morning to escape the heat of the day. If it does rain, the spadefoots emerge almost immediately and begin to move about on the desert floor, feeding on insects. Even a very light shower initiates emergence—less than 2.5 mm of rain brings the animals onto the surface. So little rain could not penetrate to the bottoms of the burrows; probably an annual cycle of activity brings the animals to the tops of their burrows at the proper time of year to await a rainfall that will make surface conditions suitable for activity.

Heavier rainfall is needed to form the breeding ponds. The first spadefoots arrive at those sites the same day the rain falls, and the first clutches of eggs may be laid that night. Amphibian eggs are damaged by high temperatures. The eggs of desert species tolerate higher temperatures than those of species from mesic (moderately moist) regions, and they also develop faster. Rate of development is an important adaptation because the eggs become increasingly resistant to damage from high temperatures as development progresses. Because they develop rapidly, the eggs of desert anurans pass through the most temperature-sensitive stages at night, and by the time the sun rises the following morning and the water in the pool starts to get hot, the eggs are well on their way to hatching.

The rapid development of desert anuran eggs continues through hatching and metamorphosis. Spadefoot toads develop to metamorphosis in 2 or 3 weeks, compared with 10 weeks or more for species that breed in permanent ponds. Despite the rapid development, groups of spadefoot tadpoles may be trapped in rapidly shrinking pools. In that situation the tadpoles gather in one area, churning up the muddy bottom with their activity. The mud they raise settles in quieter parts of the pool, leaving an excavation that collects the water in one deep spot. The reduced surface area slows evaporation, giving the tadpoles a few more hours or days to complete metamorphosis before the last of the water disappears. Under crowded conditions of this sort, individual spadefoot tadpoles of some species are likely to turn into cannibals. Their horny jaws hypertrophy, and they switch from a diet of algae to killing and eating other tadpoles. The greater energy they derive from a carnivorous diet speeds their development and increases their chances of metamorphosing in time. It seems probable that the switch from herbivory to cannibalism reflects a genetic predisposition that is activated by a particular set of environmental conditions, but what those may be is not known.

None of the adaptations of desert anurans is unique. Instead they are the maximum expression of trends seen in species from more mesic regions. The toads (*Bufo*) are an example. Different species of toads are found in habitats ranging from humid tropical lowlands through moist temperate and subarctic regions to deserts. In all of these habitats, toad eggs and larvae develop rapidly and hatch at an earlier embryonic stage than most other anurans. American toads (*Bufo americanus*) metamorphose within 6 weeks of egg laying, and are less than one twentieth the body length of an adult at metamorphosis. In contrast, tadpoles of green frogs in the same ponds take a year or more to reach metamorphosis, and are about one fifth adult size when they transform.

10.3.4 Amphibian Metamorphosis

The importance of thyroxine for amphibian metamorphosis was discovered quite by accident in the early twentieth century by the German biologist Friedrich Gudernatsch. He was able to induce rapid precocious metamorphosis in tadpoles by feeding them extracts of beef thyroid glands. Only in the last two decades have the details of the interaction of neurosecretions and endocrine gland hormones been worked out, in large part by William Etkin and his associates. Although the work has been done largely with anurans, it is probable that in broad outline the process is similar in all amphibians.

Thyroxine and related compounds (monoiodotyrosine, diiodotyrosine, and 3,3',5-triiodothyronine) are produced by the thyroid gland. In premetamorphic amphibian larvae, the levels of circulating thyroxine are quite low. The dominant hormone of premetamorphic stages is prolactin, which is produced by the pituitary gland (Figure 10–11). Prolactin stimulates growth of larva and inhibits metamorphosis.

Late in the premetamorphic period a change occurs—the hypothalamus becomes sensitive to the circulating thyroxine. The timing of this change is genetically determined, but it can probably be modified to some extent by environmental factors. Reliable seasonal cues, such as a change in day length or water temperature, may be an important part of the program that initiates metamorphosis at a season that will bring the juvenile amphibian onto land when food and moisture conditions are suitable. Superimposed on this sort of control may be characteristics of the immediate surroundings. In general these are probably responses to changes in the aquatic habitat that make it unfavorable for further larval development—rapidly falling water levels, too little food, overcrowding, and other indications that it's time to go. The shift to cannibalism by some spadefoot tadpoles caught in drying pools may be a dramatic example of control of development exerted by proximate factors.

Associated with the development of hypothalamic sensitivity to thyroxine is the organization of the median eminence. This is the portion of the brain that connects the hypothalamus to the pituitary. In premetamorphosis it is a small structure with a loose network of capillaries. In the next stage, prometamorphosis, the median eminence enlarges as the pituitary separates from the hypothalamus, and the capillary bed is organized into a portal system. The increased efficiency of the portal system improves the conduction of the neurosecretory products of the hypothalamus to the pituitary. In mammals the pituitary gland is activated by a hypothalamic neurosecretion, thyroid releasing factor (TRF), carried to it by the portal system of the median eminence; this is thought to be true of amphibians as well (Figure 19-27).

As the portal system becomes better organized, it transports more TRF to

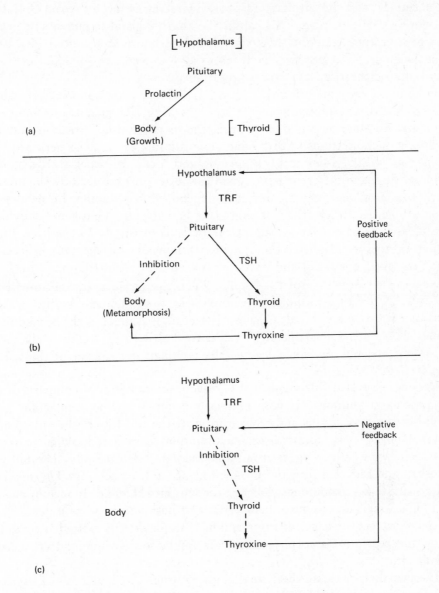

Figure 10–11. Hormonal control of amphibian metamorphosis. (a) Prolactin, produced by the pituitary, stimulates larval growth and inhibits metamorphosis. (b) At some point, which is probably determined by the interaction of genetic and environmental factors, the pituitary becomes sensitive to thyroid releasing factor (TRF) produced by the hypothalamus and transported via the portal system of the median eminence. TRF inhibits prolactin production and stimulates the body to metamorphosis. A positive feedback of thyroxine to the hypothalamus increases the production of TRF, and this results in rapidly increasing thyroxine concentrations, speeding the tempo of metamorphosis. (c) high levels of thyroxine have an inhibitory effect on the release of thyroid stimulating hormone (TSH) by the pituitary, and it is this negative feedback that returns thyroxine levels to normal concentrations at the end of metamorphosis.

the pituitary, and the pituitary responds by secreting more thyroid stimulating hormone (TSH). In turn, TSH causes the thyroid gland to increase its production of thyroxine, and the thyroxine stimulates more secretion of TRF by the hypothalamus. This positive feedback system explains the sudden increase in thyroxine synthesis in late prometamorphosis.

In addition to stimulating the release of TSH by the pituitary, TRF inhibits the production of prolactin. As a result, during late prometamorphosis the hormonal dominance shifts from prolactin to thyroxine. Prolactin had been inhibiting metamorphosis; thyroxine now stimulates it. The action of thyroxine on larval tissues is both specific and local. In other words, it has a different effect in different tissues, and that effect is produced by the presence of thyroxine in the tissue; it does not depend upon induction by neighboring tissues. The particular effect of thyroxine in a given tissue is genetically determined, and virtually every tissue of the body is involved. In the liver, for example, thyroxine stimulates the enzymes responsible for the synthesis of urea (the urea cycle enzymes) and starts the synthesis of serum albumin. In the eye it induces the formation of rhodopsin. When thyroxine is administered to the striated muscles of a tadpole's developing leg, it stimulates growth; but when administered to the striated muscles of the tail, it stimulates the breakdown of tissue. When a larval salamander's tail is treated with thyroxine, only the tail fin disappears; but thyroxine causes the complete resorption of the tail of a frog tadpole.

Metamorphosis in salamanders is relatively undramatic by comparison with the process in anurans. Extensive changes occur at the molecular and tissue level, but the loss of gills and reabsorption of the tail fin are the only obvious external changes. In contrast, the metamorphosis of a tadpole to a frog involves readily visible changes in almost every part of the body. The tail is reabsorbed and recycled into the production of adult structures. The small tadpole mouth that accommodated algae broadens into the huge mouth of an adult frog. The long tadpole gut, characteristic of herbivorous vertebrates, changes to the short gut of a carnivorous animal. Respiration is shifted from gills to lungs, and partly metamorphosed froglets can be seen swimming to the surface to gulp air.

The initial changes are the ones that are produced by low levels of thyroxine and they proceed slowly. As thyroxine levels increase the tempo of metamorphosis increases, and the final stage, metamorphic climax, takes only a few days. At this point the positive feedback of thyroxine to the hypothalamus is supplanted by a negative feedback to the pituitary. This negative feedback reduces release of TSH, and thus lowers the production of thyroxine by the thyroid gland. Circulating thyroxine drops back toward premetamorphic levels, and the metamorphosed larvae take up adult life habits.

The retention of larval characters in sexually mature individuals results from a failure of thyroxine to stimulate metamorphosis. That failure may be produced by a genetically controlled insensitivity to thyroxine, hypofunction of the thyroid gland, or environmental effects. A number of families of salamanders contain species that always retain larval characters. The mudpuppy (*Necturus*) is the most familiar example. These species cannot be made to metamorphose by treating them with thyroxine—the tissue response is absent.

In contrast, some species like the axolotl (*Ambystoma mexicanum*) rarely metamorphose in nature; however, when wild-caught axolotls are brought into the laboratory and treated with thyroxine or TSH, they readily metamorphose. Axolotls have been important animals for experimental embryology for nearly a century. Most of the animals used in these experiments are descended from fewer than 100 individuals sent to Paris in the late nineteenth century. Animals from this colony have lost the ability to metamorphose even when treated with hormones. Metamorphosed salamanders are much more difficult to care for in captivity than larval forms. As a consequence, there has been strong selection against those individuals that retain the ability to metamorphose, and the genetic capacity for metamorphosis has apparently disappeared from the colony.

The environment affects metamorphosis in several ways. Ultimately it is environmental conditions that determine the relative selective advantages of metamorphosis and paedogenesis (Section 10.1.1). More directly, low iodine concentration in the water of some western lakes may prevent synthesis of sufficient thyroxine to induce metamorphosis. Paedogenetic tiger salamanders (*Ambystoma tigrinum*) from Colorado can be induced to metamorphose by treatment with iodine. Finally, low temperatures may inhibit metamorphosis. Laboratory studies have shown that the response of larval tissues to thyroid hormones is reduced as the temperature drops and disappears entirely at 5 or 10°C.

10.3.5   Lissamphibian Life History as a Model for Paleozoic Amphibians

The reproductive modes of amphibians have clearly responded with considerable plasticity to selective forces. The great diversity of reproductive patterns among modern amphibians (which represent only a fraction of the taxonomic and ecological diversity of Paleozoic amphibians) emphasizes the caution that must be employed in extrapolating from characteristics of living amphibians to extinct ones. Although there is no direct evidence of the reproductive habits of Paleozoic amphibians, there is no reason to assume that they did not evolve at least as much diversity as modern amphibians. Speculations about the ecology of Paleozoic amphibians that assume they adhered to the "typical" amphibian pattern of laying eggs in water may be very misleading.

10.4   WATER RELATIONS OF AMPHIBIANS

Unlike the Paleozoic amphibians, which probably retained the scaly covering inherited from their rhipidistian ancestors, Lissamphibia have a glandular skin without external scales that is highly permeable to water. Both the permeability and glandularity of the skin have been of major importance in shaping the ecology and evolution of lissamphibians.

The readiness with which water moves through the skin has frequently been emphasized as a feature that limits the geographic distribution of amphibians. Terms like "evolutionary backwater" and "defeated group" have been applied to amphibians by paleontologists. These views represent a misunderstanding of amphibian ecology. It is quite true that many species of amphibians must remain in moist microhabitats because they lose water rapidly, but some amphibians (most notably certain anurans) have utilized their permeable skin in a way that has allowed them to invade deserts.

10.4.1 Permeability of Amphibian Skin

Both water and gases pass readily through amphibian skin. In biological systems, permeability to water is inseparable from permeability to gases, and amphibians depend upon cutaneous respiration for a variable but significant part of their gas exchange. The importance of cutaneous respiration has been greater than any advantage that would be derived from a reduction in the permeability of the skin. Terrestrial species of amphibians do not have less permeable skin than aquatic species—in fact, the opposite is true. Aquatic amphibians must cope with an osmotic influx of water, and decreased skin permeability appears to be one way they do it.

Although the skin permits passive movement of water and gases, it controls the movement of other compounds. Sodium is actively transported from the outer surface to the inner, and urea is retained by the skin. These characteristics are important in the regulation of osmotic concentration and in facilitating uptake of water by terrestrial species. The internal osmotic pressure of amphibians is approximately two thirds that characteristic of most other vertebrates. The primary reason for the dilute body fluids of amphibians is low sodium content—approximately 100 milliequivalents (meq) compared to 150 meq in other vertebrates. Amphibians can tolerate a doubling of the normal sodium concentration, whereas an increase from 150 to 170 meq is the maximum humans can tolerate.

A watery animal with a permeable skin seems an unlikely candidate for success in an arid habitat, and most amphibians are restricted to moderately moist microhabitats. Differences in the ability to invade dry microhabitats can be ecologically important. The lungless plethodontid salamanders are restricted to moister microhabitats than lunged salamanders, but there are differences even among the plethodontids. For example, the ravine salamander (*Plethodon richmondi*) inhabits relatively dry pockets of soil in rocky slopes, whereas the red-backed salamander (*P. cinereus*) is excluded from them by its requirement for slightly moister conditions. The slimy salamander (*P. glutinosus*) inhabits still more moist microhabitats. The California slender salamander (*Batrochaseps attenuatus*) is found in very dry chaparral in the Santa Monica Mountains where it shelters under rocks and logs and gradually follows the cracks that form in soil in the dry season down to moister levels. Among amphibians anurans have been by far the most successful invaders of arid habitats. All but the harshest deserts have substantial anuran populations, and, in different parts of the world, different families have converged on similar adaptations.

10.4.2 Turning a Permeable Skin to Advantage

The spadefoot toads (Family Pelobatidae) of North America are the most thoroughly studied desert anurans. Their reproductive adaptations have already been described (Section 10.3.3). They inhabit the desert regions of North America, including the edges of the Algodones Sand Dunes in Southern California where the average annual precipitation is only 6 cm and in some years no rain falls at all. An analysis of the adaptations that allow an amphibian to exist in a habitat like that must include consideration of both water loss and gain. We have already pointed out that the skin of terrestrial amphibians is as permeable to water as that of species from moist regions. Thus, there is no morphological or physiological reduction of evaporative water loss. A desert anuran must control its water loss behaviorly by its choice of sheltered microhabitats free from solar radiation and wind movement. Different species of anurans utilize different microhabitats—a hollow in the bank of a desert wash, the burrow of a ground squirrel or kangaroo rat, or a burrow the anuran excavates for itself. All of these places are cooler and moister than exposed ground.

Desert anurans spend extended periods underground, emerging on the surface only when conditions are favorable. Spadefoot toads construct burrows about 0.6 m deep, filling the shaft with dirt and leaving a small chamber at the bottom which they occupy. In southern Arizona the spadefoots construct these burrows in September, at the end of the summer rainy season, and remain in them until the rains resume the following July. When spadefoot toads are dug out of their burrows, some have patches of thick, black skin adhering to their bodies. This covering is not continuous over the body surface and readily flakes off. It is probably not a barrier to water movement, but only an accumulation of shed skin. Some leptodactylid frogs in Australia and South America do form cocoons of epidermis that cover them entirely and greatly reduce water loss. The aquatic salamander *Siren* also forms a cocoon of this sort and can aestivate in the mud of a dried-up pond.

At the end of the rainy season when the toads first bury themselves, the soil is relatively moist. The water tension created by the normal osmotic pressure of a toad's body fluids establishes a gradient favoring movement of water from the soil into the toad. In this situation, a buried spadefoot can absorb water from the soil just as the roots of plants do. With a supply of water available, a toad can afford to release urine to dispose of its nitrogenous wastes.

As time passes, the soil moisture content decreases and the soil moisture potential becomes more negative until it equals the water tension of the toad. At this point there is no longer a gradient allowing movement of water into the toad. When its source of new water is cut off, a toad stops excreting urine and instead retains urea in its body, increasing the osmotic pressure of its body fluids. Osmotic concentrations as high as 600 milliosmoles (mOsm) have been recorded in spadefoot toads emerging from burial at the end of the dry season. The low water potential produced by the high osmotic pressures of the toad's body fluids may reduce the water gradient between the animal and the soil so that evaporative water loss is reduced. Sufficiently high internal osmotic pressures should create potentials that would allow toads to absorb water from even very dry soil. The osmotic concentrations measured in newly emerged spade foots should be high enough to allow them to extract water from the soil in which they were buried.

The ability to continue to draw water from soil enables a toad to remain buried for 9 or 10 months without access to liquid water. In this situation its permeable skin is not a handicap—it is an essential feature of a spadefoot's biology. If the toad had an impermeable skin or if it formed an impermeable cocoon as some other amphibians do, water would not be able to move from the soil into the animal. Instead the toad would have to depend upon the water contained in its body when it buried. Under those circumstances spadefoot toads would probably not be able to invade the desert because their initial water content would not see them through a 9-month dry season.

Two other features of amphibians that assist them to colonize arid habitats are their low osmotic concentration and storage of water in the bladder. The low osmotic concentration can be considered a water reserve that an amphibian can draw upon. For example, an amphibian that starts with an osmotic concentration of 200 mOsm can lose enough water to increase its concentration by 50 percent (to 300 mOsm) and still be no more concentrated than the normal levels for most vertebrates. The ability of amphibians to withstand dehydration is related to the aridity of their habitat. Aquatic species can withstand a loss of 20 to 40 percent of their body water, but species from arid habitats can tolerate a 50 to 60 percent loss. Another amphibian characteristic that is related to the aridity of the habitat is bladder volume. The anuran kidney produces urine hyposmotic to the blood. In the bladder this urine represents a source of water that can be drawn back into the tissues to combat dehydration. In general, the more arid the habitat, the larger the bladder. Aquatic species have bladders that can hold urine equivalent to 1 to 10 percent of the body weight, whereas in desert anurans the weight of urine in the bladder can equal or exceed the body weight.

A different pattern of adaptation to arid conditions is seen in a few treefrogs. The African rhacophorid *Chiromantis xerampelina* and the South American hylid *Phyllomedusa sauvagei* lose water through the skin at a rate only one tenth that of most frogs. *Phyllomedusa* has been shown to achieve this low rate of evaporative water loss by using its legs to spread the lipid-containing secretions of dermal glands over its body surface in a complex sequence of wiping movements. These two frogs are unusual also because they excrete nitrogenous wastes as salts of uric acid rather than as urea. (See Section 7.4). This uricotely provides still more water conservation.

Adaptations for obtaining water in arid environments have received less attention than those for retaining it. Amphibians do not normally drink water. Because of the permeability of their skins, species that live in aquatic habitats face a continuous osmotic influx of water which they must balance by producing urine. Species in arid habitats rarely encounter enough liquid water in one place to drink it, and if they should find a puddle they can quickly absorb the water they need through their skins. The impressive adaptations of amphibians from arid regions are ones which facilitate rehydration from limited sources of water. One such adaptation is the **pelvic patch**. This is an area of highly vascularized skin in the pelvic area that is responsible for a very large portion of an anuran's cutaneous water absorption. Toads that are dehydrated and completely immersed in water rehydrate only slightly faster than those placed in water just deep enough to wet the pelvic area. In arid regions, water is frequently available as a thin layer of moisture on a rock, or in a very shallow puddle. Concentrating the cutaneous water uptake in the pelvic region allows an anuran to absorb this water by squatting down on it.

As one would expect, the pelvic patch is not confined to anurans from arid regions. It occurs in every family so far investigated except for the totally aquatic pipids. The rate of uptake of water through the patch is greater in species from arid habitats than those from moist ones; it takes the leopard frog (*Rana pipiens*) 48 hr to rehydrate from moist sand, whereas the red-spotted toad (*Bufo punctatus*), a desert species, takes only 6 hr. Uptake of water by toads may be facilitated by the structure of the skin outside the highly vascularized pelvic patch. A network of fine grooves draws water over the entire body surface by capillary action. It is not clear whether this water is absorbed or used for evaporative cooling.

10.5 SKIN GLANDS OF AMPHIBIANS

Aside from its permeability, the glandular nature of the skin of lissamphibians is its most striking feature. Three major types of glands are distinguished, hedonic glands (which were discussed in section 10.3.2), mucous glands, and poison glands (Figure 10–12). All play important roles in the biology of lissamphibians and are among the most important of the characteristics that have shaped their evolution.

10.5.1 Mucous Glands

Mucous glands are distributed over the entire body surface and secrete mucopolysaccharide compounds. The primary function of the mucus is to keep the skin moist and permeable. For an amphibian, a dry skin means reduction in permeability to water and gases. That, in turn, reduces oxygen uptake and the ability of the animal to use evaporative cooling to maintain its body temperature

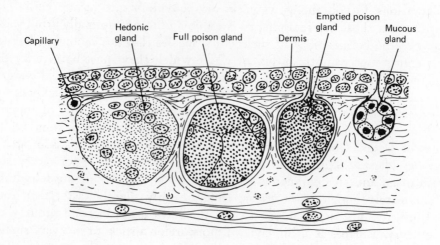

Figure 10-12   A cross section of skin from the base of the tail of a red-backed salamander, *Plethodon cinereus*. Three types of glands can be seen. (Modified from Noble [4].)

within equable limits. Experimentally produced interference with mucous gland secretion can lead to lethal overheating in frogs undergoing normal basking activity.

10.5.2 Poison Glands and Other Defense Mechanisms

Although there is some evidence that the secretions of the mucous glands of some species of amphibians are irritating or toxic to predators, the chemical defense system is located primarily in the poison glands. These glands are concentrated on the dorsal surfaces of the animal, and defense postures of both anurans and urodeles (Figure 10-13) present the glandular areas to potential predators. A wide variety of irritating and, in some cases, exceedingly toxic compounds are produced by amphibians. The compounds produced by different groups reflect their phylogenetic relationship, and skin toxins have been helpful in classification in some taxa.

Many amphibians advertise their distasteful properties with conspicuous aposematic, or warning, color and behavior. The arrow poison frogs of tropical America are among the most spectacular examples. These small frogs are only 2 to 7 cm long, but they are so brightly colored that it is virtually impossible to overlook them. Color ranges from sky blue through green to yellow, orange, red, and combinations of these colors. Unlike many amphibians, these species are active in daylight and make no attempt to conceal themselves under leaf litter. A predator that makes the mistake of seizing one is likely to spit it out because it is distasteful. The toxins in the skin may also induce vomiting that reinforces the unpleasant experience for the predator. Subsequently, the predator will remember its unpleasant experience and avoid the distinctly marked

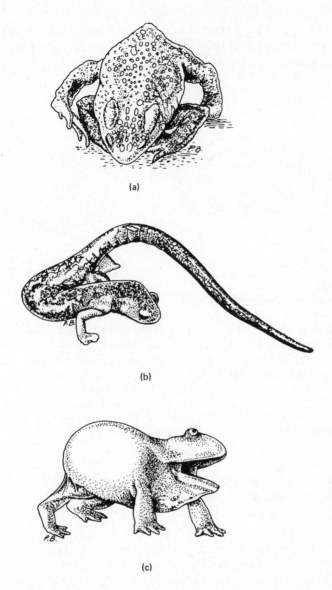

(a)

(b)

(c)

Figure 10-13 The defense postures of amphibians have the dual results of making the animal look larger and presenting to a predator the areas protected by poison glands. In toads the dorsal surface, and especially the rear of the head where the large parotoid glands are located, has the greatest concentration of poison glands and these are directed toward a predator (a). Many salamanders, like the plethodontid *Bolitoglossa* (b) have poison glands on the dorsal surfaces of the head, trunk, and tail. They present these surfaces to a predator, and may swat it with the tail when it approaches. Budgett's frog (*Lepidobatrachus*, c) from South America preys upon other frogs and is large enough to threaten predators by inflating its lungs with air and hopping forward with its mouth open. (Modified from the following sources: a from a photograph by W. S. Pitt in M. Smith [1966] *The British Amphibians and Reptiles*, 4th ed. Collins, London; b from a photograph by E. D. Brodie, Jr.; c from a photograph by G. Durrell in Cochran [1].)

animal that produced it. Some toxic amphibians combine a cryptic dorsal color with an aposematic ventral pattern. Normally the cryptic color conceals them from predators, but if they are attacked they adopt a posture that displays the brightly colored ventral surface (Figure 10–14).

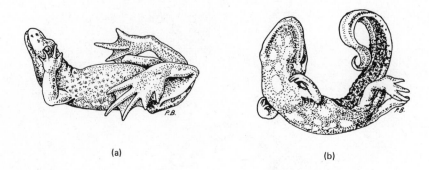

(a)

(b)

Figure 10–14. Aposematic displays of amphibians present bright colors which predators have learned to associate with the animals' toxic properties. (a) The European fire-bellied toad has a cryptically colored dorsal surface and a brightly colored underside which is displayed in the unken reflex when the animal is attacked. (b) The Hong Kong newt has a brownish dorsal surface and a mottled black and red venter, which is revealed by its aposematic display. Unlike the defensive displays illustrated in Figure 10–13, these aposematic displays present relatively poorly protected areas of the body to predators. (a, modified from H. Gadow [1909] *Amphibia and Reptiles.* Macmillan and Co., Ltd. London; b from a photograph by E. D. Brodie Jr.)

An aposematically colored species offers an opportunity for mimicry by another species that does not have such potent chemical deterrents. If the palatable species (the mimic) looks enough like the unpalatable one (the model) to convince the predators to refrain from attacking it, the mimic will derive an advantage from that resemblance. Several cases of mimicry have been described among amphibians, especially salamanders. The red eft stage of the red-spotted newt has already been discussed (Section 10.1.1). Red efts are classic examples of aposematic animals. Their color and behavior make them extremely conspicuous and they contain tetrodotoxin, a potent neurotoxin. Touching an eft to your lips produces an immediate unpleasant numbness and tingling sensation, and the behavior of animals that normally prey on salamanders indicates that it effects them the same way. As a result, an eft that is attacked by a predator is likely to be rejected before it is injured. After one or two such experiences, a predator will no longer attack efts. Some red salamanders (*Pseudotriton*) are sympatric with red efts and very similar to them in color. Edmund Brodie, Jr. found that predators that have learned by experience not to attack efts refuse to attack red salamanders, but naive predators that have not experienced red efts eat red salamanders.

Red salamanders were long considered to be palatable to predators, and thus they were thought to be **Batesian** mimics of the unpalatable red efts. Recently James Huheey and Ronald Brandon have found that red salamanders themselves have a skin secretion that is toxic. This discovery suggests that red efts and red salamanders are **Müllerian** mimics. Experience with either species of salamander should teach a predator to avoid all red salamanders in the future. Both species of salamanders would benefit from this sort of mimicry.

Although chemical defense mechanisms are extremely important to amphibians, they are not relied upon by all species. Some amphibians, ranid frogs, for example, are capable of rapid escape from predators, whereas others stand their ground and fight back. Amphibians that rely upon rapid escape appear to have a physiological specialization that distinguishes them from the more sedentary forms that rely upon poison or stand and fight. The California slender salamander (*Batrachoseps attenuatus*) attempts to escape predators by coiling its body and suddenly straightening. As it snaps straight, it propels itself into the air, moving several centimeters with each leap. More than 90 percent of the energy the animal uses for escape comes from anaerobic metabolism. Lactic acid accumulates as an end product of anaerobic metabolism, and within 1 or 2 min has reached such a high level that the salamander can no longer coil and spring.

The same pattern of high levels of activity and rapid fatigue is seen in other species that rely on flight to escape predators, including the leopard frog (*Rana pipiens*). In contrast, species that have sedentary defensive techniques are incapable of carrying on anaerobic metabolism at a high rate. As a result, a toad, for example, cannot leap away as quickly as a frog can; it just keeps moving at its usual placid hopping speed. On the other hand, because it is not producing as much lactic acid, a toad does not become exhausted. Low levels of continuous activity or brief periods of high activity are apparently alternative evolutionary specializations for amphibians. Closely related species may use different means to escape predation; the arboreal salamander (*Aneides lugubris*) relies upon its teeth to discourage predators and has a correspondingly low level of maximum activity, whereas the black salamander (*A. flavipunctatus*) runs away and has a higher capacity for anaerobic metabolism.

10.6   AMPHIBIANS AS AN EVOLUTIONARY INTERMEDIATE

In many respects the biology of living amphibians is determined by the properties of their skin. On one hand, as we have seen, the permeability of their skin limits them to microhabitats in which they can control water gain and loss; on the other hand, given the proper microhabitat, they can utilize the permeability of their skin to achieve a remarkable degree of independence of liquid water. Thus, the picture that is often presented of amphibians as animals barely hanging on as a sort of evolutionary oversight is misleading. Only a detailed examination of all facets of their biology can produce an accurate picture of an amphibian's life.

An examination of that sort reinforces the view that the skin is a dominant structural characteristic of Lissamphibia. This is true not only in terms of the limitations and opportunities presented by its permeability to water but as a result of the intertwined functions of the skin glands in defensive and reproductive behavior. The structure and function of the skin are the primary characteristics that have shaped the evolution and ecology of the Lissamphibia; they are also the major features distinguishing lissamphibians from Paleozoic amphibians. The differences between scaled and scaleless amphibians are so profound that lissamphibians provide only a very restricted insight into the lives of Paleozoic amphibians and the evolution of terrestrial vertebrates.

References

[1] Cochran, D. M. 1961. *Living Amphibians of the World.* Doubleday, New York. A lavishly illustrated semipopular review of the biology of living amphibians.

[2] Carroll, R. L. and P. J. Currie. 1975. Microsaurs as possible apodan ancestors. *Zoological Journal of the Linnean Society* 57(3):229-247. Illustrates the difficulties in trying to draw phylogenetic inferences from the amphibian fossil record.

[3] Mertens, R. 1960. *The World of Amphibians and Reptiles.* McGraw-Hill, New York.

[4] Noble, G. K. 1931. *The Biology of the Amphibia.* McGraw-Hill, New York. (Reprinted in 1955 by Dover Publications, New York.) A classic summary of amphibian biology, out-dated in some respects but still the best single source for information about many aspects of morphology and life history.

[5] Taylor, D. H. and S. I. Guttman (editors). 1977. *The Reproductive Biology of Amphibians.* Plenum Press, New York. Most recent review of various aspects of amphibian reproduction, genetics and behavior.

[6] Vial, J. L. (editor) 1973. *Evolutionary Biology of the Anurans.* University of Missouri Press, Columbia, Missouri. Various approaches to the study of amphibian evolution, paleontology, ecology, morphology and behavior.

11

Geology and Ecology
During the Mesozoic

Synopsis: The major geologic event of the Mesozoic Era was the rupture of Pangaea. Pangaea had coalesced in the Devonian, and it was not until the Jurassic, a span of over two hundred million years, that it split into two huge, but lesser continental masses—Laurasia to the north and Gondwanaland to the south. The reptiles, the subject of Chapters 12, 13, and 14, originated in the Carboniferous and were the dominant vertebrates during the Mesozoic Era. The early reptile species were widespread, until the Jurassic, for there were few barriers to their movement across Pangaea. The ends of both the Paleozoic and the Mesozoic Eras were characterized by periods of major mountain building. These orogenic forces created barriers that had profound effects on continental climates. Initially warm and moist, climatic conditions in the late Carboniferous and the Permian became more arid and cooler. Major extinctions occurred among plants and many invertebrates, as well as with some of the major groups of fishes and amphibians. In the Mesozoic Era the Triassic climate was warm and arid. In the Jurassic moisture returned, and isolation of continental floras and faunas took place because of the rupture of Pangaea. Isolation of the once contiguous terrestrial tetrapods led to new adaptive radiations, typified especially in the diverse Cretaceous ruling reptiles, the dinosaurs. Similar extensive radiations co-evolved in terrestrial flowering plants and in the pollinating insects. By the end of the Cretaceous, the ruptures of Laurasia and Gondwanaland led to the continents we recognize today. The Cretaceous, warm and moist for the most part, ended with orogenies and a cooling trend. Accompanying this climatic change was the extinction of many of the dominant reptiles, an event which made way for the radiations of birds and mammals.

363

11.1 CONTINENTAL GEOGRAPHY DURING THE MESOZOIC

Pangaea persisted from the Devonian until the end of Jurassic time—a period spanning almost 210 million years. But even following the Jurassic epoch, when Pangaea began to break apart, limited contact was maintained between the Northern Hemisphere (southern Europe) and Gondwanaland (North Africa). Plant and animal dispersals over this land route were, at least theoretically, still possible. Only in the Cretaceous did the newly forming oceans produced by the rupture of Pangaea become a barrier to dispersal of terrestrial plants and animals. During this long span of time when Pangaea remained a continental entity, the major evolutionary lines of terrestrial plants, invertebrates, and vertebrates were established. Among the vertebrates, Pangaea belonged to the reptiles, whose diversity of shapes and sizes dominates the fauna. The evolution of mammals and birds, however, is closely interwoven with the radiation of reptiles.

Geological and ecological changes surely affected each of these vertebrate classes, but often in different ways. Fossils of mammallike reptiles can be traced to the Pennsylvanian, much further back in time than the earliest known fossilized birds, which are limited to a few Jurassic finds. The origin and radiation of the various orders of birds and mammals, however, are associated largely with the Cretaceous and, especially, the Cenozoic Era. Here we place emphasis on the geology and ecology pertinent to reptile radiations that occurred in the late Paleozoic and Mesozoic Eras. Nevertheless, we must keep in mind that the physical and biotic factors that effected dramatic changes in reptiles also influenced the earliest mammalian and avian stocks. What major geological changes occurred?

In contrast to the dramatic continental movements of the early Paleozoic that led to the formation of Pangaea, little separate continental block drift took place until the Jurassic (compare Figures 8–1, 11–2, 11–3, and 11–4). Pangaea however, was not locked in position. A close examination of the figures indicates a slow northward drift of Pangaea from Devonian through Triassic time. For example, paleomagnetic data fix the position of the South Pole in South Africa in the Devonian (Figure 8–1), near the middle of Antarctica in the Carboniferous (Figure 11–1), further across the Antarctic in the Permian (Figure 11–2), on the far edge of Antarctica in the Triassic (Figure 11–3) and in the "Antarctic Sea" in Jurassic time (Figure 11–4). A similar comparison reveals that the magnetic North Pole during the Devonian was located in the ocean off Japan, and in Jurassic time within the "Arctic Ocean."

More subtle movements (rotary changes) also occurred, but the northward drift probably had the greatest effect on worldwide and local climates. The rate of northward drift was slow in terms of commonly used units (1.6 cm/yr). The total drift appears to have been slightly in excess of 30° latitude. Nevertheless, as imperceptibly slow as this northward displacement might seem, its effects were profound. For example, many areas of Pangaea that were located along equatorial latitudes in the Carboniferous were located in temperate latitudes by the Jurassic, whereas temperate locations were displaced to tropical

Figure 11-1 Continental positions during the Carboniferous. The earliest tetrapods, the ichthyostegans, are found in upper Devonian deposits of Greenland (⊕). Amphibians are found in several North American and European locations during the Carboniferous (C): labyrinthodonts, ○ = early C., ● = late C.; lepospondyls, △ = early C., ▲ = late C. The distributions closely approximate the Carboniferous palaeoequator, suggesting that the earliest amphibians, as widely believed, required moist, warm conditions. Ichthyostegan fossil sites in Greenland, during the Devonian, were close to the palaeoequator (see also Figure 8-1). The fossil distributions suggest that the early amphibian radiations may have spread eastward, southward, and westward from a center of origin located in northeastern Canada and the British Isles. Carboniferous reptiles show a similar pattern. The distribution of Carboniferous coal forests are shown by the letter C within the encircled areas. Many of the largest coal deposits coincide with the two main Devonian floral zones (Figure 8-1). The extent of glaciation in the late Carboniferous is shown by the dashed lines and the direction of glacial movement by the small arrows.

latitudes. Discrepancies remain to be resolved in this general picture, especially in the climates and biota of Antarctica. The important point is that Pangaea, although it remained a large, unified land mass through most of this time, was not static and that its drift resulted in changes in the climates of particular areas.

During late Triassic time the tectonic forces that were to rupture Panagea began to act. The first major event was the splitting of North America from South America and Africa (Figure 11-4). The Caribbean Sea and the North Atlantic Ocean resulted, although connections with Greenland and with Europe were retained. A single North Hemispheric land mass resulted, often called Laurasia. In comparison to the land units of the Northern Hemisphere today it lacked only India, which historically had long been a component of Gond-wanaland (Figure 4-3). The tectonic forces imparted a clockwise displacement. North America, for example, although moving slightly westward, also moved slightly north. At the same time Siberia and China moved south. Laurasia seemed to rotate about a focus near its contact with North Africa (Figure 11-4).

The breakup of Gondwanaland probably began in the late Jurassic with a continental separation that initially included South America, Antarctica and Australia as one unit and Africa and India as another unit. By the late Cretaceous, South America had separated from Antarctica. The splitting of Pangaea is an astonishing event, for this huge continent had persisted through 200 million years. Even more surprising is the breakup of Gondwanaland, which had maintained an identity from the Cambrian through most of the Jurassic, a period of approximately 430 million years.

The splitting of Pangaea affected not only the geographic position of the land but also produced a series of orogenic events that continue even today. For example, throughout most of the Triassic and Jurassic Periods, as well as most of the Paleozoic Era, a fairly continuous sequence of sediments was deposited in western North America within the Cordilleran geosyncline (Figure 4-2), which then represented the area now occupied by the coastal range and Sierra Nevada mountains of the United States. Near the end of the Jurassic these sediments were folded and uplifted to form the ancestral Sierra Nevadas, a range much wider and loftier then than now. In Cretaceous time a similar folding and uplifting of the midcontinental sediments of North America (and western South America) produced the Rocky Mountains and the Andes. This event, often referred to as the Laramide Revolution, closed the Mesozoic Era. Subsequent uplift and erosion in the Cenozoic produced the present altitudes of these lofty ranges. Today we realize that this Mesozoic orogeny coincides with the initial westward drift of North America and of South America as Pangaea broke apart.

As fascinating as the geological events underlying the fission of Pangaea are, it is the effects that these continental separations had on terrestrial organisms that most concern us. Immediate effects were the isolation of continents—for example, South America from Africa—and the influence of new mountain ranges on climate—for example, the Sierra Nevada and Andes ranges. Also, the separate continental movements, as with the northward drift of Pangaea

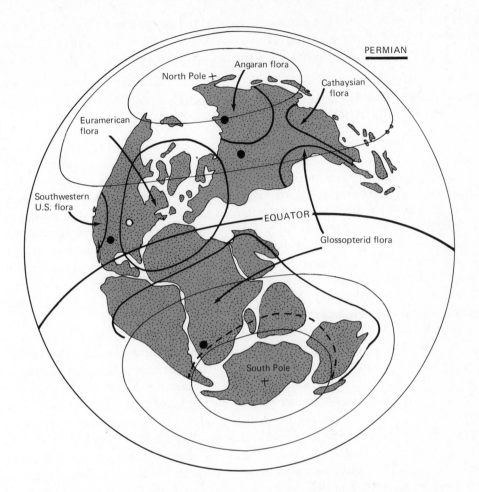

Figure 11-2 Floral regions and continental positions during the Permian. The migrating South Pole, as a result of the northward drift of Pangaea, transgressed most of Antarctica during the period from Carboniferous to Permian time. With this polar displacement, the glaciers retreated to the south in Africa, India, and South America (dashed line). By the late Permian, a glossopterid flora highly adapted to seasonally cooler and drier conditions had evolved and spread over all of Gondwanaland. In Laurasia, only the Angaran flora showed seasonal differences. Three essentially distinctive tropical floras (southwestern United States, Euramerican, Cathaysian) also existed. These tropical floristic regions show a slow transition from the Carboniferous coal type forests (spore bearing lycopods, and so on) to various gymnosperms (for example, conifers) as the Permian progressed. In Permian strata of south China and parts of Indonesia, a glossopterid flora is fossilized. This apparent paradox is explained if this region, like southern India, is considered to have been a part of Gondwanaland rather than Laurasia. Recent geological evidence supports this view (Figure 4-3).

No epicontinental seas are shown, as no single map can give a proper perspective. The Permian appears to be a period of great reduction in the extent of such seas followed by a brief reinvasion in the earliest Triassic.

Early reptiles were widespread in the Permian ($\bullet$), but seem restricted to the Euramerican floral province at their first appearance in the late Carboniferous ($\circ$).

from the Devonian into the Jurassic, also affected continental climates through latitudinal displacements. What then does the record tell us about climatic changes from the Carboniferous until the end of the Mesozoic?

11.2 CLIMATIC PRELUDE TO THE MESOZOIC— HABITATS DURING THE LATE PALEOZOIC

Extensive forests spread across Pangaea during the Carboniferous. Fossilization of these earliest forests produced major coal deposits. Indeed, the Carboniferous takes its name from the coal deposits. Giant club mosses were dominant—large arborescent lycopods, especially the genera *Lepidodendrum* and *Sigillaria*. Incredible horsetails, some 12 to 15 m tall, constituted a secondary element in these forests. Early ferns were also prevalent. By the middle Permian, however, almost all of the lycopods and horsetails had become extinct. Only a few herbaceous forms, *Lycopodium* in the club mosses and *Equisetum* among horsetails, survive today. The extinction of this extensive non-seed-bearing flora represents one of the major changes that has taken place in the earth's vegetation. What caused it?

There are no simple answers. In the late Carboniferous and the early Permian, however, a series of major glaciations took place in Gondwanaland. Ice sheets must have covered all of Antarctica, most of Australia, and at least the southern halves of India, Africa, and South America (Figures 11–1 and 11–2). In addition, extensive evaporites, deposits usually indicative of aridity, occur over large areas of the Permian landscape between 15° and 30°N paleolatitude. A broad retreat of continental seas, as a result of general uplift of the land and the containment of water in glacial ice caps, may also have influenced the demise of these forests. This extensive and general cooling in the late Carboniferous, followed by increased aridity in the Permian, suggests that many of the very water-dependent sporebearing plants could not compete with the evolving seedbearing plants.

A transitional flora is fossilized in late Carboniferous rocks from most of the land mass that comprised Gondwanaland. Represented are two major groups: (1) elements of the early Carboniferous flora and (2) a new group of plants, the glossopterids. This flora is interspersed with glacial deposits, suggesting that it could endure colder conditions. Indeed, the glossopterid flora evolved a suite of characteristics considered highly adaptive to cool climates: seasonal growth of wood and leaves, leaf clusters with short shoots, and varying degrees of seed enclosure to protect the plant embryo from desiccation. By the late Permian, when glaciation had ceased, the older elements had been replaced by a pure glossopterid flora.

Four distinctive floral regions existed in Laurasia in Permian time (Figure 11–2). Unlike the sudden appearance and explosive radiation of the glossopterids in Gondwanaland, a much slower and more orderly transition from spore bearers (for example, lycopods) to gymnosperms (for example, conifers) occurred in the Laurasian floral regions. The southwestern United States, Euramerican, and

Cathaysian (Chinese) floras possessed tropical characteristics, whereas the Angaran (central Asian) flora, as in the Devonian (see Figure 8-1), retained temperate characteristics. The virtual extinction of the non-seedbearing plants during the Permian, therefore, followed two different paths, one of which produced a uniform and widespread flora in Gondwanaland and the other a series of interrelated regional floras in Laurasia.

A surprising parallel exists among early tetrapods. Amphibians flourished in the Carboniferous and early Permian. Indeed, they dominated other terrestrial animal groups. But these amphibians were largely replaced by the early reptiles, like the captorhinomorphs and pelycosaurs, in the middle Permian. Could the same ecological conditions of the Carbo-Permian period (mainly lowered temperatures and aridity) have affected both the flora and the fauna in the same way? It is hard to be sure, but the parallel between extinctions and replacements with ecological change suggests something more than coincidence.

11.3 MESOZOIC CLIMATES AND HABITATS

In North America, the Paleozoic and Mesozoic boundary is marked by the Appalachian Revolution, a period of mountain building that initiated the folding and uplift of the present Appalachian Mountains of the eastern United States. This uplift began in late Carboniferous time and continues the middle Carboniferous orogeny that produced the Ouachita Mountains of the southwestern United States. Each of these mountain belts resulted from compressional forces produced in the coalescence of Pangaea. Because of a corresponding general and continuing uplift, by the end of the Permian all of North America was above sea level.

The Triassic climate reflects a continuation of the late Permian climate. Overall the climate was dry and perhaps warmer, although latitudinal differences and seasonal changes in climate must have existed. There is little evidence of continental glaciation, perhaps in part because the North and South Poles lay over or very close to open oceans (Figure 11-3). Widespread aridity is indicated by the abundance of red beds and dune sands, as well as a restriction in the coal deposits that were so typical of the Carboniferous and Permian periods. Thus, desert and semidesert conditions must have been prevalent in Triassic time. Nevertheless, various types of reptiles are found in Triassic deposits throughout most of Pangaea (Table 11-1). Geographic barriers must have acted only as a limited impediment to faunal movements.

Floristically, in the Triassic the seedbearing plants came into their own. Lycopods and horsetails dwindled mostly to small herbaceous forms. Important gymnosperms were conifers, cycads, and bennettitales, ginkgos and pteriodsperms (seed ferns). The unique Permian glossopterid flora of Gondwanaland persisted through the Triassic but disappeared as it was replaced by other gymnosperms of uncertain affinities (*Thinnfeldia, Dicroidium*). The dramatic

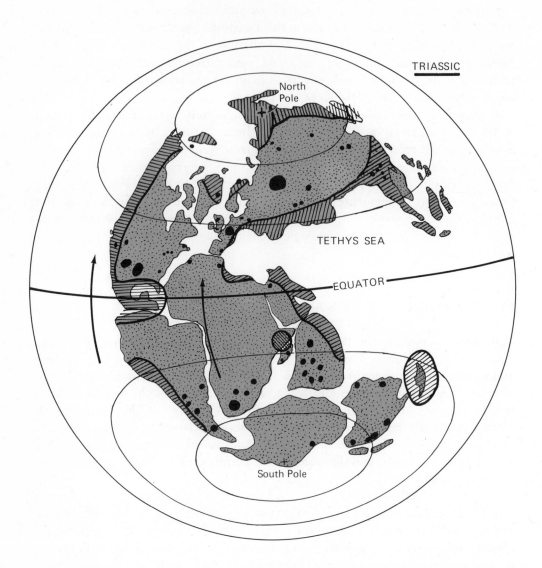

Figure 11-3 Distribution of epicontinental seas and land tetrapods during Triassic time. The relief of Pangaea throughout the Triassic increased. Epicontinental seas became limited mostly to continental margins (cross hatching). Land tetrapods were widely distributed as shown by the fossil sites (black dots; extensive sites shown by larger dots). Surprisingly few terrestrial vertebrates are fossilized in tropical latitudes (for example northern South America and Central Africa). This probably results from poor conditions for their fossilization rather than their absence from a tropical area. The existence of Pangaea is supported by the high percentage of taxa common to the various fossil sites (Table 11-1 and text for details). This high taxonomic affinity over such a wide are can be accounted for only if the various continents were confluent and intercontinental migration possible. The large arrows indicate the general northward drift of Pangea that took place from late Devonian through Triassic time.

Table 11-1. Faunal affinities of land tetrapods in the Triassic*
(Percentages are based on the number of families that each region shares.)

	Europe	Africa	Asia	India	South America
North America	88	75	44	59	56
Europe		58	80	81	71
Africa			89	75	74
Asia				41	49
India					56

Source: After Cox in Hallam (Ch. 4[1]).
*A comparison with Figure 11–3 reveals that regions with more than 50 percent affinity are geographically closer than regions with less than 50 percent affinity.

evolutionary floral changes instituted in the Permian persisted well into the Triassic with a continuing selection of adaptations fitting plants for survival in less moist and in cooler conditions.

The Jurassic Period, which ushers in the breakup of Pangaea, was characterized by a continuation of warm conditions, but was moister than the Triassic. Evaporites and red beds are confined within 30°N and S latitudes. Temperate-type coals (peat) are common only in high paleolatitudes (30–70°). Glacial deposits are unknown. Perhaps the opening of the Atlantic produced sufficient evaporation to provide a new flow of moisture to the middle of the continents (compare Figures 11–3 and 11–4). Also, shallow epicontinental seas once again covered parts of central North America and Europe. The epicontinental seas withdrew in Europe at the end of the Jurassic only to return again later in the Cretaceous.

In the early Cretaceous the climate seems to have been fairly uniform, both warm and moist, over widespread areas. Thus, subtropical organisms were fossilized as far south as 70° paleolatitude and coals are found even further south. Although continental glaciation was virtually absent throughout the Mesozoic, cooling set in late in the Cretaceous.

The Laramide Revolution which began in the late Cretaceous must have resulted in drastic but local climatic changes (Figure 11–5). The Rocky Mountains in the United States and the early Andes in South America, produced by this great period of mountain building, would have disrupted the flow of air over these continents, as they do today, and produced complex local climatic changes.

In summary, Mesozoic climates were characterized by a latitudinally widespread and prolonged warm, dry period, followed by continued warmth but increased moisture. Only toward the close of the Mesozoic did temperatures decline. During the Triassic epicontinental seas were limited, but in Jurassic and Cretaceous times they became extensive, showing withdrawal at the close of each epoch.

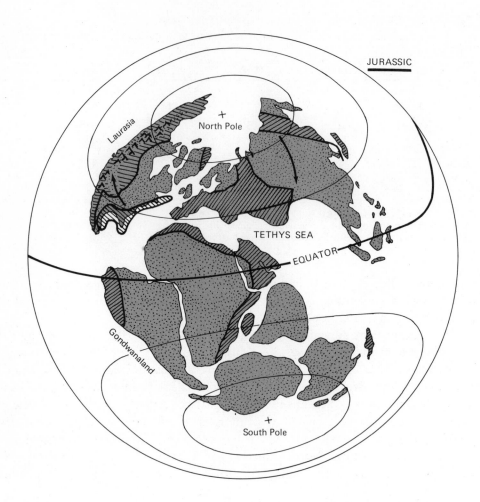

Figure 11-4 The break-up of Pangaea—Jurassic time. The initial opening of the North Atlantic Ocean occurred in Jurassic time. In some areas epicontinental seas (cross hatching) were more extensive than in the Triassic. The northward drift of land continued, but Laurasia also began to rotate (arrows), ripping North America from its contact with South America and North Africa and diminishing the size of the Tethys Sea. Toward the end of the Jurassic, compressional forces generated by the northwest drift of North America produced the ancestral Sierra Nevada Mountains (ʌʌʌ).

11.4 FLORA AND LAND ANIMALS OF THE CRETACEOUS

Floristic changes in the Cretaceous represent one of the major events in the evolution of plants. The epoch is renowned for the rapid (explosive) evolution of the angiosperms, or flowering plants. The first fossil angiosperms are found in early Cretaceous sediments. Various fossilized Jurassic plants are claimed to represent primitive angiosperms, but their identification is uncertain and their

Figure 11-5 The break-up of Laurasia and Gondwanaland—Cretaceous time. Gondwanaland ruptured into three major continental blocks, but Africa retained contact with Eurasia. General drift direction is shown by the arrows. North America retained intermittent connections with Eurasia. The Cretaceous ended in North America with the Laramide Revolution, the uplifting of the ancestral Rocky Mountains. Volcanic activity was common throughout most of the Mesozoic, especially along the western boundaries of North and South America. Movements of "Pacific" oceanic plates under continental margins released the necessary large quantities of volcanic magmas. In South America extensive volcanism and igneous intrusions at the end of the Cretaceous produced the granitic base of the Andes. Similar volcanic activities occurred along the margins of the Tethys Sea, especially in southern Europe and Middle East locations. A variety of localized orogenies, resulting from oceanic and continental plate interactions, took place. Shallow seas were extensive over many of the newly forming continents (cross hatching). The proximity of South America, South Africa, Madagascar and India probably was closer than indicated, at least during the early to middle Cretaceous. Sauropod dinosaurs of the genus *Titanosaurus* occur in upper Cretaceous sediments of Argentina, Niger and India; and species of *Laplatosaurus* occur in Argentina, Madagascar and India.

origin remains obscure. Most botanists believe that angiosperms evolved in tropical regions and then dispersed south and north to cooler regions. On this basis, according to drift theory, Africa and South America seem the most likely sites for their origin (Figure 11–4 and 11–5). The African paleobotanist Edna P. Plumstead believes that the close affinities between living Southern Hemispheric plants (African, South American, Australian) can be traced back to an origin in Gondwanaland. Further, she suggests that modern angiosperms possibly represent the culmination of the glossopterids of Permo-Triassic time. There are morphological similarities between fossil glossopterids and many living angiosperms, but in the absence of transitional fossils (Jurassic) this attractive hypothesis remains debatable.

Whatever their origin, the Cretaceous angiosperms were, relatively speaking, an instant success. What advantages did they possess over other plants? A precise answer is not available. We can only speculate from what we know about modern angiosperms in comparison to modern gymnosperms. Two major differences that may have provided angiosperms with an advantage were (1) fruits to enhance protection of the embryo and insure its dispersal and (2) improved pollination (fertilization). These differences between angiosperms and gymnosperms must have been less perfected in the Jurasso Cretaceous periods when they evolved. But even slight differences that enhance reproductive success and dispersal would be favored and rapidly fixed through natural selective forces. Only the conifers among gymnosperms appear to have withstood the evolutionary onslaught of the angiosperms. The pteridosperms, so common in Triasso-Jurassic time, became extinct, and the ginkgos and cycads were greatly reduced in numbers.

Little has been said of the evolution of terrestrial invertebrates in describing the changing earth's history. Chapter 18 considers a series of important evolutionary events that are intricately interwoven with the evolution of the mammals—the rise of the angiosperms and radiation of the insects. The earliest insects, found in late Devonian deposits, were followed in Carboniferous time by primitive silverfish, grasshoppers, and roaches. In the upper Carboniferous, flying insects such as dragonflies, stone flies, and beetles evolved. Few new orders are recorded in the Triassic. In contrast, a profusion of new groups suddenly appears in the Jurassic. Earwigs, caddis flies, dipterans (flies, mosquitos), and hymenopterans (bees, wasps, ants) all enter the fossil record. During the Cretaceous these latter groups rapidly radiated into diverse forms in parallel with the diversification of angiosperms.

11.5   MESOZOIC EXTINCTIONS

The close of the Paleozoic era, as we discussed earlier, was a period of major floral and faunal changes—replacement of spore-bearing plants by gymnosperms and of amphibians by reptiles. So also was the end of the Mesozoic, for the ruling reptiles and several invertebrate groups disappeared along with several lineages of plants. The next radiations include the birds and mammals among vertebrates and the angiosperms among plants. The cause of the late Mesozoic extinction of major reptilian taxa has been the subject of heated controversy over many years. We already know that the Mesozoic was an era of great geographic changes. Indeed, the separation of the land into our present continents was largely a Cretaceous occurrence (Figure 11-5). Did geographic isolation produce extinctions? Did the accompanying climatic changes do so? There are as many hypotheses as questions, some more plausible than others. Before dealing with the Cretaceous extinction of the ruling reptiles, let us consider what reptiles are and how they have changed from their origin in the Carboniferous to the present.

References

[1]   References 2, 5, 6 and 9 in Chapter 4 provide a variety of illustrations on the positions of the continents before and during the Mesozoic.

[2]   Colbert, E. H. 1973. *Wandering Lands and Animals*. E. P. Dutton, New York. A delightful general account of the theory of continental drift and how it has affected vertebrate evolution.

[3]   Spinar, Z. V. 1972. *Life Before Man*. American Heritage Press, McGraw-Hill, New York. A simple description of the evolution of life, this book is enjoyable for its many colored illustrations that provide a visual image of how past habitats, and the plants and animals that lived in them, may have appeared. Vertebrates are especially well treated.

12

Origin and Early Radiation of Reptiles

Synopsis: The evolution of reptiles in the late Carboniferous was coincident with a major radiation of terrestrial insects. The earliest reptiles (Table 12-1), the romeriid captorhinomorphs, were small, superficially lizardlike animals that probably fed on these terrestrial insects. Increased body size developed quickly after the initial appearance of reptiles, and the pelycosaurs of the late Pennsylvanian and early Permian were 2 to 3 m long. Pelycosaurs split into herbivorous and carnivorous lineages. One of the carnivorous groups, the sphenacodonts, appears to have given rise to the therapsids, a group still more specialized for an active terrestrial life. The therapsids radiated into a number of carnivorous and herbivorous lineages. Most of the herbivores were large animals, probably relatively clumsy and slow-moving. In contrast, the fossils of carnivorous therapsids give an impression of fleetness and agility. Physiological adaptations enhancing the ability to sustain high levels of activity probably accompanied the structural changes that we can see in fossil therapsids. It is likely that advanced forms had an insulating covering of hair and maintained their body temperatures above air temperature by metabolic heat production. The ecological role of these carnivorous therapsids was probably more mammalian than reptilian.

12.1 THE EVENTS LEADING TO THE EVOLUTION OF REPTILES

In the Upper Carboniferous there was an evolutionary event which, although it did not involve vertebrates, was to have a tremendous influence on the course of their evolution. This event was the radiation of insects in terrestrial habitats. Insects had appeared in the Devonian, and we have speculated about the role they might have played in the evolution of the earliest tetrapods. Through the Mississippian, however, the diversity of terrestrial insects appears to have remained limited, and the insects themselves may have been confined to areas

377

Table 12.1 Taxonomic arrangement of early reptiles and the
reptilelike mammals.

Class Reptilia

Subclass Anapsida

  Order † Captorhinomorpha (includes romeriids and captorhinids)
  Order † Mesosauria (*Mesosaurus* and related forms)
  † "Cotylosaurs" (a heterogeneous collection of large herbivores.
  Like the "batrachosaurs" discussed in CHapter 9, the cotylosaurs
  probably include diverse phylogenetic lineages. Procolophonids and
  pareiasaurs are included. If *Diadectes* and related forms are not
  considered batrachosaurs, they are included in the cotylosaurs.

Subclass † Synapsida

  Order † Pelycosauria
    Suborder † Ophiacodontia (relatively unspecialized carnivores)
    Suborder † Sphenacodontia (specialized carnivores including *Dimetrodon*)
    Suborder † Edaphosauria (specialized herbivores including *Edaphosaurus*
     and *Casea*)

  Order † Therapsida
    Suborder † Pthinosuchia (primitive carnivores)
    Suborder † Theriodontia (predominantly carnivorous forms including
     the relatively primitive gorgonopsids and the more advanced cynodonts,
     therocephalians, and bauriamorphs as well as the rodent-like tritylodonts)
    Suborder † Dinocephalia (very large herbivores)
    Suborder † Anomodotia (dromasaurs— small herbivores—and dicynodonts—
     large herbivores)

† = extinct.

near water. In the early Pennsylvanian the situation changed. The fossil record indicates that there was an abrupt expansion of many orders of insects, including dragonflies, stoneflies, and roaches. It is probable that the diversity of insects in the fossil record at this time reflects their spread into a variety of terrestrial habitats. The radiation of terrestrial insects was probably a response to the increasing quantity and diversity of terrestrial vegetation in the Carboniferous.

It is difficult to escape the conclusion that each of these groups, which today are often so interdependent (for example, bees seeking a food source and angiosperms requiring pollinators), rapidly exploited this new symbiotic niche to their mutual benefit. How did these changes affect vertebrates? One needs only think of how dependent many modern birds, mammals, reptiles, and amphibians are upon flowering plants and insects to realize that their mutual evolution has profoundly influenced vertebrate evolution.

Terrestrial vertebrates at that time were probably carnivorous. (No adult amphibian among living forms is herbivorous, and there is no evidence in the fossil record to suggest that Paleozoic amphibians were herbivores). Carnivorous vertebrates could not respond directly to the energy supply offered by terrestrial plants, but they could and apparently did respond to the opportunities presented by the radiation of insects. Probably for the first time in evolutionary history there was an adequate food supply to support fully terrestrial vertebrate predators.

The two forces primarily responsible for the origin and initial radiation of reptiles were directly related to the exploitation of this new energy source. One was the evolution of a more effective jaw mechanism specifically adapted to feeding on insects. The evolution of more effective jaws was accompanied by changes in body structure that permitted more effective locomotion on land. A final change, the evolution of the amniotic egg, was the definitive step that separated reptiles from their amphibian ancestors.

12.1.1 Carboniferous Amphibians and Reptiles

Several groups of vertebrates evolved specializations for terrestrial life in the Carboniferous (Figure 12-1). We have already mentioned the microsaurs and the rhachitomes (temnospondyls). In the anthracosaur lineage, a number of groups (collectively called batrachosaurs) developed extensive adaptations for terrestrial life.

Robert Carroll has stressed the importance of the small body size of early reptiles and their hypothetical amphibian ancestors. Paleozoic reptiles were almost all small animals, and the earliest fossils of the lineages we know are the smallest forms, increasing progressively in size through the Pennsylvanian and early Permian. It seems probable that small body size was an important feature in the development of reptiles in several respects. In the first place, it simplified the transition to a fully terrestrial life. The mass of an animal's body increases as the cube of its linear dimensions. A 10 cm long animal weighs only about one eighth as much as one 20 cm long. The lighter the animal, the less skeletal modification was needed to support the body weight on land without the buoying effect of water. Also it is believed that only a small animal could have first produced the amniotic egg (see section 12.2.2).

Feeding mechanisms underwent adaptive modification in early terrestrial vertebrates. These structural alterations changed the mechanical function of the jaws from quick seizure of prey to the application of crushing force to prey after capture.

In the rhipidistian crossopterygians and in Paleozoic amphibians, the jaw muscles produced their maximum force when the mouth was open. The insertion of the muscles on the posterior part of the mandible gave them leverage to produce a quick snap of the jaws that impaled prey on the enlarged palatal teeth (Figure 12-2). When the jaw was closed, the muscles could exert very little

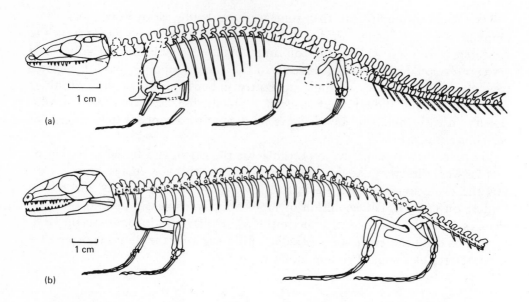

Figure 12-1 Among the anthracosaur labyrinthodonts becoming adapted for terrestrial life in the Carboniferous were the gephyrostegids, *Bruktererpeton* (a). These were small, agile animals with body lengths of 10–20 cm. The earliest reptiles known, the romeriid captorhinomorphs, *Paleothyris* (b), were very similar to the gephyrostegids and may have evolved from that group. (Modified from R. L. Carroll [1969] *Biological Review* 44:393–432.)

force because their angle of contraction was too nearly parallel to the jaw. In primitive reptiles it was apparently important to be able to apply a crushing force to prey while it was being held in the mouth, and this ability was enhanced by a progressive differentiation of the single muscle mass characteristic of fishes and labyrinthodonts into two distinct masses, the **temporalis** and the **pterygoideus.** The pterygoideus inserted on the lateral surface of the mandible and had its origin on the pterygoid. It had the leverage to produce quick jaw closure, but little force once the jaws were closed. The temporalis originated on the rear of the skull and inserted on the posterior dorsal surface of the mandible. When the jaws were closed, it ran nearly perpendicular to the jaw and thus had a great mechanical advantage and could produce the static pressure characteristic of reptilian jaws. The structure of the palate and the presence or absence of pterygoid flanges for attachment of a differentiated pterygoideus muscle are among the key characters used by paleontologists to distinguish Paleozoic amphibians from reptiles. The importance of a jaw mechanism allowing application of static pressure is emphasized by the fact that the first radiation of reptiles coincided with its development.

12.2 THE AMNIOTIC EGG

A major difference between living amphibians and reptiles is the occurrence of an **amniotic egg** in the latter group. The amniotic (or cleidoic) egg is sometimes referred to as the "land egg," but this is a misnomer. As we have seen, a large number of amphibians and some fishes have anamniotic eggs that develop quite successfully on land. Many terrestrial invertebrates also lay anamniotic eggs. Even the differences in moisture requirements of anamniotic and amniotic eggs are not great if the incubation requirements of eggs of modern reptiles can be used as a guide. These must have relatively moist conditions to avoid desiccation. Nonetheless, paleontologists have long regarded the evolution of the amniotic egg as a major evolutionary event and the definitive character that distinguishes reptiles from amphibians.

The amniotic egg, as we know it, is characteristic of reptiles, birds, monotremes, and, in modified form, of therian mammals as well. It is assumed to have been the reproductive mode of Mesozoic reptiles, and fossilized dinosaur eggs are relatively common in some deposits. An amniotic egg is a remarkable example of biological engineering (Figure 12-3). The shell, which may be leathery or calcified, provides mechanical protection while allowing movement of respiratory gases and water vapor. The albumin (egg white) gives further protection against mechanical damage and provides a reservoir of water and protein. The large yolk is the energy supply for the developing embryo. At the beginning of embryonic development, the embryo is represented by a few cells resting on top of the yolk. As development proceeds these multiply, and endodermal tissue surrounds the yolk and encloses it in a yolk sac that is part of

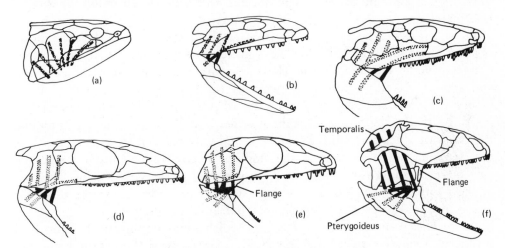

Figure 12-2 Evolution of jaw muscles in tetrapods. In fishes (a, the rhipidistian *Ectosteorhachis*) the jaw muscles form a fan-shaped array that was little changed in *Ichthyostega* (b). In the anthracosaur lineage there was increasing differentiation of the muscle mass into forward-directed components that produced a quick closure and upward-directed units that applied pressure after the mouth had closed (c, *Palaeogyrinus*; d, *Gephyrostegus*). In advanced anthracosaurs (e, *Paleothyris*), a flange of the pterygoid bone (T) provided an origin for the mouth-closing muscle (the pterygoideus), and the temporalis that produces a crushing force when the mouth is nearly closed was well differentiated. The modern *Iguana* (f) shows the same arrangement of muscles. (Modified from R.L. Carroll [1969] *Biological Review* 44:393–432.)

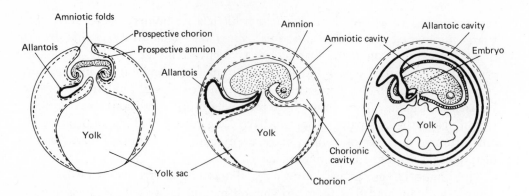

Figure 12-3 The extraembryonic membranes of the amniotic egg are formed by outgrowths from the embryo. The chorion and amnion start as a pair of two-layered folds at the anterior and posterior ends of the embryo. They meet and fuse to form the chorion, which lies just inside the egg shell, and amnion, which surrounds the embryo itself. The allantois is an outgrowth of the hindgut that eventually occupies most of the chorionic cavity. (Modified from T.W. Torrey [1962] *Morphogenesis of the Vertebrates.* John Wiley & Sons, New York.)

the developing gut. Blood vessels differentiate rapidly in the tissue of the yolk sac and transport food and gases to the embryo. By the end of development, only a small amount of yolk remains, and this is absorbed before or shortly after hatching.

In these respects the amniotic egg does not differ greatly from the anamniotic eggs of amphibians and fishes. The significant differences lie in three other extraembryonic membranes—the **chorion, amnion,** and **allantois.** The chorion and amnion develop from outgrowths of the body wall at the ends of the embryo. These two pouches spread outward and around the embryo until they meet. At their junction, the membranes merge and leave an outer membrane, the chorion, which surrounds the embryo and yolk sac, and an inner membrane, the amnion, which surrounds the embryo itself. The allantoic membrane develops as an outgrowth of the hind gut posterior to the yolk sac and lies within the chorion. It functions as a respiratory organ and as a storage place for nitrogenous wastes produced by the metabolism of the embryo.

As we have seen, the large-yolked eggs of fishes and amphibians have evolved a variety of surfaces to facilitate exchange of gases with the external environment. In most cases these surfaces are outgrowths of the body, usually the gills or tail, and are reabsorbed by the embryo at hatching. As a result, they do not provide a disposal site for waste products. In contrast, the allantois is left behind in the egg when the embryo emerges, and the wastes stored in it do not have to be reprocessed. This is one of the respects in which the amniotic egg represents an advance over anamniotic eggs. Another is the efficiency of the gas exchange system, which allows amniotic eggs to be very large. Eggs of some large birds weigh more than a kilogram. The largest terrestrial anamniotic vertebrate eggs scarcely exceed a gram in weight.

12.2.1 Origin of the Amniotic Egg

There is no direct evidence about the evolution of the amniotic egg among advanced amphibians or early reptiles. The oldest fossilized eggs, from Permian sediments, show that amniotic eggs had evolved by that time, but there is no way to tell which of the various amphibians and reptiles found with the eggs were the ones that produced them. Robert L. Carroll has suggested the following sequence of steps in the evolution of amniotic eggs:

1. Development of a terrestrial habit by the adult.
2. Initiation of internal fertilization.
3. Reduction in body size.
4. Reduction in number of eggs, with increase in size and quantity of yolk in each egg.
5. Abbreviation and later elimination of independent larval stage.
6. Laying of small anamniotic eggs on land.
7. Development of amniotic membranes.

The widespread development of stages 1–6 among lissamphibians illustrates the strength of selective forces favoring those changes.

Modern amniotic eggs are seen in their basic form in reptiles, birds, and prototherian mammals. In metatherian and eutherian mammals they appear in a modified form. In all these animals the details of embryonic development are so similar that it seems unquestionable that all are descended from a common group in which the amniotic egg had evolved. The most recent common ancestor of all those groups is found among the captorhinomorphs of the Carboniferous (Figure 13–1). Thus, it seems likely that these animals had evolved an amniotic egg and were, by the narrowest of definitions, true reptiles.

12.2.2 "How to Tell the Type of Eggs an Animal Lays by the Shape of Its Ears"

Robert Carroll emphasized the importance of small body size in relation to the evolution of the amniotic egg in a lecture with the above title, and later expanded his ideas in a symposium honoring A. S. Romer. Carroll pointed out that plethodontid salamanders that lay anamniotic eggs on land are the closest living parallels to amphibians that were evolving terrestrial habits in the Paleozoic. Among living plethodontid salamanders, the size of the eggs is directly related to the size of the adults. The largest eggs have diameters of 7 to 8 mm and are laid by salamanders with head-plus-body lengths of 70 to 80 mm. It appears that 8 mm is the largest diameter possible for a salamander egg. Apparently larger eggs do not have enough surface area to permit the rate of gas exchange required to sustain the embryo.

Carroll suggested that the same physical constraints would have applied to the anamniotic eggs of terrestrial Paleozoic amphibians, and that the maximum head-body lengths for the transitional forms between amphibians and reptiles probably did not exceed 80 to 100 mm. An examination of the fossil record does not, at first, appear to support this hypothesis. The fossils presently known are not those of transitional forms but of animals clearly on the amphibian side of the transition, the gephyrostegid anthracosaurs, or clearly on the reptilian side, the romeriid captorhinomorphs. Both groups of fossils are larger than Carroll's prediction. Adult gephyrostegids and romeriids seem to have had head-body lengths around 200 mm. Nonetheless, Carroll believes that his hypothesis is valid because there are features in the skulls of romeriids that suggest that the fossils we know had evolved from smaller ancestors.

This interpretation is based on the relationship between the size of the semicircular canals and the size of the entire skull in vertebrates. There is little difference in the size of the semicircular canals in vertebrates smaller than 2 kg. Species weighing 10 g have semicircular canals nearly as large as those of species 200 times heavier. Apparently a certain radius of curvature is required in the canals for proper function, and the radius cannot be reduced beyond that limit. As a result of the relative uniformity of the absolute size of the semicircular canals, the otic capsules that house the canals occupy relatively more space in the skull of a small vertebrate than a large one. In lizards weighing about 10 g, the semicircular canals occupy 75 percent of the width of the skull. Similar ratios are seen in small fossils.

On the basis of the relationship between the lateral extent of the semicircular canals and the width of the skull, Carroll estimated that in Paleozoic vertebrates with skulls less than 30 mm long the otic capsules would have approached the sides of the skull. Changes in the structure of the rear of the skull would be needed to accommodate the otic capsules. These changes are found in romeriids, although the fossils in which they are seen are large enough so that the otic capsules do not distort the skull. Carroll reasons that the presence of these changes in the skulls of animals that are too large to require them indicates that the romeriids we know evolved from smaller ancestors, which did have heads small enough to require structural changes to accommodate the otic capsules.

Carroll estimated that the head-body length of an animal small enough to require this rearrangement would be 80 to 150 mm. Animals that small would also have been small enough to lay terrestrial eggs. Thus, the structure of the otic region of the skull casts light on the mode of reproduction that might have been utilized by ancestors of the romeriids. The batrachosaurs do not show the key rearrangement of the rear of the skull, suggesting that unlike the romeriids their lineages had not passed through a stage in which the body size was small enough to have permitted terrestrial reproduction. These forms presumably laid eggs in water, and the young passed through an aquatic larval stage before taking up a terrestrial life as adults. Thus, Carroll concluded that the amniotic egg probably evolved only once, in an evolutionary line that ran from gephyrostegid anthracosaurs through yet-unknown intermediate forms to the romeriid captorhinomorphs.

The romeriids were small, lightly built reptiles that probably occupied much the same adaptive zone as modern lizards (Figure 12-1). They are thought to have fed on invertebrates that they caught on land, probably competing with and replacing the microsaurs by the end of the Permian. Long before that, however, the romeriids had radiated and produced a number of specialized lineages (Figure 12-4). Some of these romeriid offshoots flourished briefly but died out without leaving descendants. *Mesosaurus*, for example, was a marine reptile about a meter long with an elongated snout armed with many sharp teeth. It may have fed on small marine invertebrates. The terrestrial procolophonids and pariesaurs, probably also derived from romeriids, had short skulls and broad teeth adapted for crushing their food. In their general body form they resembled *Diadectes* as well as the herbivorous pelycosaurs that were their contemporaries (Figure 12-5).

The correct taxonomic allocation of these captorhinomorph derivatives is not clear. Even their origin from romeriids is only assumed—there are no transitional fossils known. They differ considerably from the romeriids in their heavy body form, large size (up to 3 m), and specializations for an herbivorous diet. They are often referred to as "cotylosaurs." Like the term "batrachosaurs" discussed in Chapter 9, "cotylosaurs" has been used in different senses by various paleontologists. For our purposes it is a grouping of convenience that should not be interpreted as implying that we know of a close phylogenetic relationship among the animals included.

Diadectes, which was classified as a batrachosaur in Chapter 9, is considered a reptile by many paleontologists who include it among the cotylosaurs. This uncertainty about which class a relatively well-known form belongs to illustrates the difficulty of using criteria derived from living animals to classify fossils.

Four other groups of reptiles that arose from the captorhinomorphs had greater evolutionary success and are accorded subclass rank. Three of these subclasses, the Lepidosauria, Archosauria, and Euryapsida, underwent their major radiations in the Mesozoic. These groups are discussed in Chapter 13. Before that, however, in the Permian and Triassic there was a major radiation of another subclass of reptiles, the Synapsida, that led rapidly (in geological time) through the Orders Pelycosauria and Therapsida to the Class Mammalia. For 100 million years these animals were extraordinarily successful. In the lower Permian more than 70 percent of the known reptile genera were pelycosaurs, and in the upper Permian 84 percent were therapsids.

12.3 THE PELYCOSAURS (Subclass Synapsida)

In the romeriids the skull was solid and muscles ran inside the dermal bones. This is the **anapsid** skull condition ("without openings") that gives the Subclass Anapsida its name. Although this skull structure was strong and rigid, it was also heavy, and the total mass of the jaw muscles was limited by the space that was available between the outer dermal bones of the skull and the braincase.

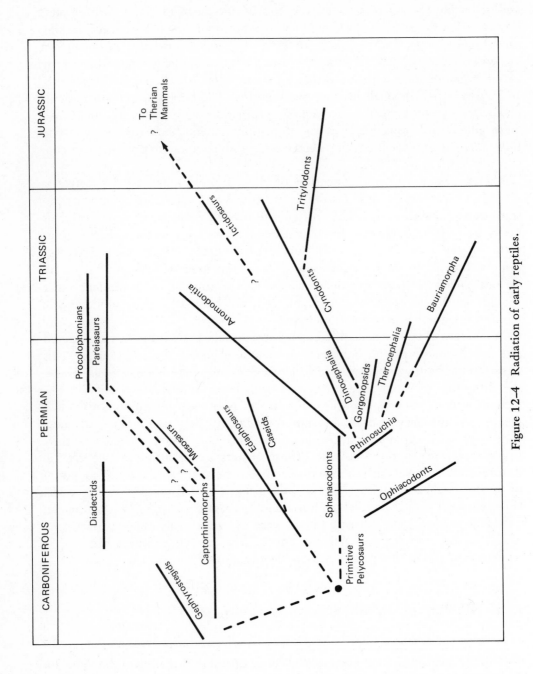

Figure 12–4 Radiation of early reptiles.

Figure 12-5 Herbivorous tetrapods of the Permian and Triassic. Early herbivorous tetrapods were large animals and their fossil remains show few signs of fleet-footedness. *Diadectes* (a) is variously classified as an amphibian or reptile. It was about 3 m long. The pariesaur *Scutosaurus* (b) was about 2.5 m long. The caseid pelycosaur *Cotylorhynchus* (c) was 2–3 m long, as was the dinocephalian therapsid *Moschops* (d). The dicynodont therapsid *Lystrosaurus* (e) was 1.5–2 m in length. (Modified from E.H. Colbert [1965] *The Age of Reptiles.* W.W. Norton, New York.)

In the other subclasses of reptiles, both problems were alleviated by the evolution of fenestrae in the sides of the skull. These openings permitted space for muscles to bulge when they contracted and at the same time lightened the skull without reducing its strength; the arches of bone that remained provided the necessary rigidity.

The position of these fenestrae is used to separate reptiles into subclasses. The Subclass Synapsida is characterized by having a single fenestra low on the side of the head. (See Figure 13-2 for details.) The most generalized synapsid reptiles are the ophiacodont pelycosaurs such as *Ophiacodon*. They differed little from the romeriid captorhinomorphs except for the presence of the temporal fenestrae. Their heads were long and slender and the upper jaws contained 20 or more small teeth. The teeth on the premaxillary bone pointed backward, and two teeth on each side of the upper jaw near the front of the maxilla were enlarged. Three groups of pelycosaurs can be distinguished (Figure 12-6). The sphenacodonts, and to a lesser extent the ophiacodonts, became increasingly specialized for predation, whereas the edaphosaurs and caseids were herbivores.

12.3.1 Carnivorous Pelycosaurs: The Sphenacodonts and Ophiacodonts

Sphenacodonts probably preyed on a variety of small and large animals, including other sphenacodonts as well as the herbivorous cotylosaurs and edaphosaurs. The progressive changes that can be seen in the structure of the skull, jaws, and teeth clearly reflect selection for effective predation. An increase in the size of the anterior premaxillary teeth can be traced from generalized forms such as *Ophiacodon, Varanops,* and *Mycterosaurus* to advanced forms like *Dimetrodon* and *Eothyris*. As the anterior teeth became larger, the posterior premaxillary teeth formed a graded series and the number was reduced until, in *Dimetrodon*, the large anterior premaxillary teeth were separated from the maxillary teeth by a gap. The enlarged teeth on the lower jaw fitted into this space when the mouth was closed.

Other changes in the form of the sphenacodont skull reflect the mechanical requirements of the specialized dentition. The enlarged maxillary teeth seen in *Dimetrodon* have deep roots in the maxilla, and in the evolution of these animals the maxilla gradually increased in depth. As the premaxilla and maxilla grew downward, the palate, which was flat in primitive genera, became arched. The significance of this arch is twofold; first, an arched palate is mechanically stronger than a flat one; second, the arch of the palate provides a space for air to pass over prey held in the mouth and reach the lungs. The arched palate of sphenacodonts was the first step in the evolution of the internal nasal passages of mammals.

The postcranial skeleton also changed in ways that appear to reflect selection for increased effectiveness in predation. The legs, although they were still held out horizontally from the body in the reptilian pattern, were longer and slimmer in the advanced sphenacodonts than in the primitive forms, suggesting increasingly active search and pursuit of prey. A remarkable feature of some sphenacodonts, including *Dimetrodon*, was the elongation of the neural spines of the trunk region. In a *Dimetrodon* that was 3 m long, the spines rose as much as 1.25 m above the vertebrae (Fig. 12.7a). The function of these enormous spines has long been debated by paleontologists, and some of their suggestions reflect more imagination than common sense. The spines have been described as secondary sex characters developed only in males, camouflage (for hiding the

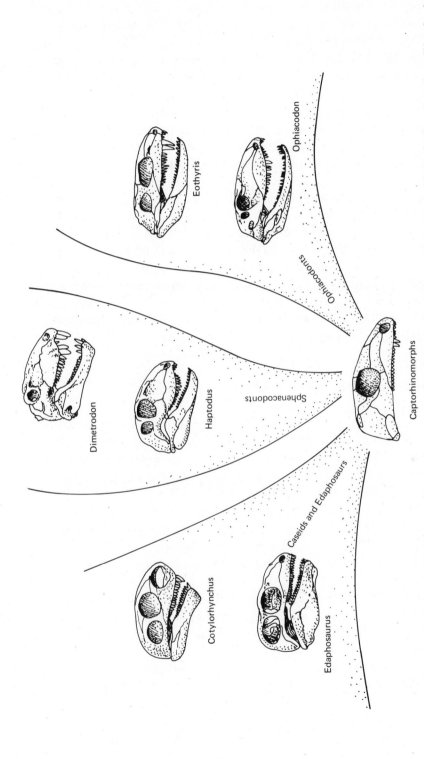

Figure 12-6 Three major evolutionary lineages of pelycosaurs can be traced by progressive modifications of the skull and teeth. All evolved from an ancestor similar to the romeriid captorhinomorph illustrated. The primary feature that distinguishes early pelycosaurs from captorhinomorphs is their synapsid temporal fenestra. Ophiacodont pelycosaurs changed relatively little from cotylosaurs. The herbivorous edaphosaurids and caseids lost the enlarged fangs in the premaxilla and maxilla and evolved a homodont dentition adapted to shearing and crushing plant material. The carnivorous sphenacodonts showed the greatest change in dentition. The initial heterodonty was greatly increased in forms like *Dimetrodon*. The large fangs were rooted in a deep maxilla and premaxilla, and the increasing depth of these bones produced an arched palate. (Modified from R. L. Carroll [1969] *Biological Review* 44:393–432 and from A. S. Romer, Ch. 15 [4].)

animals in reed beds), sails with which they were able to tack across Permian seas, or merely nonfunctional decorations which a highly successful group of animals was entitled to indulge in. Most biologists now agree that the sail supported by the elongated neural spines was a temperature-regulating device.

As ectotherms, the pelycosaurs relied upon absorbing solar energy to raise their body temperatures to activity levels. Marks of blood vessels on the spines indicate that the tissue they supported was heavily vascularized. The potential blood flow through the sail, indicated by the extent of vascularization, so greatly exceeds any reasonable metabolic requirements for such a tissue that it seems virtually certain that the animals shunted blood into or out of the sail in response to their thermoregulatory requirements. In the morning a *Dimetrodon* could orient its body perpendicular to the sun's rays and allow a large volume of blood to flow through the sail to be warmed and carry heat into the animal's body. When it was warm enough, blood flow through the sail could be restricted, and the heat retained within the body. If the animal needed to cool off, it could orient the body parallel to the sun's rays and shunt blood through the sail to lose heat by radiation and convection. Very similar circulatory mechanisms are used by living reptiles to control their rates of body temperature change (section 14.2.3). Control of body temperature would enable a predatory animal like *Dimetrodon* to extend its period of daily activity.

12.3.2 Herbivorous Pelycosaurs: The Edaphosaurs and Caseids

While the sphenacodonts and ophiacodonts evolved a body form that made them more effective predators, another lineage of pelycosaurs evolved in a very different direction to become specialized herbivores. In general body form both the edaphosaurs and caseids were similar to each other and to the herbivorous cotylosaurs and to *Diadectes* (Figure 12-5). All were heavy-bodied animals with short, sturdy legs sprawled out from the body. The skull was small in proportion to the body, and the dentition specialized for crushing vegetation rather than killing and tearing apart prey (Figure 12-6). The distinction between premaxillary incisorlike teeth and maxillary canines and postcanines, which became so pronounced in the carnivorous synapsids, disappeared in herbivorous forms. The teeth were nearly uniform in size, not as sharp as those of the carnivorous forms, and frequently broadened into crushing surfaces. Large teeth in the palate increased the surface area available for crushing plant food.

The edaphosaurids were the first radiation of herbivorous pelycosaurs in early Permian times. *Edaphosaurus* had a dorsal sail like that of *Dimetrodon*, but other edaphosaurs lacked sails. By the mid-Permian edaphosaurids had been replaced by caseids, which probably radiated later from the same stock that had given rise to the edaphosaurids. The caseids were in some respects more specialized than the edaphosaurids; the skull was short and blunt with very large nasal openings, and the legs were heavily muscled and clawed as if for digging.

Figure 12-7 In contrast to the heavily built herbivores, carnivorous pelycosaurs and therapsids were relatively slim and agile. The pelycosaur *Dimetrodon* (a) was about 3 m long. The large dorsal "sail" of *Dimetrodon* was supported by greatly lengthened neural spines. A similar structure evolved independently in some, but not all, other pelycosaurs and later in some dinosaurs. It was probably a temperature-regulating device. The theriodont therapsids were quite mammalian in posture. *Lycaenops* (b), a gorgonopsid, was about the size of a modern fox, whereas *Cynognathus* (c) was wolf-sized. (Modified from E.H. Colbert [1965] *The Age of Reptiles.* W.W. Norton, New York.)

The significance of that adaptation is unclear, because Permian plants had not evolved underground tubers and the 3 m long caseids seem too large to have been burrowing animals. Possibly the caseids ripped the exterior covering from conifers and tree ferns to reach the softer inner parts.

12.4 THERAPSIDS

From the mid-Permian to the mid-Triassic there was a flourishing fauna of mammallike reptiles and reptilelike mammals grouped under the general name of therapsids. These therapsids were derived from pelycosaurs and, like the cotylosaurs and pelycosaurs before them, radiated into herbivorous and carnivorous forms. The herbivores were a diverse group of heavy-bodied, slow-moving animals, many of which probably grazed in herds as modern herbivorous animals do. The carnivores, known collectively as theriodonts because of the similarity of their teeth to mammalian teeth, were far more progressive The pattern of increasing effectiveness of predation, which we have traced from the earliest anthracosaur labyrinthodonts through cotylosaurs and sphenacodont pelycosaurs, continued in the theriodont therapsids.

The success of therapsids was noteworthy. Late Permian fossil deposits all over the world indicate that the ecosystems of the time were based on these animals with a mixture of herbivorous cotylosaurs and herbivorous and carnivorous pelycosaurs. The remarkable similarity of faunas in different parts of the world at the end of the Paleozoic testifies not only to the continuity of land masses and climates but to the very high degree to which therapsids were successful in exploiting their environments.

The transition from the Permian to the Triassic was a period of great extinction comparable to the better-known extinction of many reptilian forms at the end of the Mesozoic. The known genera of tetrapods represented by fossils dropped from 200 in the late Permian to 50 in the early Triassic. The pattern of extinction was not regular—some old groups, such as the stereospondylous labyrinthodonts and the procolophonids, persisted, while seemingly more progressive forms disappeared. Among the herbivorous therapsids there was a considerable reduction of numbers, and only the tusked dicynodonts survived into the Triassic in abundance. Among carnivorous therapsids the most progressive groups survived the Permo-Triassic transition and evolved rapidly in the early Triassic.

The origins and interrelationships of the therapsids are a subject of considerable controversy among paleontologists. The difficulties inherent in separating parallel and convergent evolution from true genetic continuity have already been mentioned. In the case of therapsids the situation is even more complicated because they have become a battleground for proponents of monophyletic and polyphyletic origins of mammals. Most paleontologists agree that all therapsids were evolved from sphenacodont pelycosaurs, but beyond that point there are very diverse opinions. Even the origin of therapsids is in dispute—E. C. Olson has suggested that the herbivorous forms evolved from caseids.

12.4.1   Herbivorous Therapsids: Dinocephalians and Anomodonts

Many paleontologists have grouped all the herbivorous therapsids in the Suborder Anomodontia, but J. A. Hopson has pointed out that this grouping lumps together two different evolutionary trends in the jaw mechanism. He believes that those differences warrant recognition of the Dinocephalia as a suborder distinct from the Suborder Anomodontia.

The dinocephalians were huge, clumsy animals. Some, like *Moschops*, were nearly 3 m long, and more than 1.5 m at the shoulder. These animals were characterized by a sharply sloping facial region and interlocking incisor teeth that were inclined forward. The jaw articulation was a simple hinge, but as the incisor teeth swung to their closed position they interdigitated and produced a shearing force (Figure 12–8a).

The second jaw mechanism was carried to its greatest expression in the dicynodonts, and may be the basis for the success those animals enjoyed throughout the late Permian and first part of the Triassic. They radiated into a great variety of forms with a diversity comparable to that seen among grazing animals on the African plains today. Their name comes from the two tusks that were the only teeth they retained. The other teeth, in the advanced forms,

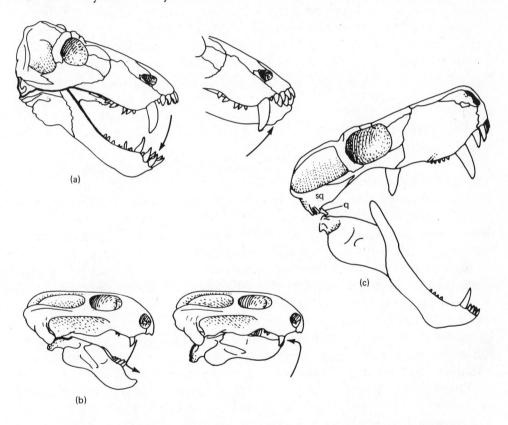

Figure 12-8 Three major divisions of therapsids are distinguished on the basis of the jaw closing mechanism. (a) The herbivorous dinocephalians had a simple hinge that was only slightly offset from the plane of the tooth row. A grinding component was introduced to the closing jaws by the interlocking, forward-pointing incisor teeth. (b) Anomodonts had and elongated articular surface. When the jaw opened it was protraced and on closing it was retracted, producing a fore-and-aft grinding motion. (c) The quadrate of theriodonts may have flexed on its articulation with the squamosal, producing a sligth fore-and-aft movement of the lower jaw. (Modified from J. A. Hopson [1969] *Annals of the New York Academy of Sciences* 167:199–216.)

were replaced by a horny, turtlelike beak. The lower jaw was capable of large fore-and-aft grinding movements as the articular surface of the lower jaw slid along the surface of the fixed quadrate (Figure 12–8b).

12.4.2 Carnivorous Therapsids: Theriodonts

Carnivorous therapsids, the Suborder Theriodontia, may be traced to sphenacodont pelycosaurs via a group of primitive therapsids, the Pthinosuchidae, found in mid-Permian deposits in Russia. The pthinosuchids show their relationship to pelycosaurs in the general features of the skull but also foreshadow the changes that are seen in theriodonts. The temporal opening of pthinosuchids is larger than that of pelycosaurs, the jaw articulation is lower and farther forward, and the enlarged caninelike teeth are single rather than paired. Hopson has suggested that a third type of jaw suspension evolved in the theriodonts (Figure 12–8c). He believes that there was a limited amount of flexibility in the articulation between the quadrate and the skull, allowing theriodonts to achieve a wider gape. In Hopson's opinion some theriodonts secondarily invaded an herbivorous niche, and in their case the movable quadrate permitted some fore-and-aft movement of the lower jaw for grinding plant material.

Two or more major evolutionary lineages of theriodonts are distinguished by paleontologists (Figure 12–9). It is among the theriodonts that the immediate ancestors of mammals are found. It is clear that a large amount of parallel evolution took place among the theriodont therapsids. The common denominator was selection for increased effectiveness as a terrestrial predator. Changes are seen in the vertebral and axial skeleton as well as the skull. The legs were longer in proportion to the body than were those of pelycosaurs, and they appear to have been held more nearly under the body in a mammalian posture instead of sprawled outward (Figure 12–7). The vertebral column was strengthened by increased ossification, and all traces of the intercentra had disappeared. The angle of articulation of adjacent vertebrae had changed so that the vertebral column formed a more arched structure than was seen in the captorhinomorphs. Advanced theriodonts probably ran like mammals with the vertebral column bending in a vertical plane and extending the reach of the legs, which moved nearly straight forward and backward. The long reptilian tail seen in pelycosaurs was considerably reduced in size in therapsids. In their general appearance, they would probably strike us as looking more mammalian than reptilian.

More detailed examination of these fossils increases the impression of a mammallike animal. The gorgonopsids were among the early theriodonts; they disappeared at the Permo-Triassic transition. The skull had developed a zygomatic arch produced, like that of mammals, by the outward growth of the squamosal and jugal bones (Figure 12–10). The arch increased the space for the temporalis muscle, which originates on the roof of the skull and inserts on the lower jaw, and provided an origin for the masseter muscle, which also inserts on

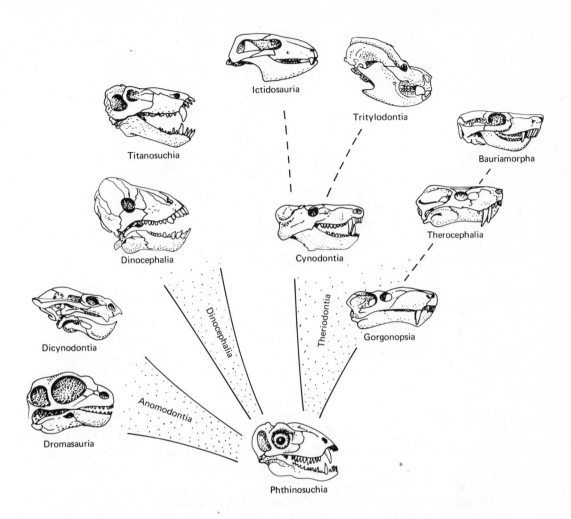

Figure 12-9 Adaptation to different diets can be traced in the skulls of therapsids. The dinocephalians and anomodonts were herbivores. Dicynodonts such as *Lystrosaurus* (Figure 12-5e) were the most successful of the herbivores. In those forms teeth were replaced by a horny beak. Most theridonts were carnivores, and the tooth morphology of advanced forms is very like that of mammals. Tritylodonts were secondarily herbivorous theriodonts with broad flat teeth and the jaw articulation offset from the plane of the tooth row. (Modified from A. S. Romer Ch. 15 [4].

the jaw. The anterior teeth on the premaxillary and maxillary bones were well developed and met equally sharp teeth on the lower jaw. Canine teeth projected below the lower jaw in some forms, and there was a series of conical cheek teeth.

Fossils of Triassic cynodonts suggest that they were still more mammallike than the gorgonopsians. In a classic paper, the South African paleontologist A. S. Brink pointed out a large number of morphological features of *Cynogna-*

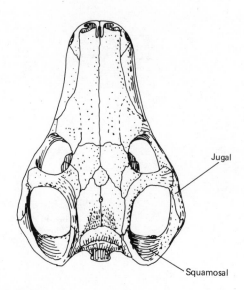

Figure 12-10 Primitive theriodonts like the gorgonopsian *Scylacops* had a well-developed zygomatic arch formed by the jugal and squamosal bones, as it is in mammals. The arch provided space for the temporalis muscle to pass inside it as well as providing an origin for the masseter muscle. (Modified from D.M.S. Watson [1951] *Paleontology and Modern Biology.* Yale University Press, New Haven.)

thus which, when they are considered together, lead many biologists to the conclusion that the advanced theriodont therapsids had achieved a structural grade at least similar to that of the living prototherian mammals, the platypus and echidna. Indeed, some paleontologists maintain that the therapsids should be placed in the Class Mammalia.

Mammals differ from reptiles in a variety of morphological, physiological, and behavioral characteristics. Different features evolved at different rates, and of course only morphological features fossilize. Conclusions about behavior and physiology are based on comparisons of fossils with living animals. Many paleontologists favor a definition of mammals based on the nature of the jaw articulation (see section 18-1). By that criterion, therapsids are reptiles. Other paleontologists have adopted a more speculative approach. They have tried to deduce as much as possible about the natural history of therapsids from the fossil material, and then decide if an animal with that sort of natural history is more mammallike than reptilelike. This is the line of reasoning that has led to the suggestions that therapsids should be classified as mammals. The problems of drawing a dividing line between reptiles and mammals will be discussed in more detail in section 18.1. For the present, one major physiological difference between reptiles and mammals must be emphasized—the distinction between **ectothermy** and **endothermy**.

The distinction between poikilotherms, (fish, amphibians, and reptiles) and homeotherms (birds and mammals) that was widely used through the middle of the twentieth century has become less useful as our knowledge of the temperature-regulating capacities of a wide variety of animals has become more sophisticated. Poikilotherm and homeotherm are terms that are supposed to describe the variability of body temperature. In practice, some "poikilotherms" live in very stable habitats and have correspondingly stable body temperatures.

Because of complications of this sort, it is very difficult to use the terms "homeotherm" and "poikilotherm" in a rigorous way. Mammalogists and ornithologists still use those terms, but biologists concerned with the temperature regulation of other animals prefer the terms "ectotherm" and "endotherm." These words were coined by Raymond Cowles whose pioneering study of reptilian thermoregulation is discussed in section 14.2.2. They are *not* synonymous with poikilotherm and homeotherm because, instead of referring to the precision of temperature regulation, they refer to the source of energy used in temperature regulation. Ectotherms rely primarily on an external source of energy. Usually this is solar radiation, but it may reach an animal by an indirect route, for example, by conduction from warm ground. Endotherms rely on metabolic heat production for temperature regulation.

Endothermy is thus a distinctive difference between reptiles and mammals, and the evolution of a capacity for endothermal heat production was a major step in mammalian evolution. It must be remembered that evolution proceeds by slow steps, and the evolution of endothermy was no exception. There would not have been a sudden change from an ectothermal reptile to an endothermal mammal. Instead there would have been a gradually increasing capacity to produce enough heat to raise the body temperature significantly. When we speak of the evolution of endothermy in therapsids, we do not mean to imply that they regulated their body temperatures continuously or precisely, merely that the capacity to produce a significant increase in the body temperature by metabolic heat production and to sustain that temperature for some period of time was present in these animals. Even a limited endothermal capacity would have produced profound changes in the natural history of the animals.

Probably the key selective force in the evolution of endothermy was a continuation of the trend toward increased predatory effectiveness. A major difference between living mammals and reptiles involves their physiological response to exercise. Mammals are capable of prolonged exertion because they can supply oxygen and metabolic substrates to active muscles. In contrast, reptiles rely upon anaerobic metabolism and become exhausted in a few minutes. This difference leads to profound differences in the ecology of reptiles and mammals that are discussed in more detail in Section 14.6. For the moment, the significant point is this: The capacity for sustained activity by mammals is achieved by an increase in the basal metabolic rate compared to that of reptiles. For a predator that runs after its prey, as the theriodont therapsids probably did, the physiological capacity to sustain activity would be an essential adjunct to the morphological adaptations for a cursorial life that we can see in the fossils. Thus, although there is no direct fossil evidence, it seems reasonable to suppose that mammallike morphology was accompanied by mammallike physiology.

The ability to maintain a high rate of activity is not the only benefit an animal can derive from a high resting metabolic rate, although in an evolutionary perspective it is probably the first feature that was selectively advantageous to theriodont therapsids. An animal that maintains a high metabolic rate is producing a great deal of heat. In the case of a reptile, which has very little insulation, the heat is lost from the body and has little effect on body temperature. The addition of an insulating layer (fat, hair, or feathers) gives an animal with a high resting metabolic rate the potential of being an homeotherm.

It is probable that insulation, in the form of hair, evolved after the evolution of a high basal metabolic rate. Insulation has no value for an animal that is not producing heat internally. In fact, it would be a disadvantage for an ectothermal animal to have hair on the body surface because it would inhibit the exchange of energy with the environment on which ectothermal animals depend. Only after there was significant heat production within the body would there be selective value in an insulative layer of hair to slow heat exchange. It is quite possible that this stage had been reached by the theriodont therapsids, and that they had achieved considerable endothermy. Although they would probably appear primitive to us in comparison to placental and marsupial mammals, they were probably the only group of vertebrates verging on endothermy in the late Permian and early Triassic. Their heat production and insulation would not have had to be very effective by our standards to have been better than that of any competing group.

12.5   THE TRIASSIC ECOSYSTEM

The weight of evidence strongly suggests that the advanced theriodont therapsids—the cynodonts and parallel lineages such as the therocephalians, diademontids, and ictidosaurs—had achieved significant endothermy. Most important, they probably occupied a mammalian position in the ecosystem of the early Triassic. In this period, for the first time in the history of terrestrial life, fossil remains indicate an ecological pyramid of numbers of the sort we are accustomed to in the modern world. In present-day ecosystems, a large base of herbivores supports a small number of carnivores. Before the Triassic, terrestrial ecosystems were characterized by inverted pyramids—large numbers of carnivores preying on small numbers of herbivores and on each other. An inverted pyramid of that sort is not compatible with a mammal-dominated ecosystem because mammals use too much energy producing heat. The appearance of a modern trophic system in the fossil record at that time reinforces the morphological evidence with some ecological data. Both sorts of evidence support the view that advanced theriodont therapsids occupied an ecological position more like that of mammals than reptiles.

In retrospect it does not seem surprising that the selection for effective predators, which so clearly shaped the morphological features of therapsids, should have been reflected in their physiology as well. What is surprising is that these mammallike animals evolved in the Permian and early Triassic and then vanished. The widespread therapsid faunas gave way to the reptilian faunas that dominated the world for 100 million years, from the late Triassic to the end of the Cretaceous. During that period reptiles were the dominant animals by any criterion one chooses—biomass, numbers of species, or ecological diversity. Mammals were present only as small rat- or mouse-sized animals. The burgeoning of reptiles in the Mesozoic is discussed in Chapter 13, and Chapter 14 considers the ecological position of modern reptiles and suggests an explanation for the successive replacements of primitive endotherms by ectotherms and those ectotherms in turn by endotherms.

References

[1]   Brink, A. S. 1956. Speculations on some advanced mammalian characteristics in the higher mammal-like reptiles. *Palaeontologica Africana* 4:77–96. A classic paper illustrating the sorts of inferences about physiology and ecology that can be drawn from the fossil material.

[2]   Crompton, A. W. and P. Parker. 1978. Evolution of the mammallike masticatory apparatus. *American Scientist* 66:192–201.

[3]   Hopson, J. A. 1969. The origin and adaptive radiation of mammal-like reptiles and nontherian mammals. In *Systematics and Comparative Neurology of Mammals.* New York Academy of Sciences, New York. Reviews the evolution of mammals from the pelycosaur stage. Emphasizes the principle adaptations of the basal stock of each of the major radiations.

[4]   Jenkins, F. A. Jr. 1970. Cynodont postcranial anatomy and the "prototherian" level of mammalian organization. *Evolution* 24:230–252. Concludes that fundamental mammalian characters were present in late Triassic therapsids.

[5]   McNab, B. K. 1978. The evolution of endothermy in the phylogeny of mammals. *The American Naturalist.* 112:1–21. Concludes that mammalian endothermy evolved in large animals, and that the small body size of Mesozoic mammals was a secondary development.

[6]   Olson, E. C. 1961. Food chains and the origin of mammals. International colloquium on the evolution of lower and unspecialized mammals. Kon. Vlaamse Acad. Wetensch. Lett. Schone Kunsten Belg. pt. 1:97–116. A careful interpretation of tetrapod evolution in terms of changing habitats and ecological conditions.

[7]   Olson, E. C. 1976. The exploitation of land by early tetrapods. In *Morphology and Biology of Reptiles*, edited by A. d'A. Bellairs and C. B. Cox. Academic Press, London.

13

Mesozoic Reptiles

Synopsis: Reptilian subclasses are defined by the patterns of temporal fenestrae in the skull. Two subclasses of reptiles, the Anapsida and the Synapsida, were discussed in Chapter 12. Both of these subclasses experienced their greatest diversity in the Paleozoic, and both left descendants in the Mesozoic. Turtles are anapsid reptiles that may have evolved from captorhinomorphs, and mammals had evolved from the synapsid lineage by the early Mesozoic.

The great burst of reptilian evolution in the Mesozoic involved three new subclasses. The subclass Euryapsida consists of four orders of marine reptiles. The ichthyosaurs (Order Ichthyosauria) evolved a very fishlike body form, whereas the nothosaurs and plesiosaurs (Orders Nothosauria and Plesiosauria) stayed closer to the tetrapod pattern. The placodonts (Order Placodontia) were mollusc-eating forms that lacked the specializations for rapid movement seen in the other orders.

The Subclass Archosauria includes the most familiar Mesozoic reptiles, the dinosaurs. "Dinosaur" is a popular term that lumps two unrelated orders of archosaurs, the Ornithischia and the Saurischia. The Order Saurischia (the lizard-hipped dinosaurs) included both carnivores and herbivores. *Tyrannosaurus rex* is probably the most familiar carnivorous saurischian, but there were also small, lightly built carnivores among the saurischian dinosaurs. *Brontosaurus* (more properly but less familiarly called *Apatosaurus*) was an herbivorous saurischian.

The second order of dinosaurs, the Ornithischia (bird-hipped dinosaurs) were all herbivores. There was a great variety of body form among the ornithischians. The ornithopods were bipedal and may have tried to outrun the large carnivorous saurischians that preyed on them. The quadrupedal stegosaurs and ankylosaurs probably depended on bony armor and tails with spikes or clubs to deter predators. The quadrupedal horned dinosaurs were the last group of dinosaurs to evolve. They may have roamed in herds like modern herbivorous mammals.

Other archosaurs include the flying reptiles (Order Pterosauria) which appear to have been covered by hair, and the phytosaurs (Order Phytosauria) that superficially resembled crocodiles. The true crocodiles (Order Crocodilia) are also archosaurs.

The Subclass Lepidosauria was the third group of reptiles that radiated in the Mesozoic. This subclass also contained a lineage of crocodilelike animals, the champsosaurs. The champsosaurs persisted into the early Cenozoic, well past the extinction of the dinosaurs and pterosaurs. The rhynchosaurs (Order Rhynchosauria) were an order of large lepidosaurs known from Triassic fossils. The most successful lepidosaurs were the squamates (Order Squamata). This group of reptiles includes the familiar lizards and snakes as well as two less familiar groups, the amphisbaenians and the tuatara. These modern reptiles are discussed in Chapter 14.

Our increasing knowledge of the behavior and physiology of living reptiles has forced reconsideration of classical views of the probable behavior and physiology of extinct reptiles. Mesozoic reptiles were probably more alert and active than paleontologists had previously assumed. Inferences drawn from indirect evidence like fossil dinosaur tracks and fossil deposits that contain the bones of different size individuals of a species suggest that some dinosaurs may have moved in herds or family groups. Behavioral complexity of that sort can no longer be considered beyond the scope of reptiles. The social and predatory behaviors of modern crocodilians and lizards are as complex as those of many mammals and birds.

Studies of temperature regulation in living reptiles has cast new light on the sorts of thermoregulatory behavior dinosaurs might have employed. Suggestions that dinosaurs were endotherms like birds and mammals appear untenable on both a physiological and an ecological basis. Instead it seems probable that in the equable climates of the Mesozoic, which were warmer and less variable than present-day climates, large reptiles would have experienced stable body temperatures as a result of their large body size. This stenothermy was achieved by utilizing external energy sources instead of metabolic energy, and consequently it would have been energetically less costly than the endothermal homeothermy of birds and mammals. The assumption that dinosaurs relied on external energy sources for thermoregulation helps to explain how they dominated terrestrial ecosystems during the Mesozoic, and why they disappeared with the advent of cooler, more variable climates in the Cenozoic.

13.1 INTRODUCTION

The Mesozoic Era, frequently referred to as the Age of Reptiles, extended for some 160 million years from the close of the Paleozoic 225 million years ago to the beginning of the Cenozoic only 65 million years ago. Through this vast period evolved a worldwide reptilian fauna that diversified and radiated into most of the adaptive zones occupied by all the terrestrial vertebrates living today and some which no longer exist (for example, the enormous herbivorous and carnivorous reptiles popularly called dinosaurs). Although the dinosaurs are the most familiar representatives of the Age of Reptiles, they are only two among many reptilian lineages.

Inevitably such a huge group of animals is complicated and confusing, not only upon first acquaintance but even after study. The situation is not helped by a plethora of inappropriate and confusingly similar names. The Eosuchia, for example, are a primitive order of reptiles that gave rise to lizards and snakes, whereas the Eusuchia are the suborder of crocodilians that contains the living crocodiles, alligators, and gharial. The suffix "suchia" means "crocodilian," but the proterosuchians and pseudosuchians are only distantly related to true crocodilians. In the discussion that follows we have tried to keep confusions of this sort to a minimum by using as few names as possible.

Parallel and convergent evolution were widespread in Mesozoic reptiles. Long-snouted fish eaters evolved repeatedly, as did heavily armored quadrupeds and highly specialized marine forms. A trend to bipedalism was general among the thecodonts, and a secondary reversion to quadrupedal locomotion is seen in many forms. Knowledge of phylogenetic relationships is in a state of flux, and the scheme outlined in Figure 13–1 will undoubtedly need revision as additional fossils are collected and analyzed. Current views of the ecology of dinosaurs are likewise undergoing a radical revision. Classic ideas have been based on a naive impression of the ecology and behavior of large living reptiles, especially crocodilians and lizards. Paleontologists have assumed that Mesozoic reptiles lived placid lives, and these ideas have colored their interpretations of posture and behavior. Recent reexamination of the morphological evidence suggests that some Mesozoic reptiles were considerably more active than had been realized, and these reevaluations have changed our views of their ecological relationships.

This chapter commences with a brief review of the phylogenetic relationships of Mesozoic reptiles and some aspects of functional morphology and major evolutionary trends. More detailed information on these topics, and additional illustrations of members of the groups discussed, can be found in the references cited at the end of the chapter. Following a consideration of some aspects of the ecology of Mesozoic reptiles, we will consider reasons for their disappearance at the end of the Cretaceous.

13.2 PHYLOGENETIC RELATIONSHIPS AND RADIATIONS OF MESOZOIC REPTILES

There are four subclasses of reptiles that originated in the late Permian (Figure 13–1). The evidence presently available indicates that each represents a separate radiation from captorhinomorphs. In addition the captorhinomorph stock persisted in the form of the turtles to create the fifth reptilian subclass. Reptiles have been classified on the basis of the temporal fenestrae (Figure 13–2). A particular pattern of lateral and/or temporal fenestrae was considered to be a conservative character unlikely to have evolved more than once and thus to be a suitable basis for phylogenetic divisions. More recent analyses have modified this view, and different authorities have different opinions of phylogenetic relationships. One possible division of reptiles into subclasses is the following:

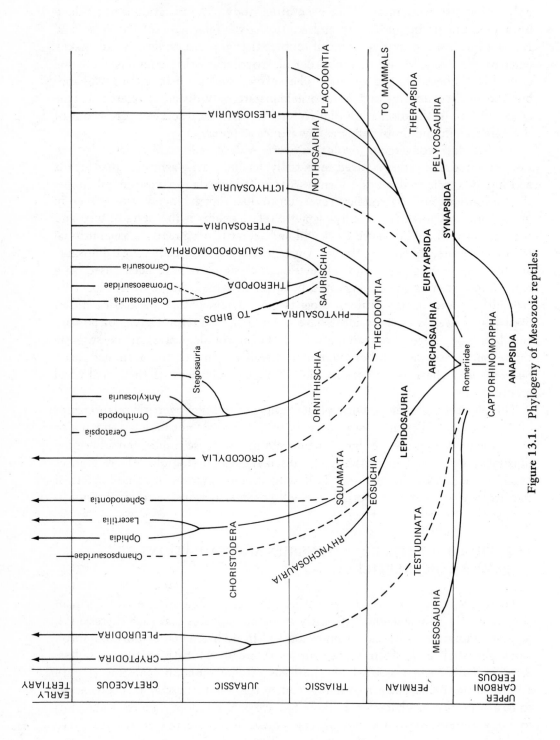

Figure 13.1. Phylogeny of Mesozoic reptiles.

Subclass Anapsida: No temporal openings. The jaw muscles extend posteriorly through the large occipital openings.

Subclass Synapsida: A single lateral temporal opening low on the side of the head bounded dorsally by the postorbital and squamosal bones.

Subclass Euryapsida: A single upper temporal opening high on the side of the head bounded ventrally by the postorbital and squamosal bones.

Subclasses Lepidosauria and Archosauria: Two temporal openings, one above and one below the postorbital–squamosal connection. Some paleontologists combine archosaurs and lepidosaurs in the subclass Diapsida.

The evolution and radiation of the synapsid reptiles, the pelycosaurs and therapsids, was discussed in the preceding chapter. Of the anapsids, which are the basal stock from which all the other reptilian groups arose, only the turtles survive.

13.2.1 Subclass Anapsida, Order Testudinata

The features that distinguish the advanced turtles are not unique to that group (parallels will be pointed out among the placodonts and ankylosaurs) but

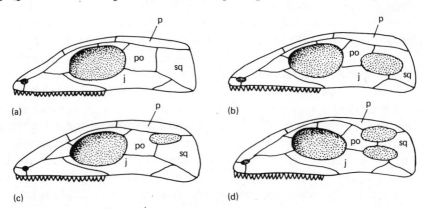

Figure 13.2. Reptilian subclasses are defined in part by the presence and position of fenestrae in the skull. (a) The subclass Anapsida (captorhinomorphs, cotylosaurs and chelonians) lacks openings in the temporal region of the skull. (b) The subclass Synapsida (pelycosaurs and therapsids) is characterized by a single fenestra bounded dorsally by the postorbital and squamosal bones. (c) The subclass Euryapsida (nothosaurs, plesiosaurs, placodonts and ichthyosaurs) has a single opening bounded ventrally by the postorbital and squamosal. (d) The subclasses Lepidosauria (snakes, lizards, amphisbaenians, and rhynchocephalians) and Archosauria (crocodilians, dinosaurs, pterosaurs, and a variety of other forms) both have two fenestrae, one above and one below the postorbital–squamosal junction. These two subclasses are often combined in the subclass Diapsida. Code: j = jugal, p= parietal, po = postorbital, sq = squamosal. (Modified from A. S. Romer, Ch. 15 [4].)

nowhere else are the modifications of the body to present an armored surface to the world so extensively developed. If turtles had become extinct at the end of the Mesozoic, they would rival the dinosaurs as objects of astonishment. It is only because we can see living turtles that we accept them as commonplace and (most inappropriately) employ them as dissection specimens to illustrate the reptilian structural grade. In fact, turtles are extensively modified in both their skeletal structure and soft anatomy and represent a very successful and highly specialized side branch of reptiles.

A turtle's shell is its most distinctive feature (Figure 13–3). The upper shell, the carapace, is formed by plates of dermal bone that are fused to the underlying ribs and vertebrae and are overlaid by horny plates of epidermal origin. The epidermal plates do not lie directly over the dermal bones but alternate with them, like bricks in a wall. The result is the same in both cases—the structure is stronger when the joints do not coincide. The lower shell, the plastron, is derived from the abdominal ribs that were widespread in many early reptiles. Parts of the shoulder girdle are incorporated into the plastron; the clavicles form the epiplastra and the interclavicle is the entoplastron. The skull is toothless and the maxillae, premaxilla, and dentaries form a beak that is covered with a horny plate.

The evolution of these features can be traced from the earliest turtle fossil yet discovered, *Proganochelys* from the Triassic. All of the shell elements known in modern turtles are present in *Proganochelys* plus some additional ones that have since disappeared. The clavicles and interclavicle were incorporated into the plastron but are still identifiable. There was a horny beak, but teeth persisted on the palate. The head and legs probably could not be drawn into the shell. In Jurassic and Cretaceous fossil turtles all teeth had disappeared, but the head still could not be withdrawn into the shell. The evolution of that defensive mechanism marked a major split in turtle evolution seen in the two suborders of living turtles, the Cryptodira and Pleurodira. Pleurodire turtles withdraw the head by bending the neck in a horizontal S. Their common name of side-neck turtles is derived from this lateral bend. Pleurodire turtles are known from fossils in the upper Cretaceous of North America, but living pleurodires are restricted to the southern hemisphere. The turtles familiar to inhabitants of the northern hemisphere all belong to the Subclass Cryptodira in which the neck is bent into a vertical S when the head is withdrawn into the shell.

The origin of turtles is completely unknown. *Proganochelys* is so specialized that it reveals little more of turtle origins than do modern turtles. The skull structure of turtles, including *Proganochelys*, can most reasonably be derived from the unspecialized romeriid captorhinomorphs, and turtles are assumed to have stemmed from that group. A fossil, *Eunotosaurus*, from the mid-Permian of South Africa has been suggested as a transitional link between captorhinomorphs and turtles, but the idea has not been received with great enthusiasm. *Eunotosaurus* showed some turtlelike features, notably a shortened trunk and ribs broadened to enclose most of the trunk. Its superficial resemblance to a turtle is unquestionable, but it probably represents a parallel or convergent

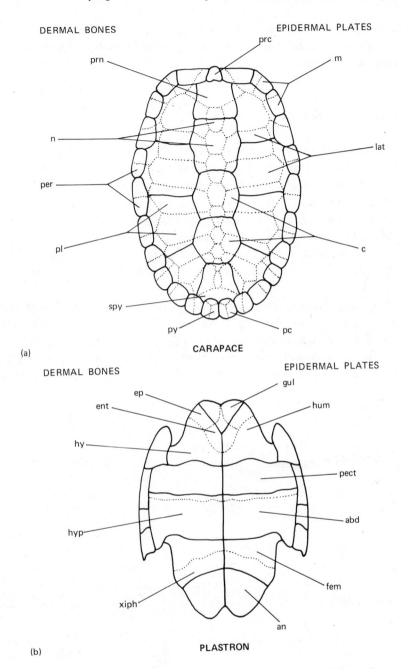

Figure 13-3 Bony shell (dotted lines) and horny plates (solid lines) of a turtle. Note that the joints between plates do not coincide with those between bones. Code for bones: ent = entoplastron, ep = epiplastron, hy = hyoplastron, hyp = hypoplastron, n = neurals, per = peripherals, pl = pleurals, prn = proneural, py = pygal, spy = suprapygal, xiph = xiphiplastron. Code for plates: abd = abdominal, an = anal, c = central, fem = femoral, gul = gular, hum = humeral, lat = lateral, m = marginal, pc = postcentral, pect = pectoral, prc = precentral. (Modified from F. M. Bergounioux in J. Pivateau Ch. 9 [4] using the terminology of A. Carr [1952] *Handbook of Turtles*. Cornell University Press, Ithaca.)

development of armor. A turtle shell is formed of dermal plates, not expanded ribs of the sort seen in *Eunotosaurus*, and there are no affinities in the skull of *Eunotosaurus* and turtles.

13.2.2 Subclass Euryapsida

This subclass contains a variety of forms that are very unlike each other. The features they share are the euryapsid skull structure, with a fenestra high on the side of the head, and extreme adaptation for marine life. The nothosaurs and plesiosaurs were fish eaters that retained a basically reptilian body form, whereas the ichthyosaurs became so highly modified that their body form was like that of a fish, although the skeleton remained distinctively reptilian. The placodonts were mollusc eaters specialized for defense rather than swift movement to capture prey. Some placodonts developed a body armor as extensive as that of turtles.

The sauropterygians appear in the fossil record in the form of nothosaurs in the Triassic (Figure 13–4a). The feet were probably webbed, but otherwise not greatly modified for swimming. The leg could be bent at the knee and the toes could flex. Nothosaurs were elongate animals with long necks (about 20 cervical vertebrae) and well-developed abdominal ribs. The nostrils were located a

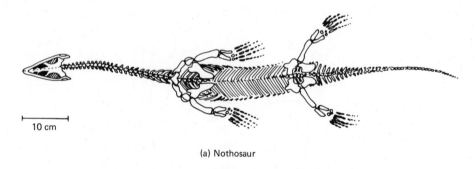

(a) Nothosaur

10 cm

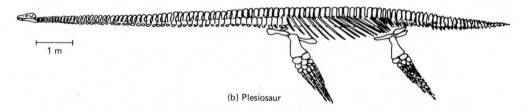

1 m

(b) Plesiosaur

Figure 13–4. (a) Except for the relatively short legs and large feet (which were probably webbed) the skeleton of a nothosaur (ventral view) differs little from that of a terrestrial reptile. (b) The plesiosaurs, in contrast, were highly specialized for marine life. In particular the girdles were reduced in size, the limbs greatly shortened, and the feet converted to paddles. (Modified from Peyer in A. S. Romer Ch. 15[4] (a) and from Welles in J. Piveteau Ch. 9[4] (b).)

short distance in front of the eyes. The plesiosaurs that appeared at the Triassic/ Jurassic transition and persisted to the Cretaceous carried the specializations we see emerging in the nothosaurs to their extreme (Figure 13–4b). Two lineages of plesiosaurs can be distinguished, and these appear to have evolved side by side, indicating a considerable ecological separation between them (Figure 13–5).

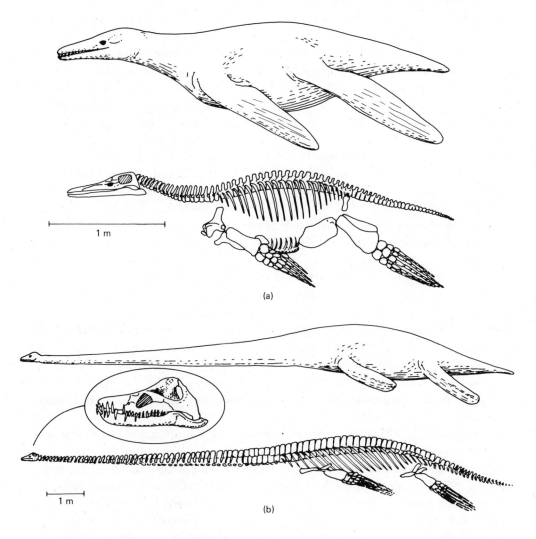

Figure 13-5 Two radically different sorts of plesiosaurs had evolved by the Jurassic. The short-necked forms like *Polycotylus* (a) had large heads with long snouts. Their paddles were large and the body was streamlined, approaching the 4:1 length:diameter ratio characteristic of fast-swimming animals. (See Figure 7-7.) The long-necked forms like *Elasmosaurus* (b) had small heads and small paddles. The neck became progressively longer during the evolution of the group, and the body shape departed increasingly from the ideal ratio of length to diameter. (Body outlines modified from D. M. S. Watson [1951] *Paleontology and Modern Biology*. Yale University Press, New Haven; skeletons modified from Andrews and from Welles in J. Pivateau Ch. 9 [4].)

One lineage comprised long-necked pleisosaurs with small heads, whereas the other contained short-necked animals with long skulls. Both had heavy, rigid trunks and appear to have swum through the water with limbs that had evolved into paddles. Hyperphalangy, the addition of joints to the toes, increased the size of the paddles. In both types of plesiosaurs the nostrils were located high on the head just in front of the eyes.

The long-necked plesiosaurs reached their zenith in *Elasmosaurus*, which lived in the upper Cretaceous. The lineage can be traced back to *Plesiosaurus* in the lower Jurassic. That form had 35 cervical vertebrae. The *Elasmosaurus* line is characterized by a progressive increase in the number of cervical vertebrae and a reduction in the size of the head. Not only did the number of cervical vertebrae increase but individual vertebrae became longer. *Microcleidus* of the middle Jurassic had 39 or 40 cervical vertebrae, whereas *Elasmosaurus* had 76. Even in the early forms the body was not well streamlined, and, as the neck became longer, the streamlining became even poorer. The size of the paddles relative to the size of the body decreased from the Jurassic to the Cretaceous. Clearly the *Elasmosaurus* line of plesiosaurs were not rapid swimmers.

Most reconstructions of plesiosaurs show them with their long necks twining in snakelike fashion as they seize fish or fight off predators. D. M. S. Watson's analysis of the cervical vertebrae of these long-necked plesiosaurs convinced him that such reconstructions are inaccurate and that plesiosaur necks were inflexible. The cervical ribs turned sharply backward and ribs on adjacent vertebrae ran parallel to each other only a centimeter apart. The intercostal muscles would have been very short and the whole neck rigid. Only in the most anterior vertebrae is there any indication of flexibility. Watson concluded that plesiosaurs did not capture their prey by sudden snakelike darts of the head, but rather by pivoting the entire body by stroking forward with the paddles on one side and backward with those on the other side. This movement would swing the head rapidly through a large arc, and both the speed of head movement and the arc in which prey could be captured would be increased as the neck lengthened. The small, vertically compressed skull and thin neck would offer little resistance to sidewards movement.

The short-necked plesiosaurs started from a form not very different from the starting point of the long-necked forms but followed a completely different evolutionary pathway leading to increasingly streamlined forms. The neck became shorter, and the paddles larger. Watson suggested that these animals captured their prey by pursuit in the manner of modern seals and sea lions. In a sense the short-necked plesiosaurs were converging on the morphological adaptations of ichthyosaurs, although they never attained the perfection of streamlining seen in advanced members of that group.

The placodonts were a group of awkward-looking, heavy-bodied animals specialized for eating molluscs (Figure 13–6). In many aspects of their life they probably resembled walruses. The anterior teeth projected forward and were probably used to nip mussels from rocks or extract clams from the sea floor. The posterior teeth were massive structures capable of crushing mollusc shells.

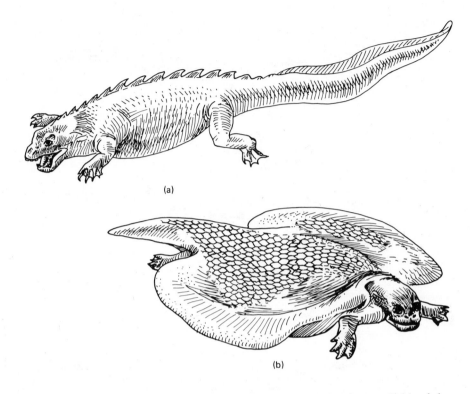

Figure 13-6. Placodonts were slow-swimming animals that probably fed on molluscs. *Placodus* (a) was relatively unspecialized, but *Henodus* (b) developed dermal armor plate almost as extensive as a turtle's. (Modified from various sources.)

The coronoid process of the lower jaw on which the adductor mandibulae muscle inserts was well developed, an unusual feature in reptiles and an indication of the strong forces the jaw could exert. *Helveticosaurus*, a placodont from the mid-Triassic, was much like a nothosaur with an elongate body and pointed teeth. The advanced placodonts *Placochelys* and *Henodus* developed a heavy dermal armor superficially similar to that of a turtle but composed of a mosaic of many small plates rather than a few large ones.

The ichthyosaurs are so very different from the nothosaurs and plesiosaurs that the euryapsid skull condition may well have evolved independently in this group (Figure 13-7). At present there is no fossil evidence. Early ichthyosaurs were elongate animals with the legs modified into paddles by hyperphalangy and **hyperdactyly** (the addition of extra digits). The vertebral column turned downward to create a reverse heterocercal tail. The outline of the upper lobe of the tail and of a fleshy dorsal fin, neither of which had skeletal support, can be seen in fossils from fine-grained sediments. The skull was elongate and the jaws contained large numbers of simple pointed teeth. Primitive ichthyosaurs were poorly streamlined, and the primary evolutionary trend was toward greater streamlining, which was presumably associated with a more active pursuit

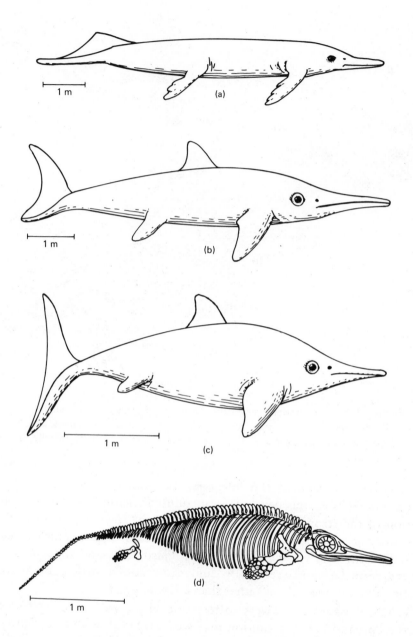

Figure 13-7. Evolutionary change in icthyosaurs led to increasingly streamlined animals. (a) *Cymbospondylus* (Triassic) was elongate with relatively small paddles, a small caudal fin, and no dorsal fin. It probably swam primarily by lateral undulation of the body. (b) *Ichthyosaurus* (Jurassic) and (c) *Ophthalmosaurus* (Cretaceous) had large caudal fins that provided the main thrust for swimming. The front flippers were enlarged to serve as stabilizers and there was a dorsal fin that reduced roll. (d) The dorsal fin and the upper lobe of the caudal fin were stiff tissue, not supported by bone. (Body outlines modified from D.M.S. Watson [1951] *Paleontology and Modern Biology.* Yale University Press, New Haven; skeleton modified from Andrews in J. Piveteau Ch. 9[4].)

of fish. The similarity in body form of advanced ichthyosaurs, sharks, and cetaceans is one of the classic examples of convergent evolution. Although the body forms of all three animals are similar, many details are different. Sharks depend on chemosensory cues and the lateral line system to locate prey, and the eyes are relatively small. Cetaceans depend upon echolocation, and also have small eyes. In contrast, advanced ichthyosaurs had enormous eyes, suggesting that they were primarily visually oriented.

13.3 THE DIAPSID REPTILES: SUBCLASSES LEPIDOSAURIA AND ARCHOSAURIA

Petrolacosaurus, a fossil eosuchian from Upper Pennsylvanian deposits in Kansas, is the oldest diapsid reptile known. It retains many of the general characteristics of the romeriids but shows new features as well. The most striking of these are the two lateral fenestrae perforating the lateral wall of the skull behind the eye (Figure 13–8). The skull of *Petrolacosaurus* is relatively smaller than that of romeriids, and the eyes are larger. The long neck and lightly built postcranial skeleton suggest that *Petrolacosaurus* may have been a more active predator than the romeriids. Robert Carroll believes that *Petrolacosaurus* and other early diapsids lived in dry areas, probably further from water than those inhabited by romeriids.

Later diapsids, eosuchians like *Heleosaurus*, had bladelike teeth set in sockets and larger temporal fenestrae than *Petrolacosaurus* (Figure 13–8). In Carroll's view an animal like *Heleosaurus* represents a potential common ancestor for lepidosaurs and archosaurs. Consequently he lumps those groups as the subclass Diapsida. Common features shared by eosuchians like *Heleosaurus* and *Youngina* and pseudosuchians like *Euparkeria* include teeth set in sockets (thecodont), and dermal armor along the vertebral column. The shape of the femur suggests that posture was upright, and rapid locomotion may have been achieved with a bipedal gait. Unlike romeriids and *Petrolacosaurus,* early archosaurs and lepidosaurs had an otic notch at the posterior margin of the skull. This notch probably supported a relatively large tympanum in life. The combination of a large tympanum and a slim stapes that would have had little inertia suggests that sensitivity to airborne sounds was important.

In contrast to Carroll's view of a common origin of archosaurs and lepidosaurs, A. S. Romer felt that separate origins of the two groups from captorhinomorphs was more likely. On this basis he treated Archosauria and Lepidosauria as subclasses. Table 13–1 uses that format because it fits the usual taxonomic arrangement of living reptiles better than a single Subclass Diapsida.

13.3.1 Subclass Lepidosauria

Champsosaurs

The eosuchians were largely replaced by the end of the Triassic by their descendants the Squamata and Rhynchosauria. The only eosuchian lineage

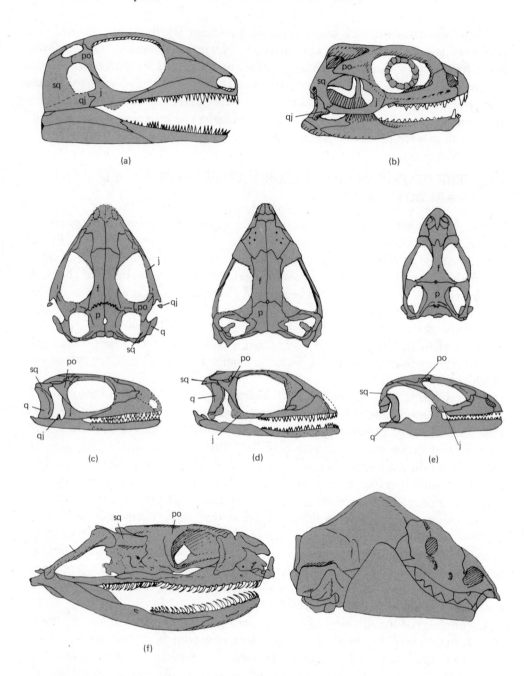

Figure 13-8 Modifications of the skull in lepidosaurian reptiles. Fully diapsid reptiles like the Permian eosuchian *Petrolacosaurus* (a) retain a connection between the quadratojugal and jugal that forms the ventral border of the lateral fenestra. This condition is maintained in the living sphenodontian, the tuatara (b). Lizards have achieved a degree of flexibility in the skull by developing a gap between the quadrate and jugal and simplifying the sutures between other skull bones. In *Paliguana* (c) from the Upper Permian or Lower Triassic, there is a narrow gap between quadrate and jugal and a complex interdigitation of the

to persist was the Suborder Choristodera (champsosaurs) which were amphibious forms with extremely long snouts specialized for capturing fish (Figure 13-13). The nostrils were located on the tip of the snout and a secondary palate was formed primarily by an ossification of the ethmoid. Champsosaurs appear in the fossil record in the Upper Cretaceous with no indication of a connection to the Permian forms. Clearly there must have been a stock of eosuchians retaining the diapsid condition without the specializations that marked the Squamata and Rhynchosauria, but we have no other fossils from that line. Champsosaurs persisted into the Eocene, longer than most Mesozoic reptiles.

Rhynchosaurs

The rhynchosaurs were large, heavily built tetrapods as much as 2 m long (Figure 13-9). The distinctive specializations of rhynchosaurs lie in the structure of the jaws and teeth. The teeth were set in deep sockets and fused at their bases (ankylothecodont dentition). The premaxilla overhung the front of the jaw to form a beaklike projection. In rhynchosaurs the beak was toothless and the maxilla carried heavy tooth plates. The cheek region was expanded, suggesting that the jaw muscles were powerful. Clearly rhynchosaurs were specialized to feed on a distinctive type of food. The body was heavy, suggesting that they were herbivores rather than predators, and one suggestion is that they fed on fruit with an outer pulpy coat that could be seized by the beak and a pit that was crushed by the tooth plates to free an edible seed. Rhynchosaurs appeared in the late Permian or early Triassic. (The earliest fossil is a primitive rhynchosaur from the early Triassic of South Africa.) They flourished during the Triassic but disappeared at the end of that period.

Origin of Squamate Reptiles

Squamate, or scaly, reptiles diverged from eosuchians in the Permian. Fossil squamates from Permian and Triassic deposits in South Africa show a combina-

---

frontal and parietal bones that would have limited movement between them. In the Upper Triassic lizard *Kuehneosaurus* (d), the amount of kinesis in the skull appears to have been increased by widening the quadrate-jugal gap and simplifying the joint between the frontals and parietals. In a modern lizard like the collared lizard (e) the quadrate-jugal gap is still broader, the median bones of the skull are fused at the midline, and the frontal-parietal joint is a straight suture that functions as a hinge. In snakes (f) still further flexibility has been achieved by the loss of a connection between the postorbital and squamosal bones and by a general loosening of connections between several other bones in the skull. Burrowing amphisbaenians (g) have moved in the opposite direction. Extensive ossification and complex sutures produce a rigid skull that can withstand the stress of tunneling. Code: f = frontal, j = jugal, p = parietal, po = postorbital, q = quadrate, qj = quadratojugal, sq = squamosal. Modified from the following: a, R. R. Reisz [1977] *Science* 196:1091–1093; b, A. S. Romer Ch. 15[4]; c, d, e, R. L. Carroll [1977] pp 359–396 in *Problems in Vertebrate Evolution.* Edited by S. M. Andrews, R. S. Miles, and A. D. Walker, *Linnean Society Symposium Series,* No. 4; f, g, C. Gans Ch. 14[3].)

Table 13-1 Taxonomic arrangement of mesozoic reptiles. The major groups discussed in text are included.

Class Reptilia
    Subclass Anapsida (captorhinomorphs, cotylosaurs, turtles—see Tables 12-1 and 14-1)
    Subclass †Synapsida (pelycosaurs and therapsids—see Table 12.2)
    Subclass †Euryapsida (marine reptiles)
      Order †Ichthyosauria (ichthyosaurs)
      Order †Nothosauria (nothosaurs)
      Order †Plesiosauria (plesiosaurs)
      Order †Placodontia (placodonts)
    Subclass Archosauria (ruling reptiles)
      Order †Thecodontia (early archosaurs)
      Order †Ornithischia (bird-hipped dinosaurs)
        Infraorder †Stegosauria (stegosaurs)
        Infraorder †Ankylosauria (armored dinosaurs)
        Infraorder †Ornithopoda (hadrosaurs and related forms)
        Infraorder †Ceratopsia (horned dinosaurs)
      Order †Saurischia (lizard-hipped dinosaurs)
       Suborder †Theropoda (carnivorous dinosaurs)
        Infraorder †Coelurosauria (Ostrich-like dinosaurs and dromaeosaurs)
        Infraorder †Carnosauria (large carnivorous dinosaurs)
       Suborder †Sauropodamorpha (herbivorous dinosaurs)
        Infraorder †Prosauropoda (primitive Triassic forms)
        Infraorder †Sauropoda (advanced Jurassic and Cretaceous forms)
      Order †Pterosauria (flying reptiles)
      Order †Phytosauria (phytosaurs)
      Order Crocodilia (alligators and crocodiles — see Table 14.1)
    Subclass Lepidosauria (scaly reptiles)
      Order †Eosuchia (early lepidosaurs)
      Order †Choristodera (champsosaurs)
      Order †Rhynchosauria (rhynchosaurs)
      Order Squamata (lizards, snakes, amphisbaenians, *Sphenodon* — see Table 14.1)

† = extinct

tion of lizard and eosuchian features. The pectoral girdle and forelimbs were lizardlike, whereas the vertebral column and pelvic girdle were more like those of eosuchians. Primitive lizards were small with head-body lengths less than 150 mm. In modern lizards and sphenodontids growth is determinate. That is, fusion of the epiphysial centers at the ends of the long bones halts growth when adult body size is reached. The structure of the long bones of fossil squamates indicates that growth was determinate in these forms as well. Robert Carroll has proposed that the small body size of primitive squamates was a specialization associated with their presumably insectivorous diet.

Suborder Sphenodontia

The tuatara, *Sphenodon*, a lizardlike reptile living now only on several islands off the coast of New Zealand, was long considered the sole surviving

(a)

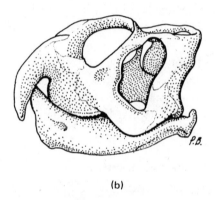

(b)

Figure 13-9 Triassic rhynchosaurs were heavy-bodied herbivores about 2 m long (a). The overhanging premaxilla formed a prominent beak (b). (Part (a) modified from E. H. Colbert [1965] *The Age of Reptiles.* W. W. Norton, New York: (b) from A. S. Romer [1956] *Osteology of the Reptiles.* University of Chicago Press, Chicago.)

rhynchosaur. Fossils very similar to the tuatara can be traced back to the Jurassic. A recent reanalysis of the structural features and especially the dentition of tuataras and fossil sphenodontids indicates that they are closely allied to squamate reptiles and not to rhynchosaurs. Sphenodontids have specialized epiphysial joints like those of lizards and share a number of other skeletal characters with primitive lizards. They may have diverged from the lineage of eosuchians that gave rise to lizards after the evolution of the specializations associated with small body size but before the appearance of changes in the skull that characterize lizards.

Modifications of the Skull in Squamate Reptiles

The fenestrae of the diapsid skull of eosuchians provided space for muscles without reducing the rigidity of the skull as a unit. The fully diapsid pattern was retained in most sphenodontids including the tuatara (Figure 13-8). A basic change in skull proportions and the mechanics of jaw function appeared in lizards. Even in primitive lizards the connection between the quadrate and jugal disappeared and the joint between the parietal and frontal bones was simplified. These changes permitted some flexing of the skull (kinesis) as prey was seized and chewed. Skull kinesis is widespread in modern lizards, and it is an important feature of their feeding mechanisms. The evolution of increasing flexibility in the skull can be traced through early lizards. The gap between the quadrate and jugal widened and the frontoparietal joint became increasingly hingelike. Additional areas of flexion evolved at the front and rear of the skull and in the lower jaws of some lizards.

In snakes the flexibility of the skull was increased still further by loss of the connection between the postorbital and squamosal bone (Figure 13-8). A further increase in the flexibility of the joints between other bones in the palate and in the roof of the skull produced the extreme flexibility of modern snake skulls. The role of skull kinesis in the evolution of snakes is discussed further in section 14.1.6.

The fourth suborder of squamates has a completely different sort of skull specialization. The amphisbaenians are small, elongate, legless, burrowing reptiles. They use their heads as rams to construct tunnels in the soil. Their skulls are heavy with rigid joints between bones. The dentition is specialized for biting small pieces out of large prey.

Mesozoic Lizards: Order Squamata, Suborder Lacertilia

Lizards radiated into a variety of specialized niches early in their development. *Kuehneosaurus* and *Icarosaurus* from the Upper Triassic had enlarged ribs that extended out from the sides of the body and supported an extension of the body wall that formed a sail with which the animal could glide through the air as do the living "flying dragons" of the genus *Draco*. We know little about Mesozoic lizards because most of the fossil deposits found to date have been marine. They provide fossils of marine forms, and those terrestrial animals large and tough enough to remain intact while they were washed out to sea, but relatively little in the way of small, fragile terrestrial animals. This unfortunate circumstance applies also to the fossils of early therian mammals which are also scarce (Chapter 15). The deposits do provide a good record of the Cretaceous marine lizards called mosasaurs. The mosasaur body form was similar to that of the short-necked plesiosaurs (Figure 13-10). The skull was elongate with many teeth in the jaws. The nostrils were situated midway between the tip of the snout and the eyes. Mosasaurs appear to have had more flexible bodies than plesiosaurs, and probably swam with serpentine lateral undulations, using the feet (which were modified into paddles) for steering. The mosasaurs are very closely related to the living monitor lizards (Family Varanidae) and differ primarily in their adaptations to marine life.

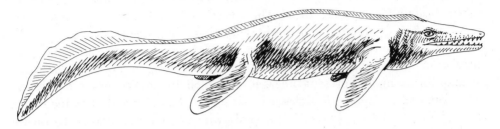

Figure 13–10. Mosasaurs were marine lizards that had a nearly world-wide distribution in the Cretaceous. Closely related to living monitor lizards, they showed a number of specializations for marine life including a tail fin and the modification of limbs into flippers. The form shown, *Tylosaurus*, was about 10 m long. (Modified from Osborn in J. Piveteau Ch. 9 [4].)

13.3.2 Subclass Archosauria

The archosaurs, or ruling reptiles, are the group that includes the enormous ornithischian and saurischian dinosaurs that dominated the world during the Mesozoic. Crocodilians are archosaur descendants, as are birds. In fact, birds are so much like archosaurian reptiles in all respects except those specifically associated with flight that they have been helpful models for paleontologists seeking to reconstruct the ecology of archosaurs.

Archosaurs are diapsid reptiles that retain the two temporal openings without loss of any of the connecting bars. In addition they developed one or two fenestrae in front of the eye (the antorbital fenestrae) and one on the side of the lower jaw at the junction of the dentary, angular, and surangular (mandibular fenestra). The legs were held beneath the body, and a tendency toward bipedalism dominated the anatomy of the group. It was reflected in structural changes in many parts of the body: The hind legs were considerably larger than the front legs. The well-developed tail may have been a counterweight, and it seems possible that the body was carried in a birdlike posture with the tail and trunk forming a horizontal beam balanced on the hind legs.

Pelvic Morphology and Dinosaur Evolution

Many of the morphological trends that can be traced in archosaur evolution are adaptations leading to increased locomotor efficiency. The two most important developments were the movement of the legs under the body and the widespread tendency toward bipedalism. Paleontologists disagree about the order in which those trends developed. Captorhinomorphs and early thecodonts had a sprawling posture like that of many living amphibians and reptiles. The humerus and femur were held out from the body and the elbow and knee were bent at a right angle. In advanced archosaurs, the legs were straight pillars held beneath the body. Was the rotation of the legs beneath the body a prerequisite for development of bipedality, or did increasing bipedality require a simultaneous rotation? The question is not trivial, because there are a number of quadrupedal archosaurs that have been assumed to have reevolved a quadrupedal stance from bipedal ancestors. If, as some paleontologists believe, bipedality was not universal among archosaurs, the quadrupedal forms may represent lineages in which bipedality never developed.

In primitive tetrapods, muscles originating on the pubis and inserting on the femur protract the leg (move it forward) whereas muscles originating on the ischium retract it (move it posteriorly). The primitive tetrapod pelvis, little changed from *Ichthyostega* through primitive thecodonts, was platelike (Figure 13-11a). The ilium articulated with one or two sacral vertebrae, and the pubis and ischium did not extend far anterior or posterior to the socket for articulation with the femur (acetabulum). The pubofemoral and ischiofemoral muscles extended upward from their origins on the pubis to the femur. (The downward force of their contraction was countered by iliofemoral muscles that ran from the ilium to the dorsal surface of the femur.) As long as the femur projected horizontally from the body, this system was effective. The pubofemoral and ischiofemoral muscles were long enough to swing the femur through a large arc relative to the ground. As the legs were held more nearly under the body, the muscles became less effective. As the femur rotated closer to the pubis and ischium, the sites of muscle origin and insertion moved closer together and the muscles themselves became shorter. A muscle's maximum contraction is about 30 percent of the resting length; thus the shorter muscles would have been unable to swing the femur through an arc large enough for effective locomotion had there not been changes in the pelvis associated with the evolution of bipedalism.

The ornithischian and saurischian dinosaurs that evolved from thecodonts carried the legs completely under the body and show associated changes in pelvic structure. The two groups attained the same mechanically advantageous end in different ways (Figure 13-11). In saurischians, the pubis and ischium both became elongated and the pubis was rotated anteriorly, so that the pubofemoral muscles ran back from the pubis to the femur and were able to protract it (Figure 13-11b). Among primitive ornithischians the pubis did not project anteriorly (Figure 13-11c). Instead the ilium was elongated anteriorly, and it seems certain that the femoral protractors originated on the anterior part of the ilium from which they ran posteriorly to the femur. This arrangement would have placed the pubofemoral muscles in a very peculiar situation: When the femur was fully retracted, the pubofemoral muscles would have run posteriorly to their insertion, and their contraction would have protracted the femur. As the femur swung forward, the angle of the pubofemoral muscles would have changed, until, as the femur passed the pubis, they would have been pulling the femur directly toward the midline of the body (adduction). Finally, as the

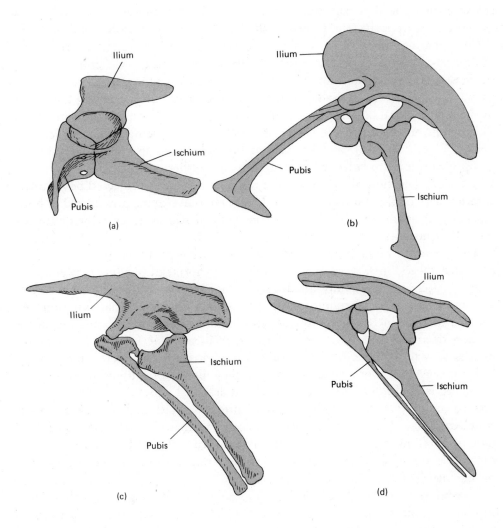

Figure 13-11. Pelvic morphology of a thecodont (a, *Euparkeria*), a saurischian dinosaur (b, *Ceratosaurus*), and two ornithischian dinosaurs (c, *Scelidosaurus*; d, *Thescelosaurus*). The evolution of an upright, bipedal posture in large archosaurs involved great changes in pelvic structure. (a) The primitive platelike pelvis that characterized tetrapods from *Ichthyostega* through primitive thecodonts was functional while the femur projected horizontally from the body, but did not provide adequately positioned sites for muscle origin when the femur was held vertically beneath the body. (b) Extension of the pubis and ischium in saurischian dinosaurs provided sites for muscle attachment fore and aft. (c) Primitive ornithischian dinosaurs achieved the same mechanical result with a forward extension of the ilium and a backward extension of the ischium. The pubis was rotated back to lie parallel with the ischium. This morphology was retained in ankylosaurs, but in other advanced ornithischians a forward extension of the pubis (sometimes called a prepubis) provided additional space for origin of femoral protractors. (a, modified from W. K. Gregory [1951] *Evolution Emerging.* Macmillan, New York; b, c, d, modified from Charig [6].)

femur swung ahead of the pubis, the pubofemoral muscles would have acted as retractors, opposing the effect of the newly developed iliofemoral muscles.

Coordinating the actions of a muscle that changes its function from protractor to adductor to retractor to adductor to protractor in the course of each stride seems a formidable bit of neurophysiological engineering, and ornithischians did not do it. Instead, the pubis rotated posteriorly, carrying the origin of the pubofemoral muscles with it and changing their action to pure retractors. This condition is seen in the pelvis of primitive ornithischians such as *Scelidosaurus* (Figure 13–11c), and appears to be maintained in the ankylosaurs, a group of advanced quadrupedal ornithischians. Other ornithischians developed an anterior projection of the pubis that ran parallel to and projected beyond the anterior part of the ilium (Figure 13–11d). This development occurred in both bipedal and quadrupedal lineages and provided a still more anterior origin for protractor muscles.

The trend towards bipedality, which was widespread if not universal in advanced thecodonts, was equally important in terms of opening new adaptive zones to archosaurs. A fully quadrupedal animal uses its forelegs for walking, and any changes in morphology must be compatible with that function. As animals become increasingly bipedal, the importance of their forelegs for locomotion decreases and the scope of the specialized functions that can evolve increases. Many of the smaller carnivorous dinosaurs that were fully bipedal had forelegs that were adapted for grasping prey. One dinosaur, the small herbivorous *Hypsilophodon*, may have been arboreal and able to climb trees to browse on leaves free from the threat of predators, although Peter Galton has concluded that *Hypsilophodon* was a cursorial animal. The ultimate specialization of forelimbs, once they have been freed from the requirements of quadrupedal locomotion, is as wings, and this occurred twice among reptiles, once in the evolution of birds and once in pterosaurs.

Bipedal animals have hind legs that are considerably longer than their forelegs, and the degree of disproportion between hind legs and forelegs is assumed to reflect the extent of bipedalism in a given species. The quadrupedal archosaurs had longer hind legs than forelegs, and this condition is thought by most paleontologists to indicate that they were secondarily quadrupedal, having evolved from bipedal ancestors.

Origin of Archosaurs

As we have implied, the origin of the archosaurs can be traced, perhaps from eosuchians like *Heleosaurus*, through the Order Thecodontia, a group of diapsid reptiles in which the teeth were set in sockets (**thecodont dentition**) instead of on the margin of the jaw bones. The thecodonts were early Triassic forms that showed little indication of incipient bipedalism, although the hind legs were slightly longer than the forelegs. *Euparkeria*, a thecodont from the lower Triassic of South Africa, was about 1 m long. It may have been moderately bipedal; the hind legs were half as long again as the forelegs but the pelvis was still close to the primitive reptilian pattern (Figure 13–11a). The skull was moderately high and slender with well-developed antorbital and mandibular fenestrae. More advanced thecodonts had a greater disproportion in the size of the forelegs and hind legs (Figure 13–12).

Figure 13-12 Archosaur ancestors. (a) The late Pennsylvanian eosuchian *Petrolacosaurus* had a generally lizardlike body form with fore and hind legs of approximately equal length. (b) In the Permian eosuchian *Heleosaurus*, the forelegs were shorter than the hind legs, suggesting that rapid movement was accomplished by bipedal locomotion. (c) The Triassic thecodont *Saltopsuchus* had a great disparity in the size of fore and hind legs and was probably entirely bipedal. (Modified from the following: a, R. R. Reisz [1977] *Science* 196:1091–1093; b, R. L. Carroll [5]; c, from Z. Burian in Z. V. Spinar [1972] *Life Before Man.* McGraw-Hill, New York.)

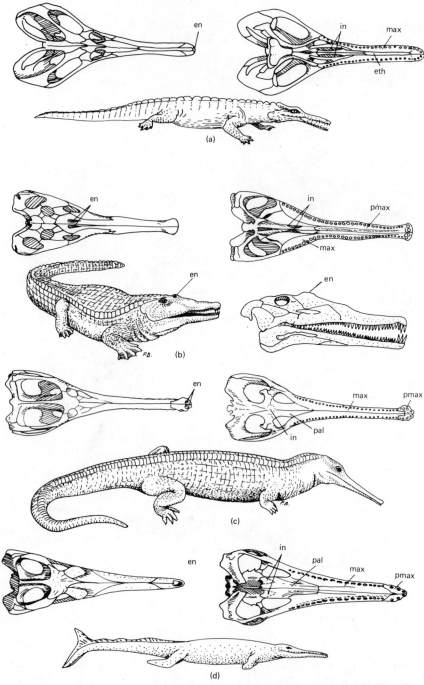

Figure 13-13. A number of specialized aquatic fish-eaters have evolved among reptiles. The plesiosaurs (Figure 13-5), ichthyosaurs (Figure 13-6), and mosasaurs (Figure 13-10) are examples, and the trematosaurs (Figure 9-15) were an amphibian example of the same specialization. In addition to the true crocodilians (Figure 14-1), crocodilelike animals evolved in both the lepidosaurian and archosaurian lineages. (a) The champsosaurs, derived from eosuchians, first appear in Cretaceous fossil beds. They survived into the Eocene. In general

Phytosaurs and Crocodilians

The thecodont stock gave rise to two independent lineages of aquatic fish eaters, the phytosaurs and crocodilians. The phytosaurs were the earlier radiation, and during the Triassic they were abundant and an important element of the riparian fauna. In contrast to crocodilians, in which the nostrils are at the tip of the snout and a secondary palate separates the nasal passages from the mouth, phytosaur nostrils were located on a boss or elevation just anterior to the eyes (Figure 13–13b). True crocodilians appeared in the Triassic and seem to have replaced phytosaurs by the end of that period (Figure 13–13c). In most respects crocodilians conform closely to the skeletal structure of archosaurs, but the skull and pelvis are specialized. Crocodilians retained the nostrils at the tip of the snout and developed a secondary palate (Figure 13–13c). The increasing involvement of the premaxilla, the maxillae, and pterygoids in the secondary palate can be traced from Mesozoic crocodilians to modern forms. Crocodilian hind legs are longer than forelegs, even in living species, and the disparity was greater in Jurassic forms, indicating an origin from a thecodont with considerable bipedality. In modern forms the legs can be brought partially under the body when a crocodile wants to move fast.

In water the heavy, laterally flattened tail propels the body, and the legs are held against the sides. In the late Jurassic, a lineage of specialized marine crocodiles enjoyed brief success. These thallatosuchians had long skulls with pointed snouts (Figure 13–13d). They lacked the dermal body armor typical of most crocodilians, and had developed a lobed tail very like that of the primitive ichthyosaurs, with the vertebral column turned downward into the lower lobe and the upper lobe supported by stiff tissue. The feet were paddlelike.

---

appearance they closely resembled crocodilians, and the nostrils were at the tip of the snout. A secondary palate was formed by the palatine bones and what is thought to be an ethmoid ossification. (b) Phytosaurs were a separate radiation from the thecodont stock that also gave rise to true crocodilians. The nostrils were located on a raised boss just anterior to the eyes. This alternative to the evolution of a secondary palate is seen in other aquatic reptiles. Both solutions place the internal nares in the posterior part of the mouth so that a fleshy valve can prevent water from entering the trachea. (c) True crocodilians replaced phytosaurs in the Jurassic. In these animals the nostrils were located at the tip of the snout and the premaxilla, maxillae, and palatines formed a secondary palate. (d) The thallatosuchians were a marine offshoot of the crocodilians that enjoyed brief success in the late Jurassic and early Cretaceous. They were the most specialized crocodilians and, in general body form, were very like primitive ichthyosaurs (Figure 13–6). Code: en = external nares, eth = ethmoid, in = internal nares, max = maxilla, pal = palatine, pmax = premaxilla. (Modified from the following sources: a. skull from A. S. Romer Ch. 15[4], restoration from J. C. Germann in B. Kurten [12]; b, skull modified from W. K. Gregory [1051] *Evolution Emerging.* Macmillan, New York, restoration from various sources; c, skull modified from A. S. Romer Ch. 15[4], restoration from various sources; (d, skull from Gregory [b] restoration from Schmidt in J. Piveteau Ch. 9[4].)

Pterosaurs

The archosaurs gave rise to two independent radiations of fliers. The birds are one of these lineages, and their similarity to archosaurian reptiles is so striking that, had they disappeared at the end of the Mesozoic, they would be considered no more than another group of highly specialized reptiles. Because they survived, they are discussed in Chapters 16 and 17. The other lineage of flying reptiles were the pterosaurs of the Jurassic and Cretaceous (Figure 13-14). They ranged from the sparrow-sized *Pterodactylus* to *Quetzalcoatlus* with a wingspan of 15 m. The wing formation of pterosaurs was entirely different from that of birds. The fourth finger was elongate and supported a membrane of skin anchored to the side of the body and the hind leg. A small splintlike bone was attached to the front edge of the carpus and probably supported a membrane that ran forward to the neck. Primitive pterosaurs had a long tail with a "rudder" on the end that was presumably used for steering; more advanced forms had lost the tail.

Flight is a demanding means of locomotion for a vertebrate, and it is not surprising that pterosaurs and birds show a high degree of convergent evolution. The long bones of pterosaurs were hollow, as they are in birds and many other archosaurs, reducing weight with little loss of strength. The sternum, to which the powerful flight muscles attach, was well developed in pterosaurs, although it lacked the keel seen in birds. The eyes were large, and casts of the brain cavities of pterosaurs show that the parts of the brain associated with vision were large and the olfactory areas small, as they are in birds. The cerebellum, which is concerned with balance and coordination of movement, was large in proportion to the other parts of the brain as it is in birds. There was a trend in pterosaurs to lose their teeth and evolve a birdlike beak. Even in their physiology pterosaurs and birds may have been very similar. Birds are endotherms, their feathers providing the insulation required to retain metabolically produced heat. Some fossils of pterosaurs in fine-grained sediments show impressions of structures that look like hairs and strongly suggest that pterosaurs, too, were endotherms. Certainly the energy requirements of flight are great enough to suggest that pterosaurs had higher metabolic rates than do living reptiles. The extent of their flight capabilities has long been debated, and most hypotheses about their ecology have been based on the assumption that they were weak fliers, perhaps depending on constant winds to produce updrafts they could use for soaring like modern eagles and vultures.

A recent aerodynamic analysis of *Pteranodon* by R. S. Stein indicates that previous views were entirely incorrect. Stein found that *Pteranodon* had a wing structure quite different from that of soaring birds but one that suggests that it was a good glider and capable of strong flapping flight. His analysis shows that a *Pteranodon* would have been quite capable of taking off with a flap of its wings and, thus, would not have been restricted to landing on cliffs and launching itself from them as some paleontologists have suggested. The peculiar crest on the head probably supported a membrane that stretched to the back. The

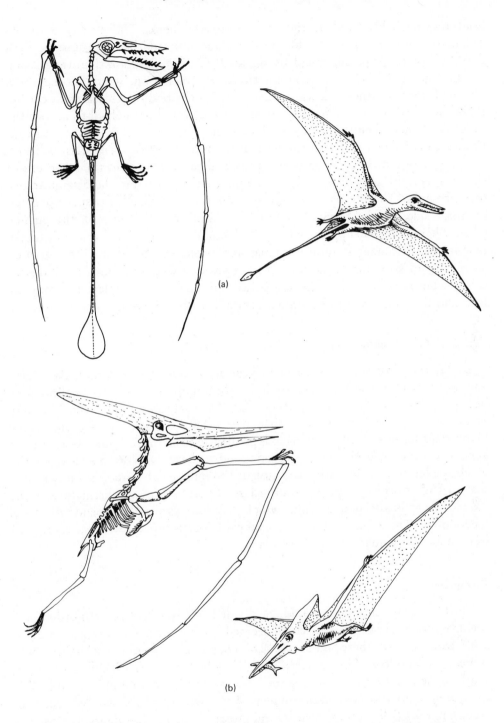

Figure 13-14.   (a) Primitive pterosaurs like the Jurassic *Rhamphorhynchus* had long tails. In some forms a flattened tip probably served as a rudder. (b) Advanced pterosaurs like the Cretaceous *Pteranodon* lacked long tails and teeth. (Skeletons from A. S. Romer Ch. 15[4] restorations modified from various sources.)

head was probably held at right angles to the neck. The structure of the articulating surfaces suggests this position, and it would be more suitable for a slow flier of the type suggested by the analysis of *Pteranodon* than the stream-lined form usually shown in reconstructions. The membrane would have exerted a lifting force when the head was turned, turning the body in the same direction. If the membrane was moderately elastic, it may also have helped to absorb the shock of impact when the beak seized a fish from the water. Stein suspects that fish were carried in a pouch at the base of the beak like that of modern pelicans.

The greatest difference between birds and pterosaurs is in the structure of the wing, and this may be a key to the persistence of birds and the disappearance of pterosaurs. The pterosaur wing was composed of a sheet of tissue supported on three sides by the forelimb, trunk, and hind limb, and free on the outside edge. There was no support within the membrane, and a tear that ran to the margin would render it nonfunctional and probably would not heal because there was no structure to hold the torn flaps of tissue together. In contrast, a bird's wing is composed of separate feathers that can be replaced if they are lost, and the wing functions even with several feathers missing.

Saurischians and Ornithischians

By far the most generally known of the archosaurs are the Orders Saurischia and Ornithischia. These groups are linked in popular terminology as dinosaurs, but in fact they are no more closely related to each other than either is to any of the other archosaurian orders that have been discussed. Both groups arose in the Triassic, apparently evolving independently from thecodonts. Both groups appear to have been primitively bipedal and to have evolved some secondarily quadrupedal forms. The early evolution of the saurischians is better known than that of the ornithischians. Two suborders of saurischians are distinguished, the Theropoda and Sauropodamorpha, which may have evolved independently from thecodonts. Thereopods were advanced carnivorous bipeds, whereas sauropods were secondarily quadrupedal herbivores.

Theropods

Until recently theropods could be divided neatly into two infraorders, the coelurosaurs and carnosaurs, which seemed to have evolved in opposite directions from a basic theropod stock. The coelurosaurs were small, lightly built animals with hollow, thin-walled bones. They had small, light skulls, large eyes, and long necks. *Coelophysis* (Figure 13–15a) was less than 3 m in total length, and *Campsognathus* was even smaller. They were probably active, cursorial predators on small reptiles, lizards, and insects. Their descendants, the ostrich-like dinosaurs (ornithomimids), carried the characteristics of the coelurosaurs to a greater degree of specialization. *Ornithomimus* of the Cretaceous was ostrich-like in size, shape, and probably in its ecology as well (Figure 13–15b). It had a small skull on a very long neck and its toothless jaws were covered with a horny

Figure 13-15. Varied types of saurischian theropod dinosaurs. Coelurosaurs were relatively small, lightly built animals. (a) *Coelophysis*, a Triassic form, shows evidence of moderate bipedality in the disproportion of fore and hind legs. (b) By the Cretaceous, coelurosaurs like the toothless *Ornithomimus* were very birdlike in structure. (c) Carnosaurs were enormous animals that preyed on even larger herbivorous dinosaurs. *Megalosaurus*, a Jurassic form, was 6 m long and some Cretaceous species were twice as large. (d) Dromaeosaurids were intermediate between coelurosaurs and carnosaurs. They were small, about 2 m, and fleet like coelurosaurs, but they had large heads and teeth. The Cretaceous *Deinonychus* had an enlarged claw on the second toe that was specialized as a weapon. (a, b, modified from various sources; c from N. Parker in D. F. Glut [1971] *The Dinosaur Dictionary*. Citadel Press, Secaucus, N.J.; d, from R. Bakker in J. H. Ostrom [1969] *Peabody Museum of Natural History. Bulletin* 30, Yale University, New Haven.)

bill. The forelegs were longer than those of coelurosaurs, and only three digits were developed on the hands. The inner digit was opposable and the wrist was flexible, making the hand an effective grasping organ for the capture of small prey. Like ostriches, ornithomimids were probably omnivorous and fed on fruits, insects, small vertebrates, and reptile eggs. Quite possibly they lived in herds, as do ostriches, and their long legs suggest that they inhabited open regions rather than forests.

Not all coelurosaurs preyed on small animals. A recently described theropod from the Gobi Desert, *Deinocheirus*, had hands more than 60 cm long that appear to have been adapted for grasping and dismembering large prey If this theropod had the same body proportions as other coelurosaurs it may have been more than 7.5 m tall, exceeding the carnosaur *Tryannosaurus rex*, previously the largest theropod known, by 2 m.

The carnosaurs are the large carnivorous dinosaurs that form the centerpiece of paleontological displays in museums. They were far more massive than coelurosaurs, but otherwise not greatly different from the basal stock of that line. The long bones were hollow, but the walls were heavy. Their skulls were large and their necks short. Among the carnosaurs there was a trend to increasing body size, probably paralleling a similar trend in the herbivorous saurischians and ornithischians that were their prey. *Megalosaurus* of the Jurassic was 6 m long, and more fully bipedal than earlier forms (Figure 13-15c). There were five sacral ribs giving increased support to the pelvic girdle, and the pubis and ischium had lost the platelike shape characteristic of thecodonts and assumed the elongate structure of saurischians. The head was large in proportion to the body and the long teeth were fearsome weapons. The Cretaceous theropods such as *Gorgosaurus* and *Tyrannosaurus* were still longer, up to 15 m in length, and stood some 6 m high. The front legs of the most specialized of these giant carnosaurs seem absurdly small; they were too short even to reach the mouth. Instead of relying on the forelegs to capture prey as coelurosaurs and ornithomimids probably did, carnosaurs appear to have concentrated their weapons in the skull. The size of the head increased relative to the body, and the neck shortened. The head was lightened by the elaboration of antorbital and mandibular fenestrae that reduced the skull to a series of bony arches providing maximum strength for a given weight. The teeth were formidable weapons as much as 15 cm long, dagger-shaped and with serrated edges.

The neat division of theropods into small, lightly built, fleet-footed coelurosaurs and enormous carnosaurs has been upset recently by the discovery of a third type of theropod that combines features of both coelurosaurs and carnosaurs. *Deinonychus* was unearthed by an expedition from Yale University in early Cretaceous sediments in Montana (Figure 13-15d). It was a small saurischian, a little over 2 m long. Its distinctive features were the claw on the second toe of the hind foot and the tail. In other theropods, the hind feet were clearly adapted for bipedal locomotion and were very similar to bird feet. The third toe was the largest, the second and fourth smaller, and the fifth sometimes disappeared entirely. The first toe was turned backward, as in birds, to provide support behind the axis of the leg. In *Deinonychus* the second toe, and especially the claw, were enlarged, and the joints of the toe were arranged so that it was held off the ground and could be bent upward even farther.

It seems very likely that *Deinonychus* used these claws in hunting, disemboweling its prey with a sudden kick. Several flightless birds including the cassowary and ostrich kick in this manner to defend themselves from predators, and a blow from the foot of either one can be lethal to a human. Although the birds kick forward, the structure of *Deinonychus'* legs suggests that it kicked backward, using the most powerful leg muscles and achieving considerably greater force than the birds. The structure of the tail was equally remarkable. The caudal vertebrae were surrounded by bony rods that were extensions of the prezygapophyses (dorsally) and hemal arches (ventrally) that ran forward about 10 vertebrae from their place of origin. The contraction of muscles at the base of the tail would be transmitted through these bony rods, drawing the vertebrae together and making the tail a rigid structure that could be used as a counterbalance or swung like a heavy stick. Possibly the tail was part of the armament of *Deinonychus*, used to knock prey to the ground where it could be kicked.

Sauropods

The advanced sauropods (Suborder Sauropodomorpha) are so different from the theropods that they might well be placed in their own order if there were not some transitional bipedal forms known. The primitive Triassic forms are grouped in the Infraorder Prosauropoda. *Plateosaurus* is an example of this group (Figure 13–16a). It was something over 6 m in length and not fully bipedal. The forelimbs were shorter than the hind limbs, but not strikingly so, and the digits on the forefeet were all well developed. The limb bones were solid, rather than hollow as they are in advanced bipeds. The neck was long and the skull comparatively small. Most prosauropods were herbivorous, but there are some poorly known forms that appear to have been carnivorous. Those were replaced by theropods by the end of the Triassic.

The Jurassic and Cretaceous forms are placed in the Infraorder Sauropoda. These were enormous quadrupedal herbivores. Their origin from forms that were at least partly bipedal is strongly suggested by the disparity in size of the forelimbs and hind limbs. The sauropods were the largest terrestrial vertebrates that have ever existed, reaching lengths of 25 m and weights of 20,000 to 50,000 kg. A newly discovered fossil sauropod from Colorado, not yet fully described or named, may have weighed as much as 90,000 kg, the equivalent of 14 elephants. *Apatosaurus* (formerly known as *Brontosaurus*), *Diplodocus*, and their relatives were massive animals with long tails and necks and very small heads (Figure 13–16b, c). The orbits were large, and the nostrils were located just in front of the eyes. The jaws were short, and the teeth were relatively small. In most forms teeth were confined to the front of the mouth. A sacral plexus (a neural relay center) in the spinal cord was several times as large as the brain and probably coordinated the movements of the legs and tail. The sauropods

Figure 13-16. The sauropod dinosaurs were quadrupedal forms that evolved from bipedal ancestors. (a) The Triassic prosauropod *Plateosaurus* was a primitive form that links bipedal ancestors with their quadrupedal descendants. With a length of "only" 6 m it was a relatively small sauropod. The advanced sauropods were quite similar to each other in their gross form but differed in their body proportions. (b) *Apatosaurus* was heavily built, and its estimated weight of 50,000 kg makes it one of the largest terrestrial animals that has ever lived. (c) *Diplodocus* was longer but more lightly built. It may have reached lengths exceeding 30 m. (a modified from R. Kane and c from C. R. Knight in E. H. Colbert [1961] *Dinosaurs*. E. P. Dutton, New York; b modified from C. R. Knight in D. F. Glut [1972]. *The Dinosaur Dictionary*. Citadel Press, Secaucus, N.J.)

were enormously heavy and the vertebrae show features that helped the spinal column to withstand the stresses to which it was subjected (Figure 13–17). The vertebrae themselves were massive, and the neural arches well developed. Strong ligaments transmitted forces from one arch to adjacent ones to help equalize the stress. The head and tail were cantilevered from the body, the vertebral column forming the compression member and the spinal ligament the tension member. The sides of the neural arches and centra had hollows in them, presumably occupied by air sacs in life, which reduced the weight of the bones with little reduction in strength. The feet of these forms were elephantlike; fossilized tracks indicate that the hind legs bore about two thirds of the body weight. Interestingly, the trackways show no tail marks, indicating that the tails were carried in the air, not dragged along the ground in the manner shown in almost all illustrations of these dinosaurs.

Ornithischians

The Order Ornithischia may have been independently derived from thecodonts but shows some convergence of body form with saurischians. The ornithischians were exclusively herbivorous, however, and radiated into considerably more diverse morphological forms. The ornithischians were never as bipedal as the saurischians, and the forelegs were never reduced to the same extent. Four infraorders are recognized:

> Ornithopoda: Bipedal forms including the duck-billed and casque-headed dinosaurs.
> Stegosauria: Quadrupedal forms with a double row of plates or spines along the back and tail.
> Ankylosauria: Heavily armored quadrupeds.
> Ceratopsia: The horned dinosaurs.

The first dinosaur fossil to be recognized as such was an ornithopod, *Iguanodon*, found in Cretaceous sediments in England (Figure 13–18a). Subquently specimens have been found in Europe and Australia. *Iguanodon* reached lengths of 10 m, although most specimens are smaller. Its name comes from the structure of the teeth, which were flat with serrated edges, very similar to the lateral teeth of living herbivorous lizards like *Iguana*. The first digit on each forefoot was modified as a spine that projected upward. These spines show a striking resemblance to the spines on the forefeet of some frogs that may be used as defensive weapons and in intraspecific encounters. One or both functions may have existed in *Iguanodon*.

Hadrosaurs and other advanced ornithopods

Advanced ornithopods are specialized forms—the hadrosaurs (duck-billed dinosaurs) and the pachycephalosaurids (casque-headed dinosaurs). As their

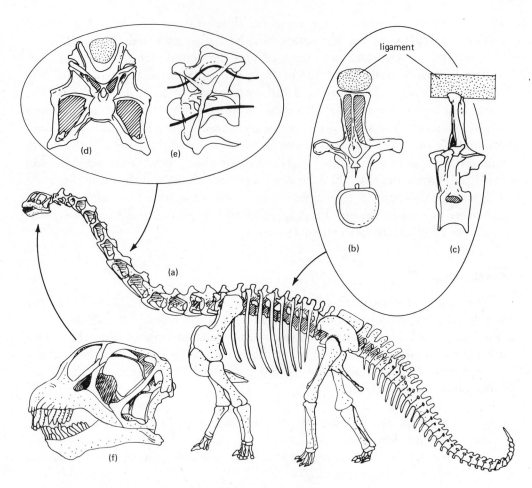

Figure 13-17. The skeletons of large sauropods like the Jurassic *Camarasaurus* (a) combined lightness with strength. Bony arches supported the animal's weight while the spaces between them were filled by a lighter tissue—possibly fat or air sacs. Vertebrae from the dorsal region (b, posterior view; c, lateral view) and neck (d, anterior view; e, lateral view) show the bony arches that acted like flying buttresses on a large building. (The black ribbons in e indicate the position of these arches.) The neural spines accomodated an enormous ligament that supported the neck and tail. The skull (f) is small in proportion to the body and the bony structure, reduced to a series of arches. (a, f, modified from Osborn and Mook in J. Piveteau Ch. 9[4]; b–e modified from E. H. Colbert [1961] *Dinosaurs*. E. P. Dutton, New York.)

name implies, duck-billed dinosaurs had flat snouts with a ducklike bill projecting from the upper jaw (Figure 13-18b). These were large animals; some reached lengths in excess of 10 m and weights greater than 10,000 kg. They seem to have been semiaquatic; the forefeet were webbed and the tail flattened. The anterior part of the jaws was toothless, but there was a massive battery of teeth posteriorly. On each side of the upper and lower jaws were four tooth rows, each containing about 40 teeth packed closely side by side to form a massive

Figure 13-18. Ornithopod dinosaurs. The ornithischian dinosaurs show greater morphological diversity than the herbivorous saurischians. They appear to have been derived from bipedal ancestors, and the group known as ornithopods retained that posture. *Iguanodon* (a) was a generalized ornithopod, whereas the duck-billed hadrosaurs (b, *Hadrosaurus*) may have been semiaquatic. The pachycephalosaurs (c, *Pachycephalosaurus*) had a very thick bony dome on the skull. Probably males stood head to head and tried to shove each other backward during agressive encounters. (Modified from various sources.)

tooth plate. Several sets of replacement teeth lay beneath those in use, so a hadrosaur had several thousand teeth in its mouth of which several hundred were in use simultaneously. The throat was quite small. What sort of food requires a horny bill for its capture, a massive battery of teeth to chew it, and a small throat to swallow it? E. H. Colbert has suggested that hadrosaurs fed in shallow swampy water, taking in huge mouthfuls of plant material that carried small vertebrates, invertebrates, and molluscs along with it. This material was ground to a pulp by the teeth, and then the excess liquid allowed to escape through grooves on the inner surface of the bill while the food was retained and swallowed.

Three lineages of hadrosaurs are distinguished by paleontologists—flat headed, solid crested, and hollow crested (Figure 13–19). In the solid-crested forms the nasal and frontal bones grew upward meeting in a spike that projected over the skull roof. In the hollow-crested forms a similar projection was formed by the premaxillary and nasal bones. In *Corythosaurus* those bones formed a helmetlike crest that covered the top of the skull, whereas in *Parasaurolophus* there was a long, curved structure that extended over the shoulders. Although the crests of the solid-crested forms contain only bone, the nasal passages run through the crests of the hollow-crested forms. Inspired air traveled a circuitous route from the external nares through the crest to the internal nares, which were located in the palate just anterior to the eyes.

J. A. Hopson has suggested that these bizarre structures were associated with species-specific visual displays and vocalizations. The crests probably supported a frill attached to the neck, which could have been utilized in lateral displays analogous to the displays of many living lizards that have similar frills. Hopson suggested that in the noncrested forms the nasal regions were covered by extensive folds of fleshy tissue that could be inflated by closing the nasal valves. Analogous structures can be found in the inflatable probocises of elephant seals and hooded seals. In the seals, and perhaps in the dinosaurs, the inflated structures are resonating chambers used to produce vocalizations.

The pachycephalosaurs are among the most bizarre ornithopods known (Figure 13–18c). The postcranial skeleton conformed to the bipedal ornithopod pattern, but on the head an enormous bony dome thickened the skull roof. The bone was as much as 25 cm thick in a skull only 60 cm long. The angle of the occipital condyle indicates that the head was held so that the axis of the neck extended directly through the dome. Small bony spines projected from the dome, and the entire structure suggests a greatly enlarged version of the heads of some large lizards, such as the Galapagos marine iguana and the rhinoceros iguana of the Carribean. Male lizards press their heads together and wrestle during territorial disputes; possibly the pachycephalosaurs used their bony heads in the same way.

Armored Dinosaurs—Stegosaurs and Ankylosaurs

The stegosaurs, known as plated dinosaurs, were a group of secondarily quadrupedal herbivorous ornithischians that flourished in the Jurassic. *Stego-*

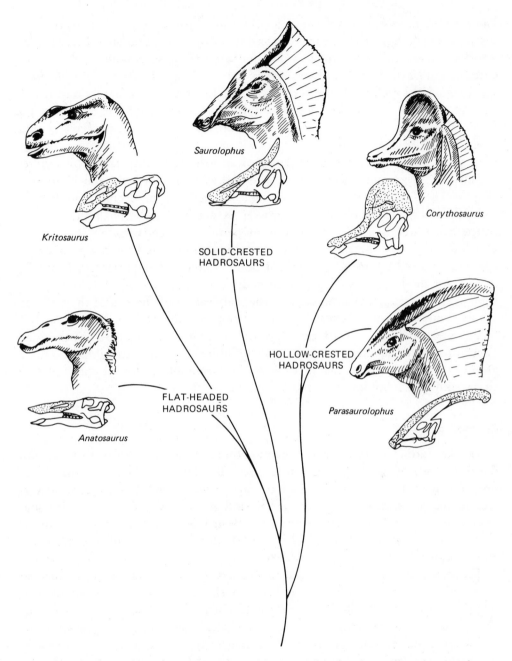

Figure 13-19. The bizarre development of the nasal and maxillary bones of some hadrosaurs gave their heads a superficially antelopelike appearance. In the flat-headed and solid-crested forms the nasal passages ran directly from the external nares to the mouth. In the hollow-crested forms, as a result of the involvement of the premaxillary and nasal bones (indicated by shading) in the formation of the crest, the nasal passages were diverted up and back through the crests before they reached the internal nares. (Modified from various sources, especially E. H. Colbert [1945] *The Dinosaur Book.* American Museum of Natural History, New York, and [1961] *Dinosaurs.* E. P. Dutton, New York.)

saurus is the most familiar of these, a large animal (up to 6 m in length) with its forelegs much shorter than hind legs, a legacy of its bipedal ancestors (Figure 13-20a). The skull was small, as in the saurischian sauropods, and a sacral enlargement of the spinal cord was several times the size of the brain. The body was protected by a double series of leaf-shaped plates set alternately on the left and right sides of the vertebral column. Two pairs of spikes on the tail made it a formidable weapon. Related forms had various combinations of plates and spikes on the back and tail.

In the Cretaceous a different group of armored quadrupedal herbivores replaced the stegosaurs. These were the ankylosaurs, or armored dinosaurs (Figure 13-20b). Ankylosaurs were large animals, 5 m or more in length. Dermal armor was developed to a greater extent in these animals than in any of the other archosaurs; a mosaic of bony plates enclosed them in a shell nearly as solid as that of a turtle. Even the head was armored, and the temporal openings were closed by an overgrowth of bone. Large spines projecting from the margins of the body must have made it difficult to attack an ankylosaur, and some forms were further protected by the elaboration of the end of the tail into a huge bony mass with coossification of the caudal vertebrae forming a stiff "handle." In all, the tail of these animals resembles nothing so much as an enormous medieval mace. Seventy-five million years later the glyptodonts, a group of South American mammals related to armadillos, evolved armor plating that strikingly resembles that of ankylosaurs.

Ceratopsians

The last group of dinosaurs to evolve, the ceratopsians (horned dinosaurs), did not appear until the Upper Cretaceous. By this time the easy transit from one continent to another that had characterized much of the Mesozoic was coming to an end. Ceratopsians are found in western North America and eastern Asia (which were connected by a Bering Sea bridge), but they were excluded from the rest of the world by the shallow seas that divided both North America and Eurasia in the late Mesozoic.

The distinctive features of ceratopsians are the frill over the neck, formed by an enlargement of the parietal and squamosal bones, and the development of one or more horns on the head (Figure 13-20c, d). The protective aspects of the frill have long been emphasized, but mechanical analysis of the jaw structure and of evolutionary changes in the frill suggest that it originated as an attachment site for jaw muscles and any protective function was secondary. The horns—a pair located over the eyes, or a single horn on the nose, or both—were undoubtedly defensive and backed by the weight of a dinosaur as much as 10 m in length must have deterred even the large carnosaurs. The dentition of ceratopsians was as specialized as their jaw muscles. The front of the mouth was occupied by a toothless beak, and in the rear was a battery of teeth. Because the jaw articulation was at the level of the tooth row, the teeth closed with a shearing bite. That morphology is more characteristic of carnivores than herbivores. The specializations of the jaw mechanism of ceratopsians suggest that they fed on some food not available to other reptiles, and their great abundance indicates that there was plenty of it. John Ostrom has suggested that the fibrous leaves of palms and cycads were the staple of the ceratopsians' diet.

13.4 ECOLOGY OF DINOSAURS

Many of our ideas about the sorts of lives dinosaurs led have persisted with little change from the nineteenth century. Although the amount of information available on such topics as biomechanics, paleoclimatology, and the ecology of living reptiles has increased by several orders of magnitude in the last century, it is only quite recently that paleontologists have returned to an examination of dinosaur fossils armed with this new perspective.

Classic views of the lives of dinosaurs were extrapolated from a superficial view of the biology of large living reptiles, especially crocodilians and large lizards. These animals are usually seen in zoos where they are well-fed and isolated from the stimuli they experience in their normal habitats. Zoo animals present an exaggerated impression of the lethargy of modern reptiles, and they are quite unsuitable as models for Mesozoic reptiles.

The Mesozoic world was quite different from the modern world, and the roles of Mesozoic reptiles were correspondingly different from those of modern reptiles. Reptiles were the dominant terrestrial vertebrates of the Mesozoic, and they undoubtedly lived more active lives and filled more adaptive zones than modern reptiles do in the mammal-dominated terrestrial ecosystems of the Cenozoic.

Recently some paleontologists have begun to compare the anatomy of dinosaurs with the anatomy of large mammals and birds and to speculate about ecological and behavioral similarities on that basis. At the same time, studies of the behavior of modern reptiles under field conditions have shown that many have much more complex behavior patterns than had been suspected from observations of captive animals. These lines of reasoning have led to the conclusion that dinosaurs were probably considerably more active than had been suspected and that they probably had an intraspecific social organization at least as complex as that of modern crocodilians (see section 14.6.1). Peripheral information about ecology and behavior of dinosaurs is provided by material such as fossilized stomach contents of dinosaurs and fossilized impressions of dinosaur footprints.

13.4.1 Ecology of Sauropods

The large sauropod dinosaurs, *Apatosaurus, Diplodocus*, and related forms, were the largest terrestrial animals that have ever lived. The largest of them may have reached body lengths exceeding 30 m and weights of 90,000 kg. (For com-

Figure 13-20. The remaining ornithischian dinosaurs were quadrupedal, but the disproportion between fore and hind legs suggests that they were derived from bipedal ancestors. Unlike the ornithopods, all the remaining ornithischians were armored. (a) The stegosaurs had a double row of plates and spikes on the back and tail. These were not fused to the vertebrae; consequently, they fell off as a stegosaur corpse decayed and were fossilized separately. Early restorations

parison, an elephant is about 5 m long and weighs 5000 kg.) From the earliest discovery of their fossils, paleontologists doubted that such massive animals could have walked on land and felt they must have been limited to a semiaquatic life in swamps. Mechanical analysis of sauropod skeletons does not support that conclusion. As we pointed out in section 13.3.2 the skeletons of the large sauropods clearly reveal selective forces favoring a combination of strength with light weight. The arches of the vertebrae acted like flying buttresses on a large building, while the V-shaped neural spines held a massive and possibly elastic ligament that helped to cantilever the head and tail. In cross section the trunk was deep, shaped like the body of a terrestrial animal such as an elephant rather than rounded like that of the aquatic hippopotamus. The tail of the sauropods shows none of the lateral flattening that is characteristic of tails used for swimming. Instead it was round in cross section and terminated in a long, thin whiplash. It fits our ideas of a counterweight and defensive weapon far more closely than the image of an organ of propulsion.

Fossil trackways of sauropods indicate that the legs were held well under the body; the tracks of the left and right feet are only a single foot width apart. Analysis of the limb bones suggests that they were held straight in an elephant-like pose and moved fore and aft in a parasagittal plane. No other leg morphology is possible for a large animal because of the mechanical properties of bone. A bone is far less resistant to bending forces exerted at an angle to its axis than it is to compressional forces exerted along the axis. As the body increases in size, the mass grows as the cube of the linear dimensions, but the cross-sectional area of the bones increases as the square of the linear dimensions. The strength of bone is roughly proportional to the cross-sectional area. As a result, as the body size of an animal increases, the strength of the skeleton increases more slowly than the stress to which it is subjected. One evolutionary response to this is a disproportionate increase in the diameter of bones—an elephant skeleton is proportionately larger than a mouse skeleton. Another response is to transform bending forces to compression forces by bringing the legs more directly under the body. In a large animal, such as an elephant or a sauropod dinosaur, the legs are not only held under the body but the knee joint tends to be locked as the animal walks. This produces the ambling locomotion familiar in elephants, and it is likely that sauropods walked with an elephantlike gait with their heads and tails held well off the ground.

---

of *Stegosaurus* placed the plates in a single row in the dorsal midline. Current opinion holds that there were two alternating rows of plates. (b) Ankylosaurs replaced stegosaurs in the Cretaceous. These animals were heavily armored. Dermal ossifications formed a shell nearly as solid as that of a turtle, and spines provided further protection. In some species the tail ended in a clublike mass of bone. The horned dinosaurs (ceratopsians) were the last group to evolve. Early forms like *Protoceratops* (c) had a large "frill" over the neck that provided increased area for the attachment of jaw muscles. Some later forms like *Triceratops* (d) had evolved long horns for defense. (Modified from various sources.)

Sauropod teeth are usually described as being small and weak. Certainly they were small in proportion to the size of the body, as was the entire skull. In absolute terms, however, they were neither small nor weak. They were larger than the teeth of modern browsing mammals, and there is no reason to believe that plant material was tougher in the Mesozoic than it is today. Sauropods had only anterior teeth, similar in position and shape to the incisor teeth of cattle and horses. There were no molariform teeth at the back of the jaw to crush the ingested plant material. This function may have been served by gastroliths, stones that are deliberately eaten by an animal and retained in some part of the gut where they crush food. Crocodilians use gastroliths in this fashion, as do birds, and accumulations of gastroliths have been found near fossil sauropods.

An important recent discovery is the fossilized stomach contents of a sauropod dinosaur found in Jurassic sediments in Utah. The fossil fragments are chiefly sections of twigs and small branches about 2.5 cm long and 1 cm in diameter. The fragmented and shredded character of the woody material indicates that even without molariform teeth the sauropod had some method of crushing its food. This discovery appears to confirm the view of sauropod ecology that was developed from study of the skeleton and analysis of plant fossils found in association with sauropod fossils. Sauropods were fully terrestrial animals that occupied open country with an undergrowth of ferns and cycads and an upper story of conifers. They were no doubt preyed upon by the large theropod carnivores and sought escape in flight or defended themselves by whipping their tail. Their long necks probably enabled them to graze from tree tops, and it may be significant that Mesozoic conifers bore branches only near the tops of the trees, far out of reach of any but a very long-necked dinosaur. The adults' necks were long enough to reach branches more than 10 m above the ground, and it has been suggested that they could stand on their hind legs, using the long tail as a counterweight, to reach even higher.

Sauropods might have been an important force shaping the landscape and preventing ecological succession from transforming open countryside to dense forest. As such, their presence could have been important in creating and preserving suitable habitat for other species, such as the cursorial coelurosaurs. Like many modern herbivorous reptiles, the sauropods may have been omnivorous and opportunistic in their feeding, taking whatever was readily available, including carrion. The fossilized stomach contents mentioned previously contain traces of bone as well as a tooth from the contemporary carnivorous dinosaur *Allosaurus*.

Fossil trackways reveal a few details from which glimpses of dinosaur behavior can be reconstructed. A famous trackway found in Texas, parts of which are now on display at the University of Texas at Austin and in the American Museum of Natural History in New York City, shows the footprints of a sauropod trailed by a large theropod. In some cases the theropod had stepped into the footprints of its potential prey. Unfortunately the trackway ended before the results of this Cretaceous drama were revealed. Some insight into the intrapsecific behavior of sauropods may be revealed by a series of tracks found in early Cretaceous sediments at Davenport Ranch in Texas. These reveal the passage of a herd of 30 brontosaurlike dinosaurs that passed by some 120 million years ago. A group of 30 individuals moving in the same direction at the same time would be remarkable in most living reptiles, but the brontosaur tracks suggest that this is what happened. Instead of moving as a casual group, the brontosaurs seem to have had an organized social structure. The largest footprints are on the outside of the trackway, and the smallest in the center. It appears that the adults served as guards on the outside of the herd while the more vulnerable juveniles were kept sheltered in the middle. This type of behavior is seen in mammals, but is unknown in modern reptiles, although protection of juveniles by adults is widespread in crocodilians.

13.4.2 Ecology of Ornithopods

Ornithischians were more diverse morphologically than sauropods. Indeed, one of the surprising things about sauropods was their basic similarity—all were enormous, long-necked browsers. Among ornithischians four major types are known and different specializations can be distinguished within those groups.

The specializations of hadrosaurs have already been discussed in some detail. These semiaquatic dinosaurs were probably good swimmers, as shown by the laterally flattened tail, but the strong hind legs and pelvis indicate that they were capable of active terrestrial excursions as well. They probably relied primarily upon running to escape carnivores if they were attacked on land. They were contemporaries of the ceratopsians, but whereas ceratopsians were upland inhabitants that were barred from invading eastern North America by the shallow sea that cut across the middle of the continent (Figure 11–5), the more aquatic hadrosaurs waded or paddled their way to the east coast.

Ankylosaurs and stegosaurs were specialized for a more sedentary life. Body armor gave considerable protection from the attacks of carnivores, and in both groups the tail could be used as a defensive weapon. The mosaic armor of ankylosaurs very nearly approached that of turtles in its solidity, and the heavy body set on short legs must have been very difficult for a predator to overturn. Spines projected from the side of the body of some ankylosaurs, and the macelike tail of others would have been a formidable weapon. Probably an ankylosaur curled itself into a ball while resisting attack to protect its underside, exposing only the heavily armored dorsal surface to a predator. The armored cordylid lizards of Africa use that sort of defensive behavior very successfully, as do armored mammals like armadillos and pangolins. There is little in the structure of the skeleton of either ankylosaurs or stegosaurs to indicate that they led an active life, and no fossil trackways are known that imply a social structure like that of sauropods.

The ceratopsians were probably the most active of the ornithopod dinosaurs. They evolved from bipedal forms like the ornithopod *Psittacosaurus* but the ability to move bipedally probably did not persist past the earliest genera. With increasing body size a quadrupedal gait was adopted. *Protoceratops* was only 2 m long, but later forms approached 10 m in length and weights of 10,000 kg. One paleontologist has proposed that these enormous animals "had the ability to gallop at speeds up to 30 mph." If that is true, they would have been the Cretaceous equivalent of a runaway locomotive; however, other biologists doubt that they could have attained anything approaching those speeds. Much of the leg movement of galloping mammals comes from rotation of the scapula, and ceratopsian scapulas seem incapable of the rotation needed to gallop. An ambling gait with the massive body supported on relatively straight legs was probably characteristic of ceratopsians as it was of sauropods.

13.4.3 Ecology of Theropods

The carnivorous saurischian dinosaurs, like the herbivorous forms, present an impression of relatively small morphological diversity. The major evolutionary lineages have already been traced—the coelurosaurs that gave rise to the ostrichlike ornithomimids, the giant theropods, and the newly discovered deinonychosaurs that combined the small size of coelurosaurs with the predatory habits of the theropods. The ornithomimids parallel flightless birds so closely in morphology that it seems safe to assume that they lived essentially the same life in the Cretaceous that ostriches, rheas, and related forms live now. The appearance of increasingly cursorial forms that relied upon running to escape predators may indicate that habitats became increasingly open during the Mesozoic. The increase in body size seen in theropods almost certainly was a response to selective forces produced by the concomitant increase in body size of their herbivorous prey. It was advantageous for the herbivores to be large because it reduced the risk of predation by carnivores, and it was advantageous for carnivores to be large because it enabled them to prey on the large herbivores. This sort of selection can continue until the disadvantages of large body size outweigh the benefits to be gained from further increase, but there is no evidence that theropods had reached that point. The Cretaceous forms were larger than their Jurassic ancestors.

There was probably a trend to increasingly specialized methods of prey capture. This is suggested by the reduction of the size of the forelimbs in theropods that presumably indicates an increased reliance upon the teeth for both seizing and killing prey. The deinonychosaurids relied instead upon fleetness of foot to capture more active prey and possibly used the rigid tail to knock it to the ground before disemboweling it with the heavy claws on the hind feet. The theropods, too, may have used their heavy tails to try to overturn prey, especially when attacking armored forms like stegosaurs and ankylosaurs.

13.4.4 Other Terrestrial Reptiles of the Late Mesozoic

There is a tendency to look upon the Jurassic and Cretaceous as the Age of Dinosaurs and to forget that there was a very considerable nondinosaur terrestrial fauna. To a certain extent the dinosaurs do form a separate unit. The large theropods were the only animals capable of preying upon adults of the large herbivores. Nonetheless, there must have been significant interaction between dinosaurs and nondinosaurs. So far as we know, all dinosaurs reproduced by laying eggs, and dinosaur eggs were a food source that could be exploited by small predators. The Nile monitor lizard today is an important predator on eggs of crocodiles, and it is likely that Cretaceous monitor lizards had a similar taste for dinosaur eggs.

One of the distinctive features of dinosaurs is their large size. The smallest dinosaur skeletons known indicate a total length of about a meter, and these may represent juveniles. Even a meter is large in comparison to living reptiles; most lizards are smaller than 50 cm, and only crocodilians approach the bulk of even moderate-size dinosaurs. There were, of course, adaptive zones available for smaller vertebrates in the Jurassic and Cretaceous, and these were filled as they are now by squamate reptiles, turtles, amphibians, and mammals. Among these were the animals that might have stolen eggs from dinosaur nests and served as food for juvenile dinosaurs.

Although dinosaurs are so impressive and distinctive that there is a tendency to think of them as inhabiting a world of their own, they were in reality a part of an ecosystem that included some vertebrates that would not appear strange to us today. Living families of amphibians and reptiles are known from Mesozoic fossils, and Mesozoic mammals, although they are quite distinct from living forms, were superficially similar in appearance to small mammals like shrews, opossums, and rodents. The striking difference in the Mesozoic world was the absence of large mammals, and the occupation of that adaptive zone by large reptiles. Figure 13–21 shows the relative abundance of different groups of vertebrates at two Cretaceous fossil localities. The Lance locality appears to have been a wooded swampy habitat with large streams and some ponds. The Bug Creek Anthills locality was probably downstream from such a swamp, in the delta of a major waterway. In both localities dinosaurs are a minor component of the community in terms of species diversity, although one dinosaur is the equivalent of a great many smaller animals in terms of biomass.

13.5 WERE DINOSAURS ENDOTHERMS?

Because living reptiles are ectotherms and dinosaurs were unquestionably reptiles, it has been assumed that they too were ectotherms. A classic series of experiments carried out by E. H. Colbert, R. B. Cowles, and C. M. Bogert in the 1940s used alligators as models for dinosaurs. The experiments indicated that an ectothermal dinosaur could maintain a very stable body temperature by behavioral means because its large mass gave it great thermal inertia. Nonetheless, the suggestion that dinosaurs were endothermal is made periodically.

The most recent presentation of this idea by Robert Bakker draws on three main categories of information: (1) Similarities between dinosaurs and birds

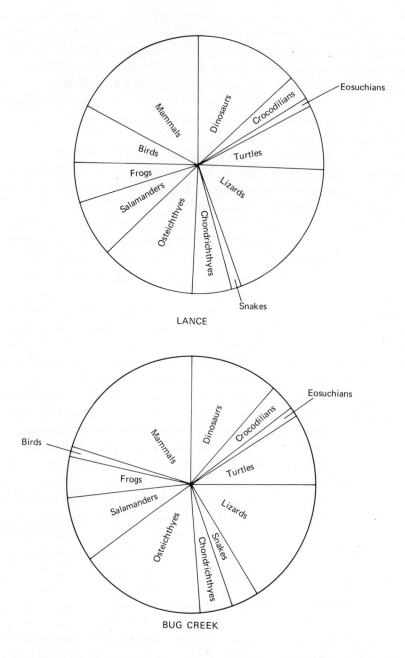

Figure 13–21. Relative abundance of genera at late Cretaceous fossil sites. Although dinosaurs were the most spectacular terrestrial vertebrates during the Mesozoic, they were surrounded by a large assortment of animals that were not very different from living species. Reptiles comprise more than 40 percent of the genera at both the Lance and Bug Creek localities, but only a quarter of the reptiles were dinosaurs. The remainder were snakes lizards, turtles, crocodilians, and one genus of eosuchian. Mammals were well represented at both sites. (Calculated from data in Estes and Berberian [8].)

and mammals in body form and bone structure; (2) the presence of dinosaur fossils in far northern localities, and (3) ratios of carnivorous to herbivorous dinosaurs in fossil deposits.

Some paleontologists appear to have been attracted by the popular interest in dinosaurs that these proposals have stimulated. In contrast, biologists familiar with the physiology and ecology of reptiles have generally rejected the idea that dinosaurs were endotherms. Point-by-point rebuttals by critics have disproved Bakker's main contentions. For example, Bakker placed great emphasis on a presumed similarity in the microscopic structure of the bones of dinosaurs and mammals. Both, he stated, have systems of Haversian canals in the bone. (In living vertebrates these canals are part of the system that transports calcium and phosphorus between the skeleton and other tissues.) Marianne Bouvier reviewed vertebrate bone structure and concluded that the similarities between dinosaur and mammalian bone cited by Bakker are misleading. Haversian bone systems are found in some modern reptiles and in other fossil reptiles besides dinosaurs, and Haversian bone is absent from some mammals and birds. This bone structure appears to be related to the mechanical strength of bone and to an animal's growth rate and body size, but not to endothermy or ectothermy. Likewise, the skeletal similarities of dinosaurs and large mammals cited by Bakker are more likely to reflect convergent morphological adaptations to similar mechanical stresses than shared endothermy. A fully erect posture, for example, is the only one a large animal can assume because bone has limited resistance to the bending stress that would be imposed by a sprawling posture. The vacuities in the vertebral columns of large sauropod dinosaurs, which Bakker compares to those of birds, appear to have evolved as structures that reduce weight with little loss of strength. The arched structure of dinosaur skulls, especially of the advanced theropods, has already been mentioned as another example of this sort of weight reduction.

Mesozoic climates were more equable than modern climates, and the continents have moved away from the equator since the Mesozoic. The northern deposits that yield fossils of allegedly endothermal dinosaurs also contain fossils of modern reptiles like crocodilians and turtles. We know that these modern reptiles are ectotherms and occur only in warm climates.

Ecological projections based on predator/prey ratios led Bakker to conclude that dinosaurs were endotherms because the ecological pyramid of numbers assumed the shape typical of mammal-dominated ecosystems rather than one that might be expected from a reptilian system. Projections of this sort are complicated by the vagaries of fossilization, and by the presence in the ecosystem of many species of vertebrates besides the dinosaurs Bakker considered. The number of fossils of a species in a deposit tells something about how likely that species was to die in a place where it would be fossilized but nothing about the relative abundance of different species in the area. For example, the scarcity of fossils of animals from highland habitats is one of the major reasons for apparent gaps in the fossil record. James Farlow concluded that uncertainties of this sort made it impossible to draw conclusions about predator/prey ratios in Mesozoic deposits. Richard Tracy applied Bakker's line of reasoning to data on garter snakes and their prey and found that it could lead to the conclusion that garter snakes are endotherms!

A variety of theropods and ornithopods had elaborate structures on their backs that appear to have been heat exchangers. The Cretaceous ornithopod *Ouranosaurus*, closely related to *Iguanodon*, had a dorsal "sail" like that of the pelycosaur *Dimetrodon* that was supported by elongate neural spines. A Cretaceous carnosaur, *Spinosaurus*, had a similar dorsal sail. In a *Spinosaurus* 11 m long the dorsal spines supporting the sail reached the height of 2 m. The dorsal plates of *Stegosaurus*, the familiar ornithopod from the Jurassic of North America, also may have been heat exchangers. Examination shows that they were extensively vascularized and could have carried a large flow of blood to be warmed or cooled according to the needs of the animal. *Kentrosaurus*, the African counterpart of *Stegosaurus*, had much smaller dorsal plates than *Stegosaurus* and the plates extended only from the neck to the middle of the back. Here the double row of plates was replaced by a double row of sharp spines that ran down the tail and appear to have had a primarily defensive function rather than a thermoregulatory one. It is frustrating to be unable to compare the thermoregulatory behaviors of the two kinds of stegosaurs in a controlled experiment.

The widespread occurrence of heat exchangers in dinosaurs widely separated phylogenetically and in space and time suggest that ectothermy was characteristic of both saurischians and ornithischians. The dorsal sails and plates of the dinosaurs, like the sails of the pelycosaurs, could have facilitated warming in the morning and late afternoon if the dinosaur oriented its body sidewards to the sun so that the sail intercepted a maximum amount of solar radiation. In midday the animal could face directly into the sun, exposing only the narrow dorsal edge of the sail to the sun's rays, and lose heat to the environment by radiation and convection.

The debate about dinosaur physiology that has centered on the question of endothermy vs. ectothermy provides imaginative examples of the use (and misuse) of paleontological and ecological data. Several recent papers on this subject are listed at the end of the chapter. References to earlier papers can be found in the literature citations of the papers listed.

To say that dinosaurs were not endotherms is not to say that they did not maintain stable body temperatures, however, because the two processes are not the same. Endothermy refers to the maintenance of stenothermy by high levels of metabolic heat production, but that is only one way to achieve a stable body temperature. Deep sea fish are ectothermal, for example, but, because the water they inhabit changes temperature less than 1°C in a year, their body temperature is very stable. Because of their size, dinosaurs achieved considerable stenothermy even in a variable environment. The large mass of the body could exchange energy with the environment only through the relatively small surface area, and the process was slow. A 1°C change in the body temperature of an adult sauropod weighing 50,000 kg represents the gain or loss of 40,000 kcal of energy. As a reptile's body size increases, its rate of heat production and of heat loss per unit mass of tissue both decrease. George Bartholomew and Vance Tucker measured these rates in monitor lizards of different sizes and found that the decline in the rate of heat loss was slightly greater than the decline in heat production. As a result of that relationship, an increasing amount of metabolic heat would be retained in the body as the mass of a reptile increased. Bartholomew and Tucker extrapolated that relationship and found that a mass of nearly 200 million kg (220,000 tons) would be needed to achieve a body temperature 7°C above ambient temperature. That mass is 4,000 times the mass of the largest dinosaurs. "Obviously," they concluded, "endothermy evolved some other way."

Another approach to the analysis of dinosaur body temperatures was used by James Spotila and his associates. They based computer similations of dinosaur body temperatures on extrapolations from alligators. Their simulations showed that a relatively small dinosaur, one with a body diameter of 1 m, would maintain a body temperature between 28.5 and 29.6°C in an environment with a daily fluctuation between 22 and 32°C. A larger dinosaur would show proportionally less variation in body temperature, and a smaller one more variation. The time constant of temperature change is a convenient way to express the lability of body temperature in a way that allows different-sized animals to be compared. The time constant is a mathematical term defined as the time required for an animal's body temperature to complete 63.7 percent of the change from its original equilibrium value to its new equilibrium when the external temperature changes. Time constants calculated for reptiles of various body sizes are shown in Table 13–2. They range from 12 min for a small lizard (2 cm body diameter) to more than 5 days for a large dinosaur (200 cm body diameter). The exact values, of course, depend on the assumptions one makes about the thermal properties of dinosaur body tissue and the external conditions, but the most extreme reasonable variation in those factors changes the time constants by only a factor of 2.

The computer simulations indicate that dinosaurs could have achieved many of the benefits of stenothermy without the energy cost of endothermy as long as the daily and seasonal temperature variation was within limits that resulted in tolerable body temperatures. In an equable climate, dinosaurs had the best of both worlds—the advantages of stenothermy without the cost of endothermy.

Table 13–2 The calculated relationship between temperature lability and body size in reptiles

Body diameter (cm)	2	10	20	100	200
Time constant* (hr)	0.22	2.04	5.30	49.0	127

*The time constant is the period required for the body temperature to complete 63.7 percent of the change from one equilibrium temperature to a new equilibrium temperature when the external temperature changes. (Based on Spotila et al. [13].)

13.6   THE END OF THE AGE OF REPTILES

After thriving and dominating the terrestrial habitat for 130 million years, dinosaurs and their contemporaries, the ichthyosaurs, plesiosaurs, and pterosaurs, disappeared abruptly near the end of the Cretaceous. The term "abruptly" must be understood in the context of geological time. Intervals of 100,000 years cannot be defined in late Cretaceous sediments, so a process that seems sudden from our perspective probably stretched over several hundred thousand or even millions of years. Nonetheless, the shift from a reptile-dominated ecosystem of the Mesozoic to the mammal-dominated system of the Cenozoic is striking.

Biologists have struggled for years to explain the rapid change in faunal composition that occurred at the end of the Mesozoic, and many hypotheses, although of doubtful biological value, are classic in their own right. The catastrophism school of thought has enjoyed some vogue with the view that the giant reptiles were wiped out by some geological or cosmic disaster. There are various candidates for the disaster. One suggestion is that the moon was scooped from the Pacific Ocean basin to the accompaniment of earthquakes, tidal waves, and volcanic eruptions. The explosion of a supernova, showering the earth with radiation, is another proposal that has been advanced to explain the extinction of dinosaurs. It would be a surprisingly selective sort of catastrophe, however, that would exterminate solely large reptiles, leaving birds, mammals, small reptiles, and amphibians unharmed and leaving no trace in the geologic record.

Proponents of the idea of an epidemic that wiped out dinosaurs have difficulty explaining how it could have affected animals as diverse as ichthyosaurs and pterosaurs while leaving other reptiles unharmed. A variation on that theme that has a certain charm is the idea that the spread of flowering plants in the Cretaceous reduced the abundance of conifers and cycads. These plants contain irritating chemicals. No longer ingesting these chemicals, the theory runs, the herbivorous dinosaurs died of constipation, and the carnivores that preyed on them starved when their food source became extinct. Among other shortcomings, that theory does not account for the disappearance of marine or flying reptiles. Racial senescence, a vague teleological concept that some biologists resort to in situations where reason fails, does not supply a satisfactory answer, especially considering the success of the amphibians and reptiles that did not become extinct. The idea that mammals exterminated the dinosaurs by eating their eggs does not account for marine or flying reptiles nor the forms that did survive, and there is no obvious reason why dinosaurs and mammals would coexist from the Triassic to the Cretaceous and then become incompatible.

Because of the worldwide scale of the late Cretaceous extinctions and the wide variety of animals affected, explanations based on broad-scale changes in climate have been proposed repeatedly. "Climatic change" is a vague term, and various authors have suggested that the climate became hotter (or colder) or wetter (or drier). Recently D. I. Axelrod and H. P. Bailey attempted to reconstruct the probable climate of the Mesozoic by analyzing the climates in which descendants of Cretaceous flora still occur. The primary characteristic of areas in which ancient groups of plants—cycads, tree ferns, ginkos, redwoods, araucarias, and others—are now most successful is equability. Temperatures change very little on a daily or annual basis in these areas, although the mean temperature is high in some areas and relatively low in others. Axelrod and Bailey concluded that the Mesozoic was characterized by extremely equable temperatures with very little variation from day to night, from summer to winter, and from north to south. Casper, Wyoming, is an important dinosaur fossil locality, and a reconstruction of its Mesozoic temperature regime indicates that the long-term temperature extremes probably lay between 11 and 34°C, with a mean of 22°C. Clearly the climate has changed since then: The present temperature extremes at Casper are −27 to +40°C, with a mean of 8°C.

Geologic evidence indicates that climates did become both cooler and more variable at the end of the Cretaceous. A general uplift of land surfaces drained the shallow inland seas that had made the continents' surfaces archipelagos during the Mesozoic. Greater elevation of the land, greater differences in elevation from place to place, and the absence of the inland seas that buffered temperature change would all increase the daily and seasonal variation in temperature. If large reptiles depended on body temperature stability derived primarily from their enormous body size, a small increase in temperature variation would be very deleterious to them. The time constant for temperature change in a dinosaur with a body diameter of 1 m is 2 days. That is long enough to withstand a moderate day-night temperature fluctuation, but not enough to stay warm during several months of cool weather in the winter. The appearance of significant seasonal temperature variation could create a situation in which a dinosaur could not function, and if the increased variation was accompanied by a general decline in the average temperature, the situation would be even worse.

How does this explanation handle the problems that have stymied other explanations—the diversity of animals affected and the survival of some? It does so promisingly, although there are many questions still left unanswered. Ichthyosaurs, mosasaurs, and plesiosaurs may have suffered primarily from the sudden disappearance of the shallow seas. Reduction of suitable habitat and increasingly variable temperatures combined with a decline of about 5°C in sea temperatures at the end of the Cretaceous appear to be sufficient explanation for the extinction of large marine reptiles.

The hypothesis of climatic change marked primarily by increased variability and only secondarily by a reduction of the annual mean temperature can also account for the survival of small reptiles. These forms have considerable ability to regulate their body temperatures (see section 14.2). They can bask in the sun to warm up and retreat to shade to avoid getting too hot. Both of these techniques are denied to a large reptile because of its size. Temperature change is slow in a large reptile, so even an entire day of basking produces relatively little change in the body temperature of a dinosaur. There are few places a large animal can take shelter when the sun gets too hot. A 1 m long lizard can find shade under a bush or in a burrow, but a 25 m long sauropod needs a small forest for shelter. The only large reptiles that survived the Mesozoic, crocodilians and some turtles, support the hypothesis because they occur only in the most equable existing climates. Crocodilians are primarily tropical, with a few sub-tropical forms, and are always associated with water, which buffers temperature change and provides a retreat from temperature extremes. Small turtles penetrate far north and south, but large forms occur only near the equator. Again, the presence of water as a temperature buffer is important. Sea turtles spend virtually their entire lives in water and the large terrestrial turtles occur on equatorial islands.

Although we can only speculate about the details and wonder why particular suborders and families persisted longer than others, the hypothesis of increasing climatic variation at the end of the Cretaceous provides a context in which physical and biological stresses can be interwoven to provide a plausible explanation of why certain types of reptiles became extinct whereas others did not. As will be pointed out in section 14.6, this hypothesis also presents a reasonable explanation for the sudden diversification of mammals following 130 million years of relative obscurity.

References

[1] Axelrod, D. I. and H. P. Bailey. 1968. Cretaceous dinosaur extinction. *Evolution* 22:595–611. Proposes worldwide climate change as mechanism of dinosaur extinction.

[2] Bakker, R. T. 1971. Dinosaur physiology and the origin of mammals. *Evolution* 27:170–174. One of the first "hot blooded dinosaur" papers. References to other papers by Bakker can be found in the bibliography of this paper and papers listed below by Bennett and Dalzell, Bouvier, Charig, Farlow, and Feduccia.

[3] Bennett, A. F. and B. Dalzell. 1973. Dinosaur physiology: a critique. *Evolution* 27:170–174. Calls attention to errors in Bakker's reconstruction of dinosaur physiology.

[4] Bouvier, M. 1977. Dinosaur haversian bone and endothermy. *Evolution* 31:449–450. Points out errors in Bakker's interpretation of the microstructure of dinosaur bone.

[5] Carroll, R. L. 1976. Eosuchians and the origin of archosaurs. In *Essays in Honour of Loris Shano Russell*, edited by C. S. Churcher. Royal Ontario Museum Misc. Publ.

[6] Charig, A. J. 1972. The evolution of the archosaur pelvis and hindlimb: an explanation in functional terms. In *Studies in Vertebrate Evolution*, edited by K. A. Joysey and T. S. Kemp. Winchester Press, New York.

[7] Charig, A. J. 1976. Dinosaur monophyly and a new class of vertebrates: a critical review. In *Morphology and Biology of Reptiles*, edited by A. d'A. Bellairs and C. B. Cox. Academic Press, London. Concludes that there is no basis for Bakker's grouping of the saurischian and ornithischian dinosaurs with birds as a class Dinosauria.

[8] Estes, R. and P. Berberian. 1970. Paleoecology of a late Cretaceous vertebrate community from Montana. *Brevoria* (Museum of Comparative Zoology, Cambridge, Mass.), Number 343. A report on an unusually complete analysis of a fossil locality.

[9] Farlow, J. O. 1976. A consideration of the trophic dynamics of a late cretaceous dinosaur community (Oldman Formation). *Ecology* 57:841–857. Discusses the diffi-

culties of attempting to infer information about biomass of species from the relative numbers of fossils in a locality. Concludes that the relative numbers of carnivorous and herbivorous dinosaurs suggest that dinosaurs had the low metabolic rate expected of ectotherms.

[10] Feduccia, A. 1973. Dinosaurs as reptiles. *Evolution* 27:166–169. Presents an alternative view to the "hot blooded dinosaur" theory.

[11] Hopson, J. A. 1975. The evolution of cranial display structures in hadrosaurian dinosaurs. *Paleobiology* 1:21–43. Uses analogies with living reptiles and mammals to infer social behavior of hadrosaurs.

[12] Kurten, B. 1967. *The Age of the Dinosaurs.* McGraw-Hill, New York.

[13] Spotila, J. R., P. W. Lommen, G. S. Bakken, and D. M. Gates. 1973. A mathematical model for body temperatures of large reptiles: implications for dinosaur ecology. *The American Naturalist* 107:391–404. A mathematical model of dinosaur body temperatures indicates that they would have achieved considerable stenothermy because of their large size.

[14] Stein, R. S. 1975. Dynamic analysis of *Pteranodon ignens*: a reptilian adaptation for flight. *Journal of Paleontology* 49:534–548. An aerodynamic analysis of an advanced pterodactyl, using models in a wind tunnel.

[15] Stokes, W. L. 1964. Fossilized stomach contents of a sauropod dinosaur. *Science* 143:576–577. One of the few bits of direct evidence about dinosaur food habits.

[16] Swinton, W. E. 1970. *The Dinosaurs.* John Wiley, New York. A detailed review of dinosaur phylogeny.

14

Modern Reptiles

Synopsis: Living reptiles include representatives of three subclasses, all of which can be traced from the Mesozoic. Turtles (Subclass Anapsida, Order Testudinata) have changed little since the Triassic. The shell that encloses a turtle's body is the distinctive feature of the order. Turtles are very specialized morphologically for life within a rigid shell and there is relatively little further morphological specialization within the group. Terrestrial tortoises with their high, domed shells and sturdy, elephantlike feet are at one extreme of turtle morphology and ecology. The opposite extreme is represented by the stream-lined sea turtles with their legs specialized as flippers.

Crocodilians (Subclass Archosauria, Order Crocodilia) are a second group of living reptiles that have changed little since the Mesozoic. Crocodilians are the largest living reptiles, and probably have the most complex social behavior in the Class Reptilia. Both males and females of some species assist hatchlings to get out of the eggs and to reach the water. Young American alligators remain with their mother for two years after hatching. Crocodilians are a morphologically conservative group. Differences in body size appear to be related to the size of the bodies of water a species inhabits. Differences in the shape of the snout are associated with specialized dietary habits.

More than 95 percent of the approximately 6000 species of living reptiles belong to the Order Squamata (Subclass Lepidosauria). Four living suborders of squamates are recognized and two of these, the snakes and lizards, account for 98 percent of the species in the order. Lizards (Suborder Lacertilia) are the most abundant and diverse squamates. Most lizard species are small (80 percent have adult body weights of 20 g or less) but there are a number of medium to large species. The largest species is the Komodo monitor lizard that reaches a weight of 75 kg and preys on deer and even water buffalo. Leglessness has evolved repeatedly among squamates, and snakes (Suborder Serpentes) are extremely specialized legless derivatives of the squamate stock. Snakes differ from legless lizards in a number of morphological features. The most important differences are the adaptations of the skulls of snakes that allow them to swallow prey with a diameter greater than the diameter of the snake's head. Unlike lizards, snakes are relatively large vertebrates; nearly 75 percent of snake species have body weights greater than 20 g.

Two small and relatively poorly known suborders of squamates complete the roster of living reptiles. The amphisbaenians (Suborder Amphisbaenia) are burrowing reptiles found in tropical regions. Morphological specializations for different methods of burrowing can be distinguished and the most specialized forms are found in the most compact soils. The tuatara (Suborder Sphenodontia) has only recently been added to the squamates; it was previously classified with the rhynchosaurs (Chapter 13). The tuatara is a morphologically primitive squamate that has changed little since the Mesozoic. Its distribution is limited to the New Zealand region.

Reptiles are ectotherms, and many species have an elaborate repertoire of behavioral and physiological thermoregulatory mechanisms that enable them to maintain their body temperatures within a range of a few degrees during periods of activity. The range of body temperatures maintained during activity (the Activity Temperature Range) is characteristic of a particular species and is part of its ecological niche. Physiological and biochemical adjustments to a species' Activity Temperature Range can be demonstrated experimentally.

As ectotherms, reptiles interact with their physical and biological surroundings differently from birds and mammals. Many features of reptilian behavior and physiology that have traditionally been regarded as primitive are more reasonably considered specializations associated with ectothermy. An important feature of ectothermy is the energetic efficiency it permits. Because food is not burned to produce heat in reptiles as it is in mammals and birds, reptiles can use a higher proportion of the energy they ingest for growth and reproduction. The efficiency of biomass production (growth of individuals and production of young) by reptiles is nearly 10 times as great as that of birds or mammals.

14.1   INTRODUCTION

Although the reptile fauna of the Cenozoic is far less extensive than that of the Mesozoic, living reptiles exhibit a tremendous ecological and morphological diversity. In size they range from geckos, which may be mature at body lengths of 2 cm, to crocodilians, which reach 7 m in length, and snakes 9 m long. While the Asian "flying lizards" of the genus *Draco* glide through the air by spreading their elongated ribs to form an airfoil, legless amphisbaenians burrow in the earth. Turtles plod along protected by their shells as they have since the Triassic, and the tutara is very like its Mesozoic relatives.

The greatest diversity of reptile life is found in the tropics. As one moves away from the equator diversity decreases, and large species in particular disappear. Deserts support extensive lizard faunas even in temperate regions. The ranges of some species of lizards and snakes extend northward to the edge of the Arctic Circle in Europe and others range south to the tip of South America.

Table 14-1 Partial Classification of living reptiles.

Subclass Anapsida
  Order Testudinata (turtles)
    Suborder Cryptodira
      Family Testudinidae (tortoises)
              Emydidae (land and pond turtles)
              Chelydridae (snapping turtles)
              Kinosternidae (mud and musk turtles)
              Trionychidae (soft-shelled turtles)
              Cheloniidae (sea turtles)
              Dermochelyidae (leatherback turtle)
    Suborder Pleurodira
      Family Chelidae (snake-necked turtles)
Subclass Archosauria (ruling reptiles)
  Order Crocodilia
      Family Alligatoridae (alligators and caimans)
              Crocodilidae (crocodiles)
              Gavialidae (gharials)
Subclass Lepidosauria (scaly reptiles)
  Order Squamata
    Suborder Sphenodontia (tuatara)
    Suborder Amphisbaenia (amphisbaenians)
    Suborder Lacertilia (lizards)
      Family Agamidae (Old-World lizards)
              Iguanidae (New World lizards)
              Chamaeleonidae (true chameleons)
              Varanidae (monitor lizards)
              Helodermatidae (Gila monster and Mexican beaded lizard)
              Teiidae (whiptail lizards, ameivas and tegus)
              Lacertidae (Old World and Asian lizards)
    Suborder Serpentes (snakes)
      Family Boidae (boas and pythons)
              Colubridae (non-venomous and rear-fanged snakes)
              Elapidae (cobras, mambas, and kraits)
              Hydrophiidae (sea snakes)
              Viperidae (true vipers and pit vipers)

14.1.1 Order Crocodilia (Crocodilians)

In many respects crocodilians are the living reptiles most like Mesozoic forms, and they have been used as models for dinosaurs in attempts to analyze the ecology of extinct reptiles. All crocodilians are specialized for an aquatic life (Figure 14-1). The nasal openings are at the tip of the snout, and a secondary palate carries the air passages posteriorly to the rear of the mouth. A flap of tissue arising from the base of the tongue can form a watertight seal between the mouth and throat. Thus, a crocodilian can breathe while only its nostrils are exposed without inhaling water.

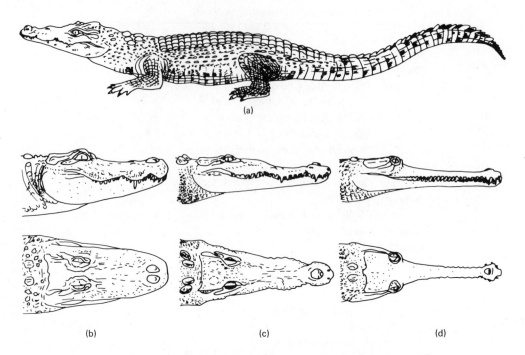

Figure 14-1 Modern crocodilians differ little in outward appearance from each other or from Mesozoic forms. The greatest interspecific variation is in head shape. Alligators and caimans are broad snouted forms with varied diets. Crocodiles include a range of snout widths. The widest crocodile snouts are almost as broad as those of most alligators and caimans, and these species of crocodiles also have varied diets that include turtles, fish, and terrestrial animals. Other crocodiles have very narrow snouts, and these species are primarily fish eaters. The species illustrated are near the middle of the range for crocodiles. The gharial is a specialized fish eater with a very narrow snout. (a, Cuban crocodile; b, Chinese alligator; c, American crocodile; d, gharial. (Modified from H. Wermuth and R. Mertens [1961] *Schildkröten, Krokodile, Brückenechsen.* Gustav Fischer Verlag, Jena.)

Systematists divide crocodilians into three families: The Alligatoridae includes the two species of living alligators and the caimans. With the exception of the Chinese alligator, the Alligatoridae is solely a New World group. The American alligator is found in the Gulf Coast states, and several species of caimans range from Mexico to South America and through the Caribbean. Alligators and caimans are fresh-water forms, whereas the Family Crocodilidae includes species like the salt-water crocodile that inhabits estuaries, mangrove swamps, and the lower regions of large rivers. This species occurs widely in the Indo-Pacific region and penetrates the Indo-Australian archipelago to northern Australia. In the New World, the American crocodile is quite at home in the sea and occurs in coastal regions from the southern tip of Florida through the Caribbean to northern South America.

The salt-water crocodile is probably the largest living species of crocodilian. Until recently, adults may have reached lengths of 7 m. Crocodilians grow slowly once they reach maturity, and it takes time to attain large size. In the face of intensive hunting in the last two centuries, few crocodilians now attain the sizes they are genetically capable of reaching. Not all crocodilians are giants; there are a number of small species that are specialized for life in small bodies of water. The dwarf caimans of South America and the dwarf crocodile of Africa live in swift-flowing streams. They have more extensive dermal ossifications than most crocodiles, and this body armor probably protects them from injury when they are swept against rocks by swift currents.

The third family of crocodilians, the Gavialidae, contains only a single species —the gharial, which once lived in large rivers from northern India to Burma. This species has the narrowest snout of any crocodilian; the mandibular symphysis (the fusion between the mandibles at the anterior end of the lower jaw) extends back to the level of the 23rd or 24th tooth in the lower jaw. A very narrow snout of this sort is a specialization for feeding on fish that are caught with a sudden sidewards jerk of the head. We have already called attention to the evolution of similar skull shapes in a variety of Mesozoic animals, including trematosaurs, phytosaurs, and the short-necked pleisiosaurs. Alligators and most caimans stand at the opposite end of the scale of head width from gharials. Most species are broad-snouted and quite generalized in their feeding habits, eating fish, turtles, and a variety of semiaquatic birds and mammals. Crocodiles bridge the gap between alligators and the gharial. The broadest-snouted crocodiles are very like alligators in morphology and feeding habits, whereas the false gharial approaches the head shape of the true gharial and feeds on fish, as do some caimans.

14.1.2 Order Testudinata (turtles)

Turtles found a successful approach to life in the Triassic and have scarcely changed since. The shell, which is the key to their success, has also limited the diversity of the group. For obvious reasons there are no gliding turtles and few are very arboreal. Shell morphology reflects the ecology of turtle species (Figure 14–2). The most terrestrial forms, the tortoises of the Family Testudinidae, have high domed shells and elephantlike feet. Smaller species of tortoises may show adaptations for burrowing. The gopher tortoises of North America are an example. The forelegs are flattened into scoops and the dome of the shell is reduced. The Bolson tortoise, recently discovered in northern Mexico, constructs burrows a meter or more deep and several meters long in the hard desert soil. These tortoises bask at the mouths of their burrows, and when a predator appears they throw themselves down the steep entrance tunnels of the burrows to escape just as an aquatic turtle dives off a log. The pancake tortoise of Africa is a radical departure from the usual tortoise morphology. The shell is flat and flexible because its ossification is much reduced. This turtle lives in rocky foothill regions and scrambles over the rocks with nearly as much agility as a lizard. When threatened by a predator, the pancake tortoise

Figure 14-2 Turtles come in a variety of shapes that reflect the demands of their habits and habitats. Tortoises are terrestrial animals; most have rigid domed shells (a). When a tortoise withdraws into its shell, the heavily scaled forelegs are drawn across the anterior opening and the soles of the feet protect the rear. The flat, flexible shell of the African pancake tortoise (b) allows it to wedge itself in cracks between rocks to escape predators. Box turtles (c) are terrestrial, and the anterior and posterior ends of the plastron can be raised to cover the openings when the turtle retreats into its shell. Aquatic turtles that swim actively through the water like sliders (d) are moderately streamlined and have large webbed feet. The aquatic softshell turtles (e) have achieved still greater agility. Turtles that crawl on the bottom of a body of water are less streamlined than those that swim. The mud turtles (f) can close their shells like box turtles. Snapping turtles and the alligator snapper (g) rely on large size (as much as 60 cm and nearly 100 kg for the alligator snapper) to protect them.

There are no truly terrestrial pleurodire turtles comparable to the cryptodire tortoises and box turtles. Some species of pleurodires like the African pond turtle (h) move overland from pond to pond. The African pond turtle, like the box and mud turtles, can raise the ends of the plastron to close off the openings

of its shell. There are a number of pluerodires like the snake-necked turtle (i) with extremely long necks. The common name for pleurodires, side-necked turtles, comes from the lateral S-shaped bend with which the neck is withdrawn into the shell. The matamata (j), a South American pleurodire, is the most bizarre fresh-water turtle. The flaps of tissue on the head and neck not only obscure the outline of the turtle but some of them are also sensitive to the minute water movements produced by nearby fish.

The most specialized aquatic turtles are cryptodires, the sea turtles. In true sea turtles like the loggerhead (k) the limbs have evolved into flippers. The leatherback turtle (l) is even more specialized for aquatic life—the shell has been reduced to a series of bony platelets embedded in the tough skin. (a–d, g, h, j, k modified from H. Wermuth and R. Mertens [1961] Schildkröten, *Krokodile, Brückenechsen*, Gustav Fischer Verlag, Jena; e, 1 modified from R. C. Stebbins [1954] *Amphibians and Reptiles of Western North America*. McGraw-Hill, New York; f, modified from A.C.L.G. Günther [1885–1902] *Biologia Centrali-Americana, Reptilia & Batrachia*, Taylor & Francis, London; i, from J. de C. Sowerby and E. Lear [1872] *Tortoises, Terrapins, and Turtles Drawn from Life*. Henry Sotheran, Joseph Baer & Co., London, reprinted in 1970 by the Society for the Study of Amphibians and Reptiles.)

crawls into a rock crevice and uses its legs to wedge itself in place. The flexible shell presses against the overhanging rock and creates so much friction that it is almost impossible to pull the tortoise out.

Less specialized terrestrial turtles have moderately domed **carapaces** (upper shells) like the box turtles of the Family Emydidae. This is only one of several kinds of turtles that have evolved flexible regions in the **plastron** (lower shell) that allow the front and rear lobes to be pulled upward to close off the openings of the shell. Aquatic turtles have low carapaces that offer little resistance to movement through water. The Family Emydidae contains a large number of pond turtles, including the painted turtles and the red-eared turtles usually seen in pet stores and anatomy and physiology laboratory courses.

The snapping turtles (Family Chelydridae) and the mud and musk turtles (Family Kinosternidae) prowl along the bottom of ponds and slow rivers and are not particularly streamlined. The mud turtle has a hinged plastron, but the musk and snapping turtles have very reduced plastrons. They rely upon strong jaws for protection. A reduction in the size of the plastron makes these species more agile than most turtles, and musk turtles may climb several feet into trees, probably to bask. If a turtle falls on your head while you are canoeing, it is probably a musk turtle.

The soft-shelled turtles (Family Trionychidae) are specialized for fast swimming. The ossification of the shell is greatly reduced, lightening the animal, and the feet are large with extensive webbing. Soft-shelled turtles lie in ambush for prey partly buried in debris on the bottom of the pond. Their long necks allow them to reach considerable distances to seize the invertebrates and small fish on which they feed.

The two suborders of living turtles can be traced through fossils to the Mesozoic. The **cryptodires** (*crypto* = hidden, *dire* = neck) retract the head into the shell by bending the neck in a vertical S-shape. The **pleurodires** (*pleuro* = side) retract the head by bending the neck horizontally. All of the turtles discussed so far have been cryptodires, and these are the dominant suborder of chelonians. Cryptodires are the only turtles found in most of the northern hemisphere, and there are aquatic and terrestrial cryptodires in South America and terrestrial ones in Africa. Only Australia has no cryptodires. Pleurodires, which are found only in the southern hemisphere, are very much less diverse than cryptodires. There are no terrestrial pleurodire turtles even in Australia, which has no land turtles at all.

The snake-necked pleurodire turtles (Family Chelidae) are found in South America, Australia, and New Guinea. As their name implies, they have long, slender necks. In some species the length of the neck is considerably greater than that of the vertebral column. These forms feed on fish which they catch with a sudden forward stroke of the head. Other snake-necked turtles have much shorter necks. Some of these feed on molluscs and have enlarged palatal surfaces used to crush shells. The same specialization for feeding on molluscs is seen in certain cryptodire turtles.

The most specialized feeding method among turtles is found in a pleurodire, the matamata of South America. Large matamatas reach shell lengths of 40 cm. They are bizarre-looking animals. The shell and head are broad and flattened, and numerous flaps of skin project from the sides of the head and the broad neck. To these are added trailing bits of sloughed skin and adhering algae. The effect is exceedingly cryptic. It is hard to recognize a matamata as a turtle even when you see one in clear water sitting on the slate bottom of an aquarium; against the mud and debris of a muddy river bottom they are practically invisible. In addition to obscuring the shape of the turtle, the flaps of skin on the head are sensitive to minute vibrations in water caused by the passage of a fish. When it senses the presence of prey, the matamata abruptly opens its mouth and expands its throat. Water rushes in carrying the prey with it, and the matamata closes its mouth, expels the water, and swallows the prey. Alone among turtles, the matamata has lost the horny beak that other turtles use for seizing prey or biting off pieces of plants.

The marine turtles are cryptodires. The Families Cheloniidae and Dermochelyidae show more extensive specialization for aquatic life than any fresh-water turtle. All have the forelimbs modified as wings with which they "fly" through the water. The largest marine turtle, the leatherback, reaches shell lengths of more than 2 m and weights exceeding 500 kg. This is a pelagic turtle that ranges far from land, feeding on seaweed, invertebrates, and possibly small fish. The dermal ossification has been reduced to bony platelets embedded in the skin. The largest of the sea turtles of the Family Cheloniidae is the loggerhead, which once reached weights exceeding 400 kg. Sea turtles, like crocodilians, have suffered so much predation by humans that now they seldom, if ever, grow as large as they did three centuries ago.

14.1.3 Order Squamata (The Scaly Reptiles)

The Order Squamata (Subclass Lepidosauria) includes four suborders—the lizards (Lacertilia), snakes (Serpentes), amphisbaenians (Amphisbaenia), and the tuatara (Sphenodontia). There are about 6000 living species of squamates and about half of them are lizards. Snakes make up the vast majority of the remaining species. There are fewer than 200 species of amphisbaenians and the sphenodontians are represented by a single species, the tuatara.

Suborder Sphenodontia (The Tuatara)

Tuataras are found only on a few small islands off New Zealand. It is likely that until recently they inhabited the larger islands as well, but the advent of man and his associates—cats, dogs, rats, sheep, and goats—exterminated them there. Access to the islands where tuataras occur is controlled by the government of New Zealand (as well as being physically difficult because the islands rise as sheer cliffs from the sea) and the chances for the tuatara's survival in its last stronghold appear reasonably bright at present.

Tuataras reach a maximum length of about 60 cm. They are nocturnal, and in the cool, foggy nights that characterize their habitat they cannot raise their body temperatures by basking in the sun as most reptiles do. Body temperatures near 11°C have been reported for active tuataras, and these are extraordinarily low for reptiles (see section 14.2). During the day tuataras bask in the sun and raise their body temperatures. Tuataras feed largely on invertebrates with an occasional gecko or baby seabird for variety. The pace of their lives is slow, even by reptilian standards. One specimen lived for 50 years in captivity, and the maximum life span may be even greater. The eggs are laid in spring and require more than a year to develop.

Suborder Amphisbaenia

The amphisbaenians are a small group of extremely fossorial reptiles (Figure 14–3). Herpetologists set them apart from snakes and lizards as a separate suborder of reptiles because of the structure of their skulls and particularly because they possess a single tooth in the midline of the upper jaw—a feature unique to this group among vertebrates. The median tooth is part of a specialized dental battery that makes amphisbaenians formidable predators capable of subduing a wide variety of invertebrates and small vertebrates. The upper tooth fits into the space between two teeth in the lower jaw, and the combination forms a set of nippers capable of excising a piece of tissue from a flat surface too large for the mouth to engulf as a whole.

The burrowing habits of amphisbaenians make them difficult to study, and even the taxonomy of the order is poorly understood. There are four families of amphisbaenians, and three major functional categories can be recognized in the largest family. Some of the species have blunt heads; the rest have either vertically or horizontally wedged snouts. In many places representatives of the three types occur sympatrically and share the subsoil habitat. The unspecialized blunt-headed forms live near the surface where the soil is relatively easy to tunnel through, and the specialized forms live in deeper, more compact soil. The geographic range of the unspecialized forms is greater than that of the specialized ones, and in areas in which only a single species of amphisbaenian occurs it is a blunt-headed species.

The relationship between unspecialized and specialized burrowers is puzzling—one would expect that the specialized forms would replace the unspecialized ones, but this has not happened. The explanation appears to lie in the conflicting selective forces on the snout. On one hand it is important to have a snout that will burrow through soil, but on the other hand it is also important to have a mouth capable of tackling a wide variety of prey. The specializations of the snout that make it an effective structure for burrowing appear to reduce its effectiveness for feeding. The range of prey sizes that the specialized forms can handle is probably less than that available to the unspecialized forms. Thus, in loose soil where it is easy to burrow, the blunt-headed forms have an advantage

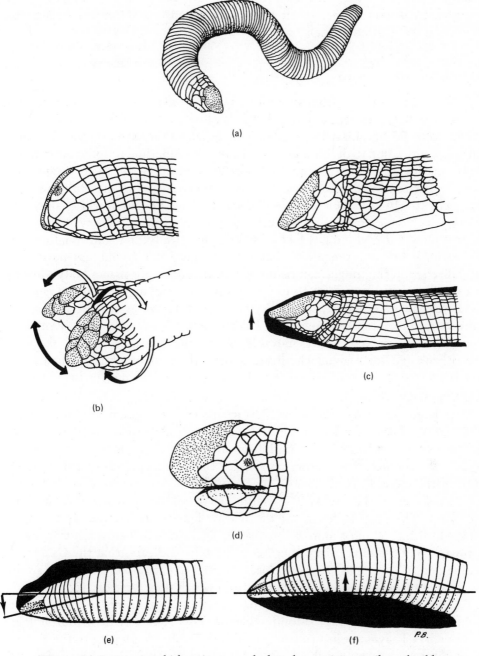

Figure 14-3 (a) Amphisbaenians are legless burrowing reptiles. (b) blunt-snouted forms burrow by ramming their heads into the soil to compact it. Sometimes an oscillatory rotation of the head with its heavily keratinized scales (stippled) is used to shove material from the face of the tunnel. (c) Shovel-snouted forms ram the end of the tunnel, then lift the head to compact soil into the roof. (d) Wedge-snouted forms (seen here in lateral view) ram the snout into the soil, then use the snout (e) or the side of the neck (f) to compress the material into the walls of the tunnel. (a, d original; b, c, e, f modified from Gans [3].)

because they can eat a wide variety of prey. Only in soil too compact for a blunt-headed form to penetrate do the unspecialized forms find the balance of selective forces shifted in their favor.

Suborder Lacertilia (Lizards)

Lizards range in size from diminutive geckos only a couple of centimeters long to the Komodo monitor lizard that is close to 3 m long at maturity and weighs some 75 kg. Lizards are adaptable animals that have occupied habitats ranging from swamp to desert and even above the timberline in some mountains. Many species are arboreal, and the most specialized of these are frequently laterally flattened and often have peculiar projections from the skull and back that help to obscure their outline (Figure 14-4). The Old World chameleons (Chamaeleonidae) are the most specialized arboreal lizards. Their **zygodactylous** (*zygo* = joined, *dactyl* = digit) feet grasp branches firmly, and additional security is provided by the prehensile tail. The tongue and hyoid apparatus are specialized, and the tongue can be projected forward more than a body's length to capture insects that adhere to its sticky tip. This feeding mechanism requires good eyesight and especially the ability to gauge distances accurately so that the correct trajectory can be employed. The chameleon's eyes are elevated in small cones and are independently movable. When the lizard is at rest, the eyes swivel back and forth giving the lizard a view of its surroundings. When an insect is spotted, both eyes fix on it and a cautious stalk brings the lizard within shooting range.

Most large lizards are herbivores. The iguana (Family Iguanidae) are arboreal inhabitants of the tropics of Central and South America. Large terrestrial iguanas occur on islands in the Caribbean and the Galapagos Islands, probably because the absence of predators has allowed them to spend a large part of their time on the ground. A number of species of lizards live on beaches, but very few actually enter the water. The marine iguana of the Galapagos Islands is an exception. The feeding habits of the marine iguana are unique. It feeds on seaweed, diving 10 m or more to browse on algae growing below the tide mark. Comparison of the molecular structures of proteins indicates that the Galapagos marine and land iguanas are more closely related to each other than they are to the iguanas of mainland South America from which they were derived.

An exception to the rule of herbivorousness in large lizards is found in the monitor lizards (Varanidae). Monitor lizards show relatively little morphological diversity. They range in size from small forms only 60 cm long to the Komodo monitor, which may reach a length of 3 m and weigh 75 kg. Some elongate species are specialized for arboreal life, and a few species are quite aquatic and have laterally flattened tails. Varanids are active predators that feed on a variety of vertebrate and invertebrate animals including birds and mammals. Few lizards are capable of capturing and subduing birds and mammals, but varanids have morphological and physiological adaptations that make them effective predators. The Komodo monitor lizard is capable of killing adult water buffalo, but its normal prey is deer and feral goats. Its hunting method are very similar to those employed by mammalian carnivores. In the early morning a Komodo monitor waits in ambush beside the trails deer use to move from the hilltops, where they browse at night, to the valley where they rest during the day. If no deer pass the lizard's ambush, it moves into the valleys, systematically stalking the places that deer are likely to be found. This purposeful hunting behavior is in strong contrast to the opportunistic seizure of prey that characterizes the behavior of most reptiles.

Leglessness has evolved repeatedly among lizards, and every continent has one or more families with legless, or nearly legless, species. Leglessness in lizards is usually associated with life in dense grass or shrubbery in which a slim, elongate body can maneuver more easily than a short one with functional legs. Some legless lizards crawl into small openings among rocks and under logs, and a few are fossorial (*fossor* = a digger).

Suborder Ophidia

Functionally as well as phylogenetically, snakes are extremely specialized legless lizards. They appear to have reached this specialization from a fossorial stage. Differences in the eyes of lizards and snakes provide evidence of that transition. In lizards the eye is focused by distorting the lens, thus changing its radius of curvature; in snakes the lens is moved in relation to the retina to bring objects into focus. The morphology of the retina of snakes differs from that of lizards. There is no fovea centralis, the retinal cells lack colored oil droplets, and there is a unique ophidian double cone. These differences in the eyes are interpreted as evidence that snake ancestors passed through a stage in which they were so specialized for burrowing that the eyes had nearly been lost. Among the most specialized living burrowing snakes and lizards, the eyes are very reduced and probably capable of little more than distinguishing between light and darkness. When snakes reentered an above-ground niche, the eye reevolved but the structural details of the original lizard eye were not exactly reduplicated.

Snakes are more specialized for legless life than are legless lizards. Snakes have 160 to 400 vertebrae. The movements of adjacent vertebrae in snakes are limited by an additional pair of articulations, the zygosphenes, that have evolved on the ventral surfaces of the centra. Legless lizards do not have zygosphenes. Instead, the rigidity required to transmit forces from one part of the body to another is provided primarily by heavy dermal scales, frequently including dermal ossifications. This armor plating makes legless lizards far more rigid than snakes. Both legless lizards and snakes crawl readily with lateral undulations in grass clumps or brush (see Figure 14–5). Legless lizards lack the suppleness that allows snakes to conform to small irregularities in the ground, and consequently they cannot crawl as readily on smooth surfaces as snakes.

Snakes range in size from diminutive burrowing species that feed on termites and grow to only 20 cm to the large constrictors that approach 10 m. The

Figure 14-4 Lizards have evolved different body forms in different habitats, and parallel and convergent evolution are widespread. Some of the most conspicuous lizards belong to the New World Family Iguanidae and the Old World Family Agamidae. Swifts (a) are small, generalized iguanids, whereas the very similar japalures (b) are among the swifts' agamid counterparts. Both families include a number of specialized species. The spiny-tailed iguanas (c) and the mastigures (agamids, d) are both herbivores that live in arid and semiarid regions. (The posterior portion of the iguana's tail has been broken off and regenerated.) The horned lizards (e) and the Australian agamid moloch (f) are flat, spiny lizards that feed primarily on ants.

Geckos are small to medium-sized lizards often found in or near human habitations (g). Because of their close association with humans, many species have been accidently transported to widely distant parts of the world where they have established new populations. Chameleons (h) are highly specialized for arboreal life. The feet are modified to grasp twigs and most forms have a prehensile tail.

(e)   (f)

(g)   (h)

(i)

(j)

Legless lizards have evolved in several families (i). Unlike snakes, in which the tail is rarely as much as a quarter of the total length, legless lizards may be two thirds tail. The monitor lizards (j) are voracious predators on small vertebrates. Except for a modest amount of lateral flattening of the tail in aquatic forms, they show little morphological diversity. (Sources of illustrations: a, e modified from R. C. Stebbins [1954] *Amphibians and Reptiles of Western North America*, McGraw-Hill, New York; b, modified from various sources; c modified from A.C.L.G. Günther [1885–1902] *Biologia Centrali-Americana, Reptilia & Batrachia*. Taylor & Francis, London; d, f, g, i original; h modified from H. Gadow [1909] *Amphibia and Reptiles*. Macmillan & Co., London; j, from a photograph in Schmidt and Inger [8].)

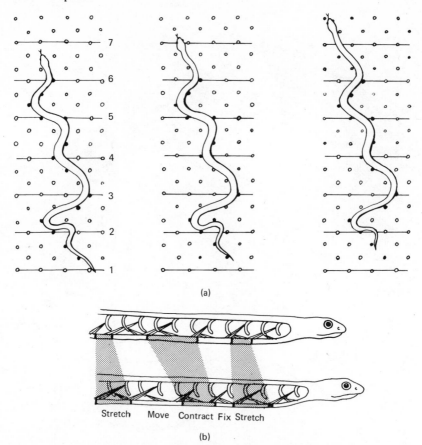

(a)

Stretch   Move   Contract Fix Stretch

(b)

Figure 14-5 Snakes commonly employ four types of locomotion.

(a) In curvilinear or serpentine locomotion the body is thrown into a series of curves. The curves may be irregular, as is shown in the illustrations of a snake crawling across a board dotted with fixed pegs. Each curve presses backward; the pegs against which the snake is exerting force are shown in solid color. The lines numbered 1 to 7 are at 3-in. intervals, and the snake's position at successive 1-sec intervals is shown. Serpentine locomotion is used by legless lizards, but they are less efficient than snakes on smooth surfaces.

(b) Rectilinear locomotion is used primarily by heavy-bodied snakes. Alternate sections of the ventral surface are lifted clear of the ground and pulled forward by muscles that originate on the ribs and insert on the ventral scales. The intervening sections of the body rest on the ground and support the snake's weight. Waves of contraction pass from anterior to posterior, and the snake moves in a straight line. It is a slow means of progression, but effective even when there are no surface irregularities strong enough to resist the sideward force exerted by serpentine locomotion. Because the snake moves slowly and in a straight line, it is inconspicuous, and rectilinear locomotion is used by some snakes when stalking prey.

(c) Concertina locomotion is used in narrow places such as rodent burrows that do not provide space for the broad curves of serpentine locomotion. A snake anchors the posterior part of its body by pressing several loops against the walls and extends the front part of its body. When the snake is fully extended, it forms new loops anteriorly and anchors itself with these while it draws the rear end of the body forward.

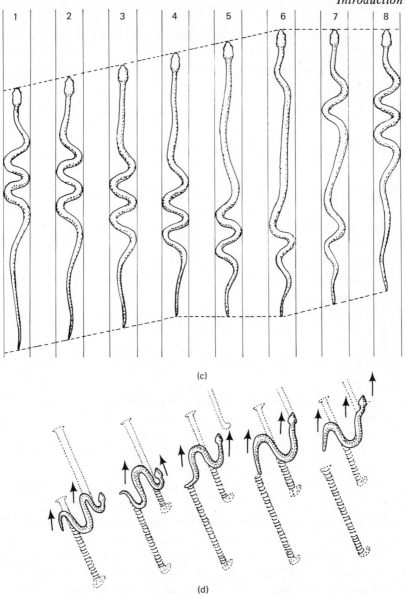

(c)

(d)

(d) Sidewinding is used primarily by snakes that live in deserts where wind-blown sand provides a substrate that slips sidewards during serpentine locomotion. A sidewinding snake raises its body in loops, resting its weight on two or three points, which are the only parts of the body in contact with the ground. The loops are swung forward through the air and placed on the ground, the points of contact moving smoothly along the body. Force is exerted downward; the lateral component is so small that the snake does not slip sidewards. The imprint of the ventral scales can be seen in the tracks. Because the snake's body is extended nearly at a right angle to its direction of travel, sidewinding is an effective means of locomotion only for small snakes that live in barren habitats. (a, drawn from a photograph in J. Gray [1946] *Journal of Experimental Biology* 23:101–120; b, modified from H. W. Lissman [1950] *Journal of Experimental Biology* 26:368–379; c, modified from C. Gans [3]; d, modified from W. Mosauer [1932] *Science* 76:583–585.)

anaconda of South America is considered the largest living species of snake, and the reticulated python of southeast Asia is nearly as large. Not all boas and pythons are large—a number of secretive fossorial and semifossorial species are considerably less than 1 m long as adults.

14.1.4 Evolutionary Origin and Specializations of Snakes

Snake skeletons are delicate structures that do not fossilize readily. In most cases we have only vertebrae and as a result there is little information to be gained from the fossil record about the origin of snakes. The earliest fossils known are from Cretaceous deposits and seem to be related to boas. Colubrid snakes, the family that includes about two thirds of the living species, are first known from the late Oligocene, and venomous snakes (elapids and viperids) appeared during the Miocene.

Snake body form is highly specialized, and relatively little additional morphological specialization is associated with even very divergent ecological niches (Figure 14-6). If the proportions of a moderately active terrestrial snake such as a garter snake (*Thamnophis sirtalis*) or ringelnatter (*Natrix natrix*) are used as a basis for comparison, one can say that the very active terrestrial snakes such as whip snakes (*Masticophis*) or sand snakes (*Psammophis*) are more elongate and have large eyes indicating the importance of vision in their hunting. Burrowing snakes are short and stout-bodied with wedge-shaped heads, small eyes, and short tails. The head shape assists in penetrating soil, and a short body and tail create less friction in a burrow than would the same mass in an elongate body. Arboreal snakes are extremely elongate. Their length distributes their weight and allows them to crawl over even small twigs without breaking them. Most aquatic snakes show no pronounced morphological changes associated with their aquatic habits, but there are a few specialized types.

The elephant trunk snakes of the Indo-Australian region are one of the most specialized aquatic forms. In these snakes the large ventral scales (**gastrosteges**) have been lost and the entire body is covered by granular conical scales. When an elephant trunk snake is at rest, the skin appears too large for the body and lies in loose folds. When the snake swims, the body is laterally compressed forming an efficient swimming structure. The sea snakes (Family Hydrophiidae) are a group related to the cobras and kraits (Family Elapidae). Sea snakes are characterized by extreme morphological adaptation for aquatic life: The tail is laterally flattened into an oar, the gastrosteges are reduced or absent, and the nostrils are located dorsally on the snout and have valves to exclude water. The lung extends back to the cloaca and apparently has a hydrostatic role as well as a respiratory function. Oxygen uptake through the skin during diving has been demonstrated in both sea snakes and elephant trunk snakes. Some sea snakes return to land to lay eggs, but others are so specialized for marine life that they are helpless on land and are viviparous.

Origin of Snakes from Lizards

There are a number of habitats in which legs are not particularly useful to a small predatory reptile. Dense vegetation entangles the legs as an animal tries to draw them forward, and in small openings, such as cracks in rocks, there is no space to use the legs. Probably the initial radiation of snakes took advantage of just these sorts of microhabitats. Modern lizards show an array of elongate forms with reduced legs, and Mesozoic lizards were probably at least equally diverse. Before an animal can dispense with its legs, an alternative mode of locomotion is necessary. Many elongate lizards fold their legs against the body and progress by lateral undulations when they are in a hurry, and the lizards that gave rise to snakes probably conformed to that pattern.

For a number of reasons, herpetologists suspect that snakes have their closest affinities to the platynotan lizards. This superfamily includes the monitor lizards, the Gila monster, and a Bornean lizard known as the earless monitor (*Lanthanotus*). It is *Lanthanotus* that gives us the best idea of what a mid-Mesozoic snake ancestor may have looked like. The earless monitor is a small (40 cm), elongate, secretive semiaquatic forest dweller that folds its legs against its body and uses lateral undulations when it moves rapidly, although the legs are well developed. Among the anatomical features it shares with snakes are the loss of the upper temporal arch, the absence of a pineal foramen, the incipient enclosure of the brain by downward projections of the frontal bones, and a moveable joint in the lower jaw between the dentary and the posterior bones of the jaw. *Lanthanotus* itself is, of course, too late in time to be a snake ancestor, but it provides an idea of the sort of animal that might have given rise to snakes in the Mesozoic.

The specializations of snakes compared to legless lizards reflect two strong selective pressures—locomotion and predation. Elongation of the body is characteristic of snakes. Legless lizards retain regional differentiation in the vertebral column corresponding closely to that seen in lizards with legs, but snakes have lost that differentiation and developed new sorts of vertebral differentiation associated with serpentine life. The reduction of the body's diameter associated with elongation has been accompanied by some rearrangement of the internal anatomy of snakes. The left lung is reduced or entirely absent, the gall bladder is posterior to the liver, the right kidney is anterior to the left, and the gonads may show similar displacement.

Legless reptiles face problems in capturing and swallowing prey. The primary difficulty is not the loss of limbs, because few reptiles use the legs to seize or manipulate food. The difficulty stems from the elongation that is such a widespread characteristic of legless reptiles. As the body lengthens, the mass is redistributed into a tube with a smaller diameter. As the mouth gets smaller, the maximum size of the prey that can be eaten also decreases, and a legless reptile is faced with the difficulty of feeding a large body through a small mouth. Legless lizards and amphisbaenians bite pieces out of prey, whereas snakes have evolved morphological adaptations that permit them to engulf prey

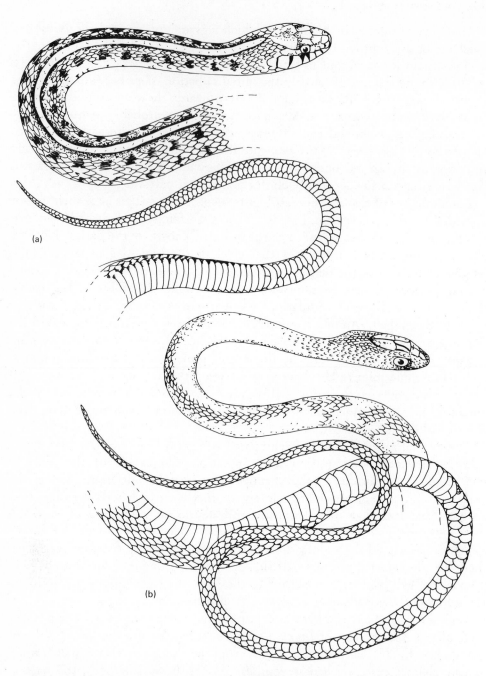

(a)

(b)

Figure 14-6 The body form of even a generalized snake like the garter snake (a) is so specialized that little further morphological specialization is associated with different ways of life. Active, visually oriented terrestrial snakes like whipsnakes (b) are longer and slimmer than generalized snakes and have larger eyes. (c) Arboreal snakes are longer still. (d) Burrowing snakes, at the opposite extreme, are short and smooth-scaled with blunt heads, reduced eyes, and very short tails. (e) Vipers, especially African forms like the Gaboon viper, are heavy-bodied with

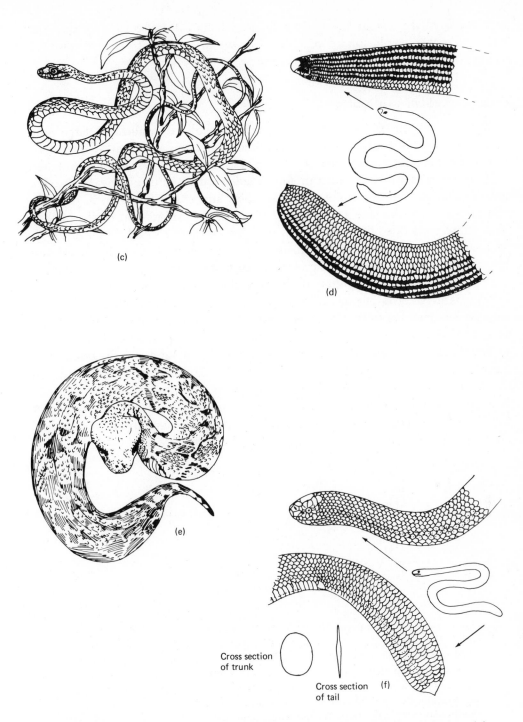

(c)

(d)

(e)

Cross section
of trunk

Cross section
of tail

(f)

broad heads. (f) Sea snakes have short, laterally compressed tails. (a, b, d, and f are from *Iconographie Générale des Ophidiens* by Georges Jan and Ferdinand Sordelli, a classic herpetological work published in Milan between 1860 and 1881; c, from H. Gadow [1909] *Amphibia and Reptiles*. Macmillan & Co., Ltd. London; g, from various sources.)

considerably larger than the body diameter. This difference is probably the key to the great evolutionary success of snakes in contrast to the very limited success of legless lizards and amphisbaenians.

Feeding Specializations of Snakes

The first step in the evolution of a skull capable of swallowing large prey may have been liberation of the mandibular tips from a firm attachment. In lizards the mandibles are joined in a symphysis, but in snakes they are attached only by muscles and skin and can spread sideward and move forward or back independently. Loosely connected mandibles and extensible skin in the chin and throat allow the jaws to be spread to allow the widest part of the prey to pass ventral to the articulation of the jaw with the skull (Figure 14-7). This change means that the interquadrate distance is no longer the sole determinant of the maximum size of prey that can be swallowed.

In addition to the flexibility of the jaws that results from the absence of a bony symphysis, the entire snake skull is much more flexible than a lizard's (Figure 14-8 and 14-9). Carl Gans has proposed that three changes could account for the difference: (1) A general loosening of articulation between several bones including the premaxillary, prefrontal, and palatal bones. (2) Simplification of the dorsal attachment between the maxillae and palatines and the nasal capsule to permit a lateral rotation of the tooth-bearing bones. (3) Increased motility of the basipterygoid articulations permitting the pterygoids to follow the laterorostral movements of the quadrate-mandibular articulation.

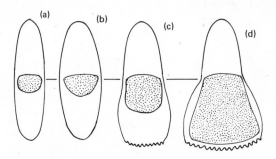

Figure 14-7 Evolution of the ophidian mechanism of prey ingestion. The mouth is shown diagramatically in a 180° gape; horizontal lines indicate the articulation of upper and lower jaws. The shaded area shows the cross section of the largest object that can be swallowed. (a) Symphysis between the mandibles firmly fixed and no cranial kinesis—for example, a tuatara. (b) Mandibular symphysis fixed, but limited spread at the quadrates as in most lizards. (c) Loss of mandibular symphysis frees the tips of the quadrates as in most snakes. (d) Very elastic tissues allow the tips of the mandibles to spread farther than the quadrates. This mechanism allows the greatest diameter of the prey to pass beneath the quadrates, so the distance between them no longer sets the maximum diameter of prey that can be swallowed. (Modified from C. Gans [1961] *American Zoologist* 1:217–227.)

The postorbital-squamosal arch may already have been lost in snake precursors, as it has been in *Lanthanotus*. That change frees the quadrate to move on its articulation with the supratemporal. Elongation of the supratemporal and simplification of its articulation with the skull to allow some movement in effect adds another joint to the lower jaw.

The snake skull contains eight links with joints between them that permit rotation. This number of links gives a staggering degree of complexity to the movements of the ophidian skull, and to make things more complicated the linkage is paired; each side of the head acts independently. Furthermore, the pterygoquadrate ligament and quadratosupratemporal ties are flexible. When they are under tension they function as rigid members, but when they are relaxed they permit sidewards movement as well as rotation. All of this results in a considerable degree of three-dimensional movement in a snake's skull (Figure 14-9). When the mouth opens, the pterygoids are pulled forward and upward. This movement is transmitted through the ectopterygoids to the maxillae. The prefrontal and premaxilla are bent upward, and the anterior teeth point outward, When the mouth closes these movements are reversed. The effect is to embed the teeth in the prey and draw it into the mouth. When the mouth is opened the teeth are disengaged and move forward to take a new grip. In the lower jaw a similar outward rotation can occur, and it is enhanced by the flexibility of the jaw at the intramandibular joint.

Swallowing movements take place slowly enough to be observed easily (Figure 14-10). A snake usually swallows prey headfirst, probably because that approach presses the limbs against the body out of the snake's way. Small prey may be swallowed tailfirst or even sideward. The teeth of one side of the jaw are anchored in the prey and the opposite jaw is advanced, partly by the movements described previously that move the tooth-bearing bones forward and partly by pivoting the head. The teeth that were engaged in the prey are lifted from it by protraction of the palatomaxillary arch, and the head is pivoted in the opposite direction to advance the teeth over the prey. The mandibles are protracted independently or together and grip the prey ventrally. Once it has reached the esophagus, the prey is forced toward the stomach by contraction of the body musculature. Frequently the neck is bent into a sharp loop just anterior to the prey and the loop travels posteriorly, forcing the prey into the stomach.

In a sense the two major selective forces operating on snakes work at cross purposes. On one hand there are advantages to repackaging the body's mass into a long, slim shape capable of crawling into small holes in pursuit of prey. On the other hand the skull is specialized for swallowing large prey. It is possible for a snake to swallow prey too large for it; the lump the prey makes in the body interferes with locomotion and may obstruct circulation and inhibit the peristaltic movements of the gut that are necessary to pass the prey through the digestive system. Snakes sometimes regurgitate large meals, but there are cases in which snakes even in the wild state have died after swallowing very large prey. Apparently a snake needs a certain amount of discretion; its jaws are capable of engulfing prey that will make a bulge that may prove lethal.

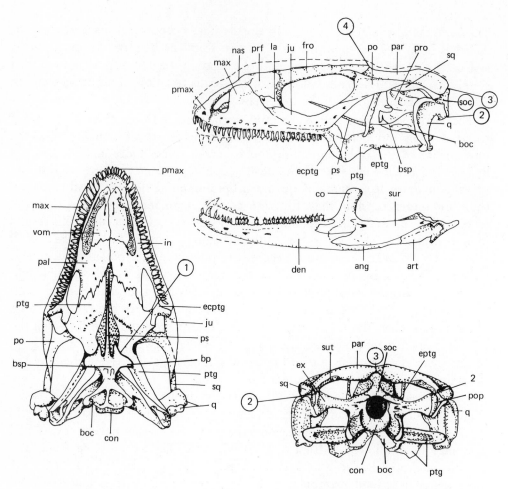

Figure 14-8 Kinesis in a lizard skull. Considerable movement of different parts of the skull is achieved in lizards such as the spiny iguana by four flexible joints. (1) The pterygoids slide forward on the basipterygoid pushing the snout upward as the mouth is opened (dashed line). Their movement is facilitated by reciprocal movement in other parts of the skull. (2) Between the paraoccipital process and the parietal, supratemporal and quadrate. (3) Between the parietal and supraoccipital. (4) Between the frontal and parietal. Flexibility at the quadrate-articular joint and between the dentary and angular bones confers additional gape on the lower jaw. When the mouth closes these movements are reversed and the snout flexes downward (dashed lines). Code: ang = angular, art = articular, boc = basioccipital, bp = basipterygoid process, bsp = basisphenoid, co = coronoid, con = occipital condyle, den = dentary, ecptg = ectopterygoid, eptg = epipterygoid, ex = exoccipital, fro = frontal, in = internal naris, ju = jugal, la = lacrimal, max = maxilla, nas = nasal, pal = palatine, par = parietal, pmax = premaxilla, po = postorbital, pop = paraoccipital process, prf = prefrontal, pro = prootic, ps = parasphenoid, ptg = pterygoid, q = quadrate, soc = supraoccipital, sq = squamosal, sur = surangular, vom = vomer. The kinetic joints are identified by numbers: (1) between basipterygoid process of basisphenoid and pterygoid. (2) Between paraoccipital process and parietal, squamosal and quadrate. (3) Between supraoccipital and parietal. (4) Between frontal and parietal. (Modified from Bellairs [1].)

The majority of snakes have unspecialized prey-catching techniques. They merely seize prey and swallow it alive as it struggles. The risk of damage to the snake during this process is a real one, and various features of snake anatomy seem to have evolved to give some protection from prey. The frontal and pareital bones of a snake's skull extend downward, entirely enclosing the brain and shielding it from the protesting kicks of prey being swallowed. Hypophyses project downward from the vertebrae in the neck to give additional protection to the spinal cord from bulky and possibly struggling prey. Probably the size of prey that can be attacked by snakes without a specialized feeding mechanism is limited by the snake's ability to swallow the prey without being injured in the process.

Constriction and venom are specialized feeding mechanisms that permit a snake to tackle large prey with little risk of injury to itself. Constriction is characteristic of the boas and pythons as well as a number of colubrid snakes. Despite travelers' tales of animals crushed to jelly by a python's coils, the process of constriction involves very little pressure. A constrictor seizes prey with its jaws and throws one or more coils of its body about the prey. The loops of the snake's body press against adjacent loops, and friction prevents the prey from forcing the loops open. Each time the prey exhales, the snake takes up the slack by tightening the loops slightly, and in a few minutes the prey has been suffocated. Seldom are any bones broken in the prey, and when that does occur it is the result of the prey's struggles rather than the force the snake has exerted.

The use of venom to kill prey is a still safer method for a snake. Enlarged teeth (fangs) on the maxillae have evolved in a variety of snakes (Figure 14–11). Three major categories of venomous snakes are recognized. **Opisthoglyph** (*opistho* = behind, *glyph* = hollowed) snakes have one or more enlarged teeth near the rear of the maxilla with smaller teeth in front. In some forms the fangs are solid; in others there is a groove on the surface of the fang that may help to conduct saliva into the wound. The rear-fanged condition has evolved repeatedly in the Family Colubridae, and there can be considerable variation in this character even among closely related forms. The ringneck snakes (*Diadophis*) of western North America are robust animals with well-developed opisthoglyph dentition. In eastern North America, ringneck snakes are smaller and have little or no development of fangs. Several African and Asian opisthoglyphs can deliver a dangerous or even lethal bit to large animals including humans, but their primary prey is lizards or birds, which are often held in the mouth until they stop struggling and then swallowed.

Proteroglyph snakes (*proto* = first) include the cobras, mambas, and coral snakes in the Family Elapidae and the sea snakes (Hydrophiidae). Recent studies indicate that the coral snakes of the New World had a separate evolutionary origin from the cobras and mambas of the Old World, and thus the proteroglyph dentition of these two lineages represents convergence. The hollow fangs of the proteroglyph snakes are located at the front of the maxilla, and there are often several small, solid teeth behind the fangs. The fangs are

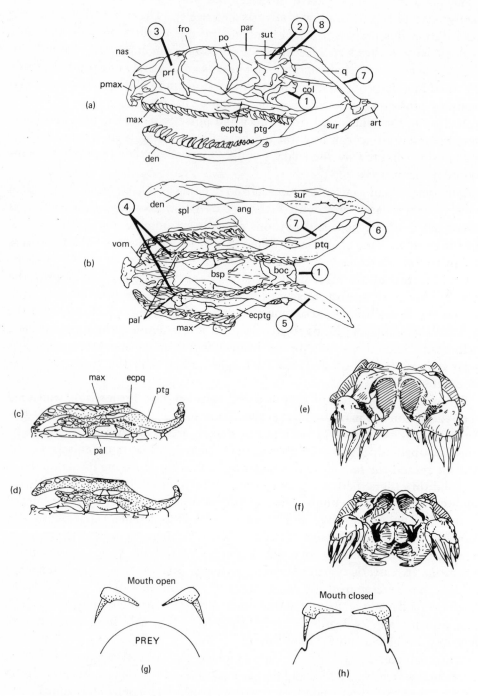

Figure 14-9 Skull of a snake—lateral (a), ventral (b, c, d) and frontal (e, f) views. (a, b, colubrid snake; c-f python.) Kinesis in a snake skull is very much more complex than that of a lizard. A snake skull contains eight moveable links: (1) braincase, (2) supratemporal, (3) prefrontal, (4) palatine, (5) pterygoid, (6) pterygoquadrate ligament, (7) quadrate, (8) quadratosupratemporal tie. This

permanently erect and relatively short. **Solenoglyph** (*solen* = pipe) snakes include the pit vipers of the New World and the true vipers of the Old World. In these snakes the hollow fang is the only tooth on the maxilla which rotates so the fang is folded against the roof of the mouth when the jaws are closed. This folding mechanism permits solenoglyph snakes to have long fangs that inject venom deep into the tissues of the prey. The venom first kills the prey, and then speeds its digestion after it has been swallowed.

It is probable that there is more convergent evolution within the three major categories of venomous snakes than is yet realized. African mole vipers (*Atractaspis*) were traditionally considered primitive solenoglyphs, but recent studies suggest that they represent an independent evolution of a viperlike fang mechanism from a colubrid ancestor. In mole vipers, the fangs project out of the sides of the mouth when they are erected. A mole viper jerks its head rapidly from side to side when it is molested, hooking at its attacker with its fangs instead of striking or biting as other venomous snakes do. Harry Greene examined the stomach contents of mole vipers in museum collections and found that many individuals had eaten several small rodents, all the same size, in one meal. He suggests that a mole viper feeds by crawling into a rodent nest, striking sidewards with its fangs until it has bitten all the babies, and then swallows the entire litter. Probably future studies that combine functional morphology with ecology and behavior will reveal additional examples of independent origins of specialized feeding habits and venom delivery mechanisms among snakes.

Snakes that can inject a lethal dose of venom into their prey have evolved a very safe prey-catching method. A constrictor is in contact with its prey while

---

linkage system is paired; that is, the left and right sides of the skull operate independently. Furthermore, the pterygoquadrate (6) and quadratosupratemporal (8) links are flexible. When they are under tension they are rigid, but when they are relaxed they permit movement in three dimensions. There is also greater flexibility in the lower jaw of a snake than a lizard, and the tips of the mandibles are joined by elastic tissue instead of a firm symphysis. When the mouth is opened (c = closed; d = opened) the pterygoid is protracted (moved anteriorly). That motion causes the palatine to turn outward, and the force is transmitted to the maxilla by the ectopterygoid. The tooth-bearing surfaces of the palatine and maxilla are rotated about their long axes so that the teeth point outward when the mouth is opened [compare e (mouth closed) and f (mouth opened)]. When the mouth closes, the movements are reversed and the teeth impale the prey (g, h). Similar movements, controlled independently on each side of the head and combined with independent movements of the mandibles disengage and engage the teeth and pull prey into the mouth during swallowing (Figure 14–10). Code: ang = angular, art = articular, boc = basioccipital, bsp = basisphenoid, col = columella, den = dentary, ecptg = ectopterygoid, fro = frontal, nas = nasal, pal = palatine, par = parietal, pmax = premaxilla, po = postorbital, prf = prefrontal, ptg = pterygoid, q = quadrate, spl = splenial, sur = surangular, sut = supratemporal, vom = vomer. (a, b, modified from C. Gans [1961] *American Zoologist* 1:217–227; c–h, modified from T. H. Frazetta [1966] *Journal of Morphology* 118:217–296.)

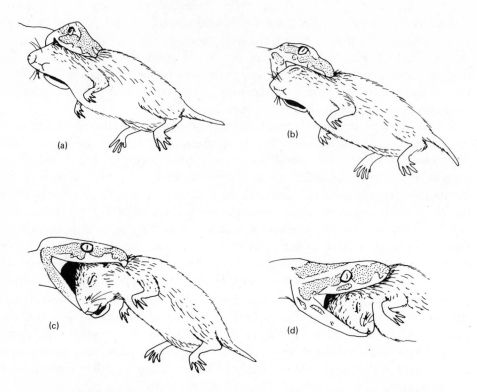

Figure 14-10 Snakes use a combination of head movements and protraction and retraction of the jaws to swallow prey. (a) Prey grasped by left and right jaws at the beginning of the swallowing process. (b) The upper and lower jaws on the right side have been protracted, disengaging the teeth from the prey as shown in Figure 10-9f. (c) The head is rotated counterclockwise, moving the right upper and lower jaws over the prey. The recurved teeth slide over the surface of the prey like the runners of a sled. (d) The upper and lower jaw on the right side are retracted, embedding the teeth in the prey and drawing it into the mouth. Notice that the entire head of the prey has been engulfed by this movement. Next the left upper and lower jaws will be advanced over the prey by a clockwise rotation of the head. The swallowing process continues with alternate left and right movements until the entire body of the prey has passed through the snake's jaws. (Modified from R. H. Frazetta [1966] *Journal of Morphology* 118:217-296.)

it is dying and runs some risk of injury from the prey's struggles. A solenoglyph snake needs only to inject venom and allow the prey to run off to die. Later the snake can follow the scent trail of the prey to find its corpse. This is the prey-capture pattern of most of the vipers. Several features of the body form of vipers suggest that the relative safety of their hunting methods allows them to attack larger prey in relation to their own body size than nonvenomous snakes can. Many vipers, like rattlesnakes, the fer de lance, and especially the African puff adder and Gaboon viper, are very stout snakes. The triangular head shape that is usually associated with vipers is a result of the

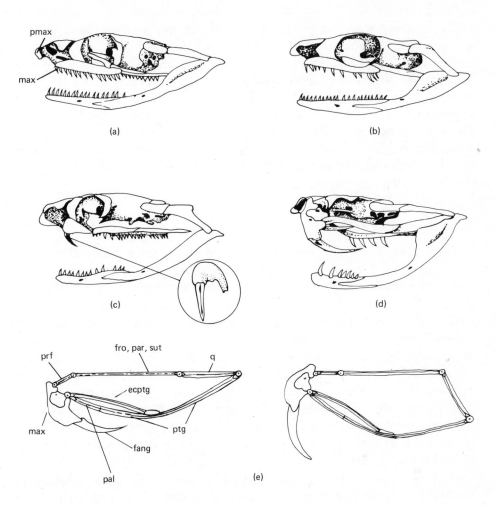

Figure 14-11 Four major types of dentition are distinguished in snakes. (a) Aglyphous snakes have no enlarged teeth. (b) Opisthoglyph snakes have enlarged, sometimes grooved, teeth near the posterior end of the maxilla. This condition is widespread among snakes usually considered nonvenomous and the number, position, and morpholoy of the enlarged teeth are variable. (c) Proteroglyph snakes, such as cobras, mambas, and sea snakes, have relatively short, permanently erect hollow fangs at the front of the maxilla. The maxilla may bear some small solid teeth in addition to the fang. (Inset: a frontal view of a cobra fang.) (d) Solenoglyph snakes have long hollow fangs that are the only teeth on the maxilla. The maxilla rotates, folding the fang against the roof of the mouth when the mouth is closed. This condition is characteristic of vipers and pit vipers. (e) The fangs of vipers are erected when the mouth is opened by a mechanism similar to the kinesis snakes use for feeding. The forward movement of the pterygoid is transmitted to the maxilla via the ectopterygoid causing it to rotate. Abbreviations as in Figure 14-8. (a–d modified from G. Jan and F. Sordelli [1860–1881] *Iconographie Generale des Ophidiens,* Milan; e modified from L. M. Klauber [1956] *Rattlesnakes.* University of California Press, Berkeley.)

outward extension of the rear of the skull, especially the quadrates. As a result of these morphological features, vipers can swallow large prey. The wide-spreading quadrates allow passage through the mouth, and even a large meal makes little bulge in the stout body, and thus does not interfere with locomotion. Vipers have specialized as relatively sedentary predators that can prey even on quite large animals. The other family of terrestrial venomous snakes, the cobras, mambas, and their relatives are primarily slim-bodied snakes. There is one exception to the generalization that elapids are slim-bodied. This is the death adder of Australia (*Acanthophis*). Vipers are absent from Australia, but the death adder is an elapid that has evolved a typical viperid body form— the body is short and stout, and the head broad.

Modifications of the Venom Apparatus for Defense

There is no doubt that venom evolved in snakes as an aid to capturing and subduing prey. There are a large number of snakes usually considered harmless that have a portion of the superior labial gland differentiated into "Duvernoy's gland." This gland contains serous protein-secreting cells and its secretions are mildly toxic. In humans there is a considerable amount of individual variation in sensitivity to the venoms of these "harmless" snakes; some people have little or no reaction to a bite, whereas others may suffer considerable local swelling and discomfort. It is possible, although not demonstrated, that the venom may be particularly toxic to the usual prey of these snakes which frequently feed on arthropods, fishes, amphibians, or other reptiles. Delivery of the venom is a slow process; usually the snake bites and retains its grasp on the prey until the venom takes effect. In some cases that may be 5 hr or more. In contrast, prey bitten by a rattlesnake usually dies in less than 5 min.

Once the venom-injecting apparatus had evolved to the point at which a bite produced an immediate unpleasant sensation, it could serve a defensive function in addition to prey capture. This stage has been reached by the solenoglyph snakes (vipers) that bear long erectile fangs on the maxillae, by the proteroglyph snakes (elapids) that have short, permanently erect fangs on the maxillae, and by some opisthoglyphs (colubrids) with enlarged teeth at the rear of the maxillae (Figure 14–11).

Snake venom is a complex mixture of enzymes and specific toxins, the exact composition of which varies along phylogenetic lines and within species. There may be differences between the venom of juveniles and adults, or in a single individual at different times of the year. Frequently there is geographic variation in venom components within a species. Table 14–2 shows the distribution and function of a number of components of reptilian venom.

A generation ago it was common to divide snake venoms into two categories— hemotoxic and neurotoxic. Hemotoxic venoms, considered to be typical of vipers, caused tissue destruction and widespread hemorrhage. Neurotoxic venoms, typical of elapids and sea snakes, produced death by paralysis of the respiratory muscles. As Table 14–2 shows, that view was oversimplified. Both

Table 14–2   Components of reptile venoms

Compound	Occurrence	Effect
Proteinases	All venomous reptiles, especially viperids	Tissue destruction
Hyaluronidase	All reptiles	Increases tissue permeability, hastens spread of the other venom constituents through tissues
L-Amino acid oxidase	Many snakes, but absent from some vipers, elapids, and sea snakes	Attacks a wide range of substrates, causes great tissue destruction
Cholinesterase	High in elapids, may be present in sea snakes, low in viperids	Unknown, it is *not* responsible for the neuromuscular effects of elapid venom
Phospholipases	All venomous reptiles	Attacks cell membranes
Phosphatases	All venomous reptiles	Attacks high energy phosphate compounds like ATP
Basic polypeptides	Elapids and sea snakes	Blocks neuromuscular transmission

viperid and elapid venoms contain both neurotoxic and hemotoxic elements, and it is not possible to account for the function of every component that can be identified in snake venom. Within limits, the actions of different venoms can be explained on the basis of their constituents. For example, a bite from a viper causes massive tissue destruction. Depending on how rapidly the venom is absorbed into the circulation, the destruction may be primarily near the site of the bite or distributed through the entire body.

Because of the great activity of the proteinases in viper venom, a bite is likely to produce massive tissue destruction near the site of the bite, and this frequently necessitates amputation. The discovery that cholinesterase in elapid venoms is not the component that produces neuromuscular paralysis is puzzling, and its function is yet to be discovered. Elapid bites produce less local necrosis than viperid bites, but their systemic effects are more pronounced. Loss of consciousness and cessation of respiration or of heart activity are not uncommon. Even the opisthoglyph snakes, not usually considered dangerously venomous, can deliver lethal bites. A noted herpetologist, Karl P. Schmidt, died in 1957 from the bite of a boomslang, an African opisthoglyph.

The use of venom for defense has involved certain behavioral and morphological adaptations. A warning behavior that allows a venomous snake to deter potential predators before they actually come to blows gives the snake additional protection. Venomous snakes have evolved a number of these mechanisms. The best known is the rattle of rattlesnakes (Figure 14–12). This unique structure is formed by the retention of a small, dumbbell-shaped piece of skin each time the epidermis is shed. The interlocking segments rub against each other,

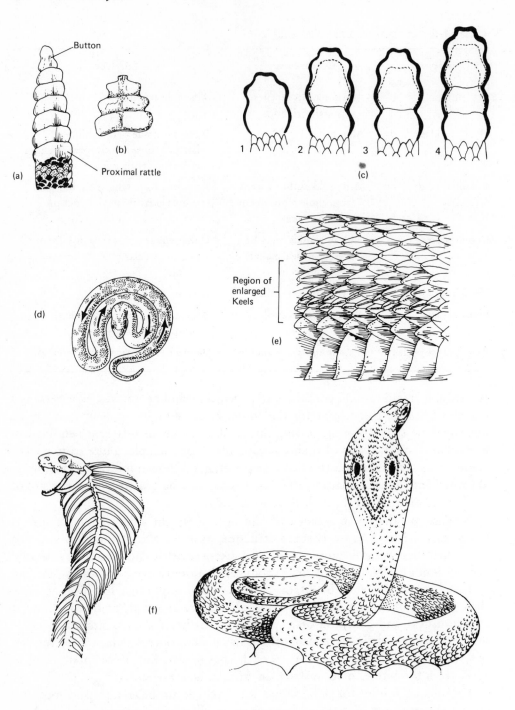

Figure 14-12 Venomous snakes have evolved a number of aposematic structures and behaviors that warn potential predators of the snakes' noxious qualities. (a) The rattle of a rattlesnake is composed of a number of two-, three-, or
four-lobed segments (b) that interlock loosely and produce a distinctive buzzing
sound when the tail is vibrated. A rattlesnake is born with a "prebutton" that
adds a second lobe and becomes a "button" shortly after birth (c1). Sub-

and when the snake vibrates its tail they produce a distinctive buzz. A similar sound is produced by some African snakes by a completely different mechanism; several rows of scales low on the side of the body have their enlarged keels broken into projecting points. In its defensive display, the snake throws its body into a tight S-shaped coil and keeps the loops of the body moving. As adjacent loops rub against each other, the clicking of the points on the scales sums into a hissing noise.

Many snakes expand their necks in threat displays, and in cobras a large horizontal expansion is produced by spreading the ribs in the neck sidewards to form the hood. The snake lifts the anterior third of its body off the ground, and a 2-m-long cobra with its hood spread, weaving from side to side and hissing, is an impressive sight. The ultimate in adaptation of the venom-delivery system for defensive purposes has probably been achieved by the spitting cobras. The name is a misnomer, the snakes do not spit; rather, they can spray their venom from the fangs with sufficient force to cause the stream to break into fine droplets that travel 2 m or more. This is a defensive mechanism that is particularly effective against large mammals and birds. The snake appears to aim for the attacker's eyes, and the venom is rapidly absorbed and causes immediate pain and blindness that will probably become permanent unless prompt treatment is given. As little as one drop of venom from a spitting cobra applied to the eye of a white rat caused death in 20 min, but venom from a viper applied in this way had no apparent effect. Furthermore, venom from a spitting cobra appeared to be absorbed more rapidly through mucous membranes than venom from a nonspitting species of cobra, indicating that venom of the spitting cobra has undergone biochemical evolution that has made it more effective as a deterrent to predators.

---

sequently a new segment is added to the rattle every time the snake sheds its skin (c2–c4). Many nonvenomous snakes vibrate their tails when they are threatened; in dry vegetation the sound produced is very similar to a rattlesnake's buzz. An Old World viper produces a similar buzzing noise by throwing the body into a series of loops that press against each other as they move in opposite directions. Heavily keeled scales on the lower part of the side of the body rub against scales in the adjacent loop to produce the sound. (d) Nonvenomous egg-eating snakes mimic the vipers and have evolved similar behavioral and morphological adaptations. (e) The enlarged scales of an egg-eating snake. (f) The cervical ribs of cobras are elongated. When a cobra is at rest the ribs extend backward along the neck, but when the snake is alarmed the ribs are protracted to produce the familiar hood of a cobra. Some cobras have distinctive markings on the dorsal surface of the neck that are displayed when the hood is spread. Many nonvenomous and mildly venomous snakes spread their necks either horizontally like cobras or vertically. As the scales are spread apart, bright colors may be revealed on the underlying skin. (Sources of illustrations: a–c, modified from L. M. Klauberg [1955] *Rattlesnakes.* University of California Press, Berkeley; d, e modified from Gans [3]; f drawn from a photograph in R. L. Ditmars [1933] *Reptiles of the World.* Macmillan, New York; and modified from F. Werner [1922] *Die Lurche and Kriechtiere von Alfred Brehm.* 2nd ed. Bibliographic Ins., Leipzig.)

14.2   REPTILIAN THERMOREGULATORY MECHANISMS

From the time of Aristotle onward reptiles and amphibians have paradoxically been called "cold-blooded" although they were thought to be able to tolerate extremely high temperatures. Salamanders frequently seek shelter in logs, and when a log is put on a fire, the salamanders it contains may come rushing out. Observations of this phenomenon gave credence to the belief that salamanders live in fire. In the first part of the twentieth century biologists were using similar lines of reasoning. In the desert lizards often sit on rocks. If you approach a lizard it runs away, but touching the rock shows that it is painfully hot. Clearly, the reasoning went, the lizard must have been equally hot. Biologists marvelled at the heat tolerance of reptiles, and statements "documenting" this are found in the authoritative textbooks of the period.

A study of reptilian thermoregulation by Raymond Cowles and Charles Bogert published in 1944 demonstrated the falsity of earlier observations and conclusions. They showed that reptiles regulate their body temperatures with considerable precision, and the level at which the temperature is regulated is characteristic of a species. The implications of this discovery in terms of the biology of amphibians and reptiles are still being explored. Virtually everything a reptile does is related directly or indirectly to its body temperature and the activities required to regulate its temperature. We have already pointed out that the geographic distribution of reptiles is related to environmental conditions. Outside of the tropics, lizards (which rely largely upon basking in the sun) are most abundant in desert areas where sunny days are the rule. Snakes depend less upon basking and are as abundant in warm cloudy regions as in sunny ones.

The microhabitat requirements of reptiles also reflect their thermoregulatory activities. Activity temperature is one of the niche dimensions that must be considered in an ecological comparison of two reptiles. Behavior is related directly to body temperature because reptiles are sluggish at low body temperatures and incapable of capturing prey or avoiding predators. Their sluggishness at low body temperatures is reflected in the flight distance of *Anolis* lizards; when they are cold they do not let a predator get as close to them before they run away as they do when they are warm and can run faster.

Another relationship between body temperature and defensive behavior occurs in geckos. Many lizards, geckos among them, are able to sacrifice their tail to a predator if that is the only way they can escape. The process is called autotomy. Some of the tail vertebrae have built-in fracture planes. When a lizard's tail is seized by a predator, the lizard breaks the tail at one of these autotomy planes. Sphincter muscles close off the blood vessels so there is little loss of blood, and the lizard runs away leaving part of its tail in the predator's mouth. The broken piece of tail twitches and jumps about, presumably distracting the predator while the lizard makes its escape. The lizard regenerates the lost part of its tail but the vertebrae are replaced by a cartilaginous rod without autotomy planes, and future tail loss is most likely to occur closer to the lizard's body than the original loss. Thus, the closer to the tip of the tail each break occurs, the more often during its lifetime a lizard will be able to use this escape mechanism. Furthermore, growing a new tail is a large investment of energy. For both of these reasons it is to the lizard's advantage to lose as little of its tail as possible each time. When a warm gecko's tail is seized in a pair of forceps, the gecko breaks it off just above the forceps' grasp and keeps its loss to a minimum. When a cold lizard is treated the same way, however, the tail breaks much closer to the body. Perhaps a predator that gets a large portion of the lizard's tail is delayed longer and as a result the cold lizard has more time to escape.

The traumatic experiences of a reptile's life are not the only ones that are intimately related to thermoregulation; such day-to-day features as diet are also involved. The desert iguana occurs in desert regions of California, Arizona, and northern Mexico. It is the most thermophilic of North American lizards, maintaining a body temperature between 40 and 42°C during activity. It is active during the middle of the day in deserts, when other animals are sheltering from the heat. As a result the desert iguana avoids competition from other lizards and the threat of predation. On the other hand, at midday insects have also sought shelter from the sun, and desert iguanas have adopted a primarily herbivorous diet probably because plants are the food most readily available to them.

14.2.1 Energy Exchange Between an Organism and Its Environment

A brief discussion of the pathways by which thermal energy is exchanged between a living organism and its environment is necessary to understand the thermoregulatory mechanisms employed by terrestrial animals. There are several pathways by which that energy is either gained or lost by an organism, and by adjusting the relative flow through various pathways an animal can warm, cool, or maintain a stable body temperature.

Figure 14–13 illustrates pathways of thermal energy exchange. In the case of an ectotherm, the sun is the source of virtually all the energy used to regulate body temperature; however, solar energy can reach the animal in several ways. Direct solar radiation impinges on an animal when it is standing in a sunny spot. In addition, solar energy is reflected from clouds and dust particles in the atmosphere and from other objects in the environment and reaches the animal by these circuitous routes. The wavelength distribution of the energy in all these routes is the same—the portion of the solar spectrum that penetrates the earth's atmosphere. About half this energy is contained in the visible wavelengths of the solar spectrum (400 to 700 nanometers) and the other half is in the ultraviolet and infrared regions of the spectrum.

There is an important exchange of energy in the infrared. All objects, animate or inanimate, radiate energy at wavelengths determined by their absolute temperatures. Objects in the temperature range of animals and the earth's surface (roughly −20 to +50°C) radiate in the infrared portion of the spectrum. Animals continuously radiate heat to the environment and receive infrared radiation from the environment. Thus, infrared radiation can lead to either heat gain or loss, depending on the relative temperature of the animal's body surface and the environmental surfaces and on the radiation characteristics of the surfaces themselves.

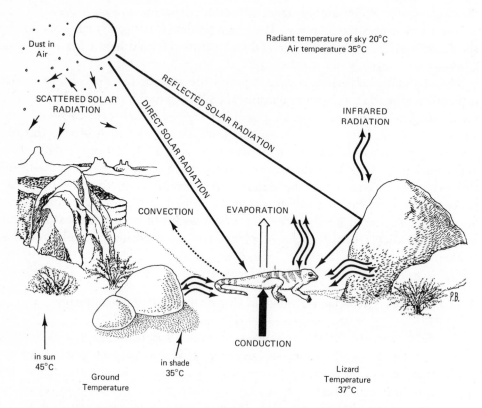

Dust in Air

SCATTERED SOLAR RADIATION

REFLECTED SOLAR RADIATION

DIRECT SOLAR RADIATION

Radiant temperature of sky 20°C
Air temperature 35°C

INFRARED RADIATION

CONVECTION

EVAPORATION

CONDUCTION

in sun 45°C

in shade 35°C

Ground Temperature

Lizard Temperature 37°C

Figure 14-13 There are many pathways of energy exchange between a terrestrial organism and its environment. These are illustrated in simplified form by a lizard resting on the floor of the desert arroyo. Solar energy reaches the animal directly and by reflection from dust in the air or objects in the environment. An animal also exchanges energy with its surroundings by conduction and convection. In the example shown, the ground on which the lizard rests is warmer than the lizard whereas the air is cooler. As a result, the lizard gains energy by conduction from the ground and loses it by convection to air. Evaporation of water from the skin and respiratory passages is another pathway of energy loss. Energy exchange in the infrared is more complicated because it depends on the relative temperatures of the lizard and different objects in its environment. The lizard is cooler than the sunlit rock, so it receives more energy by radiation from that source than it loses to it. The lizard is warmer than the shaded rock arroyo, however, so it has a net loss of energy in that radiative exchange. The radiative temperature of clear sky is about 20° C, so the lizard loses energy to the sky. Small adjustments of posture or position change the magnitude of the various routes of energy exchange and give the lizard considerable control of its body temperature.

Heat is exchanged between objects in the environment and the air via convection. If an animal's surface temperature is higher than air temperature, convection leads to heat loss; if the air is warmer than the animal, convection is a route of heat gain. In still air convective heat exchange is accomplished by convective currents formed by local heating, but in moving air forced convection replaces natural convection and the rate of heat exchange is greatly increased.

Conductive heat exchange resembles convection in that its direction depends on the relative temperatures of the animal and environment. Conductive loss occurs between the body and the substrate where they are in contact. It can be modified by changing the surface area of the animal in contact with the substrate and by changing the rate of heat conduction in the parts of the animal's body that are in contact with the substrate.

Evaporation of water occurs from the body surface and from the pulmonary system. Each gram of water evaporated represents a loss of about 585 cal (the exact value changes with temperature). Evaporation almost always occurs from the animal to the environment, and thus represents a loss of heat. The inverse situation, condensation of water vapor on an animal, would produce heat gain but it rarely occurs under natural conditions.

Metabolic heat production is the final pathway by which an animal can gain heat. Among ectotherms metabolic heat gain is usually trivial in relation to the heat derived directly or indirectly from solar energy There are a few exceptions to that generalization, and some of them are discussed later. Endotherms, by definition, derive most of their heat energy from metabolism but their routes of energy exchange with the environment are the same as those of ectotherms and must be balanced in order to maintain a stable body temperature.

14.2.2 Behavioral Thermoregulation of Reptiles

R.B. Cowles and C.M. Bogert's work concentrated upon behavioral thermoregulatory mechanisms of reptiles, and their pioneering study has been extended by other biologists. Lizards show better-defined thermoregulatory behavior than other reptiles, and most studies have centered on them. Cowles and Bogert studied wild reptiles in the field as well as individuals confined in large pens built in the desert. They correlated particular behavior patterns with specific body temperatures and were able to define eight categories of temperature to which the reptiles showed characteristic behavioral reactions:

1. The **lethal minimum** is the low temperature that disrupts the physiology of an animal and produces death after a short exposure.
2. The **critical minimum** is an ecological lethal temperature rather than a physiological end point. At the critical minimum, the animal is incapacited by cold and cannot move. At this temperature reptiles cannot escape enemies or move to a warmer place.
3. The **voluntary minimum** is the lowest temperature at which a reptile

attempts activity. Below the voluntary minimum the animal retreats to shelter and becomes inactive.

4. The **basking range** is the temperature region in which a reptile's efforts are devoted to warming itself to body temperatures at which normal activity is possible.

5. The **normal activity range** is the temperature region in which a reptile carries out its full repertoire of activities—feeding, courtship, territorial defense, and so on. For many lizards it is as narrow as 4°C, but for other reptiles it may be as broad as 10°C. For convenience in comparing different species it is common to use the **eccritic** temperature which is the mean of all temperature measurements in the normal activity range. The significance of the eccritic temperature has often been misunderstood. The average of temperature measurements in the activity range is a convenience for biologists, not a biologically significant value for reptiles. A reptile's thermoregulatory efforts are directed to staying within the normal activity range. The precise temperature it maintains within this range may depend upon internal conditions. For example, several reptiles have been shown to maintain higher body temperatures within the activity range when they are digesting food than they do when fasting. Reproductive state and also the season of the year can produce shifts within the normal activity range.

6. The **maximum voluntary tolerance** is the highest temperature at which a reptile continues to be active. When its body temperature exceeds that level, it retreats to shade or underground and waits for things to cool off.

7. The **critical maximum** is the high temperature at which an animal is incapacitated by heat. Unable to move, it cannot escape from conditions which, if prolonged, will lead to its death either from further heating or because a predator finds the helpless animal. This is an ecological lethal in the same sense that the critical minimum was an ecological rather than physiological end point.

8. The **lethal maximum** is the high temperature that does irreversible physiological damage and kills the animal. It is a physiological end point, as is the lethal minimum.

The behavioral mechanisms involved in temperature regulation are quite straightforward and are employed by insects, birds, and mammals (including humans) as well as reptiles. Movement back and forth between sun and shade is the most obvious. Early in the morning or on a cool day, reptiles are basking in the sun whereas in the middle of a hot day they have retreated to shade and make only brief excursions into the sun. Sheltered or exposed microhabitats may be sought out—in the morning when a lizard is attempting to raise its body temperature it is likely to be in a spot protected from the wind. Later in the day when it is approaching the upper end of its normal activity range it may climb into a bush or onto a rock outcrop where it is exposed to the breeze and its convective heat loss is increased.

The amount of solar radiation absorbed by an animal can be altered by changing the orientation of its body with respect to the sun, the body contour, and the skin color. All of these mechanisms are used by reptiles. An animal oriented perpendicular to the sun's rays intercepts the maximum amount of solar radiation, and one oriented parallel to the sun's rays intercepts minimum radiation. Reptiles adjust their orientation to control heat gained by direct solar radiation. Many reptiles are capable of spreading or folding in the ribs to change the shape of the trunk. If the body is oriented perpendicular to the sun's rays and the ribs are spread, the surface area exposed to the sun is maximized and heat gain is increased. Compressing the ribs decreases the surface exposed to the sun and can be combined with orientation parallel to the rays to minimize heat gain. Horned lizards provide a good example of this type of control. If the surface area that a horned lizard exposes to the sun directly overhead when it sits flat on the ground with its ribs held in their resting position is considered to be 100 percent, the maximum surface area the lizard can expose by orientation and change in body contour is 173 percent and the minimum is 28 percent. That is, the lizard can change its radiant heat gain by more than sixfold, solely by changing its position and body shape.

Color change can further increase a reptile's control of radiative exchange. Lizards darken by dispersing melanin in melanophore cells in the skin and lighten by drawing the melanin into the base of the melanophores. The lightness of a lizard affects the amount of solar radiation it absorbs in the visible part of the spectrum, and changes in heating rate (in the darkest color phase compared to the lightest) are from 10 to 75 percent.

Lizards can achieve a remarkable independence of air temperature as a result of their thermoregulatory capacities. Lizards occur above the timberline in many mountain ranges and during their periods of activity are capable of maintaining body temperatures 30°C or more above air temperature. While air temperatures are near freezing, these lizards scamper about with body temperatures as high as those of species that inhabit lowland deserts.

The repertoire of thermoregulatory behavior seen in lizards is greater than that of other reptiles. Turtles, for example, cannot change their body contour or color, and their behavioral thermoregulation is limited to movements between sun and shade and in and out of water. Crocodilians are very like turtles, although young individuals may be able to make minor changes in body contour and color. Most snakes cannot change color, but some rattlesnakes lighten and darken as they warm and cool.

14.2.3 Physiological Control of Body Temperature

A new dimension was added to studies of reptilian thermoregulation in the 1960s by the discovery that reptiles can use physiological mechanisms to adjust the rate of temperature change. The original observations, made by George Bartholomew and his associates, showed that several different kinds of large lizards were able to heat faster than they cooled when exposed to the same differential between body and ambient temperatures. Subsequent studies by other investigators extended these observations to turtles and snakes. From the animal's viewpoint, heating rapidly and cooling slowly prolongs the time it can spend in the normal activity range.

Fred White and his colleagues have demonstrated that the basis of this control of heating and cooling rates lies in changes in peripheral circulation. Heating the skin of a lizard causes a localized vasodilation of dermal blood vessels in the warm area. Dilation of the blood vessels, in turn, increases the blood flow through them, and the blood is warmed in the skin and carries the heat into the core of the body Thus, in the morning when a cold lizard orients its body perpendicular to the sun's rays and the sun warms its back, local vasodilation in that region speeds transfer of the heat to the rest of the body.

The same mechanism can be used to avoid overheating. The Galapagos marine iguana is a good example. Marine iguanas live on the bare lava flows on the coasts of the islands. In midday, beneath the equatorial sun, the black lava becomes extremely hot—uncomfortably if not lethally hot for a lizard. Retreat to shade of the scanty vegetation or into rock cracks would be one way the iguanas could avoid overheating, but the males are territorial and these behaviors would mean abandoning their territories and probably having to fight for them again later in the day. Instead, the marine iguana stays where it is and uses physiological control of circulation and the cool breeze blowing off the ocean to form a heat shunt that absorbs solar energy on the dorsal surface and carries it through the body and dumps it out the ventral surface.

The process is as follows: In the morning the lizard is chilled from the preceding night and basks to bring its body temperature to the normal activity range. When its temperature reaches this level the lizard uses postural adjustments to slow the increase in body temperature, finally facing directly into the sun (parallel orientation) to minimize its heat load. In this posture the fore-part of the body is held off the ground. The ventral surface is exposed to the cool wind blowing off the ocean, and a patch of lava under the animal is shaded by its body This lava is soon cooled by the wind. Local vasodilation is produced by warming the blood vessels. It does not matter whether the heat comes from the outside (from the sun) or from inside (from warm blood). Warm blood circulating from the core of the body to the ventral skin warms it and produces vasodilation increasing the flow to the ventral surface. The lizard's ventral skin is cooler than the rest of its body—it is shaded and cooled by the wind, and in addition it loses heat by radiation to the cool lava in the shade created by the lizard's body In this way the same mechanism that earlier in the day allowed the lizard to warm rapidly is converted to a regulated heat shunt that rapidly transports solar energy from the dorsal to the ventral surface, and keeps the lizard from overheating. In combination with postural adjustments and other behavioral mechanisms, such as the choice of a site where the breeze is strong, the physiological adjustments allow the lizard to remain on station in its territory all day.

In the afternoon the situation reverses. As the sun goes down the lizard begins to cool. First the body temperature cools until the blood is no longer warm enough to produce vasodilation in the ventral region, and the heat shunt stops. As long as the sun warms the dorsal surface, vasodilation there increases heat absorption. When the dorsal surface cools below the temperature that produces vasodilation, all circulation to the body surface is restricted, and the lizard cools slowly because the transfer of heat from the warm core to the cool surface is limited by the reduction of peripheral circulation. These changes are accompanied by the appropriate behavioral changes in orientation and posture. Finally, as the sun goes down, marine iguanas collect in groups of 20 or more that form piles of lizards. This behavior, in effect, creates a "super-lizard" with a more favorable surface/mass ratio than an individual animal and thus reduces the fall in temperature overnight.

The behavioral and physiological thermoregulatory mechanisms of reptiles are intimately intertwined. Although we have tried to simplify our presentation by discussing them separately, it is essential to realize that neither behavioral, nor physiological, nor morphological thermoregulatory mechanisms function by themselves. They are used in combination, and they have evolved in combination. The thermoregulation of a reptile (and, as we will see later, a bird or mammal) involves all of these mechanisms simultaneously.

14.3 ROLE OF THERMOREGULATION IN THE LIFE OF REPTILES

The elaborate thermoregulatory mechanisms employed by reptiles have been described in some detail because observations of the behavior of reptiles in the field indicate that thermoregulatory activities occupy a considerable portion of an animal's time. Less obvious but just as real are the limitations that the need for thermoregulation sets on a reptile's life. Certain species are excluded from certain habitats, for example, because it is impossible to thermoregulate. Similarly, in temperate regions the activity season lasts only during the months that are warm and sunny enough to permit thermoregulation. During cool months reptiles become inactive. In contrast, many fish and amphibians are eurythermal. They function well at high temperatures in summer and at low temperatures in winter (see section 7.5.4). What advantages do reptiles gain from thermoregulation that make the consequent limitations on their activities acceptable?

14.3.1 Physiological Adaptations to the Normal Activity Range

The advantages associated with a stable body temperature have been mentioned in Chapters 7, 12 and 13. An organism's tissues are the site of a tremendous variety of biochemical reactions proceeding simultaneously and dependent upon one another to provide the proper quantity of the proper substrates at the proper time for reaction sequences. Because each reaction has a different sensitivity to temperature, it is difficult to evolve a regulatory system for all these processes that works under a variable temperature regime. The more the range of temperature can be restricted by thermoregulatory mechanisms, the more precise the integration of reaction sequences can be. Birds and mammals achieve this restriction of temperature variation by using metabolic heat to maintain high and reasonably constant body temperatures (see section 18.2.3) whereas reptiles (and to a lesser extent amphibians) depend on balancing heat uptake and loss.

There are really two aspects of temperature regulation—temperature stability and the level at which temperature is regulated. A reasonably constant body temperature allows integration of complex reaction sequences whether the temperature itself is high or low. This degree of regulation is achieved passively by a large variety of animals—for example, aquatic animals living in stable temperature regimes and very large terrestrial animals like dinosaurs. A high body temperature is a second adaptation. Chemical reactions proceed faster at higher temperatures than at low ones. (Reaction rates approximately double with a 10°C increase in temperature.) The advantages an animal can derive from maintaining a high body temperature will be discussed in Chapter 18.2.3 in the context of endotherms. In this chapter we will concentrate on the advantages of minimizing the variation in body temperature during activity.

If the temperature stability achieved by reptilian thermoregulatory processes is important to the animal, one would expect to find that the internal economy of a reptile functions most efficiently at its activity temperature. Indeed, this is the case. Examples of physiological processes that work at maximum efficiency in the normal activity temperature range can be found at the molecular, tissue, and whole-animal levels of reptiles. Some enzymes, lactic dehydrogenase and myosinadenosine triphosphatase (myosin ATPase) for example, have temperature optima that correspond to the activity temperature of the particular species tested. The enzymes' activities are maximal at those temperatures and decline at higher or lower temperatures.

The oxygen transport characteristics of blood of lizards spanning a wide range of activity temperatures has evolved in such a manner that oxygen delivery to the tissues is greatest at the activity temperature. When the resting oxygen consumption of a reptile is measured, one frequently finds a plateau in the region of the activity temperature range in which the oxygen consumption is nearly independent of body temperature (Figure 14–14). Above and below that range oxygen consumption increases sharply with increasing body temperature, but in the normal activity range the animal's energy production (and hence activity) is not sensitive to temperature.

Finally, the maximum oxygen consumption during strenuous activity increases with increasing temperature and, in many species reaches a peak in the normal activity range. As a result, the aerobic scope (the difference between resting and maximum oxygen consumption) is greatest in the activity temperature range. This means that an animal's ability to increase its oxygen consumption to sustain activity (catching food, escaping a predator, defending a territory) is greatest in its normal activity range. (It must be noted that reptiles depend primarily on anaerobic metabolism for high levels of activity. In contrast to aerobic metabolism, anaerobic metabolism is quite insensitive to body temperatures and produces a set amount of energy over a broad range of body temperatures. The total energy production—aerobic plus anaerobic—is

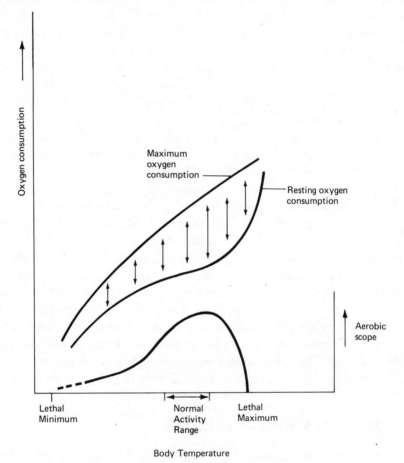

Figure 14-14 Among ectotherms, the rate of oxygen consumption is profoundly influenced by temperature. The maximum oxygen consumption that can be sustained during activity rises nearly linearly as body temperature increases. The resting oxygen consumption frequently shows a plateau of relative temperature independence near the activity temperature range and is very temperature-sensitive at temperatures higher or lower than that range. The aerobic scope is the difference in oxygen consumption between resting and maximum levels. In the upper graph it is shown by the length of the arrows; the lower graph plots aerobic scope as a function of body temperature. The lethal maximum and minimum are the temperatures at which the resting oxygen consumption would equal the maximum consumption possible. The normal activity range is usually only a few degrees below the lethal maximum.

thus greatest in the normal activity range, and the proportion of the total energy contributed by anaerobic metabolism is minimum.)

14.3.2 Ecological Adaptations of the Normal Activity Range

In some species of reptiles the normal activity range is important in defining the ecological niches they occupy. More than 20 species of the small arboreal lizard *Anolis* occur in Cuba, but only one species on Dominica. In Cuba the five most common species divide the habitat with minimum overlap (Figure 14–15). Perch height, the diameter of trees on which they occur, and the type of habitat (deep forest, forest margins, or open areas) separate the species. Each species has a characteristic body temperature; the forest dwellers have the lowest temperatures and the savanna forms the highest. Different populations of each of the five species in different areas have very similar mean body temperatures. The range of mean body temperatures among populations of any one of the species is only 2 to 3°C. The total range for the five species is 7.5°C.

On Dominica *Anolis oculatus* occupies the full range of habitats occupied in Cuba by the five different species. The range of mean body temperature in seven populations of *A. oculatus* from different habitats was 7.5°C. This is identical to the range of mean body temperatures for the five Cuban species combined. In other words, in the absence of competing species, *A. oculatus* has been able to occupy the full range of habitats that the five Cuban species divide among them, and this eurytopy (broadening of the niche) has been accompanied by eurythermy (broadening of the normal activity temperature range). A similar eurythermy has been described in other species of *Anolis* in which populations living in shaded forest areas do not raise their body temperatures above air temperature but other populations living nearby at the edge of the forest do bask in the sun and maintain high body temperatures. Clearly reptiles have a considerable degree of flexibility in their temperature relations.

14.4 WATER RELATIONS OF REPTILES

Reptiles have a reputation for being highly tolerant of arid conditions. Probably their evolutionary position as the earliest vertebrates to invade upland habitats has caused biologists to place more emphasis on the relatively low rates of cutaneous water loss of reptiles than those rates really deserve. The rate of evaporative water loss through the skin of reptiles is about the same as the rate of loss through mammalian skin. Thus, reptiles have no advantage over mammals in arid habitats on that basis, and we must look elsewhere for an explanation of the abundance of lizards and the scarcity of mammals in the desert during daytime. Three characteristics of reptiles are particularly important preadaptations for water conservation in arid habitats: their low metabolic rates, excretion of nitrogenous wastes as uric acid and its salts, and the presence in many taxa of an extrarenal pathway of salt excretion.

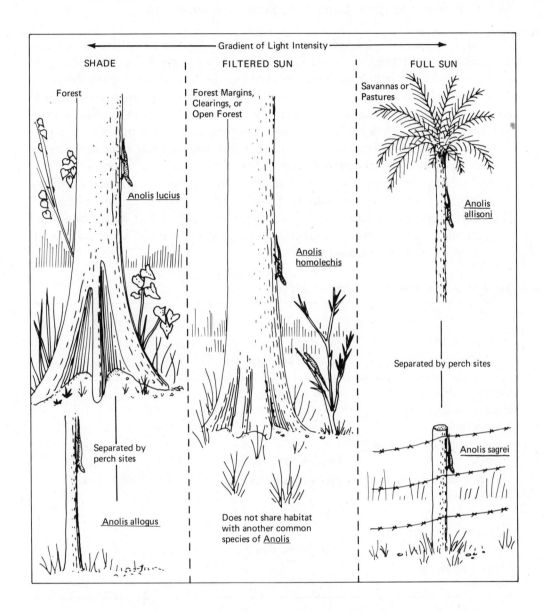

Figure 14–15 The five most common species of *Anolis* lizards on Cuba partition the habitat in several ways. Two species occur in deep shade in forests, one in partial shade in clearings and at the forest edge, and two in full sun. In the forest the two species are separated by their choice of perch sites: *Anolis lucius* perches on large trees up to 4 m above the ground, whereas *A. allogus* rests on small trees within 2 m of the ground. *A. homolechis*, which does not share its habitat with another common *Anolis*, perches on both large and small trees. The species that live in open habitats are also spearated by perch site: *A. allisoni* perches more than 2 m above the ground on tree trunks and houses whereas *A. sagrei* perches below 2 m on bushes and fenceposts. (Based on R. Ruibal [1961] *Evolution* 15:98–111.)

14.4.1   Evaporative Water Loss: Respiratory and Cutaneous

Inhaled air must be humidified in the respiratory passages so that it is saturated with water by the time it reaches the lungs. When the air is exhaled, much of the water vapor is lost with it. Because reptiles have lower metabolic rates than mammals, the volumes of air passing through the respiratory system in a given time are also lower and respiratory water loss is correspondingly low. The water saving achieved by the low metabolic rates of reptiles is a major factor in their water economy, but low metabolic rates are not an adaptation to conserving water—these metabolic rates are characteristic of reptiles in both moist and dry habitats. There is an adaptive reduction in skin permeability, however. Reptiles from dry habitats have lower rates of cutaneous water loss than those from moist habitats (Table 14–3). The species that are active in the desert during the day face the greatest evaporative stress and have the lowest rates of evaporative water loss.

Table 14–3 Evaporative water loss of reptiles

Species	Habitat and habits	Cutaneous $mg \cdot cm^{-2} \cdot day^{-1}$	Respiratory $mg \cdot g^{-1} \cdot day^{-1}$
		Rates of water loss	
Caiman (*Caiman crocodilus*)	Tropical, aquatic	32.9	9.6
Gecko (*Sphaerodactylus macrolepis*)	Tropical, nocturnal	9.4	—
Night lizard (*Lepidophyma smithi*)	Tropical, cave dweller	6.9	—
Green iguana (*Iguana iguana*)	Tropical, arboreal	4.8	3.4
Carolina anole (*Anolis carolinensis*)	Subtropical, arboreal	4.6	—
Side-blotched lizard (*Uta stansburiana*)	Semi-desert, diurnal	2.4	—
Chuckwalla (*Sauromalus obesus*)	Desert, diurnal	1.3	1.1

From P. J. Bentley [1976] Osmoregulation. *In The Biology of the Reptilia.* C. Gans and W. R. Dawson (eds.). Academic Press, New York. G. K. Snyder [1975] *J. comp. Physiol.* 104:13–18. W. J. Mautz, unpublished data.

14.4.2   Urinary Water Loss

Most reptiles (and birds as well) are **uricotelic**; they excrete nitrogenous wastes primarily as uric acid or its salts. Uric acid is only slightly soluble in water, and as a consequence it readily precipitates in urine, forming a whitish semisolid mass. This mass is a complex compound that includes sodium, potassium, and ammonium salts of uric acid. As these substances precipitate from solution, water is left behind and can be reabsorbed from the urine. Water freed by the precipitation of urates is reabsorbed by the blood, used to produce more urine, again freed by precipitation of urates, and so on. This recycling of water is important for reptiles because their kidneys are not able to form urine with a higher osmotic pressure than that of their blood plasma.

Not all of the ions in urine are precipitated with the urates, some remain in solution. There is active transport of sodium and potassium ions from urine in the bladder or cloaca back into the blood. At first glance, this process does not appear to make sense. Why should a reptile expend energy in the bladder or cloaca to bring back into its blood the same ions it used energy to remove from its blood in the kidney? The answer to the paradox lies in the third water-conserving feature of reptiles, extrarenal salt excretion.

14.4.3 Reptilian Salt Glands

In at least three groups of reptiles (lizards, snakes, and turtles) there are some taxa that possess glands specialized for the selective transport of ions out of the body. Salt glands are most widespread in lizards, where they have been found in five families. In all these families it is the lateral nasal gland that excretes salt. The secretions of this gland are emptied into the nasal passages, and a lizard expels them by sneezing or by shaking its head. In two families of specialized aquatic snakes, the sea snakes (Hydrophiidae) and the elephant-trunk snakes (Acrochordidae), the posterior sublingual gland secretes a salty fluid into the tongue sheath from which it is expelled when the tongue is extended. Finally in sea turtles (Cheloniidae) and in the diamond back terrapin, an emydid turtle which inhabits estuaries, a lachrymal gland secretes a salty fluid around the orbits of the eyes. (Photographs of nesting sea turtles frequently show clear paths streaked by tears through the sand that adheres to the head. These tears are the secretions of the salt glands.) Clearly salt glands have evolved independently at least three times in reptiles, because the three groups with salt glands utilize different glands. Independent evolution of salt glands within some of the three groups is also likely To make things more confusing, many birds also have functional salt glands, and these glands appear to be homologous with the nasal salt glands, but are clearly derived independently from the evolution of salt glands in lizards.

Despite their different origins and locations, the functional properties of reptilian salt glands are quite similar. They secrete fluids containing primarily sodium or potassium cations and chloride or bicarbonate anions in high concentrations. The total osmotic pressure of the salt gland secretion may reach 2000 mOsm/liter—more than six times the osmotic concentration of urine that can be produced by the kidney. This efficiency of excretion is the explanation of the paradox of active uptake of salt from the urine. As ions are actively reabsorbed, water follows passively, so an animal recovers both water and ions from the urine. The ions can then be excreted via the salt gland at much higher concentrations with a proportional reduction in the amount of water needed to dispose of the salt. Thus, by investing energy in recovering ions from urine, a reptile with salt glands can conserve water by excreting ions through the more efficient extrarenal route. The combination of low metabolism, uricotely, and extrarenal salt excretion has allowed lizards to adapt readily to life in arid habitats.

14.5   SOCIAL BEHAVIOR AND SPECIES RECOGNITION

As we mentioned in Chapter 13, there is apparently less social behavior in most reptiles than there is in many mammals and birds. Some species seem to show virtually no social behavior, in others males may engage in ritualistic combat during the breeding season but ignore each other at other times of the year. Territoriality is quite common among lizards and crocodilians. Usually it is the males that are territorial, but there are exceptions. Some species hold territories only during the breeding season, and in these cases it is presumably females that are the resource in short supply. Other species hold territories year-round, and in these cases food is probably the limiting resource.

14.5.1   Species-specific Displays

Among territorial species of reptiles the displays used to advertise the presence of a territorial individual have a species-specific component that allows conspecifics to recognize each other. Snakes probably rely primarily upon scent to identify the species and sex of other snakes. Male snakes in a number of families engage in a type of wrestling match inappropriately called the combat dance. The snakes intertwine the foreparts of their bodies and try to wrestle each other to the ground (Figure 14–16). When one snake has pinned the other, the loser crawls away. The snakes do not bite each other during these encounters; like most intraspecific combats these contests minimize the risk of injury.

The South American forest tortoise (*Geochelone denticulata*) and the red-legged tortoise (*G. carbonaria*) show a similar sort of behavior. During the breeding season male tortoises move actively about and challenge any object they see that might be another tortoise. The challenge is conveyed by moving the head to the side and back to the midline. In the forest tortoise this is a smooth, continuous movement, but in the red-legged tortoise it is divided into a series of short jerks. When a male tortoise is challenged in this manner, it responds with the head movement typical of its species. If the challenger is answered by the "wrong" head movement, indicating that it has challenged a male of the other species, both individuals lose interest and move off. If the response to the challenge indicates that the two tortoises are males of the same species, they fight by ramming their shells together and trying to overturn one another.

After the fight the winner wanders on and the loser, after turning himself right-side up, scurries away. A female tortoise gives no response to a male's challenge, so when the challenged object does not bob back, the male walks up to it and investigates. These tortoises do not appear to have very keen vision, and the object often turns out to be a rock or clump of grass. If it is a female tortoise, the male sniffs her cloacal region and if she is the same species as the male he begins to court her. Some lizards, like snakes, are primarily oriented to scent and show little species-specific behavior or color. Other lizards, particularly the iguanids of the New World and the very similar agamids of the Old World, are primarily visually oriented and have bright colors that can be displayed in species-specific behavior patterns. Because humans are visually oriented animals,

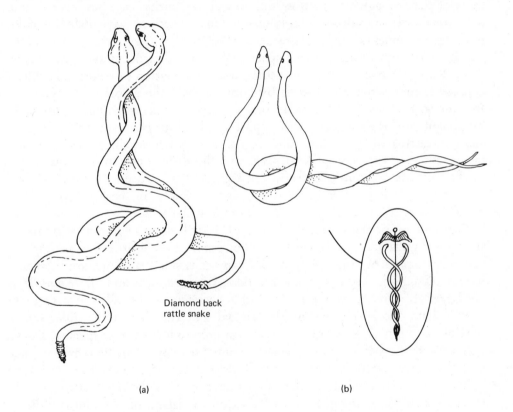

Diamond back
rattle snake

(a)   (b)

Figure 14-16 Combat "dance" of male snakes. The word "dance" is a misnomer that derives from early observers who regarded this behavior as a part of courtship. Later more careful observation revealed that the snakes involved were always males. The snakes do not bite each other, but try to pin their opponent to the ground. (a) Combat posture of the western diamondback rattlesnake. (b) Combat posture of the Aesculapian snake of Europe. This species of snake was associated in Roman mythology with Aesculapias, the god of healing, and was kept in his shrines. The lyre-shaped configuration formed by the foreparts of the bodies of the two snakes is probably the source from which the caduceus was derived. (modified from C. M. Bogert and V. D. Roth [1966] *American Museum Novitates* no. 2245.)

we readily perceive visual displays and this aspect of lizard behavior has received a great deal of attention.

In most cases the bright colors that are used to identify a lizard's species and sex are concealed most of the time. Many lizards have contrasting markings on the ventral surface, which are inconspicuous except when the lizard flattens its body to display them. In other lizards the gular fold under the throat is brightly colored and can be protruded by extending the hyoid apparatus to create a very conspicuous colored signal.

14.5.2 Species Recognition and Redundancy

One of the striking features of the species-specific displays of vertebrates is the fact that enough information is conveyed to identify a species several times over. In lizards the information is encoded in a variety of ways including color patterns, movements, and the site chosen for display. Figure 14–17 shows the colors of the gular fans of eight species of *Anolis* that occur sympatrically in Costa Rica. No two of them have the same combination of colors, and thus it is possible to identify any species by knowing only the colors of its gular fan. In addition, each species has a behavioral display that consists of raising the body by straightening the forelegs (called a pushup), bobbing the head, and extending and contracting the gular fan. The combination of these three sorts of movements allows a very complex display, and the **display action patterns** (DAP) of the Costa Rican *Anolis* are shown in Figure 14–17. Again, no two DAPs are the same, so it is possible to identify a species solely on the basis of its DAP without knowing anything about the colors of its gular fan.

Examination of the lizards in Figure 14–17 will show a wide variation in the size of the gular fan in relation to the body size of the lizard; in *A. pentaprion* and *A. cupreus* it is large; in *A. tropidolepis* it is small; and in the others it is intermediate. The body size itself varies—*A. pentaprion* and *A. tropidolepis* are large; *A. humilis, A. limifrons,* and *A. sericeus* are small; the others are intermediate. Both the size of the animal and the size of the gular fan in relation to the body size are cues that can help to identify the species. Dorsal pattern and the perch from which the animal displays provide more information. In short, every time an animal displays it provides many times as much information as is needed to identify its species. Calculations based on eight species of *Anolis* that occur at La Palma on the island of Hispaniola indicate that enough information is present in their displays to allow at least 500 species to be individually identified.

There are fewer than 500 species of *Anolis* in the world. Why do eight sympatric species on one small island need so much redundancy in their species-specific displays? Part of the answer probably lies in the fact that not all the channels of information function at once. For example, in a forest a lizard may be visible only as a silhouette against a patch of light. In that situation colors would not be visible, but the body and gular fan outline and the DAP would identify the species. Conversely, in dense vegetation it might be impossible to see anything except the color of the gular fan, and so on.

Redundancy presumably ensures that the message gets through even if many of the channels on which it is broadcast are blocked. This hypothesis permits predictions that should be tested in the field. For example, species that occur sympatrically with many other species should have more complex DAPs and color patterns than species that occur alone or with only a few other species. Similarly, a species with a broad geographic range that comes into sympatry with a large number of other species in different parts of its range might have more redundancy than a species with a restricted distribution, or it might show geographic variation. Finally, species in habitats where vision is restricted, such as deep forest, might require more redundancy than species that live in open habitats.

14.6   REPRODUCTION AND CARE OF YOUNG

Reptiles have evolved a large variety of reproductive modes, each advantageous in a particular situation. The terms "oviparous," "ovoviviparous," and "viviparous" have been used in an attempt to distinguish different reproductive modes. Unfortunately the terms have been applied loosely, and reptiles themselves do not fall conveniently into three clear-cut categories, so the result has been more confusion than enlightenment. In this section we will distinguish only between oviparity, meaning that the female lays eggs, and viviparity, meaning that she produces living young.

14.6.1   Turtles and Crocodilians

All turtles and crocodilians are oviparous. Female turtles use their hind legs to excavate a nest in sand or soil. The number of eggs in a clutch ranges from four or five in small species to more than 100 in the largest sea turtles. In some species with broad geographic ranges, the reproductive pattern shows adaptation to local climatic conditions. For example, painted turtles (*Chrysemys picta*) are distributed in North America from Canada to the southern United States. In Wisconsin an individual female lays two clutches each containing about 11 eggs, whereas in Louisiana where the annual period of activity is much longer each female lays four or five clutches with four eggs in each. The total number of eggs produced in a year is approximately the same in both regions, but the southern populations take advantage of the longer activity season by not putting all their eggs in two baskets. When a nest is lost, either to a predator or by flooding, all the eggs are likely to be killed. A single nest discovered by predators would eliminate half of a female's annual reproductive effort in the Wisconsin population but only a quarter or a fifth of that of a turtle in Louisiana.

Turtles exhibit no parental care of either the nest or young. In some cases interactions between the young may be essential to allow them to leave the nest. Sea turtle nests are quite deep, the eggs may be buried 50 cm beneath the sand,

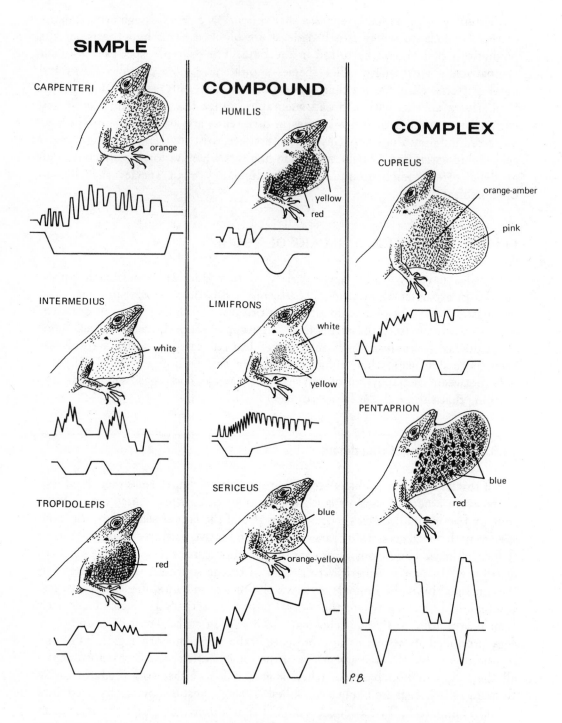

Figure 14-17 Species-specific displays of *Anolis* lizards. Many characteristics of a lizard enter into its display. Eight sympatric species of *Anolis* from Costa Rica can be separated into three groups on the basis of the size and color pattern of their gular fans. *Simple* fans are unicolored. *Compound* fans are bicolored,

and the hatchling turtles must struggle upward through the sand to the surface. After several weeks of incubation the eggs all hatch within a period of a few hours, and the hundred or so baby turtles find themselves in a small chamber at the bottom of the nest hole. Spontaneous activity by a few individuals sets the whole group into motion, crawling over and under each other. The turtles at the top of the pile scrape sand from the roof of the chamber as they scramble about, and the loosened sand filters down through the mass of baby turtles to the bottom of the chamber.

Periods of a few minutes of frantic activity are interspersed by periods of rest, possibly because the turtles' exertions reduce the concentration of oxygen in the nest and they must wait for more oxygen to diffuse into the nest from the surrounding sand. Gradually the entire group of turtles moves upward through the sand as a unit until it reaches the surface. As the baby turtles approach the surface, high sand temperatures inhibit further activity, and they wait a few centimeters below the surface until night when a decline in temperature triggers emergence. All of the babies emerge from a nest in a very brief period, and all of the babies in different nests that are ready to emerge on a given night leave their nests at almost the same time because their behavior is cued by temperature. The result is a sudden appearance of hundreds or even thousands of baby turtles on the beach, each one crawling toward the ocean as fast as it can.

Simultaneous emergence is a very important feature of the reproduction of sea turtles, because the babies suffer severe mortality crossing the few meters of beach and surf. Terrestrial predators await their appearance—crabs, foxes, raccoons, and other predators gather at the turtles' breeding beaches at hatching time. Some of the predators come from distant places to the beaches at hatching time in order to prey on the baby turtles. In the surf, sharks and bony fish patrol the beach. Few, if any, baby turtles would get past that gauntlet of predators if it were not for the simultaneous emergence that brings all of the babies out at once and temporarily swamps the predators. Since any babies that appear either early or late are virtually certain to be eaten, it is clear that selection has strongly favored precise timing.

In contrast to the absence of parental care among turtles, all crocodilians probably provide some protection of the nest and in some species quite extensive

---

and *Complex* fans are bicolored and very large. Fan color alone distinguishes all of these eight species individually—no two have the same combination of colors. Fan shape plus fan size/body size ratio also distinguishes several species individually. The display action patterns (DAP) are also species-specific. The DAP of each species is represented graphically beneath the drawing of the lizard. In the convention used to record DAPs, the horizontal axis is time (the duration of these displays is about 10 sec) and the vertical axis is vertical height. The upper line represents up and down movements of the head, and the lower line represents downward extensions of the gular fan, which are coordinated with head movements. No two species have the same DAP. (Modified from A. A. Echelle *et al.* [1971] *Herpetologica* 27:221–288.

care of the young has been described. Female crocodilians build one of two types of nest. Species that occur in marshy habitats, like the American alligator, build elevated nests by dragging together accumulations of vegetation. Other species, like the American crocodile, nest where the ground is high enough to dig a nest in the soil without danger of its flooding. Guarding of the nest by the female has been described or inferred in several species and may be universal. The female's presence may deter predators, and if the nest is threatened the female may attack. A recent example of this behavior by an alligator in Ever- glades National Park was described in a herpetological journal: "At the site discussed here, visitors were kept 5 m from the nest by barricades. On one occasion, however, a man left the roadbed and went down to the nest. In this instance the guarding female chased him up the bank and 30 m along the road before stopping. Whether this action was an attack or prolonged bluff could really have been decided only if the man had stopped and opened himself to attack."

The female American alligator bites the sides off the nest to release the young when she hears them squeaking. She then picks the babies up in her mouth, a few at a time, and carries them to the water where she releases them. This process is repeated until all of the hatchlings have been carried from the nest to the water. In some species of crocodilians one or both of the parents may gently crush eggshells with their teeth to help the young escape. Young crocodilians stay near their mother for a considerable period—for two years in the American alligator—and may feed on small pieces of food the female drops in the process of eating. They are also capable of catching their own food and are not dependent upon the female for nutrition. The distress squeak of a baby crocodilian stimulates both adult males and females to attack a predator. A female American alligator has been observed to rescue babies giving the distress call by picking one up in her mouth and carrying it to the water, and then returning to pick up a second baby. The parental care displayed by crocodilians appears to be as complex and widespread as that of birds, and studies of other reptiles are revealing more complex behavior than had been suspected. These discoveries discredit the hypothesis that behavior patterns inferred for some dinosaurs were too complex to have existed in ectothermal animals.

14.6.2   Snakes and Lizards

It is among the squamate reptiles that the distinction between oviparity, ovoviviparity, and viviparity is blurred. In general reproductive modes follow taxonomic lines, but there are numerous exceptions. One of the characteristics that distinguishes pythons (Family Boidae, Subfamily Pythoninae) from boas (Subfamily Boinae) is that pythons lay eggs whereas boas are viviparous.

As one travels north or south from the equator, the proportion of viviparous species of snakes increases. A female snake that retains eggs in her body can bask in the sun, thus raising the temperature of the eggs and speeding their development. The increased rate of development becomes increasingly important as one moves into cooler areas where warm microhabitats in which to deposit the eggs are harder to find. For the same reason there is an increase in the proportion of viviparous species as one moves upward in elevation. In Arizona, for example, the several species of horned lizards that live in the desert flatlands are oviparous, but the mountain species is viviparous. Not many snakes or lizards guard the eggs or nest site, although there are some that do. If a snake belongs to a species that is relatively immune from predation because it is large or venomous, the eggs may have a greater chance of survival if they are retained in the female until they hatch. This line of reasoning may explain why most vipers are viviparous.

14.7   THE ROLE OF ECTOTHERMAL VERTEBRATES IN TERRESTRIAL ECOSYSTEMS

In Chapter 12 we called attention to the seemingly paradoxical replacement of therapsids—which appeared to have reached a prototherian structural grade—by the reptilian archosaurs as the dominant members of the terrestrial fauna during the Jurassic and Cretaceous. Not until the Paleocene did large mammals appear and take over the herbivorous and carnivorous niches occupied by dinosaurs. Biologists' interpretations of that paradox vary. (Even among the authors of this book there are different opinions.) Some paleontologists feel that replacement of mammals by reptiles is so unlikely an event that our conclusions about the physiological and ecological characteristics of therapsids must be incorrect. Despite the gross as well as detailed skeletal similarities of therapsids and mammals, this view maintains that therapsids were physiologically and ecologically more reptilian than mammalian, and they were replaced by other reptiles which, in their turn, were replaced by mammals at the end of the Mesozoic when mammalian physiology had evolved. Other biologists find the evidence that therapsids had a capacity for endothermy comparable to that of prototherians too strong to dismiss.

Terrestrial endotherms and ectotherms play very different roles in an ecosystem. Ectotherms have low metabolic rates and low energy requirements. The vertebrate ectotherms in a terrestrial ecosystem generally consume far less than 1 percent of the net primary production of the ecosystem. In contrast, endotherms have high metabolic rates, and consume larger portions of the primary production of terrestrial ecosystems.

Ectotherms and endotherms differ also in the allocation of energy they ingest. (Table 14.4). Endotherms expend more than 90 percent of the energy they take in to produce heat to maintain their high body temperatures. Less than 10 percent, often as little as 1 percent, of the energy they assimilate is available for net production (that is, increasing their species' biomass by growth of an individual or production of young). Ectotherms do not rely on metabolic heat

Table 14-4 Efficiency of biomass production by tetrapods. These are net production efficiencies calculated as $\dfrac{\text{calories produced}}{\text{calories assimilated}} \times 100$.

Endotherms		Ectotherms	
Species	Production efficiency	Species	Production efficiency
Kangaroo rat (*Dipodomys merriami*)	0.8	Red-backed salamander (*Plethodon cinereus*)	48
Least weasel (*Mustela rixosa*)	2.3	Mountain salamander (*Desmognathus ochrophaeus*)	76–98
Field mouse (*Peromyscus polionotus*)	1.8	Panamanian anole (*Anolis limifrons*)	23–38
Meadow vole (*Microtus pennsylvanicus*)	3.0	Side-blotched lizard (*Uta stansburiana*)	18–25
Red squirrel (*Tamiasciurus hudsonicus*)	1.3	Hognosed snake (*Heterodon contortrix*)	81
Savannah sparrow (*Passericulus sandwichensis*)	1.1	Python (*Python curtus*)	6–33
Marsh wren (*Telmatodytes palustris*)	0.5	Adder (*Vipera berus*)	49
Average of 19 species	1.4	Average of 12 species	50

From a summary presented in Pough [7].

the solar energy they use to warm their bodies is "free" in the sense that it is not drawn from their food. Thus, most of the energy they ingest is converted into the biomass of their species. Values for net productivity of amphibians and reptiles are between 30 and 90 percent. As a result, a given amount of chemical energy "invested" in an ectotherm produces a much larger biomass return than it would have from an endotherm. A study of salamanders in the Hubbard Brook Experimental Forest in New Hampshire showed that, although their energy consumption was only 20 percent that of the birds or small mammals in the watershed, their productivity was so great that the annual increment of salamander biomass was equal to that of birds or small mammals.

In exchange for using most of their energy to produce heat, endotherms achieve a greater independence of environmental constraints than do ectotherms. Endotherms can be active on cloudy days, in cold habitats, and during the winter—all situations in which activity of terrestrial ectotherms is severely limited or entirely eliminated. Furthermore the greater concentrations of mitochondria found in endotherms allow them to produce more energy per unit weight of body tissue than an ectotherm and thus to be more active. Indeed, we have suggested that it was the importance of high levels of activity to predatory therapsids that provided the selection that led to the evolution of endothermy (Section 12.5.2 and 18.1).

Independence of environmental conditions is a valuable feature in climates that have characterized the Cenozoic, and its value is amply demonstrated by the current predominance of mammals and birds compared to reptiles and amphibians in terrestrial habitats. Mesozoic climates were characterized by extreme equability (Chapter 11). The average temperature was several degrees warmer, and seasonal and latitudinal variation was slight. Under those conditions, computer simulations have shown that large reptiles could have maintained very stable body temperatures solely by virtue of their great thermal inertia (Section 13.4). They had no need to expend their ingested chemical energy to produce heat, and as a result they could have put that energy into producing more biomass. It takes as much energy to maintain a single 1-kg mammal for a day as it does to maintain twelve 1-kg reptiles, and a herd of 10 or 11 cow-size dinosaurs could be supported for the energy cost of maintaining a single cow. (Table 14.5).

This is the sort of competition that an endothermal therapsid may have faced at the end of the Triassic. Reptiles would have been capable of maintaining stable body temperatures without a great investment of energy and consequently would have been able to multiply their own species' biomass many times faster than therapsids could. Therapsids were presumably capable of achieving and maintaining higher levels of energy production during activity than predatory reptiles could, but in the Triassic there would be little significance to that difference because the prey was reptilian. A reptile can prey on other reptiles as well as a mammal can. Probably therapsids received little benefit from the energy they invested in a high metabolic rate. Under those circumstances, replacement of mammalian therapsids by the increasingly agile reptilian archosaurs is plausible.

Table 14.5   Daily energy cost for reptiles and mammals of the same body mass. [It costs more energy to maintain a given mass of tissue in an endotherm (mammal) than it does in an ectotherm (reptile).]

Body Mass	*(kcal Energy per animal per day)* Mammal	Reptile	Ratio	*Mammal / Reptile*
10 g	2.5	0.2		13:1
100 g	13.8	1.1		13:1
1 kg	77.8	6.5		12:1
10 kg	437.8	38.5		11:1
100 kg	2,461.7	226.7		11:1
1,000 kg	13,843.2	1335.0		10:1
10,000 kg	77,846.0	7861.4		10:1

Calculations are based on weight-metabolism relationships for reptiles at 30°C (Bennett and Dawson. [1976] in *Biology of the Reptilia*, vol 5, edited by Carl Gans and W. R. Dawson, Academic Press, New York) and placental mammals (Schmidt-Nielsen [1975] *Animal Physiology*. Cambridge University Press, New York).

This view may also explain the persistence of a few mammals through the Mesozoic and their radiation and replacement of reptiles in the Cenozoic. Large reptiles, as we have said, have great thermal inertia and their temperatures change slowly. The time constant of temperature change for even a moderate-sized dinosaur (of 1 m body diameter) is so great that there is very little fall in body temperature during the night (Table 13-2). Thus, large reptiles should not have been dependent upon the sun to allow them to be active. Small reptiles—juvenile dinosaurs and lizards—change temperature rapidly and would have cooled off at night. They probably relied upon behavioral and physiological thermoregulation in the Mesozoic just as modern forms do, and their activity was probably limited to daytime. This limitation of activity in small reptiles probably provided the only niche that mammals were able to exploit during the Mesozoic. Investment of energy in endothermy allowed mammals nocturnal activity that was not possible for small reptiles. Mesozoic mammals were rat-sized or smaller and probably were nocturnal. They could have preyed upon small reptiles at night while the reptiles were lethargic (see section 18.1).

The change in climate that marked the end of the Mesozoic altered the competitive balance between reptiles and mammals. In the slightly cooler and much more variable climate of the Cenozoic, investment of energy in endothermy became profitable. The resulting freedom from environmental constraints allowed mammals and birds to exploit niches unavailable to reptiles. Among reptiles large body size changed from an advantage to a handicap. In the equable climate of the Mesozoic large body size meant that an animal never got too hot or cold, but in the cool, variable Cenozoic climate it prevented large animals from getting warm. The largest living reptiles are limited to equatorial habitats where water buffers temperature changes and provides a retreat from temperature extremes. As one moves away from the equator or to higher altitudes, the number of species of reptiles and their average body size both decrease. In this context large, tropical reptiles like giant tortoises and crocodilians can plausibly be regarded as survivors of the Mesozoic Age of Reptiles. Considering all modern reptiles in that light is not justified because it overlooks the specializations that have made reptiles successful in the Cenozoic world. The most abundant and diverse modern reptiles, the lizards and snakes, are those that exploit the advantages of ectothermy. It is these advantages that have allowed squamate reptiles to occupy adaptive zones that are not available to birds or mammals.

References

[1]   Bellairs, A. 1970. *The Life of Reptiles.* Universe Books, New York. A readable review of most aspects of reptile biology.

[2]   Cowles, R. B. and C. M. Bogert. 1944. A preliminary study of the thermal requirements of desert reptiles. *Bulletin of the American Museum of Natural History*, vol. 83, article 5, pp. 261–296. (This important reference has been republished by The Society

for the Study of Amphibians and Reptiles in their "Miscellaneous Publications" series [1974].) This is a classic paper that served as the basis for most studies of reptilian thermoregulation.

[3]   Gans, C. 1974. *Biomechanics: An Approach to Vertebrate Biology*. J. B. Lippincott, Philadelphia. Most of the examples are drawn from reptiles, and snakes in particular are considered in detail. This is one of the few sources of information about amphisbaenids outside of scientific journals.

[4]   Gans, C. and others. 1969– . *The Biology of the Reptilia*. Academic Press, New York. This continuing series of volumes contains review papers covering a wide range of topics, especially morphology, ecology and physiology.

[5]   Heath, J. E. 1965. Temperature regulation and diurnal activity in horned lizards. *University of California Publications in Zoology* 64(3):97–136. An outstanding account of behavioral thermoregulation in a lizard.

[6]   Mertens, R. 1960. *The World of Amphibians and Reptiles*. McGraw-Hill, New York. A readable well illustrated account of reptilian biology.

[7]   Pough, F. H. 1980. The advantages of ectothermy for tetrapods. *The American Naturalist* , vol. 115.

[8]   Schmidt, K. P. and R. F. Inger. 1957. *Living Reptiles of the World*. Doubleday, New York. A largely taxonomic account of living reptiles with excellent photographs.

15

The Geology and Ecology of the Cenozoic

Synopsis: The origins and, especially, the early evolution of birds and mammals were as much a consequence of the geological events of the late Mesozoic and early Cretaceous as they were the results of competition with the Mesozoic ruling reptiles. Fossils of ancient birds are scant, but fossil mammals present us with a reasonable picture of how geologic events probably influenced their radiation. The central event was the Jurassic rupture of Pangaea initially isolating the huge Northern land mass of Laurasia from its southern counterpart, Gondwanaland. By the early Cretaceous the North Atlantic Ocean and the Caribbean Sea were formed as North America separated from South America and Europe. By the Middle Cretaceous, India and Africa had split away from the remainder of Gondwanaland. South America, Antarctica and Australia, however, retained land connections until the early Cenozoic.

This sequence of events led to isolation of the earliest mammals, and presumably also of birds, on the different drifting land masses. Various theories concerning the types of mammals trapped by these isolations and how the isolations have led to the distributions of modern mammals have been proposed. Three major faunistic assemblages are usually recognized: a Laurasian groupin that includes all the northern continents and Africa; a South American grouping; and an Australian grouping. Successful evolutionary theories must account for the presence of essentially a pure placental mammalian assemblage in Laurasia, a mixed placental and marsupial assemblage in South America, and a virtually pure marsupial assemblage in Australia. The modern concepts of continental drift have begun to unravel the bases of these unusual distributions of mammals.

15.1 CENOZOIC TIME—THE AGE OF MAMMALS

The Cenozoic era covers the last 65 million years, a period when the geography and faunas of the world again changed and began to assume their present characters. The kinds of organisms and the relationships between the

515

continents we observe today are, of course, also the result of what occurred prior to Cenozoic time. But a great deal of vertebrate evolution has occurred over the last 65 million years, especially in endothermic vertebrates—the mammals and birds. It is appropriate to examine the continental movements and climates in relationship to the geologic periods covering the evolution of birds as well as mammals. However, the fossil record for birds is still so poorly known that relationships are somewhat unclear. In contrast, the fossil record of mammals is far better. As a result, we shall emphasize in this chapter how late Mesozoic and especially Tertiary geologic changes have affected mammalian evolution, keeping in mind that birds were evolving simultaneously. What the sparse fossil record tells us about birds will be examined in Chapter 16.

The Cenozoic era is often called the "Age of Mammals". Following their origin in the Triassic and humble persistence through the Mesozoic, mammals exploded in the Cenozoic and evolved the astonishing array of forms we know today. Extinction of the ruling reptiles in late Cretaceous time opened a vast series of niches for the mammals to occupy and to exploit. They did not "wait" long. In excess of 150 genera of fossil mammals are recognized in Paleocene rocks. The American vertebrate paleontologist, George Gaylord Simpson, compiled one of the first lists of the Cenozoic mammals that numerically documents their expansion during the Tertiary epochs. Using more recent data we have repeated his work (Figure 15-1). Although the rate of appearance of new genera increased throughout the Tertiary, there were two peaks in total number of genera—the first in the Eocene, the second in the Miocene-Pliocene. The peaks reflect what has long been accepted, that the radiation of mammals has occurred in two waves: an early Tertiary radiation in which a series of rather primitive mammals evolved, and a later Tertiary radiation in which the primitive forms were replaced by more modern mammals. Professor Simpson's analyses are compilations for all Cenozoic mammals. They approximate what occurred in the northern continents (that is, Laurasia) and in South America. They cannot be applied to Africa and Australia because very early Cenozoic mammals have yet to be found on these continents.

Today the mammals of the northern continents, of Africa, of South America, and of Australia are strikingly different. They were also distinct in the early Cenozoic. These four geographic groupings fall into three different faunal lineages: (1) a Laurasian fauna, here taken to include the northern continents plus Africa (often considered a separate Ethiopian fauna), (2) a South American fauna, and (3) an Australian fauna. The isolation of these three regions is reflected in the distinctive composition of the basic types of mammal species present. In Laurasia and Africa the first Cenozoic mammals were predominantly eutherians (placentals; see Chapter 18.1). They remain so today. In South America the early Cenozoic saw a mixed fauna, balanced between eutherians and metatherians (marsupials). Australia began with a metatherian lineage that has been upset only slightly by later arriving eutherian migrants. What produced these distinctive faunal differences?

They have primarily resulted from the evolution of the three stocks in isola-

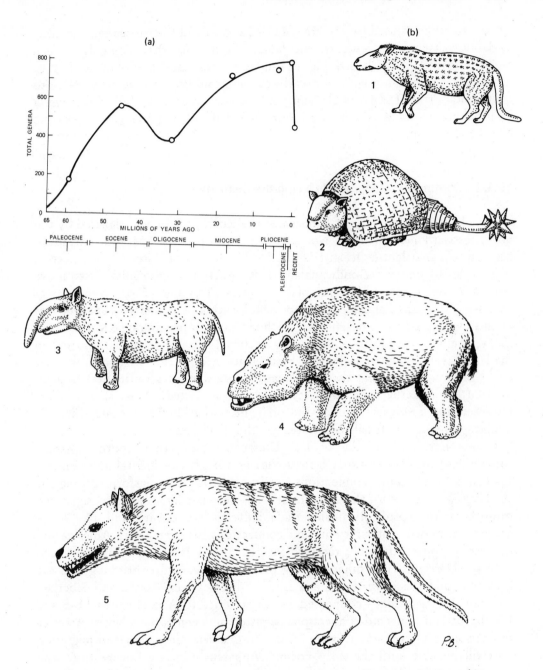

Figure 15-1 (a) Increase in mammalian genera during the Cenozoic. Note the bimodal peak in total genera. (b) Some extinct mammals of the Cenozoic. (1) *Phenocodus*, a Paleocene and Eocene herbivorous condylarth of Europe and North America. (2), (3), and (4): South American mammals evolved on that continent during its long isolation. (2) One of the armored glyptodonts, mid-Eocene to Pleistocene edentates. (3) Elephantlike *Pyrotherium*, a late Eocene amblypod. (4) Rhinoceroslike *Toxodon*, a Pliocene and Pleistocene notoungulate. (5) *Andrewsarchus*, a late Eocene condylarth from Mongolia that appears to have been carnivorous. Not to the same scale.

tion, a situation caused by the breakup of Pangaea and the subsequent splitting of Gondwanaland. In spite of the faunal differences, the mammals in each region provide us with striking examples of convergent evolution. Let us examine mammalian origins and early distributions to see how they relate to continental drift. The basic elements of this story are simple, but it has several interlocking plots. We begin by examining drift during the later Mesozoic and the Cenozoic.

15.1.1 Continental Geography During the Cenozoic

Recall that the breakup of Pangaea in the early Jurassic was signaled by the northwestward movement of North America to open the ancestral Caribbean Sea and North Atlantic Ocean (Figure 11-4). Somewhat later in the Jurassic, rifts began to open in Gondwanaland (Figure 11-4). India split off to move northward on a separate oceanic plate (Figures 11-5 and 15-4), eventually to coalesce with Eurasia and occupy its present location. The inertial consequences of this event produced the Himalayas, the highest mountain range in the world today. Actually, mild orogenic activity occurred as early as the late Eocene in the Himalayan region of India, but reached its peak in the Oligocene. Recent investigations suggest that the Himalayas represent an intercontinental orogeny uplifted before actual land connections with Asia existed. As we shall see, this interpretation resolves some of the difficulties that India presents when attempting to relate its fossil vertebrates to other land masses.

It was not until middle to late Cretaceous time that South America-Antarctica-Australia separated from Africa as a single continental unit (Figures 11-4 and 11-5). Land connections persisted across this platform during the Paleocene and, perhaps, the early Eocene. Somewhat later Australia separated from Antarctica, probably during the middle to late Eocene and, like India, drifted northward. Intermittent land connections between South America and Antarctica apparently were retained via the Scotia Island arc until recently.

In the Northern Hemisphere during the Cretaceous and continuing into the Paleocene and the early Eocene a land bridge between Alaska and Siberia—the trans-Bering Bridge—connected North America to Asia (Figures 11-5 and 15-2). Intermittent land connections persisted between eastern North America and Europe via Greenland (Figure 15-2). In Laurasia, then, overland migration was still possible until the early Eocene. Apparently in late Eocene time, and definitely in the Oligocene, the land bridges were submerged (Figure 15-3). However, they emerged again in the Miocene.

These massive continental rearrangements also modified the oceans. The Atlantic Ocean widened as did the southern oceanic connection with the newly forming Indian Ocean (Figure 15-2). The northwestward drift of North and South America diminished the size of the Pacific Ocean. The creation of new seas and the reduction in size of older seas disrupted oceanic circulation patterns and, as a result, changed climates. The equatorial heat input from the sun

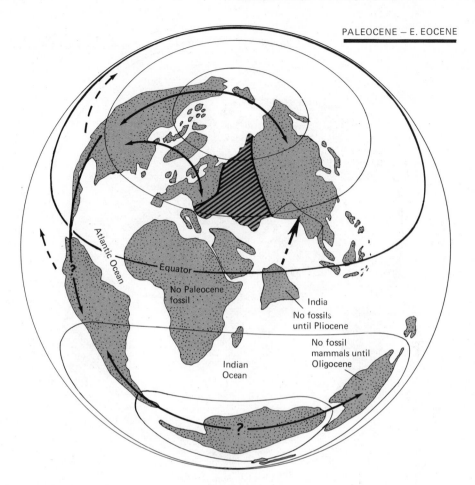

Figure 15-2 Continental positions in the late Paleocene and the early Eocene. Shaded areas are land, hatched areas epeiric seas. Cenozoic fossil mammals are known in Laurasia and South America from the beginning of the Paleocene, in Africa from the Eocene, in Australia from the Oligocene, and in India from the Pliocene. Solid arrows are major land bridges. Dashed arrows show direction of drift.

would presumably have been transported more readily to polar seas in the single huge ocean that surrounded Pangaea than in the more restricted ocean basins of today (Figures 11–3 and 15–4). On the other hand, the breakup of Pangaea increased maritime borders and arranged continental interiors so they were closer to the air-moistening seas. This would favor more equable continental climates than existed in central Pangaea, because water evaporated over the sea and returned as rain on nearby land masses.

Whatever the effects of the changing seas and continental land areas, paleo-climatologists conclude that the general warm and humid conditions typical of the Jurassic and early to middle Cretaceous changed toward the end of the Cretaceous when widespread moderate cooling took place. The Paleocene continued this trend, but was followed by warming in the Eocene. From the

Figure 15-3 Continental positions in the late Eocene and the Oligocene. Shaded areas are land. Solid arrows stress migratory routes for mammals, cross marks on arrows represent occasional or chance dispersal routes. Dashed arrows indicate direction of continental movement.

Oligocene to Pliocene, general cooling continued and was capped in the Pleistocene by a period of major glaciation. Overall, the Cenozoic climate tended to become cooler and habitats changed. Paleobotanical studies regularly show that forest boundaries were displaced southward in North America during the Cenozoic.

15.2 EARLY TERTIARY MAMMALS IN LAURASIA

The first Tertiary mammals are found in early Paleocene rocks in North America. There were three major mammalian types (Figure 18-1). First, primitive pouched mammals or marsupials (metatheria), omnivorous and arboreal and not unlike modern opossums, existed. A second important primitive group were small placentals (eutheria), mostly insectivorous mammals (leptictids), much like shrews. From these early insectivores the greatest variety of living mammals have evolved. Another group were also placentals (deltatherids), not too far removed from the insectivores but with more diverse feeding habits. Each group persisted to give rise to modern mammalian orders. A third group, the multituberculata (named for their multicusped teeth), were also small mammals with nontherian teeth but rodentlike dental arrangements. They probably filled the ecological niche the rodents occupied after multituberculate extinction later in the Cenozoic (Fig. 18-1).

All four mammalian types also occurred earlier in late Cretaceous rocks. The characters that separate the placental leptictids and deltatherids (slight differences in dentition), however, are difficult to distinguish in the earlier Cretaceous fossils of North America. Nevertheless, in North America eutherian and metatherian fossils are recognizable as separate mammalian lineages. The nontherian multituberculates can be traced back to the middle Jurassic where they occurred with the pantotheres, the mammalian group that most vertebrate paleontologists believe gave rise to the placental and marsupial stocks in the very early Cretaceous.

If the fossil locations of these general groupings of mammals are examined from the early Cretaceous to the early Paleocene, an unusual biogeographic distribution unfolds. In the early Cretaceous rocks of England, Mongolia, and the famous Forestberg site in Texas, relatively unspecialized therians are found. In association with these early therians are several nontherian mammals characteristic of Jurassic time: triconodonts, symmetrodonts, and multituberculates. At Forestberg, however, differentiated therians are found and represent the earliest known marsupials and possibly also placentals. Did the metatheria and eutheria originate in North America or, at least, the Western Hemisphere? As so often happens, the paleontological record is not yet, and perhaps never will be, complete enough to provide a total answer. But it is likely that marsupials originated in or were limited to the Western Hemisphere.

A gap follows in the fossil record, for mammals are not found again in Cretaceous rocks until the end of the period. Very little has been found in European rocks so far—only a single tooth of questionable identity (possibly eutherian). In Mongolia and North America, however, fossils of late Cretaceous age are abundant. At both locations the triconodonts and symmetrodonts are absent. Only the multituberculates prevailed. These were diverse and many genera are recognized. The older nontherian mammals were disappearing, probably the result of competition with the more recent therians. Primitive insectivores were diverse and common in Mongolia and North America. In the Mongolian rocks an insectivorous form, *Zalambdalestes*, had several rabbitlike characteristics and probably was ancestral to that group. Deltatherid mammals with long canines (indicative of carnivorism) are found at each site. In fact, the fossil mammals found in the Mongolian and North American rocks are surprisingly similar. Their close relationship indicates that migration of land animals between east Asia and western North America was possible in the

Cretaceous. The existence of such a connection is also supported by the distribution of Cretaceous dinosaurs, the suborder Ceratopsia (Chapter 13), and the reconstructed location of continents (Figures 11–5 and 15–2).

In late Cretaceous and Paleocene time, a shallow epicontinental sea separated Europe from Asia (Figure 15–2). As a result, faunal interchange probably was limited and a lack of close mammalian faunal relationship between these regions is to be expected.

In late Cretaceous deposits, marsupials are found only in the North American rocks; they have not been found in either Mongolia or Europe. Apparently the broad land bridges that allowed intercontinental migration of the eutherians and multituberculates effectively stopped the marsupials. Why did these ancient land bridges act as selective filters, allowing only certain types of mammals to cross? No one is sure. It is true, however, that these land bridges occurred at high latitudes. Possibly the marsupial niches were so restrictive that they were unable to adapt or to find adequate shelter. Cooler boreal conditions may have excluded the marsupials or, alternatively, appropriate foods to sustain their passage may have been unavailable in these regions. We can really only guess. That this long journey from western North America to Mongolia would be executed quickly is unlikely. More probable would be a dispersal dependent on the limited territorial explorations of generations of these small mammals. This would involve immigration rates measured, at best, in kilometers per generation, not hundreds of kilometers.

This type of slow generational dispersal is made especially vivid by the recent movements of the opossum northward in eastern North America (Figure 15–4). Opossums on the northern limits of the species' distribution frequently suffer frostbite of ears and tail, and it may be that low temperatures per se are limiting their northward spread. Of course it would be possible for a chance dispersal to occur. But a unique event in which a male-female pair or even a pregnant female is transported over several thousand miles is highly improbable. Such rare chance dispersals have been referred to as "Sweepstakes Dispersals" by Simpson. As we shall find out, sweepstakes dispersals do occur, but they naturally require proper conditions at the new habitat if they are to produce a successful colony. If a chance dispersal of early marsupials from North America into Asia did occur, it did not succeed.

The important early Paleocene deposits of North America provide a continuous depositional sequence from the late Cretaceous into Tertiary time. A distinct change in the mammal fauna is observed. The marsupial fauna which was common and diverse in the late Cretaceous, become impoverished. In contrast, the eutherian fauna became very rich. The earliest Paleocene insectivores show morphological signs of radiations important in more recent faunas (primates, bats, and so on). Among the deltatherids were carnivorous forms that led to the more advanced Paleocene creodonts, and an herbivorous group that gave rise to the later condylarths (Figure 15–1 and section 18.2). Fossil indications of a change in faunal balance suggest that the marsupials were unable to compete effectively with the eutherians in North America. If so,

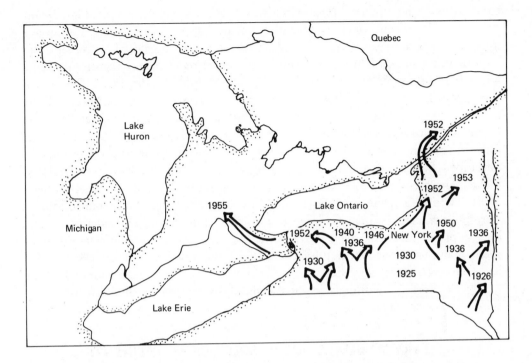

Figure 15-4 Northward invasion of the opossum (*Didelphis virginianus*) in eastern North America. Capture dates are placed over localities from which specimens were first obtained.   Direct introductions into northern areas made during the early 1900s were always unsuccessful. Subsequently, naturally occurring invasions into these same northern areas by expanding, locally adapted populations have been very successful. (Data from W. J. Hamilton 1958 *Cornell Univ. Agric. Exp. Sta. Mem.* 354:1–48, and R. L. Peterson 1966 *The Mammals of Eastern Canada,* Oxford University Press, Toronto.)

why were they so successful in North America in the Cretaceous and then suddenly much less successful in the Paleocene?

One explanation hinges on the possibility that the earliest eutherians originated in Asia and first radiated there, the marsupials in North America. The evolution of each group may have taken place isolated from that of the other. Filtering of early eutherians into North America from Asia across the Bering Bridge would have been slow and against competition offered by less specialized theria, as well as the early marsupials, all of which were small, shrewlike, and probably similar in requirements. Professor Simpson argues that the sudden increase in eutherians in North America occurred over too short a period of time to be explained by an *in situ* radiation of eutherians. He makes the persuasive point that the sudden appearance of new eutherians upset the harmonious community balance that had slowly evolved between the marsupials and the other nondifferentiated Cretaceous mammals of North America. This argument implies, however, that marsupials cannot compete with placentals for the same or closely related niches. As we shall see, this competitive "inferiority" to placentals has often (but not always) led to extinction of marsupials.

By late Paleocene and the Eocene a wide variety of different eutherians inhabited Laurasia (section 18.2). Most forms can be traced from the leptictid insectivores from which, for example, bats and primates clearly arose. Rodents were first common in the early Eocene and quickly displaced the multi-tuberculates. From the deltatherids arose the carnivorous creodonts and the herbivorous condylarths (Figure 15-1). All possessed large canines, but the cheek teeth of the herbivorous forms were modified for crushing plant materials. Although most of the early forms were clawed, some of the condylarths developed hoofs on their toes, suggesting that they filled an early "ungulate" niche. The creodonts and condylarths were initially of modest size, but some of the late Paleocene and Eocene forms—the uintatheres and pantodonts—were huge placentals, many as large as a modern rhinoceros. The paleontological record places many of these early eutherians over much of Laurasia. Most of these mammals were extinct by the end of the Eocene, although some (for example, creodonts) persisted through the Miocene. The establishment of the modern orders of mammals in Oligocene-Miocene time spelled the doom of these early eutherians.

15.3 TERTIARY MAMMALS OF THE SOUTHERN CONTINENTS

So far we have examined temporal changes only in mammals of the Northern Hemisphere. Did mammalian evolution follow a similar or a different course in the southern continents? Most of us are aware that the modern mammalian fauna of Australia differs from the faunas of Asia, Europe, Africa, and North America. The mammals of South America also differ, but less so than the Australian mammals. Obviously mammals must have followed different evolutionary paths in these two regions of the world. The differences reflect the long-term isolation of South America and Australia from Laurasia and Africa. In the Triassic, Pangaea existed and the first mammals (evolved from the therapsid reptiles) found worldwide migration possible. In fact, most land tetrapods were widely distributed (Table 11-1). During the Jurassic, the pantotheres evolved as Pangaea began to rupture, but widespread dispersal over land still remained possible (Figure 11-4). It was only in the Cretaceous that continental isolation became a reality (Figure 11-5). Continental isolation, therefore, coincided with the evolution of the therian mammals and their dichotomy into marsupial and placental lineages.

15.3.1 Cenozoic Isolation of South America

A sequence of exotic mammalian fossils are preserved in South America, mostly in Paleocene through Pleistocene rocks of Patagonia. As a result, vertebrate paleontologists have an excellent record of the types of mammals that existed there through the Cenozoic.

Only two basic groups seem to have been present in the early Paleocene—primitive marsupials and primitive placentals. The former were represented by small opossumlike creatures, the latter by two divergent stocks, one leading to edentates (sloths, anteaters, armadillos) and the other leading to a variety of bizarre hoofed animals (liptoterns, Figure 19-10; notoungulates and so on, Figure 15-1). Strangely, in South America the predatory carnivore role was not filled by placentals, as happened with Laurasia's creodonts, but rather by the marsupials. A diverse group, the "borhyaenids," evolved into formidable predators (Figure 19-9). By the miocene wolflike borhyaenids roamed South America and by the Pliocene saber-toothed marsupials stalked prey (Figure 19-11). The borhyaenids are one of the classic cases of convergent evolution, for, even though they were a very different subclass of mammals than those on any other land mass, they evolved grossly similar morphologic types that assumed the same ecological role—that of top carnivore.

The early placentals of South America also evolved a diverse group of omnivorous forms, the edentates (section 19.1.2 and Figures 15-1 and 19-1). This assemblage was without parallel in the rest of the world. In contrast, the herbivorous mammals of South America were derived from primitive condylarths, a group that also occupied Laurasia. Nevertheless, their lineage is distinct, since these early Tertiary condylarths evolved in complete isolation from those in Laurasia. There is a question, in fact, as to the actual affinities of the South American condylarths. Some suspect they were derived from early primitive (shrewlike) eutherians isolated in South America during the early Cretaceous when Pangaea first split. Others favor a view that they arrived somewhat later and in a more advanced (condylarth) eutherian condition by chance dispersals during the late Cretaceous or the Paleocene. The isolation of these "condylarth-like" mammals in South America produced a strange array of hoofed forms that usually are arranged into four groupings. Two of these taxonomic units, the litopterns and the notoungulates (Figures 15-1 and 19-10), gave rise to mammals that paralleled in morphology and habits the ungulates of Laurasia, the perissodactyls and artiodactyls. The third group, the pyrotheres (Figure 15-1) were large, unusual creatures that occupied a niche similar to the elephants and their relatives in Laurasia. The final group, the astrapotheres, were also large animals with unusual canine teeth. Their ecological role, in the sense of its convergence with Laurasian mammals, is unclear.

Three additional placental groups found their way into South America in the early Tertiary. Rodents are present in early Oligocene deposits, and these must have provided the stocks from which the capybara, guinea pig, porcupine, and chinchilla evolved. There are no earlier indications of rodents in South American rocks. It is likely, therefore, that this group reached South America by chance dispersal—perhaps by rafting on floating trees. South America also has an unusual group of monkeys. These monkeys, unlike the monkeys in Africa and Asia, possess a prehensile tail. The presence of three rather than two premolar teeth, a dental arrangement present only in the more primitive anthropoids of the Old World is also important. The earliest fossil monkeys in

South America occur in very late Oligocene rocks. Like the rodents they must have slipped into South America by chance. The procyonids (the third placental group in South America) dispersed during the Miocene and gave rise in South America to the modern coatis.

The rodents, monkeys, and procyonids were all relatively small placentals, conditions more favorable to successful sweepstakes dispersal by rafting. Further, in the Eocene deposits of North America, fossil rodents and primates have been found with close affinities to the early South American forms. It requires little imagination to envision the movement of the first of these differentiated placental stocks down island chains from North America and their chance dispersal into South America on floating trees and other debris. This vast array of diverse mammals was isolated from the rest of the world from late Cretaceous until Pliocene time, when the Isthmus of Panama rose above sea level and bridged North and South America.

The Panamanian land bridge was to lead to the extinction of much of the South American mammalian fauna. Migration of North American placentals into South America was rapid. The borhyaenids were quickly replaced by placental carnivores. The varied hoofed mammals of South America also gave way to competition from the grazing and browsing placentals from the north— deer, tapirs, and camels. Amongst the edentates, a few survived the arrival of the North American mammals. Only the armadillo and smaller sloths have survived to the present. The migration was not all one way, however. The large edentates, the glyptodonts and ground sloths, moved into North America. Like a variety of other mammals, however, they were probably extinguished in the Pleistocene by prehistoric man (Chapter 21). Armadillos and porcupines have been the most successful invaders from South America.

By and large, however, the Panamanian land bridge spelled the doom of the ancient mammals of South America. Again, as in the extinction of the North American Cretaceous marsupials (Section 15.2), we see an example of placentals displacing marsupials when they are suddenly brought together and compete. Of the entire marsupial fauna, only the Virginia opossum persists in North America and a few forms in South America (like the mouse opossum, *Marmosa*; the yapok, *Chironectes*; the opossum rat, *Caenolestes*). These marsupials share in common a small and generalized body structure, perhaps factors to which they owe their survival.

The long isolation of South America, although providing time for natural selection to produce a fascinating assemblage of unusual mammals all adapted to their particular situation, nevertheless produced a fauna that was limited in its adaptive plasticity. Why were the waves of North American placentals such effective competition for the resident mammals of South America? No doubt any answer must involve a higher grade of intelligence, more effective repro- duction, and more efficient feeding mechanisms on the part of the North American invaders. Most importantly, the mammals of Laurasia did not evolve in total isolation from each other. To the contrary, intermittent intercontinental migration continued through most of the Tertiary. As a result, the Laurasian mammal fauna evolved under the continuous onslaught of invading competitors from the far reaches of Laurasia (North America, Europe, Asia) and Africa; Figures 15-2 and 15-3.

In a sense the larger land mass of Laurasia provided a greater proving ground in which natural selection could mold and more finely hone the adaptive features of the mammals evolving there than was possible in South America. Thus, the fitness of the mammals in these two areas, although adequate in their respective homelands, when brought together largely favored those from the larger land mass. There is an important evolutionary lesson here—total isolation produces lineages with different morphological solutions to similar conditions. However, heightened interlineal competition, due to moderate discontinuous isolation, results in a highly refined evolutionary product. Thus, the effectiveness ("superiority") of the North American mammals compared to the South American forms is attributable not only to the larger land mass of Laurasia and of Africa in which they evolved but also to the fact that partial and not total isolation occurred. Competitively superior adaptations evolved as the North American and Old World faunas intermittently mixed through the Cenozoic.

15.3.2 Cenozoic Isolation of Australia

The mammalian fauna of Australia is unique. From its beginnings it has been composed almost entirely of marsupials and not placentals. Only late in the Cenozoic did limited groups of eutherians enter Australia. For reasons unknown, primitive placentals did not enter Australia. Our present knowledge of continental locations during the Mesozoic and Cenozoic allows us to guess why this happened (Figures 11-4, 11-5, and 15-2). There are several distinct time periods and different routes that may explain the entrance of marsupials into Australia.

1. Late Jurassic—Early therians arrived in Antarctica-Australia where the marsupials subsequently evolved as autochthonous (of local origin) forms.
2. Early to middle Cretaceous—Early marsupials arrived in Australia from northern regions and then radiated in isolation.
3. Paleocene—Marsupials entered Australia from South America.
4. Eocene—Chance dispersal of marsupials into Australia.

Each view (with variations) has had proponents. The first assumes a southern center of origin for marsupials, possibly Antarctica, and dispersal to South and North America and to Australia. Since the first mammals suspected of being marsupials occur in North America in the early Cretaceous, their origin in southern regions must have taken place by then or earlier in the late Jurassic. But why then have Cretaceous fossil marsupials not turned up in South America? They certainly are abundant in Cenozoic rocks! Another question arises—why are fossil marsupials unknown in Africa? If the marsupials originated in the late Jurassic in southern Gondwanaland, they should have spread into Africa, unless a barrier existed. The separation of the South America-Antarctica-Australia platform from Africa, which was final by middle Cretaceous (Figure 11–5), was underway and may have presented such a barrier. In Antarctica today massive ice sheets cover the continent and extreme cold deters the collection of fossils. The lack of knowledge is due mostly to the inaccessibility of fossils, not necessarily to their absence. Recent attempts in Antarctica, made by Professor Colbert and his colleagues, have met with stunning success with respect to Mesozoic reptiles but not mammals.

An interesting variation that resolves the absence of Cretaceous South American fossils considers marsupials as polyphyletic, one stock originating in North America and another in a southern region. Few paleontologists accept this view, since the modern American opossum and some of the small Australian dasyurid marsupials appear to have very close affinities.

The second suggestion of an early to middle Cretaceous arrival from the north is plausible because marsupials were established in North America and were expanding rapidly by the middle Cretaceous. To reach Australia at that time, a land route over South America and Antarctica may have existed (Figure 11–5). Again one wonders why Cretaceous fossil marsupials have not been found in South America. Also, why did only marsupials make it to Australia? Early Cretaceous nondifferentiated therians, and possibly even primitive eutherians, were present in North America. Yet they are unknown in South America. However, if eutherians originated in Asia and metatherians in North America in the early Cretaceous, as some suggest (section 15.2), the absence of early primitive eutherians in Australia can be resolved by their "too late" Upper Cretaceous arrival in North America over the trans-Bering Bridge. By the Upper Cretaceous, Antarctica straddled the South Pole (Figure 15–5) and not the more temperate latitude it occupied earlier in the Cretaceous and the late Jurassic (Figure 11–4). Any animals attempting to enter Australia across Antarctica would have encountered severe climatic difficulties. The marsupials therefore may have slipped into Australia just before the Antarctic became polar—a position it has retained through the Cenozoic.

The third concept requires that marsupials entered Australia from South America when Antarctica was in a polar position (Figure 15-2). Since eutherians were also present in South America, we must again consider why they did not make it. It seems unlikely that the only marsupials would manage to cross the severe Antarctic Bridge. Nevertheless, the singular presence of marsupials suggests that chance somehow played a large part in the origin of the Australian mammals.

The fourth suggestion rests on the concept that marsupials crossed into Australia by rafting over oceans in the late Paleocene or Eocene. The absence of eutherians is attributed purely to chance. This view is championed by Professor Simpson. The source of these rafted marsupials, Simpson suggests, was Southeast Asia. Perhaps this was the case, but the fossil evidence is sorely deficient. Fossil marsupials are known only from South America, North

America, Europe, and Australia. In Australia, fossils are unreported prior to the Oligocene, and by that time they were already diverse.

In summary, there is too little information to settle the question of exactly how the marsupials got to Australia. The answer will depend on additional fossil evidence about early mammals from Australia and, especially, from Antarctica, of which we know nothing at present. The information about continental separations does, however, provide considerable perspective to the question: it dictates in time and space whether a given concept is credible or not.

However the marsupials reached Australia, once there they enjoyed the privileges and liabilities of long-term isolation. As a result, a radiation of diverse types followed. The familiar kangaroos and wallabies, the koala, bandicoots, and the recently extinct Tasmanian wolf evolved to fill a variety of niches with food habits ranging from complete herbivority to carnivorism (Figure 3-4). During this period Australia drifted northward in isolation from an Eocene location entirely below 30°S latitude (Figure 15-2) to its present position between 10 and 45°S latitude.

First in the Miocene, and on several occasions since then, by chance the placental rodents have founded colonies in Australia. No doubt they were rafted from Southeast Asia. The probability of successful rafting increased as Australia continued to drift northward. The rodents have differentiated into a variety of types that fill ecological niches occupied elsewhere by such familiar forms as mice, voles, kangaroo rats, squirrels, and rabbits. The only other major group of placentals that reached Australia was the bats, an achievement obviously attributable to flight. Endemism has evolved in Australian bats, but as one might expect, it is less prevalent in these highly mobile and constantly intermingling forms than in the much less mobile Australian rodents. These later Cenozoic incursions have probably had little overall effect on the Australian marsupials. A far greater threat has been the prehistoric invasion by man and dog, which already has led to the extinction of the Tasmanian wolf and the grave endangerment of numerous other forms.

15.3.3 Africa—An Example of Periodic Isolation

In the Cretaceous, the Tethys Sea presumably separated Africa from the large northern continental complex of Laurasia (Figure 11-5). Africa, therefore, like South America and Australia, was isolated. In the Mesozoic rocks of Africa, as in North America, Europe and Asia, there is abundant evidence of a major reptilian fauna (Table 11-1). The early Jurasso-Cretaceous mammals of North America and Europe seem to have been present in Africa too, but as yet few fossils are known. The only Mesozoic mammals found to date in Africa are the Triassic docodont, *Erythroptherium*, and the upper Jurassic pantothere, *Brancatherium*. An enormous gap exists, for it is not until the late Eocene that fossil mammals are again encountered in Africa. But how surprising these

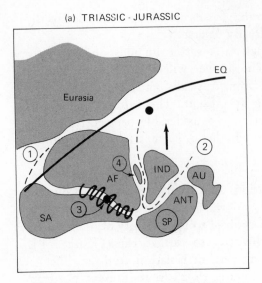

(a) TRIASSIC - JURASSIC

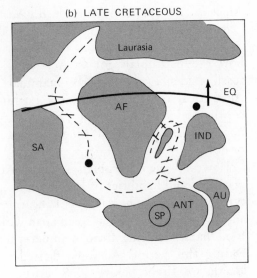

(b) LATE CRETACEOUS

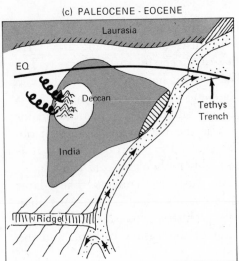

(c) PALEOCENE - EOCENE

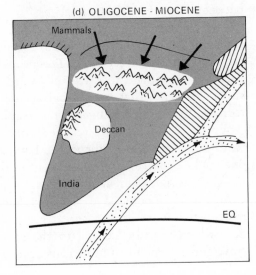

(d) OLIGOCENE - MIOCENE

Figure 15-5 The positions of India during the Mesozic and Cenozoic. (a) The rifts that ruptured Gondwana separated India by middle Jurassic. One, two, and three refer to the position and chronology of these rifts. (b) A suggested chronology of the breakup of Gondwana was (1) India, (2) SA + Ant + Au, (3) Madagascar from Africa. (c) India contacted the Tethys Trench in its northward movements. The Deccan plateau formed in passing over an Indian Ocean hot spot. No mammalian fossils are known. (d) Formation of the Himalayas from the Indian plate overriding the Tethys Trench and later collision with Laurasia. The Himalayas apparently acted as a prepositioned barrier to the infiltration of Laurasian mammals until the Paleocene. EQ = Equator, SP = South Pole, black circles = volcanic hot spots.

mammals are!—the moeritheres, the first proboscideans; the hyracoids, the early conies; the sirenians, the first sea cows (Figure 19-9).

These early distinctive mammals suggest that Africa during the Paleocene and Eocene was effectively isolated from Eurasia, in spite of its close proximity (Figure 15-2). But in the Oligocene, Miocene, and Pleistocene the faunal barriers eased periodically and mixing occurred. In the early Oligocene the earliest mastodons are found in Egypt, along with the endemic hyracoids and sirenians, and soon, also present are carnivorous creodonts and herbivorous anthracotheres from Eurasia. By Miocene time most of the very primitive mammalian groups of Laurasia were extinct. Their replacements are familiar to us for, by and large, they represent the mammalian families present today. Of importance here, however, is the mixing of the autochthonous mammals and immigrants in Africa and Eurasia. Rhinoceroses, cats of varied types, pigs, and other ungulates crossed into Africa. In the opposite direction the mastodons entered Eurasia and even made their way to North America across the trans-Bering Bridge, which had again been uplifted. An important group of immigrants into Eurasia was the monkeys and apes, also autochthonous African mammals.

These faunal mixings were never complete, however. The filter between Africa and Eurasia was selective, allowing only some to pass. As a result, African mammals today represent an intermediate blend—largely endemic, but with many close affinities to the mammals of Laurasia.

15.3.4 India—A Cenozoic Mystery

Like the Antarctic, India presents an evolutionary puzzle, for as yet fossil mammals are virtually unknown in Cenozoic rocks of the Indian peninsula. Cenozoic rock formations from the Paleocene to the Pliocene have yet to yield fossil mammals. Nor, for that matter, have Jurasso-Cretaceous mammals been found. What past events lurk behind this mystery? Perhaps early mammals were never in India. The isolation of India in the Jurassic could explain this, but one would expect at least Jurassic reptiles, and possibly even mammals, to be found. A single genus of a large sauropod dinosaur is known in Central India. It seems that even in Africa the earlies mammals were sparsely represented (section 15.3.3). Did a Mesozoic barrier between Africa and India act as an effective filter, stopping the entrance of early mammals? At present we do not know.

Colbert's studies of the Triassic reptile *Lystrosaurus* and the vertebrate fauna associated with it, however, show that migration between Africa, Antarctica, and India was possible in the Triassic. *Lystrosaurus* does occur in all these regions. The Indian vertebrate faunal association, according to Colbert, is less diverse, suggesting that the entrance of Triassic vertebrates into India was selective. This finding implies that few, if any, early mammals were in India when it was isolated from Africa.

Land connections with Laurasia brought about by the northward drift of India did not occur until Oligocene-Miocene time. However, with the approach of India toward Asia in the Eocene some infiltration of Laurasian mammals into India must have occasionally occurred. Why then are their fossils not found in peninsular India? Perhaps if they did disperse, the Himalayas, which are now believed to be an intercontinental cordillera, effectively acted as a barrier against immigration into peninsular India. This, incidentally, might explain the presence of Eocene and later fossil mammals to the north of the Himalayas but not to the south. These ideas are summarized in Figure 15-5.

A second possibility is that early Cretaceous mammals were present in India, and as probably happened in Africa, South America, and possibly Australia, they underwent radiation in isolation during most of the Cenozoic. Because of the fortuitous nature of fossilization, they may not have been preserved, or at least they have not yet been found. Indeed, in all of the southern continents it is only in South America that abundant fossil mammals have been found in Cenozoic rocks older than Eocene age. Yet if a unique fauna evolved in India we would expect some living relicts to still remain as in South America. There are none!

Madagascar represents a fascinating exception to the general lack of early Cenozoic mammalian fossils in the southern continents. There, fossils occur in Paleocene rocks and more or less continuously to the present. Madagascar has enjoyed a degree of isolation from Africa since the early Cenozoic. Yet a variety of different mammals exist in most of its Cenozoic rocks. For example, Paleocene rocks reveal that primitive insectivores were present. This is astonishing because similar insectivores are yet to be found in Africa. In the Eocene primitive lemurs are found, in the Oligocene civetlike carnivores, in the Miocene rodents, and the hippopotamus as late as the Pliocene. Most of these forms must have arrived by chance dispersal, for the fossils and the modern fauna possess a high degree of endemism and indicate the island's long isolation. The fauna is unusual and depauperate—for many common African and Laurasian mammals are absent. At present Madagascar actually has a relict fauna protected largely by the absence of modern, native carnivorous predators.

The fossil mysteries of Africa, India, Australia, and Antarctica remain. But active search continues, and the results should prove fascinating, for they will fill the gaps in our knowledge concerning the breakup of Gondwanaland and the relative importance of continental drift to the evolution of Cenozoic vertebrates.

The evolution of mammals into widely diverse groups, as discussed in this chapter, seems to stem from a series of isolations. As Pangaea ruptured, the continents and their faunas drifted apart, only to partly coalesce again into new associations. It is in this same geological scenario that birds evolved. Given the simultaneous evolution of birds and mammals it is unfortunate that only the mammals have left us much evidence of their history in the rocks. Birds, too, are diverse and we can glean much about their lives and possible evolution from an examination of their scant fossil record and from living birds, the subject of Chapters 16 and 17.

References

See references in Chapters 4, 8 and 11 for geological works.

[1] Clemens, W. A. 1968. Origin and early evolution of marsupials. *Evolution* 22:1–18. A concise essay that traces the possible ancestry of marsupials to the late Cretaceous genus *Alphadon* that, presumably, was derived from *Pappotherium*, a well-known middle Cretaceous therian from Texas.

[2] Clemens, W. A. 1970. Mesozoic mammalian evolution. *Annual Review of Ecology and Systematics* 1:357–390. A presentation of the view that mammals in the late Triassic constituted two basic groups, the prototheria and theria, each of which may have been derived from a common mammalian ancestor. If so, mammals must be considered monophyletic.

[3] Hopson, J. A. and A. W. Crompton. 1969. Origin of mammals. In *Evolutionary Biology*, edited by T. Dobzhansky, M. K. Hecht and W. C. Steere, vol. 4, Appleton-Century-Crofts, New York. A presentation of the view that the various Jurassic orders of mammals were monophyletic, and evolved from a single lineage of therapsids in the Triassic.

[4] Romer, A. S. 1966. *Vertebrate Paleontology*. Third edition.. University of Chicago Press. A magnificent reference book to fossil vertebrates. The data on which Figure 15-1 was based are largely derived from this book.

[5] Simpson, G. G. 1959. Mesozoic mammals and the polyphyletic origin of mammals. *Evolution* 13:405–444.

[6] Simpson, G. G. 1965. *The Geography of Evolution: Collected Essays*. First edition. Chilton Books, Philadelphis.

16

The Origin of Birds and
the Function of Feathers

Synopsis: Birds are feathered archosaurs that evolved as part of the great reptilian radiation of the Mesozoic. The earliest known avian fossil, *Archaeopteryx* from the late Jurassic, already possessed fully differentiated feathers but differs from modern birds in possessing teeth, an elongate tail of caudal vertebrae, functional toes and claws on its wings, and other primitive traits. According to one theory *Archaeopteryx* was an arboreal climber and glider. Another view holds that it was a cursorial predator that used its "wings" to capture prey. In either case, *Archaeopteryx* probably was not a direct ancestor of modern birds.

The derivation of feathers from reptilian scales is clouded by the absence in the fossil record of intermediate structures. Most ornithologists agree that feathers evolved in association with endothermy as an insulative cover and that special wing and tail feathers evolved later in association with flight. Five basic types of feathers are recognized: (1) contour feathers, (2) semiplumes, (3) downs, (4) bristles, and (5) filoplumes. Feathers are excellent insulation because they trap dead air around a bird's body. By varying the depth of this layer of dead air space birds achieve a high degree of thermoregulation over a wide range of environmental temperatures.

The necessary modification for flight are of two sorts: weight-reducing adaptations and power-promoting adaptations. Birds are capable of flight because of feathers, endothermy, wings and associated pectoral muscles, hollow bones, a special respiratory system, and a large, powerful heart. The avian wing acts as a variable airfoil and propeller. In flapping flight most of the lift is provided by the secondaries of the inner wing, whereas movement of the primaries of the outer wing generates most of the thrust.

There are many subsidiary functions and adaptations associated with feathers. These include physical colors and iridescence, water repellency and adsorption by the plumage, display and concealment, mechanical sound-production, and sound-muffling in flight. The evolution of feathers—structurally and functionally the most complex derivatives of the vertebrate integument—has been the key to adaptive radiation and ecological success of birds.

535

16.1   INTRODUCTION: BIRDS AND REPTILES COMPARED

The famous nineteenth century defender of Darwinian evolution, Thomas Henry Huxley, wrote that birds are nothing more than "glorified reptiles." While feathers absolutely distinguish birds from all other vertebrates, their plumage is about the only morphological feature that is conspicuously different from their reptilian relatives. Even without fossils, it would be easy to establish a relationship between reptiles and birds, especially with the archosaurian crocodiles and alligators. Huxley, in fact, was so impressed by their many similarities that he placed birds and reptiles together in his classification scheme as a Class, Sauropsida.

Many of the most basic similarities are skeletal, but not all. The following list includes the main characteristics shared by birds and some reptiles, especially the archosaurs:

1. Skull and neck hinged by a single occipital condyle.
2. Lower jaw with several bones and articulating with skull by the quadrate and articular bones.
3. Single sound-transmitting bone in middle ear, the *columella*.
4. Uncinate processes on ribs (some reptiles only, for example, tuatara).
5. Intertarsal ankle joint (birds and archosaurs).
6. Hollow, pneumatic bones (most birds, some reptiles).
7. Epidermal scales (legs and feet of birds).
8. Relatively few skin glands.
9. Respiratory air sacs (similar structures in turtles, some lizards).
10. Pecten in retina (birds and many lizards).
11. Nucleated red blood corpuscles.
12. Cleidoic, amniotic egg with shell; egg tooth and special hatching muscle (birds and some reptiles, especially turtles).

The main differences between birds and reptiles relate to avian adaptations involving flight and endothermy (homeothermy). Both of these fundamental avian functions are closely dependent upon the structure of feathers. The evolution of feathers as a special epidermal derivative and the modification of feathers for various functions are, therefore, key processes in the adaptive radiation and ecological successfulness of birds as a class. Perhaps in no other major group of vertebrates has evolutionary success been so determined by the appearance of a single structural feature as in the case of birds with their highly modifiable feathers.

Birds have high, usually constant body temperatures around 40 to 41°C. This high body temperature is determined in part by superior insulative qualities of feathers and in part by a high rate of resting metabolism and heat production. Generally speaking, the resting metabolic rates of birds average some five to ten times the metabolic rates of reptiles at the same body temperature. Most biologists agree that feathers evolved first as an effective and variable insulation in association with the evolution of endothermy in the ancestors of birds, and that the modification of certain feathers on the fore-limbs and tail for flight surfaces was a later, secondary adaptation.

In any case, the 8700 living species of birds, and the tens of thousands of extinct ones that have lived at some time since the first birds emerged from their reptilian ancestry in the Jurassic, attest to the adaptive value of feathers as a skin covering.

16.2   ORIGIN AND EVOLUTION OF BIRDS

As we have already shown in Chapter 13, the Mesozoic was the period of major radiation of the various reptile groups, particularly the archosaurian assemblage. By the Jurassic, land-based and aquatic niches for amniotes were pretty well packed, and the most exploitable unfilled ecological opportunities for potential evolving types lay in the possibilities inherent in locomotion through the air. At least three different groups of diapsid reptiles experimented with volant adaptations. The dawn lizards, which were the first tetrapods to become airborne, represent an early attempt in the Triassic to exploit aerial niches by volplaning or gliding on a pair of sails supported on modified ribs, much like the adaptation found in the modern day East Indian lizards of the genus *Draco*. It was, however, the pterosaurs and the birds that developed true wings and flapping flight. Probably because of their relatively large size, the pterosaurs were primarily gliders and possessed only limited flapping flight capacity. They seem to have been restricted mostly to a pelagic, marine existence; it was left to the birds to capitalize on powered flight by flapping wings and large pectoral muscles.

16.2.1   Archaeopteryx Fossils: The Oldest Known Birds

The oldest known birds from the Jurassic are about 140 million years old and are even more reptilelike than modern birds. Our knowledge about these birds is based on one fossil impression of a feather and on five fossilized skeletons with associated feather impressions (Figure 16-1). If it were not for the certain association of feathers with these skeletons, they would no doubt be classified as archosaurian reptiles. They differ enough from modern birds to be separated in classification as a subclass Archaeornithes, Order Archaeopterygiformes. The name *Archaeopteryx* means "ancient wing." All other birds, including other fossil forms, are placed in the subclass Neornithes (new birds).

Table 16-1 lists the main anatomical features of *Archaeopteryx* with notations on their phylogenetic significance. The thecodont and archosaurian affinities of *Archaeopteryx* are prominent, but the most significant features of these fossils are their feathers. These structures are already fully developed and

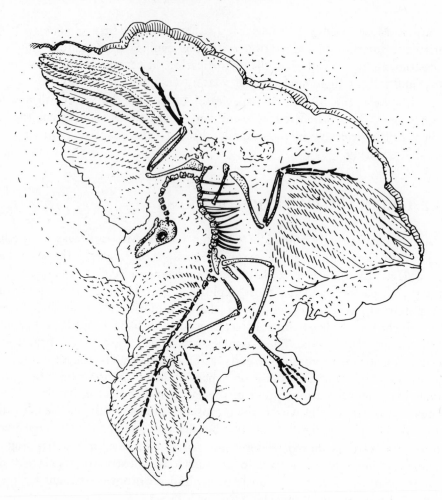

Figure 16-1 *Archaeopteryx lithographica*, sketch of fossil impression showing skeleton and feathers. Note apparent absence of a sternum, lack of fusion in bones of forelimbs, and the long, free tail with paired feathers on each side.

functionally diversified in *Archaeopteryx*, indicating a long prior history of evolution. In addition to a complete covering of body feathers, the wing feathers are well differentiated into an outer series of primaries on the hand bones and an inner series of secondaries along the outer arm, essentially as in modern birds. The tail, on the other hand, is peculiar in that there are 15 pairs of rectrices, or tail feathers, arranged dichotomously along the sides of the sixth through twentieth caudal vertebrae, a condition found in no other bird living or fossil.

Unfortunately there are no fossils intermediate in character between *Archaeopteryx* and any of the immediate reptilian groups from which it might have evolved. The details of the pelvis could suggest a tie with the ornithischian dinosaurs (see Chapter 13), but it is more likely that these similarities result from convergence associated with the independent evolution of bipedalism. The limb bones are more like those of bipedal saurischians, particularly the

Table 16-1. The Main Anatomical Characteristics of *Archaeopteryx*

Anatomical character	*Phylogenetic remarks*
Large eye orbit with sclerotic ring	Found in some modern birds and thecodont reptiles like *Euparkeria*
Teeth in sockets of both jaws	Basic thecodont reptilian character
Fenestra in lower jaw posterior to dentary bone	Basic archosaur character; reduced in modern birds
10 cervical vertebrae, 6 with ribs	Reptilian; modern birds have 13–25, none with ribs
19–20 trunk vertebrae; 6 sacral vertebrae, only 5 fused into synsacrum	Reptilian; more fusion in modern birds
20 free caudal vertebrae, no pygostyle	Reptilian; modern birds have only pygostyle
Vertebral centra biconcave or "amphicoelous"	Primitive vertebrate character; modern birds with "heterocoelous" centra
None of bones hollow	Primitive; modern birds have pneumatic bones
None of wing bones fused	Reptilian; modern birds with fused carpometacarpus
Hand reduced to 3 digits with external, functional claws	Common in bipedal Mesozoic reptiles; external claws vestigial in some modern birds
Two clavicles fused into *furcula*	Present in modern birds
Sternum without keel	Condition found in some flightless modern birds, e.g. "ratites"
Abdominal ribs	Reptilian; lacking in modern birds
Feet with 4 digits, three forward, one back (hallux)	Many reptiles and modern birds
Intertarsal joint with fusion of leg bones	Many bipedal reptiles and modern birds; archosaur trait
Feathers, both "flight" and contours well developed	Also in modern birds

relatively small coelurosaurs (theropods), many of which were fast, cursorial predators that apparently ran down their prey, which was grasped by well-clawed hands. The pterosaurs had wings, but these reptiles were virtually contemporary with *Archaeopteryx* and have a fundamentally different body plan for flight. They clearly represent an independent evolution of winged flight and can in no way be considered the ancestors of birds.

The more distant ancestry of birds probably lies in some early group of bipedal, diapsid thecodonts. The pseudosuchians in the Lower Triassic are sufficiently generalized to be good ancestors, and some of them show tendencies in the direction of *Archaeopteryx*. The skull of *Euparkeria* has often been compared with the skull of *Archaeopteryx* and with that of a modern bird (Figure 16-2). Although a plausible case can be made for *Euparkeria* or some similar type of pseudosuchian as the remote ancestral stock from which birds arose, a gap of some 100 million years separates the pseudosuchians from the first known birds. This gap can be accounted for in two ways: first, many slow changes were required to achieve the avian condition; and second, it is likely that the first "birds"—that is, the first archosaurs to possess feathers— antedate *Archaeopteryx* by many millions of years. Their fossilized remains simply have not been found. Recent studies by the paleontologist John Ostrom indicate that *Archaeopteryx* evolved from a line of reptiles that also gave rise to the theropod dinosaurs, and he places *Archaeopteryx* close to the coelurosaurs.

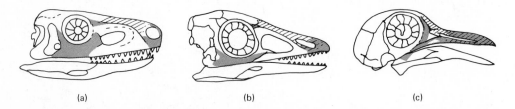

(a)   (b)   (c)

Figure 16-2 Comparison of reptile and bird skulls to show evolutionary stages. A. The pseudosuchian reptile, *Euparkeria capensis* from the Triassic of Africa. B. *Archaeopteryx lithographica*, Upper Jurassic of Bavaria. C. The common pigeon, *Columba livia*, Recent. Dark areas are the premaxilla, lacrimal, and jugal; diagonal shaded, the nasals. (Redrawn and modified from Heilmann, G. [1926] *The Origin of Birds*. Appleton-Century, New York.)

16.2.2 Cretaceous Fossil Birds

The next fossil birds known are from the Lower Cretaceous (Figure 16-3), about 10 million years after the appearance of *Archaeopteryx*. This is not a very long period of evolutionary time, and yet the Cretaceous birds are almost fully modern in every respect, except for the presence of teeth in the jaws of a few species. They all have fused carpometacarpals in the wings; a large, platelike sternum, usually keeled; a pygostyle instead of free caudal vertebrae; and other features of modern birds. No fossil has yet been found that is in any way intermediate in character between *Archaeopteryx* and the early Cretaceous birds. All these facts indicate that *Archaeopteryx* represents a side-branch of avian evolution and that this genus of birds was not directly ancestral to the modern birds that we know.

The Cretaceous was an important period for the evolution of birds (Chapter 15). There are about 35 known species, which are placed in 19 genera, 13 families, and 8 orders, indicating considerable diversity. Six of the orders

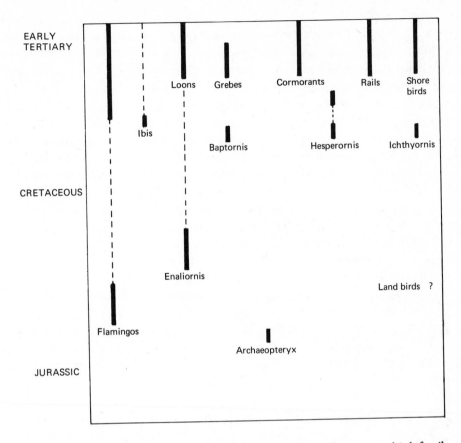

EARLY
TERTIARY

Loons   Grebes   Cormorants   Rails   Shore
birds

Ibis

Baptornis   Hesperornis   Ichthyornis

CRETACEOUS

Enaliornis   Land birds   ?

Flamingos   Archaeopteryx

JURASSIC

Figure 16-3 Temporal occurrence of major groups of Mesozoic bird fossils. The span of each epoch is given in million of years. Broken lines show gaps in the record. (Modified from Brodkorb [1].)

and two of the families present in the Cretaceous are represented by species living today.

We know that aquatic birds had evolved early in the period and that they underwent several radiations before the end of the Cretaceous. Presumably land birds also existed at the same time, adapting and evolving in association with the new biomes created by the diversification of flowering plants in the extensive tropical and subtropical forests of the late Mesozoic. Unfortunately, we have no direct proof as yet. No avian fossils have been discovered among the terrestrial faunas of the Cretaceous.

The earliest known Cretaceous bird (genus *Gallornis*) is a flamingolike species that is usually placed in the Suborder Phoenicopteri with modern flamingos. Other of these aquatic forms are clearly related to such groups as the ibises (Ciconiiformes), loons (Gaviiformes), grebes (Podicipediformes), cormorants (Pelecaniformes), rails (Gruiformes), and shorebirds (Charadriiformes).

Primarily owing to the enthusiasm and imagination of an American paleontologist named O. C. Marsh, it used to be thought that all Mesozoic birds possessed teeth. This idea was based in part on the incorrect association of teeth-bearing reptile jaws with bona fide avian skulls and postcranial skeletons and in part on the assumption that Mesozoic birds *should* possess teeth even when no fossil evidence existed to support the notion. Students of fossil birds now agree that the only birds other than *Archaeopteryx* certainly known to have possessed teeth are the gull-like birds called *Ichthyornis* and the peculiar, flightless, loonlike birds in the genus *Hesperornis* (Figure 16–4). The latter were large birds up to 2 m in length with vestigial wings often reduced to just the humerus. They swam and dived with powerful feet and legs set far back on the body like loons. They had conical teeth in both jaws, apparently adapted for catching fish, and, owing to the lack of a mandibular symphysis and the presence of a transverse joint between the splenial and angular jaw bones, the lower jaws could bulge out, allowing these birds to swallow prey larger than the circumference of their undistended mouths.

The general characterization of the known avifauna at the end of the Cretaceous is that of an aquatic assemblage of birds distinctly modern in appearance and including loons, grebes, cormorants, a pelicanlike genus, flamingos, ibises, rails, and sandpipers. Conspicuously absent from this assemblage are other groups of aquatic birds that become prominent in the early Tertiary—petrels, ducks, cranes, gulls, and auks. The terrestrial avifauna of the Cretaceous remains completely unknown, but it must have been at least as diverse as the aquatic one.

Figure 16–4 Reconstruction of the large, diving bird, *Hesperornis*, of the Cretaceous. Note teeth in jaws; also there is some evidence that the feet were lobed like those of grebes rather than fully webbed. (Modified from Gleeson, in Welty [11].)

16.2.3 Evolution of Modern Orders and Families in the Cenozoic

Most, if not all, of the orders of birds had evolved by the end of the Eocene some 40 to 50 million years ago (Figure 16-5). Groups definitely associated with deposits of Eocene age include the ostriches (Struthioniformes), the rheas (Rheiformes), the recently extinct elephant birds of Madagascar (Aepyornithiformes), the penguins (Sphenisciformes), the tube-nosed aquatics (Procellariiformes), the waterfowl (Anseriformes), the fowls (Galliformes), owls (Strigiformes), the kingfisherlike birds (Coraciiformes), the woodpeckers (Piciformes), and the perching birds (Passeriformes). The diurnal birds of prey (Falconiformes) had already appeared in the Paleocene. Five other orders first appear in the fossil record around the end of the Eocene or the beginning of the Oligocene: the doves and pigeons (Columbiformes), the cuckoos (Cuculiformes), the goatsuckers (Caprimulgiformes), the swifts (Apodiformes), and the trogons (Trogoniformes), so that by the beginning of the Oligocene at least 26 orders of birds had become differentiated.

There were two major radiations of avian families during the Tertiary. The Eocene was the epoch of greatest importance for the diversification of birds. More families that are still surviving arose in that period than at any other time, 20 for certain and eight other likely ones. Most of these families consist of additional kinds of water birds and nonpasserine forest dwellers. A second radiation of 20 families occurred in the Miocene and included a few additional kinds of water birds but mostly land-dwelling forms and particularly passerines that were adapted to drier, less forested environments. Most families of birds had evolved by the end of the Miocene, and the world avifauna had a distinctly modern look by that time. Many still-existing genera and some species were present by the Pliocene.

16.3 STRUCTURE AND EVOLUTION OF FEATHERS

As the main covering of avian skin, feathers are structurally and functionally the most complex derivatives of the integument to be found among the vertebrates. They develop from invaginated pits or follicles in the skin, and they are generally arranged in distinct tracts or **pterylae**, which are separated by patches of unfeathered skin, the **apteria** (Figure 16-6); in some species, such as the ratites, penguins, and mousebirds, the feathers are more uniformly distributed over the surface of the skin.

For all their macroscopic complexity, feathers are remarkably simple and uniform in chemical composition. More than 90 percent of the substance of a feather consists of beta keratin, a protein structurally related to the keratin of reptilian scales and to the hair and horn of mammals. Keratin is extremely resistant to proteolytic enzymes and is otherwise a strong and resilient material well suited for its role as the basic component of feathers. About 1 percent of a feather consists of lipids, about 8 percent is water, and the remaining fraction consists of small amounts of other proteins and pigments such as melanin.

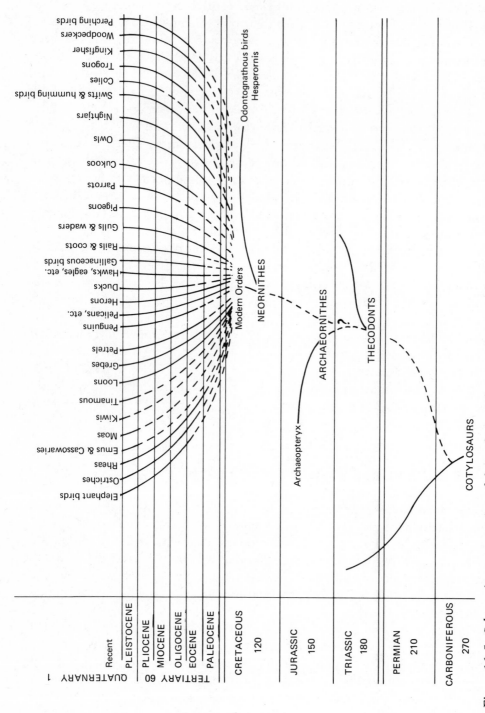

Figure 16–5 Schematic representation of the evolution of avian orders. Solid lines indicate the known fossil record. Dashed lines show postulated extensions and relationships.

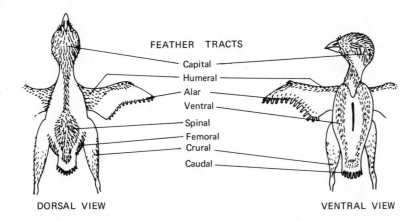

FEATHER TRACTS
Capital
Humeral
Alar
Ventral
Spinal
Femoral
Crural
Caudal

DORSAL VIEW　　　　　　　　　　　　VENTRAL VIEW

Figure 16-6　Pterylography of a typical songbird. (Redrawn from Welty [11].)

16.3.1　Basic Types of Feathers

Ornithologists usually recognize five basic structural types of feathers, although intermediate conditions exist. The five types are (1) **contour feathers**, including typical body feathers and the flight feathers (**remiges** and **rectrices**), (2) **semiplumes**, (3) **down feathers** of several sorts, (4) **bristles**, and (5) **filoplumes**.

The structural details of a typical feather are most easily seen by examining a contour feather (Figure 16-7). The short, tubular base of the feather shaft is the **calamus**, which remains firmly implanted within the follicle until molt occurs. Distal to the calamus there is a long, tapered **rachis**, which bears closely spaced side branches called **barbs**, the lowermost of which externally mark the division between calamus and rachis, as does a pit on the underside of the shaft, the **superior umbilicus**, a remnant of the open, upper end of the tube of epidermis that forms the growing feather. The rachis is further distinguished from the calamus by its more solid, pith-filled construction. The **inferior umbilicus** at the bottom of the calamus is the opening which is formed by the drying of the pulpy core and the separation of the feather from its papilla after growth is complete.

The barbs on either side of the rachis constitute a surface called a **vane**. The two vanes of a contour feather may be symmetrical or asymmetrical; the flight feathers are typically asymmetrical. The proximal portions of the vanes of a feather have a downy or plumulaceous texture, being soft, loose, and fluffy. This part of the contour feathers and feathers that are completely downy give the plumage of a bird its excellent properties of thermal insulation. The more distal portions of the vanes have a pennaceous or sheetlike texture, firm, compact, and closely knit. This is the exposed part of a contour feather that provides an airfoil, protects the downy undercoat, sheds water, reflects or absorbs solar radiation, and may have a visual or auditory communicating function, as will be discussed later.

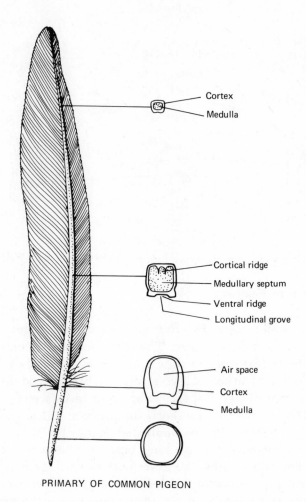

PRIMARY OF COMMON PIGEON

Figure 16-7 Typical contour feather (wing quill), showing main structural features. (Redrawn from Pettingill [7] and Thomson [10]).

The barb itself consists of an axis called the **ramus** and numerous closely spaced branchlets, the barbules (Figure 16–8). The **barbules** are essentially stalks one cell thick that are differentiated to some extent from base to tip. Downy or plumulaceous barbules have a short, straplike base, whereas the distal segment, or **pennulum**, is a long, simple, jointed stalk. Pennaceous barbules, on the other hand, are highly differentiated and bear several kinds of projections termed **barbicels**. The barbules facing the proximal end of the feather have a prominent flange along the dorsal edge of their bases, whereas the barbules facing the distal end of the feather possess little or no dorsal flange but instead have many well-developed, hooklike barbicels or **hamuli** (singular, **hamulus**). The distal barbules of each barb cross over the proximal barbules of the next higher barb on the shaft, and their hamuli grasp the dorsal flanges of the proximal barbules (Figure 16–9). These are the structural details that maintain the pennaceous character of the feather vanes. They are arranged in such a way that any physical disruption to the vane is easily corrected by the bird's preening behavior, in which the barbs are realigned by the bird's drawing its slightly separated bill over them.

Contour feathers also often bear outgrowths on the underside attached to the rim of the superior umbilicus. These structures are termed **afterfeathers** and may consist of a fringe or cluster of barbs around the umbilicus or of an

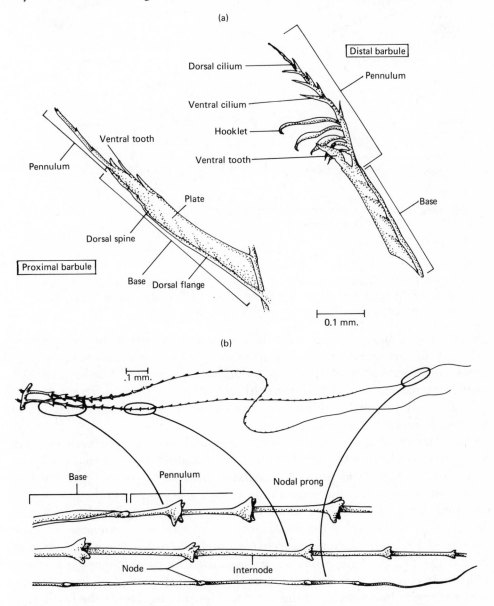

Figure 16-8 Types of barbules. (a) Pennaceous barbules from contour feather of domestic fowl. (b) Plumulaceous barbules from downy base of pigeon feather. (Redrawn from Lucas and Stettenheim [4]).

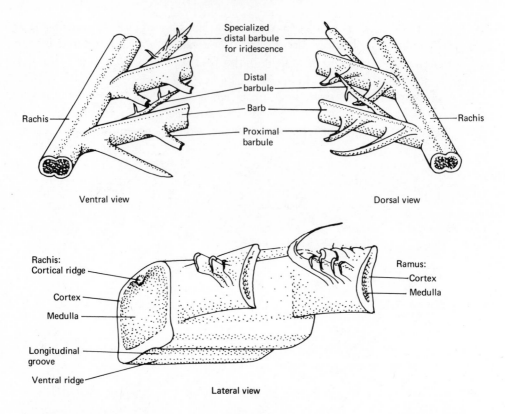

Figure 16-9 Detailed structure of a typical barb from a pennaceous feather, showing the hook and groove arrangement for binding adjacent barbs together. (Based on Thomson [10] and Pettengill [7]).

aftershaft with barbs on each side. In most cases, the latter appears as a small auxiliary feather attached onto the main one, but in some species, notably the emus and cassowaries of Australia and New Guinea, the aftershaft virtually duplicates the main feather. Afterfeathers vary not only in different parts of a bird's body but also in different taxa of birds, being absent in some groups and highly developed in others.

The **remiges** (wing feathers, singular **remex**) and **rectrices** (tail feathers, singular **rectrix**) are large, stiff, mostly pennaceous contour feathers that are highly modified for flight. Adaptations associated with their use in flight are both gross and microscopic, and they are most highly developed in the primaries of the outer wing, the feathers that are involved in producing the propulsive force for flight. For example, the distal portions of the outer primaries in many species of birds are abruptly tapered or notched, so that when the wings are spread the tips of these primaries are separated by conspicuous gaps or slots (Figure 16-10). This condition reduces the drag of the wing and, in association with the marked asymmetry of the outer and inner vanes, allows the feather tips to twist as the wings are flapped and to act somewhat as individual propeller blades. The pennaceous barbules of flight feathers are generally larger and have more highly developed barbicels than those of typical body feathers. A frequent

Figure 16-10 Flight silhouette of the California condor, showing extreme development of slotting in the outer primaries, an adaptation for static soaring.

adaptation involves **friction barbules**, distal barbules that carry special lobate, dorsal barbicels, which occur on those parts of flight feathers that overlap when a wing or tail is spread (Figure 16-11). By rubbing against the rami of overlapping vanes, they help to keep the feathers from slipping too far apart.

Ear coverts are modified contour feathers that surround the external ear openings of most birds. Their vanes have an open, pennaceous texture resulting from the wide spacing between stiff barbs and the small, simple structure of the barbules. They apparently function to screen the ear opening, while allowing access of sound waves to it, and may also be involved in special hearing abilities in some species such as the nocturnal owls.

Semiplumes are feathers intermediate in structure between contour feathers and down feathers. They combine a large rachis with entirely plumulaceous vanes and can be distinguished from down feathers by the fact that the rachis is longer than the longest barb. The barbules are plumulaceous, and an afterfeather occurs if such a structure is also present on the contour feathers (Figure 16-12). Semiplumes occur along the margins of contour feather tracts and in tracts by themselves, where they are mostly hidden beneath the contour feathers. They provide thermal insulation and help to fill out the contours of a bird's body.

Down feathers of various types are entirely fluffy feathers in which the rachis is shorter than the longest barb or entirely absent. Natal downs (**neossoptiles**) provide an insulative covering on many birds at hatching or shortly thereafter. Their barbules are plumulaceous but smaller than those of adult feathers, and the tips of their terminal barbs lack barbules, a characteristic by

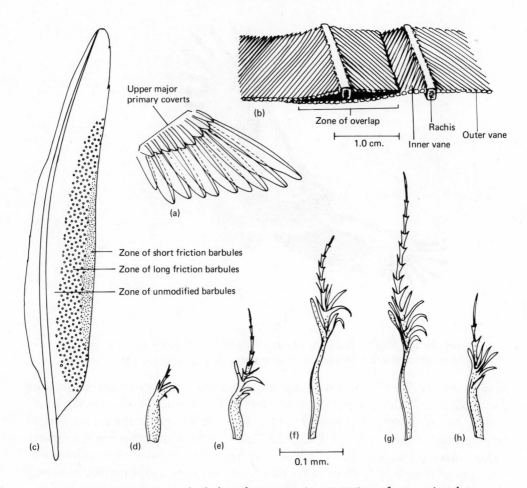

Figure 16-11 Friction barbules of pigeon's wing. (a) View of outer wing showing slots and areas of overlap between adjacent primaries, (b) two primaries viewed from distal end to show overlap, (c) primary showing friction zones, (d) lateral view of unmodified distal barbule from inner vane of a primary, and (e to h) special friction barbules. (Redrawn from Lucas and Stettenheim [4]).

which they can be distinguished from the downs of later plumages. Natal downs usually precede the development of the first contour feathers, semiplumes, and down feathers associated with apteria (the spaces between the contour feather tracts); there can be two or three generations of natal down, depending on the species. Definitive downs are those that develop with the full body plumage. They always have a true calamus, although the rachis may be very short and tapered, and an afterfeather often occurs in species in which this structure is present on the body contour feathers. Uropygial gland downs represent a specialization associated with the large sebaceous gland found at the base of the tail in most birds. The papilla of the gland usually bears a tuft of modified, brushlike down feathers that aid in transferring the oily secretion from the gland to the bill, the instrument used to apply the oil as a waterproof dressing to the plumage.

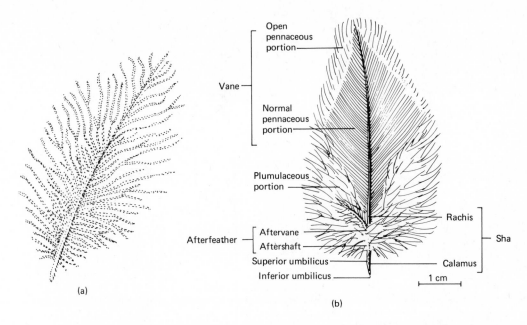

Figure 16-12   Semiplume (a) and body contour feather (b) compared.

Powder feathers, which are difficult to classify by structural type, represent another modification of basic feather structure found in many species of birds. They produce an extremely fine, white powder composed of granules of keratin about 1 micrometer (μm) in diameter. The powder, which is shed into the general body plumage, is nonwettable and is therefore assumed to provide another kind of waterproof dressing for the contour feathers. Powder feathers most often have the structure of downs, and are often termed "powder downs," but in some birds powder is also produced by special semiplumes and by the modified basal down portions of contour feathers. Depending on the species, powder feathers may occur among the ordinary downs and contour feathers, in distinct patches by themselves, or both in patches and intermingled among ordinary feathers. In species such as herons and bitterns, which shed copious amounts of powder, the powder producing feathers grow continuously.

Bristles are specialized feathers with a stiff rachis and barbs only on the proximal portion or none at all (Figure 16-13). They often grade structurally into adjacent contour feathers and occur most commonly around the base of the bill, around the eyes, as eyelashes, and in some birds, on the head or even on the toes. The distal rachis of most bristles is dark brown or black, owing to a dense concentration of melanin granules. The melanin pigment not only colors the bristles but also adds to their strength and resistance to wear and photochemical breakdown. Bristles and structurally intermediate feathers called "semibristles," function in part to screen out foreign particles from the nostrils and eyes of many birds, in part as tactile sense organs, and possibly as aids in the aerial capture of flying insects, as in the case, for example, of the long bristles at the edges of the jaws in nightjars and flycatchers.

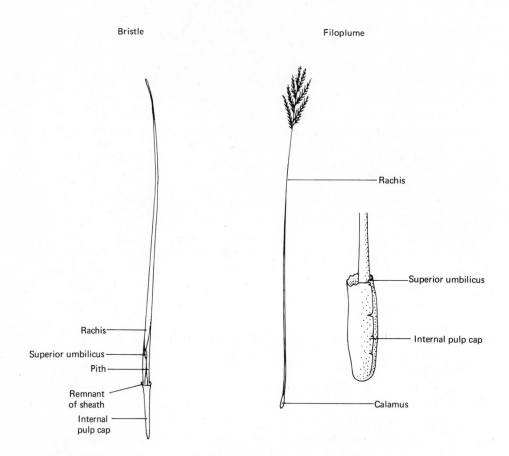

Figure 16-13 Comparison of a bristle (left) and a filoplume (right). (Redrawn from Pettingill [7].)

Filoplumes are fine, hairlike feathers that consist of a thin shaft with a few short barbs or barbules at the tip. They are the least variable of feathers and do not grade into other types. Although a calamus is present, it often breaks off when a filoplume is pulled out of its follicle because the rachis is constricted at its base. In some birds such as cormorants and bulbuls, the filoplumes grow out over the contour feathers and contribute to the external appearance of the plumage, but usually they are not exposed.(Figure 16-13).

It used to be thought that filoplumes are degenerate contour feathers, but more recent information about them indicates that they are highly specialized sensory structures that aid in the operation of other feathers. Filoplumes have numerous free nerve endings in their follicle walls, and these nerves connect to pressure and vibration receptors around the follicles. Apparently the filoplumes transmit slight vibration or movement of the contour feathers to these receptors. Sensory input of this sort probably plays a role in keeping the contour feathers in place and adjusting them properly for flight, insulation, bathing, or display.

16.3.2   Origin of Feathers from Reptilian Scales

Feathers are considered to be homologous with reptilian scales and to have evolved from them, but there is not much direct proof of homology and only some speculations as to the selective forces that might have transformed scales into feathers. Except for their biochemical similarities and the fact that feathers and reptilian scales both develop from similar embryonic germs consisting of dermis and epidermis, there is little else upon which to base their homology. It has also been suggested that the main part of a contour feather is homologous with the dorsal part of the reptilian scale and that the afterfeather is the homolog of the ventral half, whereas the underside of feather vanes corresponds not to the underside of scales but to the innermost layer of scale epidermis. Feathers are molted as individual structures from their follicles, whereas the reptiles that are most familiar, the snakes and lizards, molt the cornified layer of their skins entire; however, the crocodilians, closest living relatives of birds, and some turtles, replace the outer layers of their scales individually, and it is possible that this was a general archosaurian characteristic, or at any rate that it was characteristic of the archosaurian precursors of birds. Also, some pseudo-suchians have been described as having a paired row of epidermal scales along the back, twice as long as they are broad and with slight sideways ribbing, a detail suggesting a trend in the modification of scales toward a precursor of feathers. Little more can be said about homology between feathers and scales.

There is no agreement as to whether the first feathers were downlike or more like typical contour feathers. All we know is that *Archaeopteryx* had fully developed contour feathers; no structures intermediate between feathers and scales have been discovered in the fossil record. Most authorities on the subject agree that feathers first evolved as an insulative body covering in association with the development of endothermy in the preflying ancestors of birds, but a few have argued that they first originated from elongated scales on the arms as primitive aids in gliding flight and that a general body covering for insulation was a later development. The basis for this idea is that the body contour feathers have a rachis and double vanes like the flight feathers, and there is a question why the body feathers should have this complex structure if they were the first to evolve under selective forces relating to thermoregulation. Another idea is that reptilian scales first evolved into elongate, finely branched structures that functioned as radiant heat-shields, and these intermediate scales later evolved into contour feathers. It is an interesting fact, however, that birds like the ratites and penguins, which have lost the power of flight, have body feathers that are not highly pennaceous and contour-like, suggesting that selective forces associated with flight are responsible for the evolution of this type of feather. All such theories about the origin of feathers remain speculative in the absence of fossil structures intermediate between reptilian scales and feathers; however, because insulation is the most generalized function of feathers in all known birds, living and fossil, the most *reasonable* explanation for their evolutionary origin lies in the selective advantages of plumage in relation to endothermy.

16.4   FEATHERS AND THERMOREGULATION

Birds, like mammals, are homeotherms that regulate their high body temperatures over an amazingly wide range of environmental conditions by elaborate mechanisms for precisely balancing heat loss and heat gain. Heat production by metabolism is one aspect of this regulation. Not only is the resting metabolic heat production of a bird many times greater than that of a reptile, but a bird can also change the intensity of its heat production by varying metabolic rate over a wide range. In this way, it can maintain a constant high body temperature by adjusting heat production to equal heat loss from its body under different environmental conditions.

Heat is, of course, a byproduct of all oxidative metabolic reactions taking place in an organism. Birds produce their endothermic heat in several ways. One obligatory fraction of their heat production derives from the so-called basal or resting metabolic rate, which is the sum of all the oxidative chemical reactions required for the maintenance of living tissues and organs when the bird is at rest. Another is the heat increment of feeding, often called the **specific dynamic action** or effect of the food ration. This added heat production after ingestion of food apparently results from the energy requirements for assimilation and varies in amount depending on the type of foodstuff being processed. It is highest for a meat diet and lowest for a carbohydrate diet.

The voluntary actions of skeletal muscle produce large amounts of heat in birds, especially during flight, which can result in a heat production 15 or more times greater than at rest. This muscular heat can be advantageous for balancing heat loss in a cold environment, or it can be a problem requiring special mechanisms of dissipation in hot environments that approach or exceed the bird's body temperature. Finally, there is an important regulatory mechanism of heat production involving **shivering**, the generation of heat by muscle fiber contractions in an asynchronous pattern that does not result in gross movement of the whole muscles. Mammals also possess mechanisms for nonshivering, regulated thermogenesis, but such mechanisms have yet to be demonstrated in birds.

Since birds for the most part live under conditions in which ambient temperatures are lower than the regulated body temperatures of the birds themselves, heat loss to the environment is a more usual circumstance of their lives than heat gain, although heat gain can be a major problem in deserts. Control of heat loss to a cold environment is, therefore, one of the most important regulatory functions of an endotherm, and birds employ their plumage in a very effective way as an insulation against heat loss.

16.4.1   The Insulative Qualities of Feathers

Man has long known the practical value of feathers as an insulative material. Centuries ago the Eskimos learned to make a very light-weight but warm parka from the skins of birds such as eider ducks, geese, and other kinds of water birds. These garments were such an effective insulation against the loss of body heat that they could not be worn outdoors during periods of activity but were reserved for use when sitting or sleeping in unheated quarters during the long winter nights of the far north. Even today, with the development of so many synthetic materials, feather comforters and eider down sleeping bags are still much in demand.

Pound for pound feathers are a better insulative material than mammalian hair or almost any other natural or synthetic material. The reason can be understood by comparing the thermal conductivities of some common materials and substances listed in Table 16-2. Copper, for example, has one of the highest thermal conductivities known and is therefore a very poor insulator. A rod of copper with a cross-sectional area of 1 cm^2 conducts 0.99 gram calories (g cal) of heat per linear centimeter per second per degree difference along that centimeter. By contrast, a column of still air conducts only 6×10^{-5} g cal. Eider down approaches this value more closely than any of the common insulating materials.

Any material that traps a lot of dead air is a good insulator against conductive heat transfer. The structure of feathers, especially the downs, semi-plumes, and downy parts of contour feathers, are admirably designed to hold dead air, and this is the reason why a bird's plumage is such a good insulating cover.

Table 16-2. Comparison of Thermal Conductivities of Some Common Materials

Substance	*Conductivity* g cal cm^{-2} cm^{-1} sec^{-1} $^{\circ}$C^{-1}
Copper	0.99
Glass	0.0025
Water	0.0014
Soft wood	0.0009
Cotton wool	0.0004
Sheep's wool	0.00025
Eider down	0.00016
Still air	0.00006

Without feathers a bird's body would conduct heat away from its surface at about the same rate as water, but the insulative properties of a feathery covering reduce this rate by a factor of 10 or more between the actual body surface and the exterior surface of the feathers, depending on the temperature difference between the body and external feathering. Thus, the mechanism of conductive insulation relies upon an outer shell of a relatively nonconducting material that covers an inner core or body with a relatively high conductivity.

Two further points need to be noted. Because of its heterogeneous structure and various physiological functions, such as blood flow, a living body transfers heat in a more complicated way than the simple physical conductivity through a homogeneous, nonliving substance. Secondly, conduction is only one of three modes of heat transfer. Heat may also move in or out of a bird's body by radiation or by convection. As insulation against radiative and convective heat transfer is just beginning to be understood in animals, here we restrict our attention to conductive heat transfer.

16.4.2 Ecological Evidence for the Insulative Value of Feathers

The insulative qualities of feathers are sufficient to keep even some very small birds protected against subarctic winter cold. Some species of chickadees, redpolls, and creepers, which weigh only about 10 g, manage to survive quite well through the northern winter, during which they must face subfreezing temperatures for many days on end. Although it is true that they can seek out microclimatic conditions that are more moderate than those indicated by standard meteorological measurements, even so they frequently have to exist under conditions in which the difference in temperature between the core of their bodies and the ambient air is more than 70°C. The Emperor Penguin, weighing about 30 kg, is so well insulated that it can safely incubate its egg during the middle of the Antarctic winter when the chill factor (the combined effect of temperature and wind) is equivalent to −50 or −60°C or lower.

Aside from this sort of distributional evidence, there are other kinds of correlational evidence for the role that feathers play as insulation. Birds existing in cold climates have heavier, denser plumage with more down than birds of comparable size in the tropics. Many species of north temperate zone birds show seasonal changes in the insulative quality of their plumages, the winter plumage being heavier and downier than the summer one. Young birds frequently are covered with dense down during the early stages of development when their powers for generating heat are still weak; later, the down is shed and replaced by contour feathers.

16.4.3 Quantitative and Experimental Evidence

Direct measurements of the insulative value of plumage or fur in terms of heat loss per unit of surface area for a given temperature difference between the inside and the outside of a living animal are extremely difficult to make. A group of researchers at Point Barrow, Alaska, approached this problem by removing pelts from dead birds and mammals. They carefully applied the fresh skins, with feathers or fur attached, to the surfaces of a hot plate, the temperature of which could be regulated by varying the electrical input. When the insulated hot plate was placed in a controlled temperature chamber, the amount of energy (in the form of electricity) required to maintain a constant "body temperature" in the hotplate could be measured for any given temperature difference between the inside and the outside.

Their methods proved better for fur than for plumage (see section 20.2.1), because it is difficult to obtain a natural arrangement of feathers on a dead pelt. Also, a living bird has a high degree of control over the position of its feathers. By a complicated neuromuscular arrangement interconnecting adjacent feather follicles, birds can fluff out their feathers to a maximum extent, thereby increasing the insulative qualities of their plumage in the cold, or they can depress their feathers very close to their bodies and thereby decrease the thickness and insulation of their plumage. The influence of ambient temperature on feather fluffing has been well demonstrated by some studies on ring doves (Figure 16-14), and the ability to vary the insulative qualities of plumage allows many birds to adjust their body heat loss rather precisely for thermoregulation over a range of ambient conditions.

Most experimental measures of the influence of plumage on heat loss are indirect and involve the metabolic rate and body temperature of the bird. Metabolic rate is commonly measured as oxygen consumption per unit of body weight per unit of time (ml O_2 × hour/mass), and the body temperature is measured electrically by some kind of implanted temperature sensor such as a thermistor. When simultaneous measures of the core body temperature and metabolic rate are made by these or similar methods, characteristic curves in relation to environmental temperatures are obtained for different species of

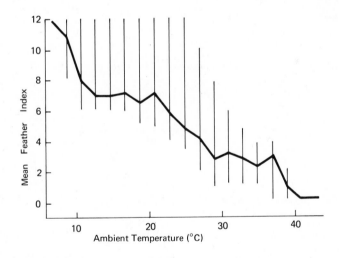

Figure 16-14 Influence of air temperature on the ptilomotor control of feathers in the Ring Dove. The feather index on the ordinate is based on the sum of the following three scores for six different regions of the dove's body: 0 = "sleeked," 1 - "normal," and 2 = "raised." The index provides a quantitative scale ranging from 0 to 12. Vertical lines represent the range of values for several birds. Some of the birds have all of their feathers raised (index = 12) below 25° C. (Modified from McFarland and Baher, in Farner and King [1974] *Avian Biology.* Vol. 4, Academic Press, New York).

birds and mammals under "resting" conditions. Most of them conform to the generalized scheme shown in Figure 16–15.

For each species there is a definable range of environmental temperatures (t_2 to t_5) over which the core body temperature can be regulated through appropriate physiological adjustments of heat loss or endothermic heat production. Above this range the animal's ability to dissipate heat by evaporation is overpowered, and both the body temperature and metabolic rate increase as air temperature increases. A positive feedback system results in explosive heat death at t_6, which is actually a time-dependent variable and not a fixed temperature point on either the ambient or body temperature scales. At environmental temperatures below the range of homeothermy, the animal's ability to generate more endothermic heat to balance heat loss is exceeded, body temperature falls, metabolic rate declines, again as a positive feedback effect of the lowering body temperature, and cold death results at t_1, which is also a time-dependent variable like t_6. Values for the core body temperature of birds at t_1

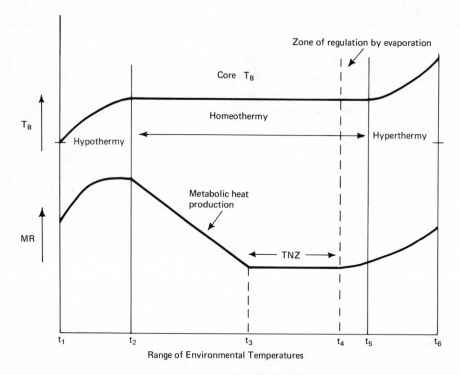

Figure 16–15 Generalized pattern of changes in body temperature and metabolic heat production of an endothermic homeotherm in relation to environmental temperature. Core T_b is the normothermal body temperature and varies somewhat for different mammals and birds (see Table 18–3). t_1 = temperature for hypothermic death; t_2 = incipient lower lethal temperature; t_3 = lower critical temperature; t_4 = upper critical temperature; t_5 = incipient upper lethal temperature; t_6 = temperature for hyperthermic death; TNZ = thermoneutral zone. (Based on various sources, especially W. S. Hoar [1966] *General and Comparative Physiology*. Prentice-Hall, New Jersey.)

are usually in the range of 15–25°C, although there are exceptions that go much lower, and values for t_6 lie in the range of 46 to 48°C. Large species, because of their more favorable surface to mass ratio, usually have a wider range between t_2 and t_5 for homeothermy than small species; well insulated species have lower values for t_2 and t_3 than poorly insulated ones, whereas thinly feathered species tend to achieve a higher value for t_5 than densely feathered ones, particularly when evaporation from the skin is an important mechanism for heat dissipation at high temperatures.

The thermoneutral zone is the range of ambient temperatures within which the bird's resting metabolic rate remains basal and temperature regulation is effected solely by varying the rate of heat loss (conductance) from the body surface to the air. Most of this regulation is brought about by changing the insulative qualities of the plumage (fluffing or sleeking), but postural adjustments and changes in the peripheral vascularity (vasoconstriction or dilation) of exposed parts of the body (feet, legs, face) both influence the total conductance of heat from the surface. The greater the ability of a bird to increase its insulation and reduce conductance by other avenues, the lower the value of t_3 or the **lower critical temperature**. Once these physical limits have been reached, the bird must increase its metabolic rate to compensate for increased heat loss in order to continue thermoregulating; however, the slope of the increase in metabolic rate between t_2 and t_3 is influenced by the conditions of minimum conductance achieved by the bird at the lower critical temperature. Thus, both the lower critical temperature and the slope of metabolism below the thermoneutral zone are fairly good indirect indicators of the insulative qualities of a bird's plumage, because insulation accounts for most of the reduction in conductance.

Figure 16–16 shows two generalized curves of metabolic rate in relation to ambient temperature. These curves can be used to illustrate the results from

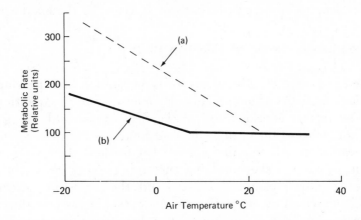

Figure 16–16 Schematic curves of metabolic rate in relation to ambient temperature drawn to illustrate the influence of feathery insulation on heat production. Curve "a" represents a poorly insulated bird; curve "b" a well insulated one. See text for examples and discussion.

several types of experiments that demonstrate the importance of feathers for insulation. There is a genetic anomaly among domestic breeds of chickens known as the "frizzle fowl," so named because of a mutation that results in abnormal formation of weak, stringy barbs. Feathers composed of such barbs have poor insulative value compared to the feathers of a normal fowl. As revealed by metabolic measurements, the frizzle fowl has a higher lower critical temperature and a steeper rate of increase in metabolism below thermoneutrality (curve *a*) than a normal fowl (curve *b*). Similarly, if all the feathers are plucked from a duck or a pigeon, such a nude bird will conform to curve (*a*) in Figure 16–16, whereas its intact, feathered counterpart will conform to curve (*b*). If two similar sized species are compared, the one from a warm climate will be represented by curve (*a*), the one from a cold climate, by curve (*b*). Curve (*a*) could also represent the metabolic pattern of a blue jay in summer plumage, and curve (*b*), the same bird in winter dress.

These examples are sufficient to illustrate the importance of feathers in thermoregulation, and they provide comparative information for the very similar kinds of mechanisms that have been independently evolved by mammals (section 20.2).

16.5   AVIAN FLIGHT

Flapping flight has evolved four different times during the course of evolution. Among the arthropods, many kinds of insects fly by flapping movements of stiff, membranous wings, and analogous modes of locomotion have appeared in three separate groups of vertebrates—pterosaurs, birds, and bats. As has been discussed in Chapter 3, the wings of these different vertebrates represent examples of convergent evolution, and the actual structural details of the wing design are quite different in the three groups. Only birds employ a complicated series of overlapping epidermal derivatives (feathers) as the main wing surface.

16.5.1   Evolutionary Origin of Flight in Birds

What were the selective advantages for the evolution of wings and flight in the "proavian" ancestors of birds? To look for answers to this question we need to give some further consideration to the first known bird, *Archaeopteryx*, and to its life habits as they can be inferred from its structure. *Archaeopteryx* was fully feathered, and this fact means that feathers themselves must have evolved much earlier in the nonflying ancestors of birds, probably under the influence of selective pressures associated with the acquisition of endothermy in this clade of archosaurs. It seems reasonable, then, to assume that the evolution of avian wings and flight began in a line of well-feathered, insulated, homeothermic, highly active, bipedal animals.

What kind of a bird was *Archaeopteryx* in life? It is usually considered to have been an arboreal climber, jumper, and glider with limited powers of flapping flight allowing it to extend the distance it could travel through the air between trees. It scampered about bipedally on the branches of trees, aided by grasping toes and a reversed hallux, by the grasping claws on the leading edges of its wings, and by a long, balancing tail. Most reconstructions and pictures of *Archaeopteryx* are based on this conception.

Ornithologists have often compared *Archaeopteryx* with the hoatzin and other arboreal species with limited powers of flight, such as chachalacas and turacos (Figure 16–17, Table 16–3). The young hoatzin has been particularly attractive to adherents of this arboreal conception because it climbs about in the branches aided by functional clawed fingers on its wings; its claws, however, are deciduous and do not occur on the adult.

The arboreal theory of the origin of avian flight has been the most commonly held one. According to this theory, the proavian ancestors of *Archaeopteryx* were tree-climbers, which jumped from branch to branch and from tree to tree much like some squirrels, lizards, and monkeys do. Under selection pressures favoring adaptations that increase the distance and accuracy of travel between trees, structures that would aid in gliding by providing some surface area for

Figure 16–17 Reconstruction of "arboreal" *Archaeopteryx* compared with hoatzin, chachalaca, and touraco.

Table 16–3. Ratios of Lengths of Wing and Leg Bones in *Archaeopteryx* and Similar Modern Birds (percent)

Species	Ulna: Humerus	Carpomet.: Humerus	Dig II, phal 2: Humerus	Femur: Tibiotarsus	Tarsomet.: Tibiotarsus
Archaeopteryx, I	87	43	24	73	49
Archaeopteryx, II	93	46	26	73	55
Turaco (*Crinifer concolor*)	89	28	22	67	56
Chachalaca (*Ortalis leucogastra*)	97	51	19	72	64
Hoatzin (*Opisthocomus hoazin*)	107	56	25	74	60

Modified from Brodkorb [1].

lift would be favored. As a result, contour feathers on the arms elongated into primitive flight feathers, and the evolution of volant forms passed from gliding stages through intermediate stages, such as *Archaeopteryx*, in which gliding was aided by weak flapping flight, to fully airborne flapping fliers.

One problem that has never been satisfactorily explained by the arboreal theory is selection for bipedalism in an arboreal habitat. A bipedal stance and locomotion are not very adaptive for arboreal life, unless the forelimbs have also been specially modified, as for brachiation in primates. The grasping claws on the wings might have evolved in response to such needs of a bipedal animal moving into the trees; but most of the early thecodonts that could have been ancestral to birds were already bipedal, ground-dwelling runners, and it seems certain that bipedal locomotion preceded the evolution of flight. In fact, it seems most plausible that even the initial modifications of the forelimbs for flight did not occur until an obligate bipedal condition had been achieved and the forelimbs had been completely freed of any involvement in quadrupedal locomotion and weight support.

If, as seems reasonable from the fossil record of the thecodonts and the structure of *Archaeopteryx* itself, the ancestral lineage giving rise to birds consisted of bipedal, cursorial forms, is it necessary to invoke arboreal selection pressures at all for the evolution of avian flight? Another much criticized theory but one as old as the arboreal theory, in fact, postulates that flapping flight evolved directly from ground-dwelling, bipedal runners.

According to this "cursorial theory" the proavians were fast, bipedal runners that used their primitive wings as planes to increase lift and lighten the load for running. In a later development, the wings were flapped as the animal ran to provide additional forward propulsion, much as a chicken flaps across the barnyard to escape from a dog. Finally, the pectoral muscles and flight feathers became sufficiently developed for full powered flight through the air.

There are quite a few skeletal details to indicate that *Archaeopteryx* was derived from a cursorial ancestor and that it was itself mainly ground-dwelling. The pelvis is very similar to that of the bipedal, cursorial ornithischian dinosaurs and certainly must represent a convergence for postural support and the accommodation of muscles for similar locomotion. The hindlimb and hindfoot show many cursorial adaptations that have almost exact parallels in the smaller saurischian theropods, especially the coelurosaurs, whose skeletons reveal them to have been fast-running, bipedal carnivores. These features include (1) a vertical, parasaggital position of the limbs, (2) a fused metatarsus, (3) an intertarsal joint, and (4) a tridactyl foot with digit 1 reduced, directed to the rear, and slightly elevated, providing a digitigrade stance common in ground-runners. A decumbent hallux, as found in perching birds, would be the expected condition if avian evolution had been strongly under the influence of selective forces associated with arboreal conditions for a long period prior to *Archaeopteryx*. Further, the claw-bearing ungual phalanges have a structure more nearly like that of terrestrial forms than either arboreal or predatory forms (Figure 16–18), all of which indicates that powerful opposition of the hallux against the three forward toes was not important for *Archaeopteryx* either for branch-grasping or for prey-grasping.

Many have asserted that the long tail and the three claws on each wing of *Archaeopteryx* argue against the cursorial theory and for a gliding origin, because such a tail would have been dragged on the ground and quickly worn to a frazzle, and because no one has believed that the claws on the wings could be used for a function other than clinging onto branches. But all of the bipedal dinosaurs had long tails, which indeed are a necessary counterweight for the bipedal stance of a cursor and which they no doubt held up off the ground when running. Also a number of ground-dwelling birds, such as the roadrunner, the secretary bird, and several species of pheasants, have long feathery tails, which they manage to keep intact while running through deep grass, brush, and cactus. Finally, the coelurosaurs had forelimbs with three long-clawed fingers that were apparently used for snatching prey on the run.

The cursorial theory in its original form fails, however, in physical and mechanical terms as an explanation, because flapping is not an effective mechanism to increase running speed. Maximum traction on the ground is required for acceleration or high speed from leg movements, and this requirement can only be met by solid contact with the feet on a firm substrate. Planing with primitive wings would have reduced traction, and the push from the small surfaces of the protowings probably would not have compensated the loss in speed from the hindlegs, much less added to acceleration.

A recently identified specimen of *Archaeopteryx* misidentified as a pterosaur for over 100 years has revealed some previously unknown details of the hand and has led to a plausible modification of the cursorial theory. Some elements of the manus are extremely well preserved in this specimen and show the actual horny claws on digits 1 and 3, a preservation that is unique among Mesozoic fossil vertebrates (Figure 16–19). These claws look more like the talons of a predatory bird than like devices for clinging to tree trunks, and the

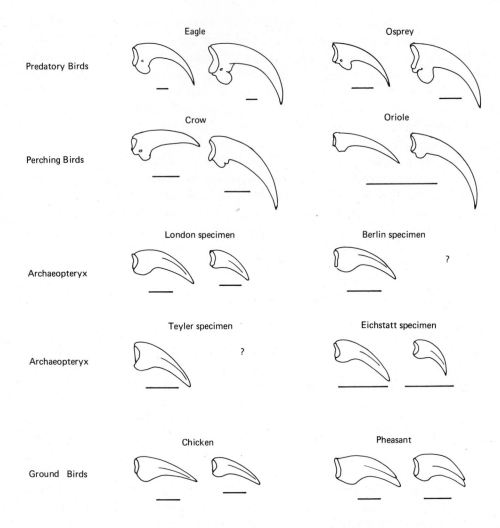

Figure 16-18 Comparison of the ungual phalanges of *Archaeopteryx* foot with those of a modern arboreal, terrestrial, and predatory bird. Horizontal lines under each are 1 cm. (Redrawn from Ostrom [5].)

well-developed flexor tubercles on the unguals of these claws (Figure 16–20) support the idea that they had a powerful grasping action.

The striking similarities in morphology between the hand, metacarpus, forearm, humerus, and pectoral apparatus of *Archaeopteryx* and those of several coelurosaurs are evidence of a similarity in biological roles for both— a powerful grasping function for predation. (They may also be evidence of a closer phylogenetic relationship than previously thought.) Although bearing feathers, the forelimb of *Archaeopteryx* has not been structurally much modified from the skeletal condition of these small theropods, and it differs from all other known birds in lacking a number of features that are critical for powered flight (fused carpometacarpus, restricted wrist and elbow joints, modified coracoids, and a platelike sternum with keel). Thus, the entire pectoral append-

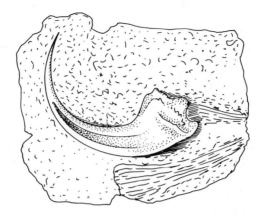

Figure **16-19** Detail of finger with claw on the Teyler Museum specimen of *Archaeopteryx*. Straight length of claw is 10.5 mm. (Drawn from photograph by Ostrom [5]).

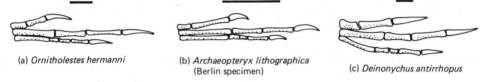

(a) *Ornitholestes hermanni*   (b) *Archaeopteryx lithographica* (Berlin specimen)   (c) *Deinonychus antirrhopus*

Figure **16-20** Comparison of the wing claws of *Archaeopteryx* with predator claws of coelurosaurs. Horizontal lines are 1 cm. (Redrawn from Ostrom [5].)

age of *Archaeopteryx* appears to have been better adapted for predation than for flight.

From these considerations, John Ostrom has postulated that enlargement and elongation of the contour feathers on the proavian forelimb might have been related to this predatory function, natural selection favoring those individuals with improved prey-catching ability. According to this view, the incipient "wings" of the proavians evolved first as snares to trap insects or other prey against the ground or to knock them down out of the air, making it easier for them to be grasped by the claws and teeth. These structures subsequently became further modified into flapping appendages capable of subduing larger prey and coincidentally aided in leaping attacks on that prey. It is interesting to note that some modern birds—geese are a well known example—flail their enemies on the ground by striking out with their wings, and many insectivorous and raptorial birds "flash" their wings at potential prey, an action that apparently either startles or confuses the prey and makes them more vulnerable to capture.

The known structural features of *Archaeopteryx* are consistent with those of a ground-dwelling, bipedal, running predator—rather like the modern roadrunner—which was not capable of powered flight other than fluttering leaps after aerial prey. In support of the hypothesis that *Archaeopteryx* could fly, their primaries had the same asymmetrical arrangement of outer and inner vanes as found in modern flying birds (Figure 16-7b). Sustained, flapping flight eventually emerged from this preadaptation for it (Figure 16-21).

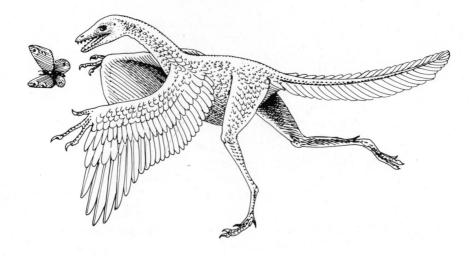

Figure 16-21   Reconstruction of *Archaeopteryx* as a ground-running predator.

16.5.2 Modification of Form and Function for Flight

To one who is familiar with many avian species, birds may appear quite varied in their shapes and sizes. In some ways they are variable; there are many different kinds of feet and beaks adapted for different modes of locomotion and feeding (see Chapter 17), but the structural design of birds is far more uniform than that of either mammals or reptiles. Take body size for example: The largest living bird is the flightless ostrich weighing around 144 kg, and the largest known to have lived, one of the elephant birds, weighed an estimated 450 kg; one of the smallest is the scintillant hummingbird, which averages 2.3 g. The ostrich is some 66,000 times heavier. If one considers only flying species, the largest volant bird known was one of the giant condors, *Teratornis incredibilis*, which had a wingspan estimated to be 5 m and a possible body weight of some 20 kg, roughly 10,000 times larger than the hummingbird. By contrast, the largest mammal, the blue whale weighing 136,200 kg, is 59 million times heavier than the pigmy shrew at 2.3 g. Or, if one restricts his comparison to quadrupedal, terrestrial mammals, then the elephant weighing in at some 2000 kg is roughly a million times larger than the shrew.

Clearly birds have not been able to reach the large body sizes of mammals, and presumably there must be physical and physiological limits to the maximum size attainable by flying birds. These limits are determined by surface-to-volume ratio, capacity to do work, which can be expressed as pectoral muscle mass, attainable flight speed, wing-loading, and other aerodynamic properties. It is interesting to note, however, that the largest known pterosaur had a wingspan estimated to be 15.5 m, based on the size of its humerus, whereas the largest flying mammal, a Javanese flying fox (Megachiroptera), attains a wingspan of less than 1.6 m and a body weight of about 2.5 kg.

The actual limitations on attainable body size have been different for the three different groups of flying vertebrates, and it is not possible, really, to say how closely these achieved sizes have explored the physical and physiological limits of a particular design (pterosaurian, avian, chiropteran), and how much they may have been determined by ecological limitations (other volant competitors, predators, availability of exploitable food). It seems doubtful that *Teratornis incredibilis* and the other large Pleistocene condors could have evolved in the absence of the North American mammalian megafauna or a comparable source of large carcasses upon which to feed, and it certainly was no accident that these large condors died out synchronously with the large ungulates.

The structural uniformity of birds is seen even more clearly if their body shapes are compared with those of mammals or reptiles. The design of nearly every species of bird is determined by requirements for flight, and even those species that have become secondarily flightless retain many features of their volant ancestors. The uniformity of birds is reflected in the requirements for sound aerodynamic design, and no successful volant species can deviate very far from these requirements.

Birds fly as the result of several evolutionary inventions—feathers, endothermy, wings and associated pectoral muscles, hollow bones, a special respiratory system, and a large, powerful heart being some of the most important avian features for flight. These modifications for flight can be considered under two basic categories: weight-reducing adaptations and power-promoting adaptations (Table 16–4).

Table 16–4. Avian Adaptations for Flight

Weight-reducing adaptations	*Power-promoting adaptations*
Thin, hollow bones	Homeothermy and high metabolic rate
Extremely light feathers	Heat-conserving plumage
Elimination of teeth and heavy jaws	An energy-rich diet
Elimination of some bones and extensive fusion of others	Rapid and efficient digestion
A system of branching air sacs	Highly efficient respiratory system
Oviparity rather than viviparity	Air sacs for efficient cooling during muscular activity associated with flight
Atrophy of gonads in nonbreeding periods	Breathing movements synchronized with wing beats
Eating concentrated foods	Large heart and rapid high pressure circulation
Rapid and efficient digestion	
Excretion of uric acid instead of urea	

Modified from Welty [11].

Despite the fact that birds must have larger and stronger pectoral and pelvic girdles and appendages than quadrupedal mammals, so that the full requirements for locomotion can be supported by the wings alone or the legs alone, the avian skeleton is much lighter in relation to total body weight than that of a mammal of similar size; for example, the skeleton of a pigeon is about 4.4 percent of body weight, whereas that of a laboratory rat is 5.6 percent. A frigate bird with a 1.5 m wingspread has a skeleton that weighs only 114 g, less than the weight of its feathers.

Hollow tubes have greater strength than solid rods of the same mass, and the avian skeleton has taken advantage of this principle to produce a supporting structure that is not only light-weight but also very strong and elastic, necessary characteristics in an air-frame subject to the stresses of aerial maneuvers. Internal trusslike reinforcements in some of the hollow bones of birds further add to their strength. This type of reinforcement is particularly well developed in the avian skull, and in some of the hand bones of large, soaring birds, in which they have an arrangement like the Warren truss used in aircraft and bridge design (Figure 16–22). Birds have further decreased weight in the head region through the loss of teeth: the skull and teeth of a fox comprise about 1.25 percent of its body weight, whereas the skull and horny beak of a golden eagle are only 0.2 percent of its total weight. Fusion of bones and loss of some bones also aid in weight reduction and increased strength.

The feathers themselves are not only light but extremely strong and resilient for their weight. The specific gravity of a feathered Mallard is 0.6, whereas that of a plucked duck is 0.9. By comparison, the specific gravity of most mammals and other vertebrates is close to 1.0. The lower value for a plucked bird results from the extensive system of air sacs that take up much of the body volume inside (Figure 16–23).

Modification of some internal organ systems also contributes to weight reduction. Birds excrete uric acid in a semi-solid paste and, therefore, do not possess a urinary bladder or other storage organ for the waste products of nitrogen metabolism and excretion. No bird is viviparous, and in most species there is only one (left) ovary and oviduct. The gonads of both males and females atrophy during the nonbreeding period of the life cycle. Also most birds consume food of high caloric value; their digestion and elimination are rapid; and they readily store extra calories in the form of fat, which gram for gram has the highest energy content of any biological product.

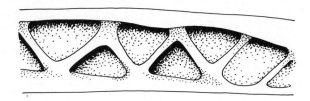

Figure 16–22 Internal struts in the hollow metacarpal of a vulture. (Redrawn from Welty [11].)

Power-promoting adaptations are equally important for flapping flight. Since the rate of biochemical reactions roughly doubles for each 10°C rise in temperature, the high body temperatures (about 40–41°C) of birds allow them to operate their metabolic machinery at a fast speed (section 7.5.2). Most of the related functions in Table 16–4 are involved in the rapid and efficient conversion of chemical energy (from food or body fat reserves) to kinetic energy or mechanical work, which is performed by powerful muscle action. The pectoral muscles provide the motive force for flapping flight. In a strong flying bird, such as a pigeon or falcon, the breast muscles make up more than 30 percent of the body weight, whereas gliding and soaring species that flap less usually have reduced keels on their sterna and pectoral muscles that are less than 30 percent of body weight. There is also a big demand on the circulatory system to move oxygen and other needed materials rapidly to sites of high metabolic activity and to remove carbon dioxide and other wastes; therefore, the avian heart has evolved into a large and powerful muscle with rapid contractions (Table 16–5). In both birds and mammals, the smaller the species is, the larger its relative heart size and resting heart rate; but gram for gram of body weight birds have larger and faster hearts than mammals.

The avian respiratory system not only allows for efficient one-way movement of air through the lungs, in contrast to the tidal flow in mammals, but the extensive air-sac system apparently allows for a powerful cooling mechanism to dissipate the large amounts of heat generated by the pectoral muscles and heart during flight (Figure 16–23).

Successful flight also demands a high degree of neuromuscular coordination and navigational ability. Birds have relatively large brains, comparable in size to those of rodents and much larger than the brains of lizards. The rather large avian forebrain and cerebellum are supplied with numerous association centers, especially centers concerned with coordinating muscle actions. As a bird relies mainly on visual information for maneuvering through the air, the optic lobes are large, as are the eyes, so much so that the eye sockets meet in the center of the skull. The sense of smell, on the other hand, is not highly developed in most birds, and the olfactory lobes of the brain are correspondingly reduced in size.

Table 16–5. Comparison of Heart Sizes and Rates in Some Vertebrates.

Species	Heart as percent body weight	Heart beats per minute
Boa constrictor	0.31	20
Bull frog	0.32	22
Man	0.42	78
Domestic dog	1.05	140
Turkey vulture	2.07	301
Common crow	0.95	342
House sparrow	1.68	460
Ruby-throated hummingbird	2.37	615

Modified from Welty [11].

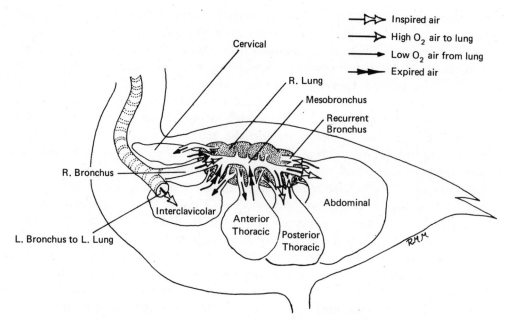

Figure 16-23 Diagram of the avian respiratory system to show the relation of lungs to the connecting air sacs in a bird's body. Only the respiratory system of the right side is shown.

16.5.3   Aerodynamics of Avian Wing Compared to Fixed Airfoils

Unlike the fixed wings of an airplane, the wings of a bird function both as an airfoil (lifting surface) and as a propeller for forward motion. The avian wing is admirably designed for these functions, consisting of a light, variable airfoil for contact with the air (Figure 16-24). The primaries, inserted on the hand bones, do most of the propelling when a bird flaps its wings, while the secondaries along the arm provide lift. Experimental removal of flight feathers from the wings of doves and pigeons show that when only a few of the primaries are pulled out the bird's ability to fly is greatly altered or entirely prevented; however, even when as much as 55 percent of the total area of the secondaries has been removed, the bird can still fly.

A bird also has the ability to alter the area and shape of its wings and their positions with respect to the body for the purposes of maneuvering, changing direction, landing and taking off, and changing the aerodynamic properties of its wings to meet the requirements for flight under various circumstances. Moreover, a bird's wing is not a solid structure like a conventional airfoil (airplane wing) but allows the flow of some air through and between the feathers. Other parts of the body surface play a more passive role in flight, but the tail feathers (rectrices) are spread or moved in various ways to assist in braking for landing, turning, or to increase area.

Obviously the functioning of a bird's wing in flight—even in nonflapping flight—is vastly more complex than the functions of a fixed wing on an airplane

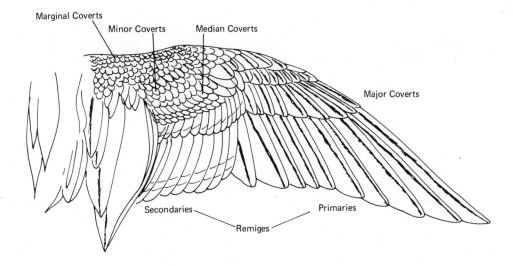

Marginal Coverts
Minor Coverts
Median Coverts
Major Coverts
Secondaries
Remiges
Primaries

Figure 16-24 Dorsal surface of a bird's wing, showing arrangement of the various feather tracts that produce a variable airfoil. (Redrawn from Thomson [10].)

or glider. Nevertheless it is instructive to consider a bird's wing in terms of the basic performance of a fixed airfoil. Although a bird's wing actually moves forward through the air, it is easier in some ways to think of it as stationary with the air flowing past, as would be the case in an experimental wind tunnel. The flow of air produces a force, which is usually termed the **reaction.** It can be resolved into two components, the **lift,** which is a vertical force equal to or greater than the weight of the bird, and the **drag,** which is a backward force opposed to the bird's forward motion and to the movement of its wings through the air. (Figure 7-2).

Consider now what happens when a drop of water falls through the air. The friction and pressures of the resisting air mold it into a tear-drop or fusiform shape. The reason is that this shape has the least resistance to the surrounding air, the least *drag* in aerodynamic terms. The same principle applies to any body moving in air or water and accounts for the evolution of fusiform bodies in aquatic vertebrates and birds, as we have already considered in Chapter 7.

When the leading end of a symmetrically streamlined body cleaves the air "head on" it thrusts the air equally upward and downward, reducing the air pressure equally on the dorsal and ventral surfaces. No lift results from such a condition. There are two ways to modify this system to generate lift. One is to increase the **angle of attack** of the airfoil, and the other is to modify its surface configuration.

When the contour of the dorsal surface of the wing is convex and the ventral surface is concave (a **cambered foil**) the air pressure against the two surfaces is unequal because the air has to move farther and faster over the dorsal convex surface relative to the ventral concave surface (Figure 16-25). The result is reduced pressure over the wing or lift. When the lift equals or exceeds the bird's body weight, it becomes airborne.

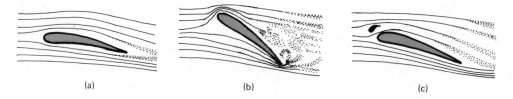

(a)   (b)   (c)

Figure 16-25 Diagram of air flow around a cambered airfoil. At a low angle of attack (a), the air streams smoothly over the upper surface of the wing and creates lift. When the angle of attack becomes steep (b), air passing over the wing becomes turbulent, decreasing lift enough to produce a stall. A wing-slot (c) helps to prevent stalling turbulence by directing a flow of rapidly moving air close to the upper surface of the wing. (Redrawn from Welty [11]).

If the leading edge of the wing is tilted up so that the angle of attack is increased, the result is also increased lift up to an angle of about 15°, the **stalling angle**. This lift results less from an increase in pressure below the airfoil than from a decrease in pressure over the dorsal surface.

Increasing the angle of attack also increases the drag. A comparison of an inclined flat plane, a cambered plane, and a streamlined airfoil (Figure 16–26) reveals that the cambered airfoil has the greatest **lift to drag ratio** (L/D ratio) and is therefore the most efficient design for flight. This shape is rather easily evolved from a bird's forelimb, the thick leading edge being filled out by the bones and muscles, while the thin trailing edge is made up of the backward projecting flight feathers, the whole surface being further streamlined and cambered by the patagium and the various rows of upper wing and under wing coverts (Figure 16–24). Drag is also influenced by a number of variables such as the weight and size of the bird, its body shape, the shape of its wings, the texture of its surface, and by air disturbances around the wings and body.

If the smooth flow of air over the wing becomes disrupted, the air flow begins to separate from the wing resulting from the increased air turbulence over the wing. The wing is then stalled. Stalling can be prevented

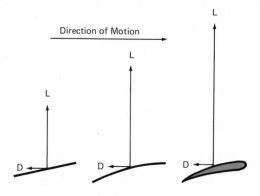

Figure 16-26 Comparison of lift (L) and drag (ᴅ) for an inclined plane, a cambered plane, and a streamlined airfoil. (Redrawn from Welty [11].)

or delayed by the use of slots or auxilliary airfoils on the leading edge of the main wing. The slots help to restore a smooth flow of air over the wing at high angles of attack. The bird's **alula**, or bastard wing, functions for this purpose, particularly during landing or taking off (Figure 16-27). Also, the primaries function as a series of independent, overlapping airfoils, each tending to smooth out the flow of air over the one behind.

At a given angle of attack, lift increases with increased air speed. Conversely, for lift to equal weight and for the bird to remain airborne, the angle of attack must increase as air speed decreases. The speed can fall until the stalling angle is reached. This is the **stalling speed**, the lowest speed at which flight can be maintained for a given airfoil and wing loading.

Another characteristic of an airfoil has to do with wing tip vortexes—eddies of air resulting from outward flow of air from under the wing and inward flow from over it. This is so-called **induced drag**. One way to reduce the effect of these wing tip eddies and their resulting drag is to lengthen the wing, so that the tip vortex disturbances are widely separated and there is proportionately more wing area where the air can flow smoothly. Thus, long, narrow wings are more efficient than short, broad ones, but the former require more support from the skeleton and are less maneuverable. The ratio of length to width is called the **aspect ratio**. Long, narrow wings have high aspect ratios and correspondingly high lift to drag ratios. High performance sailplanes and albatrosses,

Figure 16-27 A landing bird, showing maximum extension of the alula beyond the leading edge of the main wing. (Based on a photograph of a Robin in Welty [11].)

for example, have aspect ratios of 18:1 and *L/D* ratios in the range of about 40:1 (not actually measured for the albatross).

Wing loading or **span loading** is another important consideration. This is the weight of the bird divided by the wing area. Induced drag is also proportional to wing loading. The lighter the loading, the less power needed to sustain flight. There is a general relationship with body size: Small birds have lighter wing loadings than large ones; but wing loading is also related to adaptations for powered versus soaring flight. The comparisons in Table 16-6 point out both of these trends. The hummingbird, barn swallow, and mourning dove have lighter wing loadings than the larger peregrine, golden eagle, and mute swan; yet the 3-g hummingbird, a powerful flier, has a heavier wing loading than the more bouyant, sometimes soaring barn swallow, which is more than five times heavier; similarly the rapid stroking peregrine has a heavier wing loading than the larger, often soaring golden eagle.

The conditions for stable, level flight are summarized in Figure 16-28. At a given air speed, the system is in equilibrium when the lift is equal to weight, and the forward propulsion (P) is equal to the total drag (D_t) on the wings and

Table 16-6. Wing Loading in Some North American Birds

Species	Weight (g)	Wing area (cm²)	Grams per cm² wing
Ruby-throated hummingbird	3.0	12.4	0.24
House wren	11.0	48.4	0.23
Black-capped chickadee	12.5	76.0	0.16
Barn swallow	17.0	118.5	0.14
Chimney swift	17.3	104.0	0.16
Song sparrow	22.0	86.5	0.25
Leach's petrel	26.5	251.0	0.11
Purple martin	43.0	185.5	0.24
Redwinged blackbird	70.0	245.0	0.29
Starling	84.0	190.3	0.44
Mourning dove	130.0	357.0	0.36
Pied-billed grebe	343.5	291.0	1.20
Barn owl	505.0	1683.0	0.30
Common crow	552.0	1344.0	0.41
Herring gull	850.0	2006.0	0.42
Peregrine falcon, female	1222.5	1342.0	0.91
Mallard	1408.0	1029.0	1.40
Great blue heron	1905.0	4436.0	0.43
Common loon	2425.0	1358.0	1.80
Golden eagle	4664.0	6520.0	0.71
Canada goose	5662.0	2820.0	2.00
Mute swan	11602.0	6808.0	1.70

Modified from E. L. Poole [1938] *Auk* 55:511-517.

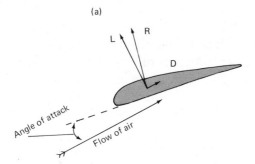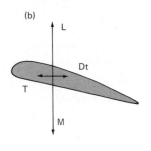

Figure 16-28 Conditions for a stable glide and for stable level flight. L = lift, D = drag, and R = resultant force, and T = Thrust. In (a) the force(s) acting on the wing during a stable glide (usually during descent) is represented by R, which vectorially can be divided into L, a force in opposition to gravity, and D, a resistance force in opposition to the direction of motion. The scheme in (b) shows the conditions during stable level flight, when L equals body mass (M) and the total drag (D_t) equals the forward propulsive force (N) or the thrust (T) generated by wing motion. (Modified from Borwn [2]).

body. Thus, the reaction (R) operating on an airfoil is proportional to the square of its velocity through the air, to the wing loading, and roughly to the angle of attack up to the stalling angle of about 15°.

16.5.4   Gliding Flight

All birds can stretch their wings and glide, and some species depend primarily on this mode of flight to soar on rising air currents (see Chapter 17 for details on soaring adaptations). The conditions of a stable glide, one in which air speed remains constant in still air, are like those for level flight, except that the flight path is inclined downward at some angle that is determined by the lift to drag ratio (Figure 16–28). The lift is inclined ahead of the vertical axis of the airfoil to produce a force in the direction of flight that is equal to the total drag (Section 7.1). Obviously a glide can continue without loss of air speed only by loss of altitude, in which case the drag multiplied by distance traveled along the glide path will equal the loss of potential energy, which is the weight multiplied by the loss in height:

$$\frac{\text{weight}}{\text{rate}} = \frac{\text{distance moved}}{\text{height lost}}$$

The higher the *L/D* ratio, the smaller the glide angle has to be for a given air speed, and the lower the **sinking speed**, the height lost per unit time, and consequently the farther a bird can move horizontally per distance dropped. Since the *L/D* ratio is importantly influenced by the aspect ratio of the wing and by size and weight of the flying body, it can be expected that different species of birds will have vastly different gliding performances, depending on their sizes and the shapes of their wings.

Some studies of different kinds of birds have been carried out in wind tunnels or by observing the actions of wild birds in free flight in comparison with a sailplane flying under the same conditions, in order to determine the basic performance of birds during gliding. The **glide polar** is a curve showing the relationship between sinking speed and forward speed through the air and is a convenient way of summarizing the gliding performance of a bird or aircraft. Figure 16–29 shows the main characteristics of the glide polar. The forward speed at which loss of height is minimal, the speed for minimum sink (V_{ms}), is somewhat higher than V_{min}. Still higher is the speed for best glide ratio (V_{bg}), which can be determined by drawing a tangent to the curve from the origin of the graph. By flying at V_{bg} a bird or aircraft covers the greatest horizontal distance for a given loss of height. Figure 16–30 presents generalized glide polars for several species of birds and compares them with the performances of the astomite model airplane and the SHK sailplane.

Because of its very much larger size and wingspan, the sailplane has a far superior L/D ratio than any bird that has been studied, approaching 40:1 at air speeds of 20 to 25 m/sec. Some vultures approach the L/D ratio of the sailplane, and it can be estimated on the basis of known aspect ratios and flying speed that albatrosses probably have L/D ratios in the range of 30:1 to 40:1. Again, because of its size and weight, the sailplane has to maintain a faster forward air speed than the birds, and its minimum gliding speed is around 18 m/sec, well above the speed for best glide ratios of most birds. Birds have much lower stalling speeds and optimum air speeds for gliding because of their lighter wing loadings (again primarily a function of size). In general, soaring birds, such as vultures and albatrosses, have gliding performances superior to birds like pigeons and falcons that rely mainly on flapping flight.

As a mode of locomotion through air, gliding flight can function in two ways. One is to cover distance over ground to a given destination along a single glide

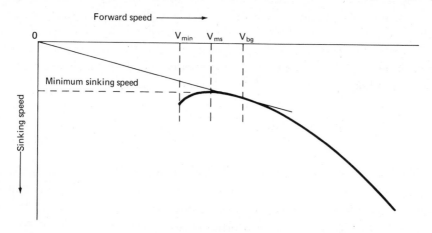

Figure 16-29 Main parameters of the "glide polar." The minimum gliding or stalling speed (V_{min}) is a function of the maximum lift coefficient, body mass, and the wing area. (Redrawn from Pennycuick [6].)

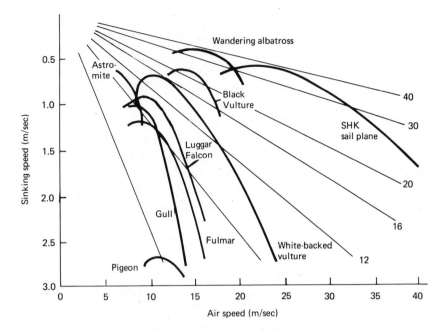

Figure 16-30 Generalized glide polars for several species of birds and aircraft. Diagonal straight lines represent points with the L/D values indicated. The curve for the albatross is speculative and based on known air speeds and L/D values approximated from aspect ratios and body weights. (Modified after V. A. Tucker and G. C. Parrott [1970] *J. Exp. Biol.* 52:345–367, and after Penny-cuick [6]).

path, as when a vulture launches from the top of a cliff in still air and glides to some point on the ground below. The other is to remain aloft by soaring, that is, by gliding in air that is rising faster than the sinking speed of the bird. Any aircraft or bird accomplishes these two goals to the degree that is determined by its design and prevailing atmospheric conditions. A design that is optimal for one goal is not optimal for the other, and this is the basic meaning of the different glide polars shown in Figure 16–30.

Birds have lower *L/D* ratios and lower velocities than the sailplane, and therefore their performances are inferior when the goal is to achieve maximum distance along a glide path in still air, although again birds that are primarily adapted for soaring have better performances than birds that rely on powered flight. When static soaring is the goal, however, then some birds are as good or better than the sailplane. The best avian soarers have lighter wing loadings, lower stalling speeds, and lower sinking speeds than the sailplane, and for these reasons birds can frequently rise in upward moving air when a sailplane cannot do so. The sailplane has two disadvantages compared with the birds: because of its heavier wing loading, it can only gain altitude in air that is rising considerably faster than that required to lift a bird; because of its larger size, it has a much larger turning radius than a bird and consequently it cannot maneuver as easily to take advantage of small volumes of rising air (thermals).

16.5.5 Flapping Flight

Flapping flight is remarkable for its automatic, unlearned performance. A young bird on its maiden flight uses a form of locomotion so complex that it defies precise analysis in physical and aerodynamic terms. The nestlings of some species, such as hornbills, some owls, swifts, swallows, some parrots, woodpeckers, petrels, and others develop in confined spaces (burrows, cavities) in which it is impossible for them to spread out their wings and practice flapping before they leave the nest. Despite this seeming handicap, many of them are capable of flying considerable distances on their first flights. A young African bank martin (swallow), for example, flew continually for 6 min the first time it left its nest. Young whale birds and diving petrels (Procellariiformes) have been observed to fly as far as 10 km the first time out of their burrows. On the other hand, young birds reared in more open nests frequently flap their wings vigorously in the wind for several days before flying—especially large birds such as albatrosses, storks, vultures, and eagles. Such flapping may help to develop musculature, but it is doubtful whether these birds are "learning how to fly"; however, a bird's flying abilities do improve with practice for a period after it leaves the nest.

There are so many variables involved in flapping flight that it becomes difficult to understand exactly how it works. A beating wing is flexible and porous and yields to air pressure, unlike the fixed wing of an airplane. Its shape, expanse, wing loading, camber, sweepback, and the position of the individual feathers all change remarkably as a wing moves through its cycle of locomotion (Figure 16–21a). Furthermore, the frequency and amplitude of the beats change in velocity and angle of attack during a single cycle. This is indeed a formidable list of variables, and it is no wonder that flapping flight has not yet fully yielded to explanation in aerodynamic terms; however, the general properties of a flapping wing can be described.

We can begin by considering the flapping cycle of a small bird in flight. According to the forces diagrammed in Figure 16–28, a bird cannot continue to fly straight and level unless it can develop a force or "thrust" to balance the drag operating against forward momentum. The flapping of the wings, especially the wing tips (primaries), produces this thrust, while the inner wings (secondaries) do not move very fast and continue to act as an airfoil generating lift as when the bird is gliding (Figure 16–31 b and c). It is easiest to consider the forces operating on the inner and outer wing separately. Mos of the lift, but also most of the drag, comes from the forces acting on the inner wing and body of a flying bird. The forces of the wing tips derive from two motions that have to be added together. The tips are moving forward with the bird, but at the same time they are also moving roughly downward relative to the bird (Figure 16–31d). The wing tip would have a very large angle of attack and would stall if it were not flexible. As it is, the forces on the tip cause the individual primaries to twist as the wing is flapped downward and to produce the forces diagrammed in Figure 16–31c. The forces acting on the two parts of the bird combine to produce the conditions for equilibrium flight shown in Figures 16–28b and 31e.

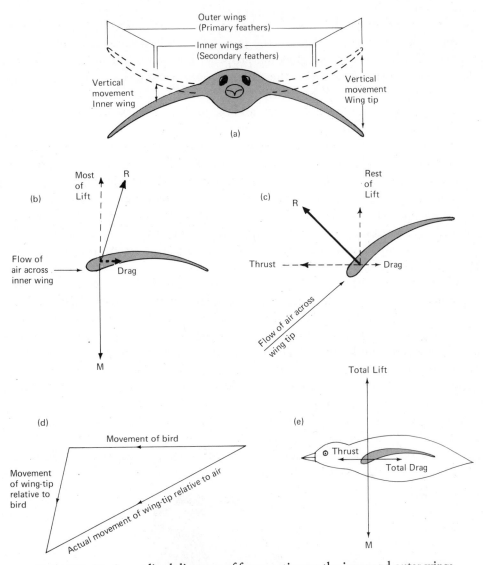

Figure 16-31 Generalized diagrams of forces acting on the inner and outer wings and the body of a bird in flapping flight. (a) Upstroke (dotted lines) and downstroke positions of the inner and outer wings reveal that vertical motion is mostly applied at the wing tips. Thus, the inner wing (area of secondaries) functions as if the bird were gliding to generate the forces shown in (b), while the outer wing generates the force shown in (c). Most of the lift to counter the force of gravity in the body mass (M) is generated by the inner wing and body. Canting of the wing tip (the primaries) during the downstroke (c) produces a resultant force (R) that is directed forward to produce thrust. The relative movement of the wing tip to the air is affected by the forward motion of the bird through the air (d). As a result of this motion and canting of the wing tips during the downstroke the flow of air across the primaries (c) is different than the flow of air across the secondaries and the body (d). When flight speed through the air is constant the forces acting on the inner wing and the body (b) and on the outer wing (c) combine to produce a set of summed vectors where thrust exceeds total drag and lift at least equals the body mass (e). (Modified from Thomson [10]).

As the wings move downward and forward on the downstroke, which is the power stroke, the trailing edges of the primaries bend upward under air pressure, and each feather acts as an individual propeller biting into the air and generating thrust. During this downbeat, the thrust is greater than the total drag, and the bird accelerates. In small birds, the return stroke, which is upward and backward, provides little or no thrust and is mainly a passive recovery stroke. The bird decelerates during the recovery stroke.

Contraction of the **pectoralis major**, the large breast muscle, produces the downstroke during level flapping flight, whereas the upstroke results from the reaction of the air on the wing as modified by the relaxing and controlling movements of the pectoralis major. During the upstroke, also, the primaries twist and open in a venetian-blind manner that allows them to slip through the air with little resistance (Figure 16–32). This twisting of the primaries in the down stroke and upstroke results from the asymmetry of the leading and trailing vanes.

In larger birds with slower wing actions, the time of the upstroke is too long to spend in a state of deceleration. A similar situation exists when any bird takes off: it needs thrust on both the downstroke and the upstroke. Thrust on the upstroke is produced by bending the wings slightly at the writs and elbow and by rotating the humerus. This movement causes the upper surfaces of the twisted primaries to push against the air and to produce thrust, as their lower surfaces did in the downstroke (Figure 16–32). In this type of flight, the wing tip describes a rough figure-of-eight through the air.

This powered upstroke results from the contraction of the **supracoradoideus**, a deep muscle underlying the pectoralis major and attached directly to the carina of the sternum. In most species of birds the supracoracoideus is a

Figure 16–32   Drawing from high speed photograph of a chickadee, showing the twisting and opening of the primaries during the upstroke in flapping flight. Further forward rotation of the wrist would give the twisted primaries a positive angle of attack on the upstroke with a resultant production of thrust. (From a photograph in Welty [11]).

relatively small, pale muscle with low myoglobin content, easily fatigued. In species that rely more on a powered upstroke, either for fast, steep takeoffs or for hovering, as in hummingbirds, or for fast aerial pursuit as in swifts, the supracoracoideus is relatively larger. The ratio of weights between the pectoralis major and the supracoracoideus is a good indication of a bird's reliance on a powered upstroke; such ratios vary from 3:1 to 20:1. The total weight of the flight muscles is also indicative of the extent to which a bird depends upon powered flight. Strong fliers like pigeons and falcons have breast muscles comprising more than 30 percent of body weight, whereas in some owls, which have very light wing loadings, the flight muscles make up less than 10 percent of total weight.

16.6 OTHER FUNCTIONS OF FEATHERS

Although feathers have evolved primarily as epidermal coverings that function as a highly effective and variable body insulation as part of the basic regulatory mechanism for endothermy and secondarily as structures on the wings and tail for flight, they also serve many other important roles in the lives of birds. These subsidiary functions are usually associated with some adaptive modification of the basic feather structure, either at the gross morphological level or at the microscopic level of the individual barbules, as the following examples illustrate.

16.6.1 The Color of Feathers

The coloration of feathers often plays a role in social communication (visual releasers), in concealment from predators, and in radiative heat exchange with the environment, and feathers have consequently become modified in various ways that produce different kinds of colors and patterns of greys, depending upon the visual capabilities of the organism viewing them.

There are two main sorts of pigments found in feathers. One class, the melanins, produce black, dull yellow, red, and brown colors, and they occur in the form of discrete microscopic particles that are nonsoluble in organic solvents. The other class of pigments, the lipochromes, are alcohol-soluble carotenoids that produce yellow, orange, and red colors, and rarely, also violet, blue, and green. Feather carotenoids are all of a particular biochemical type known as xanthophylls.

Most of the colors of feathers are influenced in part by physical structure, and some colors reflected by feathers are almost pure structural colors. Structural colors result either from interference phenomena, which produce shimmering, iridescent colors that change in hue with the angle of view, or from the scattering of short wavelengths of light by particles smaller than the wavelength of red light (less than 0.6 μm) lying within the feather structure

(Figure 16–33). Since only the shorter wavelengths of incident light can be reflected by such a physical system, the perceived color of such feathers is blue, regardless of the angle of view, so long as the light is reflected from the feather, not transmitted. This phenomenon is a special case of Tyndall scattering, which is responsible for the blue appearance of the sky.

Iridescent colors that change in hue with the angle of view, as in a soap bubble, result from a physical phenomenon called "interference." Interference occurs when different wavelengths of light are refracted and reflected from the barbules in such a way that certain wavelengths mutually intensify each other, while others mutually interfere and cancel each other out. Refraction, of course, occurs when light passes from a medium of one refractive index into a medium of a different refractive index, as from air with a refractive index of 1 to water with a refractive index of 1.33, and the degree to which the light waves are refracted depends on the difference between the two refractive indexes.

The colors actually perceived from iridescent feathers depend very much upon the position of the feather surface in relation to the incident light (sun) and on the position of the viewer in relation to the reflecting surfaces. Thus, the barbular angle, the vanular angle, and the reflecting angle all play a role in determining the appearance of iridescent feathers. In general, the smaller the vanular angle, the more nearly the two vanules lie in the same plane, the less important are the relative positions of bird, sun, and observer for the perceived color. The same wavelengths are reflected at many different angles. The large vanular angle and barbule angle of the feathers in the gorget of hummingbirds, an area of the plumage that is highly colored for intraspecific social communication, greatly restricts the position of perceiving iridescence to within a few degrees, head on. This limitation may be an adaptation to avoid detection by predators. The more cryptic, greenish iridescent feathers on the back have barbules with upper surfaces that are crescentric rather than sharply

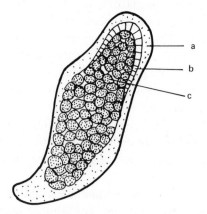

Figure 16-33 Cross-section of the ramus of a barb from a blue jay feather. Zone (a) is a transparent cortex; (b) is a thin layer of transparent, colorless cells with large, central, gas-filled vacuoles; (c) is a region of closely packed cuboidal cells with many small vacuoles and melanin granules which specifically reflect blue light. (Modified from Pettengill [7].)

V-angled. As a consequence the same wavelengths of light are reflected at all angles.

The structures responsible for iridescence are present in specialized barbules that have become flattened for part of their length and twisted at right angles so that one of the flat sides faces the viewer. The specialized flattened part may be the basal section, as in hummingbird feathers, or the pennulum, as in sunbird feathers. The torsion of the barbules is also accompanied by loss of the hooklets and flanges in the flattened section, resulting in some reduction in the mechanical strength of the vane. For this reason, fully iridescent colors are not found on flight feathers, although the speculum on the secondaries of some ducks approaches full iridescence.

It used to be thought that all iridescent colors resulted from interference in the thin, outer layer of the barbule keratin, while the melanin pigment deeper inside the barbule functioned only to absorb certain wavelengths. Recently deeper lying structures have been found to play a major role in the interference phenomena that produce iridescent colors in hummingbirds, glossy starlings, and presumably also in other groups such as the sunbirds. In the case of hummingbirds, near the surface of the flattened basal portion of the barbule are three or more thin layers consisting of a mosaic of oval platlets, each about 2.5 μm long and 1 μm wide. Each platelet contains dozens of tiny air bubbles encased in a matrix, probably melanin, having a refractive index of 2.2. The color reflected back from such a physical arrangement varies with the size of the air bubbles and the thickness of the layers of platelets. In the case of the African glossy starlings, the special color-producing barbules contain numerous air-filled tubes of melanin, each about 0.27 μm in diameter and 1.6 μm long and with walls about 0.066 μm thick. These tubes, which are packed in five rather precise layers—much like cigars in a box—produce brilliant reflected blue-violet and coppery colors as a result of the repetitive refraction and reflection of the incident light striking these layers.

The dull, velvety appearance of some ornamental feathers is the physical reverse of iridescence. The barbules of such feathers bear many vertically directed, pigmented bristles that adsorb rather than reflect light waves.

16.6.2 Water Repellancy and Adsorption

Aquatic birds can be placed at a disadvantage in terms of thermoregulation and buoyancy if water penetrates all the way through their feathers to the skin, and various adaptations have evolved to prevent water from penetrating into the air space of the downy layer underlying the contour feathers on birds that characteristically spend a good part of their lives in water. In part water repellancy is brought about by dressing the feathers with preen gland oil, but it is likely that special surface features of the feathers themselves are involved in holding and distributing the oil for maximum repellancy. In addition there is evidence that the physical arrangement of the barbs and barbules—particularly their spacing and angular relationships on the more distal and outer portions of ventral contour feathers on several kinds of aquatic birds—result in hydrophobic or anti-water-absorbing surfaces. Some aquatic birds, such as grebes and loons, also have special coiled barbules (modified pennaceous types) attached to the barbs on the outer parts of ventral contour feathers, and these coiled barbules may actually adsorb and hold a film of water by interfacial tension in the outer parts of the belly feathers (see Figure 16–34). If so, the water trapped by the barbules could act as a surface skin against the air interface in the downy layer, thereby preventing by surface tension further penetration of water. Sandgrouse and a few other birds use their feathers to adsorb and carry water from one place to another (see section 20.4.3).

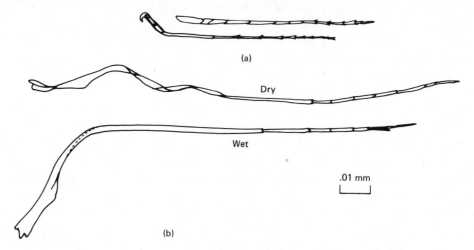

Figure 16–34 Specialized coiled barbules that hold water by interfacial tension. (a) Barbule from distal portion of belly feather of a grebe. (b) Barbule from proximal part of belly feather of a sandgrouse. Dry barbules are coiled; wet barbules are extended.

16.6.3 Display and Concealment

Feathers have become modified—often in highly conspicuous ways—both in shape and in coloration for many kinds of social communication involving visual releasers or signals and displays, highly ritualized movements or postures involving feathers. Sexual dimorphism is one aspect of these social functions. Males are often more highly colored than females or have feathers that have become structurally modified as sexual adornments. These features of the male plumage may function in sex recognition and attraction of females, or they may be used in displays that serve the function of bringing the female into sexual readiness. The colorful speculum on the secondaries of drake ducks, the red epaulets on the wings of male red-winged blackbirds, and the reddish or golden crowns on kinglets are familiar examples of specialized areas of the plumage involved in sexual behavior and display, as are the elaborate plumes found in birds of paradise and egrets, or the highly modified tail coverts of the peacock. Many of these same or other features are also used in territorial and agonistic displays. In all of these cases in which feathers are involved in visual communication, special modifications in color, shape, or movement have been employed in a variety of combinations to impart signal functions.

Conspicuous or aerodynamically cumbersome feathers can make a bird more vulnerable to capture by a visually guided predator, and the bright colors and special adornments of the breeding season are often discarded for a more sober, even cryptic appearance during the rest of the year. Thus, the African standard-wing nightjar has specially elongated and flagged number two primaries (Figure 16–35) that he uses in courtship-flight displays but that probably also slow him down and make him easier for an aerial predator like a falcon to capture. As soon as courtship is over, he bites off the projecting parts of these feathers, leaving the stubs in the wings. The pattern of molting his wing feathers is so arranged that the number two primaries are not replaced until just the right time for them to grow back in and be full length for the next breeding season. This adaptation involves a departure from the usual molting sequence found in the female and in all other species of nightjars.

Other species, especially ground-dwelling and ground-nesting ones that are vulnerable to visual predators, have evolved feathers with special colors, patterns, and textures that function to conceal the bird from its enemies. Ground-dwelling birds often match the general color of their habitat and hide effectively by remaining motionless and blending in with the background when threatened by a predator. The ptarmigan is white in winter and matches its snowy background, but it molts into a mottled brown and cryptic plumage during the spring and summer matching the tundra vegetation in which it lives. The females turn brown first and begin sitting on their nests, situated on snow-free tundra, when the males are still white and very conspicuous in their courtship and territorial dress. When a hunting gyrfalcon appears on the scene, the males seek out the last small patches of snow on which to hide from the predator. Many species and races of African larks match very precisely the soil color of their

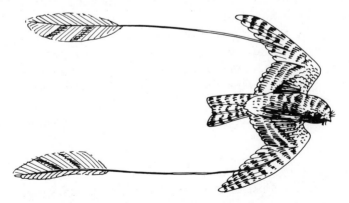

Figure 16–35 Standard-wing Nightjar, showing the elongate and flagged second primaries used in aerial courtship display. (Redrawn from Welty [11].)

habitats: On red, sandy soils there are red larks; on whitish limestone soil, there are pale colored larks; and on some black lava soils, there are black larks. In regions where the different soil types come together, each species of lark stays on that soil which its own color most closely resembles, as though the birds have some instinctive affinity for the soil type on which they are best concealed.

Counter-shading, disruptive patterns, and mimicry of inanimate objects such as rocks or dead branches are other forms of crypsis that find expression in the colors, shapes, and textures of feathers. Feathers have been almost endlessly modified both for conspicuousness in display or for concealment against predators, and often both features are combined in the same plumage, the cryptic feathers overlying and concealing the bright display feathers, which are revealed only by spreading and erecting particular feather tracts on brief occasions requiring use of the signal or display.

16.6.4   Sound Production and Sound Muffling

Although birds are usually thought of as being among the most vocal of animals, they also use a variety of nonvocal sounds in social communication or in threats against other species. Again feathers play an important role in sound production in some species and may be specially modified for this purpose.

Sounds are often produced incidentally to the beating of a bird's wings in flight, and only slight modification in the shape of the primaries or tail feathers will produce the characteristic whistling sounds made by certain kinds of ducks, doves, bustards, and hummingbirds when they fly. Such sounds may be used in territorial advertisement or as interindividual location signals among birds flying in flocks at night or in heavy fog. Other species have undergone more specific modifications of their flight feathers to produce mechanical sounds used in displays. The outer three primaries of the American woodcock are specially narrowed and stiffened so that, when spread apart during aerial courtship, they vibrate to produce a characteristic "winnowing" sound. Snipes of the genus *Gallinago* have modified their outer tail feathers in the same way; the pintail snipe has 27 tail feathers, more than double the usual number in birds, and the outer eight pairs are specially narrowed and stiffened for sound production in courtship flight. Among the tropical American manakins (Pipridae) one finds not only narrowed and stiffened primaries involved in display sounds but also secondaries with unusually thickened, clublike shafts, which apparently act somewhat like castenets to produce sounds when the wings are moving (Figure 16–36).

Other species, including goatsuckers, owls, doves, and larks, clap their wings together in flight, producing characteristic sounds associated with courtship or territorial defense. Most of these birds strike their wings together over their backs—for example, doves and pigeons—but the short-eared owl has a beautiful aerial display in which it claps its fully extended wings under its body. As it

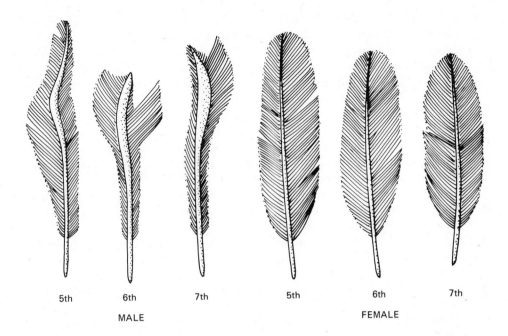

Figure 16-36 Secondaries of the South American manakin, *Allocotopterus deliciosus*, showing the special thickenings of the feather shafts of the male, structures that produce sounds when the wings are moved in display. (Based on Welty [11].)

does so it falls through the air, producing a muffled but surprisingly penetrating drum-like sound that can be heard from half a mile away over the tundra. The ruffed grouse fans its wings rapidly while perched on a log and produces the characteristic "drumming" of this species in the spring, a sound which is not made by striking the log as once was thought.

Owls (Strigiformes), goatsuckers (Caprimulgiformes), and some other nocturnal species have special modifications of their flight feathers that muffle the sounds produced in flight. Presumably this adaptation allows predatory species to slip up on unsuspecting prey without being heard. In any case sound-muffling is most highly developed in nocturnal owls, in which the flight feathers have a soft, velvety texture on their dorsal surfaces. This type of surface is produced by special "pubescent" barbules that project soft hairlike tips at right angles to the main surface of the feather, thereby altering the characteristics of the flight feathers so that they do not vibrate to produce sound waves when the bird moves through the air.

References

[1] Brodkorb, P. 1971. Origin and evolution of birds. In *Avian Biology*, edited by D. S. Farner and J. R. King, vol. 1. Academic Press, New York. A technical and detailed account of known fossils. Particularly interesting for its comparisons of *Archaeopteryx* with modern arboreal species that are presumed to be similar in form and habits.

[2] Brown, R. H. S. 1961. Flight. In *Biology and Comparative Physiology of Birds*, edited by A. J. Marshall, vol. 2. Academic Press, New York.

[3] Calder, W. A. and J. R. King. 1974. Thermal and caloric relations of birds. In *Avian Biology*, edited by D. S. Farner and J. R. King, vol. 4. Academic Press, New York. A highly technical, up-to-date treatment with a particularly interesting discussion of thermal conductance and insulation.

[4] Feduccia, A. and H. B. Tordoff. 1979. Feathers of *Archaeopteryx*; asymmetric vanes indicate aerodynamic function. *Science* 203:1021–1022.

[5] Lucas, A. M. and Pr. R. Stettenheim. 1972. *Avian Anatomy: Integument*. Agriculture handbook No. 362, U.S. Dept. of Agriculture, Washington, D.C. A detailed atlas of feathers and feather structures.

[6] Ostrom, J. H. 1974. *Archaepteryx* and the origin of flight. *Quarterly Rev. Biology* 49:27–47. Most up-to-date review of subject, emphasizing the author's novel ideas on the selective forces molding the early avian forelimb into a "wing."

[7] Pennycuick, C. J. 1975. Mechanisms of flight. In *Avian Biology*, edited by D. S. Farner and J. R. King, vol. 5. Academic Press, New York. A highly technical update of this subject, requiring considerable knowledge of mathematics and physics.

[8] Pettingill, O. S. 1970. *Ornithology in Laboratory and Field*. Fourth edition. Burgess Publishing Co., Minneapolis.

[9] Regal, P. J. 1975. The evolutionary origin of feathers. *Quarterly Review of Biology* 50:35–66.

[10] Stettenheim, P. R. 1972. The integument of birds. In *Avian Biology*, edited by D. S. Farner and J. R. King, vol. 2. Academic Press, New York.

[11] Thomson, A. L. (editor). 1964. *A New Dictionary of Birds*. McGraw-Hill, New York. The standard work of reference on birds. Subjects arranged alphabetically as in an encyclopedia.

[12] Welty, J C. 1975. *The LIfe of Birds*. Second edition. W. B. Saunders, Philadelphia. A popular text on birds with easily read, informative chapters on feathers, flight, and evolution.

17

The Diversity of Locomotion and Feeding in Birds

Synopsis: There are about 8,700 living species of birds and many fossil forms, classified in more than 30 orders and 160 families; more than half are perching birds in the order Passeriformes. Within the limits set by their basic adaptations for flight, birds have diversified adaptively in many features of their morphology, physiology, and behavior. Adaptive radiation and convergence among birds are particularly well illustrated by adaptations associated with locomotion and feeding.

Unlike most other tetrapods, birds usually possess two distinct modes of locomotion, one associated with the hind limbs, the other with the forelimbs. The avian hind limbs are adapted for pedal locomotion—walking, running, swimming—whereas in the great majority of birds the forelimbs have been modified into wings. There are four basic modes of avian flight: forward flapping flight, hovering by flapping, dynamic soaring, and static soaring. The structure and shapes of bird wings have been modified to conform with the aerodynamic requirements for these general modes, as well as for special styles of flight. Many birds have *elliptical wings*, which are adapted for maneuvering in forest and woodland habitats or for fast takeoff. Others possess *high speed wings*, which are long and pointed and adapted for sustained fast, flapping flight. *High aspect ratio wings* are long, narrow, and flat and have evolved as the design best suited for dynamic soaring. *Slotted high lift wings* are broad with wide spacing between the tips of the primaries; this form of wing has evolved in association with static soaring. The different species of flying birds show many variations and special modifications of these basic types of wing. Terrestrial adaptations of the avian hind limbs may be associated with walking or running, supporting heavy bodies, hopping, perching, climbing, wading in shallow water, or for support on insubstantial surfaces, such as snow or sand. The hind limbs and feet have also frequently been modified as swimming structures. The feet of aquatic birds are either webbed or lobed and function as paddles, but the most extreme specializations for paddling are seen among foot-propelled diving birds, such as loons, grebes, cormorants, and some ducks.

Other submarine swimmers such as penguins and auks use their wings as flippers. In the most specialized cases, the wings are no longer capable of functioning in flight.

Birds eat a wide variety of organic matter, especially energy-rich foods such as seeds, nuts, and animal bodies. The main types of avian feeders are: (1) fish and aquatic invertebrate feeders, (2) aquatic filter-feeders, (3) carnivorous or predatory birds, (4) carrion eaters, (5) insectivores, (6) pollen and nectar feeders, (7) fruit eaters, (8) seed eaters, and (9) grazers and browsers. Many species employ more than one type of feeding, especially at different seasons, and some are so specialized in food habits that they cannot be categorized easily. Each type of feeding is associated with a particular structure of the beak, features of the digestive system, and locomotor capabilities that have evolved as adaptations that promote individual survival, for example the hooked beaks and powerful taloned feet of predatory birds or the long beaks and tubular tongues of flower-probers. Frequently such adaptive traits have evolved convergently in taxa that are only distantly related to each other (owls and hawks; sunbirds and hummingbirds).

Specialist feeders range from such bizarre species as the African honeyguides that eat beeswax, which is digested by symbiotic intestinal bacteria, to some Galapagos finches that have developed the habit of biting the bases of growing feathers on nesting boobies and taking the blood that oozes from the wound.

17.1 INTRODUCTION: ADAPTIVE RADIATION AND CONVERGENCE IN BIRDS

As discussed in the previous chapter, birds have been abundant and diverse since the late Cretaceous, and new forms have continually evolved since then. At present there are about 8700 species in the world, representing some 28 orders and 157 or more families, whereas at least six orders and 40 families of birds are known only from fossils (Table 17–1). According to some estimates, on the order of 100,000 species of birds have probably evolved and become extinct since the Jurassic. As the appearance of new landforms, climatic changes, the evolution of new plant life, and the extinction of other animals (including birds) created unoccupied habitats and niches for birds, the evolutionary results of genetic changes, reproduction, and differential survival of individuals have tended continually to fill up the vacant ecological possibilities for birds with new forms. As a consequence of this adaptive radiation, or exploitation of ecological opportunities, birds live in a great variety of ways and in almost all parts of the world.

The adaptive and evolutionary possibilities for birds are, however, not without limitations. We have already considered some of the restraints on body size and shape imposed by the primary requirements for flight, noting for instance that birds have not been able to attain body sizes as large as those of some reptiles and mammals. There are other features of the basic avian organization that also set limits on the extent of adaptive radiation in this class. All birds are oviparous and, with a few exceptions, obligate brooders of their eggs; for these reasons none lead fully aquatic lives as certain mammals (whales, dugong) and the ovoviviparous sea snakes do. Although some birds nest in burrows that they

Table 17-1. A Classification of Birds to the Level of Suborders*

Class Aves

Sublcass †Archaeornithes
    Order †Archaeopterygiformes (*Archaeopteryx*)
      (1 family, 1 genus, 1 or 2 species; Jurassic, Bavaria)
Subclass Neornithes
  Superorder †Odontognathae
    Order †Hesperornithiformes (*Hesperornis*)
      (2 families, several genera and species; Cretaceous, North America)
  Superorder Palaeognathae (ratites)
    Order Tinamiformes (tinamous)
      (1 family, 9 genera, 42 species; Neotropics)
    Order Rheiformes (rheas)
      (2 families, 1 fossil; 2 living genera, 2 species; South America)
    Order Struthioniformes (ostriches)
      (1 family, 1 species; Africa)
    Order Casuariiformes (emus and cassowaries)
      (3 families, 1 fossil; 2 living genera, 5 species; Australia, New Guinea)
    Order †Aepyornithiformes (elephantbirds)
      (1 family, 2 genera, 7 species; Madagascar)
    Order Dinornithiformes
      Suborder †Dinornithes (moas)
        (2 families, 7 genera, 28 species; New Zealand)
      Suborder Apteryges (kiwis)
        (1 family, 1 genus, 3 species; New Zealand)
  Superorder Neognathae
    Order Podicipediformes (grebes)
      (1 family, 5 genera, 18 species; worldwide)
    Order Sphenisciformes (penguins)
      (1 family, 6 genera, 17 species, southern oceans)
    Order Procellariiformes (albatrosses, shearwaters, petrels)
      (4 families, 23 genera, about 95 species; all oceans)
    Order Pelecaniformes
      Suborder Phaethontes (tropicbirds)
        (1 family, 1 genus, 3 species; pantropical oceans)
      Suborder Pelecani (boobies, gannets, cormorants)
        (4 families, 1 fossil; 7 genera, 42 species; virtually worldwide)
      Suborder Pelecanoidea (pelicans)
        (1 family, 1 genus, 8 species; all continents)
      Suborder Fregatae (frigatebirds)
        (1 family, 1 genus, 5 species; pantropical oceans)
      Suborder †Odontopterygia
        (2 families of fossil forms)
    Order Anseriformes
      Suborder Anhimae (screamers)
        (1 family, 2 genera, 3 species; South America)
      Suborder Anseres (waterfowl)
        (1 family, 45 genera, 150 species; worldwide)

Table 17-1.   A Classification of Birds to the Level of Suborders (*Cont.*)

Order Phoenicopteriformes (flamingos)
   (6 families, 5 fossil; 3 living genera, 6 species; all continents but Australia)
Order Ciconiiformes
   Suborder Ardeae (herons and bitterns)
      (1 family, 32 genera, 66 species; worldwide)
   Suborder Balaenicipites (whale-head stork)
      (1 family, 1 species; Africa)
   Suborder Ciconiae (storks, ibises, spoonbills)
      (3 families, 32 genera, 50 species; pantropical and north temperate)
Order Falconiformes
   Suborder Cathartae (condors)
      (2 families, 1 fossil; 5 living genera, 7 species; New World)
   Suborder Accipitres (hawks, eagles, kites, etc.)
      (3 families, 66 genera, 219 species; worldwide)
   Suborder Falcones (falcons and caracaras)
      (1 family, 10 genera, 61 species; worldwide)
Order Galliformes (upland gamebirds and fowl)
   (5 families, 95 genera, 275 species; worldwide)
Order Gruiformes
   Suborder Ralli (rails)
      (2 families, 1 fossil; 52 living genera, 138 species; worldwide)
   Suborder Heliornithes (sungrebes)
      (1 family, 3 genera, 3 species, Neotropics, Africa, so. Asia)
   Suborder Rhynocheti (kagus)
      (1 family, 1 species; New Caledonia)
   Suborder Eurypygae (sunbitterns)
      (1 family, 1 species; Neotropics)
   Suborder Mesoenatides (roatelos)
      (1 family, 2 genera, 3 species; Madagascar)
   Suborder Turnices (buttonquail)
      (2 families, 3 genera; 15 species; Old World tropics; Australia)
   Suborder †Gastornithes
      (2 families of fossils, including *Diatryma*)
   Suborder Grues (cranes; limpkins; trumpeters)
      (4 families, 1 fossil; 6 living genera, 18 species; worldwide)
   Suborder Cariamae (seriemas and phororhacids)
      (6 families, 5 fossil; 2 living genera, 2 species; South America)
   Suborder Otides (bustards)
      (2 families, 1 fossil; 11 living genera, 24 species; Old World)
Order Charadriiformes
   Suborder Charadrii (plovers, sandpipers, and relatives)
      (13 families, 1 fossil; about 49 genera, 187 species; worldwide)
   Suborder Lari (gulls, terns, and relatives)
      (3 families, 20 genera, 90 species; worldwide)
   Suborder Alcae (Auks, murres, puffins)
      (2 families, 1 fossil; 13 living genera; 21 species; northern oceans)

Table 17-1. A Classification of Birds to the Level of Suborders (*Cont.*)

Order Gaviiformes (loons)
 (1 family, 1 genus, 4 species; Holarctic)
Order Pteroclidiformes (sandgrouse)
 (1 family, 2 genera, 16 species)
Order Columbiformes (doves, pigeons, dodos)
 (3 families, 2 fossil; 59 living genera, 306 species; worldwide)
Order Psittaciformes (parrots)
 (1 family, 80 genera, 339 species)
Order Cuculiformes
 Suborder Musophagi (turacos)
 (1 family, 6 genera, 20 species; Africa)
 Suborder Cuculi (cuckoos and relatives)
 (1 family, 38 genera, 130 species; worldwide)
Order Strigiformes (owls)
 (3 families, 1 fossil; 26 genera, 146 species; worldwide)
Order Caprimulgiformes
 Suborder Steatornithes (oilbirds)
 (1 family, 1 species; South America)
 Suborder Caprimulgi (goatsuckers and relatives)
 (4 families, 23 genera, 94 species; worldwide)
Order Apodiformes
 Suborder Apodi (swifts)
 (3 families, 1 fossil; 29 living genera, 80 species; worldwide)
 Suborder Trochili (hummingbirds)
 (1 family, 123 genera, 331 species; New World)
Order Coliiformes (mousebirds)
 (1 family, 1 genus, 6 species; Africa)
Order Coraciiformes
 Suborder Alcedines (kingfishers, todies, motmots)
 (3 families, 21 genera, 101 species; worldwide)
 Suborder Meropes (bee-eaters)
 (1 family, 7 genera, 24 species; Old World tropics and subtropics)
 Suborder Coracii (rollers and hoopoes)
 (5 families, 9 genera, 24 species; Old World tropics and subtropics)
 Suborder Bucerotes (hornbills)
 (1 family, 12 genera, 46 species; tropical Africa and Asia)
Order Piciformes
 Suborder Galbulae (jacamars, barbets, honeyguides, toucans, etc.)
 (5 families, 34 genera, 178 species; pantropical)
 Suborder Pici (woodpeckers and relatives)
 (1 family, 38 genera, 213 species; worldwide except Australia)
Order Passeriformes (perching birds)
 Suborder Eurylaimi (broadbills)
 (1 family, 8 genera, 14 species; Africa, India, Borneo)
 Suborder Furnarii (woodhewers, antbirds, tapaculos)
 (3 families, 137 genera, 529 species; Neotropics)

Table 17-1.   A Classification of Birds to the Level of Suborders (*Cont.*)

Suborder Tyranni (cotingas, manakins, tyrant-flycatchers, etc.)
   (5 families, 175 genera, 539 species; Neotropical to North America)
Suborder Menurae (scrub-birds and lyrebirds)
   (2 families, 2 genera, 4 species; Australia)
Suborder uncertain
   Xenicidae, New Zealand wrens, 3 genera, 4 species
   Pittidae, pittas, Old World tropics, 1 genus, 23 species
   Philepittidae, asities and false-sunbirds, Madagascar, 2 genera, 4 species
Suborder Passeres (songbirds)
   (46 families, 1 fossil; more than 570 genera and 4,050 species; worldwide)
Fossils of Uncertain Taxonomic Position
†*Ichthyornis* spp, toothed birds from Upper Cretaceous, Kansas and Texas

---

† = extinct

*Modified from R. W. Storer [1971]. In Farner and King [2].

excavate in the ground, none has become truly fossorial in habits—again, a not uncommon mode of life for some amphibians, reptiles, and mammals. Feathers would be an ill-suited epidermal covering for a fossorial animal, and the special avian construction of the pectoral and thoracic skeletal elements could not easily be modified into the kinds of supports needed to resist the physical stresses and pressures associated with fossorial locomotion and movement under ground.

Within the range of possibilities set by their primary and ancestral (hence, genetically least modifiable) adaptations for flight, birds have diverged in many features of their morphology, physiology, and behavior. Birds of prey, for example, have specializations in their eyes for resolution of objects at a great distance and for both binocular and lateral vision (two fovea in each retina), in their wings for fast pursuit, in their legs and feet for seizing and killing prey, and in their beaks for tearing and breaking up carcasses into pieces that can be swallowed. Special patterns of predatory behavior have evolved along with these morphological adaptations, as have special physiological processes, such as those associated with digestion of meat. All phases of a bird's life—reproduction, escape from enemies, adjustment to climatic conditions, care of plumage, and all other activities necessary for maintenance of the individual—reflect morphological, physiological, and behavioral adaptations for a given set of living conditions.

Another feature of the adaptive trends and variations among birds is convergence, which we have considered generally in Chapter 3. Many taxa of birds (families, subfamilies, genera, even a few orders) have remained confined to specific regions of the world owing to barriers against dispersal or to competitive exclusion. Because very similar habitats occur in disjunct regions, distantly related species often converge on similar niches in separate parts of the world where the environments and ecological opportunities are essentially alike; consequently they become similar in form and function. There are many examples among birds: The small auks (Alcidae, Charadriiformes) of the northern oceans and the diving petrels (Pelecanoididae, Procellariiformes) of the southern oceans [and the very parallel case of the great auk (Alcidae) and the penguins (Spheniscidae)], the Old World vultures (Aegypinae, Accipitridae, Falconiformes) and the New World vultures or condors (Cathartidae, Falconiformes), and the many shrikelike passeriform species found in such geographically diverse families as Laniidae (true shrikes), Formicariidae (antbirds), Cotingidae (cotingas and bellbirds), Vireonidae (vireos), Muscicapidae (Old World flycatchers and babblers), Cracticidae (Australian butcherbirds), Vangidae (vanga-shrikes), and others.

In the following sections we survey some of the avian adaptations that show divergent and convergent evolutionary trends. Adaptations associated with locomotion and feeding are best known in birds, but these evolutionary trends are also evident in other aspects of avian adaptation to the earth's environments.

17.2 DIVERGENCE AND CONVERGENCE IN LOCOMOTOR ADAPTATIONS

Avian locomotor adaptations are associated primarily with the wings and legs and secondarily with the tail, which may function for steering and support in flight and as a proplike support for arboreal climbers. Birds have become adapted and specialized in various ways for flight, walking and related bipedal terrestrial movements, wading in water and related movements, climbing, swimming on the surface of water, and swimming under water.

17.2.1 Flying Adaptations

Flight patterns vary greatly among the different avian species, and even the same individual bird often employs several modes of flight, depending on atmospheric conditions and on how it holds and uses its wings in the air. Four basic modes of avian flight can be distinguished: (1) forward flapping flight, (2) hovering by flapping, (3) dynamic soaring, and (4) static soaring.

The shapes of bird wings reflect adaptation for these basic modes, as well as for special styles of flight associated with particular niches. Wings may be large or small in relation to body size, resulting in light wing loading or heavy wing loading (refer to Table 16–6), they may be long and pointed, short and rounded, highly cambered or relatively flat, and so forth.

Depending upon whether the bird is primarily adapted as a powered flier or as a soaring form, the various segments of the wing are lengthened proportionally to different degrees. For example, hummingbirds (Trochilidae, Apodiformes), which have very fast, powerful wing beats, requiring maximum propulsive force from the primaries, have hand bones that are longer than the forearm and upper arm combined, and only six or seven secondaries in the inner wing. In the frigatebirds (Fregatidae, Pelecaniformes), marine species with long, narrow wings adapted for both powered flight as well as for gliding and soaring, all three segments of the forelimb are about equal in length. The soaring albatrosses (Diomedeidae, Procellariiformes) have carried lengthening of the wing to the extreme found in birds; in them the humerus or upper arm is the longest segment, and there may be as many as 32 secondaries in the inner wing (Figure 17–1).

Ornithologists recognize four structural and functional types of wings (Figure 17–2). Birds that are adapted to enclosed habitats—forests and woodlands—where they have to maneuver around obstructions in their flight space, have **elliptical wings**. These wings have a low aspect ratio (length to width), tend to be highly cambered, and usually have a high degree of slotting in the outer primaries. These are structural features generally associated with slow flight, reduction of tip vortex induced drag, and a high degree of maneuverability, although some species with elliptical wings, notably upland gamebirds (Galliformes), have fast take-off speeds but do not maintain fast sustained flight for long. Elliptical wings have evolved independently in many different groups of forest and woodland species: most gallinaceous birds, many doves (Columbiformes), some cuckoos (Cuculiformes), woodpeckers (Piciformes), and songbirds (Passeriformes), especially in the family Corvidae (crows, jays, and magpies).

Many birds that attack prey in flight, that make long migrations, or that have a heavy wing loading in relation to some other aspect of their lives—such as diving—have **high speed wings**. Such wings have a moderately high aspect ratio, taper to a pointed tip, have a flattish profile (little camber), and often lack slots in the outer primaries. In flight they show the sweepback attitude and fairing

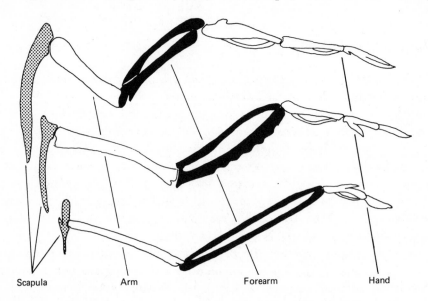

| Scapula | Arm | Forearm | Hand |

Figure 17-1 Comparison of wing bones of hummingbird (top), frigatebird, and albatross (bottom) drawn to same size to show relative proportions of distal, middle, and proximal elements.

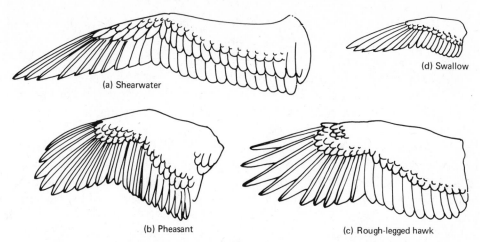

(a) Shearwater

(d) Swallow

(b) Pheasant

(c) Rough-legged hawk

Figure 17-2 Comparison of four basic types of bird wings. (a) High aspect ratio. (b) Elliptical. (c) Slotted high lift. (d) High speed. (Redrawn and modified from R. T. Peterson [4].)

to the body of jet fighter-plane wings. All really fast flying birds have converged on this form of wing. Familiar examples from diverse orders include: loons (Gaviiformes), waterfowl (Anseriformes), shorebirds (Charadriiformes), many pigeons and doves (Columbiformes), falcons (Falconiformes), swifts (Apodiformes), and, independently, two families of passerines, the true swallows (Hirundinidae) and the wood swallows (Artamidae).

Many other species of birds have wing shapes that are in some degree intermediate between elliptical and high speed.

Seabirds, particularly those like albatrosses and shearwaters that rely on dynamic soaring, have **high aspect ratio wings**, long, narrow, flat wings lacking slots in the outer primaries. Some albatrosses have aspect ratios of 25:1 and lift to drag ratios comparable to high performance sail planes (see Figure 16–30). Dynamic soaring is only possible where there is a pronounced vertical wind gradient, the lower 15- to 20-m layer of air being slowed by friction against an irregular surface, such as a wave-disturbed ocean. Further, dynamic soaring is only advantageous in regions where winds are strong and persistent, such as in the latitudes of the "Roaring Forties," where, in fact, most albatrosses and shearwaters are found. Starting from the top of the wind gradient, the albatross glides downwind with great increase in ground speed, and then, as it nears the surface, it turns and gains altitude into the wind. Because the bird flies into wind of increasing speed as it rises, its loss of air speed is not as great as its loss of ground speed, and consequently the albatross does not stall until it has mounted back to the top of the wind gradient where the air velocity again decreases. At that point, the bird turns downwind, repeating the cycle; and it can continue to remain aloft without flapping so long as it "drifts" with the wind.

This method of soaring is only possible for a heavy bird with great inertia. The larger albatrosses weight 10 to 12 kg and have wing spans greater than 3 m, making them among the largest flying birds. They have a rather heavy wing loading and must fly at high speeds. Because their pectoral muscles are relatively small and weak, they can only take off by running or paddling and flapping into a strong wing, or by launching forth from the brink of a precipice.

The **slotted high lift wing** is a fourth type, which is associated with static soaring, typified by vultures, eagles, storks, and some other large birds. This wing has an intermediate aspect ratio between the elliptical wing and the high aspect ratio wing, a deep camber (at rest), and marked slotting in the primaries. When the bird is in flight, of course, the camber disappears and, in fact, the tips of the primaries turn markedly upward under the influence of air pressure and body weight. Static soarers remain airborne mainly by seeking out and gliding in air masses that are rising at a faster rate than the bird's sinking speed. Hence, a light wing loading and maneuverability (slow forward speed and small turning radius) are advantageous. Broad wings provide for the light loading, and short wings give the low inertia required for quick, subtle response to small variations in air currents. A bird cannot soar in tight spirals at very fast speed, but to fly slowly with enough lift to prevent stalling requires a high angle of attack. The deeply slotted primaries apparently make the combination of low speed and high lift possible. The distal, emarginated portion of each primary acts as a separate high aspect ratio airfoil set at a high angle of attack. Such a design greatly reduces wing tip vortex disturbances, although the high angle of attack does increase drag. The latter is not, however, a big problem for *slow* flying birds, since drag increases as the square of air speed.

There are two sorts of rising air currents that static soaring birds use: obstruction currents and convection currents or thermals (Figures 17–3 and 17–4). In regions where topographic features and meteorological factors provide these atmospheric conditions, static soaring is an energetically cheap mode of flight. By soaring rather than flapping, a large bird the size of a stork can decrease by a factor of 20 or more the energy required for flight per unit of time, whereas the saving is only one tenth as much for a small bird like a warbler. It is little wonder, then, that most large land birds perform their annual migrations by soaring and gliding as much of the time as possible, and some condors and vultures cover hundreds of kilometers each day soaring in search of food.

There are many variations and specializations within these four basic types of wing. A brief description of the flying adaptations in hummingbirds and in owls (Strigiformes) serves to indicate the extremes of specialization. Hummingbirds can hover and also fly backwards. We have already noted the drastic modifications in the wing structure of hummingbirds, most of the skeletal structure consisting of hand bones, and most of the flight surface being made up of primaries, the forearm and upper arm being very short. In addition, there is little articulation in the elbow and wrist joints, so that the entire wing is essentially an inflexible framework that can be moved very freely and in almost any direction at the shoulder joint. Thus, the entire wing functions essentially as a variable-pitch propeller. As the bird hovers, with head up and tail down, the wings beat in a horizontal plane, both the upstroke and the downstroke being powered (Figure 17–5). As a consequence the supracoracoideus muscle

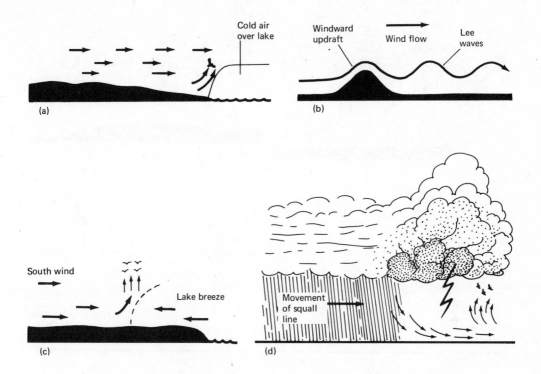

Figure 17-3 Types of obstruction currents frequently used by soaring birds. (a) Shoreline updraft along a lake; (b) upward deflection of wind over a mountain with lee waves; (c) updraft at the boundary between prevailing land wind and a lake breeze; (d) updraft in front of a moving squall line, creating highly favorable fast moving conditions for soaring. (Mofidifed from J. R. Haugh [1972] *Search* 2(16):1–59.)

is large in relation to the pectoralis, and the entire flight muscle mass is large in relation to overall body size, comprising 30 percent of total body weight. Hovering requires very high energy utilization because no lift is generated by forward momentum through the air, but, despite its energetic cost compared to other modes of flight, it is adaptive in allowing the hummingbird to exploit an energy-rich source of food (flower nectar). Also, when required to do so, some hummingbirds can compensate for a high energy utilization during activity by becoming torpid and drastically reducing energy use at night (Figure 17–6).

Owls have extremely large wing surfaces for their weights. As a consequence of their very light wing loadings, they are able to fly very slowly with a slow wing beat. Their breast muscles are smaller in proportion to body size than in any other group of birds, and their energy requirements for flapping flight must be very greatly reduced over that of similar sized birds. Being mostly nocturnal predators, owls rely more on stealth and silent approach to capture prey by surprise than do the diurnal birds of prey and, consequently, do not have to invest so much in powerful energy-demanding flight muscles.

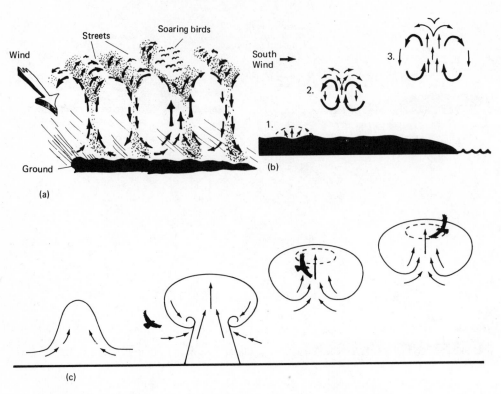

(a)

(b)

(c)

Figure 17-4 Types of convection currents used by soaring birds. (a) Thermal "streets" — under certain conditions masses of air containing convection currents align horizontally into zones of rising and sinking that are parallel to the wind direction. Under such conditions birds can soar either upwind or downwind in the rising streets. (b) A typical thermal bubble begins as a rising column of warm air heated by the surface (1), then becomes undercut by cold air (2), and continues to rise with convectional circulation (3 and 4). (Modified from J. R. Haugh [1972] *Search* 2(16):1–59 and from Peterson [4].)

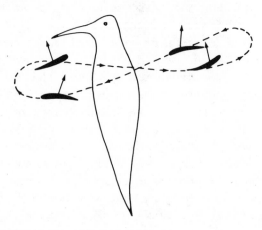

Figure 17-5 Diagrammatic representation of hummingbird in hovering flight. (Redrawn from Brown, in A. J. Marshall [1967] *Biology and Comparative Physiology of Birds*. Vol. II. Academic Press.)

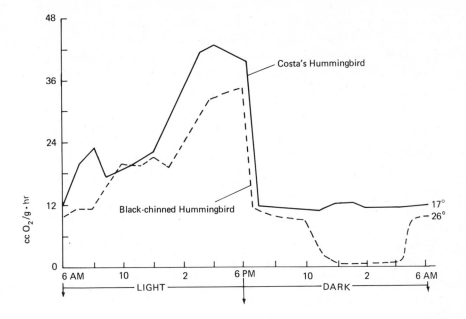

Figure 17-6 Diurnal cycle of torpor in hummingbirds as indicated by changes in oxidative metabolism. The black-chinned hummingbird weighing 2.8 gm shows torpor at night at an ambient temperature of 26° C, while the Costa's hummingbird weighing 3.4 gm does not, even at 17°C. The former saves energy equivalent to about 20 percent of the total daily energy expenditure of a similar sized hummingbird that does not become torpid at night. (From R. Lasiewski [1963] *Physiological Zoology* 36:122–140.)

17.2.2 Adaptations of the Hind Limbs for Terrestrial Locomotion

Unlike most tetrapods, birds usually are constructed for two or more different modes of locomotion—bipedal walking or swimming with the hind limbs and flying with the forelimbs. Terrestrial adaptations of the avian hind limbs may involve walking or running, supporting heavy bodies, hopping, perching, climbing, wading in shallow water, or for support on insubstantial surfaces.

It is instructive to consider cursorial adaptations first, since birds presumably evolved from bipedal cursors and because the principles of cursorial adaptation have been well worked out in the quadrupedal mammals. Modifications usually associated with running in quadrupeds are (1) a progressive increase in the lengths of the distal limb elements relative to the proximal ones, (2) a decrease in the area of the foot surface that makes contact with the ground, and (3) a reduction in the number of toes. All three of these cursorial trends are expressed to varying degrees in running birds; however, problems of balance are more critical for bipeds than for quadrupeds, and these problems have restricted the evolution in bipeds of some cursorial adaptations found in quadrupeds.

Since the center of gravity must lie over the feet of a biped in order for it to maintain balance, a reduction in the surface area in contact with the ground, as expressed by the length or number of toes, can be achieved only by some sacrifice in stability. No bird has reduced the length and number of toes in contact with the ground to the extent that the hooved mammals have, but the large, fast running ostrich (Struthioniformes) has reduced the number of toes on each foot to two, and many other cursorial species, flightless as well as flying ones, have only the three forward-directed toes in contact with the ground (Figure 17–7). The only fast-running birds with four toes in contact with the ground are the zygodactylous roadrunners in the Order Cuculiformes, and the secretary bird (Falconiformes). Moreover, long-legged birds with short toes must have their tarsometatarsi and tibiotarsi nearly equal in length in order for the center of gravity to remain over their toes when they bend their legs to squat down or to get up (Figure 17–8).

Well-coordinated running birds range in size from very small, around 20 g, to the largest living bird, the ostrich. Large species such as the various ratites, bustards (Otididae, Gruiformes), and some gamebirds (Galliformes) have long legs, take long strides, and can run relatively fast (60 to 70 km/hr)-some of them for long distances, especially the flightless ratites, which are the most specialized for cursorial locomotion. Small species have short legs, which they move very rapidly for short distances only; none of the very small runners is flightless. They run quickly and then "freeze," counting on cryptic plumage to avoid detection by visually guided predators, but, when closely pressed, they

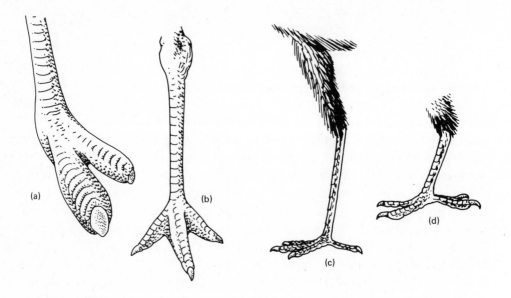

Figure 17-7 Examples of avian feet showing various adaptations for terrestrial locomotion. (a) Ostrich, with only two toes. (b) Rhea, with three toes. (c) Secretary Bird, with typical avian foot. (d) Roadrunner, with zygodactyl foot. Not drawn to scale.

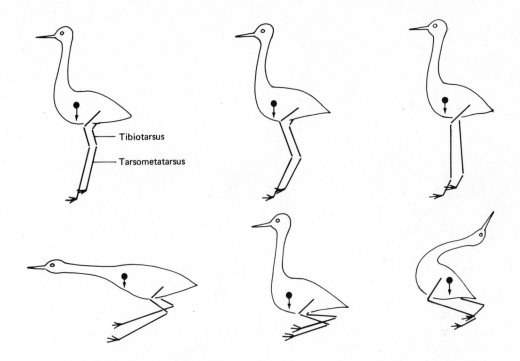

Figure 17-8 Arrangement of avian leg skeleton to accomodate center of gravity when the legs are bent for sitting down or standing up. The center figure shows the advantage of having the tibiotarsus and tarsometatarsus about the same length in order to maintain the body's center of gravity over the feet, as contrasted with conditions in which the tibiotarsus (left) or the tarsometatarsus (right) is relatively short. (Modified from Storer, [1971] in Farner and King [2].)

take flight to escape. Plovers and many sandpipers (Charadriiformes) exemplify the small, short-legged runner as they move with the wave action up and down the beach, and so do many larks (Alaudidae, Passeriformes) when they forage for wind-blown seeds and insects over the desert sands.

While graviportal adaptations—large, heavy leg bones arranged in vertical columns as supports of great body weight—are well demonstrated in some large mammals, such as elephants and rhinoceroses, no surviving birds show marked adaptations of this kind. Some of the large, flightless rails (Gruiformes) and the kiwis (Apterygidae, Dinornithiformes) of New Zealand show tendencies toward graviportal development in the legs, but such adaptations were most fully expressed by some of the large, flightless terrestrial birds of earlier times. The extinct elephantbirds (Aepyornithiformes) of Madagascar and moas (Dinornithiformes) of New Zealand were large, flightless, graviportal herbivores that evolved on oceanic islands in the absence of large carnivores and survived there until contact with primitive man in the post-Pleistocene period (Figure 17-9); the giant, carnivorous *Diatryma* (Gruiformes) and related species were successful continental forms in the Americas until the appearance of large mammalian carnivores. Among the large, flightless terrestrial birds, only some of the

(a)   (b)   (c)

Figure 17-9   Graviportal birds, (a) Elephant Bird of Madagascar and (b) moa of New Zealand, compared with cursorial ostrich (c). Not to same scale.

fleet-footed cursors like the rheas (Rheiformes) and ostrich have managed to survive in the presence of large placental carnivores.

Hopping, a succession of jumps with the feet moving together, is a special form of pedal locomotion found mostly in perching, arboreal birds, as it is primarily an adaptation for moving about on the branches and limbs of trees or bushes. It is most highly developed in the Order Passeriformes, and only a few nonpasserine birds—penguins, some members of the Cuculiformes, Coraciiformes, and Piciformes regularly hop. Many passerines can only hop rather than walk, and, even when they are on the ground, they must move by hopping. Some groups of passerines have evolved a more terrestrial mode of existence, and these birds possess a walking gait as well as an ability to hop. Larks (Alaudidae), pipits (Motacillidae), starlings (Sternidae), and grackles (Icteridae) are examples. Interestingly enough, the separation between walking and hopping passerines does not always follow strictly along family lines; for example, in the family Corvidae, the ravens, crows, and rooks are walkers, whereas the jays and magpies are hoppers.

Bipedal, ricochetal hopping has evolved as a terrestrial mode of locomotion several times among mammals (see Chapter 3), but an exact counterpart appears to be absent among birds, perhaps because the problems of balancing involved in this form of locomotion cannot be satisfactorily overcome by the avian tail. Some of the pittas (Pittidae, Passeriformes) and the peculiar African "rockfowl" (*Picathartes*, subfamily Timaliinae, Passeriformes) do progress overland by a series of rapidly repeated bounds described as "kangaroolike," but no careful analysis of their movements have yet been made (Figure 17–10).

Figure 17-10 The African bare-headed "rockfowl" *Picathartes gymnocephalus*, which hops "kangaroo-like" from rock to rock.

The best adapted avian foot for perching on branches is one in which all four toes are free and mobile, of moderate length, the hind toe being well developed, lying in the same plane as the forward three, and opposable to them. Such a foot produces a firm grip around the branch and is highly developed in the passerine birds. The zygodactylous condition, with two toes forward and opposable to two extending backward, provides an even firmer grip, but seems to be better adapted for birds like parrots and woodpeckers that climb on branches rather than for hopping and perching birds.

The tendons involved in flexing the toes of a perching bird show special modifications that physically lock the foot in a tight grip so that the bird does not fall off its perch when it reaxes or goes to sleep. These plantar tendons, which insert on the individual phalanges of the toes, slip in grooves and sheaths that are positioned anterior to the knee joint and posterior to the ankle joint in such a way that the tendons tighten and flex the toes around the perch just by the weight of the resting bird; thus, muscular contraction to hold the toes closed is not required. Furthermore, the tendons lying underneath the toe bones have hundreds of minute, rigid, hobnail-like projections to mesh with hard ridges on the inside surface of the surrounding tendon sheath, thereby locking the tendons in place in the sheaths and further helping to hold the toes in their grip around the branch by the opposing weight of the bird (Figure 17–11).

Birds that wade are adapted for walking at various depths and over various types of bottoms. Every shoreline community has its complement of short-legged, intermediate, and long-legged waders, each exploiting a somewhat different wading and feeding niche. Small sandpipers run along just at the edge of tideline ripples, while slightly larger species wade out into the receding water, and still larger species such as godwits wade in 10 to 15 cm of water and probe

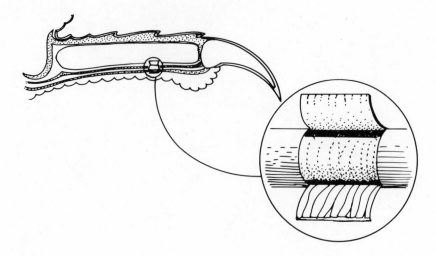

Figure 17-11 Locking mechanism in the tendons underlying the toes of perching birds. The sleeping bird's weight, by pressing down through its toes on the underlying tendons, locks them in a rachet-like mechanism in their sheaths, where numerous small projections on each tendon engage the ridges on the adjacent wall of the sheath. (Redrawn from Welty [6].)

for food with correspondingly long beaks. A similar stratification of species can often be seen in mixed species aggregations of herons in a southern swamp such as the Florida Everglades (Figure 17-12). If the bottom substrate is soft mud, wading birds require feet with longer toes, as seen in herons, or with partial webbing between short toes (flamingos, avocets, and others).

Birds that walk on insubstantial surfaces, such as herons, usually also have long toes. Ground-dwelling passerines such as larks, pipits, longclaws, and longspurs, which spend much time on loose, sandy soils, have elongated claws, especially that of the hallux. A similar condition is found in some of the coucals (Cuculiformes) living in African swamps, but the jacañas or lilly-trotters (Charadriiformes), which walk on floating vegetation, have developed the longest toes and claws relative to their body size of any birds (Figure 17-13). Certain sandgrouse (Pteroclidiformes) have especially short and broad claws and densely feathered toes, the whole foot forming a broad base for the bird when walking in loose sand. The northern grouse (Galliformes) have evolved very similar modifications of their feet for walking on snow.

Some birds have very poorly developed pedal locomotion on land, owing to their short legs, and can do little more than waddle, and most of these either take to the water or to the air when speed or movement over distance is required. Highly aerial species such as swifts, swallows, and some goatsuckers use their feet for little more than clinging to a perch, whereas very specialized aquatic forms such as loons and grebes have their legs positioned so far back to maximize propulsion through the water that they cannot walk upright on land. Penguins do normally stand and walk upright, but their legs are so short

Figure 17-12 Ecological distribution of mixed species aggregation of foraging herons as determined by water depth in relation to leg length. From left to right: green heron, little blue heron, reddish egret, common egret, great blue heron.

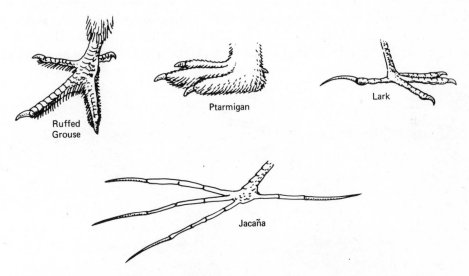

Figure 17-13 Modifications of avian feet for walking on insubstantial surfaces, such as snow (ruffed grouse, ptarmigan), sand (lark), or lily pads (jacaña). Note that elongation of the claw of the hallux is a frequent adaptation for the latter conditions.

that they cannot move very fast. They have developed a special mode of tobogganing locomotion over snow and sometimes use their modified wings (flippers) as well as their legs for propulsion, one of the few cases in which the forelimbs are used by birds for nonaerial locomotion on land.

17.2.3 Adaptations for Climbing

Birds climb about on tree trunks or other vertical surfaces by using their feet, tails, beaks, and exceptionally, their forelimbs. Several distantly related groups of birds have independently acquired similar adaptations for climbing and foraging on vertical tree trunks; they can be divided into two types—those that use their tails as supports and those that do not. Woodpeckers, (Picidae, Piciformes) woodhoopoes (Phoeniculidae, Coraciiformes), woodcreepers (Dendrocolaptidae, Passeriformes), treecreepers (Certhiidae, Passeriformes), and some ovenbirds (Furnariidae, Passeriformes) are examples of birds that use their tails. Typically such species begin foraging near the base of a tree trunk and work their way vertically upward, head first, clinging to the bark with their strong feet mounted on short legs. The claws on their anterior toes are strong and curved as an aid for grasping bark, but that of the hallux is not, and often the hallux itself is reduced or absent. Since the tial is used as a prop to brace the body against the powerful pecking exertions of head and neck, the pygostyle and free caudal vertebrae in these species are much enlarged to form the necessary support for attachment of the tail feathers, which are themselves greatly modified for strength and stiffness. A similar modification of the tail for the same function is found in certain swifts that climb on cave walls and inside chimneys. As typically happens in such a case of convergent evolution, the strengthening and stiffening of the tail feathers—superficially and functionally so similar—have actually been achieved by different structural modifications of the feathers in different taxa of wall and trunk-climbing birds (Figure 17–14).

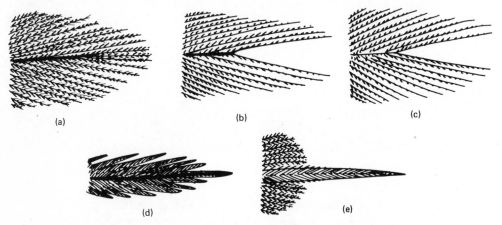

Figure 17-14 Structural variations in the central rectrices of various birds that use their tails as props for support on vertical surfaces. Tails are shown in dorsal aspect. (a) Bewick's wren (Troglodytidae) shows unspecialized structure of the tail feather not used for support. (b) Brown creeper (Certhiidae). (c) Downy woodpecker (Picidae). (d) Barred woodcreeper (Dendrocolapidae). (e) Vaux's swift (Apodidae). Barbules are represented diagrammatically: functional interlocking barbules by small lines whose tips cross; non-functional barbules by lines that do not meet. (Modified from Storer [1971]. In Farner and King [2].)

Nuthatches (Sittidae, Passeriformes) and similarly modified birds climb on trunks and rock walls in both upward and head-downward directions while foraging, and in these species, which do not use their tails for support, the claw on the hallux is larger than those on the forward-directed toes and is strongly curved. The oxpeckers, a specialized group of starlings (Sturnidae), forage for ticks by climbing about on the backs and sides of African big game mammals; they have strongly curved claws with sharp but short tips for clinging onto the hides of their mammalian symbionts without piercing the skin. The oxpeckers also have stiff tails for bracing.

A few groups of birds use their hooked beaks as well as their feet in climbing from branch to branch. This adaptation is most fully developed in the parrots (Psittaciformes) but also occurs in such diverse forms as the crossbills (*Loxia*, Passeriformes), and in the peculiar African mousebirds (Coliiformes), which in addition have all four toes directed forward and long, stiff tails.

Although a few nestling birds may clamber about with their wings, only the young hoatzin (Opisthocomidae) of South America has evolved a special modification of the forelimbs for climbing. The first and second digits on the wing bear large, functional claws, which are moved by special muscles, and these claws are used for grasping onto branches after the young hoatzin leaves the nest. Later in life the claws fall off, and the wing of the adult assumes a fully feathered, typical avian condition. Even so, the adults are weak fliers, and they continue to use their wings in a crude way to help in climbing among the dense branches of their tropical, swampy habitat, often breaking their primaries while doing so.

17.2.4 Adaptations for Swimming on the Surface

During the course of vertebrate evolution various groups of amniotic tetrapods departed from their ancestral mode of terrestrial life by evolving adaptations that allowed them to become aquatic. Although no birds have become fully aquatic, like the ichthyosaurs and cetaceans, there are at least 390 species in nine orders that are adapted for swimming. Nearly half of these aquatic species not only swim on the surface but also dive and swim under water for food or escape. The others either feed on the surface or plunge into the surface water from the air.

Modifications of the hind limbs are the most obvious avian adaptations for swimming, although other changes involve widening of the body to increase stability in water, dense plumage for buoyancy and insulation, a large preen gland, the oil from which waterproofs the plumage, and structural modifications of the body feathers that also retard penetration of water to the skin (Chapter 16). For maximum mechanical advantage in paddling, the legs need to be positioned near the rear of a bird's body, where the mass of leg muscles interferes least with streamlining and where the best control of steering can be achieved; however, these aquatic advantages can only be achieved at some sacrifice of efficient locomotion on land, as previously noted. Avian legs are basically articulated so as to thrust backward easily, and in aquatic species the tarsometatarsi are laterally compressed so that they meet the least resistance when moved through water.

To function as efficient paddles, the feet of aquatic birds are either webbed or lobed (Figure 17–15). Webbing between the three forward toes has been independently acquired at least four times in the course of avian evolution, probably more: in the ancestors that gave rise to the waterfowl and flamingos, in the petrel-penguin lineage, in the gull-auk and perhaps loon lineage, and presumably in the extinct Ichthyornithiformes. Totipalmate webbing of all four toes has apparently evolved only once, in the pelicans and their relatives (Pelecaniformes). Lobing on the toes of aquatic birds has also evolved convergently in several phylogenetic lines: the grebes (Podicipediformes), certain rails, for example, coots (Rallidae, Gruiformes), the sungrebes or finfoots (Heliornithidae, Gruiformes), the phalaropes (Phalaropodidae, Charadriiformes), and presumably in the extinct Cretaceous Hesperornithidae.

There are two different types of lobed feet. The grebes are special in that the lobes on the outer sides of the toes are rigid and do not fold back as the foot moves forward (Figure 17-16). For an efficient recovery stroke with minimum drag, a grebe has to rotate its foot 90° so that the inner side points forward, the toes with their lobes slicing through the water like knife blades. A mechanically simpler and energetically less costly mechanism for the recovery stroke occurs in all the other lobe-footed swimmers. The lobes are structured as flaps that fold back against the toes during forward movement through water but flare open to present a maximum surface on the backward power stroke. Whether there is some adaptive advantage over this method in the grebe's more complicated solution to the same problem has not been determined, but presumably the grebe might have a more effective power stroke with its rigid lobes.

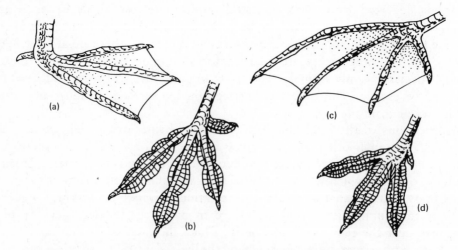

Figure 17-15 Webbed and lobed feet of some aquatic birds. (a) Duck, (b) Coot (c) Cormorant (totipalmate webbing). (d) Phalarope.

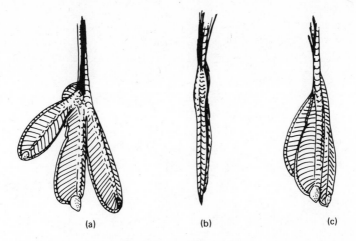

(a)   (b)   (c)

Figure 17-16 Detail of grebe's foot and method of rotation to allow for efficient recovery. (a) position of toes for backward power stroke in side view. (b) front and (c) side views of foot in recovery stroke. (Modified from Peterson [4].)

17.2.5 Locomotor Adaptations for Diving and Swimming Under Water

Diving and swimming under water usually require even greater specializations of limb and body structure for birds than swimming on the surface does. The transformation from a surface swimming bird to a subsurface swimmer has occurred in two fundamentally different ways: either by further specialization of a hind limb already adapted for swimming or by modification of the wing for use as a "flipper" under water. Highly specialized, foot-propelled divers have evolved independently in the grebes, cormorants (Phalacrocoracidae, Pelecaniformes), loons, and in the extinct Hesperornithidae, including some flightless forms in all these families except the loons. Wing-propelled divers have evolved in the Procellariiformes (the diving petrels, Pelecanoididae), in the Sphenisciformes (penguins), and in the Charadriiformes (auks and related forms, Alcidae). Only among the waterfowl (Anatidae) are there both foot-propelled and wing-propelled diving ducks, but none of these species is as highly modified for diving as specialists such as the loons or auks. Interestingly, the water ouzels or dippers (Cinclidae) are passerine birds that dive and swim under water with great facility using their wings, but they lack any obvious specialization of structure for doing so.

Highly specialized foot-propelled divers show the following adaptations: (1) a long and extremely narrow pelvis, (2) a short femur with a hingelike, double articulation at and above the acetabulum, (3) a long tibiotarsus with a long cnemial crest, (4) union of the leg with the body musculature nearly to the ankle joint, and (5) a laterally compressed tarsometatarsus (Figure 17-17). A close examination of the structural details reveals the similarities in these

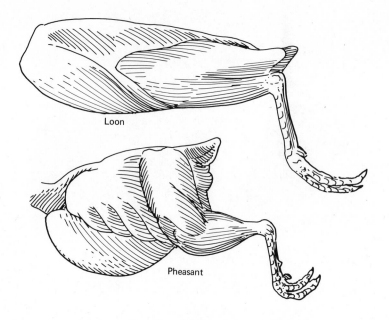

Loon

Pheasant

Figure 17-17 Comparison of the skinned bodies of a pheasant and a loon to show how the latter's leg musculature has become incorporated into the body mass. (Redrawn from Storer [1971] in Farner and King [2].)

features in the three groups of divers are the result of convergent evolution. For example, the specially elongated cnemial crest, to which the powerful extensor muscles of the leg attach, derives in different ways from bony elements in the three groups. In the *Hesperornithidae* it consists only of the patella; in loons it forms as an extension of the tibia, the patella being small and embedded in the tendons of the knee; whereas in the grebes the cnemial crest is compounded from a large patella and an extension of the tibia.

Flightless wing-propelled divers like the penguins and the recently extinct great auk apparently represent extreme specializations in parallel evolutionary transitions from Southern and Northern Hemisphere ancestral forms that originally used their wings only for flight, through intermediate states in which the wings were used both for aerial and underwater flight (Figure 17–18). Although fossils are not available to show the actual evolutionary changes, known forms can be arranged in a sequence of adaptive stages that suggest how evolution may have proceeded in these cases (Figure 17–19). The intermediate states in the evolutionary sequence, represented by the present-day auks and diving petrels, are adaptive compromises between the selective pressures favoring adaptation of the wing for flight in air and those favoring adaptation of the wing as a flipper for locomotion in water. Once the balance has been tipped in favor of loss of aerial flight, natural selection can act to modify the wing as an increasingly efficient flipper operated by powerful breast muscles; aerial constraints on wing loading and body size are also removed, so that some wing-propelled divers have been able to evolve to rather large body

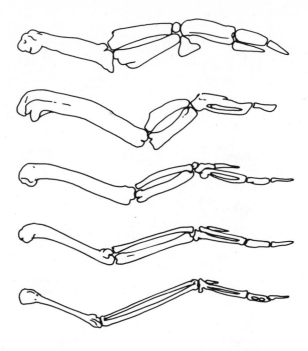

Figure 17-18 Suggested series of adaptive stages through which the penguin flipper may have evolved. From bottom to top: wing skeleton of a gull (*Larus*), razor-billed auk (*Alca*), great auk (*Pinguinus*), Lucas auk (†*Mancalla*), and penguin (*Spheniscus*). (Redrawn from Storer [1971] in Farner and King [2].)

sizes, up to an estimated 100 to 150 kg in the case of some fossil penguins. In contrast, when foot-propelled divers become flightless, their wings and pectoral muscles tend to degenerate, as seen in the flightless grebes, the flightless cormorant of the Galapagos, and in *Hesperornis*.

We have considered in some detail the locomotor adaptations for diving and swimming under water, but other morphological, behavioral, and physiological modifications are also important. Buoyancy is one problem; oxidative metabolism under water is another. Diving birds can overcome buoyancy by reducing the volume of their air sacs, decreasing the pneumaticity of their bones, by expelling air from their plumage before submerging, and, in the case of some penguins, by swallowing small stones that function as ballast. Some divers have also evolved heavier, less pneumatic bones than typical fliers have. As in the case of other diving air-breathers, birds greatly constrict their peripheral blood flow, reduce heart rate, and otherwise greatly lower their metabolic demands for oxygen while under water. They also tend to have a large blood volume with a high oxygen-carrying capacity and muscles that are especially rich in myoglobin; they have a high tolerance for carbon dioxide levels in blood; and they can obtain considerable energy from anaerobic metabolism—all adaptations that allow for extended diving times, which normally range from about 1 to 3 min in birds, with a maximum recorded survival time of 15 min.

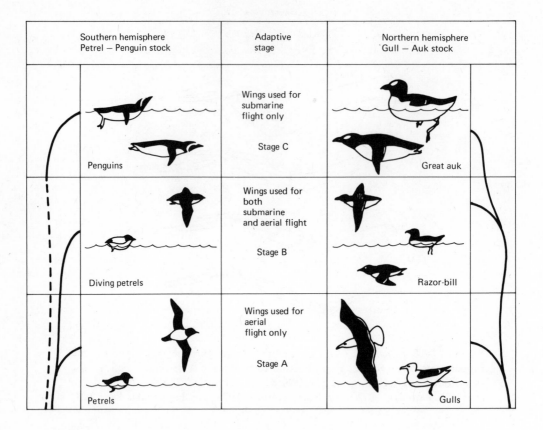

Southern hemisphere Petrel — Penguin stock	Adaptive stage	Northern hemisphere Gull — Auk stock
Penguins	Wings used for submarine flight only Stage C	Great auk
Diving petrels	Wings used for both submarine and aerial flight Stage B	Razor-bill
Petrels	Wings used for aerial flight only Stage A	Gulls

Figure 17-19   Postulated sequence of evolutionary stages from a flying ancestor to flightless divers in a Southern Hemisphere lineage (Procellariiformes-Sphenisciformes) and Northern Hemisphere group (Charadriiformes). (Adapted from Storer [1971] in Farner and King [2].)

17.3   DIVERGENCE AND CONVERGENCE IN FEEDING ADAPTATIONS

Many forms of organic matter serve as food for birds, although most species eat energy-rich substances—seeds, nuts, and particularly the bodies of other animals. Consequently, birds show a great variety of morphological, physiological, and behavioral adaptations for feeding on these diverse sources of food. Modifications of the beak for different modes of feeding are especially conspicuous and well studied—to a lesser extent modifications of the feet are also involved—but other specializations of the digestive tract (crop, gizzard, caecum, intestine) are equally important, as are associated behaviors that involve the locomotor appendages and sense organs.

Major types of avian feeders are (1) fish and aquatic invertebrate feeders, (2) aquatic filter-feeders, (3) carnivorous or predatory birds (that is, those that eat other tetrapods, especially birds and mammals), (4) carrion eaters, (5) insectivores and terrestrial invertebrate-feeders, (6) pollen and nectar feeders, (7) fruit eaters, (8) seed eaters, and (9) grazers and browsers. Many species employ more than one type of feeding, especially at different times of year, and some species are so highly specialized in their food habits that they do not fit into any major category.

17.3.1 Fish and Aquatic Animal Feeders

The problems involved in obtaining food from aquatic environments, in which birds cannot live continually or reproduce, have led to a variety of different evolutionary solutions; many are associated with the locomotor adaptations for swimming and wading that we have already discussed in the previous section. Birds can obtain aquatic animal foods from the surface, at various depths in the water below the surface, and from the bottoms of streams, lakes, and marine bodies. Figure 17–20 shows some typical examples of the main methods employed by aquatic feeding birds.

A few species of aquatic birds obtain some of their food by pursuit and capture in the air or by "piracy." Frigatebrids (Fegatidae, Pelecaniformes)

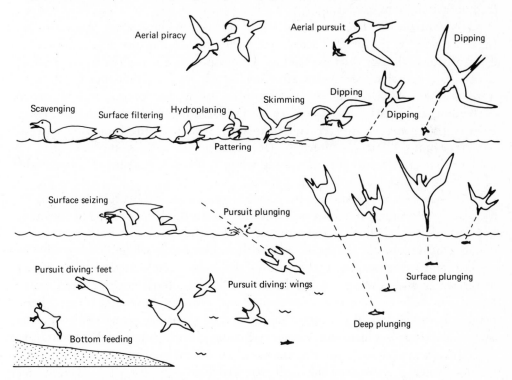

Figure 17-20 Schematic representation of aquatic modes of feeding in birds, in water too deep for wading. See text for details (Modified from Ashmole [1971- in Farner and King [2].)

capture flying fish in the air, and a number of species such as fulmars (Procellariidae), skuas and jaegers (Stercorariidae, Charadriiformes), some large gulls (Lariidae, Charadriiformes), and the fishing eagles (*Haliaeetus* spp, Falconiformes), as well as frigatebirds, frequently obtain food by forcing other birds to disgorge or drop their prey.

More frequently birds obtain food in flight by capturing prey at or just under the surface of the water. Only the specialized ternlike birds called skimmers (*Rhyncops* spp, Laridae) regularly use the technique in which the bird flies with the lower mandible immersed in water to make contact with and then seize individual prey, a mode of feeding that is only effective in calm water but does have the advantage of permitting location of prey by touch and making capture possible in turbid water or at night. "Dipping" involves picking up food in the beak from the surface or just below while the bird is in flight; it is the principal method used by many terns (Laridae, Charadriiformes) and small petrels (Hydrobatidae, Procellariiformes and others) and occasionally by other aquatic species, such as boobies (Sulidae, Pelicaniformes), frigatebirds, and gulls; however, the hydrobatid petrels feed mainly by "pattering," using their feet in addition to their wings to maintain a precise height above uneven water surfaces, a method that allows the birds to pick small prey from the surface in rapid succession. Some of the larger petrels, some terns, and some gulls also feed occasionally by pattering.

Many aquatic birds feed mainly while settled on the surface of the water. Birds as diverse as the large albatrosses and the small phalaropes (Phalaropodidae, Charadriiformes) use "surface seizing" or grasping individual items of food in the beak, as do some species of waterfowl (Anatidae) and other species of freshwater feeders. "Scavenging" on sick, injured, or dead animals is another mode of feeding frequently employed by surface-floating birds, such as the giant fulmars (*Macronectes* spp, Procellariiformes), several of the fulmarine petrels, and large gulls. "Surface-filtering" (see Section 17.3.2) is another form of feeding that grades into surface-seizing when it involves capture of single items of food along with water that is expelled.

Birds feed below the surface by diving from the air into water ("plunging"), by actively swimming under water, either by foot-propulsion or wing-propulsion, by wading into shallow water and extending the head and neck below the surface, by extending the head and neck under water while floating on the surface, and exceptionally in the case of the water ouzels or dippers by diving under water and walking on the bottom. Plunging birds such as terns, gulls, the brown pelican, kingfishers (Alcedinidae, Coraciiformes), and some others penetrate the surface by a body length or less ("surface plunging"); whereas others, such as boobies and gannets (Sulidae), with dense bodies penetrate several meters below from the momentum of their dives to capture prey ("deep plunging"). Most of the birds that use their wings for locomotion under water obtain their food by "pursuit diving", departing from the surface by diving and pursuing prey under water, as do many of the foot-propelled divers. Such birds actively chase after fishes, cephalopods, crustaceans, and other actively moving animals at various depths in the water, or they dive to the bottom to feed on shelled mollusks and other bottom-dwelling organisms in relatively shallow water. The latter group includes some of the diving ducks (eiders, old-squaw, scoters), some cormorants, and auks (*Cepphus* spp, Alcidae). Those that feed on shelled mollusks have especially enlarged and muscular gizzards which crush and grind up the shells.

Many kinds of aquatic birds capture and eat fish, and the shapes and sizes of their beaks (and tongues) are often finely adapted for particular methods of seizing, holding, and ingesting their prey (Figure 17–21). The skimmers have already been mentioned, and their beaks are perhaps the most specialized for a particular method of capturing fish. Most fish eaters seize their prey between the elongate upper and lower mandibles in a forcepslike action, and many of them have special structures that hold onto slippery bodies. The extinct hesperornithids had rows of true teeth for this function, and other extinct forms evolved toothlike structures constisting of a bony core covered with a horny overgrowth of the epithelium. The bills of mergansers, fish eating ducks, have hooklike structures along their edges derived from the rostral lamellae that characterize the beaks of other ducks and waterfowl. Serrations in the cutting edge of the bill occur in many fish eaters such as the tropic birds (*Phaethon* spp, Pelecaniformes), boobies (*Sula* spp, Pelecaniformes), anhingas (*Anhinga* spp, Pelicaniformes), and some auks (razorbill and puffins).

Still other types of fish eaters have predaceous, terminally hooked beaks for capture (some penguins, cormorants), although a very similar beak shape in the albatross and their relatives is not associated with a predominant diet of fish but rather with cephalopods and other marine invertebrates. Others like the pelicans have distensible throat pouches that aid in capturing fish. A few birds, anhingas and the western grebe, actually spear fish on their long, swordlike beaks. The anhingas and herons also have a special muscular and bony mecha-

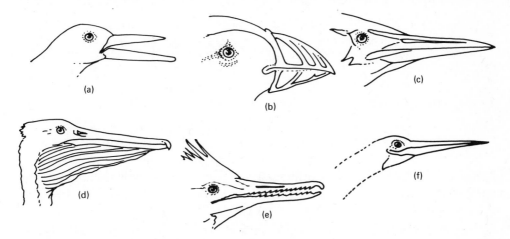

Figure 17–21 Adaptive modification of beaks used to catch fish. (a) skimmer. (b) puffin. (c) heron. (d) Pelican. (e) merganser. (f) anhinga. (Modified from various sources).

nism built into their elongate necks, making it possible for them to extend the withdrawn head and neck very rapidly to strike fish under water. The herons have a further adaptation for fishing in that, from their standing position in water, they have the ability to use binocular vision below their long beaks without having to move their heads.

Some birds of prey have feet specially adapted for catching fish (Figure 17–22). The widely distributed osprey or fish hawk (*Pandion haliaeetus*) dives from the air and catches fishes by plunging feet-first into the water and seizing a fish in its talons. The scales on the plantar surfaces of its toes bear spicules that aid in holding slippery fishes, and the outer toe is reversible, so that prey can be held in a zygodactylous grip. The Neotropical fishing buzzards of the genus *Busarellus* have similar spicules on the pads of their feet but are not as highly specialized for fish catching as the osprey. The sea eagles and fish eagles of the genus *Haliaeetus* are even less specialized in foot structure but are nevertheless very adept at catching fish from surface waters; they also feed on birds, mammals, and carrion. The Asiatic fishing owls of the genus *Ketupa* feed mainly on fish and differ from other closely related owls in having unfeathered tarsi and spicules on the soles of their feet.

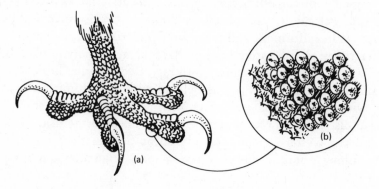

Figure 17–22 Foot structure in the fish-catching osprey. (a) Reversible zygodactyl foot; (b) detail of spicules on foot pads for holding on to slippery fish. (Redrawn from Storer [1971] in Farner and King [2].)

17.3.2 Aquatic Filter-Feeders

Filter-feeding by birds involves the intake of numerous small food particles along with water or other materials such as sand or mud that are separated from food and expelled. Visual or other sensory detection of each bit of food does not occur. Small animals and plants, such as diatoms and algae, and organic detritus are types of food usually consumed by filter-feeders.

Several types of filtering mechanism and techniques have been evolved by birds. Some of the marine fulmarlike birds (*Fulmarus, Daption, Thalassoica*) have wide-spreading lower mandibles, distensible gular pouches, and an associated muscular system for creating suction by which zooplankton can be filtered from the surface of the water. Some storm petrels may also use suction to obtain animal oils from slicks on the surface. Pelicans use their very large gular pouches like a dip-net when feeding on small shoaling fish, and this mechanism can be considered a type of filtering, too. The hydroplaning behavior of the specialized petrels called "prions" (*Pachyptila*) is a special type of filtering in which the bird rests on its breast on the surface, holding its wings outstretched and its bill under water, and propels itself with its feet. Since the flow of water through the mouth for filtering is produced by the bird's own motion, this method of feeding is analogous to a towed plankton net. The straining of small items through a series of lamellae in the bill by muscular exertion of a broad, fleshy tongue against the palate is the method of feeding used by flamingos and many ducks (Figure 17–23). The lesser flamingo sweeps its beak through the water near the surface, filtering out fine blue-green algae and diatoms, whereas the greater flamingo, which has a coarser filtering apparatus, feeds in the bottom mud, taking invertebrates and organic matter, from which water, sand particles, and other unusable materials are separated.

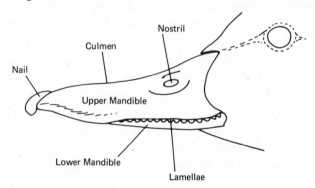

Figure 17-23 Beak of a duck, showing lamellae along the side of the horny covering (culmen) of the upper mandible. These structures are involved in the typical anatid method of filtering food from water.

17.3.3 Carnivorous or Predatory Birds

What is a "flesh-eating" or carnivorous bird? In a general sense any animal that kills and eats other animals is predatory and carnivorous, but we humans have always been expecially intrigued by animals that kill and eat other species closely related to themselves. Haunted as we are by our own ancestral inclinations to hunt and kill, birds that prey on other birds, mammals that stalk and catch other mammals, somehow seem more predatory to us than those that eat animals far different from themselves phylogenetically. In fact, no precise definition can be given for a carnivorous bird, and while we have terms like "insectivore" and "piscivore" to designate certain categories of predators, there is no comparable term for species that feed to a large extent on tetrapods, especially mammals, birds, and reptiles. The term "carnivorous" usually suggests such a predator, but many of these birds include fishes, insects, and other invertebrates in their diets too. Some are generalists, taking a wide array of sizes and species from many higher taxa, whereas others are specialists restricted to preying on a few closely related species, or exceptionally, to a single species of prey.

The diurnal birds of prey, hawks, eagles, falcons (Falconiformes), and the nocturnal birds of prey or owls (Strigiformes) are the most highly adapted predators among birds, but many other avian orders and families contain species that have specialized to some extent for feeding on terrestrial vertebrates. Frigatebirds or man-of-war birds often snatch young of other sea birds in their beaks as they fly by nesting colonies; the skuas and jaegers (Stercorariidae) and some of the large gulls (*Larus* spp) feed on rodents (lemmings) and small birds including penguin chicks, during their nesting seasons, which are spent primarily in association with Arctic or Antarctic terrestrial ecosystems, whereas during the non-breeding period these birds are mostly pelagic and coastal in existence. Carnivorous species are also found scattered through other such diverse nonpasserine orders as the Gruiformes (South American seriemas, Cariamidae), Cuculiformes (roadrunners and coucals), and the Coraciiformes (some kingfishers, some rollers, and some motmots); in all these cases, snakes, lizards, and occasional amphibians are the forms most often preyed upon. The large and diverse order of perching birds (Passeriformes) contains several families in which some species have become predatory to a degree on terrestrial vertebrates, including birds and mammals. The most highly specialized in this respect are the true shrikes or butcherbirds of the genus *Lanius* (Laniidae), but there are many other shrikelike forms that are also predatory, in particular some of the African bush-shrikes (*Malaconotus* spp, Laniidae), and the Australian butcherbirds (*Cracticus* spp, Cractididae). Less specialized but nevertheless highly capable predators are also found among the Corvidae, especially some of the jays, magpies, and ravens, and in some other passerine families.

The predatory mode of feeding involves hunting, capturing, killing, and eating prey which usually is too large to be swallowed whole. The various carnivorous birds have become specialized to varying degrees and adapted in different ways to perform these five functions.

The method of hunting (pursuit) adopted by a predatory bird depends upon its locomotor adaptations (see section 17.1); on its daily pattern of activity (whether diurnal, nocturnal, or crepuscular); and on the specialization of its sense organs for detecting its prey. Most avian predators are aerial, dirunal, and visual hunters; but some are mainly terrestrial cursors, and some, especially the owls, are nocturnal and use auditory cues as well as vision for locating prey in darkness (Figure 17–24). A bird's habitat also strongly influences its hunting behavior.

Aerial pursuit can be divided into several patterns. Aerial hunters may launch the attack from a position in the air or from a stationary perch. Most species that hunt in open habitats such as savannas, prairies, or marshes search for prey by flying and begin their attacks in the air. Large raptors like eagles and the

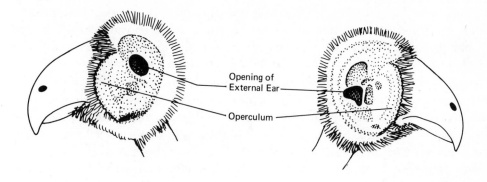

Figure 17-24 Asymmetric shape and position of right and left ear openings of an owl (*Asio*) aid in acoustic location of prey by exaggerating differences in time, loudness, and phase of sound waves reaching the two ears. (Modified from Welty [6].)

broad-winged hawks or buzzards (*Buteo* spp and relatives) typically soar on thermals at heights of 100 to several hundred meters looking for suitable quarry—mostly mammals or other ground-dwelling prey—moving beneath them; when a vulnerable animal is spotted, the raptor folds its wings close to its sides, dives or "stoops" at a steep angle of attack, and closes on its intended victim with the great speed developed by the momentum of its dive. The bird-hunting falcons attack flying prey in the same manner (Figure 17-25), stooping almost vertically from a great height and either striking their prey as they dive past or more usually pulling out behind the fleeing bird and overhauling it from the speed gained in the dive.

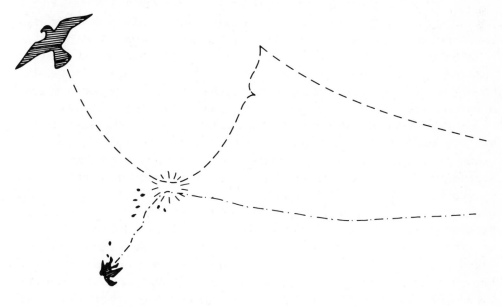

Figure 17-25 Diagram of the flight path of a stooping falcon and its prey.

Several unrelated predatory birds have the ability to maintain a more or less stationary position in the air by facing into the wind and hovering on beating wings for fractions of a minute to search the ground or water below before dropping down. Well-known examples include the osprey (*Pandion* sp), the small falcons called kestrels or "wind-hovers," certain kites and buzzards (for example, white-tailed kite and rough-legged buzzard), some owls (for example, short-eared owl), the jaegers (*Stercorarius* spp), and some of the shrikes (*Lanius* spp). The white-tailed kites (*Elanus* spp) combine hovering with a peculiar, controlled parachutelike descent on wings stretched out in a "V" angle above the body to capture mice and other small animals on the ground in dense grass. Other airborne predators such as the harriers (*Circus* spp) and some open-country owls, like the short-eared owl, quarter back and forth only a few meters above the ground and drop suddenly down onto prey caught unaware of the predator's approach.

Some birds of prey attack flying birds, bats, and large insects by direct, flapping pursuit through the air. This method is particularly characteristic of some of the strong, fast flying falcons (gyrfalcon, peregrine, merlin, bat falcon); however it is also used by the short-winged hawks or accipiters, but usually beginning from a perched position, and by some owls such as the screech owls (*Otus* spp) and pygmy owls (*Glaucidium* spp). Even some of the relatively unspecialized avian predators can catch other birds by direct pursuit through the air, for example, the glaucous gull, parasitic jaeger, the northern raven, and the great grey shrike.

In wooded or forested habitats, predatory birds usually hunt by perching and waiting for an unsuspecting animal to reveal itself or to move into a vulnerable position ("still hunting"). In fact, many of the open-country hunters already mentioned also hunt in this fashion at the edge of woodlands or from isolated trees and poles. More often, though, the true woodland and forest predators hunt by stealth from a concealed position in a tree, from which they make a short dash after their intended victims. Most of the bird-eating hawks (genus *Accipiter*) hunt in this fashion, as do many other diurnal birds of prey ranging in size from pygmy falcons (weighing 50 g) to forest eagles (for example, harpy eagle, crowned eagle, monkey-eating eagle) weighing 5 kg or more. Most owls and shrikes, and nearly all of the unspecialized woodland predators— kingfishers (Daceloninae, Alcedinidae), rollers, coucals, jays—use some variant of this method, which is highly successful in habitats with dense cover.

A number of predatory birds in grasslands and semideserts have become essentially terrestrial and catch most of their prey by running the animals down or chasing them into holes, crevices, or crannies where they can be caught. No owls are known to hunt in this way, but isolated examples can be found in several other orders. The cranelike secretary bird of Africa is a prime example among the Falconiformes; it runs after reptiles, small mammals, occasional birds, and large insects on the ground. To a lesser extent some other falconiform birds hunt the same way, for example, the chanting goshawks (*Melierax* spp) of Africa and the savanna hawk (*Heterospizias* sp) of South America.

The seriemas (Gruiformes) of the Chaco region of South America are probably the surviving representatives of the large and powerful flightless carnivorous birds of earlier periods, the Oligocene giants, *Brontornis* and *Phororachos*, and the still earlier diatrymas of the Eocene. Other ground-dwelling predators of open landscapes include the roadrunners (*Geococcyx* spp, Cuculiformes), mainly feeders on reptiles.

A few forest-dwelling species also pursue some of their prey on the ground or by using their long legs and feet to reach for animals hiding in holes in the ground, in hollows of trees, and crevices in rocks. Even the large and powerful goshawk, which is primarily an aerial chaser, will run into rabbit burrows to catch its prey or chase after animals in thick brush by running on the ground. This type of hunting is more highly developed in the long-legged Neotropical forest falcons called micrasturs (Falconidae). Two aberrant species related to the harriers, called crane hawks or harrier hawks (*Geranospiza* sp of South America and *Polyboroides* sp of Africa), secure much of their prey from cracks in rocks, tree-holes, burrows, or birds' nests, a mode of hunting for which their long legs, small feet, and peculiar intertarsal joints, which can be flexed forward and backward, are special adaptations.

Carnivorous birds actually catch and kill their prey either with their feet or their beaks or a combination of both. The more highly adapted ones in the Falconiformes, Strigiformes, and Laniidae have evolved special structural and behavioral modifications for catching and killing. Most falconiform and strigiform birds catch and kill with their feet, which tend to be large and powerful relative to body size and are equipped with long, sharp talons for holding and puncturing the prey. The foot of a goshawk, for example, shows extreme specialization for this mode of catching and killing (Figure 17–26). Its long but

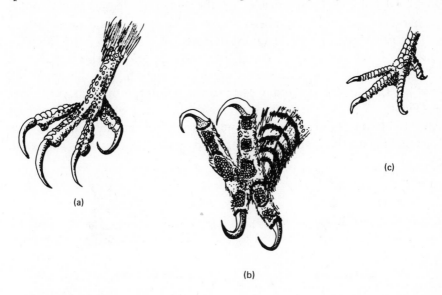

Figure 17-26 Feet of goshawk (a) and great horned owl (b) compared with foot of a comparable sized fowl (c).

powerful toes are adapted for grasping its quarry, and the long, daggerlike talons are the lethal instruments by which the prey is killed, by a powerful spasmodic gripping that forces the talons through the skin and body wall of the prey and into its vital organs. Only after the animal stops struggling and is limp in the hawk's feet does the raptor begin to use its beak to feed. Most owls catch and kill in the same way, and their zygodactylous toes and needle-sharp talons are positioned at the moment of capture so as to maximize the area of contact with the prey, a feature that is particularly advantageous when used in conjunction with acoustical location of the prey, which is less accurate than visual location (Figure 17–26). Some other less specialized predatory birds also catch some prey in their feet—for example, some shrikes and some corvids, particularly ravens—but the killing is done primarily or exclusively with their beaks, and most of the catching is also by use of their beaks.

The falcons (Falconidae) are specialists in catching and killing. In many instances they catch and kill with their feet, although their talons are not as well adapted for killing as those of most accipitrine birds of prey, eagles, and owls, nor do they possess the same powerful spasmodic grip that these other predators have in their feet. When attacking prey in a "stoop," falcons often use their feet to strike and stun their victims, holding their feet in a loose fist and using their momentum to deliver a powerful blow. Sometimes the back talon on the hallux also gashes the prey open as it rakes past. The blow may or may not kill the prey but is often sufficiently incapacitating to permit capture on a second attack. In any case, once a falcon has its feet securely around the prey, either on the ground or in the air, it automatically reaches down with its beak and bites into the neck until the cervical vertebrae have been disarticularted before beginning to tear into the prey with its beak to feed.

This neck-biting is a uniform behavior in all true falcons (genus *Falco*) and is probably characteristic of all members of the Falconidae (good observations on forest falcons, *Micrastur* spp, laughing falcons, *Herpetotheres* spp, and for some caracaras, *Polyborinae*, are lacking). In the true falcons this behavior is associated with a well-developed tomial "tooth" on each side of the upper bill subterminal to the curved tip and with a corresponding notch in the lower mandible. (Figure 17–27). These structural modifications of the beak appear to aid in disarticulating the cervical vertebrae of prey, but they may also serve as devices for breaking up the long bones of appendages before they are swallowed.

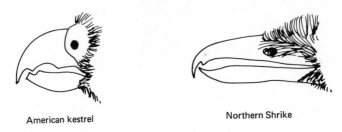

American kestrel   Northern Shrike

Figure 17–27 Tomial "teeth" of falcon (*Falco*) and shrike (*Lamius*). (From T. J. Cade [1967] *The Living Bird* 6:43–85.)

The other group of predatory birds in which neck-biting is known to be highly developed consists of the true shrikes (*Lanius* spp), and their beaks show very parallel modifications of the cutting edges. Like falcons, shrikes also break up long bones in their beaks before swallowing them. The African bush-shrikes (*Malaconotus*) and the Australian butcherbirds (*Cracticus*) probably kill vertebrate prey the same way, even though they do not have the toothlike structures of the true shrikes. Some owls such as the barn owls (*Tyto*) and the great horned owls (*Bubo*) also frequently bite into the necks of their prey immediately after capture, although they are perfectly capable of killing animals with their feet.

All of the less specialized predatory birds (those in orders other than Falconiformes and Strigiformes) catch and kill their prey primarily or exclusively with their beaks, by jabbing, pinching between the mandibles, shaking, or pounding the captured animal against the ground, a rock, or the branch of a tree until it succumbs from generalized injuries. Gulls and jaegers, kingfishers and rollers, roadrunners and coucals, and the jays, crows, and ravens all catch by grabbing their prey in their beaks and kill by some variation or combination of these methods.

The falconiform and strigiform raptors, with their specially adapted feet, typically hold their prey in their talons and tear the carcass into bite-sized pieces with their powerful hooked beaks, although owls swallow much of their prey entire, as they have especially large gapes. Some of the unspecialized forms, such as jays and ravens, use their feet for holding to some extent too, but most do not. In most cases the unspecialized predators simply swallow their prey intact, a fairly easy task with elongate animals such as lizards and snakes. Some jaegers and gulls—usually members of a pair or siblings—engage in a "tug of war" over large prey, such as an adult male lemming, and tear it into smaller pieces by pulling against each other with their beaks. The shrikes and certain other "butcherbirds" impale their prey on thorns, barbed-wire fences, or broken-off branches and then tear the carcass apart with their beaks.

The diurnal birds of prey (Falconiformes) have large saclike crops developed in the esophagous in which quantities of food can be stored, but most of the other predators do not have crops. The stomachs of predatory birds are large, saclike structures into which powerful proteolytic enzymes and hydrochloric acid are secreted to aid digestion. There is little muscular movement for mechanical processing of the food, as in the gizzards typical of the domestic fowl and seed eaters. In some species not only is all of the meat and fat digested and dissolved but also most of or all of the bones; in other species little of the bone is dissolved; and no species appears to be capable of digesting the epidermal derivatives—hair, scales, feathers, and horny structures. Consequently all predatory birds (fish eaters and insectivores too) regurgitate these undigestible materials periodically in the form of compact pellets, which often accumulate under nests or favorite perches and, when carefully analyzed, can serve as a useful index of the food habits of the species (Table 17-2).

Among the Falconiformes, several clades have independently evolved similar

Table 17-2. Analysis of 136 Pellets from Five Northern Shrike Nests at Lake Peters, 1959*

Species of Prey	Number pellets	Percent frequency	Minimum individuals	Percent total individuals
Water Pipit	1	0.74	1	0.46
Redpoll	5	3.68	5	2.29
White-crowned Sparrow	3	2.21	3	1.38
Lapland Longspur	6	4.41	6	2.75
Snow Bunting	1	0.74	1	0.46
Unidentified bird remains	52	38.24	1	0.46
Total birds	68	50.00	17	7.80
Alaska vole, adults	12	8.82	12	5.50
Alaska vole, juveniles	15	11.03	15	6.88
Tundra vole, adults	3	2.21	3	1.38
Tundra redback vole, adults	4	2.94	4	1.83
Tundra redback vole, juveniles	6	4.41	6	2.75
Brown lemming, adults	1	0.74	1	0.46
Greenland collard lemming, adults	1	0.74	1	0.46
Unidentified microtine remains	71	52.21	?	———
Total microtines	113	83.08	42	19.27
Bumblebees	62	45.59	67	30.73
Wasps	7	5.15	12	5.50
Grasshoppers	1	0.74	1	0.46
Ground beetles	62	45.59	79	36.24
Other insects	16	11.76	?	———
Total			218	

*From T. J. Cade [1967] *The Living Bird* 6:43–86.

specializations for catching snakes, including venomous species. Their adaptations involve heavy scalation on the feet and legs to protect against strikes by the quarry and short, rough, extremely powerful toes to grip and hold this type of prey. These raptors catch the snake close behind its head with one foot and around the trunk by the other and then bite off and discard the head before swallowing the rest of the snake whole, unless it is very large. Biting off the head evidently is an adaptation to avoid contact with venomous fangs; it remains to be learned whether raptors can distinguish between venomous and nonvenomous forms. Also, these birds often harass the snake first by jumping around it and flailing with their wings, forcing the snake to strike repeatedly until exhausted before grabbing. Such adaptations occur in the Old World and African snake eagles (Circaetinae, Accipitridae), in the African bateleur (*Terathopius ecaudatus*), in the Neotropical laughing falcons (*Herpetotherinae*, Falconidae), and in less specialized form in some other large species. Some owls also catch snakes, but none appears to have evolved special adaptations for this type of quarry.

A few raptors are narrowly specialized in their feeding habits. An extreme example is the snail kite or Florida Everglades kite (*Rostrhamus* sp), whose narrow, sickle-shaped beak is adapted for extracting snails from their shells (Figure 17–28). They feed almost exclusively on a single genus of fresh-water snails (*Pomacea* spp); once the snail has relaxed in the kite's foot, the bird inserts its needle-like upper bill past the operculum to destroy the collumelar muscle that holds the snail inside its shell. Obviously the survival of such a specialized predator is closely tied to the welfare of the snails; and when alterations in water level or other environmental changes occur to the detriment of the snail populations, the kites also diminish.

Figure 17-28 Beak of the snail kite (*Rostramus*), adapted for removing snails from their shells.

17.3.4 Carrion Eaters

The bodies of dead animals provide food for certain birds such as vultures (Aegypiinae, Accipitridae), condors (Cathartidae), some eagles, caracaras (Falconidae), and even some storks, the large marabous and adjutants (*Leptoptilos* spp). The basic similarities, independently evolved, in these carrion feeders are (1) naked heads and necks or much reduced feathering in these areas, especially immediately back of the beak and around the face, presumably an adaptation that reduces the chances of befouling the plumage with the juices, blood, and grease from the carcass, and (2) in the case of falconiform scavengers, nonpredaceous feet more adapted for walking on the ground than for holding quarry, especially well developed in the condors.

Most are large, soaring birds that spot the carcasses from a great height by keen vision, although some of the cathartids also use smell at closer range. Variations in body size and in beak size and shape are specializations for feeding on different sized carcasses or on different parts of the same carcass. For example, the large, heavy beaked species (lappet-faced vulture, white-headed vulture) are the ones that first open up a carcass of large animals like ungulates because they are capable of tearing through the skin, whereas the long-necked griffons are more specialized to reach deep into the body cavity once the animal has been opened up. In the East African savannas where many large ungulates die, carrion-eating niches are highly diversified, and as many as nine species can be found feeding on the same carcass (Figure 17-29). Less specialized carrion feeders include gulls, some petrels, some hawks, ravens and other corvids, and even some song birds, which occasionally glean suet and scraps of meat from large mammal carcasses in winter. One highly specialized carrion-feeder is the lammergeier or bearded vulture, which feeds on the marrow of long bones from the carcasses of large mammals. It flies into the air with a bone, or a tortoise or other hard object, and drops it on the rocks below, where it is shattered apart and can then be eaten.

Figure 17-29 Heads of East African vultures, each adapted for feeding in different ways at the same carcass. (a) Lappet-faced vulture. (b) Griffon. (c) Hooded vulture. (d) Egyptian vulture.

17.3.5 Feeders on Insects and Terrestrial Invertebrates

Insects, other terrestrial arthropods, and certain annelid worms constitute a vast array and mass of food objects for birds, and it has been estimated that as many as 60 percent of all birds are insectivorous to some degree. Birds obtain insects and other small invertebrates in the air (flying insects), from the foliage of vegetation, from stems and trunks, from rocks, from the surface of the ground, and from subterranean habitats; there is virtually no habitat-niche on earth where an insect can be completely safe from the foraging ability of some kind of bird or another, so completely have birds diversified in their foraging abilities to exploit insects as food.

An informative way to approach the diversity of foraging adaptations among insectivorous birds is to group them by functional similarities. Groups of birds that exploit the same types of food species in a similar way have been called **guilds** by R. B. Root. Although some species overlap between guilds, the following major groups find representatives in various families and orders of birds.

1. **Gleaners** are birds that pick visually located insects from foliage, trunks, rocks, and ground by a quick peck with their beaks from a perched position. They often use rapid peering from spot to spot as a means of locating insects. This method of capture is the most widespread and generalized among insectivorous birds, and it is probably the mode of feeding from which most of the other ways of feeding on insects have been derived.

2. **Flycatchers** are birds that sally forth from a perched position to capture visually located insects at some distance away in the air and then return to a perch to launch the next attack.

3. **Aerial sweepers** are birds that fly high and fast through swarms of flying insects (so-called "aerial plankton") and make repeated captures in their beaks while on the wing, perhaps in part by random contact with the prey but certainly also in many cases guided by visual detection of individual prey. There are various intermediate modes of hunting between flycatchers and sweepers, and it is likely that sweepers have evolved from flycatching ancestors.

4. Specialization in a different way, probably also from gleaning ancestors, involves **probers**, birds that poke their beaks into existing holes, cracks, or other enclosed cavities in plants, rocks, or ground or probe directly into the soil, for hidden insects, their eggs, larvae, and pupae. **Gapers** are specialized ground probers that poke their closed beaks into the ground and then open the mandibles against the resistance of the soil, thereby creating an opening that can be explored by eye and tongue for edible objects.

5. **Wood-drillers** are birds with powerful neck muscles and chisel-shaped beaks that are used to excavate holes, usually into dead wood, to expose hidden insects; the **bark-scalers** are a variant of this general

guild, which is principally defined by the ability of the birds in it to excavate or otherwise physically remove portions of woody plants to get at hidden prey. Probably wood-drillers have evolved from gleaners through intermediate forms, such as the bark-scalers.

6. **Diggers** are birds that excavate into the soil or forest litter with beaks or feet to expose subterranean items of food. Their origins are multiple, and they probably have been derived from both gleaners and probers.
7. Various hawks, falcons, and owls catch insects in their feet, either on the ground or in the air; they can be termed **grabbers**. It must be remembered that none of these species is exclusively insectivorous.
8. Certain cursorial birds run after insects disturbed in flight and catch them immediately as they alight or sometimes in the air; these birds can be grouped together as **ground-chasers**.

Detailed analyses of most guilds of insect-eating birds have not been carried out, and here we can summarize only some of the obvious morphological and behavioral characteristics that are associated with the principal insect-eating guilds and mention some of the groups of birds that share these features.

Gleaners may seek their food primarily in foilage, on the branches and trunks of trees, on rocks, or on the ground. The American wood warblers (Parulidae) are typical foilage-gleaners, as are their ecologically close counterparts, the Old World warblers (Sylviinae, Muscicapidae), and many other passerine birds in wooded or forested environments throughout the world. Typically they have rather short, thin, pointed bills adapted for picking up stationary insects. Chickadees and tits (Paridae), to some extent nuthatches (Sittidae), and a few woodpeckers (Picidae), and some other birds glean from the branches and trunks of trees and bushes. Only a few kinds of birds regularly glean from rocks—some of the nuthatches, New Zealand wrens (Acanthisittidae), some accentors (Prunellidae), some pipits (Motacillidae), and some thrushlike birds (Turdinae, Muscicapidae). Gleaners from the ground are quite varied in size as well as beak shape, and range from small passerines from various families (Alaudidae, Prunellidae, Motacillidae, and others), to medium sized hornbills (Bucerotidae) with outlandishly enlarged, scimitar-shaped beaks, with which they adroitly pick up harvester termites and other small insects from the ground.

The flycatchers are just as varied as the gleaners and show just as many examples of parallel evolution of similar behavior and structure in different taxa. The tyrant flycatchers (Tyrannidae) of the New World and the muscicapid flycatchers of the Old World are a prime example of independent evolution of the same adaptations for the flycatching guild, but many other groups of birds contain flycatchers too. The beaks of these birds tend to be broad, slightly hooked, and often with long bristles at the gape, all structural adaptations associated with catching insects in the air.

A few groups of flycatching birds have evolved specializations for eating bees, wasps, and other stinging hymenopterans, whereas others have developed an ability to select between nonvenomous castes (drones) and venemous forms (workers), apparently by differences in the sounds they produce in flight. The bee-eaters (Meropidae, Coraciiformes), jacamars (Galbulidae, Piciformes), and shrikes (Laniidae) all feed regularly and extensively on a variety of bees, wasps, and hornets. After capturing and killing a stinging insect in their beaks, bee-eaters and shrikes move the prey laterally through the bill until the beak comes into contact with the tip of the insect's abdomen; several bites are delivered to this region, causing any venom in the poison gland to be expressed out as a droplet on the sting itself. Then the bird vigorously wipes the sting against a branch several times and swallows the insect. This is obviously adaptive behavior, for bee venom is highly toxic. As little as 0.60 mg/100 g of body weight kills birds and mammals. The average volume delivered by a bee, 0.2 mg, is sufficient to kill a small bird, if stung in the act of swallowing the insect.

The diurnal sweepers include the swifts (Apodidae), swallows (Hirundinidae), and the Australian wood swallows (Artamidae, Passeriformes), fast flying birds with long, pointed wings. Their nocturnal counterparts are the somewhat owl-like nightjars or goatsuckers (Caprimulgiformes), with soft plumage and a silent, slower flight; the largest members of this order, the potoos (*Nyctibius*) and the frogmouth (*Podargus*), are not aerial sweepers. Typically these sweepers have a short, weak beak with a very wide gape and expanded palatine bones (Figure 17–30) to act as a brace against the impact of a flying insect when it strikes the mouth, but a few forms have longer, stouter beaks that are used to snap up insects in flight, very much as the flycatchers do.

Probers typically have elongated beaks that are often curved and are inserted into cracks, crevices, and holes on the branches and trunks of trees, dead wood, rocks, soil, or into flowers to obtain hidden insects (Figure 17–31). They occur in a wide variety of nonpasserine and passerine families (Table 17–3). Probing as a general feeding technique is involved in at least three different guilds, ground probers, tree and rock probers, and flower probers. Most of the latter

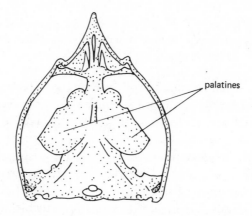

palatines

Figure 17-30 Palate of a caprimulgiform (potoo) showing enlarged palatines that protect the roof of the mouth against the impact of large insects caught in the air. (Drawn from a photograph in Storer [1971] in Farner and King [2].)

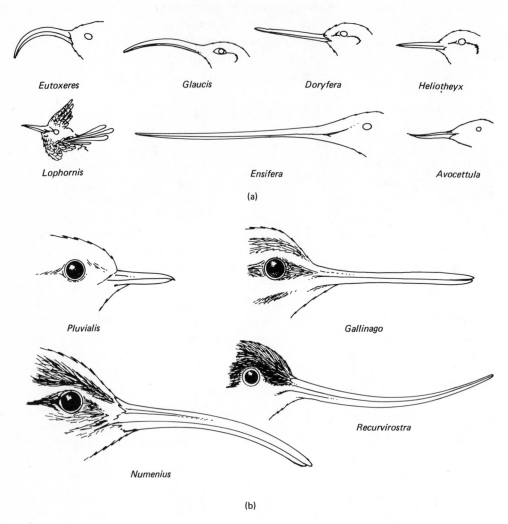

Figure 17-31 A variety of probing beaks. (a) a group of nectar-feeding hummingbirds. (b) a group of sand and mud-probing shorebirds. (From Storer, [1971] in Farner and King [2] and Thomson [5].)

are principally specialized for nectar feeding and secondarily for insect eating. Some of the Old World starlings (Sturnidae) and the New World meadowlarks, grackles, and relatives (Icteridae) are gapers (specialized ground probers); they have their eyes set forward in the skull, enabling them to see into the opening made by the bill, and very powerful muscles for raising the upper mandible and for lowering the lower one against the soil after insertion of the closed beak.

The main group of wood-drilling and bark-scaling birds are the woodpeckers (Picidae), with chisel-shaped beaks, long, protrusible, barbed tongues (Figure 17-32), and specially strengthened skulls to withstand the hammering shocks produced by drilling. Such types range in size from the small downy woodpeckers in the genus *Dendrocopus* to the crow-sized ivory-billed woodpecker

Table 17-3 Avian Probers

Taxonomic group	Food habits and substrate
Nonpasserines	
Kiwis (Apterygidae)	Worms and grubs in soil located by smell
Painted snipes (Rostratulidae)	Semiaquatic habitats
Sandpipers (Scolopacidae)	Semiaquatic habitats
Stilts and avocets (Recurvirostridae)	Semiaquatic habitats
Hummingbirds (Trochilidae)	Nectar and insects in flowers
Hoopoes and woodhoopoes (Upupidae)	Ground, tree trunks, and branches
Woodpeckers (Picidae)	Tree trunks, branches, dead wood, combined with "drilling" or "scaling"
Passerines	
Woodcreepers (Dendrocolaptidae)	Mainly trunks and branches in neotropical forests
Ovenbirds (some genera Furnariidae)	Trunks and branches in Neotropical forests
False sunbirds (*Neodrepanis*, Philepittidae)	Nectar and insects in flowers
Huia, wattlebirds (Callaeidae)	Female probes, male chisels trunks and dead wood
Some babblers (Timaliinae)	Trunks, branches, flowers
Thrashers (Mimidae)	Soil, ground insects, but mainly diggers
Honeyeaters (Meliphagidae)	Mainly nectar, some insects from flowers
Sunbirds (Nectariniidae)	Nectar and insects from flowers
Hawaiian honeycreepers, part (Drepanididae)	Nectar and insects from flowers
Neotropical honeyeaters (various genera, family affiliations uncertain)	Nectar and insects from flowers
Creepers (Certhiidae)	Trunks, branches of trees
Darwin's Woodpecker finch	Tool-prober in holes on branches

(*Campephilus*). Some other birds also drill in wood or scale bark, especially some of the woodcreepers (Dendrocolaptidae, Passeriformes), nuthatches (Sittidae), and even some chickadees (Paridae).

Diggers range in size from gallinaceous birds like turkeys (Meleagrididae), which scratch for both plant and animal foods under the surface of the soil, to small passerines. Either the feet or the beak, or both in combination, may be used for digging. Gallinaceous birds and some others typically scratch the ground with one foot while standing on the other. Passerine birds that hop rather than walk scratch by jumping up and kicking the leaf litter or loose soil with both feet simultaneously. Modifications of the beak for digging are numerous and are often associated with probing. Thrashers in the genus *Toxostoma* demonstrate varying degrees of specialization in the bill for digging,

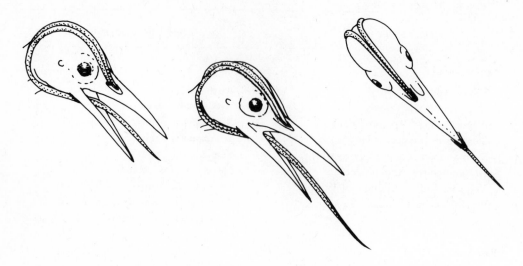

Figure 17-32 Hyoid apparatus of a woodpecker, showing mechanism for protrusion of the tongue beyond the tip of the mandibles. The tongue itself is about the length of the bill but can be extended well beyond the tip of the beak by muscles that move the elongate hyoid apparatus forward from its resting position around the head. (Modified from Peterson [4].)

from the eastern brown thrasher (*T. rufum*), which obtains most of its food on the surface by gleaning with a relatively short bill, to western species such as the California thrasher (*T. redivivum*), which excavate holes in soil by rapid sidewise movements of their long, curved bills, which are also used for probing.

17.3.6 Pollen and Nectar Feeders

Most but not all of the pollen and nectar feeders are probers with long bills and tongues. In addition to those forms listed in Table 17-3, certain parrots (Psittacidae) called lories specialize in feeding from flowers, as do some members of several other passerine families, the white-eyes (Zosteropidae), leafbirds (*Chloropsis* spp), and flowerpeckers (Dicaeidae). In the most specialized pollen and nectar feeders the tongues are tipped with brushlike structures, which collect the food, or the tongues are tubular (the tubes being formed in different ways in the different groups), or both conditions are combined together. Even within one closely related group, such as the hummingbirds, beaks are highly variable from species to species in relative length, shape, and curvature, and in many instances the bill shape appears to be adapted to particular kinds of flowers, which are quite often pollinated by the birds. A striking adaptation exists in the Neotropical honeycreepers of the genus *Diglossa*; these birds hook the curved upper mandible around the base of the corolla of a tubular flower and then pierce the sharp, straight lower mandible through the corolla, making a hole through which the tubular tongue works and obtains nectar.

17.3.7 Fruit Eaters

Fruit eating species are also widely distributed among both nonpasserine and passerine families and include cassowaries, emus, pigeons, parrots, plantain eaters, some cuckoos, mousebirds, barbets, toucans, hornbills, contingas, some tropical tanagers, some flowerpeckers, some icterids, waxwings, and other species on a less regular basis. Most of these species feed insects to their young. Bill shapes and sizes vary enormously among the different fruit eaters and may usually be molded by selective forces other than those associated with eating fruit. Specialization for this diet occurs mainly in the digestive tract, which tends to be short and tubular and without a well-developed stomach. The food is passed through the digestive tract extremely rapidly; only the pulp of the fruit is digested and usually only partially.

17.3.8 Granivorous Birds

Seeds are a rich source of nutrients for birds and are often available at seasons of the year when insects and other foods are scarce. They range in size from tiny grass and forb seeds less than a millimeter in greatest dimension to some very large nuts and pits with dimensions of 50 mm or more. Consequently, seed-eating birds show a great variety of adaptations for exploiting seeds as a source of food.

Seed eating birds can be divided into three functionally different groups: (1) those that eat the seeds and nuts entire, including the outer husk or shell, and depend upon powerful gizzards for grinding up and mechanically processing the food, (2) those that hammer seeds and nuts open with their beaks and then pick out the meat, and (3) those that have evolved special beaks, palates, and jaw muscles for husking and cracking open seeds before they are swallowed. To the first group belong such birds as the ratites, tinamous, gallinaceous birds, pigeons, sandgrouse, and the South American seedsnipes (Thinocoridae, Charadriiformes). The gizzards of some of these species, particularly nut-feeders like the turkey, can exert extremely impressive mechanical forces to crack open nuts. A turkey's gizzard can grind up two dozen walnuts in as little as 4 hr and can crack up hickory nuts that require 50 to 150 kg of pressure to break under experimental conditions. To the second group belong various corvids (jays, nutcrackers, ravens, crows), chickadees, nuthatches, and some woodpeckers. The titmice, jays, and ravens hold the seeds in their feet and crack them open with blows from their beaks, whereas the nutcrackers, nuthatches, and some woodpeckers wedge hard seeds into crevices or holes in bark, wood, or rock and then open them with their beaks.

Perhaps the most highly specialized seed eaters are those that husk, crack, or crush, or shear seeds by mandibulating them in their beaks. Many parrots feed on seeds, and almost everyone has watched the mechanical efficiency with which the Australian budgerigar or shell parakeet husks the small seeds on which it feeds. Other species of parrots use their large bills to crack open or shear hard-coated nuts, which they often hold in one foot. Various taxa of "finches" are examples of an adaptive type that has evolved independently several times among the passerine birds. They have relatively short, heavy, arched beaks operated by powerful jaw muscles.

Frequently the cutting edge of the upper bill is sharp and fits into a groove in the lower bill (Figure 17–33), or there may be other structures lingually located in the palate or floor of the mouth that function to help hold or to crush seeds (Figure 17–34). Bill size correlates generally with the size of seeds fed upon, so that in regions where several kinds of seed eating birds occur sympatrically there is usually a gradation of body and bill sizes that are associated with average differences in the kinds of seeds eaten by the different species. The crossbills (genus *Loxia*) are a group of boreal finches specialized for feeding on the seeds of conifers. The crossed bill is adapted for holding apart the scales of cones while the bird inserts its tongue and removes the seed.

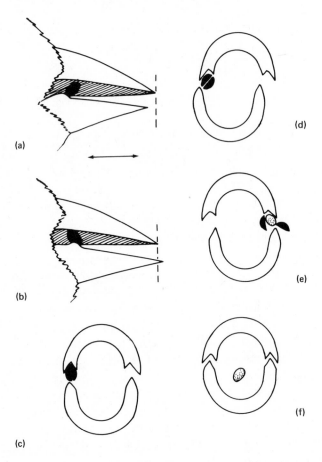

Figure 17-33 Diagram of the cutting edges of the upper and lower bills of a finch, showing jaw movements (a and b) that position a seed in preparation for husking (c through f). (Redrawn from Welty [6].)

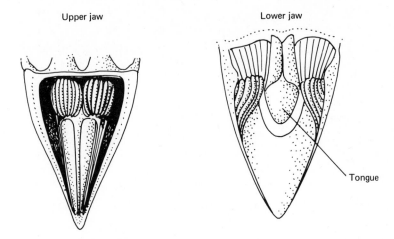

Figure 17-34 Upper and lower jaws of hawfinch, showing the ridged anvil-like structures for crushing nuts and fruit pits. This bird can crush olive pits that require a force of 48 to 72 kilograms per cm^2. (Redrawn from Welty [6].)

17.3.9   Grazers and Browsers

Only a few kinds of birds feed extensively on grass, leaves, buds, and twigs. Some of the geese and gooselike waterfowl crop grass on land. The peculiar South American hoatzin feeds mainly on the tough rubbery leaves, flowers, and fruits of arum plants and the white mangrove. Most of the mechanical processing of this food for digestion occurs in the hoatzin's crop rather than in its gizzard, which is reduced and simplified. The unique crop, on the other hand, is large, with thick muscular walls and a horny internal lining, adaptations for breaking up its vegetable food. The New Zealand kakapo or owl parrot (*Strigops*) extracts the juices from leaves, twigs, and young shoots by chewing on them without detaching them from the plant, or it fills its large crop with browse and retires to a roost, where it chews up the plant material, swallowing the juices and ejecting the fiber as dry balls.

The holarctic grouse (Tetraoninae) browse extensively on the buds, leaves, and twigs of willows and other plants of low nutritive value. They have extremely large caeca or storage sacs associated with the small intestine, in which large volumes of vegetable materials are held and are digested by symbiotic bacteria, a system analagous to that found in ruminating mammals (Figure 17-35).

A few birds also eat roots and tubers. Some, like certain pheasants, dig into the soil with their beaks, whereas others scratch with their feet. Some of the cranes, geese, and ducks are also adapted for obtaining the roots, rhizomes, and bulbous parts of aquatic plants.

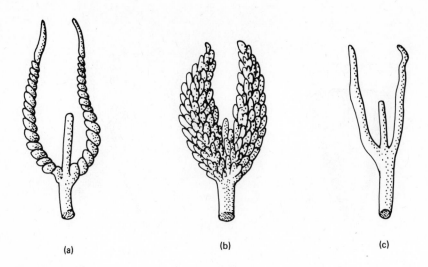

(a)   (b)   (c)

Figure 17-35 Comparison of the caecae of three avian species to show variations. (a) ostrich. (b) tinamou. (c) domestic fowl. (Modified from Farner and
King [1972] *Avian Biology*, Vol. (II [1972], Academic Press, N.Y.

17.3.10 A Few Odd Specialists

Certain species of birds feed on rather surprising materials. The African
honeyguides (Indicatoridae, Piciformes), which are known for their habit of
leading men and other mammals to beehives, eat beeswax as well as the larvae
and pupae and have been kept alive in captivity for as long as 32 days on an
exclusive diet of wax. Only a few kinds of animals have the enzymes necessary
to break down wax into its constituent fatty acids for digestion, and the honey-
guide has symbiotic intestinal bacteria that digest the wax it eats.

One of the Galapagos finches (*Geospiza difficilis*) obtains some of its food by
biting the bases of growing feathers on boobies (*Sula* sp) and eating the blood
that oozes out. A number of birds eat the dung of mammals. In the Arctic, the
ivory gull feeds on the dung deposited on the ice by polar bears, walruses, and
seals. Puffins and petrels eat whale dung floating in the water, and certain
vultures and kites feed extensively on the excrement of humans and dogs
around native villages.

Albatrosses are able to feed with impunity on stinging jellyfish, and a number
of land birds can eat berries and fruits that are poisonous to man.

References

[1]   Austin, O. L., Jr. 1961. *Birds of the World*. Golden Press, New York. This is a
       popular and informative account of the bird families of the world, well illustrated in
       color by A. Singer.

[2]   Farner, D.S., and J.R. King (editors.) 1971. *Avian Biology*, vol. 1. Academic Press, New York. This is a technical book; chapters by R.W. Storer on adaptive radiation, N.P. Ashmole on sea bird ecology, D.L. Serventy on adaptations of desert birds, and G. Orians on behavioral ecology are particularly relevant to this chapter.

[3]   Grossman, M. L., and J. Hamlet. 1964. *Birds of Prey of the World*. Clarkson N. Potter, New York. This is a popular account of the biology of the Falconiformes and Strigiformes, with many photographs and drawings.

[4]   Peterson, R. T. 1978. *The Birds*. Second edition. Life Nature Library. Time, Inc., New York. This is an updated version of a popular summary of avian biology by a well-known bird artist and bird watcher, with many photographic examples of avian adaptations for locomotion and feeding.

[5]   Thomson, A. L. (editor.) 1964. *A New Dictionary of Birds*. McGraw-Hill, New York.

[6]   Welty, J. C. 1975. *The Life of Birds*. Second edition. W. B. Saunders, Philadelphia.

18

Origin and General Characteristics of Mammals

Synopsis: Five distinctive mammalian orders were established by the early Jurassic. These groups led to the evolution of the living prototheria (monotremes), the metatheria (marsupials) and the eutherians (placentals). Although the origins of the prototheria are unknown, the marsupials and placentals arose during the Cretaceous as contemporaries of the reptiles. The first mammals were small and likely to have been insectivorous. Because of their endothermic abilities to maintain a relatively constant internal body temperature, they were able to be active at night and, especially, to remain active under cooler conditions. Endothermy is an expensive mode of living, and homeotherms require a greater amount of food than do their ectothermic counterparts. Nevertheless, there are advantages to endothermy: thermal constancy of the internal milieu provides a setting for the efficient biochemical catalysis of organic substances; rapid and consistent cell responses, especially of nerves and muscles; and a considerable independence from changes in environmental temperatures.

Homeothermy when maintained at a relatively high temperature eases the problem of thermal regulation because excess heat from endothermic processes can be dissipated via radiation to the environment. Water is an essential ingredient to active life; cooling in a hot environment requires the evaporation and loss of body water. Setting the body temperature as high above ambient temperatures as possible, therefore, minimizes water loss. The cost of endothermy can be considered by balancing what is known as an energy budget. Vampire bats provide an excellent example revealing that their geographic distribution is limited by the balance between energy needs and the coolness of the environment.

Mammalian endothermy is reinforced by at least two unique morphological structures, the integument and the kidney. In addition to contributing vital thermal insulation by means of hair and fat, the integument provides mammals with unique cooling and chemical signaling mechanisms. The basic structure

641

of the integument is also modified to produce claws, nails, hoofs, and horns and it provides for the development of antlers. The kidney, although less varied than the integument within the Class Mammalia, is just as distinctly mammalian. The elegant structural complexity of the mammalian kidney is the basis of its dynamic, powerful regulation of body fluid composition. The kidney provides for a precise control of the internal milieu within which the other complex systems of a mammal can effectively and efficiently function.

18.1   MAMMALIAN BEGINNINGS

The mammals represent a distinctive segment of living vertebrates, for they alone widely exploit the resources of earth, from pole to pole, mountain top to deep sea, and even the night sky. Woven into the fabric of this class is an adaptive plasticity not matched in other vertebrates and epitomized in humans. Mammals were derived from reptiles during the Triassic and, probably avoiding direct conflict and competition with the ruling and mostly diurnal reptiles, occupied a different niche, the night. Several specific requirements were necessary to be dominant at night. Paramount among these was the regulation of a high body temperature. This is not a simple adaptation but affects every system in the body. The most readily apparent mammalian alteration is in the hairy integument, of great importance to the efficient maintenance of a high internal body temperature. In addition to the evolution of mammalian hair, the integument has provided the morphological basis for unique features which contribute to homeostasis, communication, and nourishment of the young. Although the rudiments of the refined homeostasis of mammals are seen in their thermoregulation and the morphology of their integument, mammalian kidneys are of key importance in maintaining an internal milieu of extraordinary constancy. As with the mammalian integument, the work performed by the mammalian kidney can only be explained by examination in detail of kidney structures and associated functions. In this chapter we survey mammalian beginnings, general characteristics of living mammals, examples that illustrate the advantages and disadvantages of being a mammal, and the structure and function of the integument and kidney.

Living mammals separate into three taxa: the monotremata or **Prototheria** the marsupials or **Metatheria**, and the placentals or **Eutheria** (Table 18–1). These taxa include two reproductive groupings: the prototheria, which lay reptilelike eggs, and the theria, which are viviparous. The origin of the monotremes is uncertain, for fossils older than the Pleistocene are unknown.

The earliest therians are fossilized in early Cretaceous rocks of Asia and North America (Chapter 15). During the late Cretaceous both therian lines, the marsupials and placentals, became common. Earlier in the Mesozoic, however, several distinct fossil mammals are recognized. Was mammalian evolution monophyletic or polyphyletic? Current interpretations suggest the differences between the early Mesozoic mammals were sufficient to define at least two or

Table 18–1. Classification of Mammals with Orders and Approximate Number of Species of Living Forms.

Classification	*Major examples*
Class Mammalia	
Subclass Prototheria	
Infraclass †Eotheria	
Order †Triconodonta	
Order †Docodonta	
Infraclass †Allotheria	
Order †Multituberculata	
Infraclass Ornithodelphia	
Order Monotremata (6 species)	Spiny anteaters, duck-billed platypus
Subclass Theria	
Infraclass †Trituberculata	
Order †Symmetrodonta	
Order †Pantotheria	
Infraclass Metatheria	
Order Marsupialia (242 species)	Opossums, bandicoots, koala, wombats, kangaroos, wallabies
Infraclass Eutheria	
Order Insectivora (406 species)	Hedgehogs, moles, shrews
Order Dermoptera (2 species)	Flying lemurs
Order Chiroptera (853 species)	Bats
Order Primates (166 species)	Lemurs, monkeys, great apes, man
Order Edentata (31 species)	Anteaters, sloths, armadillos
Order Pholidota (8 species)	Pangolins
Order Lagomorpha (63 species)	Pikas, rabbits, hares
Order Rodentia (1687 species)	Rodents
Order Cetacea (84 species)	Baleen whales, toothed whales, propoises
Order Carnivora (284 species)	Dogs, bears, racoons, weasels, hyaenas, cats, sea lions, walrus, seals
Order †Condylarthra	
Order Tubulidentata (1 species)	Aardvark
Order Proboscidea (2 species)	Elephants
Order Hyracoidea (11 species)	Hyraxes
Order Sirenia (5 species)	Dugongs, manatees
Order Perissodactyla (16 species)	Horses, tapirs, rhinos
Order Artiodactyla (171 species)	Swine, hippos, camels, deer, giraffe, antelope, sheep, goats, cattle

† = extinct.

After Hopson, [1970] *J. Mam.*, 51(1):1–9; T. A.Vaughan [1978] *Mammalogy*. Second edition. Saunders, Philadelphia; E. P. Walker et al. [1975] *Mammals of the World*. Third edition. Johns Hopkins University Press, Baltimore.

even three major clades (Figure 18–1). Whether the mammals are monophyletic or polyphyletic is therefore a matter of when each of the well-recognized groups of Jurassic mammals passed through the reptilian-mammalian "boundary." If the phylogeny in Figure 18–1 is correct, mammals are polyphyletic because they derive from at least two Triassic reptilian stocks, the ictidosaurs and the cynodonts.

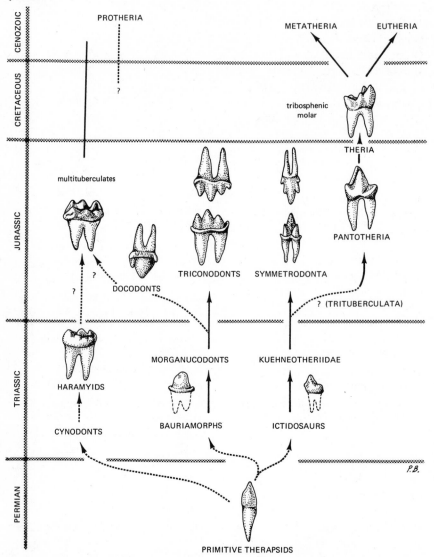

Figure 18-1 Hypothetical phylogeny of Jurassic and Cretaceous mammals from Triassic mammallike reptiles. Although tooth morphology is not the sole basis for the construction of such phylogenies, the durability of dental tissues has made such remains of paramount importance in mammalian paleontology. Shown are lateral and occlusal views of typical cheek teeth. The (generally hypothetically reconstructed) roots are unshaded and indicate whether the tooth is an upper or lower tooth by their orientation. Note that five distinct orders of Jurassic mammals can be recognized.

Before considering what these earliest mammals were like, let us define the characters that set modern mammals apart from modern reptiles (Table 18–2). Endothermy, stabilizing the internal milieu, is the most obvious and important single character setting living mammals and birds apart from living reptiles. Many of the other characters of mammals, such as insulative hair and a diaphragm to enhance breathing, relate to the high metabolism associated with endothermy. Mammary glands for postnatal feeding of the young are another mammalian distinction. Unfortunately endothermy and lactation do not fossilize.

To identify the first mammals and to seek their origins from ancient reptiles we must rely on characters that fossilize—bone. Using the criteria that define and separate living mammals and reptiles, fossils that possess a jaw with only a dentary bone hinged to the skull by a squamosal-dentary joint are mammals (Figure 3–6). The secondary palate, a double occipital condyle, and diphy-odont teeth (two sets of teeth) are also fossilizable mammalian characteristics. Otherwise similar fossils lacking these skull characters are reptiles. There are difficulties in applying this simplistic definition of a mammal, however. Many Triassic therapsids had already developed some but not all of these mammalian characters. For example, the Upper Triassic *Diarthrognathus* possessed small articular-quadrate hinge joints for the jaw as well as the typical mammalian dentary-squamosal hinge. Was *Diarthrognathus* a mammal or a reptile? From the jaw structure and tooth arrangements, it is likely *Diarthrognathus* fed more like a mammal than a reptile. It probably was endothermic and may even have possessed hair. But on the basis of structures that fossilize, *Diarthrognathus* is classified as a reptile (Chapter 12.3.2).

Table 18–2. Reptilian versus Mammalian Characteristics

Reptilian	*Mammalian*
Teeth usually single-cusped, continuously replaced	Teeth usually with distinctive cusps, diphyodont
Jaw articulation: quadrate-articular	Jaw articulation: squamosal-dentary
Several bones in lower jaw	Dentary only
Internal nares open near front of mouth (primitive forms)	Secondary palate
Single occipital condyle	Double occipital condyle
Phalangeal formula 2-3-4-5-3	Phalangeal formula 2-3-3-3-3

Ectotherms
  Low metabolic rates
  Poor insulation
	Endotherms
	  High metabolic rates
	  Good insulation (hair usually present)

Diaphragm absent	Diaphragm present
No mammary glands	Mammary glands
Heart generally three-chambered, blood exits via three arteries	Heart four-chambered, blood exits via two arteries

In despair, perhaps, some paleontologists have suggested that the Class Mammalia should include all of the presumably endothermic therapsids. Using somewhat similar arguments Robert T. Bakker at Johns Hopkins would classify dinosaurs with the birds, the basis of the inclusion being his belief that dinosaurs were endotherms (see Section 13.4). At present the fossil evidence seems far too flimsy to seek a monophyletic origin of mammals in groups where the lineages are unclear and, especially, where evidence for endothermy is speculative. Clearly different reptilian species of Triassic age, which may or may not have been endotherms, had a suite of characters that became more mammallike as the Triassic unfolded. A transition from reptile to mammal was taking place, and parallel evolution occurred in different reptilian lines. In this sense mammals represent a "grade" derived from two or more reptilian clades (Figure 18–1).

Fossil morganucodonts (probably triconodonts) and kuehneotherids (probably symmetrodonts) from the late Triassic were definitely mammals. The five orders of Jurassic age (Figure 18–1) firmly establish mammals in a variety of distinct life styles. What were these Jurassic mammals like? Almost all of the Mesozoic mammals were small, not larger than rabbits. Their fossils, as a result, tend to be disarticulated and consist primarily of teeth, parts of lower jaws, occasional skulls, and few parts of the axial skeleton. To mammalian paleontologists, the morphology of the teeth, which fossilize well, is critically important in defining the earliest mammals.

To a nonpaleontologist it seems ludicrous that a new fossil mammal can be defined, let alone named, on the basis of a single minute fossil tooth. Nevertheless many mammals have been defined on just that, a single tooth. The molar teeth (Chapter 19) of these mammals, however, were not simple conical single-rooted pegs, as in their reptilian ancestors. They had various complex cusps and were double or sometimes triple rooted. The changes in dentition, as well as in jaw structure, were important in improving the efficiency of feeding. Early Triassic cynodonts showed incipient molar cusping, but the Jurassic mammalian orders (Figure 18–1) were established on distinctive types of molar teeth. Tooth morphology is complex and very difficult to visualize; however, it is basic to appreciating the life styles of these early mammals.

In the triconodonts three aligned, conical cusps, the middle the largest, projected from each molar (Figure 18–1). Because the inner faces of the cusps of the upper molar slipped down over the external cusp faces of the lower molar, a cutting-shearing action occurred between the molars. A food bolus could be repeatedly processed by cutting it into smaller and smaller bits, the cheeks retaining the food within the mouth but external to the teeth. The docodonts, which are of less certain affinities (perhaps modified triconodonts), possessed only two cusps separated on a broader molar base. The upper and lower molar cusps interdigitated and must have led to some mashing of food caught between them. The multituberculates are distinctive because their cheek molars were broad and usually possessed many rounded cusps. These no doubt were effective in mashing food, especially when the upper and lower molars were ground against one another. Paleontologists suspect that the multituberculates were herbivores and filled a niche similar to that now occupied by the rodents.

K. H. Kermack at University College, London, and Z. Kielan-Jawarowski of the Institute for Paleozoology in Poland have suggested that the multituberculates and triconodonts, along with the living monotremes, form a related group because the lateral wall of the braincase is formed by the large petrosal bone. In contrast, this skull bone is not well developed in the closely related symmetrodonts and theria. Instead, the alisphenoid bone forms much of the lateral braincase. This dichotomy in early mammalian evolution is supported by the presence of *triangular based*, three-cusped molars in both symmetrodonts and the pantotheres (Figure 18-1), and their absence in the other groups. In the symmetrodonts the cusps of the upper and lower molars rigidly interlocked, preventing horizontal sliding movements between them. Each cusp had two faces, which sheared against the cusp faces of the opposing teeth. Symmetrodont cusps elongated during the evolution of the group until the mid-Cretaceous, when symmetrodonts died out. The pantotheres also had triple-crowned molars, but like the apparently unrelated docodonts the teeth widened basally to produce flat grinding areas. In addition, up to six transversely oriented shearing faces characterize this tooth. Thus, Jurassic mammals are believed to have represented three distinct feeding "guilds"; the triconodonts occupied a strictly carnivorous predatory niche, the symmetrodonts and pantotheres an insectivorous niche, and the multituberculates an herbivorous niche. Since the docodonts preceded the multituberculates (Figure 18-1), Kermack and Kielan-Jawarowski suspect they were ancestral to them, representing an intermediate stage between carnivorism and herbivority.

The therian molar, termed a tribosphenic molar, differs from the ancestral pantothere molar by a much enlarged talonid process on the posterior margin of the lower molars (Figures 18-1 and 18-2). The talonid process contains a basin which when occluded by an upper molar receives the latter's inner cusp, the protocone. Together they mash and grind the food caught between them. By mid-Cretaceous both marsupials and placentals had well-developed tribosphenic teeth. The cutting and grinding molars present in the Cretaceous therians quite likely were highly versatile and adaptive. It is probably not coincidence that the tribosphenic molar evolved in parallel with the rapid coevolution between angiosperms and their pollinating insects during the Cretaceous. Of the five Jurassic mammal orders, only the herbivorous multituberculates and the originally insectivorous therians persisted into the Cenozoic. Possibly the tricondonts and symmetrodonts were too limited by their dentition to compete with the therians.

How did the multituberculates and therians persist through the Cretaceous when a diverse herbivorous and carnivorous dinosaur fauna dominated the land? These early mammals were small and probably nocturnal whereas the vast majority of Cretaceous reptiles probably were diurnal. Indeed, a highly developed sense of smell and the newly transformed ear bones, (Chapter 3), which improved hearing, must have aided early mammals greatly in occupying

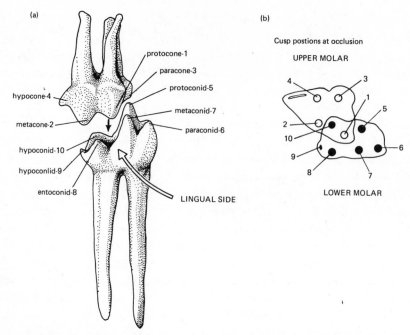

Figure 18-2 Tribosphenic molar morphology. (a) three-dimensional view of a left upper and lower molar of the modern opossum as seen from lingual (tongue) aspect. (b) a standard diagram showing how the molar cusps interdigitate when occluded.

a nocturnal niche (Chapter 19). Endothermy must also have been a benefit, permitting feeding activity during cool nights. Nor is it beyond credibility to suggest that the coincident evolution of angiosperms and pollinating insects in large part shaped therian survival and evolution. During the day pollinators flew, flitting from blossom to blossom, collecting energy that would have essentially been unavailable to a small terrestrial mammal. During the night pollinating insects settled and became subject to an active, furry, terrestrial predator small enough to find insects in their nests. To be sure, the evolution of the first therians was more complex than we have described. But not until the early Paleocene, when the diurnal dinosaurs were long extinct, did the theria evolve large diurnal forms (Chapter 15).

18.2 RECENT MAMMALS: CHARACTERIZATION OF MAJOR GROUPS

Vertebrates with elevated, constant body temperatures have high energy requirements (see Chapters 7 and 16). Much of mammalian evolution and the diversity of living forms can be attributed to trophic specializations that favor the evolution of increased efficiency in energy acquistion and the utilization of new sources of energy. The classification of modern mammals into orders is based primarily on trophic adaptations. At higher levels of classification, modern mammals are divisible into three major groups, separated on the basis of reproductive biology (Table 18–1).

The Prototheria, or monotremes, survive as about six species of primitive forms isolated in Australia and New Guinea. They are grouped in two sub-orders—the spiny anteaters and the duck-billed platypuses. Perhaps they evolved in the Mesozoic from therapsid reptiles distinct from those which gave rise to the other mammals, but no Mesozoic fossils have yet been discovered. Prototherians share a number of characters with reptiles and birds that other living mammals lack. Prototherians lay eggs that are incubated and hatched outside the reproductive tract of the females. Nevertheless, echidnas and platy-puses are hairy, endothermic, milk-producing vertebrates with a single bony element in each ramus of the lower jaw, and thus qualify as mammals.

The remaining two groups of living mammals are more closely related but have had separate evolutionary histories since the earliest Cretaceous. The 250 or so species of Metatheria or marsupials are distinguished by their short gesta-tion periods, tiny, feebly developed offspring, and (in many) the presence of a protective pouch, the **marsupium**. This pouch lies over the mammary glands of the female and the young crawl into it immediately after birth to feed and complete development. The marsupials are restricted to the Australian region and the New World. Ecologically they represent almost as diverse a group of animals as do the remaining mammals, particularly if the variety of recently extinct forms is considered.

The eutherians, or placental mammals, remain the dominant division of the class Mammalia, as they have throughout the Cenozoic. Today about 3800 species inhabit the earth and are found on every continent and in every sea. The eutherians are born at a more advanced state of development than are the marsupials. Most have spent a considerably longer time in the womb than would newborn marsupial species with the same adult body size. Many are able to run or swim by their mother's side within minutes of their birth, and even those species with a long period of external development are weaned at an earlier age than are marsupial young.

These differences between the reproductive patterns of metatheria and eutheria apparently do not reflect a superiority of one mode over the other but differences in adaptation to the stresses of the environment. Eutherians pro-duce advanced young with superior survival potential, but at high and pro-longed cost to the mother. The metatherian female invests much less before the birth of her young and can quickly recycle her reproductive apparatus to produce subsequent broods if appropriately stimulated by environmental con-ditions (Chapter 19.3.1).

Nevertheless, eutherians are by far the dominant mammals of recent times. A close examination of Table 18–1 not only confirms this numerical superiority but provides insight into the pattern of eutherian diversity. Most mammalian species are contained in about a half dozen eutherian orders. Over one third of the living eutherian species are insectivores and bats. Because of their small size, nocturnal and retiring nature and habitats which are often difficult for humans to explore, these mammals are not well known to the general public. Most often they are found as lifeless carcasses along basement walls, their presence indicated by furrowed lawns, or they are associated with mystery and the occult. Mammalogists, however, have uncovered a wealth of fascinating information about them. The Insectivora, which feed on the flesh of soft bodied invertebrates or pierce the exoskeleton of insects to eat their soft parts, provide us with models of the life of the earliest mammals. The winged Chiroptera generally feed on flying insects or on fruit and nectar. They have superb specializations which challenge our understanding and have led researchers to significant advances in the areas of biophysics, acoustics and neurobiology.

By far the most familiar of mammals people encounter in the wild are species of Rodentia. These small, gnawing, plant eating eutherians are in number of species and, certainly, in number of individuals the most successful of mammals. They are also humanity's greatest mammalian competitors and constant, if uninvited, companions.

In contrast, domesticated mammals are from among the species of plant eating Artiodactyla, upon whom we depend for much of our animal food, and the vertebrate eating Carnivora, from which we gain cooperative assistance, protection and endless hours of enjoyment. The many species of wild artiodactyls and carnivores excite the human imagination and provide us with some of the most important illustrations of the evolution of mammalian life and the interactions of environment and vertebrate populations.

Finally, two orders of mammals, although unknown to most people in the wild, are of great interest to almost every person. The fish, squid and plankton eating Cetacea include the largest vertebrates ever known to have lived on earth and encompass a suite of unique characters that inspire wonder and admiration. The Primates with their generally catholic tastes in food are man's closest relatives. As such they are a perennial attraction and an important guide to understanding our own aptitudes and deficiencies. In an age of unparalleled technology and scientific knowledge we are increasingly dependent upon primates for use in the vital medical research which daily touches our lives.

Although we will draw examples from other orders in our discussion of mammals, these six orders—Insectivora, Chiroptera, Primates, Rodentia, Carnivora and Artiodactyla—deserve and receive most of our attention. They will be considered as we examine the benefits, challenges and consequences of regulating body temperature, and the structure and function of the integument and the kidney.

18.3 THE REGULATION OF BODY TEMPERATURE

The essence of being a mammal or bird is expressed in endothermic regulation of constant and high body temperatures, that is, endothermal homeothermy. This "warm-bloodedness" opened nocturnal niches to the earliest mammals, for in equable climates of the Mesozoic, the ability to be active at night was probably the key factor that allowed them to persist in an ecosystem dominated by reptiles. Preadaptations to homeothermy in fishes, reptiles, and birds have already been discussed. What are the advantages of regulating body temperature (T_b) at a high level? How do mammals regulate body temperature? What might the ecological consequences of endothermal homeothermy be?

18.3.1 Advantages of a High Body Temperature (T_b)

Mammals and birds have high metabolic rates, at least six times the SMR of poikilotherms (Figure 7–21). Although it is energetically expensive, there are benefits to regulating T_b at a high rather than a low level. The biochemistry of vertebrates involves thousands of interacting enzyme-catalyzed reactions, most of which are temperature sensitive (Figure 7–24). A constant internal temperature is required to obtain maximum chemical coordination among these reactions. In addition, the higher the T_b the more rapid the response of cells to organismal needs. Certainly complex systems like the CNS, although capable of acting at very cold temperatures (as in arctic fishes), function more effectively at high temperatures. As one example, neurotransmitters like acetylcholine and norepinephrine act by diffusing from their site of release across the synaptic junction to the postsynaptic receptor surface. Because diffusion is a physical process its rate increases as temperature increases. A high T_b therefore enhances the rate of information processing, a competitive advantage too often neglected when considering the success of mammals and birds. Very rapid responses can be vital in catching prey and avoiding predators. Reptiles enjoy the same neurological benefits when they are warm, but very rapid responses on cool nights can occur only in endothermal homeotherms. In addition, muscle viscosity declines at high temperatures. This reduction in internal friction results in more rapid, forceful contraction and faster response times.

Endothermal homeothermy provides an additional advantage not enjoyed by ectotherms: great independence of environmental temperatures, or ambient temperature (T_a). In particular, endothermy allows animals to be active at night when solar energy is absent, and in cold climates where solar energy is not sufficient to warm an ectothermal vertebrate (Chapter 20).

18.3.2 Mechanisms of Temperature Regulation

In general the T_b of mammals and birds is at or above 36°C (Table 18.3). Why was T_b set so high in the evolution of homeothermy? Certainly the influences of a high T_b on activity and reaction times was beneficial and these are part of the reason. The mechanisms used to regulate T_b are important too, for they impose physiological and ecological limitations on homeotherms. Because it is easier to balance heat loss against gain in a cool environment $(T_b > T_a)$ than in a warm environment $(T_b < T_a)$, high internal body temperatures are ecologically beneficial. Heat flows from higher to lower temperatures,

Table 18-3. Representative Body Temperatures at Ambient Temperatures between 20° and 30° C.

Group	Normothermal body temperature* (°C)
Monotremes**	28–30
Marsupials	33–36
Placentals	36–38
Birds	40–41

*At rest and without thermal stress.
**The placental sloths also show body temperatures in this lower range.

thus when T_b is greater than T_a excess body heat can be dissipated freely to the environment. If T_a is greater than T_b, excess body heat can be lost only by costly metabolic processes. To understand this more fully let us consider how a mammal regulates its temperature as T_a varies below and above T_b.

When the environment is cooler than the body, heat loss may be increased by increasing blood flow to the skin. Areas of skin with reduced fur (feet, ears, abdomen, and so on) are highly vascularized and excess heat is quickly dissipated across these surfaces. When the environmental temperature falls, several physiological responses take place, usually in sequence, all of which lead to heat conservation and maintenance of T_b. Peripheral vasoconstriction reduces skin temperatures and, therefore, the rate of heat loss. Piloerection (section 18.4.2) increases fur thickness and the amount of insulation. Finally, shivering, an asynchronous contraction of skeletal muscles, begins and increases metabolic heat production.

When T_a exceeds T_b, the temperature gradient is reversed and heat flows into the animal; unless checked, T_b will rise. This is a much more difficult problem for a vertebrate. Erecting the fur increases insulation and reduces the rate of heat gain across the skin, but metabolic heat production causes the body temperature to rise. Mammals and birds must resort to the evaporation of body water to combat the rise in T_b. Its high heat of vaporization makes water evaporation an effective cooling mechanism. Many mammals cool themselves by sweating: secreting water through pores onto the body surface. Other mammals, devoid of sweat glands, evaporate water across the nasal, lung, and oral surfaces. They pant, ventilating rapidly, when warm. Panting, however, requires muscular activity that increases metabolic heat production. This heat compounds the problem and requires further loss of water.

Most habitats seldom have air temperatures that exceed 35 to 40°C. Even the tropics have average yearly temperatures below 30°C. Only in open situations where solar radiation impinges directly on a mammal or a bird can the environment reverse the normal heat gradient. Thus the high body temperatures maintained by mammals ensure that in most situations the heat gradient is from animal to environment. Still higher body temperatures, around 50°C for example, could ensure that mammals were always warmer than their environment. There are definite upper limits to the body temperatures that are feasible, however. Many proteins denature near 50°C. During heat stress, some birds and mammals may tolerate body temperatures of 45–46°C for a few hours. Only some bacteria, algae, and a few invertebrates exist at higher temperatures. This is another case in which the direction of vertebrate evolution has been established by a balance between biotic needs and physicochemical realities.

Although endotherms have achieved independence from the constraints imposed by environmental temperatures, the mechanism used has, in its turn, imposed a new set of limitations upon them. Because of the high metabolic rate needed to maintain a high body temperature, endotherms consume energy much faster than ectotherms. Energy comes from the food animals eat; therefore, endotherms must eat more, and more frequently, than ectotherms. Well-known examples illustrate these differences. Shrews may starve to death if they cannot consume their own weight in prey daily, but snakes may fast for several months after capture and emerge from their fast little the worse for wear.

18.3.3 Ecological Consequences of Endothermy: Cost-Benefit Analysis

A clearer understanding of the costs, the ecological and evolutionary effects, of endothermy can be obtained by constructing energy budgets for a particular species. An energy budget, like a financial budget, shows income and expenditure using units of energy as currency The energy costs of different activities and alternative behaviors can be evaluated. The importance of adaptations that allow a species to increase energy consumption or reduce energy expenditure become clear.

Recent studies of vampire bats by Brian McNab have revealed a clear-cut relationship between energy intake and expenditure and the species' geographic range. These bats of the suborder Microchiroptera inhabit the neotropics and are specialized to feed exclusively on blood (Chapter 19.1.2). If one is willing to be immunized against rabies before starting to work with the bats, and can stand the smell of their caves, vampires are ideal for calculations of energy budgets. Their daily pattern of activity is simple: they spend about 22 hr in their caves, fly out at night to a feeding site, and return after they have fed. Typically, a vampire flies about 10 km round trip at 20 km/hr to find a meal. Thus, a bat spends half an hour per day in flight. The remaining hour and a half outside the cave may be spent in actual feeding.

The vampire's food is as convenient for energetic calculations as its daily schedule, because of the relatively constant caloric content of blood.

I = ingested energy (blood). The blood a bat drinks must provide the energy needed for all of its life processes: maintenance, activity, growth, and reproduction.

$E =$ excreted energy. As in all animals, not all of the food ingested is digested and taken up by the bat. The energy contained in the feces and urine is lost.

$I - E =$ assimilated energy. This is the energy actually taken into the bat's body.

$M =$ metabolism. This can be subdivided into M_i (metabolism while the bat is inside the cave) and M_o (nonflight metabolism while the bat is outside the cave).

$A =$ the cost of activity, a half hour of flight per day.

$B =$ the biomass increase. This term is the "profit" a bat shows in its energy budget. It may be stored as fat or used for growth or for reproduction (production of gametes, growth of a fetus, or nursing a baby).

In its simplest form, the energy budget is

$$\text{energy in} = \text{energy out} \pm \text{biomass change}$$

The biomass term appears as ± because an animal metabolizes some body tissues when its energy expenditures exceed its energy intake. This is what every dieter hopes to do in order to lose weight.

Translating this general equation into the terms defined gives

$$I - E = M_i + M_o + A \pm B$$

All these terms can be measured and expressed as kilocalories per bat per day (kcal/bat · day). These calculations are based on McNab's studies and apply to a Brazilian vampire bat weighing 42 g.

Ingested energy: In a single feeding a vampire can consume 57 percent of its body weight in blood, which contains 1.1 kcal/g. Thus the ingested energy is

$$42 \text{ g} \times 57\% \times 1.1 \text{ kcal/g blood} = 26.3 \text{ kcal}$$

Excreted energy: A vampire excretes 0.24 g urea in the urine plus 0.95 g of feces daily. Urea contains 2.5 kcal/g and the feces contain 5.7 kcal/g. Thus the excreted energy is

$$0.24 \text{ g urea} \times 2.5 \text{ kcal/g} + 0.95 \text{ g feces} \times 5.7 \text{ kcal/g} = 0.6 \text{ kcal} + 5.4 \text{ kcal} = 6.0 \text{ kcal}$$

Assimilated energy: The energy the bat actually assimilates from blood equals the energy ingested (26.3 kcal) minus that excreted (6.0 kcal).

Thus a vampire's energy "income" is 20.3 kcal/day

$$26.3 \text{ kcal} - 6.0 \text{ kcal} = 20.3 \text{ kcal}$$

Metabolism: In a tropical habitat, 20°C is a reasonable approximation of the temperature a bat experiences both inside and outside the cave. While at rest in the laboratory at 20°C a vampire's metabolic rate is 3.8 ml O_2/g · hr. As a result, the terms for metabolism can be calculated and converted to calories using the caloric equivalent of oxygen (4.8 cal/ml O_2).

$$M_i = 42 \text{ g} \times 3.8 \text{ ml } O_2/\text{g} \cdot \text{hr} \times 4.8 \text{ cal/ml } O_2 \times 22 \text{ hr/day} = 16.9 \text{ kcal}$$
$$M_o = 42 \text{ g} \times 3.8 \text{ ml } O_2/\text{g} \cdot \text{hr} \times 4.8 \text{ cal/ml } O_2 \times 1.5 \text{ hr/day} = 1.1 \text{ kcal}$$

Activity: The metabolism of a bat flying at 20 km/hr is three times its resting metabolic rate (3 × 3.8 ml O_2/g · hr = 11.4 ml O_2/g · hr). The cost of the round trip from the cave to the feeding site is

$$A = 42 \text{ g} \times 11.4 \text{ ml } O_2/\text{g} \cdot \text{hr} \times 4.8 \text{ cal/ml } O_2 \times 0.5 \text{ hr} = 1.2 \text{ kcal}$$

Biomass change: The quantities calculated so far are fixed values that the bat cannot avoid. The biomass change is a variable value. If the assimilated energy is greater than the fixed costs, this energy "profit" can go to biomass increase. Fixed costs which exceed the assimilated energy are reflected as a loss of biomass. For the situation described there is an energy "profit":

$$I \quad - \quad E \quad = \quad M_i \quad + \quad M_o \quad + \quad A \quad \pm B$$
$$26.3 \text{ kcal} - 6.0 \text{ kcal} = 16.9 \text{ kcal} + 1.1 \text{ kcal} + 1.2 \text{ kcal} \pm B$$
$$B = + 1.1 \text{ kcal/bat} \cdot \text{day}$$

These calculations show that vampires can live and grow under the conditions assumed. What happens if we change some of the assumptions? Professor McNab points out that the northern and southern geographic limits of vampires conform closely to the winter isotherms of 10°C. That is, the minimum temperature outside the cave during the coldest month of the year is 10°C; the bats do not occur in regions where the minimum temperature is lower. Is this coincidence, or is 10°C the lowest the bats can withstand? Calculating an energy budget for a vampire under these colder conditions provides an answer.

Inside the bats' caves the temperature remains constant at 20°C, so only the conditions a bat encounters outside the cave are altered. Because of limitations of stomach capacity ingestion cannot increase beyond 57 percent of body weight, the value assumed in the previous calculation. Therefore we need reconsider only M_o, A, and B.

Metabolism Outside: At 10°C a bat must increase its metabolic rate to maintain its body temperature, and laboratory measurements indicate the resting metabolic rate increases to 6.3 ml $O_2/g \cdot$ hr.

$$M_0 = 42 \text{ g} \times 6.3 \text{ ml } O_2/g \cdot \text{hr} \times 4.8 \text{ cal/ml } O_2 \times 1.5 \text{ hr} = 1.9 \text{ kcal}$$

Activity: The cost of activity will not change because the metabolic rate of the bat during flight (11.4 ml $O_2/g \cdot$ hr) is higher than the resting metabolic rate needed to keep it warm (6.3 ml $O_2/g \cdot$ hr). Only the term M_o changes, increasing from 1.1 to 1.9 kcal, and the sum of the energy costs becomes 20.0 kcal/bat $\cdot$ day.

Because the assimilated energy remains at 20.3 kcal/bat $\cdot$ day, only 0.3 kcal is available for biomass increase. The assumptions in these calculations introduce a degree of uncertainty, and probably 0.3 kcal is not different from 0 kcal. In other words, at 10°C a bat uses all its energy staying alive; it does not have an energy profit for growth or reproduction and can tolerate these conditions for only a short period. The 10°C January isotherm runs south from the vicinity of Brownsville, Texas, along the Gulf Coast of Mexico and cuts across the middle of the Florida penisula. Most of the Gulf Coast of the United States is north of this isotherm and has too harsh a climate for vampires.

Additional calculations reveal more about the selective forces that shape the life of vampire bats. For example, a bat's stomach can hold a volume of blood equal to 57 percent of its body weight, but a bat cannot fly with that load. The maximum load with which a vampire can fly is 43 percent of its body weight. Before it can take off to start the flight back to its cave, therefore, a bat must reduce the weight gained from its meal. For example, if a bat weighing 42 g takes in 23.9 g of blood, it must reduce this load to 18.1 g before it can fly. Vampires do this by rapidly excreting water. Within 2 min after it begins to eat, a vampire starts to emit a stream of dilute urine. Experiments performed at Cornell University have revealed that a vampire produces urine at a maximum rate of 0.24 ml/g body weight $\cdot$ hr. Thus, in the hour and a half the bat may spend in feeding, it could excrete as much as 15 g of water—more than enough to allow it to fly.

Although rapid excretion of water solves the bat's immediate problem, it introduces another. The bat is left with a stomach full of protein-rich food that will yield a large amount of urea. To excrete this urea, the bat needs water to form urine. By the time a vampire gets back to its cave it is facing a water shortage instead of a water excess. Unlike many mammals, vampires seldom if ever drink water to compensate for water loss. Instead, they depend on blood for their water requirements. Like other mammals adapted to conditions of water scarcity, vampire bats have kidneys capable of producing very concentrated urine to conserve water (see Table 18–5). As a result of its unusual ecology and behavior, a vampire bat can be considered to live in a "desert" of its own making in the midst of a tropical forest.

As can be seen from studies of the vampire bat, consideration of the energy requirements of alternative behaviors, each of which could satisfy the needs of an organism, help us understand why vertebrates behave as consistently as they do. Application of energy-budget analysis to a variety of vertebrates is an exciting approach to behavioral evolution and ecology that has barely begun.

18.4 THE MAMMALIAN INTEGUMENT

We have established the selective advantage of having a high and constant body temperature despite its energy costs. It should have also become clear that any energy lost can be critical to the endothermic homeotherm. While behavioral microhabitat selection helps to curtail energy loss, most mammalian thermoregulation would be impossible without the characteristics and dynamic reactions of the integument. A great deal of the ability of mammals to individually and evolutionarily invade new habitats is attributable to the properties of their integument.

The integument of mammals is a single organ, the largest of the body; it includes a series of adaptive structures that are outgrowths of the skin. In mammals, as in other vertebrates, the skin functions to protect against onslaughts from, and to receive information from the outside world. The skin, its hair, glands, and sense cells interact with every other aspect of mammalian life in such significant ways that each deserves special consideration.

18.4.1 The Skin

The major divisions of the vertebrate skin (Figure 18–3) are the **epidermis** (the superficial cell layer derived from embryonic ectoderm and the **dermis** (the deeper cell layer of mesodermal origin) Both rest on the subcutaneous tissue (**hypodermis**) that overlies the muscles and bones. The uniqueness of the mammalian integument lies in its derivatives: growing, replaceable hair; lubricant- and oil-producing holocrine (sebaceous) glands; apocrine and exocrine glands that secrete volatiles, water, and ions; the unique mammary glands; scales, nails, claws, and hoofs; and horns. The integument is also a contributing factor in the growth of antlers. Not every species has each class of integumentary appendage, but the various types often occur in unrelated mammals, attesting to their broad adaptive importance.

Epidermis

The epidermis is an avascular cell layer with active and proliferating cells in its deepest part but dead cells near the surface that are regularly shed. The deep, active cells (**stratum germinativum**) obtain their metabolic needs from the underlying vascular dermis. Epithelial cell divisions at this interface force

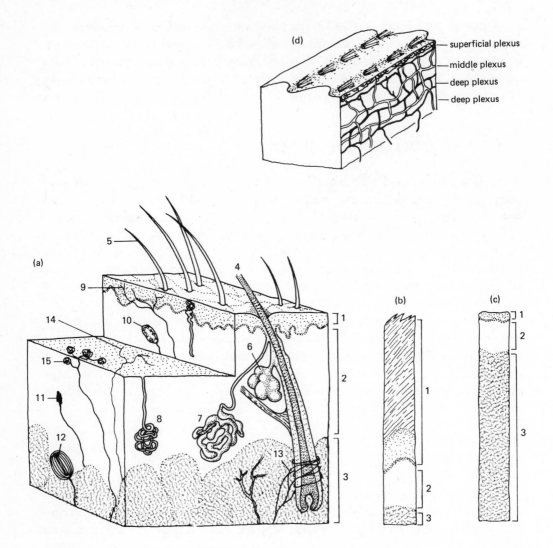

Figure 18-3 The structure of mammalian skin.

(a) Composite of skin and appendages showing: (1) layers and surface configuration of the epidermis; (2) the relatively cell-poor dermis; (3) and the hypodermis rich in subcutaneous fat cells. Appendages shown include (4) guard hairs, (5) undercoat or wool hairs, (6) sebaceous glands, (7) apocrine glands, and (8) sweat glands. Sensory nerve endings include (9) free nerve endings (probable pain receptors) and (10) beaded nerve nets around blood vessels; touch receptors in the forms of (11) Meissner's corpuscles, (12) pressure receptor Pacinian corpuscles, (13) hair movement transducing nerve terminals around the follicle and, (14) warmth and (15) cold receptors.

(b) Thick epidermis of the skin on the sole of the human foot.

(c) Thin, hairy skin and subcutaneous tissue from the human thigh.

(d) Canine skin showing typical multiple hair follicles of the undercoat, the regular texturing of the skin surface, and three of the vascular plexuses involved in thermoregulation. (Modified from various sources, especially Ham [3], and Harrison and Montagna [4].)

daughter cells toward the skin's surface. These daughter cells produce and accumulate proteinaceous keratohyalin granules and related substances, which slowly change chemically as the cell dies and is forced nearer to the skin's surface. By the time these cells reach the surface they are adherent plateletlike sacs of soft keratin, a tough, pliable hydrophobic protein.

Delicate balances between rates of cell proliferation, production of kerato-hyaline granules, "maturation," and the shedding of surface cells produce regional and species differences in skin thickness and texture. Some small rodents have exceedingly delicate epidermis only a few cells thick. Human epidermis varies from a few dozen cells thick over much of the body to over a hundred cells thick on the palms and soles. Elephants, rhinoceroses, hippo-potamuses, tapirs, and pigs were formerly classified as "pachyderms" because their epidermis is several hundreds of cells thick. The texture of the external surface of the epidermis varies from smooth (in fur-covered skins and the hairless skin of cetaceans) to rough, dry, and crinkled (many hairless terrestrial mammals). The tail of many rodents is textured with epidermal scales very like those of reptiles. Without doubt the most adaptive character of the mammalian epidermis is cohesiveness between cells during all stages of development and degeneration. This imparts marvelous characteristics to the tissue: It is living, growing, regenerative—and at the same time dry, abrasive, resistant, and expendable.

Dermis

The dermis, unlike the epidermis, is primarily composed of extracellular products. The dermis is laid down in two layers whose mechanical properties dominate and determine those of the skin as a whole. Over joints and in areas of mobility, the dermis is elastic, loose, and thin. On the back and areas of continual friction with the outside world (the soles, prehensile tails, flippers, and so on), the dermis is thick, firm, and rather immobile. Its deepest layer is rich with interlaced bundles and sheets of collagen fibers produced by sparsely scattered cells (fibroblasts). The interweaving of this dermal fabric and the amount of elastic fibers determine the extraordinary tensile strength of the mammalian skin. It is primarily this layer of mammalian skin that is used to make leather. Occasionally smooth muscle fibers occur in this area. Their contraction produces a wrinkling of the skin such as that of the nipple of the mammary glands or the male scrotum.

Twigs of blood vessels and nerves course through the lower dermis to their destination in the superficial dermis. The latter is a thin layer composed of more delicate and loosely arranged collagen and elastic fibers. It is a "reverse impression" of the complex deep surface of the epidermis (Figure 18–3a, 18–3d). Here arterioles break up to form an enormous, complex web of capillaries closely applied to the dermis-epidermis junction. These fuse again as venules, which exit from the superficial dermis often in parallel with an incoming arteriole. In addition there may occur direct connections between afferent and efferent vessels (arteriovenous anastomoses). Two or more flat networks of vessels occur at different depths within the dermis and hypodermis parallel to the body surface. These vascular networks are best developed in mammals lacking thick fur (for example, swine and humans) or in naked regions of hairier mammals (the perineum, that is, around the urogenital/anal openings; the scrotum; the nostrils, and so on). They have a special function in body cooling. The amount of blood and rate of flow within these vessels can be varied by central neural and hormonal control or directly by local temperature. Under conditions of excess heat, the vessel plexus dilates and warm blood is cooled by its close contact with the skin surface. When chilled, all but a portion of these capillaries constrict, minimizing blood flow to the skin. The close juxtaposition of afferent arterioles and efferent venules, which continue to carry small amounts of life-supporting blood to the epidermis, acts to conserve heat by countercurrent exchange (Chapter 7.6).

The dermis houses the majority of sensory structures and nerves associated with the sensations of temperature, pressure, and pain. In addition, many motor nerves run to the vascular, muscular, and secretory structures of the skin. Some free nerve endings penetrate the epithelium, but the majority terminate in the dermis as specialized end organs. Generally these organs are made up of connective tissue capsules of varying thickness (for example, Pacinian corpuscles; see Figure 18-3a).

Skin color resides in the dermis, as well as the epidermis. Some neural crest cells, the **melanocytes** (Figure 5–12), come to rest in the dermis, in the deepest layers of the epidermis, and in the interface between the dermis and epidermis. They produce granules of melanin, which may be yellow, rusty red, brown, or black in color. Through long, narrow, cytoplasmic extensions of the melanocytes, melanin is injected into adjacent cells that lack the melanin-producing enzymes. Additional hues may be produced by structural colors (Chapter 16) overlying melanin, and by the vascularization of the skin to produce shades from pink to scarlet. Skin coloration, which incorporates the vascular system, has the advantage of being rapidly variable so that red display areas may become brilliant in excited individuals.

Hypodermis

The subcutaneous tissue or hypodermis is not functionally a part of the skin but interfaces between the skin and the deep-lying muscles and bones. Collagenous and elastic fibers are prevalent, but much of the hypodermis is unstructured. This region often contains a major energy store of the body: subcutaneous fat. Fat storage reaches its maximum development in pinnipeds and cetaceans in the form of insulative blubber. Many mammals have extensive subcutaneous muscles that move the skin relative to underlying tissues. The fly-disturbing jiggle of a horse's skin is an example. In many carnivores, ungulates, and especially in man and the other higher primates, subcutaneous muscle reaches exceptional development in the facial region. Known as the platysma, this muscle can produce facial expressions, movement of the external ears, eyelid opening and closure, and the ability to purse the lips and suck—a vital attribute to early mammalian life.

18.4.2 Integumental Derivatives

The so-called "appendages" of the mammalian skin are derived from epidermis but supported, nourished, and innervated through the dermis. The primary appendage of the mammalian integument is the hair.

Hair

Hair is a unique mammalian adaptation that has no direct homolog in other vertebrates. The perplexing problem is to understand how the first hairs evolved. Although, like feathers, hair acts as insulation in thermoregulation, it is not essential for endothermal homeothermy. To provide insulation, the hairy coat of pelage must be long enough and dense enough to trap air. What gave hair selective advantage? Paul F. A. Maderson has proposed that an important function of the skin in all vertebrates is the sensory one. Tactile information about the immediate surroundings would be especially important to behaviorally thermoregulating ancestors, as well as to the earliest nocturnal, perhaps burrowing mammals. Maderson proposes that thin, rodlike elements evolved from germinal epidermal cells which stuck together to project above the surrounding unaltered epidermis as the first hairs. Basal innervation of these sparse rods, which were distributed over the body, provided tactile sensory input from bending and distortion of the rod. Therefore, like the lateral line receptors (Chapter 7.2), hairs evolved as epidermal mechanoreceptors. At some point these sensory hairs were altered to result in a localized multiplication of the units, which perhaps improved the precision of the information provided by these tactile receptors. Whatever the selective forces, Maderson argues that dense sense organs produced a pelage whose insulative properties became the focus of subsequent selection. In support of this hypothesis, hairs are richly innervated around their base even when insulation seems to be their prime function. Highly specialized tactile hairs are localized on the legs, nose, and around the mouth and eyes of many mammals. These **vibrissae**, which have bulbous complex roots, vascular beds, and a multitude of nerve endings, are extremely sensitive to deflection of the tips of their still rodlike shaft.

Although the origin of hair is obscure, a great deal is known about hair of living mammals. Pertinent features of the pelage of mammals are its growth, replacement, color, and mobility. A hair grows from a deep invagination of the germinal layer of the epidermis called the hair follicle (Figure 18–4). Hair is composed of keratin, pigments, and, in some cases, tiny encapsulated air bubbles. Although these substances are common to all tetrapods, it is their organization and chemistry that make hair a unique mammalian feature. The keratin of hair is of two types: soft and hard. **Soft keratin**, found in the core or medulla of a hair shaft, is similar to that of the epidermis. Microscopic examination of soft keratin reveals that distinct granules of its precursors form in epithelial cells and later fuse to produce a homogeneous cell mass. The soft keratin in the medulla is often interrupted by masses of trapped air bubbles. **Hard keratin** is found in the cortex covering the medulla and the scalelike cuticle or thin sheath that encapsulates the hair surface. During its development, hard keratin shows no sign of precursors. Rather, a gradual fusion of living epidermal cells into a rather hard, chemically unreactive homogeneous mass occurs. When a hard keratin cortex forms most of the material of hair, stiff hair, bristles, and quills are produced. Bird feathers and leg scales, as well as reptile scales, are hard keratin. When soft keratin is dominant, pelage that is soft and supple, such as wool and marketable furs, is produced.

The color of hair depends on the quality and quantity of melanin injected into the forming cortical cells by melanocytes at the base of the hair follicle. Medullary melanin contributes less to the overall color of the pelage than do the cortical pigments. Black hair has dense melanin deposits in both cortex and medulla; the white hair of humans has an unpigmented cortex but pigmented medulla. When the medullary portion of the hair shaft is absent, a fine often blond or reddish pelage (depending on the melanin) results. Adjacent hairs are often identical in pigmentation, but different regions of the body may produce hairs differing greatly in color and appearance. Thus, the adaptive color patterns of mammals are built up by the colors of individual hairs. Most mammals that expose themselves to light from the sky are countershaded. Many have strongly contrasting patterns visible even at low levels of illumination (e.g., skunks and many antelopes).

Because exposed hair is nonliving, it wears and bleaches. Replacement occurs by growth of an individual hair or by **molting**, in which old hairs are replaced. During active growth, human hair lengthens about one third of a millimeter per day. It finally becomes quiescent and the follicle produces a bulbous anchor that firmly locks the hair in place. Only after reactivation of the follicle and formation of a new hair is the old hair forced out of the skin. Although in some mammals, like humans, quiescent and active hair follicles occur in a mosaic over the entire body, most mammals have pelage that grows and rests in seasonal phases. Molting usually occurs once or twice a year. Only by this mechanism can mammals like the arctic fox, snowshoe hares, and ermine change from a summer camouflage of brown to a winter coat of white. Because the hairs are tightly bound to the dermis, only skins with quiescent (generally winter) pelage are of value in the fur trade.

The insulative effectiveness of hair depends upon its ability to trap air, and this is proportional to the length of the hairs. Polar and temperate mammals have longer coats in winter than in summer. The closeness of hairs is also important because it creates smaller entrapped air spaces; winter coats are "thicker" than summer pelage. A duplex pelage often occurs: guard hairs are long, cylindrical, straight shafts, whereas curly underfur or wool hairs are shorter, more numerous, and flattened. These differences in pelage have obvious seasonal adaptiveness, but thermal stresses on a mammal often cover a short time course to which morphological changes in insulation are not suited.

Most hairs are not perpendicular to the surface epithelium but angle in the same direction over a given region of the body. Attached about midway along the length of each hair follicle is a bundle of smooth muscle, the **erector pili** (Figure 18-4). This muscle pulls the hair into a nearly perpendicular orientation to the skin. The result is to thicken the pelage and thus entrap a larger volume of air. A curious side effect noticeable in near-naked mammals such as humans are the dimples ("goose pimples") on the skin's surface over the insertion of contracted erector pili muscles. Via the sympathetic nerves, cold stimulates a general contraction of the erector pili as do other stressful conditions such as fear and anger. The significance of these behaviorally induced erections of the pelage appears to lie in increasing the visible size and, therefore, the apparent strength of the excited mammal. Often such displays are limited to specific regions, such as the hackle and tail hair erections of dogs. Displays such as those of a porcupine are impressive and are clearly understood by a wide range of other animals.

Glandular structures

Secretory glandular structures of the skin develop as elaborations of the epidermis. The **sebaceous glands** open via a short duct into the neck of the hair follicles. These saclike structures often lie in the space bounded by the hair shaft, the superficial epidermis, and the erector pili muscle. When the muscle contracts the sebaceous gland is pressed between the rigid hair shaft and the keratinized epithelium, squeezing its oily contents into the follicle and thence onto the skin surface (Figure 18-4). Sebaceous glands may also be found in hairless areas of the mouth and lips, the end of the penis, and around the vulva, the nipples of the mammary glands, and the edges of the eyelids. The oils lubricate the hair and skin, giving them considerable waterproofing. Specialized sebaceous glands produce lanolin in sheep.

A second type of gland common to many mammals is the **apocrine gland**. These glands frequently open near but not into the hair follicles and also occur in skin that lacks hair. Apocrine glands are tubular epidermal invaginations that extend deep into the dermis. In addition to those differences, apocrine glands differ from sebaceous glands in the details of the secretory mechanism. The exact functions of apocrine secretions are not well understood. Although the secretion from human apocrine glands is odorless, bacteria on the body surface convert it to an odorous product. Various musk and scent glands so characteristic of mammals are thought to be enlarged, modified apocrine glands. In their development, their structure, and their mode of secretion, **mammary glands** closely resemble apocrine glands.

Mammary glands (Figure 18-5) develop as paired ridges of hypertrophied ventral epithelium of the mammalian fetus that run from the axillary to the

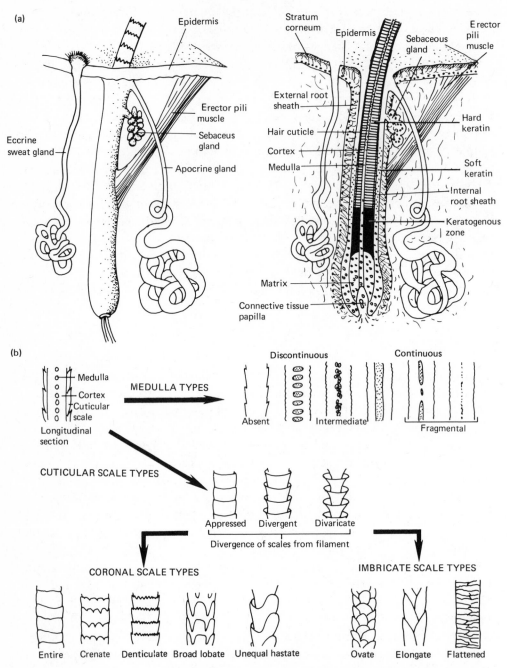

Figure 18-4 The structure of mammalian hair and associated glands. (a) Composite view of hair follicle (left) and surrounding structures with dermis dissected away and a longitudinal section (right) to show the internal structure of a typical hair. (b) Variation in internal and external structure of hairs may be used as a species specific identification character. Some of the important differences in internal structure (medulla types) and external appearance (cuticular scale types) are illustrated. (Modified from various sources, especially Ham [3], and Harrison and Montagna [4].)

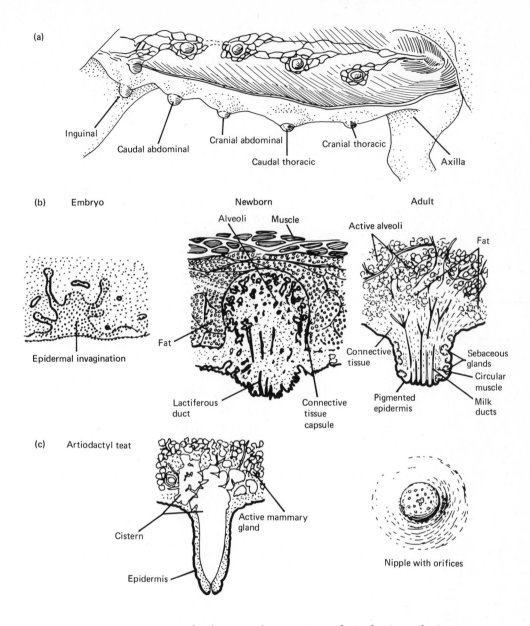

Figure 18-5 Mammary glands. (a) The positions of nipples in eutherians as illustrated by a dissection of the dog. Only the axillary mammae fail to develop in these large litter-bearing canids. (b) Stages in the development of the human mammary gland resemble those of apocrine and eccrine glands. An embryo of 25-cm length has a simple proliferation and invagination of epidermal cells (left). A new-born infant (middle) has the basic elements of a nipple, multiple lactiferous ducts and nonsecretory alveolae. The tip of the mature, lactating primate breast shows the addition of constrictor muscles, sebaceous glands, and swollen active alveolae which vent their product via multiple openings on the nipple. (c) The artiodactyl teat is designed to deliver a much larger volume of milk in a shorter time via the development of large milk-holding cisternae adjacent to the elongate teat, which has a single wide terminal opening.

inguinal regions. This mammary ridge, or milk line, differentiates into discrete localized thickenings in different positions in accordance with the species. Manatees (Order Sirenia) have one pair of axillary mammae, primates one pair of thoracic mammae, some artiodactyls have abdominal mammae, and perissodactyls have inguinal mammae. The number of mammae usually corresponds to the number of young born in a single litter. Certain opossumlike marsupials are reported to have nearly 20 mammae. In male mammals further development of the mammae does not occur after birth. In females, however, the thickened epithelial cells send numerous elongate cords that branch and proliferate into the dermis. Hollow sacs develop at the terminal ends and become continuous with channels that develop through the cords of epithelium to produce a highly branched duct system. Under proper hormonal stimulation the terminal sacs become more numerous, enlarge greatly, and begin to secrete milk. Milk is a water-based solution of proteins (primarily the phosphoprotein calcium caseinate), the unique sugar lactose, and lipids in the form of several different fatty acids combined in suspended droplets. The fat droplets often are capped by or contain cell fragments, indicating a mode of secretion similar to that of apocrine glands, which also loose portions of cells during secretion. The exact proportions of these primary constituents varies greatly from species to species (Table 18–4). In general the growth rate of a newborn mammal is positively correlated with the protein content of that species' milk. Minor components vary and may reflect elements in the maternal diet. The concentrations of these major components, including the primarily protein-bound minerals calcium and phosphorous, are far above those of the maternal blood.

Table 18–4. Composition of Milk of Various Species (in grams per liter)

Species	Sugars	Proteins	Fats	Inorganic salts
Ass	66	17	11	4
Buffalo	38	62	25	8
Camel	33	30	55	7
Cat	50	92	35	11
Cow	45	35	40	9
Dog	40	70	85	11
Elephant	72	32	190	6
Goat	47	33	40	6
Horse	60	20	12	4
Human	75	11	35	3
Pig	32	74	45	10
Porpoise	13	110	460	6
Reindeer	29	100	175	14
Sheep	50	67	70	8
Whale	4	95	200	10

Although the hormones of pregnancy stimulate the mammary glands into a physiological state capable of lactation, milk does not flow until some time after suckling has begun. Monotremes lack highly specialized openings for the mammae, but other mammals have nipples that are complexly innervated epidermal organs. Certain artiodactyls have analogous but much larger structures called teats (Figure 18-5).

Similar in structure to the apocrine and mammary glands and often grouped with them are the **sweat** or **eccrine glands**. The coiled tubules of sweat glands are never associated with hair follicles. In many mammals they are prevalent on the hairless surfaces that come in contact with the substrate: soles of feet, prehensile tails, and so on. Here their secretions, by maintaining skin pliability, increase traction and enhance sensitivity of skin pressure receptors. Most mammals, particularly those without underfur, have sweat glands widely distributed over the body surface.

Sweat glands function primarily in thermoregulation. Water secreted by the sweat gland is forced onto the skin surface by special contractile elements of the gland. Here it evaporates and cools the skin which, in turn, cools the blood flowing in the subcutaneous vascular beds. When perspiration is copious considerable sodium and chloride ions are lost and must be replaced in the diet. Some sweat glands respond to stimuli other than heat. In conjunction with nearby apocrine glands, these sweat glands contribute to odor production under conditions of stress and excitement.

Claws, Nails and Hoofs

Some integumental appendages are specialized for locomotion, offense, defense, or display. Claws, nails, and hoofs all represent accumulations of keratin protecting the terminal phalanx of the digits (Figure 18-6). Claws may be very sharp and retractable in highly specialized carnivores. Less specialized claws are nearly universal among mammals and probably gave rise to hoofs and nails. The claw aids in traction and is a major climbing adaptation in many orders of mammals. Nails evolved in primates only after the digital grip of hands and feet had been attained. The hoofs of ungulates (orders Artiodactyla and Perissodactyla) are much enlarged blocks of keratin upon which the animal rests its entire weight. They are light, resilient, and exceedingly tough structures well suited to the life of cursorial ungulates.

The cephalic horns and antlers of many ungulates are used in offense and defense. They occur in mammals whose dentition and jaw apparatus are highly modified for a herbivorous diet, which precludes biting as a primary offensive and defensive tactic. Horns are of four types (Figure 18-7a–d). The median horns of the perissodactyl family Rhinocerotidae are simple in construction and give the rhinoceros its name (Greek, *rhino* = nose, *keras* = horn). The horn of a rhinoceros is firmly attached to the skin, not to bone, and grows throughout life from a pad of epidermis. It is never shed and is composed of tubules of hard keratin fused together at their peripheries. Because the horn germinal

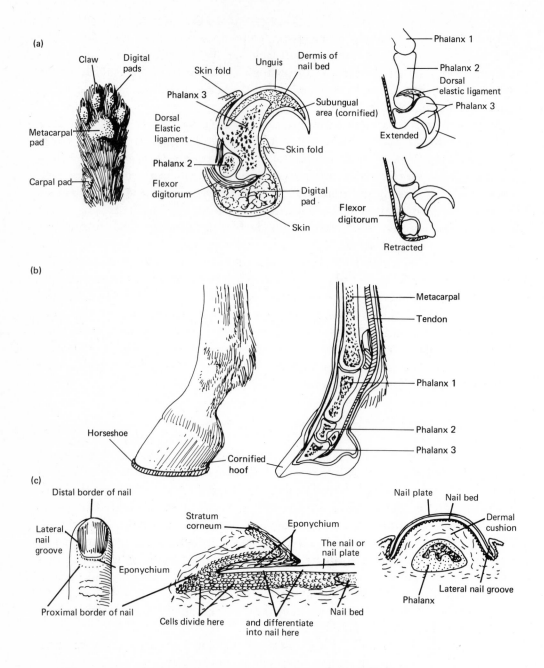

Figure 18-6 Skin appendages associated with terminal phalanges.

(a) Claws, which are considered the basic digital appendage of the skin, are illustrated by the highly specialized retractable claws characteristic of the catlike carnivores. Left: Hair and thick epidermal pads associated with the base of the claws and other pressure (friction) points constitute the integumentary structures common to many mammalian paws. Center: section of a claw shows its close relationships with the vascular, living dermis and bone of the third phalanx. Right: muscle tension on the flexor digitorum longus tendon causes the third

epidermis nearest the tip of the snout proliferates more than on the opposite edge of the pad (posterior), the rhinoceros horn grows in a sweeping posteriorly directed curve.

The horns of the only surviving member of the artiodactyl family Antilo-capridae, the North American pronghorn antelope, are of greater complexity. Like most mammalian head appendages, the horns of the pronghorns are bilaterally paired and arise from specialized epidermis that overlies spikelike projections from the frontal bones. The horns, which have been described as fused hair, grow much larger than the bony core and have a forward projection or prong as well as the erect backward curling main shaft. These horns are produced by a new growth each year, which forces the old horn to be shed. Such deciduous horns are unique among living mammals.

More typical horns are those of the large artiodactyl family Bovidae (antelopes, gazelles, buffaloes, cattle, goats, sheep, and so on). Their head appendages are supported by a process from each frontal bone, the **os cornu** (literally, "the bone of the horn"). The covering epidermis is active throughout life and produces the keratinized horn. Bovine horns are never shed nor do they ever exhibit branching; but they can be extremely complex in shape: straight, loose spirals, tight helixes, and so on. Horns are primary offensive and defensive weapons, having the same properties as claws and hoofs. Nevertheless, in some bovine genera only males have horns and in other genera the horns of males are noticeably larger than the females'. Horns, therefore, must be objects of sexual selection and have display and/or dominance functions in the male that are absent in the female.

In their early annual developmental stages, the antlers of the artiodactyl Cervidae (deer) are similar to the "horns" of the giraffe, that is, bony cores invested with hairy skin (Figure 18-7e). With the exception of female reindeer (caribou), female deer never develop antlers. The skin over the antler is covered with short, fine hair, the "velvet," and is highly vascular and innervated.

---

phalanx to rotate on the balllike extension of the second phalanx. This exposes the claw and stretches the dorsal elastic ligament. Relaxation of the flexor digitorum muscle allows the dorsal elastic ligament to retract the claw.

(b) The hoof, characteristic of ungulates, is illustrated by that of the horse. Normal appearance of the hoof and foot of a shoed horse (left). Horseshoes are used to retard wear of the hoof due to the hard substrates and heavy work man imposes on domestic horses. They may be nailed directly to the hoof, since the hoof is dead keratin. Longitudinal section of foot showing relationship of phalanges to hoof (right). The tendon is homologous to that illustrated in the cat but functions to return the hoof to a proper angle after it has been bent by the weight shifted on it during a step.

(c) The human nail is a simpler structure than either the retractable claw or the hoof, but was derived from primitive claws. Distinct regions on a nail (left) correspond to the various regional specializations of the epidermis associated with the nail (center). A cross section of the end of a finger (right) shows the nail's close association with the dermis and bone of the digit. (Modified from various sources including Ham [3].)

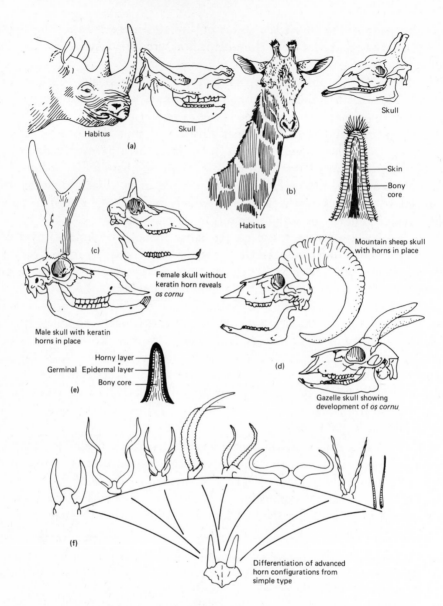

Figure 18-7   Cephalic appendages of mammals.

(a) The entirely epidermal nasal horns of a rhinoceros have no contribution from the bony elements of the skull. (b) The short, nobby "horns" of the giraffe are composed of both skin and bony contributions. The skin is scarcely differentiated from that elsewhere on the body. (c) The North American prong-horn antelope has forked epidermal horns over a simple bony projection of the frontal bone known as the *os cornu*. Unique among living mammals is the annual shedding of the fused hairlike horny sheaths. (d) The members of the artiodactyl family Bovidae have a great variety of true horns based on a common developmental plan. True horns are supported by a well developed *os cornu* illustrated by the skull of a Pliocene gazelle. Massive horns develop in the males of mountain sheep. (e) Over the *os cornu*, a nonshedding nonreplaceable horn or hard keratin grows from a living germinal epidermal layer. (f) Complex

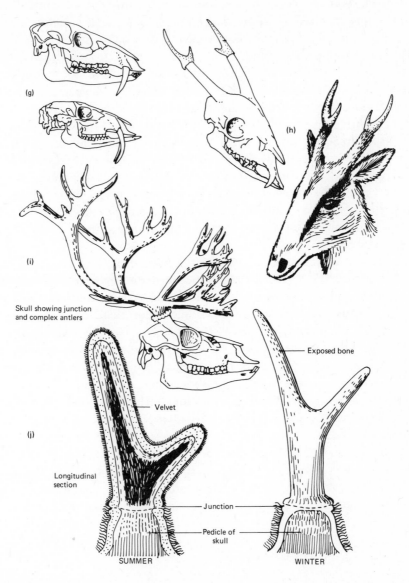

Skull showing junction and complex antlers

Exposed bone

Velvet

Longitudinal section

Junction

Pedicle of skull

SUMMER

WINTER

shapes are characteristic of many advanced antelopes of Africa and Asia. (g) Antlers are not found in primitive members of the deer family (Cervidae), as illustrated by Miocene (upper) and recent (lower) tusked forms. (h) Later cervids developed antlers in conjunction with tusks, as seen in the recent muntjac. (i) The most advanced cervids lack canines altogether, but in many species males have enormous complex antlers such as those of the caribou. Females of this species, unlike females of any other deer species, also possess antlers. (j) During their annual replacement, the integument in the form of the 'velvet' plays an important role in the nourishment and protection of the growing antler. At maturation in the early fall, the velvet dies and is shed, resulting in an entirely bony mature antler. The complex junction between the pedicle of the skull and the antler eventually weakens and the antlers are shed in the spring of each year. (Modified from various sources including Modell [10].)

Growth, maturation, and the shedding of each antler is an annual event with male deer and represents a large energy expenditure. The selective advantage of these often bizarrely magnificient appendages is puzzling. The antlers are in velvet from early to late summer and, because of their considerable vascularity and the presence of sweat glands, might be effective cooling devices. It is difficult to account for their presence only in males, however, for there is little difference in activity between males and females during summer. Antlers are used in fighting, especially in ritualized interlocked head wrestling matches during the breeding season when activities between the sexes differ most. However, at this time the vascularized "velvet" has been shed, and the antler is physiologically dead. Even as elements of strong sexual selection, antlers are of dubious value. They cost a great deal to regrow, interlocking often results in the deaths of males, and they are not absolutely essential for successful harem dominance. The resolution of this paradox may lie in their evolutionary history.

The Oligocene ancestors of the deerlike ungulates had long stabbing upper canines but no antlers. During the Miocene, forms evolved that had both stabbing canines and modest antlers, as do living Asian muntjacs. Not until the Pliocene did tuskless antlered deer become widespread. In studies of the fighting behavior of muntjacs, Cyrille Barrette has learned that tusks are used offensively, whereas antlers may be offensive when used to stab or defensive when used as shields against blows from an opponent. Males engage in head wrestling and attempt to twist and shove their opponent off balance so that the tusks can be used. Winning a tusk-and-antler fight rewards the successful male muntjac with access to receptive females and successful reproduction. A possible first stage in the evolution of antlers would be defensive shielding against tusk blows. The larger the "protoantler," the more effective the shield, and at some point in their evolution the effectiveness of antlers made tusks obsolete. However, their visual image and ability to intimidate an opponent must have lingering advantages by increasing the possessors' access to females.

18.5   THE MAMMALIAN KIDNEY

While the integument contributes to a wide variety of characteristics unique to mammalian life, the kidney has a more restricted roster of functions. However, the highly sophisticated capacity of the mammalian kidney to conserve water, rid the body of nitrogenous and other wastes, and maintain a narrow ion and acid-base variation are essential to mammalian life. Only within the constant internal milieu provided in large part by the specialized mammalian kidney can the digestive, neuro-muscular, sensory, integrative and reproductive refinements uniquely characteristic of mammals be realized. Only with the concentrating powers of the mammalian kidney could these vertebrates invade so many diverse and severe environments. Understanding the mammalian kidney is a key factor in understanding the success of mammals.

The mammalian kidney is composed of millions of nephrons, the basic micro-anatomical units of kidney structure recognizable in all vertebrates (Chapter 7.3.1, and Figure 7-19). Each nephron is composed of a **gomerulus** that physically filters the blood, producing an ultrafiltrate, and a long tubular conduit in which the chemical composition of the ultrafiltrate is altered. These tubules empty into coalescing ducts that lead to the urinary meatus (opening) and carry the final kidney product, the urine. The mammalian kidney is extraordinary because it is capable of producing a urine more concentrated than that of any anamniote or reptile and, in most cases, more concentrated than that of birds as well (Table 18-5). This ability enormously reduces water loss and the need for potable water. The ability of several lineages of mammals to highly concentrate their urine indicates a major adaptation, especially to arid habitats.

The basic problem in urine concentrattion involves removal of water from an isosmotic ultrafiltrate, leaving behind the concentrated excretory residue. Because cells are unable to transport water directly, they use osmotic gradients to "manipulate" movements of water molecules. In addition, the cells lining the nephron must actively reabsorb substances important to the body's economy from, and also secrete toxic substances into the ultrafiltrate. The single layer of cells lining the nephron differs along the length of the nephron in permeability, molecular and ion transport activity, and reaction to the hormonal and osmotic environments in the surrounding body fluids. The nephron's activity may be considered a six-step sequential process, each step localized in regions having special cell characteristics and distinctive variations in the osmotic environment.

The first step is the mechanical production of an ultrafiltrate at the glomerulus (Figure 18-8). This encapsulated knot of vessels acts like a sieve, using the hydrostatic pressure of the blood to produce an ultrafiltrate. The ultrafiltrate is isosmotic with blood plasma and resembles whole blood after the removal of (a) cellular elements, (b) substances with a molecular weight of 70,000 or greater (primarily proteins), and (c) substances between 15,000 and 70,000 molecular weights, depending on the shape of the molecule. An adult human has up to 3 million nephrons with a combined glomerular filtration surface of about 1 m^2. An average filtration rate for resting adults approximates 120 ml of ultrafiltrate per minute. Obviously a primary function of the remainder of the nephron is reduction of the ultrafiltrate volume—to excrete the 170 liters (45 gal) of glomerular filtrate produced per day is impossible!

The second step in the production of the urine is the action of the **proximal convoluted tubule** (PCT, Figure 18-9a) in decreasing the volume of the ultra-filtrate. The PCT also very effectively removes blood sugar (glucose) and other essential metabolites from the ultrafiltrate and returns them to the general circulation. The PCT cells, whose lumenal surfaces are greatly enlarged, actively transport sodium ions from the lumen to the exterior of the nephron (Figure 18-10). To maintain electric neutrality, chloride ions passively accompany the sodium ions. Water osmotically flows through the PCT cells in response to the removal of sodium chloride. By this process about two thirds of the salt and volume of the ultrafiltrate is reabsorbed in the PCT. Although it is still very

Table 18-5. Maximum Urine Concentrations of Tetrapods

Species	Maximum observed urine concentration (mOsm liter)		Approximate urine: plasma concentration ratio	
Crab eating frog (*Rana cancrivora*)		600		0.72
Pelican (*Pelecanus erythronhynchos*)		700	approx.	2
Savannah sparrow (*Passerculus sandwichensis*)		2000		4.4
Beaver (*Castor canadensis*)		520		2
Pig (*Sus scrofa*)		1100		3
Man (*Homo sapiens*)		1430		4
Bottlenose porpoise (*Tursiops gilli*)		1600–1800	approx.	5
Quokka (*Setonix brachyurus*)		2188		5–6
Hill kangaroo (*Macropus robustus*)		2730		7–8
Camel (*Camelus dromedarius*)		2800		8
White rat (*Rattus norvegicus*)		2900		8.9
Cat (*Felis domesticus*)		3250		9.9
Pack rat (*Neotoma albigula*)		4250		11 (est.)
Marsupial 'mouse' (*Dasycercus cristicauda*)	approx.	4000		12 (est.)
Kangaroo rat (*Dipodomys merriami*)	approx.	4650		12 (est.)
Vampire bat (*Desmodus rotundus*)		4650		14
Gerbil (*Gerbillus gerbillus*)		5500		14
Australian hopping mouse (*Notomys alexis*)		9370		22

Sources: W. N. McFarland and W. A. Wimsatt [1976] *Comp. Biochem. Physiol.* 28:985–1006; R. A. Macmillan [1972] *Symp. Zool. Soc. London* 31:147–174; K. Schmidt-Nielsen [1964] *Desert Animals*, Table xxvi. Oxford University Press; M. S. Gordon et al. [1977] *Animal Physiology*. Third ed. Macmillan, New York; R. L. Malvin and Rayner [1968] *Amer. J. Physiol.* 214:187–191.

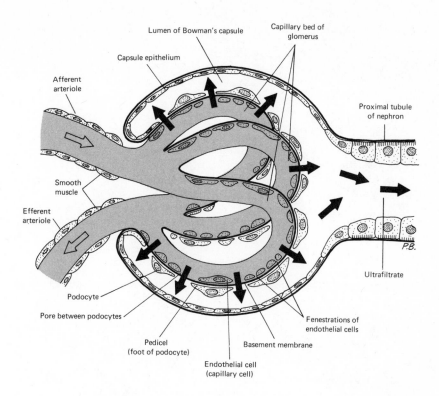

Figure 18-8 Typical mammalian glomerulus illustrating the morphological basis for its functioning. The glomerulus is a complex of capillaries embedded in a sheath of cells (podocytes) that constitute the inner wall of Bowman's capsule — the first portion of the nephron. Substances of sufficiently small size may pass from the blood through fenestrations in the capillary wall (endothelial cells) and between the "feet" (pedicels) of the podocytes to become a part of the ultrafiltrate within the nephron. The rate of glomerular filtration of a substance is dependent upon the ability of that substance to pass through the pores of the glomerulus and on the blood pressure. The hydrostatic pressure of fluids already in the nephron and the osmotic effect of substances too large to pass through glomerular pores retard filtration. Smooth muscle surrounding the afferent and efferent arterioles can restrict blood flow through the glomerulus to alter the rate of filtration. (Modified from Netter [11].)

nearly isosmotic with blood, the individual substances contributing to the forming urine's osmolarity are at different concentrations than in the blood.

The next alteration of the formative urine occurs in the descending (thin) limb of the **loop of Henle** (Figure 18-9b). The thin, nearly smooth-surfaced cells of this segment freely permit diffusion of sodium and water (Figure 18-10). Because the descending limb passes through tissues whose osmolality increases steadily along its length, water is lost from the forming urine, concentrating it. In humans the osmolality of the fluid in the descending limb may reach 1200 mOsm/liter. Mammals producing a more concentrated urine achieve corres-

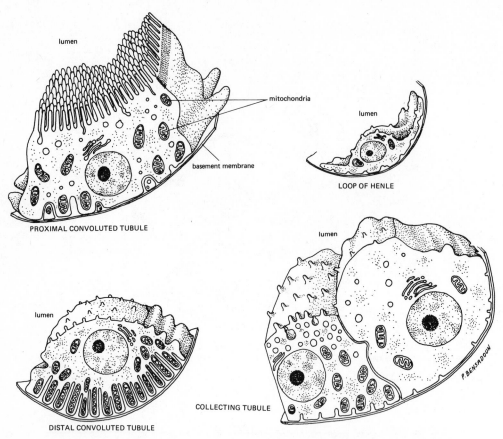

Figure 18-9   Structure of the walls of the regions of the nephron and collecting tubule. The cells of the proximal convoluted tubule (PCT) have an enormous luminal surface area produced by long, closely spaced microvilli. Not the many energy producing mitochondria. These structural features reflect the function of the PCT in rapid, massive transport of sodium from the lumen of the tubule to the peritubular space and capillaries; sodium transport is followed by passive movement of chloride to neutralize electric charge. Water then flows osmotically in the same direction. The cells of the thin segment of the loop of Henle are waferlike and contain fewer mitochondria. They function either in permitting passive flow of sodium and water (descending limb) or in active sodium removal from the ultrafiltrate, where they are impermeable to water (ascending limb). The cells of the distal segment of the nephron actively remove sodium up a concentration gradient of considerable magnitude and may or may not permit water to move osmotically depending on whether they are associated with the ascending loop of Henle (impermeable) or the distal convoluted tubule — DCT (usually permeable). These activities require considerable cellular energy at the interface between the nephron and the peritubular tissues. Large mitochondria occur here in numbers. Cells of the collecting tubule appear to be of two kinds. Most seem to be suited to the relatively impermeable, low ion pumping state characteristic of periods of sufficient body water. Other cells are mitochondria-rich and have a greater surface area. They may respond to the presence of antidiuretic hormone (ADH), triggered by insufficient body fluid, and actively exchange ions, pump urea, and freely permit the flow of water from the lumen of the tubule to the concentrated peritubular fluids. (Modified after Netter [11].)

pondingly higher concentrations of fluid in the descending limb. By this mechanism the volume of the forming urine is reduced to 25 percent of the initial filtrate volume. Although the concentration of fluid in the end of the descending limb is approximately that of the urine excreted, its volume is still large. In the human adult, for example, between 25 and 40 liters of fluid reaches this stage per day, yet only a few liters will be urinated.

The fourth step takes place in the ascending loop of Henle, which possesses cells with very large and numerous mitochondria (Figure 18-9c). The ATP produced by these organelles is utilized in actively removing sodium from the forming urine. Because these cells are impermeable to water, the forming urine volume does not decrease and it enters the next segment of the nephron hyposmotic to the body fluids. It is important to remember that the solute particles remaining within the tubule represent the residue of substances filtered from the plasma. Although this sodium-pumping, water-impermeable, ascending limb does not concentrate or reduce the volume of the forming urine, it physiologically "sets the stage" for these important processes.

The very last portion of the nephron changes in physiological character but the cells closely resemble those of the ascending loop of Henle. This region the **distal convoluted tubule** (DCT), pumps sodium from the forming urine but, unlike the ascending limb, is permeable to water. The osmolarity surrounding the DCT is that of the body fluids, and water in the entering hyposmotic fluid flows outward and equilibrates osmotically. This results in a varying fluid volume of 5 to 20 percent of the original ultrafiltrate.

The "final touch" in the formation of a scant, highly concentrated mammalian urine occurs in the **collecting tubules** (Figure 18-9d). Each collecting duct drains many nephrons. Like the descending limb of the loop of Henle the collecting ducts course through tissue of increasing osmolarity, which tends to withdraw water from them. Some sodium is removed from the forming urine in the collecting ducts, although usually in exchange for potassium, hydrogen, or ammonium ions as a component of the acid-base regulation of the body. The significant phenomenon associated with the collecting duct, and to a lesser extent with the DCT, is its conditional permeability to water. Under conditions of excess fluid intake, the collecting duct demonstrates low water permeability; only half of the water entering it may be reabsorbed and the remainder excreted. In this way a copious, hyposmotic urine can be produced by a mammal. When the mammal is dehydrated the collecting ducts and the DCT become very permeable to water and the final urine volume may be less than 1 percent of the original ultrafiltrate volume. In certain desert rodents so little water is contained in the urine that it crystallizes almost immediately upon micturition!

A polypeptide called **antidiuretic hormone**, ADH (also known as vasopressin), is produced by specialized neurons in the hypothalamus, stored in the posterior pituitary, and released into the circulation whenever blood osmolality is elevated or blood volume drops. When present in the kidney, ADH increases the permeability of the collecting duct to water and facilitates water reabsorption

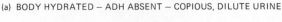

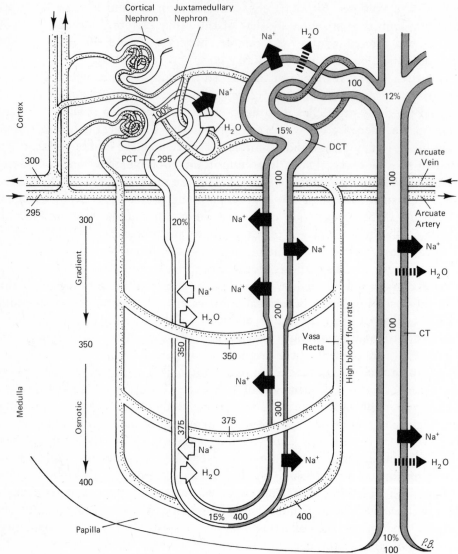

Figure 18-10 Diagram showing how the mammalian kidney produces a dilute urine when the body is hydrated and a concentrated urine when the body is dehydrated. Black arrows indicate active transport and white arrows passive flow. The number represent the approximate millios-molality of the fluids in the indicated regions. Percentages are the volumes of the forming urine relative to the volume of the initial ultrafiltrate. Dark nephron walls indicate relative impermeability to water, white walls free permeability, and interrupted walls those in which water permeability has been increased by the action of antidiuretic hormone (ADH). (a) When blood osmolality drops below normal concentration (ca. 300 mOsm for most mammals) excess body water is excreted. Excess hydration dimishes pituitary release of ADH, which renders the distal

(b) BODY DEHYDRATED — ADH PRESENT — SCANT, CONCENTRATED URINE

convoluted tubule (DCT) and the collecting tubule (CT) nearly impermeable to water. As a result of active sodium reabsorption a copious dilute urine is produced. (b) When blood osmolality exceeds normal levels water is conserved. Dehydration stimulates the release of ADH from the pituitary, which greatly increases the permeability of the DCT and CT to water. The reduction in forming urine volume in the DCT, and the enhanced osmotic flow of water into the medullary portion of the kidney from the CT produces a low volume, highly concentrated urine. During urine concentration blood flow through the vasa recta declines. As a result solutes that accumulate in the medullary interstial tissue are not rapidly flushed into the systemic circulation, and a steep osmotic gradient is formed. (Based on Netter [11].)

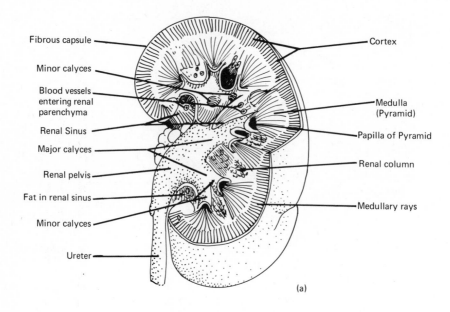

Fibrous capsule

Minor calyces

Blood vessels entering renal parenchyma

Renal Sinus

Major calyces

Renal pelvis

Fat in renal sinus

Minor calyces

Ureter

Cortex

Medulla (Pyramid)

Papilla of Pyramid

Renal column

Medullary rays

(a)

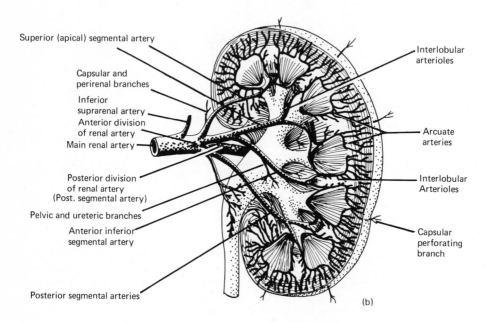

Superior (apical) segmental artery

Capsular and perirenal branches

Inferior suprarenal artery

Anterior division of renal artery

Main renal artery

Posterior division of renal artery (Post. segmental artery)

Pelvic and ureteric branches

Anterior inferior segmental artery

Posterior segmental arteries

Interlobular arterioles

Arcuate arteries

Interlobular Arterioles

Capsular perforating branch

(b)

Figure 18-11 Gross morphology of the mammalian kidney exemplified by that of a human. (a) Structural divisions of the kidney and proximal end of the ureter. (b) Renal artery and its subdivisions in relation to the structural components of the kidney. The renal vein (not shown) and its branches parallel those of the artery. (c) Enlarged diagram of a section extending from the outer

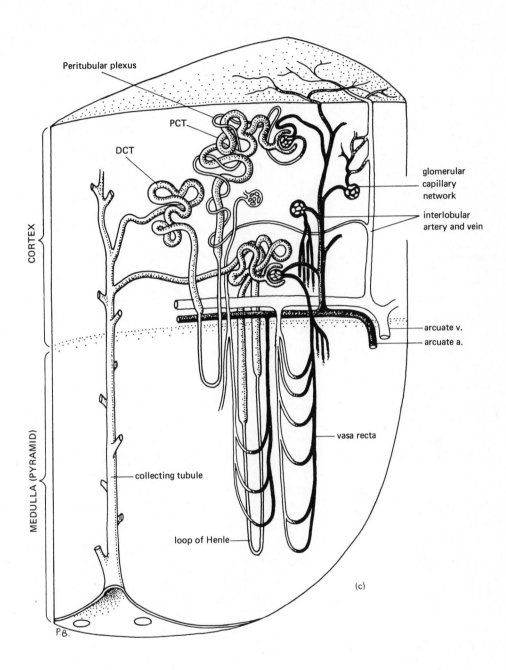

cortical surface to the apex of a renal pyramid, the renal papilla. The general relationship of the nephrons and blood vessels to the gross structure of the kidney can be appreciated in a comparison of these diagrams. (Modified after Netter [11] and Smith, H. W. [1956] *Principles of Renal Physiology*. Oxford University Press, New York.)

to produce a scant concentrated urine. The absence of ADH has the opposite effects. Alcohol inhibits the release of human ADH, induces a copious urine flow, and this frequently results in dehydrated misery the following morning.

The key to concentrated urine production clearly depends on the passage of the loops of Henle and collecting ducts through tissues with increasing osmolarity (Figure 18–10). These longitudinal osmotic gradients are formed and maintained within the mammalian kidney as a result of its structure, which sets it apart from the kidneys of other vertebrates.

The nephrons, collecting ducts, and attendant capillaries are spatially ordered within the matrix of kidney tissues (Figure 18–11). The glomerulus, PCT, and DCT of a single nephron are embedded in the superficial cortical region of the kidney. A hairpin turn along the nephron's length forms a *loop* of Henle. The descending limb of the loop departs radially from the kidney **cortex** and together with the limbs of other nephrons extends linearly into the **renal pyramid,** a conical subdivision of the **medulla.** Here the abrupt turn of the ascending limb of the loop of Henle causes it to lie parallel, and quite close, to its descending limb. Both portions of the loop of Henle are adjacent to a straight collecting duct that exits from the medulla at the tip of the papilla. Capillaries supplying this system of nephrons enter and exit the medulla only through the cortex. To do so the capillaries must also bend sharply back on themselves within the **renal pyramid.** Thus they form a series of afferent and efferent vessels called the **vasa recta,** with antiparallel or countercurrent flow. Substances at higher concentration in the ascending vessel (efferent) diffuse to the descending vessel. Sodium secreted from the ascending limb of the loop of Henle diffuses into the medullary tissues to increase their osmolarity (see preceding paragraphs). The final concentration of a mammal's urine is determined by the amount of sodium accumulated in the fluids of the medulla. Physiological alterations in the concentration in the medulla result primarily from the effect of ADH on the rate of blood flushing the medulla. When ADH is present, blood flow into the medulla is retarded. Another hormone, aldosterone, from the adrenal gland increases the rate of sodium secretion into the medulla to promote an increase in medullary salt concentration.

In addition to these physiological means of concentrating urine, a variety of mammals have morphological alterations of the medulla. Most mammals have two types of nephrons: those with a cortical glomerulus whose abbreviated loops of Henle do not penetrate far into the medulla and those with juxtamedullary glomeruli, deep within the cortex, whose loops penetrate as far as the papilla of the renal pyramid (Figure 18–11c). Obviously the longer, deeper loops of Henle experience greater osmotic gradients along their lengths. The flow of blood to these two populations of nephrons seems to be independently controlled. Juxtamedullary glomeruli are more active in regulating water excretion; cortical glomeruli function in ion regulation. Finally, some highly adapted desert rodents have exceptionally long renal pyramids. Thus, the loops of Henle and the vasa recta are extended and can produce a high differential in osmolarity from the cortical to the papillary ends. The maximum concentrations of urine measured from a given species of mammal correlate well with the length of its renal pyramids.

Some mammals have bean-shaped or lenticular kidneys (man, cat, rat); others have highly lobulate organs (cetaceans and some ungulates). Some mammalian kidneys have a single pyramid (rodents); others have fused papillae (dog) or multiple pyramids and papillae (man, pig). In spite of these gross anatomical differences, the tubular organization of the mammalian kidney is consistent. Such uniformity undoubtedly reflects the exceptional capacity of this organization to meet the ionic, osmotic, and excretory demands of diverse mammalian life styles.

From what we have described, early mammals were primarily insectivores and carnivores with a high energy demand. Their diet was rich in protein which, when metabolized, led to large amounts of urea. To void this rapidly accumulating nitrogenous waste, considerable water is required unless a means of concentrating urea is available. The unique concentrating power of the mammalian kidney may well have been an early adaptation to the accumulation of metabolic wastes from high levels of activity and the water demands of homeothermy. Once evolved, the ability to concentrate urine is a preadaptation to the invasion of very arid habitats where water to replace losses is not available.

References

[1] Barrette, C. 1977. Fighting behavior of Muntjac and the evolution of antlers. *Evolution* 31:169–176.

[2] Gordon, M. S., G. A. Bartholomew, A. D. Grinnell, C. B. Jørgensen, and F. N. White. 1977. *Animal Physiology: Principles and Adaptations*. Third edition. Macmillan, New York. Readers are referred to the chapters on thermal regulation by Bartholomew. A comparative account of how animals regulate body temperatures with emphasis on birds and mammals.

[3] Ham, A. W. 1974. *Histology*. Seventh edition. J. B. Lippincott, Philadelphia.

[4] Harrison, R. J. and E. W. Montagna. 1973. *Man*. Second edition. Appleton Century Crofts, New York. An excellently illustrated work, especially on skin structure and function.

[5] Heinrich, B. 1977. Why have some animals evolved to regulate a high body temperature? *American Naturalist* 111:623–640.

[6] Kermack, D. M. and K. A. Kermack. 1971. *Early Mammals*. Suppl. 1, Zoological Journal of the Linnaean Society of London, 50:1–203. A collection of articles on various aspects of early mammal fossils. A series of photographs are useful in visualizing the way in which the teeth of these early forms functioned.

[7] Lillegraven, J.A. 1974. Biological considerations of the marsupial-placental dichotomy. *Evolution* 29:707–722. The work presents the view that the marsupial mode of reproduction was essentially antecedent to the placentals. Those interested should also read Kirsch, J.A.: Biological aspects of the marsupial-placental dichotomy: a reply to Lillegraven, *Evolution* 31:898–900, for an interesting view that essentially suggests that both groups have evolved successful reproductive strategies, each of which has advantages and disadvantages, and that marsupials cannot be considered as antecedents to placentals only because their reproduction is considered more "primitive."

[8] Maderson, P. F. A. 1972. When? Why? and How? Some speculations on the evolution of the vertebrate integument. *American Zoologist* 12:159–171.

[9] McNab, B. K. 1973. Energetics and distribution of vampires. *Journal of Mammalogy* 54:131–144.

[10] Modell, W. 1969. Horns and antlers. *Sceintific American* 220(4);114–122.

[11] Netter, F. H. 1973. *The CIBA Collection of Medical Illustrations.* Vol. 6, *Kidneys, Ureters and Urinary Bladder.* CIBA Publications, Summit, N.J. A beautifully illustrated volume from a well-known series on human anatomy and physiology.

19

Diverse Adaptations of Mammals

Synopsis: After having examined the basic characteristics of mammals and how they contribute to mammalian life, we are prepared to turn to the adaptive radiation of modern mammals. A few selected aspects, when carefully surveyed, given an appreciation for the versatility of mammalian structure and function. The ways mammals satisfy their nutritional demands, the ways in which they sense the world around them, and the ways in which they reproduce are three areas of particular importance. Ultimately genetic information must underlie these and all other complex adaptations. Thus we consider the genetic bases of some mammalian characters, and the process by which the extensive genomes required to code for advanced vertebrate life may have evolved.

The teeth of mammals are unique and highly adaptive structures whose radiations (including remarkable convergences) are basic to mammalian evolution. Not only tooth, but also jaw, muscle, and locomotor anatomy reflect trophic adaptations. A case can be made for plant evolution having stimulated changes in herbivore evolution that, in turn, affected carnivore evolution. Similar evolutionary "advances" in mammals also occurred in the evolution of the brain, especially the cerebral hemispheres which increased greatly in size. Mammals evolved from reptiles that could process three-dimensional visual reconstructions of the world. The original sensory demands made upon a nocturnal mammal-like reptile may have acted as an evolutionary stimulus to brain enlargement. Considerations of the auditory specialists that echolocate, (bats and cetaceans), demonstrate the analytical power of the mammalian nervous system.

Reproduction stands at the base of the complex adaptations of mammals; a prolonged and plentiful energy supply to the developing mammal is essential to accomplish its complicated construction. The placenta and, subsequently, lactation provide this energy under the control of neuro-hormonal patterns of considerable complexity. Here as elsewhere there is diversity among the mammals, for the metatheria (marsupials) and eutheria (placentals) are quite different

685

with regard to reproductive patterns. Sex in its physiological aspects and sex determination are in general less diverse and plastic in mammals (although often more complex) as compared with ectothermic vertebrates. This may result from selection for rigid sex determining mechanisms in mammals as well as in birds because of the complexity of parental care of the young. The amount of DNA necessary to blueprint for mammals exceeds that in teleost fishes by as much as five times. These differences seem to have their origin in a relatively small number of genetic duplications of the total DNA. Examination of the genetic material of living vertebrates may allow us to reconstruct the history of these changes. These evolutionary windfalls rapidly increased the number of genes available for mutation while providing for the viability of the mutant.

19.1 MAMMALIAN TROPHIC BIOLOGY

As we have already seen in birds and mammals, the benefits of a constant high body temperature and a steady, predictable internal milieu are great, but so are the costs. An efficient means of obtaining and processing food to sustain high energy needs is absolutely essential to an endotherm. Mammals meet these needs in different ways from birds, and this is seen first of all in their trophic, or nutritional, adaptations. Their characteristic and highly developed trophic apparatus includes their teeth, jaws, muscles involved in chewing, and the alimentary canal.

19.1.1 Dentition

The most primitive living eutherians, the insectivores, feed upon the flesh of invertebrates (especially insects) and those vertebrates small and soft enough for them to kill. The skeletons of the earliest fossil eutherians from the late Cretaceous suggest that their feeding habits were similar. From these stocks have evolved the variety of trophic types found among the living mammals, including carnivores capable of bringing down prey many times their own size, strainers of minute floating organisms, nectar drinkers, and grazers upon dry and brittle grasses. In the evolution of mammals, as in the evolution of various reptilian and osteichthyan lines, generalized carnivores form the base of the trophic radiations that lead to various forms of herbivorism. The special dentition and digestive adaptations needed to consume plants directly have apparently locked mammalian herbivores out of further trophic diversification. One of the few exceptions is the evolution of man, whose ancestors and living relatives, the primates, are predominantly frugivores (fruit eaters) and only occasionally and opportunistically consume flesh. The effects of cultural rather than biological evolution on this dietary reversal by humans will be discussed in Chapter 22.

The eutherian placental mammals are usually divided into 17 orders, largely on the basis of dentition. Only ten of these orders contain more than a score of species (Table 18-1, page 643). Although mammalogists agree about the insectivorelike nature of the earliest eutherians, the interrelationships of the other orders is generally presented in a rather unsatisfying manner. A few orders have a fossil record indicating more recent common ancestry. But most are traced back, with increasingly vague evidence, to independent origins from generalized late Mesozoic insectivores. The only exceptions are the ungulates (orders Artiodactyla and Perrissodactyla) and their relatives the subungulates (orders Hyracoidea, Proboscidea and Sirenia). This assemblage of obligate herbivores is thought to have derived, albeit independently, from the usually hoofed but omnivorous early Tertiary order Condylarthra. In addition, the primitive ancestors of the order Carnivora are thought by many to have been related to the condylarths. Thus, we are again faced with a series of distinct phyletic lines leading back to an uncertain junction. Rather than examine the living mammals from this obscure phylogenetic pattern (Chapters 15 and 18.1), a reasonable alternative is to deal with them as ecological specialists, an approach common throughout this book.

The skulls of these larger orders of living mammals are illustrated in Figure 19-1. The arrangement in this figure illustrates a possible *functional* classification of recent eutherians based on their trophic adaptations and, especially, their teeth; it is not phylogenetic. Because digestion is basically an enzymatic chemical process, increasing the area for contact and penetration of the digestive enzymes allows rapid digestion of food. Most of the vertebrates we have studied have used their teeth to catch, grasp, or crop food but not to process it. The anterior teeth (**incisors** and **canines**) of mammals retain this basic function, but the posterior teeth (**premolars** and **molars**) are modified to **masticate** foodstuffs. During mastication, the food is repeatedly worked between interlocking **occlusion surfaces** of the upper and lower dentition by the action of the mobile, sensitive tongue and a characteristic mammalian facial structure, the cheeks. The food thus becomes a loose pulp that readily absorbs gastric juice. Archosaurs and birds show an analogous adaptation in the form of a muscular gizzard derived from the stomach. Birds periodically ingest small stones that function with the thick-walled gizzard as a mill for food mastication.

In a mammal, each tooth (Figure 19-2) is attached in its jaw socket by a type of bone known as **cement**. Cement is hard and rigid like other bone, but wears rapidly when exposed. The core of a tooth is hollow and filled with nerves, blood vessels, and the cells that produce the inner coat of the tooth called **dentine**. Dentine contains much more mineralized material and less organic matter than bone and hence is harder, heavier, and more resistant to wear than cement. The **crown**, which is that portion of the tooth exposed above the gums, is encased in a unique substance that caps the tooth and is the hardest, heaviest, and most friction-resistant tissue evolved by vertebrates. This tissue is **enamel** and, unlike both cement and dentine, is ectodermal, not mesodermal, in origin. In addition, enamel is totally acellular and thus cannot regenerate. It is nearly devoid of organic matter and is composed of large, uniformly oriented calcium phosphate crystals.

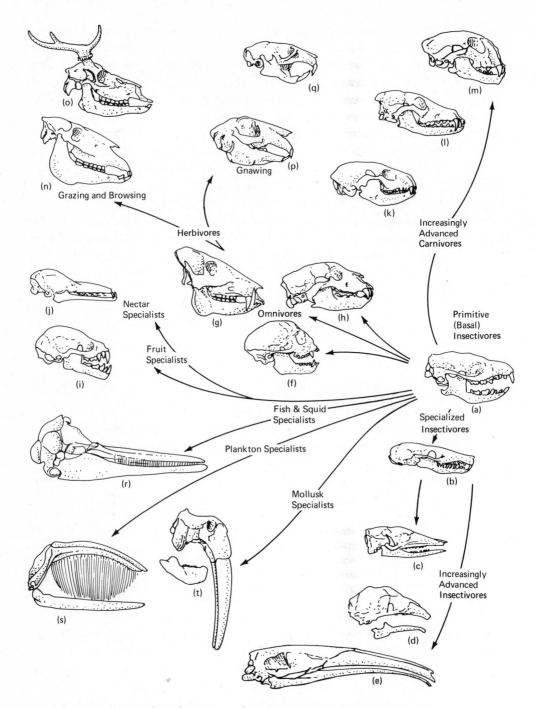

Figure 19-1 Trophic adaptations of teeth and skulls of mammals.

The earliest therian mammals had a dentition similar to that found in relatively unspecialized insectivores living today. The (a) hedgehog, *Erinaceus* has a dentition with piercing cusps on most of the teeth. More specialized is the battery of sharp teeth for piercing and holding worms and insects of (b) the mole, *Scalopus*. Mammals adapted specifically to feed on social insects in their

The teeth of a generalized mammal may be designated from anterior to posterior, on each side of each jaw, as: **incisors** (usually two to five), **canines** (never more than one), **premolars** (generally two to four) and **molars** (variable but often three). In most mammals, two sets of teeth occur during the animal's lifetime, a condition known as **diphyodonty**. The first set, although capable of dealing with the diet characteristic of the adult of the species, has fewer, smaller teeth than can be accommodated in a young animal's small jaws. The second set, when complete, is larger and appropriate to the adult jaw size. The lacteal ("milk") or deciduous dentition appears first in a generally regular order from anterior to posterior, although canines may not erupt until after the premolars. The molars are adult teeth in the sense that they function throughout life, but ontogenetically they are posterior, unreplaced members of the lacteal dentition. It is often difficult to distinguish premolars from molars in an intact skull. Collectively called **cheek teeth**, both act in mastication whereas the more anterior teeth are concerned with acquistion and bite-sized division of the food.

Because the texture and quality of mammalian foods differ enormously from species to species, the cheek teeth are variously adapted. Generalized mammalian dentitions have cheek teeth called **brachyodont** (Greek *brachys* =

---

nests show a tendency to reduction or loss of teeth and increasing snout length: (c) armadillo, *Dasypus*, (d) anteater, *Cyclopes*, and (e) giant anteater, *Myrmecophaga*.

Omnivorous mammals from many orders retain some piercing and ripping cusps in the anterior teeth but have flat broad crushing cusps posteriorly. Examples of these forms are the (f) marmoset, *Sagarinus*, (g) peccary, *Tayassu*, and, (h) bear, *Ursus*. A similar dentition with much enlarged anterior biting teeth occurs in the (i) fruit-eating bat, *Aretbius*, which bites chunks from fruit and crushes the pulp for its juices with the broad flat posterior teeth. Another bat, (j) *Choeronycterus* feeds on nectar with a long tongue and has greatly reduced dentition.

Carnivores that feed on increasing amounts of flesh, cut or sheared from the carcasses of prey, have a progressively greater development of the fourth upper premolar and first lower molar into a scissorslike pair of shearing blades, known as the carnassial apparatus. The crushing posterior molars become less important, and the carnassials enlarge in the carnivore series (k) raccoon, *Procyon*, (l) coyote, *Canis*, and (m) mountain lion, *Felis*. (See also Figure 19–3.)

Herbivores of two very different types show similarities in the great relative size of the flat grinding cheek teeth and the development of a gap between these cheek teeth and the anterior food procuring teeth. The front teeth of (n) the grazing horse, *Equus*, and (o) the browsing deer, *Dama*, clip off vegetation, which the tongue passes back to the grinding apparatus. The gnawing incisors of (p) the jackrabbit, *Lepus*, and (q) the woodrat, *Neotoma*, procure food in a very different way but show many convergent similarities of the cheekteeth. (See Figure 19–6.)

Marine habits have produced some highly specialized, unique forms such as (r) the fish trap dentition of the dolphin, *Delphinus*, (s) the toothless plankton straining right whale, *Eubalaena*, and (t) the mollusk digging and pulverizing dentition of the walrus, *Odobenus*.

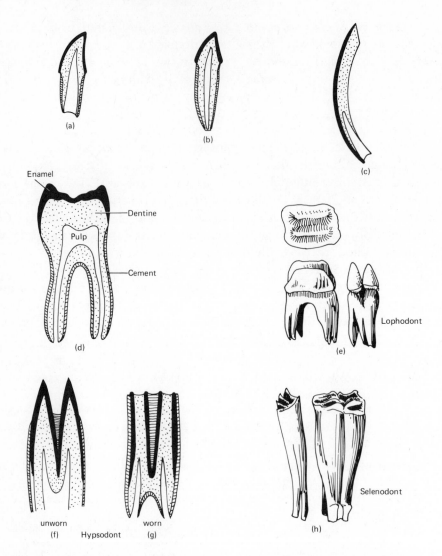

Figure 19-2 Various mammalian teeth showing structure and distinguishing characters. (a) Immature incisor tooth with wide open pulp cavity indicative of unrestricted growth. (b) Mature incisor with narrow canallike pulp cavity restricting nutrient supply and growth. (c) Incisor of the rodent, which has a persistently open pulp cavity and grows throughout life. Enamel occurs only on the outer face and wear produces a self sharpening chisel out of the tooth. (d) A typical molar tooth with multiple roots here shown in the constricted pulp cavity condition of a mature, nongrowing tooth. The low rounded cusps identify the tooth as **bunodont**. (e) If the cusps of the cheekteeth fuse into ridges, useful in grinding plant material, a **lophodont** tooth is formed. (f) As an adaptation to the wear a cheek tooth experiences from grinding a fibrous plant diet, cement may be deposited in extremely tall, **hypsodont** teeth. (g) As a hypsodont tooth wears, the various hardnesses of tooth material wear differentially to maintain an uneven grinding surface. (h) Artiodactyl hypsodont teeth with numerous ridges derived from worn crescent shaped cusps are efficient plant grinders known as **selenodont** molars.

short, *odontos* = tooth) with rectangular crowns that do not protrude much above the gums. The occlusal (grinding) surface is not smooth, but is marked by one to several sharp **cusps**, which interlock when the jaws close (occlude) (see Chapter 18). This cusp pattern is referred to as **tubercular** (Latin *tuber* = a knob or hump) and is appropriate for omnivorous and insectivorous habits because it permits both piercing and crushing of the food.

Carnivores that eat muscle and viscera must be able to shear or cut away chunks of flesh from a large carcass. Once such pieces are obtained, they are readily digested without much mastication. Hence, the occlusal surfaces are sharp and knifelike and present little grinding surface areas (Figure 19-1 l and m). The teeth of advanced carnivores are called **sectorial** (Latin *sector* = one who cuts). Sectorial cheek teeth reach their height of development in a shearing, **carnassial** apparatus of the fourth upper premolar and the first lower molar, which slide past one another like a pair of scissor blades and cut flesh most effectively (Figure 19-3).

Crushing mastication is of so little importance to cats, for example, that in the upper jaw only a vestigial first molar occurs. Dogs and their relatives are carnivores adapted to a more efficient use of a kill and in many ways are pre-adapted to the life of a scavenger. This adaptation is immediately apparent up-on examination of their cheek teeth. The anterior cheek teeth are sectorial and the carnassial apparatus occurs between the upper fourth premolar and the lower first molar, but well-developed molars with tubercular cusps follow these slicing teeth and allow dogs to crush bones for their marrow content. The structure of the teeth explains why a dog often holds a bone between its paws and gnaws with the bone far back in the corner of its mouth. When meat is on the bone, the dog does not place the bone as far back in its mouth and uses the carnassials to slice off the meat.

Diverse mammals such as bears (Order Carnivora, Figure 19-1h), swine (Order Artiodactyla, Figure 19-1g), and man and many monkeys (Order Primates, Figure 19-1f), have a more varied diet, which includes plant matter as well as animal tissue. Because of their tough cellulose cell walls, many plant tissues are resistant to efficient digestion. Grinding plant matter between rough surfaces is the standard methods of disrupting cell walls. The cheek teeth of these omnivores are brachyodont (low crowned) and **bunodont** (Greek *bounos* = a hill or mound) meaning that the cusps are rounded (Figure 19-2d). These teeth function for both animal and plant material when the former is small and soft enough to be ground (not cut) into digestible pieces and the latter is not tough grass stems and leaves but softer roots, tubers, and berries.

Tougher plant materials such as leaves, twigs, and stems require more specialized dentition. The ungulates, or hoofed mammals (Figure 19-1n and o), as well as the rodents (Figure 19-1q) and the lagomorphs (rabbits, hares; Figure 19-1p) are extremely successful herbivores and their teeth are highly modified. The tapir (Order Perissodactyla) has a **brachyodont** dentition, the cusps highly specialized into transverse, interlocking ridges that grind plant material during lateral oscillation of the jaw (Figure 19-2e). Transversely ridged

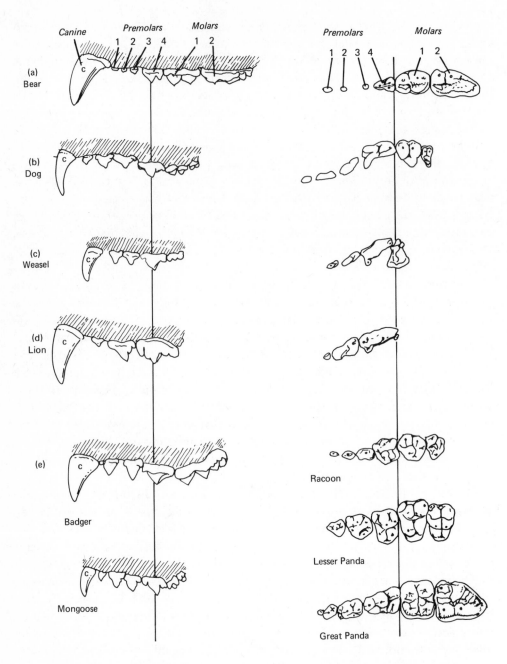

Figure 19-3 Partial dentition of the upper left jaw of various members of the order Carnivora. The vertical lines indicate the main cusp of premolar 4, the carnassial, in the lateral views and the posterior border of the carnassial in the occlusal views. A series grading from omnivore to advanced carnivore (a to d) shows how the crushing, bunodont molars are reduced and eliminated and the cusps of the fourth premolar aligned and fused to form a blade.

(a) bear, *Ursus*; (b) dog, *Canis;* (c) weasel, *Mustela;* (d) cat, *Felis.* (e) Similar dentitions may be developed independently using different elements. The posterior crushing elements of the badger, *Meles* (above), and the mongoose *Her-*

teeth are called **lophodont** (Greek *lophos* = a crest or ridge) and their development is carried to an extreme in the enormous cheek tooth surface area of the elephant (order Proboscidea).

The horse (Order Perissodactyla, Figure 19–1n), also has a lophodont dentition, but in modern species this pattern is modified in the adult by the additional development of longitudinal ridges that greatly obscure the primitive pattern. Two further specializations for herbivorism are evident in the dentition of the horse: the crowns stand quite high above the gums. The **hypsodont** condition (Greek *hyps* = high; Figure 19–2 f and g), and the occlusal surfaces of the cheek teeth are not entirely capped with enamel but show complex ridges composed of dentine adjacent to ridges of enamel and areas of exposed cement. Similar adaptations can be seen in the dentition of cattle (order Artiodactyla). In artiodactyles, however, the ridges are all derived from longitudinal cusps and are termed **solenodont** (Greek *solen* = a channel or grooved tile; Figure 19–2h). The valleys between the cusps in the developing teeth are filled with cement before they pierce the gum. As these complex teeth wear down, the different hardness of the cement, enamel, and dentine produces a rough, uneven, self-sharpening surface.

The incisors, because of their most anterior position, are used in clipping or cutting bite-sized pieces of food. In carnivores, with their highly specialized posterior carnassial apparatus, the incisor's transverse cutting ridge is present but is reduced in size and importance. Omnivores, such as the pig, and moderately modified herbivores, such as the tapir, show a more substantial incisor development coupled with an increasingly large gap between the anterior teeth and the cheek teeth. This gap places the cutting mechanism away from the face so that the snout may penetrate narrow spaces for browsing. The toothless gap is called a **diastema** and is characteristic of herbivores.

Canines are used for stabbing, slashing, holding, and manipulating food and are highly developed in carnivores. The extinct lines of saber-toothed cats and the convergent marsupial sabertooths (Figure 19–11) seem to have taken this development to an extreme. Omnivores generally have well-developed canines as do less specialized herbivores. Some ungulates have highly modified canines that are very similar in shape and function to incisors and have migrated far forward to join the incisors in a single, functional cutting edge, leaving a very large diastema in both jaws. In artiodactyls, such as cows and deer, the upper incisors are usually absent, as are the upper canines. The lower incisors and incisiform canines act against a strongly cornified (calloused) palatal plate.

---

pestes (below), are the result in the former of enlargement of the first molar and in the latter the retention of both molars 1 and 2. (f) Not all carnivores eat meat. Some are extraordinarily specialized herbivores! The series of cheek teeth seen in occlusal view shows the progressive molarization of the premolars and enlargement of the molars in the omnivorous racoon, *Procyon* (above), the herbivorous lesser panda, *Ailurus* (middle), and the bamboo eating giant panda, *Ailuropoda* (lower). (After Peyer [12].)

This apparatus, in which the very protrusible tongue plays a significant role, enables the rapid plucking of a large quantity of grass, with the tongue and lips pulling blades toward the clipping jaw tips. The food thus rapidly procured is later regurgitated, perhaps in a safe resting place away from predators, thoroughly masticated, and then returned to the stomach for the second time and final digestion. Probably of even greater adaptive significance than possible predator avoidance is the fermentation which occurs in the anterior part of the enormous, specialized four-chambered stomach of these **ruminants** (Figure 19–7). During fermentation, cellulose is digested by bacteria and converted into substances digestable by the mammalian host as well as by intestinal microbes. Later these microbes, too, are digested and absorbed by the host, enriching the quality of the originally ingested herbage.

Complex specialization of the dentition and feeding apparatus are not limited to the ruminants, however. An examination of some of the other mammals with interesting adaptations serves to emphasize how mammalian trophic structures contribute to the success of the class. We will compare the variations within a single phyletic line and between convergent forms. We will also discuss species that use their teeth for functions other than food procurement and processing, and of groups which have abandoned heterodont teeth altogether.

19.1.2 Special Dental and Trophic Adaptations

Bats

Although bats (Order Chiroptera) constitute the second largest order of living mammals (only the order Rodentia accounts for more species), we are generally unaware of their presence. The bats are yet another example of adaptive radiation. This radiation occurred after the achievement of aerial locomotion, coupled with the ability to navigate at night (thus eliminating much competition with birds). Like birds, bats possess a wide variety of specializations but remain morphologically conservative because of the strict limitations placed on a flying vertebrate. Differences in the behaviors of various bats can often be associated with differences in their diet.

The first bats, and many modern species, were insectivorous and had a primitive mammalian dentition. The teeth are characterised by having well-developed sharp cusps that can penetrate the chitinous exoskeletons of insects. These cusps are also adapted for grasping and holding large insects, such as moths. Not all bats restrict their flesh-eating habits to small, bite-sized insects. Several carnivorous forms feed on birds, smaller bats, and mice as well as insects. Both the teeth and jaws are more massive than those of insectivorous forms. The cheek teeth are broad and flat yet retain cutting cusps, much like those of the cheek teeth of the dog. Several other specialized feeding habits have evolved in carnivorous bats. Certain **piscivorous** (fish eating) bats hunt over the sea and fresh-water ponds and streams. The bats skim the surface of the water, sometimes in groups, catching small, surface-swimming fish with the long sharp claws of their feet. The fish may be eaten in flight or carried in the cheek pouches to a roost to be eaten while the bat is resting. The cusps of the cheek teeth in fish-eating bats are prominent and sharp.

Most notorious of bat feeding habits is the blood feeding (sanguivory) of the New World vampire bats (Chapter 18.3.3). The upper incisors and canines are enlarged to form a razor-sharp cutting edge. The cheek teeth are greatly reduced and without crushing surfaces. The stomach is slender and saclike. Vampire bats do not leave their caves until after dark. So far as is known, they feed only on fresh blood, alighting on or near the prospective victim. Vampires attack areas devoid or nearly devoid of hair or feathers, including the naked skin around the anus and vagina, the ears and neck of cattle and the wattle and comb of chickens. In preparation for feeding a vampire makes a quick, shallow bite removing a small plug of skin. The prey is usually undisturbed, for the bite is practically painless. Vampires never bite deeply or struggle with a victim. The tongue and a deep groove in the lower lip forms a tube. The base of the tongue is thought to produce pressure reductions in the tube, drawing blood into the mouth. The bats' saliva contains substances that inhibit clotting and the bat ingests blood until its stomach is full.

Less spectacular but equally worthy of note are the **Frugivorous** (fruit-eating bats (Figure 19–1i). The Order Chiroptera is divided into two suborders: the nearly cosmopolitan and primarily insectivorous Microchiroptera and the Old World, tropical, fruit eating bats of the suborder Megachiroptera. But some microchiropterans have also adapted a fruit eating habit, especially in the New World tropics, and the convergences between the fruit bats in the two suborders are impressive. Frugivorous bats have broad, flat, nearly cuspless molars used to crush fruit. Generally only the juices are swallowed and the pulp and seeds spit out. Another group of plant-eating bats are the nectar-feeders. Nectar-feeding bats (Figure 19–1j) resemble vampires in the fluid nature of their diet, and the cheek teeth are similarly reduced. The tongues of nectar-feeding bats, with which they probe deeply into flowers, are like those of nectar-feeding birds and marsupials such as the honey possum. Their tongues are very long, extremely protrusible, and covered to tipped with nectar-holding bristle-like papillae.

Homodonts and Toothless Mammals

We have discussed the complicated interactions of a variety of different tooth types, often within the dental battery of a single individual. Such specialized **heterodont** dentition is responsible for much of mammalian adaptive success. Nevertheless, some feeding specializations do not require highly differentiated dentition. Some of the most specialized feeding habits among mammals are associated with **homodont** dentition or even extensive reduction of teeth.

The toothed whales (Order Cetacea, Suborder Odontoceti) are a good example of mammalian homodont dentition. Most toothed whales feed on fishes, and their teeth consist of a long series of nearly identical sharp cones (Figure 19–1r). This dental pattern is convergent with the fish-trap type of dentition already described in crocodilians and holostean gars. A second type of specialized dentition in cetaceans is represented by the baleen or whalebone whales (Suborder Mysticeti; Figure 19–1s). In these animals the teeth have been replaced by an entirely different structure. Sheets of fibrous, stiff, horn-like epidermal derivative known as baleen extend downward from the upper jaw. Baleen whales are filter-feeders, straining small organisms known collectively as plankton from the water by use of their baleen sieves. The ten species of baleen whales include the largest whale (and the largest vertebrate that has ever lived), the blue whale. This giant may reach a length of 31 m and a mass of more than 160,000 kg.

Another feeding specialization little dependent upon tooth differentiation is **myremecophagy** (Greek *myremekos* = an ant; *phago* = to eat). Myremeco-phagous forms feed on ants and termites, which are generally soft-bodied, small and incapable of flight as a means of escape. Hence, in numerous unrelated mammalian groups, flat, crushing teeth or no teeth at all are to be found. These dental peculiarities are coupled with long, mobile and worm-like tongues with exceptional protrusibility. Enlarged salivary glands produce a viscous, sticky secretion which coats the tongue. Myremecophagous mammals also have elongate snouts and digging adaptations. The aardvark of sub-Saharan Africa (Order Tubulidentata) illustrates these adaptations. It has a long snout and flat, columnlike, soft (enamelless) cheek teeth. Though unrelated, armadillos (order Edentata) shows a similar condition with further reduction of the dentition and relatively weak jaws (Figure 19–1c). Finally the giant anteater, a relative of the armadillo, lacks teeth completely (Figure 19–1e).

Dental Convergence

Another example of specialized dentition demonstrates convergent evolution in two distinct clades which, because of their morphological similarity, were long considered closely related. The rabbits (Order Lagomorpha; Figure 19–1p) and the rodents (Order Rodentia; Figure 19–1q) have a greatly enlarged pair of incisors in both the upper and the lower jaw. These incisors are adapted for gnawing through hard plant coverings to reach tender material inside as well as for nibbling grasses and shrubs. The incisors of both lagomorphs and rodents continue to grow throughout life. Behind the incisors, a long diastema separates the gnawing apparatus from the plant-crushing cheek teeth. A soft fold of cheek skin is puckered inward across the diastema while gnawing and closes off the mouth from flying particles of bark and wood. The cheek teeth in both rodents and rabbits are complex and in many ways are convergent with those of herbivores such as the horse and cow.

At this point the similarity between lagomorphs and rodents abruptly ends. The architecture of the rest of the skull, axial skeleton, and the internal soft anatomy is very different; rabbits are light and delicate in appearance, rodents the size of a lagomorph are generally heavy and massive. Even the apparently similar dentition shows marked differences. The incisors of the rodent have enamel only on their anterior surfaces (the enamel is often stained brown and its limits can easily be determined by a change in tooth color). Because enamel wears less rapidly than the dentine, the teeth are self-sharpening. Rabbit incisors, on the other hand, are completely encased in enamel (except where it wears off at the tips). In addition, lagomorphs have a second set of small incisors immediately behind the first.

Paleontological evidence indicates that the rabbits and their relatives are related to the line that led to modern hoofed mammals such as the horse and cow. The origin of rodents is obscure; presumably they were derived from some insectivorous, placental stock. If an herbivorous common ancestor for such grossly different forms as ungulates and lagomorphs occurred, it may be an exception to the general principle that major radiations start from carnivorous groups.

Nontrophic Functions of Teeth

Many wild swine and some domestic types have greatly enlarged and often recurved upper and lower canines (Figure 19–1g). The shape and size of these teeth, as well as the food habits of the swine family, indicate that their canines must be functionally different from those of carnivores. Observation reveals that they are used for rooting and digging in the soil for the nutritious storage roots and tubers of plants. In addition, these tusks are larger in the males of many species than in the females. Some wild boars are famous for their upper tusks, which form complete loops or, in one species (the babirussa), pierce the upper lip, emerge on the upper side of the snout and arch toward the forehead. These tusks have lost even their digging function and must be considered sexually dimorphic characters, probably connected with aggressive and/or sexual display.

Tusks in other mammals are also modified teeth. The canines of the walrus (order Pinnipedia) are massive (Figure 19–1t). Walruses are exclusively marine, but haul out onto drifting ice or onto the shore using their tusks as levers to lift their massive bodies from the water. The great weight of the walrus skull and tusks can be energetically supported only in a marine environment and a major function of the tusks may be in social communication as well as benthic invertebrate gathering. Other well-known tusks are those of elephants. Here the incisors of the upper jaw form the tusk, which is pure dentine (ivory). Both sexes of the African elephant have tusks, but female Indian elephants are usually tuskless. The elephant's tusks are used in defense and to hold down branches pulled into reach by the trunk for browsing.

In primates, especially the various baboons and mandrills and the apes such as the gibbon and orangutan, the canines are enlarged and are larger in males than in females. By rolling back the very flexible lips, these primates present a fierce aggressive display. It has been suggested that the apparently unnecessarily large roots of human canines are a remnant of teeth that once functioned in these displays.

19.1.3   Masticatory Apparatus

Teeth, of course, are only part of a mammal's adaptation to a specialized diet. A mammal must have other skeleton-muscular adaptations allowing it to utilize its teeth effectively in mastication.

Many variations in mammalian mastication can be resolved by the study of five pairs of muscles, their sites of origin and insertion, and the articulation between the mandible and the skull. An omnivore such as man may be used to illustrate the relatively unspecialized state of these characters (Figure 19–4). The lower jaw of mammals is depressed by the paired **digastric** muscles (having two fleshy parts; Latin *di* = two, *gaster* = belly) which insert on the inner ventral borders of each side of the mandible. Each digastric passes posterior to the angle of the jaw (where the muscle constricts to a ligamentous neck) and then angles dorsally to originate in the ear region. The dual fleshy parts and angled orientation of the digastric muscles may seem a less than perfect design, but the depression of the lower jaw requires little force. The ligamentous neck occurs at the point of inflection around the angle of the jaw and resists friction and improves the mechanical advantages of the muscle fibers.

Closure of the mouth and mastication are two very different processes, and four pairs of muscles are involved in the latter. The **temporalis** originates on the skull roof and inserts on the coronoid process of the mandible. In addition another bilaterally paired muscle, the **masseter**, originates on the zygomatic arch and inserts on the posterior half of the lateral surface of the mandible. The fibers run obliquely from the zygoma posteriorly and down to the mandible. A final set of muscles of mastication is hidden from superficial view but is extremely important in adding new dimensions to the possibilities for mastication. These **pterygoideus** muscles originate on the base of the skull posterior to the palate; the fibers run laterally and obliquely to insert on the medial surface of the angle of the mandible. In man the pterygoid is divided into a lateral and a medial portion on each side. The origin of the human pterygoid complex is the base of the skull near the posterior portion of the throat. The insertion of the lateral pterygoids is very near the mandibular condyle and the jaw's articulation with the skull. Contraction of the muscles of one side pulls the mandible toward the contralateral side of the head. Acting together, they protrude the lower jaw. The pterygoid complex allows the lower jaw to be moved in a grinding rotary path relative to the upper jaw or to apply force on one side of the jaw only.

The shape of the lower jaw (mandible) and of the mandibular condyle and its fossa, as well as the orientation of the fibers in these masticatory muscles, clearly reflect adaptation to different functions. Some mammalian jaws have the condylar processes in or very near the occlussal plane (Figure 19–4). Such jaws close like scissors when the strongly developed temporalis and masseter muscles contract. The posterior teeth come into apposition before the more anterior ones. Other mammals have condylar processes high above the occlusal plane and the jaws close in a nearly parallel fashion, the occlusal surfaces approaching each other at about the same time all along the tooth row. The former orientation is ideal for slicing and cutting and is characteristic of carnivores. The latter crushes and, with the addition of lateral motion, permits grinding along the whole of the tooth row.

Because carnivores generally attack, hold, and kill with well-developed canines, it is not surprising that they usually have large temporalis muscles (Table 19–1) and prominent coronoid processes. The temporalis fibers converge on a stout tendon running anteroventrally to the coronoid process. This places maximum force near the tips of the jaw. The mandibular condyles of carnivores are cylindrical bars set deeply in skull fossae, which allow no anterior-posterior movement and very little lateral shifting of the lower jaws. These features help prevent dislocation of the jaw by struggling prey. In contrast, both the temporalis muscles (Table 19–1) and the coronoid process are reduced in highly specialized herbivores, but the ramus of the jaw is very deep to accommodate insertion of the large masseter muscles needed to apply force to the cheek teeth during mastication. The mandibular condyles of herbivores are of various shapes and orientations, permitting anterior-posterior and lateral movement. The human jaw, condyles, and musculature are intermediate between these extremes. Such generalizations, however useful, do hide very interesting functional differences within the broad dietary categories of omnivores, carnivores, and herbivores, as the variations in muscle mass seen in Table 19–1 indicate.

Although it is impossible to discuss the feeding adaptations of mammals in greater detail here, some very interesting evolutionary phenomena are beginning to be unraveled by studies of paleoecology and the adaptive evolution of entire faunas and floras. One such story is that of the coevolution of predators and prey during the changing climate of the Cenozoic.

Table 19-1. The Weights of the Jaw-Closing Muscles of Some Mammals. (Weight of each muscle as a percentage of the total)

	Temporalis	*Masseter*	*Pterygoideus*
Carnivores			
Tiger	48	45	7
Bear	64	30	6
Dog	67	23	10
Herbivore			
Zebra	10	50	40
European bison	10	60	30
Horse	11	57.5	31.5

From Alexander [1968] after Becht.

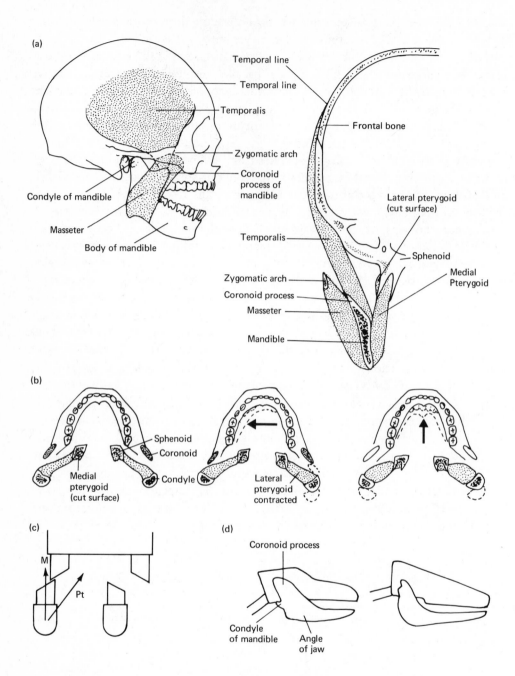

Figure 19-4 Mastication muscles and jaw modifications.

(a) The main muscles of biting and crushing are shown as they appear in the human in lateral view (left) and section of the skull and mandible at the level of the coronoid process (right). Because of the relative positions of their origins and insertions, the muscles differ in where they apply force during the bite. The **temporalis**, pulling via the coronoid process, tends to develop maximum force at the anterior portion of the jaw where the upper and lower canines and incisors

19.1.4   Coevolution of Plants and Mammals

Digestive Apparatus

The impact that grazing herbivores can have on vegetation is well known. Apparently, specialized herbivores have had as much impact on the evolution of plants as predatory carnivores have had on the herbivores themselves. Plants have reacted to the attacks of herbivores in a multitude of ways; two defensive adaptations of special interest are those of a chemical and those of a mechanical nature. Allelochemicals, or secondary compounds, are substances laid down by plants in their otherwise edible tissues that are repugnant or toxic to herbivores. Mechanical defense adaptations of plants against vertebrate herbivores include thorns, spines, and adaptations of the cell wall.

An example of mechanical defense is an excellent illustration of plant/animal coevolution. The grasses (family Gramineae) constitute a modern flowering plant assemblage of enormous importance. The 700 genera and perhaps 9000 species of grasses are not only world wide in occurrence but are also considered the zenith of flowering plant evolution. Fossil leaves that perhaps belonged to grasses are known from the Cretaceous, far back in the history of flowering

---

meet in biting. The masseter and medial pterygoid develop maximum force along the cheek teeth, which lie just anterior to their insertion on the inner and outer wall of the body of the mandible. These muscles are thus effective in crushing.

(b) A fourth pair of muscles important in the grinding portion of mastication are the lateral pterygoid muscles, which originate on the sphenoid adjacent to the medial pterygoids and insert on the inner surface of the mandibular condyle. Contraction of one of these muscles shifts the jaw laterally (middle). Contraction of both protrudes the entire lower jaw (right).

(c) Diagramatic section at the level of the cheek teeth of an herbivorous mammal showing the forces produced by the masseter (m) and medial plus lateral pterygoids (pt) in the grinding of food. Note that only one side of the jaw grinds at a time. The action of the tongue and cheeks properly positions the food bolus.

(d) The shape of the jaws is influential in producing the differing jaw actions of carnivore (left) and herbivore (right). The coronoid process is very large in carnivores and the temporalis muscle which produces the bite force of the canines is massive. (See also Figure 19–11 and Table 19–2.) The articulation of the jaw of the mandibular condyle is directly in line with the occlusal surfaces of the tooth rows. Thus, as the mouth closes the posterior teeth occlude before the anterior teeth in the manner of the blades of a scissors. This action is efficient in the uses of a carnassial apparatus (Figure 19–3). Herbivores with little forceful use of the anterior dentition have a much enlarged jaw angle for insertion of the masseter and pterygoideus muscles. (See also Figure 19–5 and Table 19–2.) The jaw articulation is above the level of the occlusal surface (as in humans) and the entire cheek dentition occlude nearly simultaneously to crush and grind. (After Alexander [1] and Campbell [3].)

plants. The first undoubted fossil grasses, however, are very modern forms from mid-Tertiary beds. It is clear that a number of adaptive features of the family must have evolved during the late Cretaceous and early Cenozoic times simultaneously with the evolution and radiation of herbivorous mammals. Many descendents of this herbivore radiation today depend on grasses. The outward simplicity of many grasses is deceptive, for their adaptations, although often derived by reduction of parts, are very specialized.

It appears from reconstruction of primitive characters of living Gramineae that the first grasses were low growing tufts. Instead of growing from the tip, grasses grow from the base of the leaf blade. A grass, therefore, elevates the oldest, nongrowing portion of its foliage, and this can be removed without stopping growth. Most grazing mammals cannot graze close enough to the ground to remove all of the leaf tissue, yet, from the earliest herbivorous radiations, the muzzle has shown elongation permitting deeper and deeper cropping of plant tufts. Grasses have repeatedly evolved nearly stemless growth forms with growing portions very close to the ground out of reach of even the most specialized herbivores.

There is other morphological evidence to suggest that early grazing mammals exerted considerable selective pressure on grasses. Two chemical adaptations highly developed in many grasses have been extremely effective against herbivores and have shaped herbivore evolution since the Miocene. Plant cells are encased in a rigid cell wall. To obtain the nutrients of the cell protoplasm, it is necessary for a herbivore to rupture this cell wall and expose the contents. Grasses have made this as unrewarding as possible by incorporating crystalline silica into the fibrous cell wall. Silica is a hard mineral and grinds the teeth to useless, flat nubbins.

Cellulose and lignin are common constitutents of all plant cell walls. Cellulose is a complex carbohydrate closely related to starch but distinctive in its indigestibility by all multicellular animals. Grasses often have greatly elongated cells and thickened cell walls, making the amount of directly digestible foodstuffs a small proportion of the total leaf. It seems improbable that vertebrates should evolve complex adaptations to enable them to feed on plants with such tooth-destroying texture and low nutritional value. Yet the orders Artiodactyla, Perissodactyla, Sirenia, Hyracoidea, Proboscidea, Rodentia, Lagomorpha, Primates, and Marsupialia contain many species that are partially or wholly adapted to grasses for their food. Why should this be so?

The whole of the Mesozoic was, by present climatic standards, warm and increasingly moist. During the Cretaceous, ferns and cycads similar to those now living in subtropical South America were distributed from the paleo-equator to nearly 90° N and S. As outlined in Chapters 11 and 15, the late Cretaceous was a period of gradual withdrawal of the shallow seas over much of the land and a decided cooling and drying of the earth's climate. Except for a brief warming trend between the Paleocene and Eocene, this deterioration of climate has continued for 70 million years. This deterioration brought about an extreme restriction of the ranges of the nonflowering plants of the Cretaceous, leaving many as relic populations in protected habitats and forcing many to extinction. The apparently more adaptable angiosperms, present since the Jurassic, radiated explosively, followed in short order by mammals. By the Miocene, even angiosperm forests had been somewhat diminished and grasslands dominated the worldwide terrestrial biota. It is clear that survival of phyletic lines would depend on adaptation to grasses as the predominant food source despite the relative difficulty in eating and absorbing them.

To deal with the thick cellulose and lignin of grass cell walls, the occlusal surfaces of the cheek teeth of mammals became enlarged and roughened (Figures 19-5 and 19-6). This change involved an overall increase in the size of the teeth, increased molarization of the premolars, and development of complex folds of enamel, dentine, and cement that results in all three substances being exposed on the occlusal surface simultaneously. To deal with the exceedingly abrasive silica content, the cheek teeth became elongate, permitting them much longer service before wearing away. Ultimately, the cheek teeth evolved either of two modes of adaptation. In some herbivores, tooth eruption patterns allow a basically diphyodont mammal to be functionally polyphydont (having several successive replacement teeth). Examples are the single functional, serially replaced teeth of Proboscideans and a similar arrangement in some Sirenians. In these mammals only a few cheek teeth are exposed above the gums at a time. As these teeth wear down under the abrasion of herbivorism they migrate forward along the jaw and are shed. New teeth erupt posteriorly to replace them. Other herbivores (many perissodactyls, artiodactyls, lagomorphs, and rodents) have persistently growing cheek teeth.

Several independently evolved clades of ruminants of the Order Artiodactyla have a complex forestomach that allows microorganisms to convert the cellulose and lignin of plant cell walls into digestible nutrients (Figure 19-7). Although the ruminants are especially worthy of note, a wide variety of herbivorous mammals, many of which feed on grasses, have evolved analogous adaptations. Kangaroos, koalas, colobine monkeys, rodents, lagomorphs, perissodactyls, and many nonruminant artiodactyls are examples. In these forms large sacculate portions of the gut, either the forestomach or a blind sac or caecum at the junction of the small intestine and colon, provide storage space for large quantities of masticated herbiage so that gut microbial symbionts can ferment cellulose to digestible end products. The nutritive nature of these fermentation products, the efficiency of such absorption, and the potential nutritive value of the feces produced by the animal depend in large part on whether the plant protoplasm was digested and absorbed before fermentation (cecal fermentation) or if only fermented or partially fermented matter is available to the mammal's digestive apparatus (forestomach fermentation).

Some small burrowing and nesting herbivores with well-developed cecal fermentation produce feces in which relatively little of the products of fermentation has been absorbed. These animals often eat some of the feces they deposit in their burrows, retrieving a part of the fermented material. Lagomorphs produce special moist fecal pellets at night, which are eaten. Young

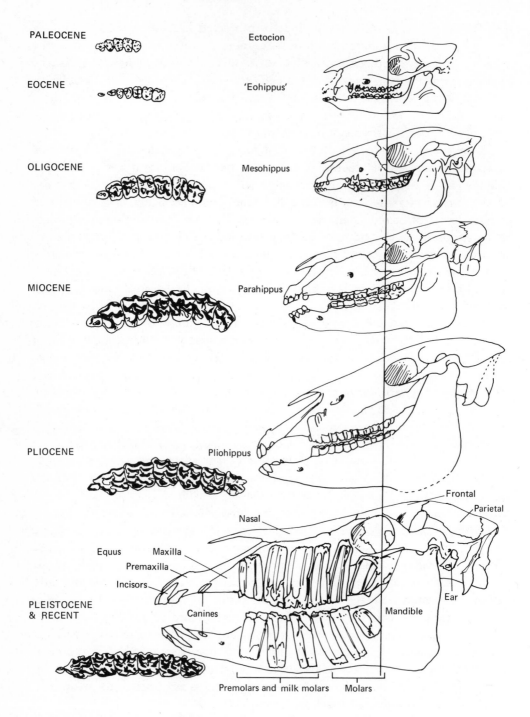

PALEOCENE — Ectocion

EOCENE — 'Eohippus'

OLIGOCENE — Mesohippus

MIOCENE — Parahippus

PLIOCENE — Pliohippus

Frontal
Parietal
Nasal
Equus — Maxilla
Premaxilla
Incisors
Ear
PLEISTOCENE & RECENT
Canines
Mandible
Premolars and milk molars   Molars

Figure 19-5 Progressive evolution of the herbivorous adaptation in horselike perrisodactyls. (Not to scale.) During horse evolution from forest browser to open plains grazer, the feeding apparatus of the skull underwent several modifications to adapt the dentition to an increasingly abrasive diet. The molars became broader and flatter, increasing their individual surface area. The premolars

koalas feed for an extended period on the feces of their mother. The end result of all gut fermentation, however, is the gleaning of more protein and energy from plant material, especially from grasses, than simple direct digestion would produce.

Cursorial Adaptations

Grasses, like most plants in seasonal or arid environments, show growth responses to rainfall. New shoots and leaves, low in cellulose and silica content, start to form shortly after the first rains. Rains over the earth's major grasslands vary from place to place and time to time. Thus, to feed on the most palatable grasses an herbivore must travel great distances to follow the rains and sprouting grasses. In leaving the cover and protection of forests, the ancestors of today's herbivores exposed themselves to easy view by predators. These factors placed a premium on efficient, rapid locomotion over the generally flat terrain of grasslands. Such locomotion is termed **cursorial** and is characteristic of most large herbivores represented today primarily by the ungulates (orders Perissodactyla and Artiodactyla).

Efficiency in cursorial locomotion correlates with a maximum length of stride, and speed is further enhanced by increasing the number of strides per unit of time. Lengthening the stride has been accomplished by lengthening of the limbs through elongation of individual bones and the assumption of **unguligrade** posture: standing and running on the tips of the toes (Figure 19-8). Increased efficiency of stride is fostered by reducing the distal mass, and therefore inertia of the limbs. The muscles of the upper part of the limb insert very close to the joint over which they act, increasing their effective speed, although reducing their power. Elastic ligaments running across the joints of the limbs are stretched when these joints are bent under the body's full weight, but snap back straightening the joint when the weight is released to return much of the energy stored in them. Behavioral characteristics also aid long distance travel. The stride length of the gait used in steady travel coincides with the natural oscillation period of a pendulum of similar dimensions to those of the animal's limb. Thus very little energy is expended in reversing the motion of the limb at the extremes of its excursion—gravity does a great deal of the work.

Such extensive locomotory adaptations have obviously had an effect on the carnivores that prey on grassland herbivores. Three families of the Order

---

became identical to the molars (molarization), thus increasing the surface area of the entire grinding battery. The occlusal surface became increasingly complex in its labyrinthine pattern of enamel, dentine, and cement (see Figure 19-2g), thus producing a self-sharpening abrasive surface. The teeth became hypsodont with long-persistent open roots for growth so that wear was effectively countered by replacement. The skull of the modern horse *Equus* is dissected around the base of the teeth to show the advanced hypsodonty. (After Peyer [12].)

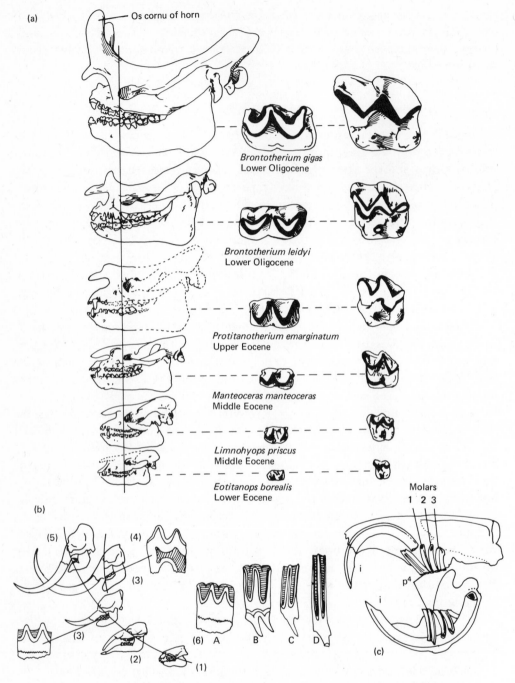

(a)

Os cornu of horn

Brontotherium gigas
Lower Oligocene

Brontotherium leidyi
Lower Oligocene

Protitanotherium emarginatum
Upper Eocene

Manteoceras manteoceras
Middle Eocene

Limnohyops priscus
Middle Eocene

Eotitanops borealis
Lower Eocene

(b)

(5)   (4)

(3)

(3)

(3)

(2)

(1)

(6) A   B   C   D

Molars
1 2 3

p⁴

(c)

Figure 19-6 Parallel and convergent evolution of herbivorous adaptations of Cenozoic mammals.   Like the horses (Figure 19-5) many herbivores evolved methods of increasing the size and/or length of function of their teeth in response to an increasingly abrasive diet.

(a) The titanotheres, culminating in the gigantic rhinoceros like *Brontotherium*, increased the individual size of the molar teeth, as indicated by the progressive enlargement of the postorbital part of the skull and the occlusal

Carnivora have members that are particularly well adapted to prey on large grassland herbivores: Hyaenidae (hyaenas), Canidae (dogs and wolves), and Felidae (cats). Both hyaenas and dogs hunt in packs and use stamina and cunning to exhaust and maim large prey. Many of the adaptations for speed and efficiency seen in ungulates have analogs in cursorial carnivores, especially the canids. Few felids run down their prey. Cats stalk by vision and, when close to their prey (usually less than 100 m), they attack in a brief burst of speed followed by a spring. Only the lion hunts in groups, and its social system is accordingly complex. The cheetah is very different from other cats and is distinct from all other carnivores in being a solitary cursorial predator. Although it rarely runs for distances greater than 0.5 km, it is the fastest terrestrial animal, having been clocked at 112 km/hr.

19.1.5 Convergent Evolution in Mammalian Feeding Types

The story of plant–herbivore–carnivore coevolution presents only a glimpse of the many interesting aspects of evolution during the "Cenozoic Age of Mammals." Recall that the earth's land masses were still intimately connected during the early Mesozoic when synapsid reptiles gave rise to a series of progressively more mammallike forms. In the late Mesozoic and early Cenozoic, the earth's terrestrial habitats were more widely separated than they have been before or since. Eurasia was divided by a great north-south epeiric sea into a European and an Asian portion. In addition, North America, South America, Africa, Antarctica, and Australia were all separate and each of these land masses had its own complement of early mammals (Figure 19–9). In spite of various and complex connections, each of these regions had its own flora and fauna, and each region showed adaptation of these elements to the worldwide cooler and drier Oligocene climate. The convergences and parallelisms between mammals that evolved on these isolated land masses is astounding. So similar are some of these forms that systematists formerly associated some, such as South America's litopterns and North America's perrisodactyls (Figure 19–10) although we now know them to belong to very different ungulate clades. As with herbivores, carnivores of extreme similarity of adaptation but wide

---

views of upper and lower molars. (b) The proboscidians reduced the number of simultaneously functioning teeth but greatly increased each tooth's size, the structural complexity of the tooth, and finally the length of time a tooth was useful by developing hypsodonty. (1) *Moeritherium*, late Eocene and early Oligocene; (2) *Serridentinus*, Miocene mastodont (= nipple tooth); (3) *Stegomastodon*, Pleistocene mastodon; (4) *Mastodon*, Pleistocene mastodon; (5) *Parelephas*, Pleistocene mammoth; (6a–d) *Elephas* Pleistocene to recent elephant species showing progressive hypsodonty. (c) Hypsodont, open-rooted (persistently growing) incisors (i), fourth premolar (p4) and molars (m1, 2, 3) of the pocket gopher, *Geomys*. (Modified after Colbert, E. H. [1969]. *Evolution of the Vertebrates*, Second Edition, Wiley and Sons, N. Y. and Peyer [12].)

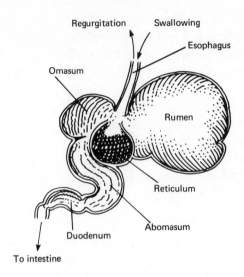

Figure 19-7 The ruminant stomach showing the complex structure for extracting the maximum in nutrients from plant material. Although all the chamber
walls are vascular and absorptive, only the abomasum produces glandular digestive secretions. This segment is homologous with the stomach of other mammals; the others are derived from the forestomach and/or base of the esophagus.
The smooth walled rumen is a fermentation chamber to which ground plant matter is added and from which partially fermented matter is regurgitated to be
chewed as the cud and then reswallowed for complete fermentation. The complex surface of the reticulum, sold in markets as honeycomb tripe, aids in straining the thoroughly fermented and liquified chyme. The omasum is also known
as the psalterium (resembling a book such as the psalter or book of psalms used
in religious service) or the manyplies. Its complexly folded wall aids in retarding
the flow of its contents and absorption of nutrients released by bacterial fermentation before the chyme reaches the true stomach or abomasum.

phyletic isolation evolved in various regions at this time. In Australia the thyla-
cinids and in South America the marsupial borhyaenids, and their derivatives
the marsupial sabertooths, paralleled the weasel, wolf, and sabertooth cats of
the Northern Hemisphere continents (Figure 19–11).

In Australia, South America, Africa, and Eurasia distinctly different forms
evolved adaptations to deal with the diverse and highly organized colonies of
termites and ants. These insects are specialized for sociality and group defense.
They often build impressive earthen nests containing thousands of individuals,
some of which are strong-jawed soldiers. Adaptations of the teeth, jaws, and
skull to myrmecophagy have already been discussed (section 19.1.2). Convergent adaptations of the forelimbs for digging (Figure 19–8) include a reduced number of enlarged digits with powerfully developed curved claws. In
many ways, adaptations of fossorial (digging) limbs are the exact opposite of
those for efficient cursorial limbs. The length of reach when digging (equivalent
to length of stride) is usually unimportant so that the limbs are generally quite

short. Power, however, is very important. Thus the muscles and their bony attachments are enlarged and arranged for maximum use of the energy expended. The muscles of the fossorial limb generally insert far from the joint over which they act, enhancing their power. The distal mass of the limb is often increased by broadening and flattening of the bony elements. A modified integument is another common feature: most forms are covered by coarse hairs, hard spines, or even overlapping scales composed of agglutinated hair. Such similarities of widely divergent phyletic lines on isolated land masses reflect the adaptive potential of mammals.

19.2   THE MAMMALIAN NERVOUS AND SENSORY SYSTEMS

In addition to the many convergences in mammalian evolution, there has been a parallel trend in brain evolution of universal importance. This trend has been the progressive enlargement and increased complexity of the central nervous system and its sensory apparatus. Unquestionably this expansion has been independent in each of the orders of mammals involved. Nevertheless, the detailed similarities in the morphological and behavioral end products are intriguing. In any attempt to unravel the story of these parallel changes we must begin by a comparative examination of the brain to determine precisely what has enlarged during the last 200 million years.

19.2.1   Brain Size and Anatomy

The mammalian brain is distinctive for its large size and the expansion of certain parts. Long ago, correlation was made between these anatomical features and alertness, responsiveness, ability to learn, and other indications of intelligence. Nevertheless, understanding how the mammalian brain functions and what forces molded its evolution has long remained uncertain.

Most vertebrate clades have evolved larger body sizes. Numerous locomotory, homeostatic, and defensive benefits derive from large size. In large animals the tension and position of additional muscle fibers must be monitored and their firing controlled; the increased skin surface brings with it more sensory organs. All these changes demand more nerve cells in the central nervous system and, especially, in the brain. Nevertheless, in vertebrates brain size does not increase in exact proportion to body mass. Rather, as body mass doubles brain mass increases by a factor of about 1.6.

To compare the brain sizes of vertebrates, which vary from the fish *Schindleria praematurus*, weighing only 2 mg, to the largest vertebrate ever to live, the blue whale *Balanoptera musculus*, weighing over 160,000 kg, it is appropriate to compare brain mass per unit of body mass. When examined in this way, the brains of living birds and mammals are approximately 15 times larger than those of other vertebrates of the same body size. To visualize the relationship between various vertebrate taxa, including extinct groups, the data for each taxon can

(a)

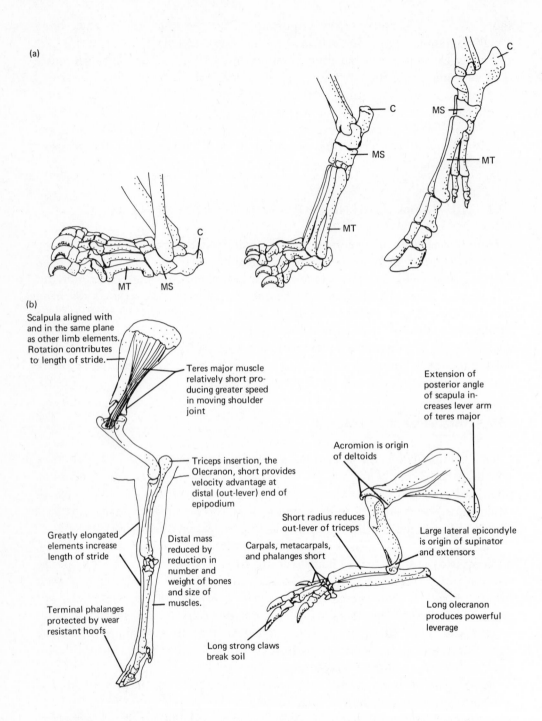

(b)

Scalpula aligned with
and in the same plane
as other limb elements.
Rotation contributes
to length of stride.

Teres major muscle
relatively short pro-
ducing greater speed
in moving shoulder
joint

Extension of
posterior angle
of scapula in-
creases lever arm
of teres major

Acromion is origin
of deltoids

Triceps insertion, the
Olecranon, short provides
velocity advantage at
distal (out-lever) end of
epipodium

Short radius reduces
out-lever of triceps

Large lateral epicondyle
is origin of supinator
and extensors

Greatly elongated
elements increase
length of stride

Distal mass
reduced by
reduction in
number and
weight of bones
and size of
muscles.

Carpals, metacarpals,
and phalanges short

Terminal phalanges
protected by wear
resistant hoofs

Long olecranon
produces powerful
leverage

Long strong claws
break soil

Figure 19-8 Contrasting adaptations of the limbs of mammals. (a) The primitive foot posture of mammals is the **plantigrade** type (left) where the entire foot skeleton complex supports the weight of the body. In the case of the hind foot illustrated support extends from the calcaneus (c) to the terminal phalanges. Several mammals, especially stealthy or moderately cursorial (running) carni-

be enclosed in a polygon (figure 19–12). Even the early mammalian orders had four to five times as much brain tissue as other vertebrate taxa of similar body size. Obviously an increase in relative brain-to-body ratios occurred in the evolution of modern mammals.

The neruoanatomy of the brain is exceedingly complex (Figure 19–13). The basic unit of neuroanatomy is the neuron. **Neurons** are made up of nerve cell bodies, which contain the nucleus and most of the metabolic machinery of the cell. Thin, sometimes very long, extensions, the **dendrites** and **axons**, extend from the cell body and transmit impulses to and from it. Axons, generally the longest and least branched of these extensions, are often encased in a fatty insulating coat, the **myelin sheath**, which increases the conduction velocity of the nerve impulse. An axon carries impulses over long distances to another nerve cell or population of nerve cells. Generally axons "in transit" from one population of cells to another are collected together like wires in a cable. Such collections of axons in the **peripheral nervous system** (PNS) are called nerves; within the **central nervous system** (CNS) they are called tracts or white matter (because of the appearance of the myelin). Nerve cell bodies are often clustered together, usually in groups with similar connections or functions, and are called ganglia (in the PNS) and nuclei (not to be confused with the cellular nucleus) and gray matter (in the CNS). The vertebrate brain is composed of five distinct regions containing both gray and white matter, each serving a different basic function.

Posteriorly, two regions differentiate from the embryonic brain region associated with the developing ear. The most posterior, the **myelencephalon**, is primarily an enlarged anterior extension of the spinal cord. It is altered in appearance by the expansion of the neurocoel cavity to form a large fluid-filled space, the fourth ventricle of the brain. The nuclei of this medulla oblongata (as the adult myelencephalon is often called) are basically continuations of the central gray matter of the spinal cord (Chapter 5). These nuclei synapse primarily with the sensory organs of the muscles and skin of the head. The myelencephalon has one distinctive set of nuclei, however: those associated with the sense organs of the inner ear. All impulses from the receptor cells of the balance (vestibular) and the hearing (cochlear) regions of the mammalian ear synapse first in these medulla oblongata nuclei.

---

vores, support their weight only on the phalanges. This (middle) digitigrade posture reduced friction and increases the length or reach of the stride by an amount equal to the length of the vertical metapodial (MT) and mesopodial (MS) elements. The ultimate in cursorial adaptations is the unguiligrade posture (right) achieved independently by numerous (primarily herbivorous) clades of mammals. Here weight is supported entirely by the hoof-clad terminal phalanx or phalanges and the effective stride includes the contribution of the second and third phalanges. (b) The skeletal elements of the left forelimb of the cursorial (running) deer and the fossorial (digging) armadillo compared. (From various sources including Hildebrand, M. [1974] *Analysis of Vertebrate Structure*, Wiley and Sons, N.Y.)

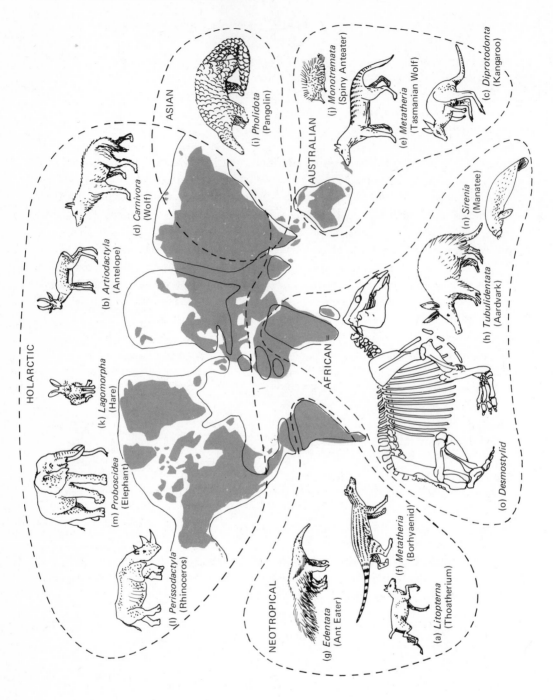

HOLARCTIC

ASIAN

(i) *Pholidota*
(Pangolin)

AUSTRALIAN

(j) *Monotremata*
(Spiny Anteater)

(e) *Metatheria*
(Tasmanian Wolf)

(c) *Diprotodonta*
(Kangaroo)

(d) *Carnivora*
(Wolf)

(b) *Artiodactyla*
(Antelope)

(n) *Sirenia*
(Manatee)

(k) *Lagomorpha*
(Hare)

(h) *Tubulidentata*
(Aardvark)

AFRICAN =

(m) *Proboscidea*
(Elephant)

(o) *Desmostylid*

(l) *Perissodactyla*
(Rhinoceros)

NEOTROPICAL

(f) *Metatheria*
(Borhyaenid)

(g) *Edentata*
(Ant Eater)

(a) *Litopterna*
(Thoatherium)

Figure 19-9   Adaptive radiation and convergence of mammals evolving in isolation during the Cenozoic.

Continental drift profoundly affected mammalian evolution by producing a maximum isolation of land masses during much of the early Cenozoic. Superimposed on a map of the *recent* relative positions of the continents (shading) are the land areas (outlines) thought to have existed during the early Eocene (see Figure 15-2). Note that these areas do not occupy positions they may have had during the Eocene. Isolating distances may be underestimated by this projection. Mammals that evolved during various periods of the Cenozoic (not just the Eocene) are illustrated with the land masses on which they probably originated. Mammals of Northern Hemisphere origin are grouped as Holarctic. The connections between land masses in this half of the earth (generally referred to as Laurasia) have been so numerous and frequent that relatively little is known about precise sites or origin.

Convergence is seen in grazing herbivores including the extinct litopterns (a) of South America (see also Figure 19-10), the holarctic artiodactyls (b), the horselike perissodactyls, and the australian diprotodonts such as the kangaroo (c). Although modes of locomotion are different, jaws, teeth and feeding mechanisms are very similar. The carnivorous mammals gave rise to similar forms such as the true wolf (d) in the Holarctic, the marsupial Tasmanian 'wolf' (e) in Australia (see Figure 19-11a), and Tasmania, and the extinct marsupial borhyaenids (f) of South America. See also Figure 19-11b for convergence in sabertooth catlike marsupials and placentals. Mammals highly specialized for feeding on social ants and termites evolved and survive in South America, Giant anteater (g); Africa, aardvark (h); Asia, pangolin (i); and Australia, spiny anteater (j); all of which are unrelated but show many morphological similarities.

Sometimes similar forms evolved under what appears to be much less isolated conditions. Thus hares and rabbits (k) evolved in the Holarctic where other gnawing herbivores, and rodents, also occurred. Other forms evolved in one area but migrated and survived only in another. Examples are the horses of North American origin now found only in Asia and Africa and the holarctic rhinoceros (1) with the same recent distribution. Unique forms also evolved on isolated land masses that seem to have had few close ecological counterparts. These include the formerly diverse proboscidians (m) of holarctic origin (see Figure 19-6b), the sirenians (n) of African origin, and the extinct desmostylians (o) which were large aquatic or marine forms thought to have evolved on the coasts of Africa but later found on the Pacific coast of North America as well. (Modified after Kurtèn, B. [1969] *Continental drift and evolution. Scientific American* 220 (3):54-64.)

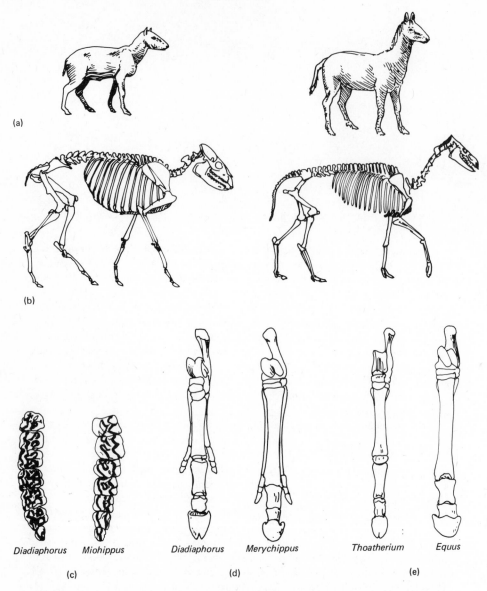

Diadiaphorus   Miohippus   Diadiaphorus   Merychippus   Thoatherium   Equus

(c)   (d)   (e)

Figure 19–10 Convergence between unrelated mammals from isolated continents to grazing and sustained running over hard ground. South American litopterns (left) and North American (holarctic) perissodactyls (right). (Not to
scale.) (a) Reconstructions of the Miocene South American litoptern *Diadiaphorus* and a late Miocene and Pliocene North American and Asian three-toed
horse *Hypohippus*. (b) Skeletons of *Diadiaphorus* and *Hypohippus*. (c) Occlusal
surface of the cheek teeth of *Diadiaphorus* and comparable developments in
perissodactyls from the mid-Oligocene to Micoene *Miohippus*. (d) Frontal view
of the left hind feet of the three-toed forms *Diadiaphorus* and *Merychippus*
(= *Protohippus*) from the mid-Miocene to the Pliocene. (e) Frontal views of the
hind foot of the Miocene one-toed litoptern *Thoatherium* and the Pleistocene
one-toed horse *Equus*. The litopterns reached a degree of specialization that the
horses did not attain until millions of years later.

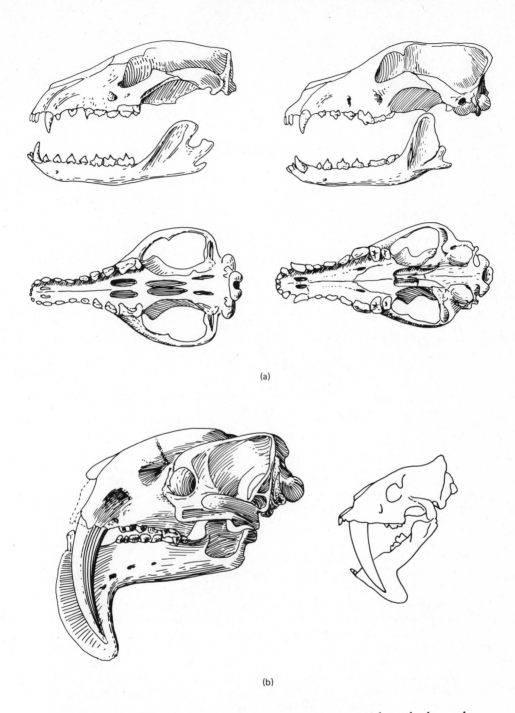

(a)

(b)

Figure 19–11 Convergence between carnivorous marsupials and placentals evolved on isolated continents. (Not to scale). (a) The skull of the recently extinct (?) Tasmanian 'wolf' *Thylacinus cynocephalus* (left) in lateral and palatal view compared with that of a greyhound *Canis familiaris*. (b) The Pliocene South American marsupial saber-tooth *Thylacosmilus atrox* (left), compared with the Eocene to Oligocene holarctic saber-tooth cat *Eusmilus* (right).

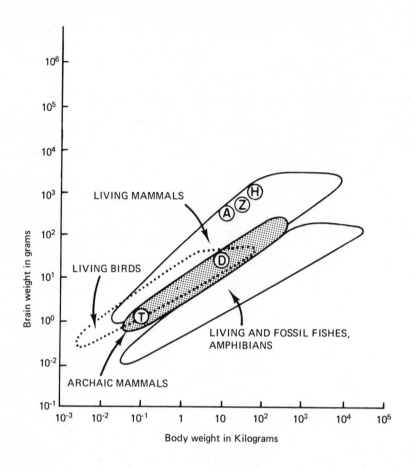

Figure 19-12 Relative Brain Weight of Vertebrates. Logarithm of brain weight plotted against logarithm of body weight based on measurements of 198 living vertebrate species and some fossils. The most extreme points representing data from a particular taxon have been connected by lines to form a polygon that includes all data points for that taxon. Data for extinct forms are estimated from endocranial cast volumes of fossils.

Living and fossil reptiles amphibians, and fishes overlap. Thus, reptiles show no trend to increase in brain size. Archaic mammals show a four- to fivefold increase in brain to body weight ratios over their reptilian forebears. Advanced living mammals have a four- to fivefold increase over their Tertiary mammalian ancestors some living mammals have retained "small" brains). Primates represent a further addition in brain matter and occupy the upper limits of the data polygon for mammals of their size. T = *Triconodon* (of the Jurrasic; D = *Didelphis* the recent Virginia opossum; A = *Australopithecus africanus*; Z = *Australopithecus boisei*; H = *Homo sapiens*. (Modified after Jerison [6].)

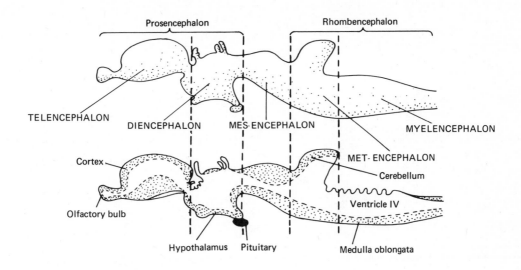

(a)

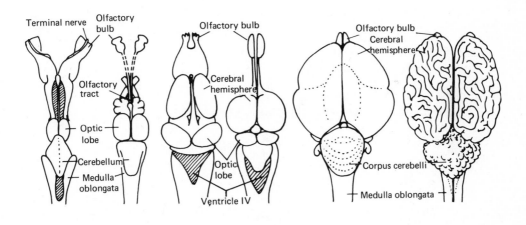

(b)

Figure 19-13 The vertebrate brain.   (a) Diagramatic developmental stage in generalized vertebrate brain showing principal divisions and structures described in text. Lateral view above; sectioned view below. (b) Representative vertebrate brains seen in dorsal view. All brains are drawn to approximately the same total length, which emphasizes relative differences in regional development. From left to right: *Scymnus*, a shark; *Gadus*, a teleost; *Rana*, a frog; *Alligator; Anser*, a goose; and *Equus*, a horse, as an example of an advanced modern mammal. (After Romer and Parsons, Chapter 1 [5].)

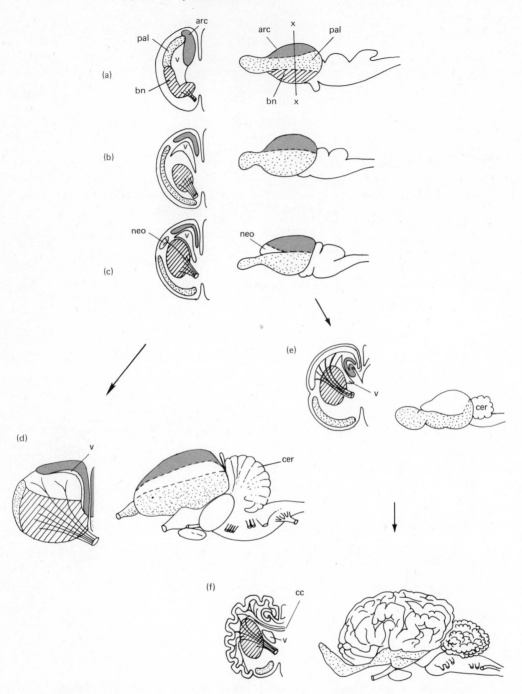

Figure 19–14 Evolution of the telencephalon. Various stages in the differentiation of the telencephalon in tetrapods in diagrammatic lateral and cross-sectional half-brain diagrams. Changes include the increase in size of the telencephalon relative to the rest of the brain, differentiation of nuclear (gray matter) areas, and appearance of new nuclear types, corticalization in which the gray matter emerges on the surface of the telencephalon forcing the white matter to a cen-

The anterior portion of the posterior region of the embryonic brain, called the **metencephalon**, develops an important dorsal outgrowth the **cerebellum**. The cerebellum coordinates and regulates motor activities whether they are "reflexive," such as maintenance of the posture, or "directed," such as escape movements. The nuclear or gray matter of the cerebellum, the **cerebellar cortex**, receives nerve impulses from the acoustic area of the myelencephalon (in particular impulses relayed from the vestibular nuclei). It also receives impulses from the complex system of muscle and tendon stretch receptors and indirectly from the skin, optic centers, and other coordinating brain centers.

The central embryonic brain region, the **mesencephalon**, develops in conjunction with the eye and plays a part in the functioning of the visual system. The roof of the mesencephalon is known as the **tectum** and receives input from the optic nerve. The floor of this region contains fiber tracts that pass anteriorly and posteriorly to other regions of the brain as well as to nuclei concerned with eye movements. A rather small region, the **diencephalon**, develops from the most anterior of the embryonic brain regions. It functions in amniotes as a major relay station between sensory areas of many modalities and the higher brain centers. A ventral outgrowth of the diencephalon contributes to the formation of the dominant endocrine organs, the pituitary gland, and the hypothalamus, which together function as the primary center for neural-hormonal coordination and integration.

---

tral position, and finally the convolution of the neopallium in advanced mammals.

(a) Primitive tetrapod with relatively large contribution of non-telecephalon brain regions to total brain weight. The telencephalon is differentiated into three deep lying (periventricular) pallial (cloak) regions: kn, the basal nuclei (= corpus striatum); pal, the paleopallium; and arc, the archipallium. Lines representing cut nerve tracts indicate an association center's connections with the brain stem. (b) Later stage, probably represented by primitive reptiles in which expansions and migrations of the paleopallium and archipallium have forced the basal nuclei to an internal position. (c) An advanced reptile stage, probably arrived at by several reptilian clades. The telencephalon now constitutes the single major portion of the brain. A new nuclear region, the neopallium (neo) has made its appearance and corticalization is well advanced. (d) In birds the paleopallium, always associated with olfaction is very small but the basal nuclei associated with innate, reflexive behaviors are enormous, being covered with a relatively thin coat of archipallium and what is thought to be neopallium. The telencephalon and cerebellum (cer) make up the major portions of the brain. (e) Primitive mammals developed in a different direction than birds. The neopallium and paleopallium are the two major divisions of the telencephalon. The archipallium is forced into a curled internal position. The majority of connections to the brain stem run directly to the neopallium unlike in primitive tetrapods, reptiles and birds where the basal nuclei serve as major associations areas. (f) In modern, advanced mammals the neopallium dominates the entire telencephalon and becomes highly folded. It communicates not only with the brain stem but also directly with its contralateral cerebral hemisphere via the corpus callosum (cc). X–X: plane of sections showing internal organziation. V: lateral ventricle. (Modified from Romer and Parsons, (Chapter 1[5].)

Finally, the most anterior region of the adult brain, the **telencephalon**, develops in association with the olfactory capsules. While functioning as the first nuclei of olfactory synapse, the telencephalon of primitive vertebrates also coordinates inputs from other sensory modalities. In mammals the primary seat of sensory integration and nervous control occurs in the telencephalon. The neuronal cell bodies in the anterior olfactory nuclei of the telencephalon are variously enlarged into masses of gray matter that are often differentiated into distinctive layers and subdivisions with complex interconnections (Figure 19-14).

Two phenomena are important in considering the evolution of the mammalian brain. First is the evolution of an external matlike covering of gray matter from the more primitive central or "buried" nuclear condition. This change places the nerve cell bodies on the outside of the brain where additions to their number will not cause disruption of the tracts leading to them. It also places the axons close to the nutritive cerebrospinal fluid and the cell nuclei adjacent to the oxygen-rich vascular meninges or nonneural coverings of the nervous system. Second is the development of a region of gray matter distinctive in its cell types and cellular organization, the **neopallium**, which dominates the entire nervous system. Birds, although as "brainy" as many mammals, have comparatively little of this type of neural tissue. Instead they show a major expansion of the basal nuclei that are related to stereotyped, reflexive reactions to stimuli. The neopallium is a site of sensory projection, of integration, of memory, and of the construction and reconstruction of past and even hypothetical patterns of sensory input. It is ultimately the seat of intelligence.

19.2.2 The Question of Brain Enlargement

Two regions of the mammalian brain show distinctive enlargement: the cerebral hemispheres and the cerebellum. Both regions are characterized by increased corticalization (thickening and differentiation of the superficial gray matter) accompanied by complex folding or convolution. This provides a maximum of surface area for nutrient and gaseous exchange between the intensely active cortical neurons and the superficial circulatory supply.

Although we now have some idea of *what* enlarged in the transition from a reptilian to a mammalian brain, we have yet to offer a reason *why* this organ changed so drastically. In a most interesting and controversial summary of his study of brains of fossil and recent vertebrates, paleontologist Harry J. Jerison offers the following hypothesis (ref. 6).

> The earliest mammals were not merely accidents in the history of life on land. Their evolution was a natural consequence of adaptations selected by niches open to them. The enlarged brain of Mesozoic mammals seems to have been a stable adaptation to an unusual adaptive zone . . . . Early mammals were specialized reptiles and, as such, could be expected to characteristically utilize sensory

modalities in a rather sophisticated three-dimensional way to gain information about events at a distance. Mesozoic mammals were the nocturnal "reptiles" of that era and their brains evolved to accommodate life in nocturnal niches in which hearing and smell, rather than reptilian vision, were distance senses.

Jerison's hypothesis leads us to examine the sensory modalities of mammals, their capacity to function as distance senses, the specialized ways in which certain mammals use these modalities, and the effects these factors had on brain evolution. Then we can return to the question of why mammals are such "brainy" vertebrates with the hope of discovering an explanation.

19.2.3 The Sense Organs

The sense organs of mammals and of vertebrates in general are classified on the basis of the type of stimulus to which they react. **Chemoreceptors** are stimulated by chemical substances. **Mechanoreceptors** are sensitive to mechanical deformation even at the level of distortion of a single cell's stereocilia. **Radio-receptors** respond variously to electromagnetic disturbances of a broad spectral range, including thermal radiation. All of these receptors are found deep within the bodies of most vertebrates as well as near or at the body's surface. Those receptive to internal stimuli are termed **interoceptors**, or **proprioceptors** if specialized to provide information on the tension of muscles or the position of joints. Interoceptors do not provide a basis for a distance sense, and we will not deal with them. **Exteroceptors** are receptive to external stimuli and may provide information on conditions at considerable distances from the organism and, in mammals, include the classic five senses of taste, touch, sight, smell, and hearing. Which of these senses might best replace the bright light vision so important to the reptilian ancestors of the mammals?

Taste

Taste, because of the location of its receptor cells and their mode of stimulation, is a poor candidate for a distance receptor capable of replacing vision. Taste buds made up of specialized sensory epithelial cells and associated nerves are restricted to the moist membranes of the walls of the mouth, the throat, and especially the tongue (Figure 19–15). Taste can be discounted as contributing significantly to the expansion of the mammalian brain. It probably was, however the major sensory modality involved in the coevolution of chemical defenses of plants and vertebrate herbivores.

Touch

At least four differently structured nerve ending–epithelium–connective tissue combinations are known to be sensitive to stimuli ranging from light touch to heavy pressure. Touch receptors may also be combined with specialized hairs,

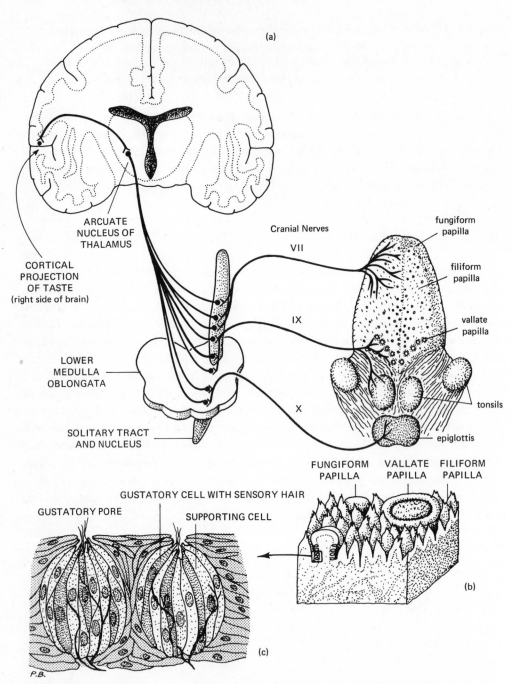

Figure 19-15 The sense of taste in mammals. (a) The tongue of a human show-ing the topography and types of lingual papillae. The distribution of innervation and route of projection onto the border of the lateral fissure of the cerebrum are shown for one side of the tongue. (b) Enlargement of region of tongue illustrat-ing details of papillae structure and position of taste buds. (c) Further enlarge-ment of taste buds on a fungiform or vallate papilla showing cellular organiza-tion and sensory "hairs." (Modified after Netter [9].)

the vibrissae on the muzzle, around the eyes, or on the lower legs. These vibrissae give early warning to nocturnal mammals of obstacles near the face and feet. The range of sensitivity of mammalian touch is nevertheless rather restricted, and touch also seems a poor candidate for a distance sense.

Vision

Vision is often considered the distance sense *par excellence* of tetrapods. Visual systems are sensitive to precisely those wavelengths of electromagnetic radiation that reach the surface of the earth with the least interference by the atmosphere. Electromagnetic radiation of these wavelengths, which we call light, have several properties that make them superior as transmitters of information over distance. Light travels great distances in air with little attenuation (absorption). Its path of travel is direct in a homogeneous medium. Light also interacts with matter in a very specific way. Thus light of different wavelengths may be absorbed, transmitted, or reflected so that the spectral (wavelength) distribution of light falling on an object is altered. These properties of light give nearly unambiguous cues to an object's gross structure, texture, and chemical composition. Another significant attribute of light is its speed. Light energy travels through most media at about 300×10^3 km/sec (186,000 mi/sec). This is instantaneous when compared to the speed of transmission and action of the vertebrate neuromuscular system. Changes in the pattern and nature of reflected or transmitted light, even at great distances, are immediately detected by the vetebrate eye.

The receptor field of the vertebrate eye is arrayed in a hemispherical sheet—the retina. Each point on the retina corresponds to specific neural connections and a different visual axis in space. Thus, a vertebrate can determine where a target is in at least two-dimensional space and whether it is stationary or moving. Because of neuronal interactions, vertebrate eyes can also detect sharp beginnings and endings of visual stimuli (ons, offs, edges) with great precision.

The eyes of mammals resemble those of other vertebrates (Figure 19–16). They consist of a light-shielded container, the sclera and choroid coats. These prevent stimulation of the eye by light from multiple directions. The eye has a variable entrance aperture, the pupil and the iris, to control the amount of light that enters. A focusing system, the cornea, lens, and ciliary body, converges the rays of light on the photosensitive cells of the retina. Because the eye is a complex and delicate organ, many of its structures are supportive, protective, nutritive and/or regenerative and not directly involved in the optics of vision.

The retina (Figure 19–17) originates as an outgrowth of the brain. The exceptional nature of the eye as an extension of the brain is not as generally appreciated as it deserves to be. For example, amphibians process distant stimuli in the elaborate neural network of the retina. The optic nerve sends signals to the brain with already encoded distinctions between predators, prey, and edges (as of lily pads). Firing patterns from the mammalian optic nerve indicate considerable retinal processing; there are approximately 100 million

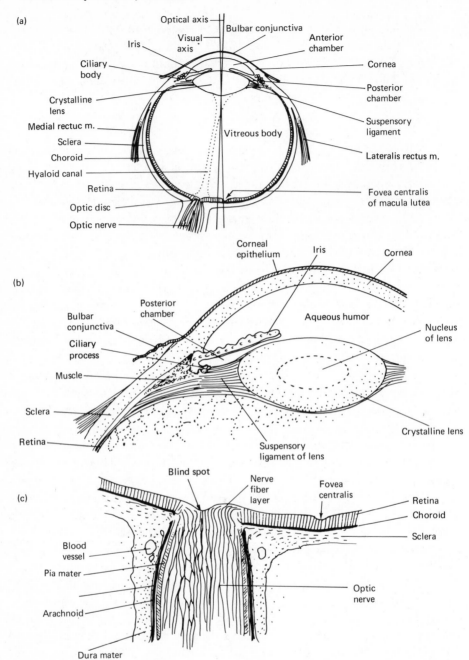

Figure 19-16 The mammalian eye. (a) The eye sectioned horizontally in the plane of the optic axis and optic nerve. In addition to the structures discussed in the text details of the anterior and fundic (deep) anatomy are shown. (b) Enlargement of the anterior portion of the eye details structures primarily responsible for image focus (cornea, lens, ciliary body) and quantitative control of light (iris). (c) Enlargement of the fundus of the eye showing the specialized sensory and neural regions found there. Note especially the continuity of the brain covering meninges (pia mater, arachnoid, and dura mater) to the protective and nutritive coats of the eye (sclera and choroid). (Modified after Crouch, J. E. [1969] *Text-Atlas of Cat Anatomy*. Lea & Febiger, Phil.)

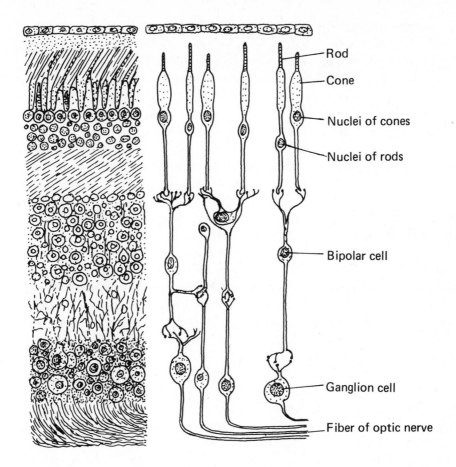

Rod

Cone

Nuclei of cones

Nuclei of rods

Bipolar cell

Ganglion cell

Fiber of optic nerve

Figure 19-17 Histological (left) and diagrammatic (right) structure of the mammalian retina as seen in man. The direction of light passage and neural conduction are indicated. The different types, great number and interconnections of nerve cells in the retina are the basis for the considerable information processing that occurs within the eye. (See text for further discussion.)

photoreceptor cells in the retina but only 1 million axons in the optic nerve. Even the rat, although not a mammal of exceptional visual capabilities, has approximately one third as many neurons in its two retinas as in its entire central nervous system.

Studies of reptiles and birds indicate that mammals evolved from, and probably competed with, vertebrates of exceptional visual capacity. Lizards and birds have better visual acuity than most mammals. In part, their ability to distinguish between two close points in space, which would appear to vertebrates with lesser acuity as a single point, is due to their nearly pure cone retinas. Cones are one of two types of light sensing cells in the vertebrate retina and are distinguished from the other type, the rods, by morphology, photochemistry, and neural connections. In addition to high visual acuity, cones are the basis for color vision.

Because of the structure of the lens and retina, vision produces a two- or three-dimensional representation of the world (Figure 19-18). Each cone on the retina corresponds precisely to a point or series of adjacent points in space. The muscles controlling the orientation of the eye and the accommodation of the lens give sufficient information for very accurate neural reconstruction of the visually perceived world. There is only one major limitation to the information-rich visual system to which mammals were heir. Cones have relatively low photosensitivity; they are at least two orders of magnitude less sensitive than rods. An eye adapted for acute vision requires daylight to function effectively, but early mammals are believed to have been nocturnal (Chapter 21). Thus nocturnal mammallike tetrapods would have required different visual systems, or even a different sensory modality, than their reptilian ancestors to obtain and process information from events at a distance.

The most obvious adaptation of a primitive mammal to the new sensory demands of a life at night would be those of a nocturnal eye. Living members of the order insectivora seem to fit most closely into the hypothetical adaptive zone of stem mammals. In addition, many otherwise specialized and advanced mammals also have a predominantly nocturnal eye. The primitive mammalian eye contains a pure rod retina. Rods are not only morphologically distinct from cones but also differ physiologically. The sum of these differences results in a cell of exceptional sensitivity to dim light. Several rods synapse with a single neural element, increasing the chances that dim light will stimulate the neuron, but achieving this increased sensitivity at the cost of visual acuity. An analogy can be made with photographic films—increased sensitivity to dim light is accompanied by increased "graininess" of the image.

The loss of acuity in a nocturnal eye is considerable. Whereas the diurnal pigeon can distinguish between two lines producing retinal images less than 1 μm apart, the rat requires a separation of retinal images of more than 20 μm. The problem can be appreciated by trying to concentrate on the form or details of objects as seen from the "corner" of our eye—peripheral rod-rich areas of the human retina. If this quality of image is in any way representative of that of early mammals, its usefulness as a distance sensor could hardly have been great.

Not all mammals are as limited as the rat, nor do humans have to use their peripheral vision. Acute vision occurs in many mammalian orders. From the variety of visual adaptations involved, it is clear that many independent alterations of the primitive mammalian eye have occurred The overwhelming majority of mammalian orders, however, appear to be color blind. Only the primates show well-developed color discrimination comparable to that found in other vertebrate classes. The cones of mammals that possess them show such distinct differences from cones of other vertebrates that they probably are analogous but not homologous structures. Mammalian cones probably evolved from pure rod retinas of Mesozoic mammals during the first major post-archosaurian radiation of mammalian types. Only recently, during the Cenozoic radiations of diurnal mammals, have retinal systems similar to those of reptiles and birds evolved in some mammals. The vast majority of mammals are not only color blind but also have rather poor acuity compared to that of the central region of higher primate retinas (the fovea). Elephants cannot distinguish between two objects if they are separated by less than 10 minutes and 15 seconds of arc on the retina. Chimpanzees, however, are representative of refined primates whose eyes only require that points in space be separated by 26 seconds of arc in order to distinguish that they are separate points and not a single one.

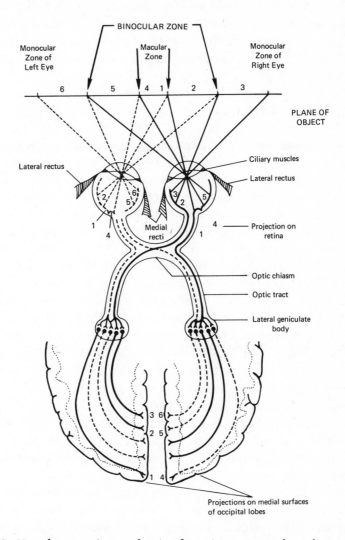

Figure 19-18 Neural connections and point for point correspondence between visual field and cerebral projections in the human brain. The one-dimensional representation of the line shown, as received by the brain, may be translated into a three-dimensional reconstruction of the visual world ("depth perception"). This is accomplished by information from the extrinsic ocular muscles on the degree of convergence or divergence of the eyes (such as from the lateral and medial rectus muscles shown) or by similar information from the ciliary muscles on the degree of lens distortion (accomodation) necessary for focusing.

Olfaction

Olfaction is generally well developed in mammals. The receptor apparatus is not drastically different from that in other macrosmatic (strongly olfactory) vertebrates. The sensory cells reside on the turbinals and nasal septum high in the nasal cavity. The lower turbinals and nasal epithelium are concerned with temperature and water regulation (Chapter 12 and Figure 19–19). Thus, a mammal must "sniff" rather vigorously to sample the air for olfactory stimuli. The olfactory cells, embedded in a supportive epithelium, have hairlike cell processes on their free end that are exposed on the moist surface of the nasal membranes. Processes of the opposite end of these cells penetrate the sieve-like cribriform plate of the cranium and synapse with the olfactory bulb of the brain.

The sensitivity of the olfactory epithelium is extraordinary. The total area of the olfactory surfaces may exceed that of the entire body surface. A single molecule can excite a receptor cell, and only a few molecules are needed to elicit a behavioral response. The neural apparatus associated with olfaction is often large, especially in primitive mammals. Even in advanced mammals olfaction plays a significant role in predator-prey interactions and is important in social-sexual contexts. Mammals, even humans with their relatively meager olfactory abilities, can distinguish thousands of odors.

A molecule of odorant probably fits more or less tightly into one or more of the different sensory pits along the sensory "hairs." Once seated, a molecule lingers, producing prolonged stimulation that allows accurate perception of the relative concentrations of stimulating substances. It also limits rapid temporal or spatial perception of the olfactory surroundings. Many mammals snort or sneeze during olfactory sensing, perhaps to flush air as completely as possible from the upper regions of their nasal cavities. The subtle differences in odor that mammals can detect are probably due to the hundreds of combinations from no more than a dozen basic odors. Such combinations would produce the individual differences in body odor that seem important in maintenance of many mammalian social groups. The physiological state of an individual is often determinable from the odors it produces especially in the feces and urine. Excretions are used by a wide variety of mammals to mark their territories or home ranges, and marking activity often occupies a great deal of a mammal's time.

Pheromones are chemical signals produced by an individual that affect the behavior and/or physiology of conspecifics. The odors of feces and urine undoubtedly have such effects. Many male ungulates sniff or taste the urine of females. This behavior is usually followed by "flaming," a behavior in which the male curls the upper lip and often holds his head high, probably inhaling. It is believed that a specialized structure of the anterior palate, the vomeronasal or Jacobson's organ, which contains branches of the olfactory nerve, is utilized during "flaming." Probably this behavior allows males to determine the stage of the reproductive cycle of a female.

All of these characteristics and phenomena point to the importance of olfaction to mammalian life. How does olfaction rate as an acute distance sense? The great sensitivity of the olfactory epithelium and the relative stability of many volatile substances suggest that stimuli can be perceived from considerable distances, but locating the source may be difficult. In still air, volatile substances diffuse radially, becoming less concentrated as the distance from the source increases. In moving air, the diffusion gradient extends downwind. In either situation, an animal using olfaction faces a challenging task in orienting and approaching the source.

Nonetheless, olfaction would have been important for early mammals. Transmission of olfactory stimuli is not impeded by obstacles, and olfactory signals can be used day and night. Because of high sensitivity, even by modest energy expenditure, a mammal can secrete a sufficient quantity of a volatile substance to be "noticed" easily above the background of environmental odors. Olfaction, despite its relative slowness and poor resolution for spatial interpretations, offers durability and an enormous variety of signals.

Audition

In contrast to the structural similarity in the sensory organs, of touch, taste, sight, and smell of mammals and other tetrapods, mammals, organs of hearing are elaborate and distinctive (Figure 19–20). The auditory apparatus of a typical mammal has three major functional units. One unit enhances the energy content and directional characteristics of sound stimuli. The second unit amplifies and transduces sound stimuli into fluid-borne analogs of the original sound. The third unit discriminates the frequency and intensity of sound and transmits this information to the central nervous system in the form of neurally encoded firing patterns.

The **pinna** (external ear) and narrowing external auditory meatus concentrate sound from the relatively large area encompassed by the external opening of the pinna to the small, thin, tympanic membrane. The pinna is unique to mammals although it may have its feathery analog in certain owls. Terrestrial mammal's auditory sensitivity is greatly reduced if the pinnae are removed. The pinna serves other functions as well. Many mammals use the position of the pinnae to indicate their emotional state, and several desert forms (best known among them are the jack rabbits) use their external ears as convective cooling devices.

The external ear is a unique mammalian character, but the middle ear, found in all tetrapods, is also highly specialized in mammals. The ear morphology that evolved in the earliest mammals greatly increased auditory sensitivity. The inner ear, where sound stimuli become encoded into neural impulses, is a series of fluid-filled membrane-lined chambers. Under most conditions the sounds tetrapods attend to are airborne. It requires considerably more energy to set the fluids of the inner ear in motion than most airborne sounds could impart directly. The middle ear receives the relatively low energy of airborne sound waves on its outer membranous end, the **tympanic membrane** or **eardrum**, and produces analogous vibrations in the fluids of the middle ear.

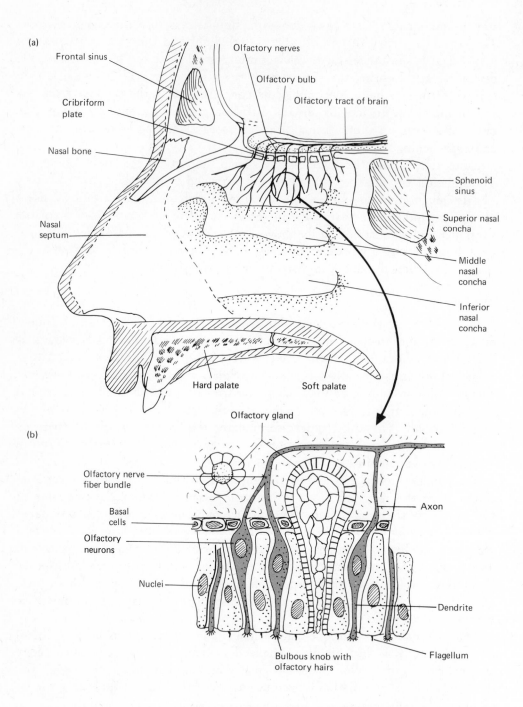

Figure 19-19   The sense of smell in mammals. (a) Cutaway diagram of saggital (medial) section of the human nasal cavity showing the distribution of the olfactory nerves and the nonolfactory nasal conchae. The cellular organization of the olfactory epithelium is enlarged. Note that the sensory cells are the actual nerve cells that penetrate the epithelium with a modified dendrite. (b) Dia-

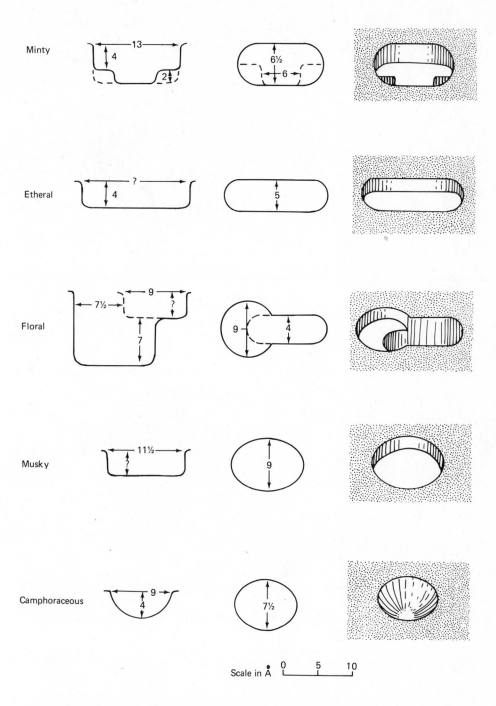

gramatic of a hypothetical basis differential sensitivity in olfaction. The three-dimensional shape of a substance determines whether or not it will lock into a sensory pit on the olfactory hairs. A sufficient number of "fits" results in sensory nerve firing. (b modified after Amoore, J. [1962] *Proc. Toilet Goods Assoc., Sci. Sect. Suppl.*, 37:1–20.)

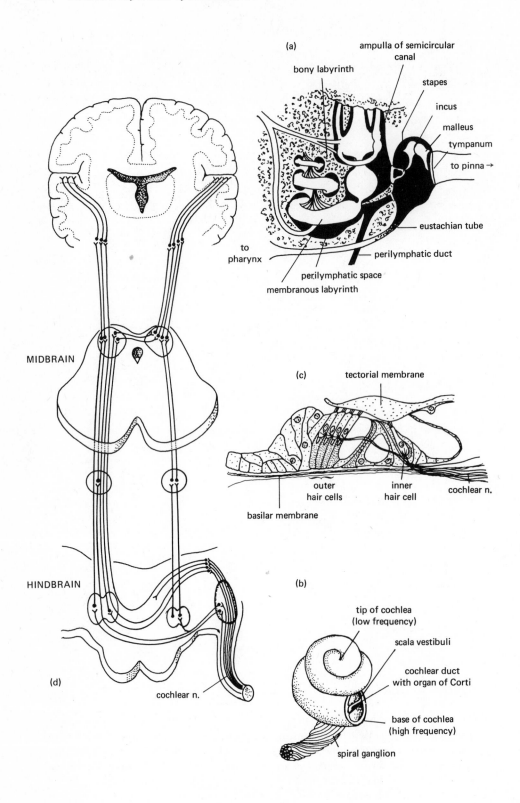

(a)

bony labyrinth

ampulla of semicircular canal

stapes

incus

malleus

tympanum

to pinna →

eustachian tube

to pharynx

perilymphatic duct

perilymphatic space

membranous labyrinth

(c) tectorial membrane

outer hair cells

inner hair cell

cochlear n.

basilar membrane

MIDBRAIN

HINDBRAIN

(d)

cochlear n.

(b)

tip of cochlea (low frequency)

scala vestibuli

cochlear duct with organ of Corti

base of cochlea (high frequency)

spiral ganglion

Two devices are used. First, the area of the eardrum is about 20 times greater than that of the oval window, which separates the air-filled middle ear from the fluid-filled inner ear. In all tetrapods the gap between the tympanum and the oval window is completed by bony elements, the ear ossicles. If mammals had a single bridging element, such as the columella of amphibians, reptiles, and birds, we would expect something on the order of a 20-fold increase in the force per unit area between sound reception at the tympanum and vibrational input at the oval window. Instead of a single bone, mammals have a chain of three ossicles linking the tympanum with the oval window. The **malleus, incus,** and **stapes** are interconnected and controlled by a series of fine muscles. Although there has been much debate about the exact nature of the ossicles' contribution, they seem to have a mechanical advantage that boosts the force of vibration on the oval window.

If the middle ear were an air-tight cavity, environmental pressure changes would interfere with auditory sensitivity. If the pressure in the environment were different from that of the middle ear, the tympanum would be distended and put under increased tension, losing some of its responsiveness to sound pressure waves. The auditory tube connecting the mouth with the middle ear alleviates this problem, but it is narrow and its response gradual. When a sound wave impinges on the tympanum, it is forced into the middle ear cavity, momentarily compressing the air in the cavity. As it moves outward the tympanum temporarily creates a reduced pressure in the middle ear. The pressure differences hinder free vibration of the tympanum and its attached ossicles by damping tympanic oscillations. Mammals have minimized these damping pres-

Figure 19-20   The sense of hearing in mammals.

(A) Diagram of the auditory apparatus. The pinna or external ear is that of a human and is simple in structure compared to that of some other mammals. The walls of the bony labyrinth are clearly separated from those of the membranous labyrinth by the perilymphatic space. In the cochlea this separation is not complete and isolates a perilymphatic space above and below the cochlear duct.

(B) Histological section of cochlea and diagrammatic section of a single turn (inset). The vestibular membrane vibrates in response to fluid movements in the perilymphatie scala vestibuli produced by stapedial excursions at the oval window. This in turn sets up vibrations in the membranes of the organ of Corti.

(c) Cellular structure of the organ of Corti. The sensory of hair cells experience distortion in their hairlike processes when relative movement between the basilar membrane and tectorial membrane occur. This distortion produces neural activity in the sensory nerve fibers.

(d) Neural pathways in audition show numerous synapses, decussations crossing overs to the contralateral side, and final dual representation in the cerebral hemispheres, which facilitate comparison of inputs from each ear. The basis for frequency discrimination lies in the relationship between a frequency's wavelength and the area of the basilar membrane maximally distorted. Hair cells nearest the base of the cochlea are distorted most by low frequencies and those near the tip by high frequencies. (d modified after Netter [9].)

sure differences between the middle ear cavity and the external auditory meatus by an enlargement of the middle ear cavity called the **tympanic bulla** (plural: *bullae*). This hollow bubble of bone creates a middle ear volume so large that the volume changes produced by oscillations of the tympanum result in insignificant pressure variations.

The auditory bullae are especially well developed in some desert rodents, including the kangaroo rats of North America and the jerboas of Africa and Asia. In these animals the bullae may constitue one third of the length of the skull and have a volume greater than that of the braincase. These nocturnal rodents are particularly sensitive to low frequency sounds, such as those made by owls in flight or by snakes moving across sand. Douglas Webster has shown that the form of these bullae is especially adapted to enhance the transmission of low frequency, low amplitude sounds from the tympanum to the inner ear.

Loud sounds occur in nature with considerable frequency, and a very strong stimulus might damage the inner ear or rupture the tympanum. Reflex contractions of certain muscles associated with the middle ear damp the excessive motion associated with loud sounds and protect the ear. This is the mechanism that allows bats to echolocate without being deafened by the emitted sounds (See Section 19.2.4).

The inner ear is phylogenetically the oldest of the three aural divisions. The earliest vertebrate fossils contain an inner ear apparatus similar to that of the living agnathans. Again, however, the complexity of the mammalian inner ear is considerably greater than that found in other vertebrates. In its simplest form the inner ear of a mammal is a cavity within extremely dense bone, the periotic or petrous region of the temporal bone. This cavity is filled with a water fluid, the perilymph, and has two flexible membrane-covered windows that face the middle ear cavity. Suspended in the inner ear cavity is a membraneous sac with walls that are lined with sensory epithelia and invested with fibers of the eighth cranial nerve (the statoacoustic, see Section 5.3.2). This sac is filled with another fluid, the endolymph, and fits loosely within the bony cavity.

Both the cavity and the membrane sac are complex three-dimensional structures as their names, the bony labyrinth and the membranous labyrinth, testify. They are better illustrated than described (Figure 19–20). Three general regions of these labyrinths can be defined: the **vestibule**, a central area adjacent to the oval window, containing some bulblike elaborations of the membranous labyrinth; a posterodorsal extension of curving tubes known as the **semicircular canals**; and the **cochlea**, an anteroventral spirally coiled tube of bony labyrinth. The cochlea is divided into upper and lower perilymph-filled canals by a loose fitting internal sleeve or tube of membranous labyrinth and endolymph.

The function of the various saclike chambers and semicircular canals is identical with that found in other vertebrates: the detection of acceleration in three-dimensional space. The cochlea is the site of origin of neurally encoded auditory signals. For our purposes it is merely necessary to note that vibrations in the perilymph initiated by the innermost of the middle ear ossicles (the stapes) at the oval window of the vestibule pass into the cochlea and travel all the way to the tip of the membranous labyrinth sleeve, around it, and back to the round window. In this long path, the pressure wave deforms the flexible endolymphatic channel, especially at points where pressure waves traveling to the tip of the cochlea are in phase with those returning from it. These sites of maximum deformation occur at different points along the length of the cochlea, depending upon the frequency of the vibrations which, of course, depend upon the frequency of sound falling on the tympanum. Arrayed along the floor of the cochlea membrane is a band of "hair cells" innervated by twigs of the auditory nerve. Above the hairlike extensions of these cells a long shelflike flap of tissue extends the length of the cochlea. When the membranous duct containing the hair cells is deformed by pressure waves in the perilymph, the "hairs" are bent by contact with the tissue flap and cause firing in their respective neural connections.

Frequency discrimination (humans can differentiate two frequencies that differ by 0.3 percent), broad sensitivity (covering 12 decibels from barely audible to painful), and directionality are thus functional characters of the mammalian ear. As a distance sense, audition has several advantages when compared with vision and olfaction. Sound is not as readily blocked by obstacles in the environment as is light, yet it is transmitted more directionally and faster than odors.

The advantages of hearing as a distance sense are emphasized by those mammals that have specialized in this mode of probing the world about them.

19.2.4 Echolocation

A variety of mammals emit sounds above 20 kHz (20,000 cycles per second) in frequency and thus above the range of normal human sensitivity (the 10 octaves between 0.002 and 20 kHz). Infant rodents emit such sounds, which seem to elicit retrieval behavior by adults. Some adult rodents, a few specialized marsupials, dermopterans, pinnipeds, many insectivores, microchiropteran bats, and odontocete cetaceans emit such sounds. The behavioral responses of these mammals indicate that they accoustically orient to the echoes of the high frequency sounds they emit. The most thoroughly studied echolocating mammals are the microchiropteran bats and the porpoises.

Bat echolocation

Because of their nocturnal and secretive habits, bats have until recently been little understood. Superstition and association of these diverse little mammals with the occult no doubt further inhibited direct observation and experimentation. Only in the last few hundred years have naturalists attempted to determine the means by which bats avoid obstacles under conditions where vision can be of little or no use.

Late in the summer of 1793, Lazzaro Spallanzani, an Italian naturalist and physiologist, then in his sixty-fourth year, visited his birthplace and summer home of Scandiano at the foot of the Appennine Mountains in Northern Italy. He had experimented with a tamed barn owl and found that the bird could navigate within a room by the light of a single candle. When the candle was extinguished, however, the owl promptly struck walls and obstacles as though it were blind. Anxious to extend this dim-light vision explanation of nocturnal object avoidance to other organisms, Spallanzani captured three bats which he took back to his room. Like the owl, the bats flew about the room by candlelight without collision. When Spallanzani blew out the candle, however, to his surprise the bat's flight did not change even though there was no light by which to see. Two years of active investigation by Spallanzani and his colleagues failed to explain how the bats navigated in the dark. Each of the senses was destroyed or occluded without seriously affecting the bats' performance. Only bats fitted with hoods struck walls and fell to the floor.

Having heard of Spallanzani's work, Louis Jurine, a surgeon and scientist in Geneva, undertook a similar series of experiments with significantly different results. Jurine observed that bats struck obstacles when their ears were plugged with paste or grease, but the same individuals flew normally when the plugs were removed. Hearing this, Spallanzani hastened to repeat tests on the auditory contribution to obstacle avoidance. He substituted pitch for the sticky plant material of his earlier experiments and other species of bats for the small one he had been using. Jurine was correct and Spallanzani elegantly extended the experimental procedure:

> I had two slender conical tubes soldered from very thin brassplate and introduced them with their thinner end in front into the ears and auditory openings of a bat. The tubes were externally covered with pitch, which served to fill up the space between tube and the deep concavity of the external ear and to attach the tube to the ear. In this way the air had no passage to the internal ear other than through the tubes. The animal (which could see) showed no influence of this impediment during its flight. But when I closed the tubes with pitch so that the air could no longer enter the auditory duct, the animal did not fly at all, or its flights were short and uncertain, and it frequently fell. This experiment which is so decisively in favor of hearing has been repeated with equal results both in blinded bats and in seeing ones.

Spallanzani also captured bats from the belfry of the cathedral at Pavia, removed their eyes, released them, and four days later captured some of these bats at the same roost. These blinded bats were as full of insects as any sighted bat, indicating that blinded bats could use some sense other than vision to capture insects on the wing and return to a home roost. Spallanzani thought that they heard the buzz of the insects' wings and that obstacle avoidance was based on a bat's ability to hear the sounds of its wings and body as they were reflected from objects. At the time of Spallanzani's studies there was a general ignorance of the properties of sound; that sound existed below and above the range of human audition was alien to scientific thought. Nevertheless, by the year of his death, 1799, Spallanzani had unequivocally confirmed Jurine's observations and extended them to a convincing hypothesis of auditory obstacle avoidance by bats.

Unfortunately this surprisingly modern, experimentally derived hypothesis was eclipsed by the scientific politics of the day. Baron Georges L. C. F. D. Cuvier, an eminent and influential French anatomist and contemporary of Spallanzani's, derided both the experiments and conclusions of the Italian. Without providing any new experimental evidence, he speculated that bats had extremely sensitive skin and, from the air stirred up during flight, could use the wing to obtain sensations of heat, cold, movement, and resistance indicating obstacles. The worldwide fame and authority of Cuvier at once brought praise for his "having brought order out of the chaotic state" of the experimental results of Spallanzani. Spallanzani's experiments eliminating the possibility of touch were never acknowledged. Spallanzani had found a bat's abilities unaltered even after coating the entire bat with as many as three coats of lacquer or when testing them in breezes that should confound such "touch" sensors.

At the beginning of the twentieth century, several researchers, often unaware of Spallanzani's work, independently concluded from experiments that ears were important for bat navigation. Without any experimental support it was also suggested that bats hear reflections of their high-pitched cries. In 1938 G. W. Pierce and D. R. Griffin, with newly developed high frequency acoustic detectors, reported that bats did emit intense "supersonic" sound synchronously with a "spitting" motion. However, in a flight room they could detect the high frequency sounds only occasionally and concluded that they were call notes or alarm signals. In the following years Griffin and R. Galambos at Harvard and S. Dijkgraaf at Utrecht independently discovered that flying bats produce ultrasonic cries that echo back from obstacles. It was also shown that bats have ears that are sensitive to these sound frequencies. They demonstrated a correlation between the rate of calls and obstacle avoidance, showing that no other sense was significant. They also rediscovered Spallanzani and, much to their credit, brought his 150-year-old work to the attention of the twentieth century. Subsequently Griffin and Galambos showed that basically the same calling behavior is used by bats for navigation and for capturing food.

Most microchiropteran bats are insectivorous, and many capture insects in flight. Although moths and large insects are acceptable, smaller insects are included in their diet—even insects the size of mosquitoes and drosophilid flies. The weight gain of bats that had been allowed to feed for a known period of time in a swarm of such tiny insects indicated that they must average as many as 10 mosquito or 14 fruit fly captures per minute. High speed photography has shown two separate catches in 0.5 second. By use of electronic transducers capable of detection of high frequency sound (ultrasonic) emissions, a three-phase hunting pattern has been defined.

The initial search phase of the little brown bat (*Myotis lucifugus*) is characterized by fairly straight flight and the emission of ten or so pulsed sounds separated by silent periods of more than 50 msec. The 10 pulses/sec are themselves about 2 msec in duration, and each pulse constitutes a downward sweep of frequencies starting at about 85 kHz and ending near 35 kHz. These bat calls are therefore frequency modulated (FM). Other bats use different pulse lengths and frequencies that vary from family to family of bats; some produce constant frequency (CF) calls that either terminate in a short downward FM sweep or, like the calls of some FM species, include several simultaneous harmonics. Despite our inability to hear their high frequency cries, many bats produce sounds of extraordinary loudness. Sound intensities higher than 200 dynes/cm (as loud as a jet airplane) have been measured 5 cm from the mouths of some bats.

The second phase begins as a bat detects an insect. Fruit flies and mosquitoes can be detected from a distance of about 1 m. The interval between pulses shortens, the silent intervals falling to less than 10 msec in *M. lucifugus*; 100 cries/sec, each lasting only 0.5 to 1 msec, are typical as the bat alters its flight path to intercept its prey.

Finally the terminal phase of the hunt is characterized by a buzzlike emission of ultrasounds. The intervals between pulses are less than 10 msec, the pulse duration is about 0.5 msec, and the frequencies drop to 25 to 30 kHz. The exact details vary from species to species, but the general behaviors are consistent. When the bat is within a few millimeters of the prey it often scoops with a wing or with the membrane between its legs and pulls the insect toward the mouth. Small insects are eaten directly, but large insects may be eaten on the wing or carried to a perch where the wings are discarded and the body eaten at leisure.

To accomplish these amazing feats, bats must hear and recognize high frequency echoes bounced off of the bodies of insects, determine target direction and distance with great accuracy, and be able to orient, approach, and capture the insect even though it may be moving. How do they do it?

The bat larynx is typically mammalian. Only in mammals is the larynx an organ capable of producing sounds of complex frequency and temporal modulation and variation. Situated just posterior to the hyoid, this box of cartilages and muscles encircles the esophageal end of the trachea. All inspired and expired air passes through its confines. The lumen of the larynx can be occluded by one or two pairs of folds of tissue that oppose each other across the tracheal lumen. The deepest of these folds form the vocal cords; they have thickened lumenal edges and considerable associated musculature. Sounds are produced by forcing air through the slit between these tensed sheets of tissue. Modulation of frequency and loudness is accomplished by alteration of the tension on the vocal cords and entire larnyx, the amount of air expelled per unit time through the structure and sometimes by extralaryngeal resonating chambers. Bats produce echolocation sounds without major modification of this basic mammalian larynx, although the whole structure is enlarged and the fleshy folds of the vocal cords are exceptionally thin. Nevertheless, the production of such brief, rapidly repeated and precisely patterned sounds seems a Gargantuan task for the tiny muscles controlling the vocal cords.

The sounds produced by the larynx are emitted either through the open mouth or the nose, depending on the family of bat. Mouth calls have a wide angle of dispersion (180° or more). The nose of those bat families that use it to broadcast is a complex structure with epidermal flaps and a nostril spacing that concentrates and focuses the sound in a narrow cone (less than 90°) in front of the bat, much like a megaphone. The calls travel through the air in radially expanding waves at 34 cm/msec. Because of the dispersion pattern, the amount of sound energy striking a target decreases as the square of the distance traveled. As a result a small object intercepts very little sound energy and thus can reflect very little. As the echo is reflected back toward the bat, its energy continues to diminish as the square of the distance.

Thus, the returned sound—despite its initial loudness—is exceedingly faint. In addition, only those wavelengths in the emitted call that are equal to or shorter than the diameter of the reflecting object will be returned. Despite these problems, bats can detect and locate remarkably small objects. Little brown bats can detect wires 1 mm in diameter from a distance of 2 m and wires of only 0.08 mm diameter from shorter distances. Even irregularities of surfaces can be located. Fish-eating bats apparently locate fish by detecting ripples on the water surface, and bats use echolocation to find the cracks in rocks that they cling to while they sleep.

Many features of an echo convey information. An object's size is indicated by the frequencies in the echo—large objects reflect longer wavelengths (lower frequencies) than small ones. The extremely high frequencies of the emitted calls are necessary to detect very small objects. The character of the reflecting surface is indicated by the character of the echo. A smooth, hard surface like the exoskeleton of a beetle returns a sharp echo, whereas a blurred echo indicates a rough surface like the body of a moth. The time required for an echo to return is directly proportional to the distance from the bat to the target, and the change in return time between successive calls can indicate the relative movement of the bat and its target. As a bat approaches a target, the call repetition rate increases, giving the bat more and more precise information about the target's location.

Several features of the morphology and neurology of the auditory system of echolocating bats contribute to the sensitivity of their hearing and their ability to process the information contained in echoes. The tympanic membranes and ear ossicles are small and light and are easily set into motion. Contraction of the middle ear muscles briefly damps the sensitivity of the ear as each cry is emitted; thus, the bat does not deafen itself. A padding of blood sinuses, fat, and connective tissue isolates the bony labyrinth of the inner ear from the rest of the skull and reduces the direct conduction of sound into the inner ear.

Perception of the direction of a returning echo is aided by the large, complex pinnae and by a neural mechanism known as contralateral inhibition. Stimulation of cells sensitive to a particular wavelength in the inner ear on one side of the head produces a transient desensitization of the cells that respond to the same wavelength in the ear on the other side of the head. The effect of that desensitization is to increase the contrast between the intensity of sound perceived by the two ears and, thus, to permit more precise determination of the direction of an echo.

One question that has long puzzled scientists is how an individual bat can discriminate between the echoes of its own call and those of other bats. When thousands of bats fly out of a roost in the evening the din must be enormous. One mechanism that probably contributes to a bat's ability to recognize the echoes of its own calls has recently been described. There is a brief neural sensitization following emission of a call to sounds of the same wavelengths that were emitted. The sensitive period begins about 2 msec after the end of the call and lasts about 20 msec. Because sound travels 34 cm/msec in air, this timing means that a bat is especially sensitive to echoes of its own call from objects at distances between $\frac{1}{3}$ and 4 m. This same mechanism probably helps a bat on its final approach to prey.

The behavior of some insects that are potential prey for bats indicates that a degree of coevolution has occurred. Some noctuid moths are sensitive to the ultrasonic sounds produced by hunting bats. Low intensity ultrasonic sounds cause the moths to fly away from the source, whereas loud sounds cause them to stop flying and drop to the ground. Some arctiid moths produce ultrasonic sounds themselves, and these sounds should be audible to bats. This moth family includes species that contain chemical compounds that make them unpalatable, and it has been suggested that their ultrasonic emissions advertise their distastefulness to bats, just as the bright colors and conspicuous displays of other insects advertise their unpalatability to predators that hunt by vision.

Cetacean Echolocation

A large number of modern mammals live in close association with water. Some are aquatic, and a few are truly pelagic. The most specialized aquatic mammals are the members of the Order Cetacea—the whales and porpoises. The terrestrial ancestors of cetaceans are unknown. Anatomical and embryological evidence suggests that they may have evolved from Paleocene sub-ungulates. The demands of aquatic life have so entirely reshaped cetaceans that there is little external indication of this relationship.

The two suborders of living cetaceans are the Mysticeti (the baleen whales) and the Odontoceti (the toothed whales). They appear to have radiated independently in the Eocene from a third suborder of primitive toothed whales, all of which are now extinct. Baleen whales, which filter small organisms from the water, are not known to use echolocation, although they produce complex vocalizations that probably are related to their complex social behavior. These are the whales whose "songs" are available on records. Calculations of the sound energy emitted suggests that their "songs" travel at least hundreds of miles through the sea. It is not inconceivable that all of the mysticete whales of a particular species in all of the oceans of the world could be in indirect vocal communication with each other.

The toothed whales share the same general modifications of body form for aquatic life that are seen in baleen whales, but are otherwise different animals. The toothed whales are much smaller than baleen whales. Although the largest species, the sperm whale, reaches a length of 18 m and a mass of 53,000 kg, most odontocetes are considerably smaller. Toothed whales are familiar to most people because many inhabit the littoral zone. There are even some purely fresh-water species of odontocetes—the Family Platanistidae has species in the Amazon and Orinoco Rivers of South America, the Ganges River in India, and Lake Tungt'ing in China. Most toothed whales eat fishes or squids, and their distributions are limited to areas of high productivity where food is abundant. Frequently, productive waters are also murky; thus many odontocetes must hunt in water in which vision is limited. Other species, like the sperm whale which feeds on squid, hunt at extreme depths where sunlight does not penetrate. Dives of 1000 m or more are probably routine for sperm whales. Odontocete evolution has solved the sensory problems in these habitats by an echolocation system as sophisticated as that of bats.

The difficulties inherent in any echolocation system are complicated for cetaceans by the acoustical properties of water. Sound travels about five times as fast in water as in air, and therefore the wavelength of any sound frequency is five times longer in water than in air. Because objects reflect only those wavelengths equal to or shorter than their diameters, cetaceans must use exceedingly high frequency sounds to produce wavelengths short enough for the detection of small objects. Not surprisingly the echolocation clicks of the bottlenose porpoise include frequencies from 20 to 220 kHz.

A second problem faced by mammals when in water is the difficulty of matching the acoustic impedance of the ear to that of water. Sound energy is not transmitted well across an air-water interface. About 99.9 percent of the energy that reaches such an interface is reflected and only 0.1 percent crosses the boundary. Terrestrial vertebrates face the same problem because the inner ear is water-filled. The middle ear converts sound energy transmitted through air to mechanical movement of bones, and thence to displacement of the fluid in the inner ear. The system does not work underwater, however, because a new water-air interface is created at the tympanic membrane and most of the sound energy is reflected from the tympanum back into the water.

A related problem is the efficiency of transmission of sound energy from water into body tissues, especially fat and muscle. A very high proportion of the acoustic energy that impinges on the body surface is absorbed and propagates through the tissues echoing back and forth to produce a diffuse buzzing noise, which conveys no directional information. Cetaceans have apparently solved these twin problems by abandoning the middle ear as a sound-receiving organ, isolating the inner ear from sound propagated through the body tissues, and conducting sound to the inner ear via a special fat body that extends from the lower jaw to the auditory bulla. The bulla is composed of extremely dense bone, which does not transmit sound readily, and is separated from the rest of the skull by soft sound-absorbing tissues. Thus the inner ear is isolated from sound approaching from other parts of the body. Kenneth Norris has proposed that the intramandibular fat body receives sound energy through an acoustic window formed by an area of very thin bone on the side of the mandible (Figure 19–21). Sound penetrates this thin bone readily and enters the fat body, which serves as a wave guide conducting the sound energy back to the bulla. The fat body terminates on the bulla itself where the bone of the bulla is extremely thin. Norris' hypothesis is that the sound a cetacean perceives is not received via the middle ear but instead comes from the lower jaw. In a test of the hypothesis, Theodore Bullock and Alan Grinnel used a speaker to beam sound energy at restricted portions of a porpoise's head while they recorded nerve cell activity in the auditory region of the brain. Sound beamed at the acoustic window of the lower jaw caused a large increase in nerve cell activity in the brain, but sound impinging on the side of the head did not.

Sound production underwater also poses problems not experienced by animals in air. Diving animals submerge with a limited supply of air, and it is probably undesirable to waste any of the air supply by bubbling air out of the mouth in the process of generating sounds. Norris has suggested that in cetaceans the site of sound production has shifted from the larynx to the complex system of nasal passages and diverticulae that extend upward from the pharynx to the "blowhole," the odontocete single external naris located on the top of the head. Air is forced back and forth in these passages to produce sound energy which is reflected forward from the broad anterior surface of the skull and probably focused by an acoustic lens formed by the oil-filled melon. This fat body is a characteristic feature of odontocete cetaceans, and it is responsible for the external shape of the head. The skull itself occupies a relatively small portion of the head. The melon of the sperm whale is the source of sperm oil that was especially valued for the clear smokeless flame it produced when burned in oil lamps. As much as 30 barrels of oil can be obtained from the melon of a large sperm whale.

Sound passes from one medium to another as it leaves a cetacean's head. Because the acoustic coupling between oil and water is good, little of the sound energy is lost by reflection as it would be in an air-to-water transmission, but the sound waves may be bent. It seems likely that the melon serves as a flexible lens, changing its shape from moment to moment to beam the sound energy in different directions. The melon of some cetaceans changes shape very conspicuously during echolocation. The echolocation sounds of the bottlenose porpoise are broad-band clicks (20 to 220 kHz). The highest frequencies in the click are emitted in a narrow beam directly forward and on a level with the animal's head. The lower frequency sounds are dispersed in wider vertical and horizontal arcs. In addition, an echolocating porpoise moves its head around as if scanning with its sound beam.

The ability of porpoises to detect objects and to distinguish between similar objects is remarkable. Many of Norris' early studies were conducted with a bottlenose porpoise named Alice. Though she was frequently described by Norris and his associates as the world's dumbest porpoise, when she was blindfolded, Alice was able to distinguish between two steel balls, one 57 mm

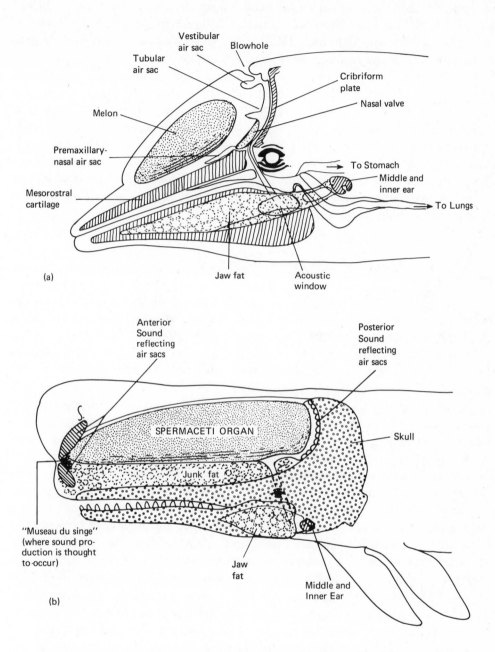

Figure 19–21 The proposed odontocete echolocation sound production and reception apparatus. Sound produced by air shuttled between air sacs through nasal valve is focused by the oil of the melon into a forwardly directed beam. Some sound may also be guided by the mesorostral cartilage. Returned sound is channeled through the mandibular (jaw) fat bodies and especially the acoustic window ('pan bone') of the lower jaw to the otherwise isolated and fused middle and inner ear. The apparatus is shown for (a) The porpoise (*Tursiops*) and (b) the sperm whale (*Physeter*). (Modified after Norris [10].)

in diameter and the other 64 mm, from a distance of 2 m in 1.5 sec. Blind-folded she could tell the difference between different species of fish or between fresh fish and day-old fish, and she was able to pick up her vitamin pill (5 mm in diameter) from the concrete bottom of her tank.

In their natural surroundings porpoises use their echolocation abilities to navigate and to find food. The low-frequency sounds that are beamed in a broad arc would reflect from large objects and are probably the important components of navigation, whereas the high frequency sounds that are con-centrated in a narrow beam are probably used to locate prey. The best sound-reflecting surface in fishes is the air bladder, and it is probably this structure that porpoises detect. It has been suggested that the reduced air bladders of some fast-swimming open-water fishes like tunas make them harder for cetaceans to detect by echolocation.

19.2.5 The Evolution of Mammalian Brain Size

Previously we asked why the earliest mammals had four or five times the brain volume of other tetrapods and more recent mammals four to five times the brain volume of the earliest mammals (Figure 19–22). Examination of the specializations of mammals and their sensory structures in particular suggests an explanation. The resulting hypothesis, although entirely speculative and admittedly oversimplified, seems plausible.

Active organisms require a three-dimensional method of perception of the world around them. In reptiles and birds, which are predominantly diurnal, vision provides this information. We have pointed out, however, that vision would not have been a totally adequate sense for early mammals, which were nocturnal. Smell and hearing appear to have been more likely senses for obtain-ing information about the world, and casts made of the inside of the brain cases of early mammals indicate that the regions associated with those sensory modalities were indeed enlarged. There is an important difference between the impressions perceived by visual means and olfactory or auditory perceptions. Provided that the eyes are placed so as to produce binocular vision, the visual impression arrives at the CNS as a three-dimensional image. Very little CNS processing is needed to perceive that one object is closer to the viewer than another. In contrast three-dimensional information about the world must be extracted from scent or sound largely by comparing the strength of a stimulus from one moment to the next, or as the head is turned from one side to the other. Thus, processing the information about its world would have required intergrative ability in the CNS of an early mammal.

This integrative ability probably had strong selective value in its own right. It should permit more complex and particularly more flexible behavior patterns. The ability to make associations between past events and a present situation is the basis of learning. This type of modification of behavior characterizes mammals and appears to distinguish their behavior from the often complex but

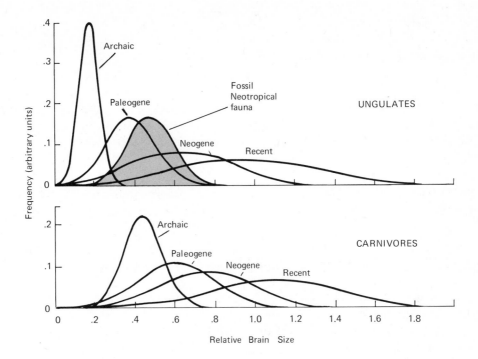

Figure 19-22 Progressive changes in the relative brain size of Northern Hemisphere mammals during the Cenozoic. The "Paelogene" includes species from the Paleocene through the Oligocene epochs, the "Neogene" species from the Miocene through the Pleistocene epochs. All archaic mammals tended to have brains smaller than more recent mammals, and archaic carnivores (perhaps because of the sensory coordination and complex behaviors required in prey capture) had larger brains than their ungulate prey. As mammals evolved, the average brain size increased as did the range of relative brain sizes. Increase in brain size may reflect the coevolution of prey and predator and the increasing complexity of defensive and offensive behaviors. The increase in the breadth of variation in brain sizes probably reflects differentiation of some ungulates and carnivores into niches eliminating the selective value of enlarged neural capacity. Such niches might include evolution of formidable size in ungulates or a scavenging habit in carnivores. Interestingly the unique mammals that evolved on the isolated South American land mass did not develop progressively larger brains. They were nearly completely replaced by late Cenozoic immigration from the northern fauna. (After Jerison [6].)

less flexible behavior of birds and reptiles. Selection for this sort of integrative ability may be the force that produced the second four to five fold increase in the brain size of mammals that occurred during the Cenozoic.

Integrative and associative capacity appears to be carried to the extreme in humans, although it would be rash to underestimate the complexity of the neural capacities of other mammals. Recent successes in communicating with chimpanzees (Chapter 1.2.2) and gorillas (F. Patterson [1978] *National Geographic Magazine*, 154(4):438–465) clearly demonstrate that many of our integrative and associative capabilities are not uniquely human. The features that we proudly consider unique attributes of human intelligence may be approached more closely by other vertebrates than we realize because these features evolved through natural selective processes that probably affected all mammals. Nobel laureate Konrad Lorenz has said:

> "I . . . consider human understanding in the same way as any other phylogenetically evolved function which serves the purposes of survival, that is, as a function of a natural physical system interacting with a physical external world."

In Chapter 22 we will examine more closely the precise physical external world of modern man's ancestors to gain insight into the evolution of our species.

19.3 REPRODUCTION IN MAMMALS

The mode of reproduction in mammals — including pregnancy, birth, and their prolonged and solicitous care of the young—is directly connected with their precise homeostasis, their adaptions to efficient food utilization, and their extraordinary neural functioning.

The structural complexity of mammals requires a great deal of time to develop. The active construction of a mammal from ultrastructure to gross morphology demands an unabating source of energy and building materials. Ready sources of energy, such as are provided by the placenta and later by lactation, are essential to mammalian development. During development the mammal is poorly adapted to its eventual role in nature. The dentition and masticatory, digestive, and locomotor skills of later life must await the time when the jaws are large enough to accommodate the appropriate teeth, the limbs of a size and strength for pursuit and escape, and so forth. Most other vertebrate young do not exist alongside their parents in the same ecological niche. The larvae of benthic fish float and feed on plankton; herbivorous lizards give rise to hatchlings that feed on insects. Many birds migrate to ensure nestlings an ample supply of insect food even though the parents may feed exclusively on fruits or grain. By a direct and continuous supply of energy to their young, mammals circumvent such drastic changes in life history and produce offspring that can assume their life-long habits as soon as they become independent. The prolonged dependency of the young mammal on the parents permits an exceptional conditioning of the nervous system. Young mammals learn feeding, escape, and other behavior patterns of great variety and flexibility during their associations with adults and litter mates.

It is not surprising that a characteristic upon which so many complex mammalian adaptations depend should itself be complex. The understanding of mammalian reproductive anatomy and physiology is not easy, but it is pivotal to understanding mammals.

19.3.1 Modes of Reproduction

Adaptations that increase reproductive success are highly favored. Because each new individual experiences natural selection, the fitness of the species, especially its reproductive fitness, is constantly readjusted. Vertebrates reproduce sexually and the means they employ to achieve fertilization and survival of the embryo are extremely varied and often elaborate, especially so in mammals.

In mammals reproductive adaptations have achieved a degree of elaboration seen in few other vertebrates. The major achievement has been the refinement of viviparity—literally defined as the bearing of living young—into a process where food for a developing embryo is supplied continuously by the mother and not from embryonic stores, such as the yolk. The refinement of viviparity in mammals therefore is associated with the evolution of the placenta. As might be expected, mammals show a gradation of adaptations from oviparity to viviparity.

Monotremes

Reproductively the platypus and echidna retain their basic reptilian heritage as egg layers (Figure 19-23a, b). The ovaries are bigger than in other mammals for they produce eggs with large amounts of yolk. As in reptiles, there are two oviducts, each opening into the urogenital sinus. The ovulated eggs are fertilized prior to entrance into the uterus. There they begin to differentiate, are coated with a leathery, mineralized shell, and are laid. Although both oviducts are present in monotremes, only the left duct functions. This is reminiscent of the situation in birds, where only the left duct functions; the right oviduct is reduced in size. Unlike birds, however, a recognizable shell gland secreting section is absent or, at least, there is no distinct separation between tube, isthmus, and uterus (Figure 19-23b).

The platypus generally lays two eggs, which are incubated in a nest. The echidna lays a single egg and carries it about in a pouch analogous to, but not homologous with, the marsupium of the metatheria. A shell tooth, like that in birds and reptiles, is possessed by each embryo to allow it to break out of its hardened embryonic home.

Maruspials

The female reproductive tract of marsupials combines structural features found in both monotremes and eutherians (Figure 19-23c, d). Following copulation, sperm pass forward from the vagina (the usual site of ejaculation of sperm in mammals) through the uterus and into the Fallopian tubes where fertilization takes place. The heavy secretions of the oviduct assist the sperm in reaching their quarry, the egg. Although the intromittent organ of male mammals, the penis, is a single organ, in many marsupials its distal end is forked to fit into each lateral vagina. As a result, following ejaculation sperm move into each lateral vagina to later enter the common confluence of the two uteri.

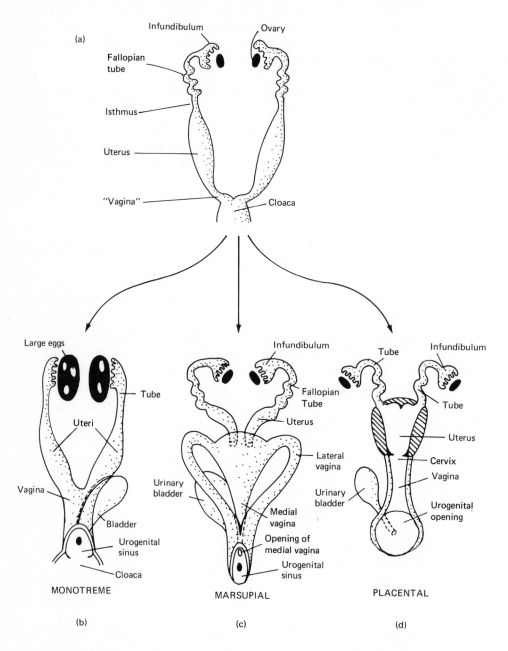

Figure 19-23 Female reproductive tracts of ancestral and major living mammals. (a) hypothetical ancestral mammal similar to living reptile; (b) egg laying monotreme; (c) marsupial showing complicated vaginal structures; (d) placental mammal as exemplified by an advanced primate.

Retention of marsupial embryos within the uteri is extremely short. Gestation takes 13 days in the opossum and only several weeks in kangaroos. The degree of prematurity at birth is probably best visualized by considering that the newborn of a large kangaroo, whose adult size equals that of man, is only 2–3 cm long. Usually the median vagina serves as the birth canal. The amount of yolk present in a marsupial egg, although considerably greater than the yolk in a placental mammal's egg, nevertheless is limited. A yolksac placenta is often formed but actual implantation, if it occurs at all, is brief. Following birth the incredibly "premature" embryos crawl from the vaginal region into the marsupial pouch (marsupium) where they grasp a mammary gland teat with their mouth. The teat swells so the baby cannot drop off. Usually more zygotes develop into embryos than there are teats. In the opossum, for example, there are about 13 nipples, but between 20 and 40 eggs are ovulated, fertilized, and gestated. This apparent waste reflects the difficulty that a newborn faces in finding and entering the marsupium and securing a teat.

Placentals

In the placentals the paired oviducts are partially fused to form a common vagina and sometimes a common large uterus, as in the human female (Figure 19–23d). In many placentals (some bats, rodents, and carnivores) the basically double uteri remain distinct as right and left halves, or are partly fused to form a T-shaped uterus. In all placentals the tubes, usually referred to as the **Fallopian tubes**, are small, short, and separate. As a rule fertilization of the egg takes place in the Fallopian tube, after the sperm travels from the internal end of the vagina, known as the **cervix**, through the uterus and up the tubes. The fertilized egg is propelled into the uterus by Fallopian contractions where it implants on the uterine wall and develops the embryonic maternal connections called the placenta. As embryonic-maternal dependency evolved in mammals, the importance of yolk diminished and, therefore, eutherian eggs are small.

Generally, the placenta of a mammal can be described either as a yolksac or a chorioallantoic placenta, or sometimes as a combination of both. The placenta is a highly vascularized and specialized extraembryonic structure, derived from the various extraembryonic membranes that evolved in the amniotic egg (Figure 19–24). In cleidoic eggs the allantois serves as a store for metabolic wastes while its vascularized membranes allow gas exchange through the shell. A placenta, as found in eutherians, can be envisioned as an internally retained cleidoic egg from which the shell has been removed and sufficient vascular connections developed between the embryonic and maternal circulations to allow for gas and nutrient exchange.

In addition, placentae are either nondeciduate or deciduate. In the nondeciduate placenta, all of the fetal and uterine membrane linings remain more or less intact throughout gestation. As a result, six distinct layers separate the fetal blood from the maternal blood (Figure 19–25). In a deciduate placenta varied degrees of erosion of the membranes occur and the number of layers

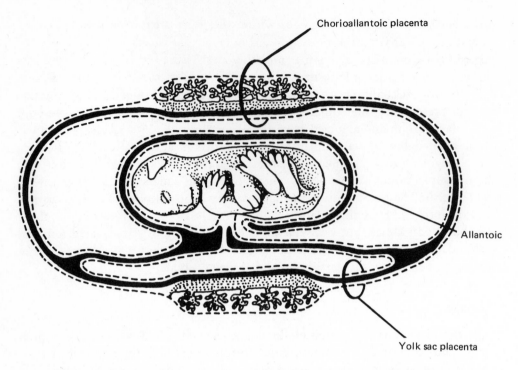

Chorioallantoic placenta

Allantoic

Yolk sac placenta

Figure 19-24   Two basic types of placental structures as seen in the cat. Both a yolk sac and chorioallantoic placenta are shown. The chorioallantoic placenta grows outward and supersedes the function of the earlier, primitive yolk sac placenta. (Modified after Ballard [2].)

decreases. In humans, only two layers remain; and in rabbits and some rodents, such as the guinea pig, fetal-maternal bloods are separated only by the fetal endothelial lining of the capillaries (Figure 19–25e). A reduction in the number and thickness of membranes between the fetal and maternal circulations increases the rate of gas and nutrient exchange.

Placentation, although a delicately balanced process, endows eutherians with specific advantages. In comparison to reptiles, birds, and monotremes, the placentals need not invest energy in laying down chemically specialized yolk for embryonic development because they can supply energy as normal blood sugars, proteins, and so on, via the placenta. Through retention, embryos are afforded the protection of the womb and develop at the high body temperatures typical of a mammal. Brooding eggs to speed embryonic development, as in birds, becomes unnecessary. Also, longer uninterrupted periods of time are available to the adult for seeking out food. The fetus can achieve a much greater size at birth and therefore a more advanced degree of development. Nevertheless, even in eutherians, adulthood comes slowly and parental postnatal nurture and protection are crucial to survival. The true value of mammalian viviparity is hard to ascribe, therefore, to any single feature. In mammals it associates best with the ideas that parents can direct more attention on fewer, larger offspring and these offspring, endowed with a highly evolved nervous system, do directly benefit from parental care by "learning."

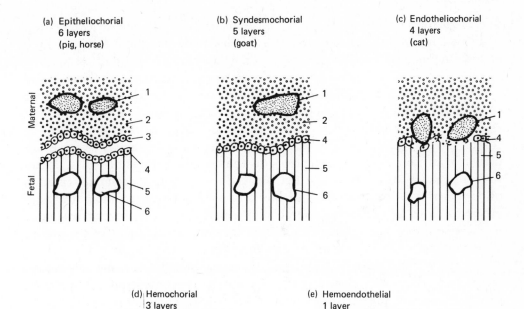

(a) Epitheliochorial
6 layers
(pig, horse)

(b) Syndesmochorial
5 layers
(goat)

(c) Endotheliochorial
4 layers
(cat)

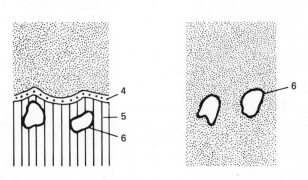

(d) Hemochorial
3 layers
(man, most rodents)

(e) Hemoendothelial
1 layer
(some rodents, rabbit)

Figure 19-25 Reduction in maternal uterine and fetal placental membranes through which nutrients and wastes must pass. (1) Maternal capillary wall; (2) maternal connective tissue; (3) uterine epithelium; (4) chorion epithelium; (5) fetal capillary wall. (Modified after Ballard [2].)

19.3.2 Hormonal control of reproduction

The sex of an individual vertebrate normally is a matter of inheritance. There are exceptions, however, and expressions of sex (as female or male) and sexual function (as fertile or sterile) can be altered by female or male hormones (estrogens and androgens respectively). In addition, reproductive cycles and their synchrony with environment are controlled by the release of sex hormones. This control is elaborate and exemplified exquisitely in mammals.

Reproductive cycles

A single reproductive cycle of the female mammal is termed **estrous**. Estrous is divided into four periods—proestrus, estrus, metestrus, and diestrus. Each period is defined by events that occur in the mature ovary. In addition, specific reproductive events are associated with each period of estrous, the durations of which vary greatly in different mammals. **Proestrus** starts each reproductive cycle, and is typified by initial growth of an ovarian follicle and its secretion of estrogen. **Estrus** begins with ovulation and modification of the follicle into the corpus luteum, which secretes progesterone. In many mammals estrus is the period of maximum female receptivity to the sexual overtures of males. **Metestrus** is a period of regression of the corpus luteum, with a declining secretion of progesterone. **Diestrus** is a variable period of quiescence before the next proestrus. In many primates metestrus is terminated by a menses, a sudden discharge of blood and cellular debris from the endometrial lining of the uterus. The reproductive cycle is therefore known as the **menstrual cycle**.

In many mammals diestrus is prolonged throughout the year, and females are receptive to males only once a year. Such species are monestrous. Polyestrous species commence proestrus following a short diestrus and are receptive several times during a breeding season or cycle continuously throughout the year. Except for the higher primates, most female mammals respond to copulation attempts by males only during estrus when ovulation takes place. The chance for fertilization of a ripe egg is thus improved.

Four hormones regulate the various periods of estrous—the anterior pituitary hormones, follicular stimulating hormone (FSH) and luteinizing hormone (LH); and the ovarian follicle hormones, estrogen and progesterone (Figure 19–26). The four hormones interact in a negative feedback loop to synchronize the various periods of estrous. Following the sequence of events from proestrus through diestrus in a polyestrous mammal, like the human female, provides insight into the complexity of vertebrate reproductive cycles. Reference to Figures 19–26 and 19–27 will greatly assist in understanding the sequence of events.

Menstruation in the human coincides with diestrus and is associated with low levels of all four hormones in the bloodstream (Figure 19–26). During this period the uterine lining is thinnest. Involved in initiation of estrous is a portion of the brain, the hypothalamus, and the pituitary gland, termed collectively the **pituitary axis** (Figure 19–27). Low estrogen in the blood is "sensed" by neurones in the hypothalamus, which respond by secreting a neurohormone (FSH releasing hormone) into the pituitary blood portal system. Upon transport to the anterior pituitary, this neurohormone influences the release of FSH into the blood. As menstruation (diestrus) proceeds, circulating FSH levels therefore increase. FSH has as its specific target the follicles of the ovary. Their growth does not proceed in the absence of FSH (Figure 19–26).

Proestrus commences as FSH induces follicle development and its secretion of estrogen. Estrogens have two primary effects in the "estrous cycle." First, they directly cause the uterine lining to grow and to enrich its vascularization

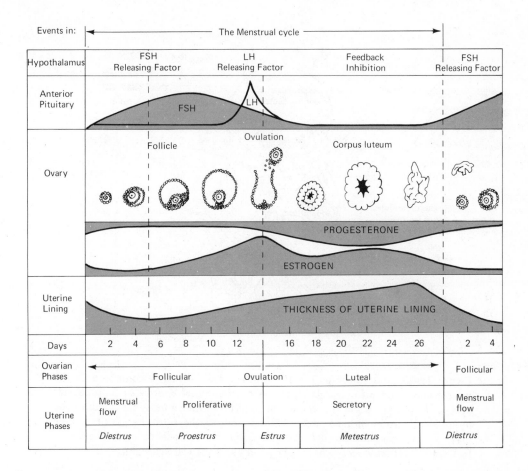

Figure 19-26 The interrelationships between reproductive structures and hormones during the menstrual cycle in the human female. Correspondence between the human menstrual cycle and the estrous cycle typical of most mammals is indicated at the bottom (uterine phases). The heighth of shading is proportional to the circulating blood levels of the indicated hormones. (Modified after Shearman, R. P. [1972] *Human Reproductive Physiology*. Blackwell Sci. Publ., Oxford).

(Figure 19–26). This is preparatory to the possible implantation of a fertilized egg. Second, the increased levels of estrogen feed back on the hypothalamus to shut off secretion of FSH releasing hormone. The higher levels of estrogens also stimulate the hypothalamus neurones to secrete a second releasing factor, LH releasing hormone (Figure 19–27). In an analogous manner to FSH releasing hormone, this neurohormone arrives in the anterior pituitary and causes release of the gonadotropic (gonad regulating) hormone LH.

At approximately this stage the reproductive cycle can be considered to pass into estrus. Follicular growth reaches the stage where ovulation occurs and the ripe ovum is shed into the coelom. LH directly stimulates the remaining follicle cells to enlarge and form the **corpus luteum**. This ovarian structure manufactures

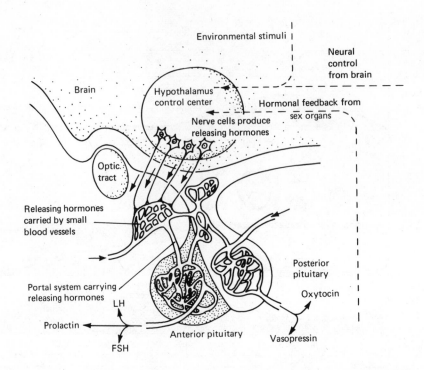

Figure 19-27 Relationships of the pituitary axis, which is composed of the hypothalmus and pituitary gland, in the neurohumoral control of reproductive cycles. Note that blood flowing to the anterior pituitary must pass through the hypophysial portal blood system. Blood flow to the posterior pituitary lobe may or may not pass through the portal system. (Modified after Shearman, R. P. [1972] *Human Reproductive Physiology*. Blackwell Sci. Publ., Oxford).

not only estrogen but also progesterone, which prepares the uterus for implantation by inducing uterine secretions and inhibiting contractions of its smooth muscle. The high levels of estrogen and progesterone feed back on the pituitary axis, inhibiting the release of FSH and LH. If pregnancy does not occur, the corpus luteum begins to degenerate, and reduced secretion of both estrogen and progesterone results in collapse of the thick uterine linings, which slough off as the menstrual flow. Metestrus can therefore be considered to end and pass into diestrus as the uterine lining begins to collapse.

In the human male both gonadotropins, FSH and LH, are secreted by the pituitary axis. Their target organs, as in the human female, are the gonads. FSH affects and stimulates the production of sperm in the seminiferous tubules of the testes. LH directly affects the interstitial cells of the testes to secrete the steroid hormone, testosterone, which along with FSH is essential to the maturation of sperm. Testosterone, like estrogen, feeds back on the pituitary axis and can reduce gonadotropic output. Despite this negative feedback, testosterone is relatively inactive in depressing the function of the pituitary, and a delicate balance is achieved: sperm and testosterone production proceed continuously. In several mammals, such as humans, a male reproductive cycle is therefore usually not described. In many mammal species, however, breeding activity and spermatogenesis are seasonal and correlate roughly with the timing of estrus.

If implantation occurs following copulation the female "estrous cycle" is interrupted and metestrus prolonged. The ovarian corpus luteum does not degenerate, but continues to produce high levels of estrogen and progesterone. Maintenance of the corpus luteum past its usual nonpregnant functional period is achieved by the secretion of HCG (human chorionic gonadotropin) from the placenta. HCG, although chemically different from LH, has a similar effect on the corpus luteum and induces continued secretion of estrogen and progesterone. Unlike LH, however, HCG secretion is not inhibited by high estrogenic levels in the blood, and continued increase in blood levels of estrogen and progesterone ensues. As the placenta grows, it becomes a new source of estrogen and progesterone (Figure 19–28). These placental secretions of estrogenic substances assure that the uterus retains its vascularity and nutrient supply to the developing embryo. As the placental source of estrogen and progesterone further increase, the levels of HCG secretion diminish and the importance of the ovary's corpus luteum in maintaining pregnancy vanishes.

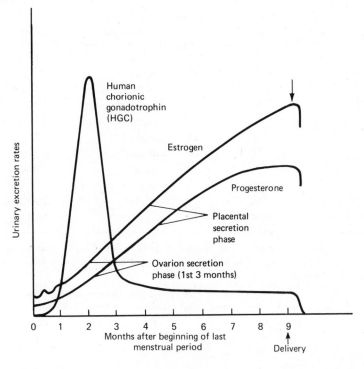

Figure 19–28 Changes in circulating hormone levels during pregnancy as evidenced by their presence in urine. During the first 3-months of pregnancy estrogen and progesterone are primarily produced by the ovarian corpus luteum; from the third month and until delivery these hormones are produced mostly by the placenta. (Redrawn from Shearman, R. P. [1972] *Human Reproductive Physiology*. Blackwell Sci. Publ., Oxford).

Just prior to delivery of the embryo (parturition) the hypothalamic peptide hormone oxytocin is secreted from the posterior pituitary gland (Figure 19-27). This hormone in combination with the high levels of placental estrogen and failing placental progesterone secretion initiate contractions of the uterus (onset of labor), which leads to expulsion of the fetus and its placenta.

In general, mammals tend to breed at times that will produce young when food is most abundant. This period often coincides with spring and early summer. As a result, mating and estrus may occur at a time of the year when environmental conditions seem less than favorable. How long an embryo is carried from fertilization to parturition, that is, the **gestation period**, varies greatly in different mammals. In general, however, larger placentals tend to have longer gestation periods than small mammals (Table 19-2). For small mammals, where gestation is short, mating can coincide with times of relative food abundance. For medium to large mammals, where gestation often exceeds 60 days or more, this is less likely and mating can occur in fall or winter to result in a summer birth period.

Table 19.2. Length of the Gestation Period and Estrous Cycle in Some Mammals (in days)

Nonprimate species	Gestation	Estrous cycle	Primates	Gestation	Estrous cycle
Mouse	20	4–5	Lemur	120–140	
Rat	22	4–5	Tarsier	180	(24)
Rabbit	63	**	Wooly monkey	139	
Cat	63	18	Rhesus monkey*	150–180	(27)
Dog	63	235	Chacma baboon*	180–190	20–36
Lion	110	21	Langur*	170–190	
Goat	150	20	Gibbon*	210	
Domestic cattle	278	21	Orangutan*	220–270	
Horse	340	20	Chimpanzee*	216–260	(36)
Dolphin	360		Gorilla*	250–290	
African elephant	660	42	Human*	267 avg.	(28)

Modified from R. J. Harrison and W. Montagna. [1973] *Man* Second edition. Appleton-Century-Crofts, New York, and Asdell, S. A. [1964] *Patterns of Mammalian Reproduction.* Second edition. Cornell University Press, Ithaca, N.Y.

*Old World monkeys and the great apes menstruate externally. Rhesus monkeys and humans show maximal behavioral changes during menstruation; other species have maximal behavioral change during estrus.

**Continuous; induced ovulation.

Lactation

One distinctive characteristic of mammals is the presence of mammary glands and suckling of the young. Following birth postnatal growth of the young is supported by the production of maternal milk in the mammary glands. This period, lacation, is a nutritional period that also is accompanied by other types of parental care. Lactation, like the "estrous cycle" and "pregnancy," is largely regulated by hormones, and also usually requires the presence of suckling young.

The high levels of circulating estrogen and progesterone during pregnancy induce enlargement of the mammary tissues and the accumulation of fats and connective tissue. Milk production does not take place until parturition has occurred. A rapid decline in estrogen and progesterone levels occurs when their source, the placenta, is expelled in the afterbirth. The hypothalamus thus released from feedback inhibition, secretes pituitary releasing hormones. In a manner analogous to FSH and LH releasing hormones, another releasing neurohormone, prolactin releasing factor, is secreted into the pituitary portal system (Figures 19-27 and 19-29). This factor instigates secretion of prolactin manufactured in the anterior pituitary gland. Prolactin directly stimulates milk production (Figure 19-29), but does not itself cause the delivery of milk. Suckling by the newborn infant stimulates nerve receptors in the nipple and this information is transmitted to the hypothalamus and on to the posterior pituitary. Oxytocin, the hormone involved in the induction of labor, is released and causes milk ejection by inducing contractions of the smooth muscles of the breast (Figure 19-29). Suckling by young maintains lactation. In many mammals, estrous is not initiated during the period of lactation for FSH secretion is low. In others, however, estrous quickly begins again. In the human female individuals vary. In some mothers menstruation is delayed for 3 to 4 months during lactation, in others the menstrual cycle commences soon after birth.

Studies of primitive hunting and gathering peoples of the Kalahari Desert of southern Africa who live in an essentially Paleolithic culture (Figure 22-15) indicate that before the rise of agriculture and animal domestication humans and human ancestors may have had less variable reproductive cycles. Although the diet of this completely nomadic culture consists entirely of nuts, wild vegetables, and meat, it is nutritionally well balanced and provides an adequate number of calories. Because the mothers have no soft foods to give to their babies, they continue to nurse them for 3 or 4 years. Continual breast feeding appears to be sufficient to maintain the neurohormonal circuit to sustain lactation. Relative to other societies the women have a low fertility; the average length of time between giving birth is 4 years.

Although the causes of this low fertility are not clear, research in American hospitals on "abnormal" menstrual cycles provides a clue. Among women of modern cultures, a certain weight loss causes a cessation of menstruation and a weight gain restores the cycle. Apparently a minimum level of stored, easily mobilized energy is necessary for ovulation (and hence a subsequent menstruation) in the human female. Women trained in ballet or competitive sports such as long-distance running and gymnastics are often non-ovulatory. Their level of body fat is usually below 15 percent of their total body weight – the apparent cessation point for ovulation.

The nearly 1000 cal/day extra needed for lactating may deplete these reserves in women living in primitive cultures to a level so low that ovulation ceases. We do know that when these nomads attach themselves to agricultural

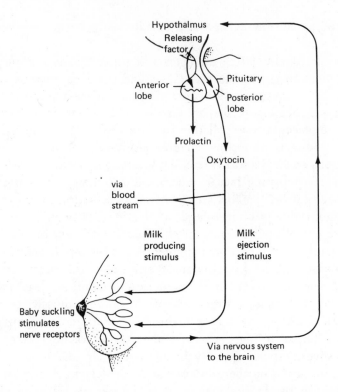

Figure 19-29 Role of the pituitary axis in lactation. Note that milk production and secretion are promoted by the anterior pituitary hormone prolactin, and milk ejection by the posterior hormone oxytocin. Both of these processes are initiated via a neural connection from the nipple to the hypothalmus, a complex refex which is stimulated by suckling. (Modified after Shearman, R. P. [1972] *Human Reproductive Physiology*. Blackwell Sci. Publ., Oxford).

and cattle-herding tribes at the periphery of the desert, drastic changes occur within a single generation. Their diet includes a great deal of cow's milk and grain, the individual becomes fatter, and women wean their babies sooner by feeding them grain meal softened with cow's milk. The birth interval drops 30 percent, as would be expected from the shortened periods of lactation and larger amounts of body fat. The children are taller, fatter, and heavier than their nomadic counterparts. Young women reach the age of menarche (first menses) earlier, and this change combined with the increased frequency of births has caused a rapid population growth in these agrarian people compared to their nomadic relatives. It is tempting to speculate that these events and their effect on the hormonal control of reproduction might reflect what took place during the Neolithic period (Figure 22–15). Prehistoric artifacts indicate that a human population expansion coincided with mankind forsaking the life of hunting and gathering and beginning to farm and to keep herds of domestic animals.

Although complex, hormonal effects and interactions in vertebrate reproductive cycles show a surprising uniformity. The gonadotropins FSH and LH are present from fish to mammal, and estrogen and testosterone are consistently associated with femaleness and maleness respectively. Different life styles and/ or evolution in isolation have led to differences in reproductive strategies and thus have emphasized different aspects of the endocrine control of reproduction. A fitting example is the reproductive difference that exists between the marsupials and the placentals.

19.3.3 Metatherians versus Eutherians

Current theories suggest that the marsupials and placentals evolved in isolation (Chapter 15). Their divergence commenced before or in the early Cretaceous. Fossil eutherians are known from Asia in that period and marsupials from North America. Gestation in marsupials seldom lasts more than 30 days. The fetus does not usually implant in the uterus until the last few days of gestation. Instead a shell-like membrane surrounds the embryo, and most of prenatal development depends upon consumption of yolk. It is a moot point whether most of marsupial reproduction is viviparous or ovoviviparous. In any case the embryo is born at an early stage of development and, unless it finds its way into the marsupium and a mammary teat, its survival is impossible. In most placentals, the embryo is born at a comparatively advanced stage of development and, although dependence upon parental nurture, care, and guidance is often prolonged, the young may be quickly weaned and begin to cope directly with the world. Humans are one of the notable exceptions to the general mammalian pattern of early weaning and independence.

Many biologists have suggested that natural selection acts most directly on factors that affect reproductive success. How reproductive success should be measured is a matter of some controversy. In Chapter 15 we presented the contention, made by a variety of biologists, that marsupials are somehow inferior to placentals because, when they are thrust together, the placentals dominate. The invasion of South America by modern placentals during Miocene-Pliocene time and the rapid elimination of the endemic marsupial fauna represents a major example. One reason proposed is that somehow the marsupial reproductive mode is inferior to the placental mode. The studies of Geoff Sharman and his colleagues in Australia, and of a growing number of other researchers, discredit this contention. Sharman, more recently John Kirsch, and others have pointed out that on the average marsupials invest less energy in reproduction than placentals.

The argument hinges on the fact that marsupial development is largely a postnatal cost (lactation), and placental development largely a prenatal cost (placentation). The latter involves a large maternal investment in intrauterine structures to assure development. When food is scarce, it is usually the mother who suffers, for the embryo's demands continue in spite of external environmental conditions. In a placental mammal, retention of the embryo is paramount to ultimate reproductive success, and natural abortion is rare. In many rodent populations, for example, limited food resources tend to restrict the initial number of zygotes implanted and not the number of fetuses aborted. In marsupials unfavorable food resources result in what might be termed "marsupium abortion," for lactation declines. The young, therefore, suffer directly and can be entirely abandoned. After parturition, placentals, of course, can also abandon young and do, but the energy already invested in reproduction is high. The marsupial reproductive plan may be more successful than that of placentals when environmental conditions at the time of birth are poor, for less energy has been invested in prenatal development. If this is a valid argument then other, nonreproductive factors must account for the so-called "superiority" of placental mammals.

Some suggest the difference may reside in a lesser intelligence of marsupials as compared to placentals. This possibility seems to be supported by data on comparative brain size, reviewed by H. J. Jerison, and studies of social organization by D. Hunsaker, J. H. Kaufmann, and others. The brains of living marsupials are only one quarter to two thirds the size of the brains of placentals with a similar body size. Primitive marsupials, such as the didelphids (New World opossums), have a solitary life style. Individuals only occasionally come together, react aggressively, and avoid further interaction. Estimates based on behavioral studies of the Virginia opossum indicate that a 3-year-old adult male may have spent less than 17 percent of his days in association with other opossums, including those spent in the mother's pouch! Even in the most socially advanced marsupials, the kangaroos, field studies indicate that only "a weak, individualistic mode of social organization" exists among adults, although mother–offspring relationships are as complex as in social placentals.

Despite a recent expansion of studies in marsupial neuroanatomy, no basis for their limited brain size nor any instructive differences in the "wiring" of the marsupial versus placental brain have been reported. The marsupial central nervous system seems to be an alternative way of organizing certain neurons for functions similar to those seen in placental mammals. Likewise, the differing modes of reproduction appear to be alternative schemes of investment of reproductive effort. Perhaps the marsupial plan offers greater adaptation to short-term unpredictabilities in the environment. Whether this reflects past climatic history on the isolated southern continents where most of marsupial radiation took place awaits further understanding of paleoclimatology. The alleged marsupial "inferiority" has yet to be explained even though faunal replacements of marsupials by placentals are well documented.

19.4   SEX DETERMINATION AND SEXUAL DIMORPHISM

Whether marsupial or placental, the long term behavioral and physiological roles of mammalian parents with respect to their offspring are precisely defined and absolutely essential for reproductive success. Only the birds have a similarly demanding parental role. A female mammal must be finely tuned physiologically and behaviorally if her investment in offspring is to survive. In species where the males participate in reproduction after insemination, they too must co-ordinate specific behaviors with their mates and offspring. Females must function as females and males must function as males if reproductive success is to sustain the genome. The determination of gender and with it future sexual roles is of the upmost importance to mammals. A similar argument can be made for birds.

The gender of most mammals is expressed by genitalia and other secondary sex characteristics and is obvious from birth. In other classes of vertebrates, however, genitalia and sexually dimorphic structures are absent or not expressed until maturity is achieved. To discuss sex determination in all vertebrates therefore we require a basic definition of sex: the female sex produces eggs; the male sex produces sperm. Egg production requires a functional ovary and sperm production a functional testis; each structure has a distinct cytological appearance. Therefore in species where sex is not declared by external sex structures, direct examination of the gonads should, and generally does, reveal the sex of an individual. However, in many vertebrates both an ovary and a testes are present in the same individual. By definition these individuals are both male and female, a seemingly bizarre condition, for in mammals individual sex is discrete and segregated into strong female and male roles. Given this wide spectrum of vertebrate sexuality, we must ask what determines individual sex. If some vertebrate individuals can function as mixed sexes, what is the purpose of rigid sex segregation?

19.4.1 Sex and Sex Chromosomes

In all vertebrates the gonad is initially indifferent and capable of producing an ovary or a testis (Figure 19–30). Individuals with both types of gonads present are termed **hermaphrodites**; individuals with either ovaries or testes are **gonochorists**. Hermaphroditism is common in lower vertebrates, especially fishes, but virtually absent among amniotes, particularly birds and mammals.

In birds and mammals, and probably most reptiles, sex reversals are atypical. A mammal's or bird's sex is genetically prescribed by the sex chromosomes. In humans, for example, zygotes of females contain 23 similar chromosome pairs, but in males one chromosome (Y) is much smaller than its mate (X). Males therefore are chromosomally XY, a condition called heteromorphic. In males the Y chromosome carries a gene dominant for maleness. Sex chromosomal anomalies are fairly common in humans. Individuals with only one X chromosome (XO) are female, although infertile. Individuals with sex chromosomal constitutions of XXY, XXXY, XXXXY, XXYY, and XXXYY are phenotypic males, but sterile. Males of XYY are fertile. The presence of the Y chromosome directs the medullary layer of the indifferent gonad to differentiate as a testis. In contrast its absence and the presence of an X chromosome causes the cortical layer to develop as an ovary (Figure 19–30).

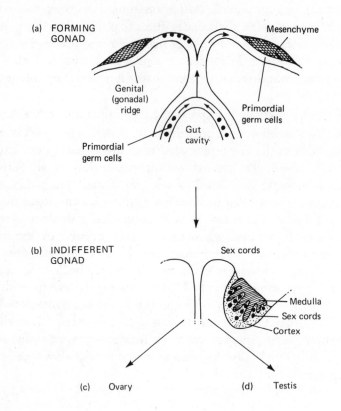

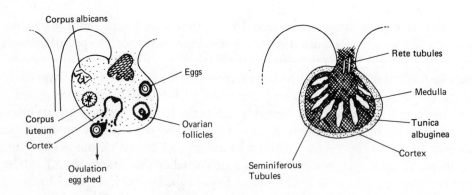

Figure 19-30 Generalized scheme of modification of the indifferent vertebrate gonad into an ovary or a testis. The gonadal structure whether primarily cortex (♀) or medulla (♂) are derived from embryonic mesoderm. (a) The primordial germ cells, which give rise to eggs or sperm, are initially located in embryonic endoderm and migrate through the mesenteries during development into the indifferent gonad. (b) The germ cells arrange themselves between the medullary and cortical tissues of the indifferent gonad. (c) In formation of the ovary the cortical tissue predominates. (d) In the testis the medullary tissue predominates.

The sex chromosomes influence only the sexual development of the gonads, that is, the primary sex characters. Precisely how the sex genes act is unclear, but once a gonad has had its primary sex declared as female or male, the sex hormones estrogen and testosterone are produced. These hormones effect the development of the secondary sex characters—all the structural and behavioral differences that exist between male and female. In humans the genitalia, breasts, hair patterns, and differential growth patterns are secondary sex characters. Horns, antlers, and dimorphic color patterns are familiar differences that we associate with sex in other mammals.

Evidence shows that sex hormones can influence an individual's secondary sex characteristics. The development of so-called freemartins in cattle provides a natural example. Twinning in cattle sometimes results in the actual mixing of fetal blood supplies. If one twin is male (XY) and the other female (XX), the female is influenced by the testosterone secretions of the developing male. At birth the female calf (a freemartin) shows a mixture of male-female traits and is infertile. The genetic male calf, however, is normal. The evidence suggests that in mammals testosterone leads to maleness, even in potentially genetic females.

In birds and reptiles heteromorphic sex chromosomes can be distinguished cytologically, but the female is the heteromorphic sex and the male the homomorphic sex. To distinguish this difference geneticists use alternative symbols: ZW for female, and ZZ for male. The dissimilar chromosome W is dominant and directs the indifferent gonad to develop as an ovary. In the cleidoic egg equivalent of freemartinism—double-yolked eggs where the embryonic blood supplies mix—femaleness dominates. ZW females are normal and ZZ males sterile intersexes. The conclusion is that estrogen effects are dominant over the effects of testosterone in these vertebrates.

In amphibians and in fishes heteromorphic sex chromosomes have seldom been demonstrated. Breeding experiments clearly indicate, however, that male- and female-determining genes are distributed over several chromosomes. Because of their morphological compatibility the "sex chromosomes" of fishes and amphibians can cross over and exchange genes, a circumstance essentially forbidden for the dissimilar sex chromosomes of mammals, birds, and reptiles. As a result, intersexuality and several types of hermaphroditism, including functional sex reversal, are widespread in anamniotes.

19.4.2 Sex Determination and Mammalian Evolution

Several questions about vertebrate sex-determining mechanisms are pertinent to a consideration of mammalian evolution and characters. Why has sex determination been fixed rigidly in birds and mammals by noncompatible heteromorphic sex chromosomes? Why is the heterogametic sex female in reptiles and birds and male in mammals?

Certainly it would be disruptive to reproduction if the activities and morphologies associated with maleness and femaleness in birds and mammals were labile. Male and female parents may play different roles in care of the young. If, for example, male baboons wavered in fending off potential predators while females hesitated to herd the young to safety, the result could reduce reproductive success. The significance of the evolution of hetermorphic sex chromosomes in birds and mammals appears to be related to the prolonged period of postnatal care provided to the young by parents. In contrast to birds and mammals, most reptiles, amphibians, and fishes usually ignore their young. In these classes, without the demands of complex postnatal care, the rigid genetic fixation of parental roles via heteromorphic sex chromosomes may be superfluous. Thus, although it is unlikely that any single factor explains fixation of sex determination, we suspect the behavioral complexity of birds and mammals is a significant component.

In addition to the complex postnatal care of the young characteristic of birds and mammals, both classes contain many species that form large groups. The social behavior of individuals in these groups often involves distinct male and female roles. Even in species that normally form only pair bonds there are usually sexually dimorphic behaviors involved in the formation and maintenance of those bonds. At least in mammals, fixed sex determination may have been essential for the exploitation of the behavioral complexity that was made possible by the increased integrative capacities associated with larger brain size (Section 19.2).

The opposite sexual expression of heterogametic individuals in mammals on one hand and birds and some reptiles on the other is intriguing. In gonochoristic vertebrates, the heterogametic sex is physiologically dominant. Carrying young *in utero* as placental mammals do, however, places strictures on the determination of sex. Ursula Mittwoch has pointed out that during pregnancy maternal estrogens can cross the maternal-fetal barrier. For a femal fetus (XX) this presents no problem, but a male (XY) would be swamped by estrogenic effects (femaleness) unless a strong antiestrogenic agent existed. Mittwoch has suggested that this crossing of the placental barrier by maternal hormones explained the dominance of testosterone in directing the primary sex differentiation of mammals. In birds the cleidoic egg isolates the embryo, whether male or female, from direct maternal influence. Genetic sex, expressed in the relative safety of the egg, could equally well be determined by female heterogamety (WZ) or by male heterogamety (XY). Exactly why female heterogamety occurs in birds in not clear from this reasoning. Perhaps it was a matter of evolutionary chance.

19.5   GENETIC EVENTS IN VERTEBRATE EVOLUTION

The increased complexity of morphological, physiological, and behavioral changes associated with vertebrates, from fish through mammal, is rooted in genetic changes—the accumulation of mutations and their tuning through natural selection. Many biologists feel that vertebrate evolution has advanced faster than can be accounted for by typical gene mutation rates. One suggestion to account for this seeming anomaly, popularized by Suhno Ohno, is based on the conclusion that the total number of gene loci in vertebrates has increased during evolution, particularly via a gene duplication process termed **tetraploidy**.

In **tetraploidy** the entire genome is doubled from a diploid to tetraploid state. Every pair of alleles therefore is complemented by an identical second pair of alleles. The importance of this excess genetic information lies in the freedom it permits for mutations to occur. Most mutations are harmful—the mutated gene's product does not function as well as the product of the normal allele. When there are two pairs of alleles for each function, one of the allelic pairs is freed to accumulate mutations that in the diploid state would have been deleterious. The other member of the allelic pair continues to perform the original function. To be retained it is only necessary that these mutations are not detrimental. These extra genes become the raw material for evolution. The potential benefit of such genetic additions that lead to increased complexity of vertebrates is the basis of Ohno's theory.

What about the factual basis of the theory? Did tetraploidy actually occur in vetebrate evolution? The answer is yes, but exactly when and in what groups it took place can only be approximated. Most geneticists agree that simpler forms of life have fewer genes than complex forms. Because genes are portions of linearly structured molecules of DNA, the number of genes in an organism is proportional to the nuclear DNA content of its cells. The number of genes in a species, its genome size, can be estimated by measuring the nuclear DNA of its cells. At best DNA is a crude index of actual number of functional genes. In many vertebrates much of the DNA is nonfunctional or represents redundant replicates of the same gene. Nevertheless, comparison of different vertebrate groups reveals increases in nuclear DNA that broadly correlate with phylogeny. Professor Ohno uses the diploid DNA content of human cells, which is 7×10^{-9} mg/cell nucleus, as a standard reference and assigns it a relative value of 100 percent. The relative nuclear DNA content of different vertebrates ranges from a low of 14 percent in some teleosts to values of 3400 percent in urodeles and dipnoans (Figure 19–31). The exceptionally high DNA content in dipnoans and urodeles represents redundant replications of single genes and is not the result of tetraploidy.

Professor Ohno thinks that tetraploidy occurred at least once and perhaps twice during vertebrate evolution. Ascidian larvae contain only 6 percent DNA, amphioxus 17 percent. Quite likely some related deuterostome underwent polyploidy in the evolution of the first vertebrate, which would have contained approximately 20 to 25 percent DNA. Living agnathans, and especially myxinoids, appear to have increased this original DNA content through replicative increases in individual gene loci, not by additional polyploid steps. In agnathans a wide variety of natural experiments in gene duplication, mutations, and recombination must have taken place, as expressed by the rapid radiation of the ostracoderms in Silurian-Devonian time (Chapter 5). Ohno postulates that the end result in some of the ostracoderms was a genome size 40 to 50 percent that of humans.

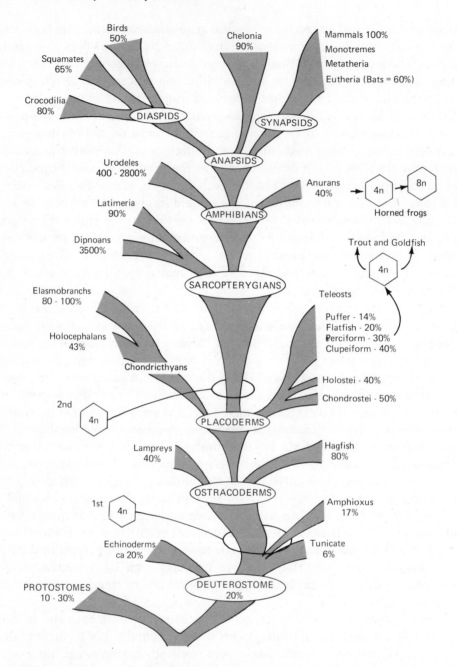

Figure 19-31 A generalized scheme of the nuclear DNA content of representative vertebrates and other animals. DNA content is expressed as a percentage of the nuclear DNA content of humans (and most mammals), which is assigned a value of 100 percent. Two major tetraploid duplications of the entire genome are postulated; the first for the stem vertebrates, the second for the sarcopterygians. Additional minor tetraploid duplications have occurred recently in fishes and amphibians. Extremely high DNA content, as measured in dipnoans and urodeles, is the result of tandem duplication of specific genes, and not of equal duplication of the entire genome. (Based upon data presented by Ohno [11].)

It is unclear whether the clade leading to teleosts underwent an additional early tetraploidy, but the retention of 48 diploid chromosomes in most fishes, which is the same as in myxinoids, suggests not. What may have occurred in chondricthyans is even more uncertain. The holocephalians have DNA values associated with the first tetraploid doubling, suggesting that no second polyploidy has occurred. In elasmobranchs the DNA index is as high as in most amniotes, and we can infer that some sort of gene duplication took place. Whether a second genome doubling occurred through polyploidy, or whether their high DNA reflects redundant gene duplications as postulated for the myxinoids is not clear. In any case at some time during the Devonian a second major tetraploidy appears to have increased the DNA content up to 80 to 100 percent that of present human levels.

Except for the dipnoans and urodeles, most living derivatives of the sarcopterygians retain evidences of the high DNA contents duplicated during the second tetraploidy (Figure 19-24). Chelonians and crocodilians have DNA indices of 80 to 90 percent, most mammals 100 percent, snakes and lizards 60 to 70 percent, and birds average about 50 percent. Anurans are highly variable in DNA content and actually provide the clearest evidence of tetraploidy in vertebrates. In several South American horned frogs, DNA content, chromosome numbers, and the karyotype suggest that a double tetraploidy has taken place in recent times (Table 19-3). Similar types of cytological evidence of recent tetraploid duplications are known in other anurans and in certain cyprinid and salmonid teleosts (Figure 19-31).

The different DNA values for vertebrates imply that extreme specialization for particular life styles may lead to a reduction in DNA. Teleosts provide a good example. Presumably early chondrosteans possessed a genome equivalent to 50 percent that of humans, produced by the first vertebrate tetraploidy. With continued mutational evolution of the actinopterygians, DNA reduction occurred through losses of the trivial replicated DNA. Thus, holosteans contain 40 percent DNA, as do the more primitive teleosts (Figure 19-31). More advanced teleosts show continuing losses to as low as 14 percent. Apparently, as these fishes have become more specialized for their particular mode of life in water, their excess DNA has been eliminated. Amniotes seem to reveal a similar

Table 19-3. Gene duplication via tetraploidy in South American horned frogs. The approximate doubling of DNA content and chromosome number (karyotype) in these closely related frogs are direct evidence for the sudden duplication of the genome of a vertebrate.

Species	2n Chromosome No.	Percent DNA
Odontophrynus cultripes	22 (2n)	70
Odontophrynus americanus	44 (4n)	125
Ceratophrynus dorsata	104 (8n)	250

pattern. Presumably starting with a genome size of 80 to 100 percent, from a possible second tetraploidy, those tetrapods specialized in flight, the birds and bats, have reduced their nuclear DNA to 50 to 60 percent. All other mammals are from 90 to 100 percent, the squamates intermediate, whereas the crocodilians and chelonians, which have been morphologically stable since the Mesozoic, retain high DNA values. Perhaps specialization for a particular life style results in reduction of the genome size to its barest essentials. Reduction of genome size may reduce the possibility of viable alteration appearing in non-functioning DNA in highly integrated systems, such as those seen in flying vertebrates. Reduction of DNA must also mean a reduction in potential adaptation. Possibly birds and bats are genetically rigidly locked into their life styles.

References

[1] Alexander, R. McN. 1968. *Animal Mechanics*. University of Washington Press, Seattle.

[2] Ballard, W. W. 1964. *Comparative Anatomy and Embryology*. Ronald Press Co., New York. A general introductory text to vertebrate anatomy with heavy emphasis on development and function.

[3] Campbell, B. G. 1966. *Human Evolution, an Introduction to Man's Adaptations*. Aldine Publishing Co., Chicago.

[4] Dijkgraff, S. 1960. Spallanzani's unpublished experiments on the sensory basis of object perception in bats. *Isis* 51:9–20.

[5] Griffin, D. R., F. A. Webster, and C. R. Michael. 1960. The echolocation of flying insects by bats. *Evolution* 8:141–154.

[6] Jerison, H. J. 1973. *Evolution of the Brain and Intelligence*. Academic Press, New York.

[7] Lorenz, K. 1977. *Behind the Mirror: A Search for a Natural History of Human Knowledge*. First American edition, translated by R. Taylor. Harcourt Brace Jovanovich, New York.

[8] Mittwoch, U. 1973. *Genetics of Sex Determination*. Academic Press, New York.

[9] Netter, F. H. 1962. *The CIBA Collection of Medical Illustrations*. Vol. 1, *Nervous System*. CIBA Publications, Summit, N. J.

[10] Norris, K. S. 1966. The evolution of acoustic mechanisms in odontocoete cetaceans. In *Evolution and Environment*, edited by E. T. Drake. Peabody Museum Centenary Celebration Volume, Yale University.

[11] Ohno, S. 1970. *Evolution by Gene Duplication*. Springer-Verlag, New York. A consideration of recent, controversial views of evolution. The work is unfortunately extremely difficult to decipher, especially for the undergraduate level student.

[12] Peyer, B. 1968. *Comparative Odontology*. University of Chicago Press. A monograph full of illustrations of extinct and living vertebrate teeth.

[13] Pond, C. M. 1975. The significance of lactation in the evolution of mammals. *Evolution* 31:177–199.

[14] van Tienhoven, A. 1968. *Reproductive Physiology of Vertebrates*. W. B. Saunders Co., Philadelphia.

20

Adaptations of Endotherms to Rigorous Habitats

Synopsis: Birds and mammals have colonized habitats ranging from the extreme polar regions to the hottest parts of deserts. It appears that the structural organization of vertebrate endotherms allows them to function in the most severe physical conditions encountered on earth. In the few areas in which endotherms are not found, their exclusion probably results from biological rather than physical factors. Endothermy is an energetically expensive way of life. If plants cannot grow, there is no energetic base to maintain consumer organisms.

There are elegant adaptations that allow specialized birds and mammals to cope successfully with extreme environments. More impressive than those adaptations, however, is the realization of how minor are the modifications of the basic bird and mammal body plans that allow these animals to endure air temperatures from $-70°C$ to $+50°C$, or water conditions ranging from continuous immersion in water to complete independence of drinking. There may be no major differences that distinguish animals from totally different habitats— an artic fox looks very much like a desert fox, and a ptarmigan from Alaska looks like a sandgrouse from the Khalahari Desert. The adaptability of these vertebrates lies in the combination of a number of minor modifications of their ecology, behavior, morphology, and physiology. A view that integrates all of these elements is needed to provide a balanced assessment of an animal in its environment.

20.1 ADVANTAGES OF ENDOTHERMY

We have emphasized some of the negative aspects of endothermy, particularly the energy cost of maintaining a high and constant body temperature. These negative aspects of endothermy are important and, unfortunately, they are often overlooked or forgotten by biologists. Being endotherms ourselves, we seldom stop to consider that ectothermy has advantages too. Despite its cost, however, endothermy gives birds and mammals specific abilities that are lacking in ectotherms. In particular, endothermy allows animals to be active at night, when solar energy is not available, or in cold climates where solar energy is not sufficient to warm an ectothermal vertebrate. We have previously emphasized the importance of nocturnality in the evolution and radiation of mammals and pointed out that, in the equable climates of the Mesozoic, the ability to be active at night was probably a key factor that allowed mammals to persist in an ecosystem dominated by reptiles (section 18.1).

20.2   ADAPTATIONS OF ENDOTHERMS TO COLD ENVIRONMENTS

Even in the moderate conditions of temperate climates, endotherms expend most of the energy they consume just keeping themselves warm, but they have proven themselves very adaptable in extending their thermoregulatory responses to allow them to inhabit even arctic and antarctic regions. Not even small body size is a insuperable handicap to life in these areas; 10-g redpolls and chickadees overwinter in central Alaska.

Aquatic life in cold regions places still more stress on an endotherm. Because of the high heat capacity and conductivity of water, an aquatic animal may lose heat at 50 to 100 times the rate it would if it were moving at the same speed through air. Even a small body of water is an infinite heat sink for an endotherm; all of the energy in its body could be converted to heat without appreciably raising the temperature of the water. How, then, do endotherms manage to exist in such stressful environments?

20.2.1   Increased Heat Production versus Decreased Heat Loss

There are potentially two solutions to the problems of endothermal life in cold environments and the special problems of aquatic endotherms in particular. A stable body temperature could be achieved by increasing heat production or by decreasing heat loss. On closer examination the option of increasing heat production does not seem particularly attractive. Any significant increase in heat production would require an increase in food intake. This scheme poses obvious ecological difficulties in terrestrial arctic and antarctic habitats where primary and secondary production is extremely low, especially during the coldest parts of the year. For most polar animals the quantities of food necessary would probably not be available, and a number of studies have shown that standard metabolic rates of most polar endotherms are similar to those of related species from temperate regions.

Lacking the ability to increase heat production significantly, conservation of heat within the body becomes the primary adaptation of polar endotherms. Pioneering studies by Per Scholander and Laurence Irving showed that the insulative value of pelts from arctic mammals are two to four times as great as those from tropical mammals. In the arctic species insulative value was closely related to the length of the fur (Figure 20–1). Small species such as the least weasel and the lemming have fur only 10 or 15 mm long. Presumably the thickness of their fur is limited because longer hair would interfere with the movement of their legs. Large mammals (caribou, polar and grizzly bears, dall sheep, and arctic fox) have hair 3 to 7 cm long. There is no obvious reason why their hair could not be longer; apparently further insulation is not needed. The insulative values of pelts of short-haired tropical mammals were similar to those measured for the same hair lengths in arctic species. Long-haired tropical mammals, like the sloths, had less insulation than arctic mammals with hair of similar length.

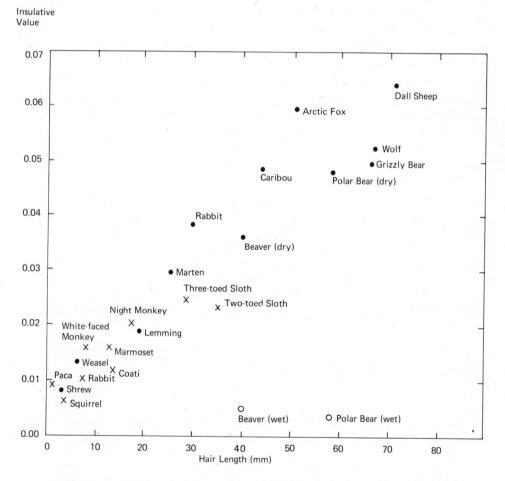

Figure 20-1 The insulative values of pelts of arctic mammals (shown by solid circles) are proportional to the length of the hair. Pelts from tropical mammals (shown by X) have approximately the same insulative value as those of tropical mammals at short hair lengths, but long-haired tropical mammals like sloths have less insulation than arctic mammals with hair of the same length. Immersion in water greatly reduces the insulative value of hair, even in such semiaquatic mammals as the beaver and the polar bear (open circles). (Modified from P. F. Scholander et. al. [1950] *Biological Bulletin* 99:237–258.)

A comparison of the lower critical temperatures of tropical and arctic mammals illustrates the effectiveness of the insulation provided (Figure 20–2). **Lower critical temperature** is the environmental temperature at which energy utilization must rise above its basal level to maintain a stable body temperature; Figure 16–15. A number of tropical mammals have lower critical temperatures between 20 and 30°C. As air temperatures fall below their lower critical temperatures, the animals are no longer in their thermoneutral zones and must increase their metabolic rates to maintain normal body temperatures. For example, a tropical raccoon has increased its metabolic rate approximately 50 percent above its standard level at an environmental temperature of 25°C.

Arctic mammals are much better insulated; even the small species like the least weasel and the lemming have lower critical temperatures that are between 10 and 20°C, and larger mammals have thermoneutral zones that extend well below freezing. The arctic fox, for example, has a lower critical temperature of −40°C, and at −70°C (approximately the lowest air temperature it ever encounters) has elevated its metabolic rate only 50 percent above its standard level. Under those conditions it is maintaining a core body temperature approximately 110°C above air temperature. Arctic birds are equally impressive. An arctic

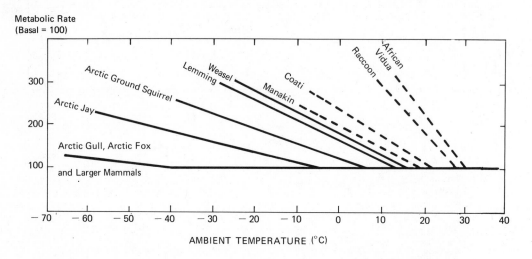

Figure 20–2 Lower critical temperatures for birds and mammals. As a result of their very effective insulation, arctic birds and mammals (shown by solid lines) have thermoneutral zones that extend below freezing in some species. For example, the rate of metabolism of the arctic jay does not rise from the basal level until ambient temperature has fallen to −5°C. Even small species like lemmings and weasels have lower critical temperatures far below those of much larger tropical mammals (shown by dashed lines). Because of their more effective insulation, arctic birds and mammals show smaller proportional increases in their rates of metabolism at temperatures below their thermoneutral zones than do tropical species. The metabolic rate of an arctic jay increases only 2.4 percent for each 10°C decline in ambient temperature whereas the rate of metabolism of a raccoon increases 12 percent for each 10°C drop in ambient temperature. (Modified from P. F. Scholander *et. al.* [1950] *Biological Bulletin* 99:237–258.)

jay has a lower critical temperature below 0°C and can withstand −70°C with a 150 percent increase in its standard metabolic rate, whereas an arctic gull, like the arctic fox, has a lower critical temperature near −40°C and can withstand −70°C with a modest increase in its metabolic rate.

Clearly hair or feathers can form a superb insulation for a terrestrial endotherm. These external insulative coverings are of limited value to aquatic animals, however, because when they are wet they lose most of their insulative value. The insulation of polar bear hair falls almost to zero when it is wet, and even seal hair loses much of its insulative value (Figure 20-1). In water fat is a far more effective insulator than hair, and aquatic mammals have thick layers of blubber. This blubber forms the primary layer of insulation; skin temperature is nearly identical to water temperature and there is a steep temperature gradient through the blubber. Its inner surface is at the animal's core body temperature.

The insulation provided by blubber is so effective that pinnipeds and cetaceans require special heat-dissipating mechanisms to avoid overheating when they engage in strenuous activity, or venture into warm water or onto land. This heat dissipation is achieved by shunting blood into capillary beds in the skin outside the blubber layer and into the flippers, which are not covered by blubber. Selective perfusion of these capillary beds enables a seal or porpoise to adjust its heat loss to balance its heat production. When it is necessary to conserve energy, a countercurrent heat exchange system in the blood vessels supplying the flippers is brought into operation; when excess heat is to be dumped, blood is shunted away from the countercurrent system into superficial veins.

The effectiveness of the insulation of marine mammals is graphically illustrated by the problems experienced by the northern fur seal (*Callorhinus ursinus*) during its breeding season. Northern fur seals are large animals; males attain weights in excess of 250 kg. Unlike most pinnipeds, fur seals have a dense covering of fur, which is probably never wet through. They are inhabitants of the North Pacific. For most of the year they are pelagic, but during summer they breed on the Pribilof Islands in the Bering Sea north of the Aleutian Peninsula. Male fur seals gather harems of females on the shore. Here they must try to prevent the females from straying, chase away other males, and copulate with willing females.

George Bartholomew and his colleagues have studied both the behavior of the fur seals and their thermoregulation. Summers in the Pribilof Islands (which are near 57°N latitude) are characterized by nearly constant overcast and air temperatures that rise to 10°C during the day. The conditions are apparently close to the upper limits the seals can tolerate. Almost any activity on land causes the seals to pant and to raise their hind flippers (which are abundantly supplied with sweat glands) and wave them about. If the sun breaks through the clouds, there is a sudden diminution of activity—females stop moving about, males reduce harem guarding activities and copulation, and the adult seals pant and wave their flippers. If the air temperature rises as high as 12°C, females, which never defend territories, begin to move into the water. Forced activity on land can produce lethal overheating.

Professional seal hunters herd the bachelor males from the area behind the harems inland preparatory to killing and skinning them. Professor Bartholomew recorded one drive that took place in the early morning of a sunny day while the air temperature rose from 8.6°C at the start of the drive to 10.4°C by the end. In 90 min the seals were driven about 1 km with frequent pauses for rest. "The seals were panting heavily and frequently paused to wave their hind flippers in the air before they had been driven 150 yd from the rookery. By the time the drive was half finished most of the seals appeared badly tired and occasional animals were dropping out of the pods [groups of seals]. In the last 200 yd of the drive and on the killing grounds there were found 16 "road-skins" (animals that had died of heat prostration) and in addition a number of others prostrated by overheating." The average body temperature of the road-skins on this drive was 42.2°C, which is 4.5°C above the 37.7°C mean body temperature of adults not under thermal stress.

Fur seals can withstand somewhat higher environmental temperatures in water than they can in air because of the greater heat conduction of water, but they are not able to penetrate warm seas. Adult males apparently remain in the Bering Sea during their pelagic season. Young males and females migrate into the North Pacific, but they are not found in waters warmer than 14 to 15°C, and they are most abundant in water of 11°C. Examples of physiological characteristics setting clear-cut geographic limits to animal distributions are relatively rare. Professor Bartholomew concluded that northern fur seals represent such a situation. Their inability to regulate body temperature during sustained activity and their sensitivity to even low levels of solar radiation and moderate air temperatures probably restricts the location of potential breeding sites and their movements during their pelagic periods. Summers in the Pribilofs are barely cool enough to allow the seals to breed there.

Another adaptation to the stress of life in cold water has been proposed by Peter Morrison and his associates on the basis of their metabolic measurements of sea otters. For convenience, the expected metabolic rate of an animal of a given size is called the **met**. Adult sea otters weigh about 30 kg. Thus they are small compared to the pinnipeds and cetaceans that share the far northern seas with them. Morrison's measurements revealed that the minimum metabolic rates of sea otters are about 2.5 times higher than would be expected from a 30 kg mammal (Figure 7–21). The observed resting rate of a sea otter therefore is about 2.5 met. This high resting metabolic rate has the effect of lowering the lower critical temperature. The lower critical temperature for an animal with the same insulation as a sea otter and a metabolic rate of 1 met is slightly below 30°C. The lower critical temperature of the sea otter is about 5°C (Figure 20–3).

Although this increase in the minimum metabolic rate does extend the thermoneutral zone to lower temperatures, it seems to be an expensive way to accomplish this end. It means that a sea otter has a continuously high energy requirement. Sea otters are known for their voracious appetites and their daily food consumption is slightly more than 20 percent of their body mass. Morrison points out that the readily available fish and shellfish upon which the

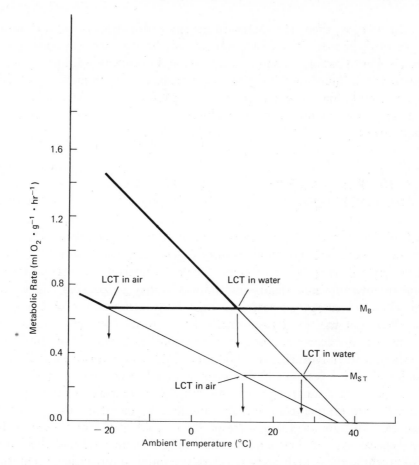

Figure 20-3 Sea otters may have adopted a physiological method of extending the thermoneutral zone to low temperatures. The metabolic rate expected for a mammal the size of a sea otter (M_{ST}) would produce a lower critical temperature of 27°C in water and 12°C in air. The actual basal rate of metabolism for sea otters (M_B) is 2.5 times the expected value. As a result of this high resting rate of metabolism, the lower critical temperatures are lowered to 12°C in water and to −20°C in air. (Modified from P. Morrison *et al.* [1974] *Physiological Zoology* 47:218–229.)

sea otters feed are very high in protein. Sea otters can be regarded as "carnivorous grazers"—that is, predatory animals that, like herbivorous grazers, are not food limited.

The cost of digesting food is called the **specific dynamic action** (SDA; see section 16.4). Because of the very high SDA associated with protein digestion, the sea otters' daily food intake must considerably exceed its energy needs. In most carnivores the SDA is reflected in a brief increase of metabolic rate immediately after a meal. In sea otters the SDA is spread throughout the day by a process not yet understood. Morrison's calculations indicate that if the SDA expected from a sea otter's diet were stretched through the day it would approximately account for the increased metabolic rate observed. In a sense, then, the increased energy production is not an extra cost for a sea otter because it is a consequence of the high protein content of the sea otter's diet. Possibly in some pinnipeds and cetaceans a similar mechanism occurs; their minimum metabolic rates are from 1.7 to 2.8 met. By coordinating the inevitable higher heat production of protein digestion to thermoregulation, these marine mammals may be able to inhabit very cold waters without metabolic stress.

20.3 TORPOR AS A RESPONSE TO LOW TEMPERATURE AND LIMITED FOOD

We have stressed the high energy cost of endothermy because the need to collect and process enough food to supply that energy is a central factor in the lives of many endotherms. In extreme situations environmental conditions may combine to overpower a small endotherm's ability to process and transform enough chemical energy to sustain a high body temperature through certain critical phases of its life. For diurnally active birds, long cold nights during which there is no access to food can be lethal, especially if the bird has not been able to feed fully during the daytime. Cold winter seasons usually present a dual problem for resident endotherms: (1) the necessity to maintain high body temperature against a greatly increased temperature difference between its internal core temperature and the ambient temperature and (2) a relative scarcity of food. As an adaptive response to such problems, some birds and mammals have evolved mechanisms that permit them to avoid the energetic costs of maintaining a high body temperature under unfavorable circumstances by entering a state of torpidity or "adaptive hypothermia."

20.3.1 Physiological Adjustments During Torpor

When an endotherm becomes torpid profound changes occur in a variety of physiological functions. Although body temperatures may fall very low during torpor, temperature regulation does not entirely cease. In **deep torpor**, an animal's body temperature drops to within 1°C or less of the ambient temperature, and in some cases (bats, for example) extended survival is possible at body temperatures just above the freezing point of the tissues. Oxidative metabolism and energy use are reduced to as little as one twentieth of the rate at normal body temperatures. Respiration is slow and irregular, and the overall breathing rate can be less than one inspiration per minute. Heart rates are drastically reduced, and blood flow to peripheral tissues is virtually shut down, so that most of the blood is retained in the core of the body.

The result is a comatose condition much more profound than the deepest sleep. Voluntary motor responses are reduced to sluggish postural changes, but some sensory perception of powerful auditory and tactile stimuli and ambient temperature changes are retained. Perhaps most dramatically, a torpid animal can arouse spontaneously from this state by endogenous heat production that restores the high body temperature characteristic of a normally active endotherm (normothermia). Some endotherms can rewarm under their own power from the lowest levels of torpor, whereas others must warm passively with an increase in ambient temperature until some threshold is reached at which arousal starts.

There are all sorts of torpor, from the deepest states of hypothermia to the lower range of body temperatures reached by normally active endotherms during their daily cycles of activity and sleep. Many birds and mammals, especially those with body weights under 1 kg, undergo **circadian temperature cycles.** These cycles vary from 1 to 5°C or more between the average high temperature characteristic of the active phase of the daily cycle and the average low temperature characteristic of rest or sleep. Small birds (sunbirds, hummingbirds, chickadees) and small mammals (especially bats and rodents) may drop their body temperatures during quiescent periods from 8 to 15°C below their regulated temperature during activity. Even a bird as large as the turkey vulture (about 2.2 kg) regularly drops its body temperature at night. When all these different endothermic patterns are considered together, no really sharp distinction can be drawn between "torpidity" and the basic daily cycle in body temperature that characterizes most small to medium-sized endotherms.

20.3.2 Body size and the Occurrence of Torpor

Species of endotherms capable of deep torpor are found in a number of groups of mammals and birds. All three subclasses of mammals include species capable of torpor. The echidna, the platypus, and several species of small marsupials possess various adaptive patterns of hypothermia, but the phenomenon is most diverse among placentals, particularly among bats and rodents. Certain kinds of hypothermia have also been described for some insectivores, particularly the hedgehog, some primates, and some edentates. Deep torpor, contrary to popular notion, is not known for any of the carnivores despite the fact that some of them den in the winter and remain inactive for long periods. Among birds deep torpor occurs in some of the goatsuckers or nightjars and in hummingbirds, swifts, mousebirds, and some passerines (sunbirds, swallows, chickadees, and others). Other species, including larger ones like turkey vultures, show varying depths of hypothermia at rest or in sleep but are not in a comatose state of deep torpor.

The largest mammals that undergo deep torpor are marmots, which weigh 5 kg or more whereas goatsuckers, the largest birds known to be capable of torpor, weigh no more than 75 to 80 g. This limitation on size reflects a balance between the energy expenditure during normothermia and torpor and the time and energy spent in entry into torpor and in arousal. Torpor is not as energetically advantageous for a large animal as for a small one. In the first place, the energetic cost of maintaining a high body temperature is relatively greater for a small animal than for a large one and, as a consequence, a small animal has more to gain from becoming torpid. Secondly, a large animal cools off more slowly than a small animal and does not lower its metabolic rate as rapidly. Furthermore, large animals have more body tissue to rewarm upon arousal, and their costs of arousal are correspondingly larger than those of small animals. An endotherm weighing a few grams, such as a little brown bat or a humming-bird, can warm up from torpidity at the rate of about 1°C per minute, and be fully active within 30 min or less depending upon the depth of hypothermia. A 100-g hamster requires more than 2 hr to arouse, and a marmot takes many hours. Entrance into torpor is slower than arousal. Consequently, daily torpor is feasible only for very small endotherms; there would not be enough time for a large animal to enter and arouse from torpor during a 24-hr period. Moreover, the energy required to warm up a large mass is very great. Oliver Pearson calculated that it costs a 4-g hummingbird only 0.114 kcal to raise its body temperature from 10 to 40°C. That is 1/85 of the total daily energy expenditure of an active hummingbird in the wild. By contrast, a 200-kg bear would require 5100 kcal to warm from 10 to 37°C, the equivalent of a full day's energy expenditure. The smaller potential savings and the greater costs of arousal make daily torpor impractical for any but small endotherms.

Medium-sized endotherms are not entirely excluded from the energetic sav-ings of torpor, but the torpor must persist for a longer period to realize a saving. For example, ground squirrels and woodchucks do not experience daily torpor but they do enter prolonged torpor during the winter hibernation when food is scarce. They spend several days at very low body temperatures (in the region of 5°C), then arouse for a period before becoming torpid again. Still larger endotherms would have such large costs of arousal (and would take so long to warm up) that torpor is not feasible for them even on a seasonal basis. Bears in winter dormancy, for example, lower their body temperatures only about 5°C from normothermal levels, and metabolic rate decreases about 50 percent. Even this small drop, however, amounts to a large energy saving through the course of a winter for an animal as large as a bear.

20.3.3 Energy Relations of Daily Torpor

Recent studies of daily torpor in birds have emphasized the flexibility and adaptive nature of the response in relation to the energetic stress faced by individual birds. Susan Chaplin's work with chickadees provides an example. These small (10 to 12-g) passerine birds are winter residents in northern latitudes where they regularly experience ambient temperatures that do not rise above freezing for days or weeks.

Chaplin found that in winter chickadees around Ithaca, New York, allow their body temperatures to drop from the normothermal level of 40 to 42°C that is maintained during the day to 29 to 30°C at night. This reduction in body temperature permits a 30 percent reduction in energy consumption. The chickadees rely primarily on fat stores they accumulate as they feed during the day to supply the energy needed to carry them through the following night. Thus the energy available to them and the energy they utilize at night can be estimated by measuring the fat content of birds as they go to roost in the evening and as they begin activity in the morning. Chaplin found that in the evening chickadees had an average of 0.80 g of fat per bird. By morning the fat store had decreased to 0.24 g. The fat metabolized during the night (0.56 g/bird) corresponds to the metabolic rate expected for a bird that had allowed its body temperature to fall to 30°C.

Chaplin's calculations show that this torpor is necessary if the birds are to survive the night. It would require 0.92 g fat per bird to maintain a body temperature of 40°C through the night. That is more fat than the birds have when they go to roost in the evening. If they did not become torpid, they would starve before morning. Even with torpor, they use 70 percent of their fat reserve in one night. They do not have an energy supply to carry them far past sunrise and chickadees are among the first birds to start to forage in the morning. They also forage in weather so foul that other birds, which are not in such precarious energy balance, remain on their roosts. The chickadees must reestablish their fat stores each day if they are to survive the next night.

Hummingbirds, too, may depend on the energy they gather from nectar during the day to carry them through the following night. These very small (4 to 10-g) birds have extremely high energy expenditures and yet are found during the summer in northern latitudes and at high altitudes.

An example of the lability of torpor in hummingbirds was provided by William Calder's and J. Booser's studies of nesting broad-tailed hummingbirds at an altitude of 2900 m near Gothic, Colorado. Ambient temperatures drop nearly to freezing at night, and under these conditions hummingbirds normally become torpid. Calder and Booser were able to monitor the body temperatures of nesting birds by placing an imitation egg containing a temperature-measuring device in the nest. These temperature records showed that hummingbirds incubating eggs did not normally become torpid at night. The reduction of egg temperature that results from the parent bird's becoming torpid does not damage the eggs, but it slows development and delays hatching. Presumably there are advantages to hatching the eggs as quickly as possible, and as a result brooding hummingbirds expend energy to keep themselves and their eggs warm through the night provided that they have the energy stores necessary to maintain the high metabolic rates needed.

On some days bad weather interfered with foraging by the parent birds, and as a result they apparently went into the night with insufficient energy supplies to maintain normothermal temperatures. In this situation the brooding hummingbirds did become torpid for part of the night. One bird that had experienced a 12 percent reduction in foraging time during the day became torpid for 2 hr, and a second that had lost 21 percent of its foraging time was torpid for $3\frac{1}{2}$ hr. Thus in hummingbirds torpor can be a very flexible response that integrates the energy stores of a bird with environmental conditions and biological requirements such as brooding eggs on a day-by-day basis. In section 20.4.5 we shall discuss a similar flexibility of torpor in relation to daily food and water requirements of small desert rodents.

20.4   ENDOTHERMS IN DESERTS

Deserts can be produced by a number of physiographic phenomena. Whatever their cause, deserts have in common a scarcity of liquid water, and that characteristic is at the root of many of the features of deserts that make them difficult places for endotherms to live. The dry air characteristic of most deserts seldom contains enough moisture to form clouds that would block solar radiation during the day or radiative cooling at night. As a result, the daily temperature excursion in deserts is large in comparison to that of more humid areas. Scarcity of water is reflected by sparse plant life and a correspondingly low primary productivity in desert communities. Food shortages may be chronic and exacerbated by seasonal shortages and unpredictable years of low production when the usual pattern of rainfall does not develop.

Not all deserts are hot. Indeed some are distinctly cold—most of Antarctica and the region of Canada around Hudson Bay and the Arctic Ocean are deserts. The low latitude deserts north and south of the Equator are hot deserts, however, and it is in these low latitude deserts that endotherms encounter the most difficult problems of desert life. Indeed, these hot, dry areas place a more severe physiological stress upon endotherms than do the polar conditions we have already discussed. The difficulties endotherms encounter in deserts result from a reversal of their normal relationship to the environment. Endothermy evolved in a situation in which an animal's body temperature is higher than the temperature of its environment. As a result, heat flow is from the animal to its environment, and thermoregulatory mechanisms achieve a stable body temperature by balancing heat production and heat loss. Very cold environments merely increase the gradient between an animal's core temperature and the environment. The examples of arctic foxes and gulls with lower critical temperatures of $-40°C$ illustrate the success that endotherms have had in providing sufficient insulation to cope with enormous gradients between high core body temperatures and low environmental temperatures.

In a desert the gradient is not increased, it is reversed. Desert air temperatures can climb to 40 or 50°C during summer, and the ground temperature may exceed 60 or 70°C. Instead of losing heat to the environment, an animal is continually absorbing heat and that heat must somehow be dissipated to maintain the animal's body temperature in the normothermal range. Physiologically it can be a greater challenge for an endotherm to maintain its body temperature 10°C below the ambient temperature than to maintain it 100°C above ambient.

Evaporative cooling is the major mechanism an endotherm uses to reduce its body temperature. The evaporation of water requires approximately 580 kcal/kg. (The exact value varies with temperature.) Thus, evaporation of a liter of water dissipates 580 kcal, and evaporative cooling is a very effective mechanism as long as an animal has an unlimited supply of water. In a hot desert, however, where the thermal stress is greatest, water is a scarce commodity and its use must be carefully rationed. Calculations show, for example, that if a kangaroo rat were to venture out into the desert sun, it would have to evaporate 13 percent of its body water per hour to maintain a normal body temperature. Most mammals die when they have lost 10 to 20 percent of their body water, and it is obvious that, under desert conditions, evaporative cooling is of limited utility except as a short-term response to a critical situation.

Unable to rely upon evaporative cooling, endotherms have evolved a number of other adaptive responses that have allowed a diverse assemblage of birds and mammals to inhabit deserts. The mechanisms they use are complex and involve combinations of ecological, behavioral, morphological, and physiological adaptations that act synergistically to enhance the effectiveness of the entire system. As a start toward unravelling some of these complexities, we can categorize three major classes of responses of endotherms to desert conditions as follows:

1. Some endotherms manage to avoid desert conditions by behavioral means. They live in deserts but are rarely exposed to the full stress of desert life.
2. Other endotherms have relaxed the limits of homeostasis. They manage to survive in deserts by tolerating greater ranges of variation in characters such as body temperature or body water content than normal.
3. Special adaptations such as torpidity in response to shortages of food or water and a reduced standard metabolic rate (and consequently a reduction in the amount of metabolic heat an animal must dissipate) are combined with some of the characters mentioned in 1 and 2 in some desert endotherms.

20.4.1 Large Animals in Hot Deserts

Large animals, including humans, have specific advantages and disadvantages in desert life that are directly related to body size. A large animal has nowhere to hide from desert conditions. It is too big to burrow, and few deserts have vegetation large enough to provide useful shade to an animal much larger than a jackrabbit. Thus a large animal cannot avoid desert conditions. On the other hand, large body size offers some options not available to smaller animals. Large animals are mobile and can travel long distances to find food or water, whereas small animals may be limited to home ranges only a few meters or tens of meters in diameter. Large animals have small surface/mass ratios and consequently absorb heat from the environment slowly. The specific heat of animal tissue is approximately 0.8 kcal/kg, and a large body mass gives an animal a large thermal inertia; it can absorb a large amount of heat before its body temperature rises.

The dromedary camel (*Camelus dromedarius*) is the classic large desert animal. Its reputation as the "ship of the desert" has lost nothing in the telling, but there are authentic records of journeys in excess of 500 km, lasting 2 or 3 weeks, during which the camels did not have an opportunity to drink. The longest trips are made in winter and spring when air temperatures are relatively low and scattered rainstorms may have produced some fresh vegetation that provides a little food and water for the camels. Dr. Gauthier-Pilters studied camels in the northern Sahara, travelling with the nomadic herdsmen. She found that in winter the camels were independent of drinking water and had to be herded continuously to keep them from straying. In summer the camels returned voluntarily to camp to be watered every 2 to 5 days.

The pioneering studies of camels' adaptations to desert life were conducted by Knut Schmidt-Nielsen and a number of associates in North Africa in the 1950s and in Australia in 1962. Subsequent work by other biologists has filled in additional details.

Camels are large animals; Schmidt-Nielsen estimated adult weights of the dromedary camels he studied to be 400–450 kg for females and up to 500 kg for males. The north African desert is cold in the winter, and the camels grow heavy coats; in summer they shed the winter coat but retain hair 50 to 60 mm long on the back and up to 110 mm long over the hump. On the ventral surface and legs the hair is only 15 to 20 mm long. The nature of the camel's adaptation to desert life is revealed by comparing the daily cycle of body temperature in a camel that receives water daily and one that has been deprived of water. The watered camel shows a small daily cycle of body temperature with a minimum at 36°C in the early morning and a maximum of 38°C in mid afternoon. When a camel is deprived of water, the daily temperature variation triples. Body temperature is allowed to fall to 34.5°C at night and climbs to 40.5°C during the day.

The significance of this increased daily fluctuation in body temperature can be assessed in terms of the water that the camel would have had to expend to prevent the 6°C rise by evaporative cooling. With a specific heat of 0.8 kcal/kg · °C, a 6°C increase in body temperature for a 500-kg camel represents storage of 2400 kcal of heat in the body. Evaporation of a kg of water dissipates approximately 580 kcal. Thus a camel would have to evaporate slightly more than 4 liters of water to maintain a stable body temperature at the nighttime level; by tolerating hyperthermia during the day it can conserve that water.

In addition to the direct saving of water not used for evaporative cooling, the camel receives an indirect benefit from tolerating hyperthermia in a reduction of energy flow from the air to the camel's body. As long as the camel's body temperature is below air temperature, a gradient will exist that causes the camel to absorb heat from the air. At a body temperature of 40.5°C the camel's temperature is equal to that of the air for much of the day, and no net heat exchange takes place. Thus, the camel saves an additional quantity of water by eliminating the temperature gradient between its body and the air. The combined effect of these measures on water loss is illustrated by data that Schmidt-Nielsen and his colleagues collected from a young camel (Table 20-1). When deprived of water the camel reduced its evaporative water loss by 64 percent and reduced its total daily water loss by half. This water economy was achieved primarily by relaxation of the limits of thermoregulation. By permitting itself to become hypothermal at night and hyperthermal during the day, the camel stored a significant quantity of heat each day that could be dissipated by conduction, convection, and radiation the following night rather than by evaporation of water. At the same time its elevated body temperature during the hottest part of the day reduced the amount of heat it gained from the hot environment.

Table 20-1. Daily Water Loss of a 260-kg Camel

Condition	Water lost (liters/day) by different routes			
	Feces	Urine	Evaporation	Total
Drinking daily (8 days)	1.0	0.9	10.4	12.3
Not drinking (17 days)	0.8	1.4	3.7	5.9

Source: K. Schmidt-Nielsen, [4].

Several morphological, behavioral, and physiological features of camel biology are integrated with this basic response and enhance the benefits a camel can gain. The thick fur on the body, for example, covers the dorsal regions and shields them from the sun. Professor Schmidt-Nielsen reported that fur surface temperatures reach 70 to 80°C, while skin temperatures are about 40°C. As a result of the insulation provided by the fur, much of the solar energy that strikes a camel's fur never reaches its skin. Furthermore, sweat glands release their secretions onto the skin surface beneath the hair. The energy that evaporates the sweat comes from the camel's metabolism. In a bareskinned animal in the sun, most of the heat that is dissipated by evaporation of sweat comes from the absorption of solar radiation by the skin. Thus, a camel's long hair allows it to make effective use of the evaporative cooling that it does employ.

Behavioral adaptations aid dehydrated camels in reducing their heat load. While it is still cool in the morning, camels lie down on surfaces that have cooled overnight by radiation. The legs are tucked beneath the body and the ventral surface, with its short covering of hair, is placed in contact with the cool ground. In this position a camel exposes only its well-protected back and sides to the sun, and places its lightly furred legs and ventral surface in contact with cool sand, which may be able to conduct away some body heat. Dr. Gauthier-Pilters has reported that camels may assemble in small groups and lie pressed closely together through the day In this posture a camel reduces its heat load by placing its sides in contact with another camel (both at about 40°C) instead of allowing solar radiation to raise the fur surface temperature to 70°C or above.

Despite the adaptations that reduce water loss, camels do continue to lose water. The conservation mechanisms of the camel Schmidt-Nielsen studied reduced its rate of water loss from nearly 5 percent of body weight per day when it was hydrated to just over 2 percent per day during dehydration. The reduced rate of water loss prolongs the period before the water loss becomes critical. In addition, camels appear to have physiological adaptations that allow them to tolerate a greater amount of dehydration than most mammals. In full sun during the summer Schmidt-Nielsen's camels lost water equivalent to more than 25 percent of their hydrated body weight without ill effect. Under the same conditions, other mammals die when they have lost half that much water.

The camel's tolerance of dehydration appears to be related to its ability to maintain its blood plasma volume near normal levels during dehydration. In most mammals dehydration affects the plasma volume disproportionately, with the result that the blood becomes increasingly viscous. Eventually the increasing viscosity of the blood prevents the circulatory system from transporting metabolic heat from the core of the body to the surface. The body temperature rises abruptly and death follows rapidly. In camels, water is lost primarily from interstitial and intercellular fluids during dehydration, whereas plasma volume is maintained. The basis of this physiological difference between camels and other mammals is not clear. Schmidt-Nielsen points out that a theoretical analysis of the osmotic and water relationships within an animal's body suggests that all animals should respond as a camel does. The difficulty lies not in explaining why a camel maintains its plasma volume but rather in explaining why other mammals do not. Whatever its physiological basis, however, the adaptive significance of a camel's tolerance of dehydration is readily apparent.

Despite their ability to reduce water loss and to tolerate dehydration, the time eventually comes when even camels must drink. These large, mobile animals can roam across the desert seeking patches of vegetation produced by local showers and moving from one oasis to another, but when they come to drink, they face a problem they share with other grazing animals: water holes can be dangerous places. Predators frequently center their activities around water holes, where they are assured of water as well as a continuous supply of prey animals. Reducing the time spent drinking is one method of reducing the risk of predation, and camels can drink remarkable quantities of water in very short periods. A dehydrated camel can drink as much as 30 percent of its body weight in 10 min. (A very thirsty human can drink about 3 percent of body weight in the same time.) Schmidt-Nielsen's measurements of dehydrated camels' body weights just before they were allowed to drink and again immediately after they had finished drinking showed that, if they had been only moderately dehydrated (moderate dehydration was defined as a loss of up to 20 percent of initial body weight), they drank enough water to restore the loss in one session. If they had been subjected to greater dehydration, they required two drinking sessions several hours apart to restore the deficit fully.

20.4.2 Other Large Mammals

Camels are not unique in their ability to rehydrate quickly. Schmidt-Nielsen and his associates also studied the wild donkeys (*Equus asinus*) that range across the arid parts of Africa and Asia. The explorer Roy Chapman Andrews reported that, in the Gobi Desert of Central Asia, wild asses were most abundant in the parts of the desert that had no water. Wild donkeys appear to use water conservation mechanisms similar to those of the camel but not quite as well developed. Like camels they tolerate dehydration of 20 percent of body weight or more while maintaining their plasma volume near normal levels and can drink enough water in a few minutes to restore their initial weights. Merino sheep, studied by W. V. McFarlane, are also able to rehydrate rapidly when allowed to drink after a period of water deprivation.

Recently Richard Taylor has investigated the temperature and water relations of several species of African antelope that live in arid grasslands or desert regions. These animals, which range in size from the 25-kg Thomson's gazelle and 50-kg Grant's gazelle (*Gazella thomsoni* and *G. granti*) to the 100-kg oryx (*Oryx besia*) and 200-kg eland (*Taurotragus oryx*) utilize heat storage like the dromedary but allow their body temperatures to rise considerably above the 40.5°C level recorded in the camel. Taylor recorded rectal temperatures of 45°C in the oryx and 46.5°C in the Grant's gazelle. (Thomson's gazelles, which do not penetrate into desert areas, began to pant at air temperatures above 42°C and used evaporative cooling to maintain body temperature below air temperature.)

Body temperatures above 43°C rapidly produce brain damage in most mammals, but Taylor observed that Grant's gazelles maintained rectal temperatures of 46.5°C for as long as 6 hr with no apparent ill effects. He has suggested that these antelope can keep brain temperature below body temperature by using a countercurrent heat exchange to cool blood before it reaches the brain. In ungulates the blood supply to the brain passes via the external carotid arteries (Figure 20-4). At the base of the brain these arteries break into a rete mirabile that lies in a venous sinus. The blood in the sinus is venous blood, returning from the walls of the nasal passages where it has been cooled by the evaporation of water. This cool venous blood thus cools the warmer arterial blood before it reaches the brain. A mechanism of this sort has been demonstrated in sheep and goats and is probably widespread.

Unlike the dromedary camel, the antelopes are apparently independent of drinking water even during summer. Taylor suggests that one of the adaptations that permits this independence is behavioral. The leaves of a desert shrub, *Diasperma*, are an important part of the diet of the antelopes. During the day, when air temperature is high and humidity low, these leaves contain about 1 percent water by weight. They are so dry that they disintegrate into powder when they are touched. At night, however, as air temperatures fall and relative humidity increases, the leaves take up water and after 8 hr, have a water content of 40 percent. By eating the leaves at night rather than in the daytime, the antelope can obtain enough water from their food to be independent of water holes.

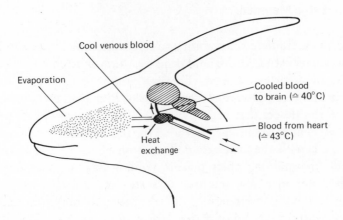

Figure 20-4 A counter-current heat-exchange mechanism that may cool blood going to a gazelle's brain. The blood supply to the brain in ungulates travels via the external carotid arteries. At the base of the brain, each of these arteries breaks into a capillary network (a *rete mirabile*) that lies within the venous sinus. This sinus is filled by venous blood that has been cooled by evaporation of water from the walls of the nasal passages. The close contact between cool venous blood and warm arterial blood permits an exchange of heat that cools the arterial blood going to the brain. In this way a gazelle may keep its brain temperature several degrees cooler than its core body temperature for long periods. (Redrawn from C. R. Taylor [1972] *Symposia of the Zoological Society of London*, no. 31, p. 225.

Large animals such as those we have discussed illustrate one approach to desert life. Too large to escape the stresses of the environment, they have adapted to them by tolerating a temporary relaxation of homeostasis. Their success under the harsh conditions in which they live is the result of complex interactions between diverse aspects of their ecology, behavior, morphology, and physiology. Only when all of these features are viewed together does an accurate picture of an animal emerge.

20.4.3   Birds in Desert Regions

Although birds are relatively small vertebrates, the problems they face in deserts are more like those experienced by camels and wild asses than like those of small mammals. Birds are predominantly diurnal, and few seek shelter in burrows or crevices. Thus, like large mammals, they meet the stresses of deserts head on and face the antagonistic demands of thermoregulation in a hot environment and the need to conserve water.

Also like large mammals, birds are mobile. It is quite possible for a desert bird to fly to a mountain range on a daily basis to reach water. For example, mourning doves in the deserts of North America congregate at dawn at water holes, some individuals flying 60 km or more to reach them. The normally high and labile body temperatures of birds give them an advantage in deserts that is not shared by mammals. With body temperatures normally around 40°C, birds face the problem of a reversed temperature gradient between their bodies and the environment for a shorter portion of each day than would a mammal. Furthermore birds' body temperatures are normally variable, and birds tolerate moderate hyperthermia without apparent distress. These are all preadaptations to desert life that are present in virtually all birds. Neither the body temperatures nor the lethal temperatures of desert birds are higher than those of related species from nondesert regions.

The mobility provided by flight does not extend to fledgling birds, and the most conspicuous adaptations of birds to desert conditions are those that ensure a supply of water for the young. Altricial fledglings, those that need to be fed by their parents after hatching, receive the water they need from their food. One pattern of adaptation in desert birds ensures that reproduction will occur at a time when succulent food is available for fledglings. In the arid central region of Australia, bird reproduction is precisely keyed to rainfall. The sight of rain is apparently sufficient to stimulate courtship, and mating and nest-building commence within a few hours of the start of rain. This rapid response ensures that the baby birds will hatch in the flush of new vegetation and insect abundance stimulated by the rain.

A different approach, very like that of mammals, has been evolved in columbiform birds (pigeons and doves) which are widespread in arid regions. Fledglings are fed on "pigeon's milk," a liquid substance produced by the crop under the stimulus of prolactin. The chemical composition of pigeon's milk is very similar to that of mammalian milk (Table 18–4); it is primarily water plus protein and fat, and it simultaneously satisfies both the nutritional requirements and the water needs of the fledgling. This approach places the water stress on the adult, who must find enough water to produce milk as well as meeting its own water requirements. The mobility of birds simplifies the problem for them, but the water stress of lactation can be very severe for mammals.

Seed-eating desert birds with precocial young, like the sandgrouse found in the deserts of Africa and the Near East, face particular problems in providing water for their young. Baby sandgrouse begin to find seeds for themselves within a few hours of hatching, but they are unable to fly to water holes as parents do and seeds do not provide the water they need. Instead, adult males transport water to their broods. The belly feathers, especially in males, have a unique structure in which the proximal portions of the barbules are coiled into helices. When the feather is wetted, the barbules uncoil and trap water (Figure 16–34). The feathers of male sandgrouse hold 15 to 20 times their weight of water, and the feathers of females hold 11 to 13 times their weight.

Tom Cade and Gordon Maclean observed sandgrouse in the Khalahari Desert of southern Africa. They found that the male birds flew to water holes just after dawn and soaked their belly feathers, absorbing 25 to 40 ml of water. Some of this water evaporates on the flight back to their nest sites, but calculations indicated that a male sandgrouse could fly 30 km and arrive with 10 to

28 ml of water still adhering to its feathers. As the male grouse lands, the juveniles rush to him and, seizing the wet belly feathers in their beaks, strip the water from them with downward jerks of their heads. In a few minutes the young birds have satisfied their thirst, and the male rubs itself dry on the sand.

The specialized structure of the belly feathers of male sandgrouse increases their water capacity fourfold compared to the belly feathers of other birds and also allows the feathers to withstand repeated wetting and stripping. In the absence of the morphological and physiological modifications of the beak and digestive system seen in pigeons, the evolution of a specialized water transportation system is the key to the success of a seed-eating bird with precocial young in desert conditions.

20.4.4 Small Mammals in Hot Deserts

Rodents are the preeminent small mammals of arid regions. It is a commonplace observation that population densities of rodents may be higher in deserts than in mesic situations. A number of features of rodent biology can be viewed as preadaptive for extending their geographic ranges into hot, arid regions. Among the most important of these preadaptations are the normally nocturnal habits of many rodents and their practice of living in burrows. A burrow provides ready escape from the heat of a desert, giving an animal access to a microenvironment within its thermoneutral zone while soil temperatures on the surface climb above 60°C. Rodents that live in burrows during the day and emerge to forage at night escape the desert heat so successfully that their greatest temperature stress may be cold. Because of the normal absence of clouds, deserts cool rapidly after sundown and many deserts are distinctly chilly at night during much of the year.

Although retreat to a burrow during the day provides direct escape from heat, it is not, by itself, a solution to the other major challenges of desert life, the chronic shortages of food and water. What the burrow does provide is the shelter and microclimate an animal needs in order to solve the other problems. We pointed out earlier that a kangaroo rat would reach its lethal limit of dehydration in less than 2 hr if it had to rely upon evaporative cooling to maintain its body temperature at normal levels during the day. By retreating into its burrow, a kangaroo rat avoids that use of water. Indeed, the water savings of a burrow probably go beyond that. As a rodent in a burrow loses water by evaporation, the air in the burrow becomes humid and may approach saturation. At the same time of day, the relative humidity of air outside the burrow may be only 20 to 30 percent. The higher humidity of burrow air reduces an animal's evaporative water loss.

A further saving is achieved in some animals (including birds and lizards in addition to mammals) by a countercurrent water recycler in the nasal passages. With his colleagues J. C. Collins and T. C. Pilkington, Schmidt-Nielsen has pointed out that the air an animal exhales leaves the nares at a temperature lower than that at which it left the lungs. That observation has important implications in terms of energy and water balance.

A brief consideration of the respiratory cycle illustrates the mechanism involved. As air is inhaled, it passes over moist tissues in the nasal passages. The nasal passages themselves are narrow and the wall surface area is large. As the air passes over these moist surfaces, it is warmed and humidified so that as it enters the lungs it is saturated with water at the animal's core body temperature. Temperature equilibration and saturation with water vapor are essential to protect the lungs, which are delicate structures that would rapidly be damaged if they were exposed to dry air. As the relatively dry inhaled air passes over the moist tissues of the nasal passages, evaporation cools the walls of the nasal passages. When the warm, saturated air from the lungs is exhaled, water from the air condenses on the cool walls. This process of evaporation on inhalation and condensation on exhalation saves water and, because a large quantity of energy goes into evaporating the water, also saves energy. In Chapter 12 we pointed out that the energetic savings are so large that it seems unlikely that an endotherm could exist without such a system, and consequently the presence of ethmoturbinal bones (which support the moist membranes in the nasal passages) in fossil *Cynognathus* is persuasive evidence that these therapsids were endotherms. It should be emphasized that this countercurrent exchange of heat and energy is not an adaptation to desert life. It is an inevitable consequence of the anatomy and physiology of the nasal passages. Indeed, Schmidt-Nielsen and his associates found that there was no consistent difference in the temperature of exhaled air in a comparison of desert and nondesert birds.

The importance of the nasal countercurrent in water recovery can be illustrated by calculations presented by Schmidt-Nielsen (Table 20–2). During the day a kangaroo rat in its burrow inhales air that we can, for convenience, say is at 30°C and 80 percent relative humidity. This air contains 24 mg water per liter of air. (Saturated air, of 100 percent relative humidity, contains 30 mg/liter.) The kangaroo rat must warm this air to core temperature (38°C) and add enough water to raise the relative humidity at that temperature to 100 percent. At 38°C saturated air contains 46 mg water per liter, so the amount of water that must be evaporated in the nasal passages is 22 mg/liter of inhaled air.

Table 20-2 Water Recycled by Nasal Countercurrent Exchange in a Kangaroo Rat

	Inhaled air				*Exhaled air*			
Conditions	*Tem-pera-ture, °C*	*RH, %*	*Water Content, mg/liter*	*Water added, mg/liter*	*Tem-pera-ture, °C*	*RH, %*	*Water Content, mg/liter*	*Water recovered, mg/liter*
Daytime in burrow	30	80	24	22	31	100	31	15
Night, on surface	15	20	2.5	43.5	14.5	100	12	34

Source: K. Schmidt-Nielson [7].

Under the conditions we have assumed, a kangaroo rat exhales air at 31°C. That is, as the air travels out through the nasal passages it is cooled 7°C by contact with the walls. The exhaled air is still saturated with water, but at 31°C saturated air contains only 31 mg water per liter instead of the 46 mg/liter it contained at 38°C. Thus, a kangaroo rat evaporates 22 mg water per liter of air to saturate the air before it enters the lungs, and recovers 15 mg water per liter of air on exhalation. The nasal countercurrent reduces its respiratory water loss by nearly 70 percent from that which the kangaroo rat would experience if the air were exhaled at core body temperature. The savings amounts to 15 mg water per liter of air under the burrow conditions we have assumed.

The water saving achieved at night when the rat is outside its burrow is still greater. The outside air is cool (15°C) and dry (20 percent relative humidity). Consequently the kangaroo rat must evaporate 43.5 mg of water per liter of air to bring it to saturation at core temperature. The increased evaporation cools the kangaroo rat's nose to 14.5 C, and air exhaled at this temperature contains only 12 mg water per liter. Thus, under nighttime conditions, a kangaroo rat recovers 74 percent of the water evaporated in the nasal passages, a saving of 34 mg water per liter of inhaled air.

Evaporation of water from the respiratory passages is one major avenue of water loss; water excreted with the urine and feces is another. Rodents in general have the ability to produce relatively dry feces and concentrated urine. The laboratory white rat, for example, can produce urine with twice the osmotic concentration humans can achieve (Table 18.5). The dromedary camel has a urine-concentrating ability approximately equivalent to that of a rat, and so do dogs and cats. In desert rodents, such as kangaroo rats, sand rats, and jerboas, urine concentrations of 3000 to 6000 mOsm/liter are commonly observed, and the current "world champion urine concentrator" appears to be the Australian hopping mouse, which can produce urine concentrations in excess of 9000 mOsm/liter. As we pointed out in Chapter 18, high urine concentrations in mammals are associated with long loops of Henle in the kidney which enhance the countercurrent multiplier function.

As a result of their low evaporative water losses and ability to concentrate urine and produce relatively dry feces, many desert rodents are completely independent of liquid water. Their water loss has been reduced to the point at which they are able to obtain all the water they need from air-dried seeds. Part of this water comes from water actually contained in the food, and part from the water that is produced as the food is oxidized (metabolic water). The amount of metabolic water produced depends on the chemical composition of the food an animal eats. Fat produces 1.1 g of water for each gram oxidized, starch yields 0.6 g water per gram, and protein yields 0.4 g water per gram when it is metabolized to urea.

The actual water content of food depends in part upon the relative humidity at which it is stored. We pointed out above that *Diasperma* leaves have a water content of only 1 percent during the day but absorb water at night when the relative humidity rises; it appears to be at night that the antelope eat them.

Seeds show a similar variation in water content with humidity. Knut Schmidt-Nielsen found that the barley he fed his kangaroo rats contained less than 4 percent water when it was kept at 10 percent relative humidity, but, when it was stored at 76 percent relative humidity, the water content rose to 18 percent. He reported that bannertail kangaroo rats (*Dipodomys spectabilis*) may store several kilograms of plant material in their burrows where it is exposed to the high relative humidity of the burrow atmosphere. The smaller Merriam's kangaroo rat (*D. merriami*) does not store much food in its own burrow, but sneaks into the burrows of bannertail kangaroo rats and may help itself to the food stored there. Gerbils in the Old World deserts also store seeds in their burrows. Air-dried seeds contained 4 to 7 percent water, whereas seeds taken from the gerbils' burrows contained 30 percent water. Thus, hoarding food in the burrow not only provides a hedge against food shortages but increases the amount of water available to an animal from the food.

20.4.5 Torpor in Desert Rodents

The significance of daily torpor as an energy conservation mechanism in small birds was illustrated earlier. Many desert rodents have the ability to become torpid. In most cases the torpidity can be induced by limiting the food available to an animal. When the food ration of the California pocket mouse (*Perognathus californicus*) is reduced slightly below its daily requirements it enters torpor for a part of the day. Vance Tucker has shown that in this species even a minimum period of torpor results in an energy saving. If a pocket mouse were to enter torpor and then immediately arouse, the process would take 2.9 hr. Calculations indicate that the overall energy expenditure during that period would be reduced 45 percent compared to the cost of maintaining a normal body temperature for the same period. In this animal, the briefest possible period of torpor gives the animal an energetic saving, and the saving increases as the time spent in torpor is lengthened.

Tucker showed that the duration of torpor is proportional to the severity of food deprivation in the pocket mouse; as its food ration is reduced, it spends more time each day in torpor and conserves more energy. Torpor induced by food deprivation has now been reported in several desert and semidesert species of white-footed mice (*Peromyscus*) by Jack Hudson and his associates. This general response probably represents an adaptation to the chronic food shortage that may face desert rodents because of the low primary productivity of desert communities and the effects of unpredictable variations from normal rainfall patterns, which may almost completely eliminate seed production by desert plants in dry years.

A possible variation of the response of daily torpor has been described by Richard MacMillen for the cactus mouse, *Peromyscus eremicus*. This species, like the others, enters torpor if its food ration is reduced below maintenance levels, but in the summer some individuals can be induced to become torpid by reducing the water available to them   Torpid cactus mice allow their body temperature to drop to the burrow temperature. Because the air in the burrow is nearly saturated with water, torpid cactus mice probably experience negligible evaporative water loss. MacMillen suggested that by remaining in their burrows and entering torpor during the driest months of summer, cactus mice are able to conserve food and water and thereby are able to inhabit regions where they could not otherwise survive in summer. Because cactus mice voluntarily stop eating when their water ration is reduced, it is not clear from MacMillen's experiments whether the shortage of water is the immediate stimulus that induces torpor in this situation. The torpor could be induced by the reduction of food intake which in turn was induced by the restricted water ration. The ecological consequences are the same in either case—the mice become torpid and conserve both water and energy.

References

[1]   Bartholomew, G. A. and F. Wilke. 1956. Body temperature in the northern fur seal, *Callorhinus ursinus*. *Journal of Mammalogy* 37:327–33.

[2]   Brown, G. W. Jr. (editor). *Desert Biology*, vol. 1 (1968); vol 2 (1974). Academic Press, New York.

[3]   Gates, D. M. and R. B. Schmerl. 1975. *Perspectives of Biophysical Ecology*. Springer-Verlag, New York.

[4]   MacMillen, R. E. 1972. Water economy of nocturnal desert rodents. In *Comparative Physiology of Desert Animals,* edited by G. M. O. Maloiy. Academic Press, New York.

[5]   Morrison, P., M. Rosenmann, and J. A. Estes. 1974. Metabolism and thermoregulation in the sea otter. *Physiological Zoology* 47:218–229.

[6]   Schmidt-Nielsen, K. 1964. *Desert Animals.* Oxford University Press. A general essay of the adapatations of vertebrates to deserts, particularly enjoyable because of the author's narrative style.

[7]   Schmidt-Nielsen, K. 1972. *How Animals Work.* Cambridge University Press. A series of selected topics from the results obtained by the author and his colleagues over many years of field and laboratory investigation.

[8]   Taylor, C. R. 1972. The desert gazelle: a paradox resolved. In *Comparative Physiology of Desert Animals*, edited by G. M. O. Maloiy. Academic Press, New York.

[9]   Whittow, C. G. (editor). *Comparative Physiology of Thermoregulation*, vol 1 (1970); vol 2 (1971); vol 3 (1973). Academic Press, New York.

21

Review of Earth History and the Geologic and Ecologic Settings During the Pleistocene

Synopsis: Throughout the 500 million years since the vertebrates first appeared in the sedimentary rocks of the late Cambrian, their evolution has been influenced by continental movements. Their initial evolution in the sea, in fresh waters, and then onto the land as the anamniotes was accomplished by the end of the Devonian and was associated with equitable tropical climates. Thus, early vertebrate fossils are mostly associated with North America, Europe and Australia because these land masses straddled the paleoequator in the early to middle Paleozoic. The rupture of Pangaea in the Mesozoic had multiple effects— it isolated floras and faunas, produced new oceans and seas, and caused generalized changes in climates on the various continents as they drifted apart, some toward the warm tropics, some toward the cooler polar regions.

In general, drift and separation of the continents led to gradual changes in climates. Dramatic and more sudden changes in worldwide climates, as exemplified by ice ages, have occurred infrequently in earth's history. Yet ice ages have had enormous effects on the evolution of vertebrates, and this is especially evident for humans, who have become the dominant vertebrate on earth from Pleistocene time until the present. The Pleistocene, which began about 1.8 million years ago, is typified by a series of climatic oscillations, in which glaciers advanced southward across the Northern continents during cooler periods, only to retreat during warmer interglacial periods. Associated with these glaciations were precipitous changes in sea level and worldwide reduction in temperature.

Both floristic and faunistic assemblages were repeatedly forced to contract their geographic ranges, a circumstance that led to the genetic isolation of populations and, therefore, an increased chance for speciation. During and since the last Pleistocene glacial episode there has been a massive extinction of large terrestrial vertebrates and, particularly, of large mammals. The cause of these extinctions is attributed to either the effects of sudden changes in climate and/or the emergence of humans as nomadic and unselective generalized predators. The controversy that revolves around this issue is an important one, for in a sense what we as humans are today is, in part, a product of what we were yesterday.

21.1 REVIEW OF CONTINENTAL DRIFT

Movements of the land over the world's surface are and have been a feature of earth's history, at least since the Pre-Cambrian. The dynamic forces that cause this drift (Section 4.2) are, seemingly, never ending. The continents still drift today—North America recedes westward and Australia northward at approximately 4 cm/year. By early Cambrian time a recognizable scene had appeared—the seas had formed, continents floated atop the earth's mantle, life had originated simply and had become complex, and even an atmosphere of oxygen had formed, signifying that the photosynthetic production of food resources had become a central phenomenon of life. The stage was set for the evolution of even more complex organisms, including the vertebrates. The direction of vertebrate evolution has been in large part molded by continental drift.

We have detailed the movements of the continents pertinent to vertebrate evolution in Chapters 4, 8, 11, and 15, and will complete the story here. Because the movements are so complex, their sequence, their varied directions, and the precise timing of the changes are difficult to summarize. By viewing the movements broadly, however, a simple pattern unfolds. Briefly stated, the pattern is described as: *fragmentation–coalescence–fragmentation.*

Expressed more completely, the continents "existed" as separate entities over 500 million years ago. In the shallow epicontinental seas of one or two of these units, destined to become North America and Europe, the ostracoderms originated. Some 300 million years ago all of these separate continents combined to form Pangaea, birthplace of the terrestrial vertebrates. Persisting and drifting northward as an entity, this single huge continent began to break apart about 100 million years ago. Its separation was in two stages, first into Laurasia and Gondwanaland, then into a series of units that have drifted and become the continents we know today (Figure 21-1).

As emphasized in the earlier chapters, continental movements have affected evolution by causing major changes in climate, in local geography, in habitat and in isolation of vertebrates on land masses and in seas of highly variable extent. Some of the major features of change (world climates, the extent of epicontinental seas, periods of major mountain building, and major glaciation) coincide with average reductions in worldwide temperatures and considerable declines in sea level. These changes are rather abrupt and occupy relatively short spans of geological time, unlike the much slower and longer lasting ecological trends associated with drift. How might these abrupt changes have influenced vertebrate evolution? This is an especially important point to consider because human culture emerged during the Pleistocene, a short period in which little continental displacement occurred. The Pleistocene, however, is characterized by precipitous changes in worldwide climate.

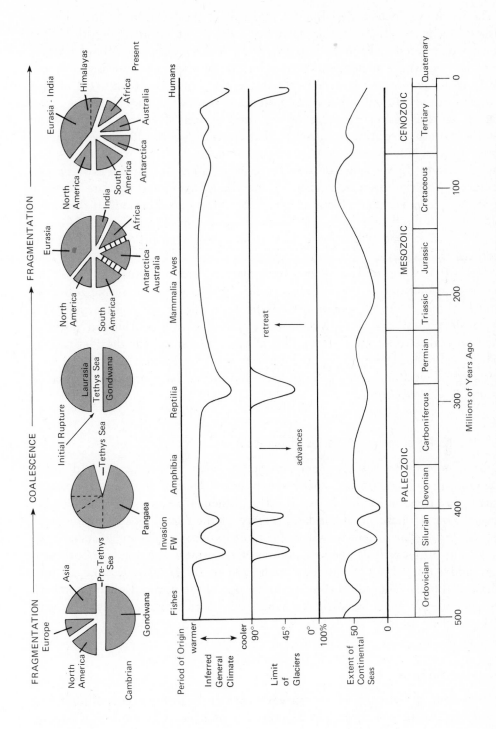

Figure 21-1 A summary of continental drift, climatic trends and the extent of Continental seas from the Ordovician to the present.

21.1.1 Periods of Major Glaciation

The fossil evidence gathered about *Homo* over the last decade continues to push our generic origin further into the past. Some anthropologists believe humans originated 3 to 4 million years ago, others only a million or fewer years ago. However ancient our generic origin, as a species we evolved within the Pleistocene and the Holocene epochs, periods that cover only 1.8 million years out of the 500 million years of vertebrate history. What a short time span that is—far less than 1 percent of the total period of vertebrate evolution! Yet the extensive continental glaciers that characterize the Pleistocene and so closely parallel human evolution are rare geoecological events. Similar extensive glacial periods are recorded only at three other times during the evolution of vertebrates—in the late Ordovician, the middle Silurian, and the late Carboniferous and early Permian (Figure 21-1). The three glacial periods of the Paleozoic were confined mostly to the southern continents (Figures 4-3, 4-4, and 11-1). Only in the Pleistocene has extensive glaciation occurred in the Northern Hemisphere.

It is often inferred that Pleistocene glaciers had a pivotal influence on vertebrate evolution. Is this inference valid? If it is valid, then what influence did the three Paleozoic glacial periods have on earlier vertebrate evolution? To consider these questions we must first know what a continental glacier is and how pervasive its effects are. This is best understood for the Pleistocene. The use of radiocarbon dating and oxygen isotope techniques, coupled with various geological and paleoclimatological indicators, have provided a detailed picture of what occurred.

21.1.2 The Pleistocene—A Series of Ice Ages

For someone who has never encountered a glacier face to face, it is difficult to appreciate its immense size and erosive power. Glaciers, therefore, are treated casually, as remote ice cubes with little influence on everyday life. Today glaciers cover 10 percent of the land surface of the earth, mostly within polar regions. In the Pleistocene they were far more extensive. Then glaciers covered not 10 but 30 percent of the land and, in North America, extended southward as a solid sheet of ice from the North Pole to 38°N latitude, that is, into southern Illinois (Figure 21-2). A similar ice sheet covered all of northern Europe. Judging from modern remnants, this Pleistocene ice mass was very thick; modern glaciers in Antarctica and Greenland are between 3000 and 4000 m thick.

Glaciers and Sea Levels

At present the volume of ice on earth constitutes about 26 million km³. During the Pleistocene there was 77 million km³ of ice. An enormous volume of water was, and still is, locked up in glaciers. Melting of the Pleistocene glaciers to their present size has caused the sea level to rise. Calculations show

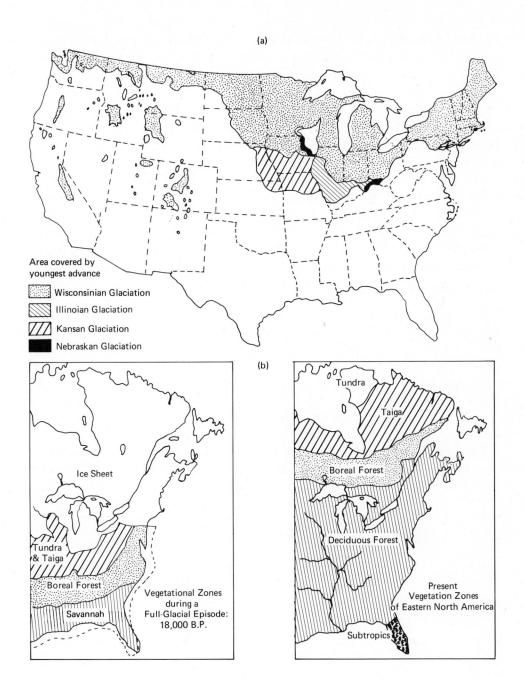

(a)

Area covered by
youngest advance

Wisconsinian Glaciation

Illinoian Glaciation

Kansan Glaciation

Nebraskan Glaciation

(b)

Ice Sheet

Tundra
& Taiga

Boreal Forest

Savannah

Vegetational Zones
during a
Full-Glacial Episode:
18,000 B.P.

Tundra

Taiga

Boreal Forest

Deciduous Forest

Present
Vegetation Zones
of Eastern North America

Subtropics

Figure 21-2 The extent of Pleistocene glaciations in the United States and effects on the position of biomes. During a representative full glacial episode (b, left) the deciduous forests native to most of eastern North America were reduced to small isolated pockets in moist regions of a savannah.

that the melt from this volume of ice is equivalent to a 200-m increase in the depth of the oceans. Isostactic movement of the continents (Chapter 4–2) has diminished the rise in sea level somewhat. The continents have lifted because melting has reduced the extent and thickness of heavy glaciers upon them. In effect, the continental crust beneath the Pleistocene glaciers has rebounded upwards as the ice melted and the ocean floors have sunk as the sea level rose. The sea level relative to a coastline actually has risen only about 140 m from its Pleistocene level; but 140 m is almost half the height of the Empire State Building. Since retreat of the last glacial maximum some 19,000 years ago, the sea surface has been rising at an average rate of 1 cm/year.

If the modern glaciers were to melt, the sea level would rise at least another 50 m. Imagine the effect! Most of the coastal cities of the world would be submerged; close to 25 percent of the current land would be beneath the sea. Such an event—catastrophic to modern civilization—has been a consequence of every past glacial period.

Glacial Episodes

Are the glaciers still receding? Some geologists and climatologists believe we are entering another glacial period. To us the important point is not whether a new Ice Age shall prevail but that glaciation is episodic. Glaciers tend to advance, to retreat, and to advance again on a cyclic schedule that is not fully understood. The episodes are clearly indicated for the Pleistocene. Four distinct glacial advances occurred (Figure 21–2a), interspersed by interglacial periods of warming and glacial retreat. However, these major glacial episodes do not represent a simple constant advance and subsequent retreat of the ice. The last glacial cycle, in fact, constituted three separate smaller cycles. Different names are applied to the multitude of episodes that occurred and even the same episode has received different names in the United States and in Europe (Table 21–1). The length of time for each cycle varied greatly. Thus, the Nebraskan episode, which introduces the Pleistocene, lasted 200,000 years, the Kansan 400,000 years, but the Wisconsin occupied only slightly more than 100,000 years.

Glaciers and World Climates

Continental glaciation is not only a local phenomenon but has a pronounced influence on world climates. For example, in the last episode of the Wisconsin glaciation the cyclonic storm belts, which today lie within the temperate region between 30° and 55° N. latitudes, were further southward; so was the temperate zone. Large biomes (distinctive floral and faunal communities) were also shifted. Conifers occurred much farther south in the United States, suggesting a generally cooler climate (Figure 21–2b). Surface temperatures of tropical seas in the Caribbean were 4 to 5°C lower than today. The climatic effects of glaciation therefore were widespread—generally, climates were cooler.

Table 21-1 Names of Glacial and Interglacial* Stages From Four Northern Hemisphere Regions (after Seyfert and Serkin 1973).

Age** (millions of years ago)	Midcontinent of North America	Rocky Mountains North America	European Alps	Northern Europe
0.03	Wisconsin—late *Interstadial II*	Pinedale	Würm—late	Weichselian *Eemian*
0.06	Wisconsin—middle *Interstadial I*	Bull Lake—late Stade	Würm—early *Riss-Würm*	Warthian *Treene*
0.12	Wisconsin—early *Sangamonian*	Bull Lake—early Stade	Riss *Mindel-Riss*	Saalian *Holstein*
0.50	Illinoian *Yarmouthian*	Sacajawea Ridge	Mindel *Günz-Mindel*	Elsterian *Cromerian* *Beestonian*
1.30	Kansan *Aftonian*	Cedar Point	Günz *Donau-Günz*	Menapian *Waalian*
1.80	Nebraskan	Washakie Point	Donau	Eburdnian

*Interglacials in italics.

**Age refers to beginning of each glacial episode in millions of years before the present.

A vivid example emphasizes this point. Many of the equatorial regions that today are moist and are occupied by lowland tropical rain forests were then arid. In the Amazon and the Congo Basins, the forests contracted into several isolated refugia with each glacial episode (Figure 21–3). During interglacial periods, the forests again spread over the basins. As a result, there was repeated isolation and then remixing of the plant and animal populations that characterized the forests. In genetic theory, isolation and remixing of populations are two of several major ingredients that bring about rapid speciation (see Chapter 2). In the refugia all populations, including vertebrates, suffered a reduction in total numbers and therefore also in their total genetic diversity. Random isolation, genetic drift, and natural selection of each species within the several isolated refugia should have led to new species. When the forests again spread, these new species mingled and some, because they were less competitive, became extinct. In other species competition forced the occupation of slightly different niches. Theoretically the number of species should have increased and, apparently, it did. Species diversity of birds in these tropical forest basins is very high (Figure 21–3). The same applies to some mammals, but less dramatically. The effect of glaciation on equatorial aridity and the resultant contraction of proper habitats is the most logical mechanism for high species diversity in these areas. Glaciation therefore can affect the lives of vertebrates everywhere, not just those in close contact with the spreading ice.

In summary, extensive continental glaciers affect many things: (1) sea level, (2) land area, (3) general climates and habitats, (4) plant and animal distribution, (5) rates of evolution, and (6) rates of extinction. How then did the Pleistocene "Ice Ages" affect mammals and, especially, humans?

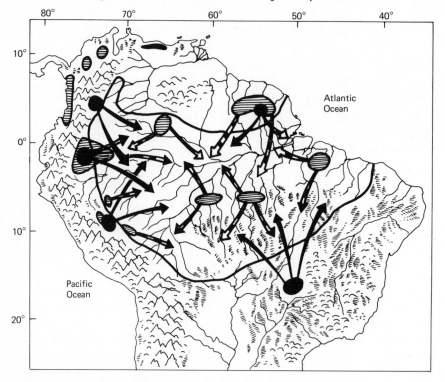

Figure 21-3 Climatic effects of pleistocene glaciation on the lowland forest vertebrates of the Amazon basin.

The black line encompasses the location from which the forests retreated because of deteriorating climate. With this retreat vertebrates were also isolated in forest regions which occurred generally in low mountain areas. The black areas represent isolation of birds (after Hoffer) and the striped areas lizards (after Vanzolini). The arrows indicate possible paths of reinvasion as the forests recovered. That this isolation and remixing of vertebrate populations could result in high species diversity is supported by Dean Amondon's estimate that 592 lowland forest bird species now exist in the Amazon forests (after P. E. Vanzolini [1972], Paleoclimates, relief, and species multiplciation in equatorial forests. In *Tropical Forest Ecosystems in Africa and South America: A Comparative Review*. Editor B. J. Meggers, E. S. Ayens and W. D. Duckworth. Smithsonian Inst. Press, Washington, D.C., 350 pp).

21.2 VERTEBRATE CHANGES DURING THE PLEISTOCENE

The number of fossilized genera of Cenozoic mammals reached a peak in the Pleistocene (Figure 15-1). Living genera that were also present in the Pleistocene (as determined by fossils) represent only 60 percent of the known fossilized Pleistocene genera. A very dramatic and recent decline in the type of mammals must have taken place. Professor Simpson and others have emphasized this decline by calculating extinction rates of different mammalian orders (Figure 21–4). Extinction seems to have occurred mostly in terrestrial mammals. Marine forms, if anything, show a decline in the rate of extinction. For all mammals, however, the rate of extinction increased from 25 genera per million years in the Miocene to 40 in the Pliocene and to over 200 in the Pleistocene.

What caused the extinctions? Some feel that the Ice Ages and the attendant climatic changes were the root cause. Others feel that the extinctions were largely caused by humans, with their newly found tools and social skills. The latter view has been pursued intensively by P. S. Martin at the University of Arizona. What types of information have led Martin to draw this conclusion? The most important are that the highest rates of extinction (1) were largely a postglacial event, (2) occurred with higher frequency in large mammals than small mammals, and (3) span but a short period of time, on the order of 1000 years.

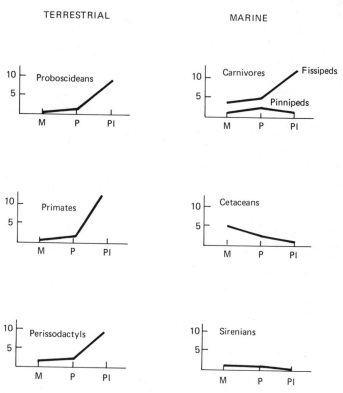

Figure 21–4 Generic extinction rates for various terrestrial and marine mammalian groups.

Numbers are the estimated number of genera which became extinct per million years. Note that the fissipeds, which are close terrestrial relatives of the marine pinnipeds, show a large increase in extinction rate, the pinnipeds do not. M = Miocene, P = Pliocene, Pl = Pleistocene.

If one scrutinizes the Pleistocene record, it is clear that high rates of mammalian extinction occurred only during the Wisconsin and, especially, in postglacial time (Figure 21-5). Examples of forms disappearing at that time are mastodons, various camels, antelopes, and ground sloths. It is Martin's contention that these large mammals were preyed upon by invading humans and, in fact, archaeological data directly support this view. Martin, and earlier A. S. Romer and others, argue that large herbivorous mammals would be more susceptible to extinction by humans than small mammals. Graphs of the changing number of genera of small versus large herbivorous and omnivorous mammals through Cenozoic time do show dramatic differences (Figure 21-6), and changes in numbers of small and large carnivores tend to follow the changes in the noncarnivorous mammals upon which they prey. But should the high extinction rate of large mammals be attributed entirely to humans?

A further relation can be shown between the arrival of prehistoric humans in different regions and the extinction of large herbivorous mammals. For example, the earliest evidence of human hunting of mammals was found in southern Eurasia and is estimated at more than 40,000 years ago. In the New World records indicate that humans were present at least 30,000 years ago. However, the hunting of mammals in the Western Hemisphere began in the colder north only when a new wave of humans invaded from Asia some 13,000 years ago. It would appear that gradually a hunting technology developed in

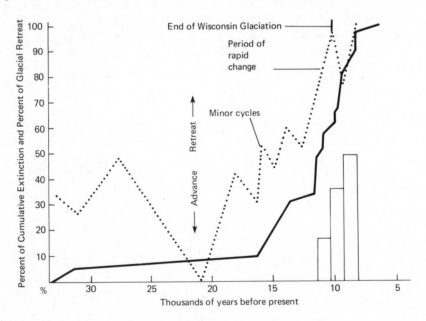

Figure 21-5 Coincidence between the extinction of North American mammals, glacial retreat and the occurrence of humans.

Solid line represents the cummulative percentage extinction of 40 mammalian species; dashed line is the percentage of withdrawal of the late Wisconsin glaciation; histograms are relative abundance of occurrence of man in North America. (After Hester, in [1]).

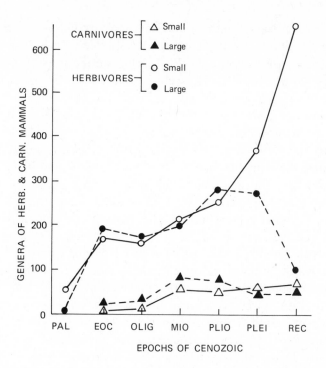

Figure 21-6 Changes in the number of genera of carnivorous and herbivorous mammals during the Cenozoic. Mammals of 20 kg adult body weight or less are considered 'small.'

Africa during the middle Wisconsin glaciation and then spread northward and burst into the Western Hemisphere as the late Wisconsin glacier retreated (Table 21-1). Although the kinds of competitive interactions may have been very different, the rapid extinctions of large mammals that followed the sudden arrival of hunting humans are reminiscent of the rapid extinction of the North American marsupial fauna during the late Cretaceous. This marsupial extinction followed the sudden entrance of Asian eutherian mammals into North America at the end of the Mesozoic (section 15.2). In both episodes of extinction it was the sudden arrival of a new competitor within the isolated American fauna that preceded mass extinctions. A large mammal extinction of similar proportion has not occurred in Africa, where over several millions of years human hunting evolved side by side with large mammals (Chapter 22.1.4). In North America mammals with slow growth and low reproductive rates, such as mammoths and rhinoceroses, were only able to persist until 11,000 years ago. Even though prehistoric humans also hunted smaller mammals—for example, rabbits, rodents, and so on—they did not become extinct. The more rapid reproduction of these small mammals (Table 19-2) sustained an effective population size for survival.

The evidence is conclusive that prehistoric humans did hunt mammals and often with devastating results. Fire, for instance, was used to stampede large herds of ungulates over cliffs to their deaths—a wasteful technique, for only a small portion of the carcasses could be utilized. Nevertheless, other scientists do not attribute the demise of the large mammals solely to human influence. In a symposium on the phenomenon of Pleistocene extinctions, B. H. Slaughter and J. E. Guilday suggest that changing climates and their influence on habitat availability were as important as humans in producing extinctions. In their view, large mammals are more susceptible to environmental stress. Severe restrictions of forage, available water, and elimination of migratory corridors to more suitable areas would affect larger mammals, with their correspondingly larger requirements, more than smaller mammals. Thus, large North American mammals may well have been on the decline when humans arrived.

Many large mammals have persisted in some regions but not in others despite prehistoric human hunting pressure. In the New World camels and horses became extinct, but the latter lived on in Europe. Others, like bison and elk, survived in the New World in spite of humans. It is difficult to resolve these two views. If stressful climates, resulting from glaciation, are the basic cause, then why are the highest rates of mammalian extinction in North America associated mostly with Wisconsin glacial and postglacial periods and not the three periods that preceded them? According to Martin, however, the rates of extinction do show a slight rise in the Aftonian interglacial period and the Kansan glaciation (Table 21-1), although not nearly so high as during the Wisconsin. There is currently no final answer; a combination of causes seems most plausible.

Another important aspect of the extinction of the large terrestrial mammals is their loss without replacement. The niches that they filled largely remain empty today. This is another way of saying that the diversity of the mammal fauna has declined and replacement, usually by more appropriately adapted species, has not occurred. Evolutionary calamities in the vertebrate assemblage of this dimension and without obvious replacements are rivaled only by the extinction of the dinosaurs in the Cretaceous and, to a lesser extent, by the reduction of amphibians and early reptiles near the end of the Permian. In each of these instances neither glaciers nor a generalized predator like humans can be invoked as causative agents. Some believe that the late Carboniferous and early Permian glaciation influenced diversification of the reptiles such as the therapsids that later gave rise to mammals (Chapters 11.2 and 12.4). Indeed, it has been suggested that some forms had evolved "incipient" homeothermy in response to the cooler climates of southern Gondwanaland (Figures 11-1 and 11-2). If so, their presumed higher levels of activity and competitiveness may have proved too much for the large Permian amphibians and other reptiles. For the moment the evidence for such competitive replacement is scant and speculations remain fragile.

In the Pleistocene, however, human technology undoubtedly hastened the extinction of the large mammals and allowed us to replace many large forms. Although no series of species has come to replace this lost megafauna of the Pleistocene, humans, using cultural and technological adaptations instead of morphological and physiological ones, have "occupied the niche" of many of the forms they eliminated in their first great population explosion in North America. Plows and grain mills have replaced grazing adaptations, or the domesticated adjuncts to human culture have been substituted for the more or less equivalent, naturally occurring forms. Today we humans continue to hunt and to take our toll of large vertebrates and to destroy natural habitats in our quest to modify the environment to our own needs. What humans are and do depends partly on what humans were and did. We are a consequence of our vertebrate heritage. To understand this heritage objectively is one of the major functions of this textbook. Let us proceed by examining humans and the vertebrates. In knowing them we may better know ourselves.

References

See references in chapters 4, 8, 11 and 15 for readings on plate tectonics and continental drift.

[1]   P. S. Martin and H. E. Wright, Jr. (editors). 1967. *Pleistocene Extinctions: The Search for a Cause.* Yale University Press, New Haven. A collection of Symposium papers presenting varied views on the extinctions of the larger mammals during the Pleistocene.

[2]   Gillespie, R., D. R. Horton, P. Ladd, P. G. Macumber, T. H. Rich, R. Thorne, and R. V. S. Wright. 1978. Lancefield swamp and the extinction of the Australian megafauna. *Science* 200:1044–1048. This recent study suggests that the extinct megafauna of Australia coexisted for over 7,000 years with aboriginal humans. The authors conclude that the model of overkill of large mammals by humans does not explain the extinction of large mammals in Australia.

[3]   Mosimann, J. E., and P. S. Martin. 1975. Simulating overkill by paleoindians. *American Scientist* 63:304–313. An interesting computer simulation of how early nomadic hunters, invading from Asia across the Bering Straits, could have in about 300 years wiped out most of the North American mammalian megafauna.

22

Homo sapiens and the Vertebrates

Synopsis: Evidence for the origin of *Homo sapiens* from apelike ancestors comes partly from the fossil record of the primates in the Cenozoic and partly from comparative study of existing monkeys, apes, and human beings. The traits that characterize the Order Primates have evolved in relation to an arboreal mode of existence, although some primates have become secondarily terrestrial, human beings most of all.

The first primates probably evolved from a line of arboreal insectivores much like the present day tree shrews (Tupaiidae). The earliest known primates appear in the late Paleocene and are basically similar in structure and closely related to the modern lemurs, potos, galagos, and other forms in the suborder Prosimii. These prosimians spread over large parts of the Old and New Worlds in a short period of geological time and differentiated into a variety of ecological forms. All were arboreal and retained certain primitive features but were otherwise identifiable as primates, particularly by their large brains in relation to body size.

By the Oligocene two quite different groups of higher primates had evolved from the prosimians—the so-called platyrrhine monkeys of the New World tropics and the Old World catarrhine monkeys and apes. The New World monkeys are considered to be more primitive but not necessarily the ancestors of the catarrhines, which include the most advanced primates. The catarrhines are characterized by nostrils set close together and opening forward and downward, by generally large body size, especially in the great apes and humans, by a marked tendency toward upright body carriage, and by secondary specializations for terrestrial locomotion. The tail is short or absent and never prehensile; and the dental formula of $\frac{2\text{-}1\text{-}2\text{-}3}{2\text{-}1\text{-}2\text{-}3}$ has been uniform in the group back to the earliest known catarrhine fossils in the Oligocene.

The anthropoid apes and humanlike species, including *Homo sapiens*, are grouped together in the superfamily Hominoidea because they are phylogenetically more closely related to each other than to any other primates. Fortunately

807

for students of human evolution, the fossil record for this clade is richer than for other lineages of primates.

While there are many morphological features that distinguish the hominoids from other catarrhines, enlargement of the brain has been the main evolutionary force molding the shape of the hominoid skull. Also, differences in cheek teeth between hominoids and Old World monkeys are paramount for the identification of fossils.

The first known hominoids occur in the Oligocene 35 million years ago. By the Miocene hominoids had diversified into a number of ecological types and had spread widely over Africa, Europe, and Asia. Among these forms were groups called dryopithecids and sivapithecids, from which evidence indicates the great apes (Pongidae) and humans (Hominidae) evolved. The first distinctively human-like fossils have been found in Pakistan, Kenya, and Turkey and date from about 17 to 7 million years ago. They have all been assigned to the genus *Ramapithecus*.

A variety of hominid fossils has now been discovered in late Pliocene and early Pleistocene deposits of Kenya and Ethiopia, some of them more than 3 million years old. Some are so similar in structure to *Homo sapiens* that they have been placed in the genus *Homo*; others with smaller cranial volumes, less upright stances, and more rugged features are assigned to the genus *Australopithecus*. As more and more of these early humanlike fossils are unearthed, it becomes increasingly difficult to make a meaningful distinction between the two genera, especially since some very *Homo* like fossils are older than some forms of *Australopithecus*, which has usually been considered ancestral to *Homo*.

Homo erectus, the renowned "Java Ape Man," ranged across Africa and Eurasia from about 1 million years to 300,000 years ago. This hominid had a brain capacity approaching the lower range of *Homo sapiens*, made stone tools, and knew the use of fire. *Homo sapiens*, the only surviving species of the family Hominidae, came into existence around 300,000 years ago, probably as a result of changes in body size and relative brain capacity in some geographically isolated population of *Homo erectus*. By 40,000 years ago the Cro-Magnon people in Europe were indistinguishable in structure from human beings as they exist today, and they had a well organized society with a rapidly developing culture. Domestication of animals, agriculture, and civilization were only a few tens of thousands of years away from human attainment.

The biological success of the human species is measured by the fact that human beings are among the most abundant of all vertebrates, by a universal distribution over the earth, and by the fact that humans derive some portion of their livelihood from every ecosystem and habitat in the biosphere. This all-pervasive status of *Homo sapiens* has had a dramatic impact on environments and on other animals, particularly on other vertebrates. Although extinction of species is a natural result of evolutionary processes through time, there is no precedent for the extent to which one species—*Homo sapiens*—has been responsible for extinctions and for the "endangerment" of many other forms of life.

The first significant human impacts on other vertebrates may have occurred in prehistoric time. Some paleoecologists think that the ice-age hunters—particulary in North America—may have been responsible for the extinction of some of the large Pleistocene mammals, such as mammoths, giant ground sloths, and others.

The influence of modern, civilized human beings on vertebrate species has occurred in three overlapping phases. During the period of exploration and discovery in the fifteenth through nineteenth centuries many island endemics were brought to extinction either directly through slaughter for food or, indirectly, as the result of exotic introductions or establishment of feral populations of domesticated animals that altered island ecosystems unfavorably for native species. Following close on the heels of the explorers, fur trappers, commercial hunters, fishermen, and other "harvesters" began exploiting wild vertebrate populations for commerce. Many vertebrate species were drastically reduced in numbers by these activities, but few were actually exterminated, because the law of diminishing returns influences profit. The most serious long-term effects of human beings on other vertebrates relate to the impact of agriculture and other land uses on natural habitats. More species will join the ranks of endangered wildlife and will become extinct in the next few decades than in all the centuries since 1600 owing to the diminution and deterioration of natural habitats resulting from the way we human beings live on the earth. Such losses will be particularly severe in the biotically rich tropical forests of South America, Africa, and Asia.

Two cardinal requirements must be met if we are to save a significant vestige of natural biotas and ecosystems intact for future generations of humans to know. These are preservation of natural habitats and ecosystems on a large and representative scale, and control of human population growth. The larger national parks and wildlife preserves of the world show what can be done in the first instance, but we need to set aside many more of these natural areas before it is too late. Ultimately and soon, human beings must learn to accept a morality of restraints—restraint on the use of resources and, above all, restraint on human procreation.

22.1 INTRODUCTION: THE BIOLOGICAL ORIGIN OF HOMO SAPIENS

Human beings are primates. Our closest living relatives are the gorillas and chimpanzees of Africa, and we have descended through a long line of intermediate forms from arboreal ancestors that lived in the early Tertiary forests 65 million years ago. What is the basis for these statements, which Darwin's Victorian contemporaries found so startling and upsetting, but which educated folk take pretty much for granted today? It is a conceptual fabric woven in part from physical evidence about the nature of humans and other vertebrates and in part from logical inference about that evidence. As the physical evidence accumulates, particularly from the fossil record, the fabric becomes more substantial and the overall pattern becomes more distinct. Our understanding of human biological origin has undergone radical change since T. H. Huxley's detailed and lucid anatomical comparisons between the human body and the great apes in the 1860s, especially in the last 10 years as more and more hominid (manlike) fossils have been unearthed in Africa, where the record indicates humans had their evolutionary origin. In the following sections we summarize the evidence and the inferences about human phylogeny and evolution.

22.1.1 Characteristics of Primates

Humans share many anatomical and other biological traits in common with animals variously called apes, monkeys, baboons, lemurs, tarsiers; and, by the principle of homology, which we discussed in Chapter 3, comparative anatomists recognized long ago that all these forms of mammals can be grouped together as an order in formal classification, the Order Primates. Some of the obvious traits that characterize primates are (1) retention of the clavicle as a prominent element of the pectoral girdle (the clavicle is greatly reduced or lost in many mammalian orders); (2) a forelimb with a shoulder joint allowing a high degree of movement in all directions and with an elbow joint permitting rotational movement of the forearm; (3) retention of five functional digits on the fore and hind limbs; (4) enhanced mobility of the digits, especially the thumb and big toe, which are usually opposable to the other digits; (5) claws modified into flattened nails (with some exceptions); (6) sensitive tactile pads developed on the distal ends of the digits; (7) reduced snout with most of the skull posterior to the orbits; (8) complex visual apparatus with forward directed eyes and highly developed binocular vision; (9) enlarged brain relative to body size, particularly the cerebral cortex; (10) only two mammary glands (some exceptions); and (11) typically only one young per pregnancy.

These are traits that obviously have evolved in relation to an arboreal way of life involving clinging to the branches of trees, jumping from branch to branch, and swinging by the forelimbs (brachiation). All of the basic primate modifications of the appendages can be seen as adaptations for arboreal modes of locomotion, as can the stereoscopic depth perception made possible by the overlapping visual fields of binocular eyes and the enlarged brain to allow for very precise neuromuscular coordination between visual perception and locomotory response. Most primates are arboreal, but some have become secondarily terrestrial, humans most of all. Even so, many of the traits that are most distinctively human derive from arboreal adaptations in the remote ancestors of *Homo sapiens*.

22.1.2 Evolutionary Trends and Diversity in Primates

Table 22-1 shows a classification of modern primates down to the level of families, and Figure 22-1 presents a simplified phylogeny in relation to the geologic time scale and the known fossil record. It is generally agreed that the first primates evolved from a line of arboreal insectivores not unlike the present day tree shrews (Tupaiidae) that live in the tropical forests of Southeast Asia. These are small, agile mammals (still classified by some taxonomists in the Order Insectivora) preferring low forest growth for the most part. They are slender in build with long bushy tails and rather short limbs, particularly the forelegs, and are more squirrellike than monkeylike (Figure 22-2). The skull and teeth show most of the unspecialized characteristics of primitive mammals, including

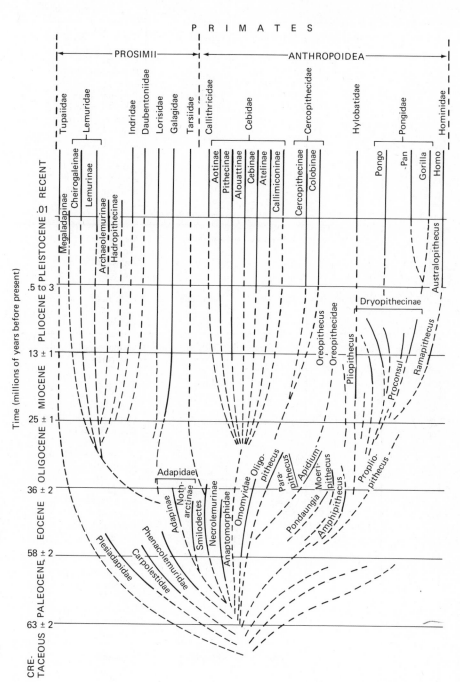

Figure 22-1 Phylogeny of main primate groups. Solid lines indicate known occurrences in time; dashed lines suggest postulated relationships and times of divergence. (Modified from Simons [10].)

Table 22-1. A Classification of the Modern Primates*

Order	Suborder	Infraorder	Superfamily	Family	Common names
Primates	Prosimii	Lemuriformes	Tupaioidea	Tupaiidae	Common Tree Shrew Smooth-tailed Tree Shrew Philippine Tree Shrew Pen-tailed Tree Shrew
			Lemuroidea	Lemuridae	Common Lemur Gentle Lemur Sportive Lemur Mouse Lemur Dwarf Lemur
				Indridae	Indiris Avahi Sifaka
				Daubentoniidae	Aye-Aye
		Lorisiformes	Lorisoidea	Lorisidae	Slender Loris Slow Loris Angwantibo Potto
		Tarsiiformes	Tarsioidea	Galagidae	Bush Baby
				Tarsiidae	Tarsier
	Anthro-poidea	Platyrrhini	Ceboidea	Callithricidae	Plumbed and Pygmy Marmosets
				Cebidae	Titi Saki Capuchin Howler Squirrel Monkey Spider Monkey Wooly Spider Monkey Wooly Monkey
		Catarrhini	Cercopithe-coidea	Cercopithe-cidae	Macaque Balck Ape Magabey Baboon Gelada Guenon Patas Common Langur Proboscis Monkey Gueraza
				Hylobatidae	Gibbon Siamang
			Hominoidea	Pongidae	Orangutan Chimpanzee Gorilla
				Hominidae	Man

*Modified from Simons, *Scientific American,* July, 1964.

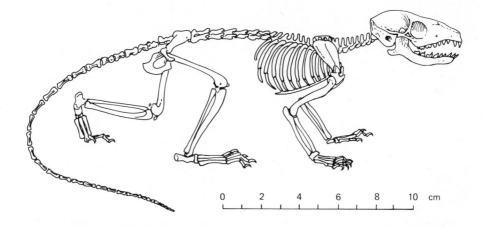

Figure 22-2   Skeleton of tree shrew (*Tupaia glis*) in typical posture with flexed limbs and quadrupedal stance. (Modified from Schultz [9].)

a long facepart projecting forward from the eyes with nostrils at the end of a snout, a large gape, eye orbits laterally situated, and the occipital condyles located at the extreme posterior of the skull with the head carried fully out in front of the vertebral column. Further, the tree shrews have lost only one incisor and one premolar from the primitive mammalian dental formula of three incisors, one canine, four premolars, and three molars, and the cheek teeth also retain the primitive tricuspid condition.

Although no fossil insectivores have yet been identified as the direct ancestors of primates, we know that they must have existed by the late Cretaceous or early Paleocene, because the first fossil primates appear in the late Paleocene and show considerable diversity in the Eocene. These fossils all belong to the Suborder Prosimii, which means that they are most closely related to modern lemurs, potos, eye-eyes, lorises, galagos, and tarsiers, and were similar to them in form and habits but at a more primitive stage of evolutionary development. The first prosimians must have been essentially like tree shrews with, perhaps, some development toward prehensile fingers and toes, opposability of the first digits, and enlarged brains.

In a relatively short period of geological time, prosimians spread over large portions of the Old and New Worlds and became differentiated into a variety of ecological forms ranging from insectivorous to vegetarian in their diets and from nocturnal to diurnal in their daily activity cycles (Figure 22-3). All were apparently arboreal and retained certain primitive features such as long, foxlike snouts and bushy tails; however, otherwise they became identifiable as primates by changes in their appendicular arrangement, the tendency for the eyes to move forward, and other basic features of the primate constellation of characters—most notably large brain capacity, for the late Eocene prosimians already possessed brains larger in relation to their body sizes than any other contemporary mammals.

Figure 22–3   Early prosimian fossils, showing variation in dentition and feeding habits: (a) *Plesiadapis*, a late Palaeocene form, was rodentlike; (b) *Notharctus* an Eocene lemur, was more omnivorous and carnivorous, while (c) *Tetonius*, an Eocene tarsier-like form, had highly specialized teeth suggesting a basically predatory mode of life, and its large eye orbits clearly indicate nocturnal habits. (Modified from Romer, Ch. 5 [7].)

The evolutionary potential and diversity of the prosimians are indicated in part by the fact that, of the 100 genera of fossil primates, over two thirds belong to the Suborder Prosimii and in part by the fact that they gave rise to the higher primates including man. Today the prosimians are restricted to parts of Africa and Southeast Asia and to the island of Madagascar, where the true lemurs and their relatives continued to maintain a diversity of forms in the absence of effective predators and competing higher primates, until true humans arrived on the scene in recent times. (Now many of the lemurs are threatened with extinction owing to man's destruction of their forest habitat.) By the Oligocene the first simians had evolved, and in most parts of the world the prosimians began to disappear in the face of the competitive superiority of the more advanced "anthropoid" monkeys and apes.

Little is known about the evolutionary transitions from the prosimians to the higher primates. Fossil primates from the critical Oligocene and Miocene periods are rare. Most of the fossils from the Miocene are, in fact, already well advanced hominoids, and no fossils intermediate in characteristics between prosimians and higher primates have yet been recovered from the Eocene or Oligocene. All we know for certain is that by the Oligocene two quite different groups of higher primates existed, the so-called platyrrhine monkeys of the New World tropics and the Old world catarrhine monkeys and apes. Both the morphological differences between these groups and their fossil histories suggest that they have evolved independently from different lineages of prosimians and that they may be no more closely related to each other than either is to surviving groups of prosimians. All known platyrrhines have been restriced to the American tropics since the first record in the late Oligocene, and almost all catarrhines with the exception of *Homo sapiens* have been restricted to the Old World for an even longer period.

The New World monkeys are usually considered to be somewhat more primitive in their features. As the group name indicates, these monkeys have flat noses, with nostrils that are wide apart and face outward. They are on average smaller than the catarrhines; there is seldom any reduction in length of the primitively long tail, which is also specialized as a prehensile organ in the Cebidae (e.g., spider monkey), and the thumb is not highly opposable to the other fingers. There are three premolars present in all cases. In addition to the typical South American monkeys, the suborder platyrrhini also includes the squirrellike marmosets (Callithricidae), in which the toes, except the hallux, have clawlike nails that function for digging into bark as aids in climbing.

The catarrhines include the most advanced primates and humans. The nostrils of these primates are close together, open forward and downward, and have a relatively smaller bony opening than in platyrrhines. There is a general evolutionary trend toward large body size, culminating in the great apes and humans. Although mostly arboreal, there is a marked tendency toward upright carriage, and secondary specializations for terrestrial locomotion, most notably in the human. The tail is often short or absent and is never used as a prehensile organ. The second premolar is always absent, giving a dental formula of: $\frac{2-1-2-3}{2-1-2-3}$; and this condition traces back as far as the earliest known fossils in the Oligocene. The group consists of the Old World Monkeys (Cercopithecidae), the gibbons (Hylobatidae), the great apes (Pongidae), and humans (Hominidae).

22.1.3　Evolution and Phylogeny of the Hominoidea

The "anthropoid" apes and humans are placed together in a superfamily, the Hominoidea, because they are phylogenetically more closely related to each other than to any other family of primates. Fortunately, also, the fossil record for this clade is better than for other lineages of primates.

The hominoids are much more diversified in morphology and habits than their nearest relatives, the cercopithecoid monkeys. There is a greater range in body size, the small gibbons being dwarfed by the huge gorillas, and the dense hair covering most of the body of the gibbons stands in marked contrast to the sparse remnants of hair in humans. The hominoids also differ in their locomotor specializations. The gibbons move through the trees most frequently by brachiation and are among the most versatile arboreal acrobats in Asia. It is interesting to note that the gibbons also possess ischial callosities on their buttocks, like those of the cercopithecoid monkeys. The larger and more sluggish orangutans rarely swing by their arms and prefer slow quadrupedal climbing among the branches of trees. The African gorillas and chimpanzees are still less adapted for brachiation and are more at home on the ground, where they most often progress quadrupedally, supported on their forelimbs by their knuckles. Although all modern hominoids can stand erect and walk to some degree on their hind legs, humans have evolved a unique, erect bipedal mode of locomotion involving specialized structure in the hind feet, thereby freeing the forelimbs from obligatory functions of support and locomotion.

Hominoids are morphologically distinguished from other recent catarrhines by a pronounced widening of the trunk relative to body length so that the shoulders, thorax, and hips have become proportionately broader than in monkeys. As a consequence the clavicles have elongated, and the iliac blades of the pelvis have broadened out. The sternum has become modified from a slender row of separate bones in all lower primates to a broad structure, the bony elements of which become fused together at some period after birth. The shoulder blades of hominoids lie over a broad, flattened back, in contrast to their lateral position next to a narrow chest in monkeys (Figure 22–4). The pelvic and pectoral girdles are relatively closer together in hominoids owing to the shortening of the lumbar region of the vertebral column with its uniquely reduced number of segments. In contrast, while the average number of sacral vertebrae is only three in the Old World monkeys, it is four or more in gibbons, five in man, and even more than five in the great apes. The caudal vertebrae have become reduced to vestiges in all recent hominoids, and normally no free tail appears postnatally. After birth, the spinal column arches toward the thoracic cavity in the hominoids and approaches the center of gravity of the body when it is held upright, an ontogenetic process that is most advanced in man but not seen at all in the lower primates. These and other anatomical specializations of the trunk are common to all hominoids and represent adaptations for maintaining the erect postures that these primates assume during sitting, climbing, and walking bipedally.

Enlargement of the brain has been the dominant evolutionary force molding the shape of the hominoid skull, which has tended to increase the neurocranial vault more than its base. The skulls of hominoids are also distinguishable from the skulls of catarrhine monkeys by their much more extensive formation of sinuses.

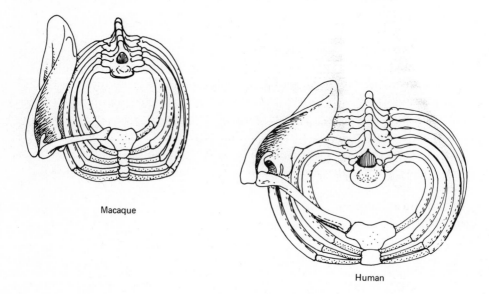

Macaque

Human

Figure 22–4 Cephalic view of chest and pectoral girdle (right half) of macaque (cercopithecoid monkey) and human (hominoid). (Modified from Schultz [9].)

As far as the identification of fossils is concerned, differences in the cheek teeth between hominoids and Old World monkeys are of paramount significance, because so many primate fossils are individual teeth, or teeth in fragments of jawbone. The Old World monkeys have molars with four cusps, one at each corner of a rectangular shaped crown; both front and rear pairs of cusps are connected by ridges. Hominoid molars have crowns with five cusps unconnected by ridges; the grooves between the cusps resemble the letter Y, with the tail of the Y pointing forward (Figure 22-5). This molar pattern is called Y-5, and its evolutionary significance lies in the fact that it is a hereditary characteristic that has persisted among all the hominoids for more than 30 million years. One problem, of course, is that highly worn teeth may not show these patterns clearly; paleontologists can argue for years about the correct taxonomic assignment of such fossils.

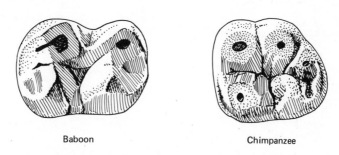

Baboon   Chimpanzee

Figure 22-5 Molar patterns of Old World monkey (baboon) and hominoid (chimpanzee) compared. (Modified from Simons [10].)

Figure 22-6 shows a phylogenetic tree of the Hominoidea, including the known fossil forms and the modern apes and humans. Note that the human, gorilla, and chimpanzee are indicated as being more closely related to each other, in terms of recency of their common ancestor, than any of them is to orangutans or gibbons. These relationships among living apes and humans, first determined on the basis of comparative anatomy by T. H. Huxley in the late 1800s, find additional support from biochemical and serological comparisons as well as from the fossil record of the earlier hominoids.

The earliest indications of the hominoid lineage come from a number of fossil teeth and jaws found in the Egyptian Fayum deposits of the Oligocene, approximately 35 million years old. Fossils assigned to the genus *Propliopithe-cus* and others of similar structure are sufficiently generalized to be considered ancestral hominoids (see Figure 22-7). Since *Oligopithecus* occurs in the same deposit, and its molars indicate that it is near the ancestral stock that gave rise to the cercopithercoid monkeys, we may assume that the major division in the catarrhine lineage between the cercopithecoids and hominoids occurred in the Eocene, but a common ancestor has yet to be found among the fossil primates of that period.

In any event, by the Miocene the ancestral hominoids had diversified into a

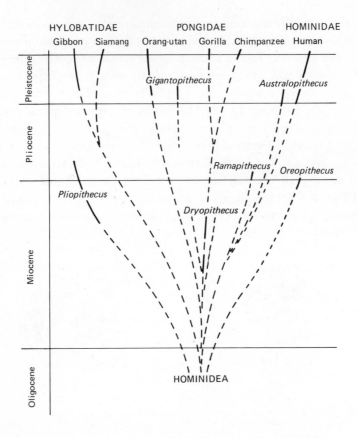

Figure 22-6 Phylogeny of the Hominoidea. (Modified from Schultz [10].)

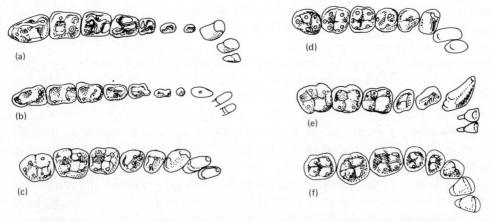

Figure 22-7 Lower left dentitions of some primates, showing evolutionary trends from (a) *Notharctus*, an American Eocene lemur and (b) *Pronycticebus*, a European Eocene lemur, to (c) *Parapithecus*, an Oligocene catarrhine to (d) *Propliopithecus*, an ancestral Oligocene hominoid, to (e) *Dryopithecus*, a Miocene hominoid, and to (f) *Homo* of middle Pleistocene age. (Modified from Romer, Ch. 5 [7].)

number of ecological types and had spread out widely over the Old World, including Africa, Europe, and Asia. One genus, *Pliopithecus* (Figure 22-8) about 20 million years old, may be ancestral to the modern gibbons, as its skull is very like that of the hylobatids; otherwise, its appendicular and trunk skeletons are generalized with the fore and hind limbs about equal in length. It was not nearly as specialized for brachiation and arboreal life as the modern gibbons or the somewhat later *Oreopithecus*, which parallels the gibbons very closely in its postcranial skeleton but has a skull with a short, flat face and other features superficially resembling a hominid (humanlike) condition (Figure 22-9). This line apparently did not survive for long in the Pliocene and may have met competition from the more advanced gibbons.

Of greater interest is a group of Miocene fossils variously referred to as dryopithecids and sivapithecids, because the evidence indicates that the great apes (Pongidae) and humans (Hominidae) derive from them. These primates occurred widely across Eurasia and Africa and have been assigned various generic names, *Dryopithecus, Sivapithecus, Gigantopithecus,* and *Ramapithecus,* and others such as *Proconsul,* which is now usually synonymized with *Dryopithecus,*

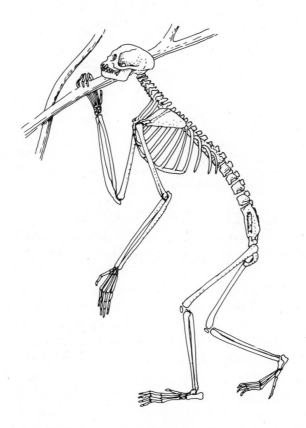

Figure 22-8   Reconstruction of *Pliopethicus*, a Miocene ape that may have been ancestral to the modern gibbons (Hylobatidae). (Redrawn from Simons [10].)

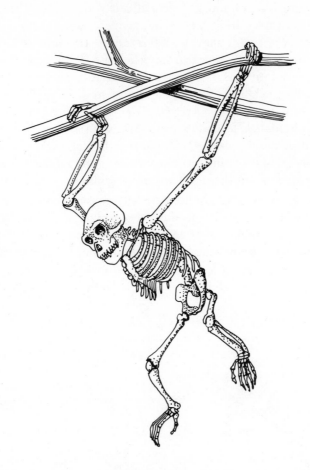

Figure 22-9   Reconstruction of *Oreopithecus*, a gibbon-like ape of the Miocene,
14 million years old. (Redrawn from Simons, [10].)

and *Kenyapithecus*, which is a synonym for *Ramapithecus*. They obviously are
all closely related, rather advanced hominoid populations, but paleoclimatic and
paleofaunistic evidence is beginning to suggest that they occupied a wide range
of ecological niches.

Some of the older Oligocene-Miocene "dryopithecines" lived in heavily forested areas of Europe and Africa and apparently never left the forests for more
open environments. The "sivapithecids," on the other hand, first appear in the
fossil record during the Miocene at a time when some of the heavily forested
areas of the Old World were giving way to mixed environments consisting of
patches of forest, savanna woodlands, and open grasslands. The kinds of
mammalian fossils with which the sivapithecids are found suggest that these
primates lived on the edge of the forests and open areas and that they derived
their livelihood from both. If true, then some of them must have been evolving
toward a terrestrial mode of locomotion, and it is among such forms that we
can expect to find the first hominid ancestors of *Homo sapiens*.

22.1.4   Origin and Evolution of the Hominidae

Humans are not only a distinct species of hominoid, but differ enough in their structure from the apes to be placed in a separate family, the Hominidae. Some of the important anatomical changes that occurred in the course of evolution from an apelike ancestor to modern humans should be reviewed at this point, particularly those that are relevant to an interpretation of hominid fossils, which, again, are mostly teeth, jaws, and skulls.

The jaw in the human clade became shorter, in association with the shortening of the muzzle, and the teeth became smaller and more uniform in size and shape. In particular the canines, which are always longer than the other teeth in apes and monkeys, came to lie in an even crown line with the incisors and cheek teeth, and the prominent diastema between the canines and incisors to accommodate occlusion of the jaws in apes has disappeared in the humans; normally all the teeth touch their adjacent members (Figure 22–10). The jaws of an ape are rectangular or U-shaped with the four incisors across one end lying at right angles to the canines and cheek teeth, which form nearly parallel lines on the two sides. The human jaw is bow-shaped, with the teeth running in a curve that is widest at the back of the mouth. The human palate is prominently arched, whereas the ape palate is flatter.

The articulation of the skull with the vertebral column (**occipital condyles**) and the **foramen magnum** shifted from the rear of the braincase to a position under the braincase, as an adaptation to balance the skull on top of the vertebral column in association with an upright, vertical posture. The braincase itself became greatly enlarged in association with an increase in forebrain size, and as it did so a prominent vertical forehead developed, in contrast to the sloping forehead of the apes. The eyebrow ridges and other "crests" on the skull became reduced in size in association with the loss of major muscles that once attached to them. The human nose became a more prominent feature of the face, with a distinct bridge and tip.

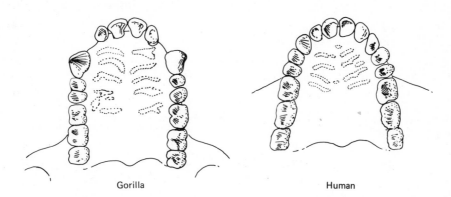

Gorilla   Human

Figure 22–10   Upper jaw of ape (gorilla) and human compared. (Modified from Schultz [9].)

The human postcranial skeleton changed in fundamental ways too, but unfortunately limb bones, girdles, and vertebrae are not often found as fossils. The most radical changes are all associated with the assumption of a fully erect, bipedal stance: the S-shaped curvature of the vertebral column, the modification of the pelvis and position of the **acetabulum** in connection with upright bipedal locomotion, the lengthening of the leg bones and their positioning as vertical columns directly under the head and trunk, with the knees as close together in a normal standing posture as are the articulations of the right and left femur with the pelvis. The feet of humans show drastic modification for bipedal locomotion, having become flattened and arched with corresponding changes in the shapes and positions of the tarsals (ankle bones) and with close, parallel alignment of all five metatarsals and digits; the big toe is no longer opposable as in apes and monkeys (Figure 22–11).

What has been the evolutionary history and origin of our distinctively human traits? As animals with a great curiosity about our ancestry, we would particularly like to know about the evolutionary events that turned the human into an obligate terrestrial biped from ancestors that clearly were arboreal and only occasionally bipedal at best, and we would like to know about the evolution of our tool-using hands and our big brain with its associated implications for intelligence, advanced social organization, and culture.

We now know a great deal more about the evolution of human physical traits than we did as recently as 10 years ago. Before Darwin's time it was commonly believed that the modern human species was only a few thousand years old, and by counting the genealogy of the Old Testament, it was possible to fix the date of creation in the Garden of Eden rather precisely. Even after the discovery of the first human fossil bones in Europe and later in Asia and Africa, it was long thought that true human beings (genus *Homo*) were a recent evolutionary innovation no more than about 500,000 years old, roughly mid-Pleistocene in origin. Recently, better methods of dating fossils and Pleistocene chronology have pushed the beginning of the Pleistocene farther back in time and have revealed that the original estimates of the age of some fossils were erroneous. Then, too, older and older fossils with hominid characteristics have been uncovered, so that it becomes more and more difficult to say when the first "true humans" appeared.

The oldest fossils that show hominid affinities are some teeth and jaw fragments from the Sewalik Hills of Pakistan, Fort Ternan, Kenya, and Turkey, dating from about 17 to 7 million years ago. They have all been assigned to the genus *Ramapithecus* and were contemporary with other Miocene forms of sivapithecids, which we have already identified as the probable ancestral stock from which the great apes and human primates evolved. The shape and relative size of the teeth are more like those of humans than like those of apes; the tooth row is more bow-shaped than rectangular or U-shaped (Figure 22–12).

The size of the teeth and jaws suggests an animal about a meter or so in stature. Nothing more can be said for certain about the structure of this group of primates. The drastic change in dentition with reduced canines and other

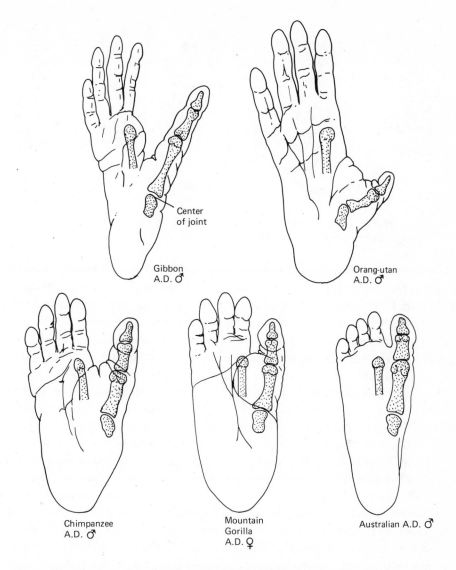

Figure 22-11 Feet of modern hominoids compared, showing some skeletal parts (metatarsals for digits I and II and phalanges of digit I) in relation to foot form. Note difference in arboreal specializations of gibbon and orangutan, and the degrees of specialization for terrestrialism in chimpanzee, gorilla, and human. (Modified from Schultz [9].)

humanlike features suggests that *Ramapithecus* may have foraged more with its hands and therefore could have been primarily bipedal. Also, all the *Rama-pithecus* fossils seem to be associated with fossils of browsing and grazing un-gulates and other mammals, indicating that these primates lived in savanna woodlands and at the edge of grasslands, and further supporting the idea that they were terrestrial or semiterrestrial in habits. The conclusion now seems inescapable that by Miocene time advanced hominoid populations widely dis-

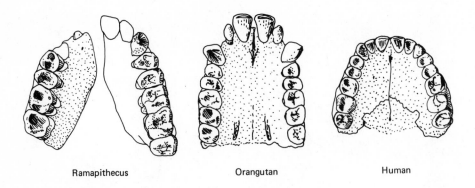

Ramapithecus   Orangutan   Human

Figure 22-12   Reconstructed jaw of *Ramapithecus* compared to jaw of ape (orangutan) and human. (Modified from Simons [10].)

tributed in Asia, Europe, and Africa had already diverged from the ancestral pongid stock and had developed important new hominid traits as adaptations for exploiting different kinds of food and ecological niches, probably involving terrestrial existence and some development of bipedal locomotion on the ground.

A gap of some 3 to 4 million years occurs in the geological record before the next hominid fossils are found in late Pliocene deposits in Kenya and Ethiopia. The most remarkable specimen to date has been scientifically described as *Australopithecus afarensis* but is known in popular literature by her nickname, "Lucy." Lucy was discovered by Dr. Don Johansen in the desert of the Afar Plain in Ethiopia not far from the Red Sea, in a deposit rather precisely dated at 3.5 million years old. She is an astonishing specimen in several respects. First of all, she is the most nearly complete pre-*Homo sapiens* specimen ever found, consisting of more than 60 separate pieces of bone, which represent portions of the skull, lower jaw, arms, leg, pelvis, ribs, and vertebrae. Secondly, she was remarkably human in structure. There is absolutely no doubt that she stood fully erect like modern woman. If she was arboreal at all, that fact is not reflected by obvious modifications of her bones. Her teeth and lower jaw are also very humanlike, but unfortunately we know little about the size and shape of her cranium. Her overall body size was small: fully grown when she died, Lucy was only about 1m tall and weighed perhaps 22 to 36 kg (Figure 22-13), but males of her species were much larger.

Lucy and her kin were contemporary with some other hominids called *Australopithecus africanus*, previously thought to be ancestral to *Homo* spp. It now seems possible, however, that Lucy's species represents the end of a line of interbreeding populations of near-*Homo* primates that stretched back into the Pliocene toward *Ramapithecus* and its relatives. Lucy's ancestors may have given rise to both the later *Australopithecus* spp and to *Homo* spp. Some of the fossil teeth and jaws of this period are also sufficiently like those of modern humans to lead some paleoanthropologists to believe that an ancestral species of *Homo* already existed then and may have been responsible for the demise of Lucy's kind.

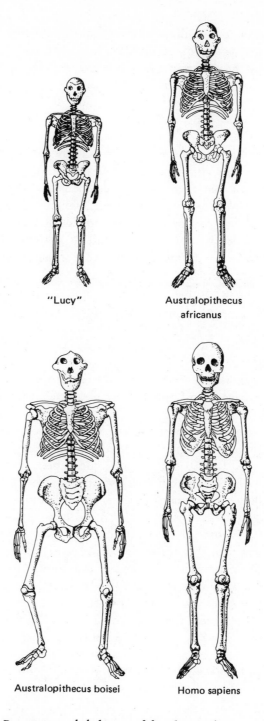

"Lucy"

Australopithecus
africanus

Australopithecus boisei

Homo sapiens

Figure 22-13 Reconstructed skeletons of four hominid species, showing relative stature and stance. Note the very close similarities between (a) "Lucy," (b) *Australopithecus africanus* ("*Homo habilis*"), and (d) *Homo sapiens*, and the divergent form of the "super-robust" *A. boisei*, (c) a specialized vegetarian. (Redrawn from Edney [3].)

The later australopithecines, which were widely distributed in East and South Africa from about 2.5 to 1 million years ago are interesting and puzzling too, because they apparently are represented by from two to four different types, depending upon how the different fossils are interpreted. In both South Africa and in East Africa there are large, *robust* forms with sagittal crests on the skulls and incompletely erect posture and smaller, more delicate, *gracile* forms, very *Homo*-like in build, with fully upright posture and lacking sagittal crests and massive eye ridges.

According to some authorities the robust forms are males, whereas the gracile forms are females of the same species; however, the weight of evidence now seems to favor the idea that there were three species, a wide-ranging and older gracile form called *Australopithecus africanus*, which was a tool-using hunter and gatherer, probably very close to the *Homo* lineage, and two robust forms, *Australopithecus robustus* of South Africa and *A. boisei* of East Africa, which were specialized terrestrial savanna-dwelling vegetarians, somewhat analogous to the forest-inhabiting gorillas. The gracile species was either competitively replaced by *Homo*, or according to some theories evolved into *Homo*; at any rate *A. africanus* disappeared from the fossil record around 2 million years ago, whereas the two robust types persisted to about 1 million years ago and were contemporary for a time with *Homo erectus*, an advanced form in the direct line of modern humans, a brainy hunter who may have killed *A. boisei* for food.

Homo erectus, originally described as *Pithecanthropus erectus*, the Java Ape Man, ranged across Africa, into Eurasia, from about 1.5 million to 300,000 years ago. This hominid had a considerably larger brain than any of the australopithecines, approaching the lower limit of brain capacity in modern humans. (For comparison: chimpanzees have a cranial capacity ranging from 280 to 450 cm^3; gorillas, from 350 to 750 cm^3; australopithecines, from 450 to 700 cm^3; *Homo erectus*, from 775 to 1100 cm^3; *Homo sapiens*, from 1200 to 1600 cm^3.) The facial features, however, remained primitive, with a massive, projecting jaw, almost no chin, large teeth, sloping forehead, prominent bony eyebrow ridges, and broad, flat nose (Figure 22–14). This hominid made stone tools and knew the use of fire and was probably the direct ancestor of *Homo sapiens*.

Homo sapiens, the only surviving species of the family Hominidae, came into existence around 300,000 years ago, probably as a result of progressive changes in overall body size and relative brain capacity in some geographically isolated population of *Homo erectus*. The first fossils, which were found in western Europe about the time Darwin was writing about his ideas on natural selection, used to be called "Neanderthal Man" (sometimes still considered a separate species, *Homo neanderthalensis*). The Neanderthal people present a curious puzzle, because we now know that a more modern form of *Homo sapiens* existed at least 200,000 years earlier. Neanderthals were about the same height as *Homo erectus*, 1.5 to 1.65 m, and they still had receding foreheads, prominent eyebrow ridges, and weak chins, but their brains were as large as modern day *Homo sapiens*. They were advanced stone tool-makers with a well-organized society and rapidly advancing culture. They and other popula-

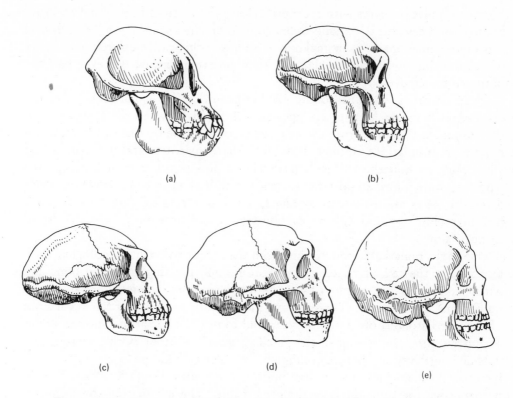

Figure 22-14 Reconstructed skulls of hominids and ancestral hominoid compared. (a) *Dryopithecus* (b) *Australopithecus africanus*. (c) *Homo erectus*. (d) *Homo sapiens neanderthalensis*. (e) *Homo sapiens sapiens*. (Redrawn from Romer, Ch. 5[7].)

tions of *Homo sapiens* quickly took over the niche occupied by *Homo erectus* and expanded it considerably, primarily by social and cultural mechanisms, only to be replaced themselves about 40,000 years ago by the still more advanced Cro-Magnon people who were the first fully modern populations of *Homo sapiens*, indistinguishable in structure from the species as it exists today.

22.1.5 Origins of Human Technology and Culture

As we pointed out in Chapter One, the human is the tool-user and tool-maker *par excellence*. Some other animals also use tools to a limited extent, usually in rather stereotyped and instinctive ways. Students of the vertebrates can provide several examples: Egyptian vultures have developed the habit of breaking open ostrich eggs by picking up stones in their beaks and dropping them on the shells; one of the Galapagos finches (Geospizinae) holds twigs and cactus spines in its beak to probe for insects in holes or under bark; and both baboons and chimpanzees use sticks and stones as weapons very much in the way that primitive humans must have done. But only humans, owning to the evolution of their manipulative hands with their precision grip and their large, contemplative brains, have been able to expand their ecological niche so far beyond the physical capabilities inherent in their makeup as to have achieved an entirely new order of relationships with the world, one that no other animal has ever had the potential to achieve.

The classical view of anthropologists has been that the use of tools led to the distinction between human and ape—that the split between the Pongidae and the Hominidae resulted from the acquisition of tool-use by one of the ancestral hominoid populations. Others now feel that environmental influences and adaptation to nonarboreal ecological niches, and possibly even birth-spacing patterns, were more important for early hominid evolution. However these divergent views are ultimately resolved, it is interesting to learn how far back human technology and culture can be traced and how they have developed through time.

One of the remarkable conclusions from the recent paleoanthropological findings is that the use of tools, chipped stones and possibly modified bone and antler, antedates the origin of the big-brained *Homo sapiens* by at least a million and a half years. There is now indisputable evidence of the occurrence of modified stone tools 2 million years old found in association with the bones of *Australopithecus africanus*, the gracile hunter. In other words, tool-use and tool-making developed before hominid brain capacity had undergone remarkable increase. The old idea that a large brain and associated high intelligence were prerequisites for tool use is no longer tenable. The use of tools by primitive hominids may, in fact, have been a major factor in the evolution of the cerebral cortex and higher intelligence, for once the use and making of tools began to favor survival, there would be a high selection pressure for neural mechanisms promoting improved crafting and use of tools. The elaborate brain of *Homo sapiens* may be a consequence of culture as much as its cause.

By at least 750,000 years ago *Homo erectus* had not only perfected stone tools considerably but had also learned how to control and use fire, as revealed by radio isotope dated hearths in caves. Fire changed human life and broadened the potentialities for coping with the world very considerably. With fire humans could cook their food, increasing its digestibility and nutritional value and preserving meat for longer periods than it would remain usable in a fresh state; they could keep themselves warm in cold weather; they could ward off predators; and they could light up the dark to see. The hearth no doubt promoted the development of social organization and allowed an opportunity for the beginning of communication through spoken language.

Neanderthal people practiced ritual burial in Europe and the Near East at least 60,000 years ago, suggesting that religious beliefs had developed by that time, and not long after the cave bear became the focus of a cult in Europe. By 40,000 years ago or a little later, Cro-Magnon people began constructing their own dwellings and living in communities. The domestication of animals and plants, development of agriculture, and the dawn of civilization were only a few tens of thousands of years away from human attainment.

22.2 THE HUMAN RACE AND THE FUTURE OF VERTEBRATES

Our human biological success as a species is reflected by our present numerical strength—we are one of the most abundant of all vertebrates—by our universal distribution over the earth, and by the fact that we derive some portion of our livelihood from every ecosystem and habitat in the biosphere and have even extended our range of activities beyond the earth (Figure 22–15). We have not achieved this all-pervasive status in the world without a very major impact on environment and other animals, and in this closing section it seems appropriate to focus attention on human ecological relationships with other vertebrates and to consider the prospects for the survival of human beings and vertebrate faunas in the closing decades of the twentieth century.

YEARS AGO	CULTURAL STAGE	AREA POPULATED	ASSUMED DENSITY PER·SQUARE KILOMETER		TOTAL POPULATION (MILLIONS)
1,000,000	Lower Paleolithic	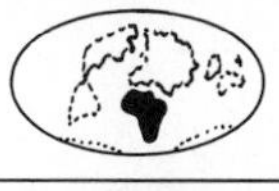	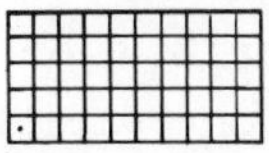	.00425	.125
10,000	Mesolithic		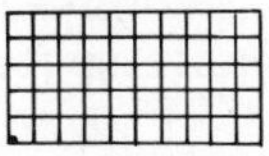	.04	5.32
2,000	Village farming and Urban		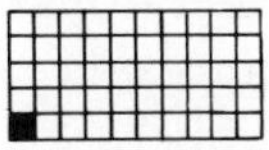	1.0	133
160	Farming and Industrial		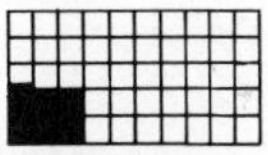	6.2	906
A.D. 2000	Farming and Industrial			46.0	6,270

Figure 22–15 Human population growth and dispersion from lower Paleolithic time through different cultural stages, projected to A.D. 2000. Several million years of hominoid evolution in Africa led to a population of *Homo sapiens* estimated at 125,000 during the middle of the Pleistocene. As the human population spread into Europe and Asia, it slowly increased to one million individuals during the next 700,000 years. By the end of the last glacial advance, about 10,000 years ago, hunting and food-gathering techniques had allowed human beings to spread over most of the earth and to reach a level of more than 5 million persons. By the time of Christ, 2,000 years ago, developing civilizations in the Old World had further increased the world population to more than 100 million, but in the New World human populations were still counted in the few millions. Not until after 1800 did the human population reach one billion individuals; by 1950 it had more than doubled again to nearly 2.5 billion; by the late 1970's it was rapidly moving toward four billion. By 2000 A.D. it could be between six and eight billion, unless catastrophe or self-control intervenes before then. Each grid unit represents an assumed population density of 1 human per km. (Based on Deevey [1]).

The preceding chapters have surveyed the great diversity of body forms and ways of living that have been evolved by the vertebrates. Some 47,000 species of living vertebrates have been described by scientists, and new species are still being discovered, especially among oceanic and tropical fishes; even in such a well-studied class as the birds an occasional new species is still found. Recently a new species of wood warbler (Parulidae) was discovered in the remnant montane forest of Puerto Rico. It has been named *Dendroica angelae*, the elfin woods warbler. This little bird has the distinction of being the first to be considered an "endangered species" simultaneously with its discovery and scientific description. Why is it endangered? Because humans have obliterated most of the forest habitat that the warbler depends upon for its existence.

Living species, of course, represent only a small fraction of the total diversity and total number of vertebrate species that have existed on earth. Since the evolution of the first ostracoderms more than 500 million years ago, hundreds of thousands of vertebrate species have no doubt evolved and then become extinct. In fact the whole history of vertebrate evolution has been characterized by the expansion and diversification of one taxonomic group of species, such as a class or order, that became successfully adapted to the prevailing conditions of a particular geological period, to be followed eventually by the extinction of many or most of its member species and their replacement by a more successfully adapted group—or by a group better adapted to new environmental conditions that slowly emerged through time. The replacement of the Mesozoic reptiles by birds and mammals in the late Cretaceous is a familiar example of this process (See Figure 22–16).

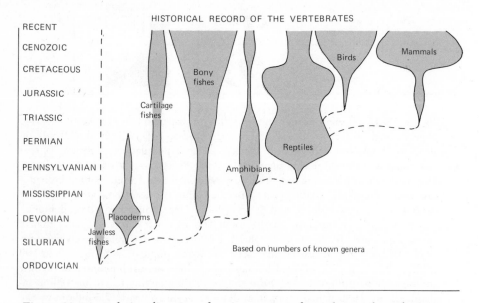

Figure 22-16 Relative diversity of major groups of vertebrates through time. The width of the pathways is relative to the number of known genera (index of diversity) at different geological periods. (Based on G. G. Simpson [1949] *The Meaning of Evolution*, Yale University Press, New Haven.)

Just as death is the natural end of the life cycle of an individual organism, so extinction of a species is a natural result of evolutionary processes. A species normally persists as long as its adaptive relationships to the abiotic and biotic environments continue to be possible, and it goes to extinction as soon as these relationships shift to the extent that its reproductive capacity is no longer able to compensate for mortality. Some students of evolution and speciation have estimated that the average survival time for vertebrate species is about 1 million years, but the standard deviation is no doubt very great, a fact that may give *Homo sapiens* some hope, as the species is now well on the way toward a million years of existence.

Currently much consideration is being given to the impending extinction of species—the so-called "endangered species"—but in our attempt to understand and correct the problems forcing some species toward extinction, we should remember that extinction *per se* is not a new or unnatural phenomenon. What is new is the extent to which one species—*Homo sapiens*—has been responsible for recent extinctions and for the general "endangerment" of many forms of life. There is no precedent in evolutionary history for this sort of dominance by one species over the welfare of all others. Past extinctions probably resulted mainly from climatic and geological changes that altered living conditions to the disadvantage of some species while favoring others. The emergence of human beings as *the dominating species* on earth adds a whole new dimension to the problems of survival for other species.

It now appears certain that *Homo sapiens*—the rational, thinking primate—more than any other animal will determine what the immediate future course for life on earth will be. Except for solar energy input and photosynthesis, human activities are now the most significant forces shaping the destiny of life on this planet, for better or for worse. For this reason, humans have some responsibilities for other forms of life, because we are the only species of vertebrate—the only known form of life—which, by virtue of versatile and cognitive brains, are capable of understanding the consequences of an action before it happens. Therein lies our success as a species; and therein lies a hope for the future of life on earth. In a little more than 1 million years of evolution, these upright, bipedal primates have reached a position in the biosphere to become the guardians of all other forms of life, *if they choose to do so*. We must choose, not only for the survival of other species but quite probably for our own survival as well.

Human dominance on earth is a curious and improbable outcome of organic evolution. It seems odd that natural selection would have so favored one species that it can now determine the survival or extinction of essentially any other forms of life and create environmental changes that formerly were only possible by the blind actions of climatic and geophysical processes of the earth and solar system. Humans can create atomic earthquakes, as on Amchitka in the Aleutian Islands; we can make great lakes where none existed before; we can change the course of rivers; we can change forests into grasslands and grasslands into deserts, but we can also reverse these processes.

All these tremendous powers that humans have over nature run contrary to biological notions about the underlying homeostatic basis of living systems. At whatever level of organization they examine, from self-replicating molecules to ecosystems, biologists have found negative feedback systems operating to hold biological functions and processes within certain set limits (Figure 22–17), much as a thermostat controls room temperature or the hypothalamus controls the body temperature of a homeotherm (see Chapter 18). According to negative feedback theory, at some point in our history some regulatory process should have been brought into play to hold humans in check and keep our population and our ecological niche in balance with the rest of nature.

How have humans broken loose from these natural, biological controls? Basically we are vertebrate predators that have gone astray through the invention of technology. Technology has allowed us to live like herbivores or omnivores with an unlimited food supply and no controlling mortality on our numbers, a sytle of life that has led to our virtually unlimited population growth for thousands of years (Figure 22–18). Even with technology, the limits of human food supply and other essential resources are now being reached because of ever-increasing numbers. What we must do is to reaffirm our basically predatory nature and come to realize that, as with any "top predator," survival of the species over the long haul depends upon self-imposed regulatory mechanisms of population control. Wolves, lions, eagles, falcons, and even territorial "songbirds" have successfully, if blindly, practiced population control for hundreds of thousands of years. Humans themselves probably practiced it in a more "primitive" period before they became captivated by modern technology (Figure 22–19; Chapter 19.3.2).

Do humans have moral obligations toward other species, as well as toward the members of our own kind? If we do, we have not exercised them much up to this time. As far as other vertebrate species are concerned, our history until very recently has been one of uninhibited exploitation for whatever purposes seemed good to us. We have slaughtered the great whales to such an extent that they are now endangered species, and we have reduced the pelagic fisheries, which numbered in the many billions of individual animals, to a remnant of their former numbers. We killed off the dodo, the greak auk, and the passenger pigeon, which numbered in the hundreds of millions of birds when John James Audubon and others first recorded the birdlife of North America.

Since 1600, there have been 200 documented extinctions among vertebrates, mostly birds and mammals; all can be attributed directly or indirectly to the activities of humans; 50 of these cases occurred in America. The current worldwide situation is deplorable. There are more than 140 officially listed endangered species of vertebrates in the United States, and there are 1000 or more in the world, about 2 percent of the living species. The best documented cases are recorded in *The Red Data Books* of the International Union for the Conservation of Nature and Natural Resources. The IUCN predicts that more species will die out between now and the year 2000 than in previous recorded history.

(a)

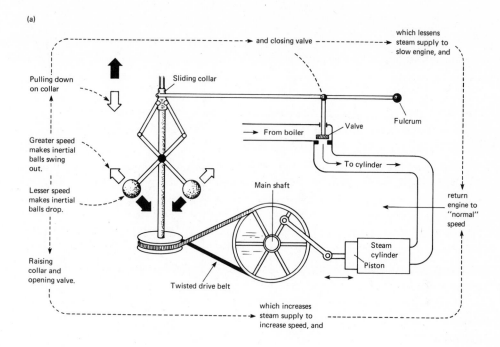

(b)

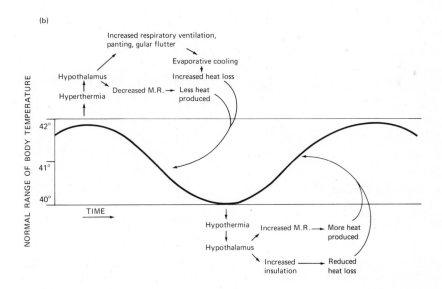

Figure 22–17 Cybernetic systems of negative feedback control. (a) governor on a steam engine. (b) body temperature control of a bird. (Based in part on Hardin [6].)

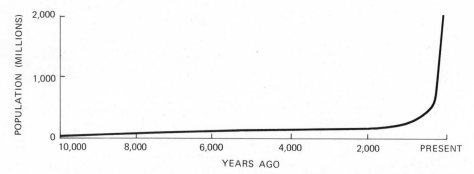

Figure 22-18 Generalized growth curve for the human population from prehistoric time to the present. (Based on Deevey [1].)

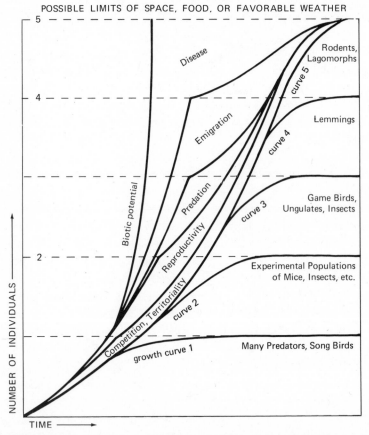

Figure 22-19 Relation between "biotic potential" or unrestrained reproduction of a population and various forms of density stabilizing and density-limiting factors. Note that the limits of population growth and stabilization are set at different "levels" of population density, depending upon which stabilizing and limiting factors a particular species population is most influenced by. Curve 1 represents the most stabilized population through time, curves 4 and 5 represent populations that tend to "cycle" from high density to low and back. The human population is currently on the steep slope of the biotic potential curve and is relatively unchecked by the limits normally set by competition, predation and disease. (Redrawn from Kendeigh [7].)

22.2.1 Primitive Human Impact on Other Vertebrates

As suggested in Chapter 21, even prehistoric humans may have caused or abetted the extinction of some of the large ice-age mammals in North America and Eurasia, although it appears likely that early human impact on vertebrate species was negligible in most cases. For one reason, there were few humans; the pre-Columbian population in the Western Hemisphere was probably not more than 10 to 15 million human beings. Moreover, primitive humans hunted with simple and inefficient weapons; but they were already by the beginning of the Pleistocene the cleverest of all predators. They knew how to use fire, how to drive ungulates over cliffs, how to set traps, and above all how to cooperate as a hunting party to obtain meat for the social group.

It is interesting to consider the moose-hunting culture of the Athabascan Indians of recent historic times as an analog of the possible relation between the wooly mammoth and humans in Alaska 10 to 15 thousand years ago during the last glacial advance. At the time the white man entered Alaska, the Kutchin tribe living in the forested interior along the major rivers, the Yukon and Tanana, depended primarily upon the moose for their food in the winter. In the summer, they fished, hunted small game, and gathered fruits and berries, but during the critical winter period they depended on the commonest large ungulate living in the region.

The Kutchin had no permanent dwellings; instead, they lived in isolated family groups in rather crude, temporary structures made of brush and animal hides. In the fall after freezing temperatures begin to prevail, the hunters would spread out from the fishing camps to find moose in the surrounding forests. Having no means to carry such a large animal, the Kutchin simply butchered each moose at the spot where it was killed and then cached the frozen meat by hanging it up in a tree. During the late fall and winter, they lived by moving their camps from one moose cache to the next. If enough moose had been killed, the group could survive through the long, cold winter.

It is tempting to speculate that the late Pleistocene human population of interior Alaska lived in much the same way by specializing on the wooly mammoth dwelling in the open grasslands and tundra. The mammoth, about the size of a modern elephant, weighed around 5000 to 7000 kg as an adult, roughly eight times the size of an Alaskan moose. Moose occur at a density of about one animal per 3 km^2, and it seems reasonable to assume that a mammoth would require roughly eight times the range of a moose, giving a density of about one mammoth per 25 km^2. The total area available to mammoths during the last glaciation in Alaska may have been around 400,000 to 500,000 km^2, so that the total population of mammoths may only have been equivalent to 16,000 to 20,000 adult animals. (Because of different size classes associated with age structure of the population, the actual numbers would have been greater than the number indicated as "adult equivalents.")

If the age structure, age of sexual maturity, and reproductive rate of mammoths were similar to modern elephants, there probably were no more than

5000 to 8000 cows of reproductive age and their annual productivity was probably not more than 500 to 800 calves. This means that primitive humans would only have had to kill a few hundred mammoths a year in order to increase mortality over natality and cause the mammoth population to go into decline. Because of its large size, it was probably profitable to continue hunting the mammoth even after it became scarce (an analogy with the great whales), so that the usual predator-prey oscillation and compensatory adjustments in numbers broke down. Once the mammoth was gone, the ever resourceful human beings merely shifted to other sources of food or emigrated into new regions where food species had not yet been exploited. Thus, a few hundred human hunters who specialized on the mammoth as a source of winter food could have been responsible for the final extinction of this proboscidean in Alaska.

Most vertebrate populations, however, were little disturbed by aboriginal human hunting, and some no doubt benefited from human practices. Moose, for example, require subclimax vegetation in the form of willows and other browse plants, and there is evidence that the Indians set fire to the spruce forests in boreal America to create favorable "moose pasturage," and similar use of fire has been employed in other parts of the world as well.

Although the exact role that primitive humans may have played in the extinction of other vertebrates will always remain speculative, the record is clear for what modern, civilized humans have done to other species.

22.2.2 Impact of Humans During the Period of Exploration and Discovery

The period of exploration and discovery, which began in the fifteenth century and ended in the nineteenth century, saw the beginning of human exploitation of other vertebrates on a large scale. The exploring ships that set out from various European ports could not carry enough provisions to last the crew for the entire trip, which might extend over several years, and consequently these exploring parties had to live "off the land" to a large extent. In particular, they used oceanic islands for provisioning and for water. On many of these islands the explorers found numerous populations of large and tame vertebrates such as tortoises, seals, and flightless birds, easily slaughtered for fresh meat. Another practice of these exploring parties, in the long run worse than slaughter for food, was to leave off domestic stock on islands for use on return trips. As a result, goats, sheep, pigs, horses, cattle, and rabbits established feral populations that destroyed the endemic vegetation and habitat for many insular land vertebrates, particularly mammals, birds, and reptiles. Accidental or careless introduction of exotic predators, for which the endemic vertebrate fauna had evolved no effective antipredator mechanisms, was another frequent result of these early visits to islands. Cats, dogs, mongoose, rats, and monkeys have all wreaked havoc on insular endemics. Captain James Cook's three famous voyages of exploration in the Pacific and Indian Oceans were probably directly or indirectly responsible for more habitat destruction and extinction on oceanic islands than any other episode in the annals of this period.

The history of extinctions among the endemic birds of the Mascarene Islands off the African coast in the Indian Ocean is a particularly dismal but revealing example of the deleterious influences of human exploration and discovery on insular faunas. The island of Mauritius was uninhabited by humans until discovered in 1505 by the Portuguese. The dodo, a large flightless bird related to pigeons, had become extinct by 1680, the victim of slaughter by sailors for food and predation on its eggs and young by introduced pigs and monkeys (Figure 22–20). A similar bird, the solitaire, had vanished on Reunion by 1750, and another, on Rodriguez, before 1800. In all, as many as 24 endemic birds have become extinct in the Mascarene Islands since 1600, more than 50 percent of the original avifauna.

A parallel case occurred in the Hawaiian Islands, where Captain Cook met his death at the hands of the Polynesians in 1779. About 60 percent of the 68 endemic species of land birds (mostly from the passerine family Drepanididae) have become extinct following the intrusion of rats, mongoose, rabbits, cats, goats, sheep, horses, cattle, and pigs, all of which have established feral populations on the islands. Of the approximately 85 full species of birds that have become extinct since 1600, no less than 77 were island endemics or were restricted to islands for nesting.

Figure 22–20 Drawing of extinct dodo, based on specimens sent to Europe from Mauritius in the 17th Century. (After Peterson, Ch. 17 [4].)

22.2.3 Commercial Exploitation of Vertebrates

The explorers were soon followed by fur trappers, commercial hunters, fishermen, whalers, plume hunters, and a host of other "harvesters" whose occupations were to make money by exploiting wild animal populations for food, leather, fur, oil, ivory, and other commercial products. These enterprises have caused great destruction among vertebrate populations but have actually brought few species to extinction. The future of the great whales, however, is doubtful owing to overharvesting by highly efficient modern techniques that continued in use through the 1960s. A few examples of heavily exploited vertebrate species serve to point out the effects of overharvesting, as well as suggesting the limits within which the wise use of renewable resources must operate.

When Vitus Bering voyaged among the islands of the Bering Sea and North Pacific Ocean he and the naturalists who accompanied him found millions of fur seals breeding on the Pribilof Islands, and great numbers of sea otters spread over an even wider range (Figure 22-21). The furs of both these marine mammals were highly valued in Europe, and soon the sealers and otter hunters began their work.

In 1885 the Russian-American Company sold 118,000 sea otter pelts, but 30 years later the number had dropped to 8000, and by 1910 only 400 could be harvested. For a time when no more pelts were received in trade, people assumed the species had become extinct, but fortunately a few pairs managed to survive in isolated parts of the range. After complete protection in North America and careful translocation of individuals from more populous islands to unpopulated ones, the otter population has slowly increased to numbers that again permit controlled harvesting, and, indeed, in some areas, like Amchitka Island, the animal shows signs of overpopulation.

Figure 22-21 Original distribution of the sea otter (open circles) compared with its present relict occurrence (closed circles). (From Ziswiler [12].)

Between 1908 and 1910 Japanese seal hunters slaughtered 4 million fur seals in the Pribilofs. Fortunately, through diplomatic negotiations undertaken by the United States, the Fur Seal Treaty of 1911 was signed by Japan, Russia, Canada, and the United States. The first international agreement to protect a marine resource, it allows for protection, management, and harvest by the United States with mutually agreed on quotas of the harvested pelts going to the respective signatories. As a consequence, the northern fur seal population rebounded in numbers and has been harvested annually on a planned and scientifically managed basis for more than 50 years. Unfortunately the southern fur seals were not so lucky, and they exist today only as remnant populations on a few islands scattered from southern California to Chile.

A similar history can be related about the northern elephant seals, which haul ashore to calve and breed on islands adjacent to southern California and western Mexico. In earlier decades they were slaughtered in the tens of thousands for dog food, until there were only a few left in the early 1930s. After protection by the governments of the United States and Mexico, the populations have increased, slowly at first and then more rapidly, until now there are more than 60,000 of these great, lumbering beasts; and they are still increasing.

These three marine mammals are good examples of how well vertebrate populations can recover from greatly reduced numbers after heavy human killing stops, so long as the environment is otherwise still supportive of the species.

A different story has to be told for the great whales. All of the large baleen whales, except the gray whale, which has responded to protection, are critically endangered as a consequence of the relentless use by the whaling nations of highly efficient modern techniques for capture and processing. The blue whale, largest of all the whales and the largest animal known to have existed, weighing as much as 25 elephants and measuring more than 25 m in length, escaped persecution during the early days of whaling, when whalers traveled in sailing ships and harpooned from open boats, simply because it was too big to be handled. The modern catcher ship, armed with cannon-launched harpoons that explode inside the animal, proved to be the nemesis of the blue whale (Figure 22–22). With the virtual extinction of the great baleen whales, the industry has turned its attention increasingly to the sperm whale, largest of the toothed whales and possessing the largest brain ever evolved. It is unclear whether the blue whale, fin-back, bowhead whale, and the two right whales can ever recover from their pathetically reduced numbers.

Humans have always regarded the seas as an inexhaustible source of food, a not unreasonable attitude in former times when the human population was small and fishing techniques were crude. Certainly some fish populations have existed in the hundreds of billions or trillions of individuals with very great productive potentials. A single female herring may lay 10 million eggs in a season. But now humans count their own numbers in the billions, and our modern methods of fishing the high seas are just as technical and efficient as those employed by the

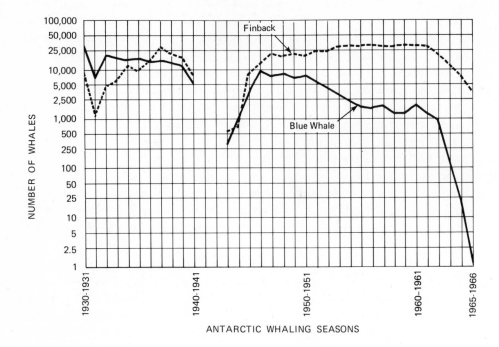

Figure 22-22 Number of blue whales and finback whales caught in Antarctic waters between 1930 and 1966. The steadily decreasing catch of blue whales beginning in the 1950's resulted in an increased catch of finbacks, which were more available. Since 1960, both species have suffered severe declines, from which they may not be able to recover. No records during years of World War II. (From McVay [8].)

whaling industry. The great cod, haddock, and herring fisheries of the Grand Banks of the North Atlantic have been drastically depleted in the twentieth century, and still no international agreements exist to regulate the take. The Pacific coast salmon and halibut fisheries have suffered from overfishing too; in addition, the salmon runs to their spawning grounds upriver have been thwarted by dams, and the spawning streams themselves have often been altered unfavorably for the fish by various kinds of "water projects" or polluted runoffs.

Humans did not stop hunting "big game mammals" with the last of the ice-age giants. They have continued to pursue all of the large ungulates and herbivores right down to the present. The history of our American bison is representative of what happened to many of these species (Figure 22-23). There probably were around 60 million "buffalo" on the American plains in 1700; they had been hunted without depletion by the Indians for thousands of years. The construction of the Union Pacific Railway in the 1860s introduced a new era of buffalo hunting. First, special hunting parties were hired to supply the railway workers with meat; Buffalo Bill Cody became famous at this time for having killed 250 bison in a single day.

The tracks of Union Pacific ran right through the heart of the bison range, and after the railway was in operation, the company encouraged buffalo shootouts from the trains, perpetrating one of the most senseless and bloodthirsty slaughters ever witnessed on the North American continent. Tens of thousands of animals were shot from the trains and left to rot along the tracks. By the 1890s only a few hundred individual bison remained in the United States, and a few hundred more of the wood buffalo variety, in Canada. Fortunately, bison reproduce well under animal husbandry, and thanks to William T. Hornaday of

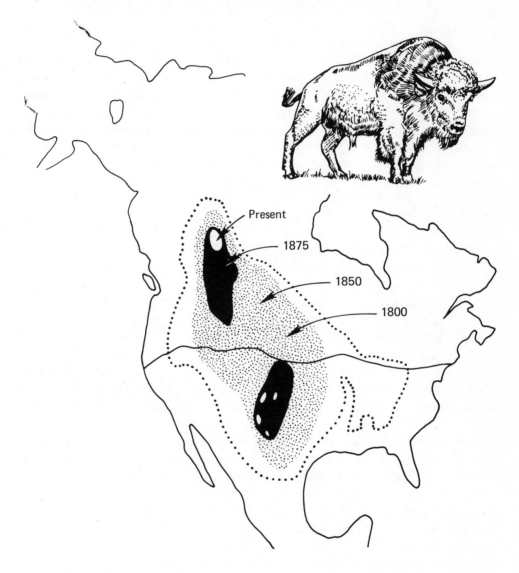

Figure 22–23 Original and present distributions of the American bison. Dotted line shows the range before 1800; stippled area, the range about 1850; the black areas, the range about 1875; and the white areas, the present distribution. (Modified from Ziswiler [12].)

the New York Zoological Society, the American Bison Society, and the governments of the United States and Canada, the species was saved through a vigorous program of captive propagation and release of animals onto protected range, such as that in the Yellowstone National Park. Today the bison exists in the tens of thousands and has even been successfully introduced into Alaska, where a related species occurred naturally during the Pleistocene.

Quite a few avian species have been overharvested for commercial purposes, and at least one, the passenger pigeon, was probably brought to extinction directly as a result of commercial hunting, although the destruction of the mast-producing hardwood forests in the eastern United States was no doubt indirectly contributory to its demise. The use of feathers and plumes in women's millinery in the late 1800s and early 1900s resulted in great destruction of birds with gaudy or fancy plumage. The African ostrich might well have become extinct as a result of this trade had it not been for the timely development of domestic breeding on ostrich farms. In 1912 more than 160 tons of ostrich feathers were sold in France alone. Bird of Paradise skins were imported in large numbers from far off New Guinea. In the Americas, plume hunters went after egrets and other showy relatives in the heron family, and at one time the American egret (common egret) was on the verge of extinction. Thanks to timely action by the National Audubon Society and a few public-spirited individuals, sanctuaries for these birds were established in the southern swamps where the birds could breed in safety from the hunters. But it was the change in women's fashions during the following World War I that really saved these birds from further wanton destruction.

At one time waterfowl were hunted for the markets, particularly in the eastern United States in areas where the birds concentrated, such as Chesapeake Bay. Often the hunters used a large bore "scatter-gun" mounted like a cannon on a swivel at the bow of an open punt boat. A single discharge could decimate an entire raft of ducks or gaggle of geese.

Roosting and nesting in dense flocks of many millions of birds made the passenger pigeon especially vulnerable to shot and nets. In 1813 John James Audubon made a 55-mile trip along the Ohio from his home to Louisville, Kentucky. Wild pigeons were emigrating through the region in numbers that stagger the imagination of a present day bird-watcher. The air was so full of birds that "the light of noonday was obscured as by an eclipse." Audubon estimated the flock to be 1 mile wide with an average density of two birds per square yard and to be moving at the rate of 1 mile/min (probably too fast an estimate); during a 3-hr period he calculated that more than 1,100,000,000 passenger pigeons must have flown by, and the flight lasted for 3 days!

Even if we reduce his estimate by half to allow for a more reasonable speed of flight, the number of pigeons must have been truly astounding. The last great nesting occurred in 1878 near Petoskey, Michigan, and covered more than 40,000 hectares of forest; many millions of birds were present. Scattered smaller nestings occurred sporadically through the latter decades of the nineteenth century, and the last bird died in the Cincinnati Zoo in 1914. The adult birds had been shot and netted by the hundreds of millions and as many squabs were removed from destroyed nests. The New York City market alone would receive 100 barrels a day week after week without a break in price. Chicago, St. Louis, Boston, and all the eastern cities, great and small, joined in the demand for pigeon flesh. Some people still speculate about the "mysterious" disappearance of the passenger pigeon—that they all drowned in a great storm over Lake Michigan or that they emigrated permanently to the Amazon Forest in Brazil— but the true story is revealed by the statistics of the marketplace and by the merciless persistence of the netters in following the pigeons from roost to roost.

In concluding this section, we should point out that in the old days—prior to the advent of the highly efficient methods of capture, transport, and processing made possible by twentieth century technology—human hunting, trapping, and fishing functioned much more like natural predation in that these activities were strongly influenced by the "law of diminishing returns." When a beaver trapper moved into a new area, at first his catch would be very high for the amount of work he put into trapping. As the beaver population became reduced by the repeated removal of trapped individuals, a good season's catch became more difficult to obtain, so that the trapper had to work harder for fewer pelts. Soon the trapping was no longer profitable, and the trapper would pick up his line and move on to another untrapped area. Relieved from the trapping pressure, the surviving beaver population could quickly build its numbers back up to a normal carrying capacity of the habitat by means of its reproductive potential. This law of diminishing returns affects human economics, and it also applies more generally to the bioenergetics of resource utilization by other animal populations in nature. It is, in fact, another example of negative feedback control. Only since the development of modern gear and transport, especially since World War II, has it become profitable for man to continue harvesting depleted populations, as particularly exemplified by the whaling industry and the marine fishing fleets.

There is nothing biologically wrong about harvesting wild animal populations for human uses, so long as we exercise the necessary wisdom and restraint to prevent overexploitation and depletion of numbers below the environmental carrying capacity *for the breeding population* of the particular species being harvested. All "healthy" animal populations produce an excess of individuals above the number that can be sustained by the ecosystem; hence, there is a necessary mortality that operates on all populations to hold each in balance with the others in its environment. This "surplus" of animals above the number required to replace losses in the breeding population provides a legitimate harvest for humans, and our biological as well as economic responsibility is to determine what share of that surplus we can take for ourselves without diminishing other animal populations over the long term, including other animals that eat the same foods we do.

As we pointed out earlier, extinction of a species or population results when the mortality of individuals is continually greater than the production of new individuals. Generally, when excessive mortality caused by humans is eliminated, even drastically reduced populations of many vertebrate species rebound to the limits imposed by factors other than human predation, as we have seen in the examples of the sea otter, bison, egrets, and other species. However, this resilient capacity of vertebrate populations has limits that vary for different species and seem to be determined by a *critical minimum population* size, below which the species cannot recover because natality has been affected independently of mortality. In such cases, even if mortality can be greatly reduced, reproduction is still insufficient to result in a net increase in numbers, and the species goes to extinction.

Highly social and gregarious species seem to have large critical minimum population sizes, whereas widely dispersed, territorial species and island endemics may have very small critical minimum numbers. The passenger pigeon is an example of the first group. After the last great nesting in 1878 in Michigan, many millions of pigeons still existed despite the slaughter that occurred there, but even though significant human depredations on the passenger pigeons stopped after that time, the remaining flocks were never again able to reproduce in sufficient numbers to replace the natural losses in the adult population, and the species had become extinct in the wild by the turn of the century, in a little more than 20 years. The little recorded information on the bird's breeding habits indicates that mating and nesting may have been initiated by social stimulation involving communal displays of large numbers of birds closely packed together. Other species seem to have such a low critical population size that the concept has no application to them. Certainly this would seem to be the case for the Laysan duck, which appears to have been reduced in 1930 to a single adult female and her clutch of fertile eggs. By the late 1950s there was again an estimated 580 to 740 ducks on Laysan Island, although the wild population has since dropped back to a lesser number of birds.

Several factors probably play a role in determining the critical minimum population size for a species, and one factor may be more important for some kinds of species, whereas another is more important for others. Much more study of this problem is needed. One possibility is that the population density becomes too low for males to have much chance to find females, a likely problem now for some of the great whales. Population density could become too low for adequate stimulation of courtship and mating, as indicated for the passenger pigeon. The population might become too small to counterbalance mortality from large numbers of surviving predators, parasites, or competitors, for example the problem Kirtland's warbler has had with nest parasitism by the brown-headed cowbird in recent years. The total population size might become too small to survive a natural disaster such as a flood, fire, or cyclone, particularly if the population is one that is localized as well as small. Remnant populations of island endemics are especially vulnerable to this factor. The last few surviving Laysan honeycreepers were blown off the island by a storm in 1923; and the same fate could be in store for the relict populations of the kestrel, pink pigeon, and parakeet on the island of Muaritius, which has a history of periodic cyclones. Reduction of fecundity or increased susceptibility to diseases and parasites owing to genetic deterioration from inbreeding are other factors to consider when a species population becomes reduced to a very few individuals; the Laysan duck may be suffering from such problems at the present time, and reduced fecundity is a frequent characteristic of highly inbred captive populations of vertebrates.

22.3.4 The Impact of Agriculture and Land Development

Human use and abuse of the land for agriculture and other purposes such as timbering, mining, and suburban living have been most detrimental to wild animal populations, primarily through the resulting destruction of natural habitat, although some species have been directly destroyed in the process of "reclaiming the land" for human uses. Few species of vertebrates have been brought to total extinction as a result of human land uses, but many have suffered major and irreversible reductions in distribution and density as a result of the way we live on the earth, and no doubt many others will join the ranks of threatened or endangered wildlife in the next few decades as a result of the diminution and deterioration of natural habitats.

Perhaps the earliest and most persistent conflict between humans and wild animals has been with other predators, particularly the large carnivores. As we have seen, early humans were efficient and destructive predators in their own right, and with some justification they regarded the carnivores as competitors for the great ungulates that they hunted for food themselves. No doubt primitive humans had to protect their kills from other predators, and they no doubt also robbed other predators of their kills, so that direct interactions between individual humans and carnivores were common in the Pleistocene period of human cultural development. From these remote and primitive times to the present, our human attitudes toward other predators have been molded, in part, by fear of the large and powerful predators and, in part, by the idea that other predators compete directly with us for available food supplies. Except for isolated, individual instances, neither belief is justified by objective and scientific analysis, but they still hold sway—almost as instincts— over many human minds in the twentieth century.

With the development of domestic livestock and pastoral nomadism as a way of life, humans began a systematic destruction of the large carnivores that has continued to the present. In North America, the history of the timber wolf, mountain lion, black bear, and grizzly bear has been one of steady retreat into more and more remote areas as cattlemen and sheep herders moved into the pasturelands of the mid-West and West. The same can be said of the lion in Africa and Asia, the tiger, and even the wildcat. Tillage of the soil and the development of intensive farming provide the final insults for large predators, as their modes of life simply are not compatible with intensive and disruptive human uses of the land. The pity is that even in their last, remote, agriculturally peripheral habitats the carnivores continue to be pursued by human agents, and governments are still persuaded to spend millions of dollars a year on misguided and unjustified "predator control programs."

Predator control usually leads to "rodent control," but rodents are much more difficult and costly to control than predators. It is wiser to "pay the cost" of the pheasants and lambs killed by coyotes and foxes in order to enjoy the greater benefits of their impact on rodent and rabbit populations.

Other vertebrates have frequently been considered to be "competitors" of humans in addition to predators. As one example, the Boer farmers who settled in South Africa regarded the existing wild ungulate herds as competitors of their cattle, sheep, and goats, and they systematically destroyed them, as they did the Cape lion. The blue buck (*Hippotragus leucophaeus*) had been exterminated by 1800; the last of the quagga (*Equus quagga*), a once common zebra, disappeared in 1878; Burchell's zebra (*Equus burchelli*) had been eliminated from all areas other than game parks by 1920; and three other ungulates, the bontebok (*Damaliscus dorcas*), white-tailed gnu (*Connochaetes gnu*), and the Cape mountain zebra (*Equus zebra*) exist today only as remnant populations in wildlife reservations.

The loss of natural environments from intensive agriculture and other disruptive uses of land and water is by far the most important factor in the overall reduction of wild vertebrate populations and will continue to be the overriding problem for the survival of many species in the closing decades of the twentieth century. The cutting of the original eastern hardwood forests in North America for timber, firewood, and agricultural clearing probably had a harmful effect on the passenger pigeon at least equal to the killing of these birds for food. The pigeons fed on the mast from oaks, hickory, and other nut-producing trees and also depended on the trees for nesting sites. As the eastern forests were demolished during the eighteenth and nineteenth centuries, the pigeons were forced westward and northward into marginal habitats, which probably were not favorable for the mass reproductive efforts that characterized the lifestyle of this species. Today much of the eastern United States is again covered by hardwood forest, but it is nearly all second-growth stands, different in relative species composition and habit from the original forest.

The prairie grasslands have been the most disrupted natural communities in North America, especially the tall grass prairie, which was ploughed under and really no longer exists. Even the short grass prairie is no longer truly "natural," because it has been invaded by many exotic species of grasses and "weeds" introduced through human agency and has been drastically altered by the overgrazing of man's domestic livestock. Now many of these western ranges are receiving additional insults from the extensive mining processes involved in extracting fossil fuels from the earth, especially strip mining for coal.

Continued growth of the human population means that more and more land must be devoted to agriculture and other disruptive uses required to sustain human beings. Thus, more and more natural habitat is lost, and there are fewer and fewer places left for wild animals to live, except for the limited number of species that can adjust to the altered conditions created by humans. Most of the agriculturally productive areas of the world have been developed, and good agricultural lands are being reduced by suburban developments, businesses, roads, airports, and artifical lakes, so that now farmers increasingly turn their attention to marginal lands. Deserts, for example, can be highly productive under irrigation, but fresh water itself is becoming an increasingly scarce commodity, and in some places continued irrigation has increased salinity of the soil to degrees that have greatly reduced the harvest. The greatest agricultural devastation in the late 1900s is taking place in the great tropical rain forests of South America, Africa, and Asia. When the forest is cut down, the soils that were under the trees do not sustain intensive agriculture for more than a few years. Once abandoned these soils are frequently so changed in composition and profile that the original forest trees cannot come back; instead there is a new edaphic climax of grasses and shrubs, which become marginal pastureland for livestock. Gone, then, is the rich diversity of tree species and associated flora and the animals that had evolved adaptations for living in these tropical forests.

There are many other human practices that threaten wildlife. In passing, we can only mention the pet trade and the traffic in wild animals for zoos and medical research programs, the introduction of exotic plants and animals, the modification of waterways and water distribution by canals and dams, and chemical pollution. We have considered enough examples in the foregoing sections to show the main kinds of effects humans have had and continue to have on other vertebrate species.

22.2.5 Human Concern for Vanishing Wildlife

Human impact on other forms of life has led to a new kind of classification of species. The categories include "rare species," "unique species," "threatened species," and "endangered species." The International Union for the Conservation of Nature and Natural Resources, in cooperation with organizations such as The World Wildlife Fund and the International Council for Bird Preservation, attempt to keep current a catalog of specific details on all rare, severely threatened, or endangered species of plants and animals. The information is published in a series of *Red Data Books*, which are arranged by taxonomic classes. Over 1000 species of vertebrates are now included in the coverage. Increasingly in recent years governments have become concerned with such species too. Public Law 93–205, the Endangered Species Act of 1973, gives authority to the U. S. Secretary of the Interior to designate endangered and threatened species and to promote ways to preserve them from further diminution and, if possible, to increase their numbers. There are also important international agreements, such as the Convention on International Trade in Endangered Species of Wild Fauna and Flora, and many of the states now also have their own endangered species laws.

By 1977, the U. S. Fish and Wildlife Service, which administers the provisions of the Endangered Species Act, had listed 146 native vertebrate species of the United States as endangered and nine as threatened; it had also listed 436 foreign species as endangered and 17 as threatened (Table 22-2). Fifty-seven recovery teams of experts had been appointed by the Director of the Fish and Wildlife Service to develop long range plans for saving endangered species in the United States; 38 "critical habitats" had been proposed for protection under the Endangered Species Act; and cooperative agreements for funding work on endangered species had been signed with 17 states. Although the executive branch of the federal government has been slow to put the provisions of the Endangered Species Act into operation, public interest in the survival of endangered species has increased greatly.

Table 22-2. Number of Endangered Species of Vertebrates Officially Listed by the United States Department of the Interior in 1977*

Taxonomic category	Number of endangered species		
	USA	Foreign	Total
Mammals	36	227	263
Birds	67	144	211
Reptiles	9	46	55
Amphibians	4	9	13
Fishes	30	10	40
Totals	146	436	582

*Data from *Endangered Species Technical Bulletin*, Vol. II, No. 6, June 1977, U. S. Fish and Wildlife Service, Washington, D. C.

22.2.6 Characteristics of Endangered Species

What are the predisposing characteristics and environmental relationships of a species that are likely to lead to its becoming an endangered species? We have already pointed out the high vulnerability of species of birds and mammals that are restricted to islands, but there are many other variables that increase or decrease the "survival potential" of a species. By comparing the characteristics of threatened and endangered species with closely related successful or "safe" species, we can begin to identify the main factors involved (Table 22-3). If we knew enough, we could probably check each species against this list of variables and come up with some sort of statistical estimate of its chances for continued survival in the modern world, but at present the best we can do is to make a qualitative appraisal for some species.

D. W. Ehrenfeld constructed an interesting description of the hypothetical *most endangered animal:* "It turns out to be a large predator with a narrow habitat tolerance, long gestation period, and few young per litter. It is hunted for a natural product and/or for sport, but is not subject to efficient game management. It has a restricted distribution, but travels across international boundaries. It is intolerant of man, reproduces in aggregates, and has non-adaptive behavioral idiosyncrasies." Although there really is no such animal, the polar bear comes close to approximating this model. Conversely, if one takes the opposite characteristics (those listed under "safe" in Table 22-2), he can develop a composite picture of a typical wild animal of the twenty-first

Table 22-3. Characteristics of Species that Decrease or Increase Survival Potential*

Endangered	*Safe*
Individuals of large size (cougar)	Individuals of small size (wildcat)
Predator (hawk)	Grazer, scavenger, insectivore, etc. (vulture)
Narrow habitat tolerance (orangutan)	Wide habitat tolerance (chimpanzee)
Valuable fur, hide, oil, etc. (chinchilla)	Not a source of natural products and not exploited for research or pet purposes (gray squirrel)
Hunted for the market or hunted for sport where there is no effective game management (passenger pigeon)	Commonly hunted for sport in game management areas (mourning dove)
Has a restricted distribution: Island, desert watercourse, bog, etc. (Bahamas parrot)	Has broad distribution (yellow-headed parrot)
Lives largely in international waters, or migrates across international boundaries (green sea turtle)	Has populations that remain largely within the territory(ies) of a specific country(ies) (loggerhead sea turtle)
Intolerant of the presence of man (grizzly bear)	Tolerant of man (black bear)
Species reproduction in one or two vast aggregates (West Indian flamingo)	Reproduction by solitary pairs or in many small or medium sized aggregates (bitterns)
Long gestation period; one or two young per litter, and/or maternal care (giant panda)	Short gestation period; more than two young per litter, and/or young become independent early and mature quickly (raccoon)
Has behavioral idiosyncrasies that are nonadaptive today (redheaded woodpecker—flies in front of cars)	Has behavior patterns that are particularly adaptive today (burrowing owl: highly tolerant of noise and low-flying aircraft; lives near the runways of airports)
Top-of-the-food-chain predator that is subject to the effects of biological magnification of chemical pollutants (peregrine)	Lower trophic level predator that feeds on less contaminated prey (gyrfalcon)

*Modified from: D. W. Ehrenfeld [4].

century. Familiar existing approximations are the herring gull, house sparrow, gray squirrel, Virginia opossum, Norway rat, and carp.

It is also instructive to take a look at a single group of closely related vertebrates to see how well the various species have managed to survive. A single order of birds, the Falconiformes, serves as a good example because many conservationists consider the birds of prey to be highly sensitive to human disturbances. The order consists of about 290 living species of hawks, eagles, falcons, vultures, and condors. Only one species has become extinct since 1600, the Guadalupe Island caracara, an endemic that lived on a small island off the western coast of Mexico. When goats and sheep were introduced onto the island in the late 1800s, the Mexican residents destroyed many caracaras because of their presumed depredations on kids and lambs, but the last ones were apparently shot by a "scientific collector" in the early 1900s. The *Red Data Book* on birds catalogs 21 rare and endangered species of Falconiformes, 14 of which are island endemics and two of which may be extinct, as no individuals have been found for a number of years. The U. S. Fish and Wildlife Service now lists six endangered North American populations—the California condor, Everglades kite, southern bald eagle, northeastern bald eagle, American peregrine falcon, and the Arctic peregrine falcon; in addition, the Hawaiian hawk is listed. Approximately 10 percent of the species in the order appear to be headed down the road to extinction, unless some actions can be taken to reverse present threats to their continued existence.

There are about five main factors involved in the high extinction potentials of falconiform birds.

1. They all carry the stigma of being *predators* and therefore harmful to man. Hunters shoot them because they kill some game birds and mammals; landowners shoot, trap, and poison them because they occasionally prey on livestock and poultry.

2. A number of species are large in body size and/or specialized in habits. The Andean condor and the California condor are Pleistocene relicts, whose large size and carrion-eating habits have subjected them to various forms of persecution by humans; not more than 20 to 40 California condors remain alive. The monkey-eating eagle of the Phillipines is another large species that has been greatly reduced by direct persecution, and the Eurasian and African lammergeyer is another. The Everglades snail kite is a food-specialist, whose existence is determined by the distribution and abundance of a single genus of snail (*Pomacea* spp); drainage of the Florida Everglades greatly decreased the snail population and, consequently, the population of kites.

3. Several species are now highly valued in the "pet trade" or for other commerical uses. Wealthy Arab falconers are prepared to pay tens of thousands of dollars for a white gyrfalcon from the Arctic or an especially handsome saker from Pakistan, and thousands of dollars apiece for some of the other species commonly used for falconry.

It is worth noting here that in 1967 records show that 74,000 mammals, 203,000 birds, 405,000 reptiles, 137,000 amphibians, and 28 million fishes were imported into the United States for keeping in captivity for one reason or another, mostly as "pets." Since that year, however, these figures have dropped owing to various laws and regulations designed to curtail the traffic in wildlife for pets, zoo exhibitions, and research.

4. Island forms always seem to be in trouble. Tameness is one problem for many kinds of island birds. The Galapagos hawk is so tame that a man can walk right up to many individuals and strike them down with a stick or club. There are estimated to be only about 200 Galapagos hawks surviving in the archipelago. The forest habitat of the Mauritius kestrel, once a common inhabitant of this western Indian Ocean island, has been reduced to only about 2000 hectares, and only about 20 individuals remain in the wild. Although strictly a lizard and small bird predator, it still suffers from its French name, *manguer de poule*, "eater of chickens."

5. A number of falconiform birds that feed at very high trophic levels of ecosystems—especially bird eaters and fish eaters—have suffered severe declines in numbers from the biological magnification of toxic chemical residues in their food species. DDT residues, acting through their biochemical influence on eggshell formation in the female's oviduct and causing the production of thin-shelled eggs that break or fail to hatch, have had particularly serious consequences on populations of the peregrine falcon, osprey, and bald eagle. More than 35 species of predatory and fish-eating birds have been measurably affected by the DDT-thin-eggshell disease since 1947, both in North America and in Europe.

In short, the result of man's influences on environment and our wanton destruction of animal life is greatly reduced diversity in faunas and floras. Increasingly we see the emergence of a relatively few, superabundant, highly successful, pestlike species that dominate ecosystems and that probably are going to interact with one another and with man's crops in boom and bust oscillations in numbers, unless sufficiently large tracts of natural ecosystems are left intact to maintain some semblance of balance.

22.2.7 Actions to Save Threatened Fauna and Flora

What can we do to prevent other species from becoming rare or endangered, or extinct? Two cardinal requirements must be met if we are to keep any significant vestige of natural biotas and ecosystems intact for future generations of humans to know: (1) Preservation of natural habitats and ecosystems on a large and representative scale and (2) control of human population growth.

The most immediately critical action, for the short term, is habitat preservation. We need to preserve relatively large blocks of natural ecosystems—on the order of 10 to 20 thousand km^2 or more—large enough for the natural communities of plants and animals to remain self-perpetuating through time. The larger national parks and game preserves of the world show what can be done when natural areas are protected against major disruptive land use by man. The Kruger National Park in South Africa, the Etosha Game Preserve in Southwest Africa—the largest wildlife sanctuary in the world—the Serengeti Park in Tanzania, Mount McKinley National Park in Alaska, and Glacier Bay National Monument, also in Alaska—these and other such areas that have been established by governments around the world show what can be done through legal protection.

The Kruger was the first game park established in Africa, and portions of it have been in preserve status since 1895; its present boundaries, which encompass more than 12,000 km^2 of lowland "bush-veld" along the border with Mosambique, have been intact since 1926. Although entirely surrounded by lands that are heavily used by humans for agriculture and stock-raising and fenced in around its entire perimeter, The Kruger has not lost a single species of plant or animal from its natural ecosystem, and it has been in existence long enough to give assurance that as long as it is properly cared for and preserved it can continue to hold its living treasures indefinitely. There one can see and study more than 44 species of fish, 29 species of amphibians, 94 species of reptiles, more than 440 species of birds, and 114 species of mammals, including 16 species of large ungulates, lions, leopards, cheetahs, hyenas, and wild dogs. There is only one thing wrong with the Kruger: it does not include the natural summer range of the migratory game, which used to move up into the mountains to the west of the park, and this oversight causes some problems with fences and with control of populations sizes.

The larger these protected natural areas are, the better the chances are for the survival of the populations of species in them. Through such legislative actions as the Wilderness Act of 1964 (Public Law 88-577) and the Alaska Native Land Claims Settlement Act of 1971, which has a provision for designating about 200,000 km^2 of Alaskan wilderness environments as new national parks, national wildlife refuges, and other national preserves, the American people still have the opportunity to keep significant fractions of their natural heritage for the future. Unfortunately, government is slow to act on specific proposals that come up for consideration under these Congressional mandates, and the people who care must remain vigilant to see that the original intent of these Acts is fulfilled before time runs out. When the Congress shirks its responsibilities, then the President can act under the executive authority given to him in the 1906 Antiquities Act to establish National Monuments.

In addition to loss of species diversity, failure to keep large portions of some natural ecosystems intact may have serious consequences on climate, composition of the atmosphere, and on the distribution and quality of water, perhaps the most limiting resource for life, as so dramatically demonstrated in recent years in the sub-Saharan regions of Africa. There is growing evidence, too, that the cutting down of the Amazon tropical forest is having an impact on rainfall patterns in that region.

Any conservation issue ultimately rests on a solution to the problems that have been created by the human population explosion, and this conclusion is certainly true for the survival of many species in the future. The human population has been increasing at an exponential rate for thousands of years, a pattern of growth that is unprecedented in other animal populations and that *cannot* continue indefinitely. With an increase in the numbers of human beings go exponential rates of increase in the use of resources of all sorts—food, fiber, fuel—and the production of "waste products" that pollute the global ecosystem, wastes from human food consumption, from use of fossil fuels, from fertilizing agricultural lands, and from synthetic chemical products (pesticides) and other sources. It is important to understand that these rates of use and waste frequently increase by exponents that are larger than the exponent of population increase. Typically in a modern technological society, use of materials increases four to five times faster than population growth; if population doubles every 30 years, the use of fossil fuels will double every 6 to 8 years, for example.

Many people still are not especially worried about human population growth or exponential rates of increase in the uses of land and resources. They believe that technology and production can keep pace with population growth. New resources of energy can be tapped, more "green revolutions" can be brought about, and more efficient occupancy of space for human dwellings can be designed. These people are still wedded to the progressivist philosophy of the eighteenth and nineteenth centuries, a philosophy that followed in the wake of the scientific and industrial revolutions. Basically this philosophy says that day to day in every way man is bettering his condition by making more and more things. It is the philosophy of the expanding economy; it is the philosophy of the TV commercials: "Progress is our most important product." "Something we do today will touch your lives tomorrow."

Other people, particularly biologists and conservationists, are not convinced that we need more technology as much as we need a new frame of reference, a new philosophy, for using technology. They look for a new morality, a new philosophical and religious world view. The morality must be one of restraint—restraint on the use of resources and, above all, restraint on human procreation. The world view should encompass a respect and a reverence for all life, not just human life.

Again, we must emphasize that there is nothing bad about technology *per se.* The same technology that can be used to destroy the world can also be used to recreate it, except that once a species is lost, it is lost forever.

22.2.8 Conclusion

As professional biologists, the authors of this text belong to many organizations and societies—The American Association for the Advancement of Science,

The Ecological Society of America, the American Ornithologists' Union, The American Society of Ichthyologists and Herpetologists, The American Society of Mammalogists, and others. Belonging to societies is one of the prestiges of being a professional scientist, and so naturally these organizations are mostly made up of other people like ourselves—middle-aged to older professionals who constitute what younger people often refer to as "the establishment." They are mostly run by full professors who have tenure and who get research grants from the National Science Foundation, National Institutes of Health, the Atomic Energy Commission (now Energy Research and Development Administration), National Aeronautics and Space Administration, Office of Naval Research, and other government funding agencies.

One of the things all of these societies do is to have annual meetings, and one of the functions of these meetings is to pass resolutions. These resolutions are usually rather bland and meaningless, accomplish little more than a formal thanks to the local committee for hosting the affair, then are promptly and deservedly forgotten.

In 1970, the American Society of Mammalogists, meeting at College Station, Texas, did something different. The assembled members addressed themselves to the pressing problems of our times and produced some rather startling statements for a professional society. The resolution by these mammalogists is so significant and still so timely that we want to end our book by quoting it in full:

> WHEREAS, mammalian populations that exceed the carrying capacity of their environments exhibit many typical characteristics prior to drastic population declines or "crashes," and
>
> WHEREAS, the world human population today is exhibiting these traits and is increasing at an exponential rate which now adds 1.3 million new people each week to the world population; and
>
> WHEREAS, incomplete figures indicate that between one and two billion people are today undernourished, and between 4 and 10 million will starve to death this year, and there is no possibility that agricultural production can be increased rapidly enough even to maintain present standards of nutrition if the population continues to increase at the present rate; and
>
> WHEREAS, modern technology depends to a large extent on nonrenewable natural resources, such as petroleum, that are rapidly being exhausted; and
>
> WHEREAS, the expansion of this technology is causing a rapidly accelerating pollution and destruction of the environment—such as air pollution which contributes to respiratory and other diseases, pollution of lakes and streams which destroys water and fishery resources, and the poisoning of the entire biosphere—including man himself—by persistent pesticides such as DDT; and
>
> WHEREAS, many renewable natural resources are being over exploited and reduced to a point far below their maximum sustainable yield; and
>
> WHEREAS, increased crowding and deprivation of large segments of the human population may be the most important factors leading to increased massive social and behavioral disruptions; and
>
> WHEREAS, in many parts of the world populations already depend upon imports of food for their survival, and the population of the United States

probably already exceeds the optimum size for maintaining a desirable standard of living in terms of aesthetic values, lack of overcrowding, preservation of open spaces, and other factors conducive to mental and physical health and well being; and

WHEREAS, all the preceding facts indicate that a massive population decline is inevitable; and

WHEREAS, actions by the responsible political leaders of the United States and of other nations to lessen the effects of this impending disaster seem to have been consistently too little and too late; and

WHEREAS, the growth of populations now makes the solution of a host of major political, social and individual problems more difficult and will eventually make satisfactory solutions impossible;

THEREFORE BE IT RESOLVED, that the American Society of Mammalogists voices its gravest concern to the President of the United States, the United States Congress, the Governors of the 50 States, officials at other levels of government, and to the people themselves, in the hope that they will assume immediately their responsibilities to take large-scale, effective, and unprecedented action to curb population growth, by promoting birth control, legalizing abortion, reducing tax incentives for natality, and by such other acts as may be needed to realize the larger goal of survival of the human species under acceptable conditions.

There outlined for all to read and understand are the self-imposed mechanisms of population control that we mentioned at the beginning of this section. To the extent that governments and individuals pay heed to its message, indeed, this extraordinary resolution is, itself, a social mechanism of birth control. Unfortunately, the authors' generation learned about the devastating effects of overpopulation, or became convinced of their seriousness, too late to do much about the population problem, except for one thing—to try to teach younger generations that it is absolutely necessary for them to do something to prevent a cataclysmic deterioration in the quality of life. While there may be honest disagreements about the particular methods to be used, the essential goal is to control the human population by humane and equitable means.

References

[1] Deevey, Edward S., Jr. 1960. The human population. *Scientific American* 203(3): 194–204.

[2] Edney, Maitland A. 1972. The missing link. *The Emergence of Man*, edited by Charles Osborne. Time–Life Books, New York.

[3] Edney, Maitland A. 1976. Three-million-year-old Lucy. Pp. 19–31 in *Nature/Science Annual*, edited by Jane D. Alexander. Time-Life Books, New York.

[4] Ehrenfeld, David W. 1970. *Biological Conservation*. Modern Biology Series. Holt, Rinehart, and Winston, Inc. New York.

[5] Fisher, James, Noel Simon, Jack Vincent. 1969. *Wildlife in Danger*. The Viking Press. New York.

[6] Hardin, Garrett. 1961. *Biology, Its Principles and Implications.* W. H. Freeman and Company, San Francisco.

[7] Kendeigh, S. Charles. 1961. *Animal ecology.* Prentice-Hall, Inc., Englewood Cliff, N.J. Cliff, N.J.

[8] McVay, Scott. 1966. *The last of the great whales. Scientific American* 215(2):13–21.

[9] Schultz, Adolph H. 1969. *The Life of Primates.* The Universe Natural History Series, edited by R. Carrington. Universe Books, New York.

[10] Simons, Elwyn L. 1964. *The early relatives of man. Scientific American.* 211(1): 50–62.

[11] Washburn, Sherwood L. 1978. *The evolution of man. Scientific American.* 239(3): 194–208.

[12] White, Edmund, and Dale M. Brown. 1973. *The First Men. The Emergence of Man,* edited by Dale M. Brown. Time-Life Books, New York.

[13] Ziswiler, Vinzenz. 1967. *Extinct and Vanishing Animals.* The Heidelberg Science Library, vol. 2. Revised in English by Fred and Pille Bunnell. Springer-Verlag, New York.

[14] Johanson, D. C. and T. D. White. 1979. A systematic assessment of early African hominids. *Science* 203:321–330.

Glossary

Technical terms, and some words with special biological meanings, are included in the glossary. Certain words have slightly different meanings in different contexts. In those cases we have limited the definition to the use of the word in this book.

abduction – movement away from the midventral axis of the body (cf. adduction).

acetabulum – a depression on the pelvic girdle that accommodates the head of the femur.

adduction – movement toward the midventral axis of the body (cf. abduction).

aestivation (estivation) – a form of torpor, usually a response to high temperatures or scarcity of water.

allochthonus – formed somewhere other than the region where found.

allopatry – the situation in which two or more populations or species occupy mutually exclusive, but often adjacent, geographic ranges.

ammonotelic – excreting nitrogenous wastes primarily as ammonia.

amniotes – those vertebrates whose embryos possess an amnion and allantois—i.e. reptiles, birds, and mammals.

amphicoelus – a vertebral centrum with both the anterior and posterior surfaces concave.

anadromus – (of fishes) migrating up a stream or river from a lake or ocean to spawn.

anamniotes – those vertebrates whose embryos do not possess an amnion or allantois—i.e. fishes and amphibians.

ankylothecodont teeth – teeth set in sockets and fused at the base; characteristic of rhynchosaurs.

aphotic – "without light"—for example, in deep sea habitats or caves.

apocrine gland – a type of gland in which only the apical part of the cell from which the secretion is released breaks down in the process of secretion (cf. holocrine gland).

aposematic – a device (color, sound, behavior) used to advertise the noxious qualities of an animal.

arcade – a curve or arch in a structure such as the tooth row of humans.

archipterygium – fin skeleton, as in a lungfish, consisting of symmetrically arranged rays that extend from a central skeletal axis.

arid – describes a habitat with a scanty supply of water (cf. mesic).

autochthonous – formed in the region where found.

autotomy – in lizards and salamanders, deliberate breaking of the tail at a specialized fracture plane within a vertebra.

benthic – living at the soil/water interface at the bottom of a body of water.

biomass – living organic material in a habitat available as food for other species.

bisymmetry – capable of being divided by a plane into mirror-image halves.

brachial – pertaining to the forelimb.

branchial – pertaining to the gills.

branchiomeric – segmentation of structures associated with, or derived from, the ancestral gill arches (cf. metameric).

carapace – a dorsal shell, as in a turtle.

catadromous – (of fishes) migrating down a river or stream to a lake or ocean to spawn.

centrum (Plural, centra) – the vertebral element formed in or around the notochord.

cephalic – pertaining to the head.

ceratotrichia – keratin fibers that support the web of the fins of chondrichthyes.

choana (plural, choanae) – internal nares.

chondrification – formation of cartilage (cf. ossification).

choroid – pigmented and vascularized middle layer of the vertebrate retina.

clade – a phylogenetic lineage of related taxa originating from a common ancestral taxon (cf. grade).

cladistic – pertaining to the branching sequences of phylogenesis.

cline – a change in a biological character along a geographic gradient.

cnemial crest – an elevated ridge on the anterior face of the tibia or tibiotarsus providing increased surface for muscle attachment.

coelom (celom) – the body cavity, lined with mesodermal tissue.

cones – photoreceptor cells in the vertebrate retina that are differentially sensitive to light of different wavelengths.

conspecific – and individual belonging to the same species as that under discussion.

cosmine – a form of dentine with branching canals characteristic of the cosmoid scales of crossopterygian fishes and early dipnoans.

countershaded – a color pattern in which the aspect of the body that is more brightly lighted (normally the dorsal surface) is darker colored than the less brightly illuminated surface. The effect of countershading is to make an animal harder to distinguish from its background.

cranial — pertaining to the cranium or skull.

detritus — particulate organic matter than sinks to the bottom of a body of water.

deuterostomy — the condition in which the embryonic blastopore forms the anus of the adult animal; characteristic of chordates (cf. protostomy).

double cone — a type of retinal photoreceptor in which two cones share a single axon (cf. cones).

durophagous — feeding upon hard material.

eccritic temperature — the arithmetic mean of body temperatures of ecto-thermal animals during their periods of activity under natural conditions.

ectoderm — one of the embryonic germ layers, the outer layer of the embryo and adult.

endentulous — lacking teeth.

embolomerous — a type of vertebral structure in which both the pleurocentrum and intercentrum form complete rings.

endemism — the property of being endemic—i.e. found only in a particular region.

endoderm — innermost of the germ cell layers of late embryos and adults.

epeiric sea (epicontinental sea) — a sea extending within the margin of a continent.

epigenetic — pertaining to the interaction of tissues during embryonic development.

epiphysis — (1) the pineal organ, an outgrowth of the roof of the diencephalon. (2) (plural, epiphyses) an accessory center of ossification at the ends of the long bones of mammals, birds, and some reptiles. When the ossifications of the shaft (diaphysis) and epiphysis meet, further lengthwise growth of the shaft ceases. This process produces a determinate growth pattern.

epiphyte — a plant that grows nonparasitically upon another plant.

euryhaline — capable of living in a wide range of salinities (cf. stenohaline).

eurythermal — capable of tolerating a wide range of temperatures (cf. steno-thermal).

eurytopy — capable of living in a broad range of habitats.

eutherian mammals — placental mammals, the infraclass Eutheria.

extraperitoneal — positioned in the body wall beneath the lining of the coelom (the peritoneum) in contrast to being suspended in the coelom by mesentaries.

fossorial — burrowing through the soil.

fovea centralis — an area of the vertebrate retina containing only cone cells, where the most acute vision is achieved at high light intensities.

furcula — the avian "wishbone" formed by the fusion of the two clavicles at their central ends.

gastrosteges — the large transverse scales on the ventral surfaces of most snakes.

geosyncline – a portion of the earth's crust that has been subjected to downward warping. Sediments frequently accumulate in geosynclines.

gestation – the period during which an embryo is developing in the reproductive tract of the mother.

gill arch – the assemblage of tissues associated with a gill; the term may refer to the skeletal structure only or to the entire epithelial muscular and connective tissue complex.

grade – a given level of morphological organization achieved independently by different evolutionary lineages–e.g. the mammalian grade (cf. clade).

gymnosperms – a group of plants in which the seed is not contained in an ovary–conifers, cycads, and ginkos.

hemal arch – a structure formed by paired projections ventral to the vertebral centrum and enclosing caudal blood vessels.

hermaphroditic – having both male and female gonads.

heterocoelus – having the articular surfaces of the vertebral centra saddle-shaped, as in modern birds.

heterosporous plants – plants with large and small spores. The smaller give rise to male gametophytes and the larger to female gametophytes. (Equivalent to protogymnosperms.)

heterotrophic – capable of using only organic materials as a source of energy.

holocrine gland – a type of gland in which the entire cell is destroyed with the discharge of its contents (cf. apocrine gland).

hydrosphere – the free liquid water of the earth–oceans, lakes, rivers, etc.

hyperdactyly – an increase in the number of digits (cf. hyperphalangy).

hyperphalangy – an increase in the number of bones in the digits (cf. hyperdactyly).

hypertrophy – an increase in the size of a structure.

hypotremate – having the main gill openings on the ventral surface and beneath the pectoral fins as in skates and rays (cf. pleurotremate).

inguinal – pertaining to the groin.

interspecific – pertaining to phenomena occurring between members of different species (cf. intraspecific).

intraspecific – pertaining to phenomena occurring between members of the same species (cf. interspecific).

isohaline – of the same salt concentration.

isostasy – a condition of gravitational balance between segments of the earth's crust or of return to balance after a disturbance.

isotherm – a line on a map that connects points of equal temperature.

leptocephalus larva – specialized, transparent, ribbon-shaped larva of tarpons, true eels, and their relatives.

lithosphere – the crust of the earth.

littoral – pertaining to the shallow portion of a lake, sea, or ocean where rooted plants are capable of growing.

lophophorate – pertaining to several kinds of marine animals that possess ciliated tentacles (lophophores) used to collect food–e.g. pterobranchs.

Mauthnerian system — paired giant neurons found in some fishes that function in a rapid escape response.

meninges — sheets of tissue enclosing the central nervous system. In mammals these are the dura mater, arachnoid, and pia mater.

mesic — describes an environment with an adequate supply of moisture (cf. arid).

mesoblast — a mesodermal cell.

mesoderm — the central of three germ layers of late embryos and adults.

mesonephric — kidney characteristic of amniotes in which the nephrons are derived from a few posterior body segments and drain via a ureter that grows from the cloaca; kidneys of anamniotes lack one or more of these characters.

metameric — pertaining to ancestral segmentation, used in reference to serially repeated units along the body axis.

metatherian mammals — the infraclass Metatheria, consisting of the marsupials.

morph — a genetically determined variant in a population.

morphotypic — a type of classification based entirely on physical form.

naris (plural, nares) — nostril.

neoteny — a condition in which a larval or embryonic trait persists in the adult body.

neural arch — a dorsal projection from the vertebral centrum that, at its base, encloses the spinal cord.

neurocranium — the portion of the head skeleton encasing the brain.

niche — pertaining to the functional role of a species or other taxon in its environment—the ways in which it interacts with both the biotic and abiotic elements.

nidimental gland — specialized segment of the reproductive tract of a female oviparous vertebrate that produces the shell surrounding a fertilized ovum.

occipital — pertaining to the posterior part of the skull.

operculum — a flap or plate of tissue covering the gills.

opisthoglyph — a type of snake dentition with one or more enlarged, solid teeth near the rear of the maxilla. The teeth may be grooved or not.

ontogenetic — pertaining to the development of an individual organism.

orogeny — the process of crustal uplift or mountain building.

osseous — bony.

oviparous — reproducing by laying eggs that hatch outside of the body of the mother.

ovoviviparous — reproducing by retaining eggs within the oviducts of the mother until hatching. Nourishment is derived from the egg yolk (cf. viviparous).

paedogenesis — a condition in which a larva becomes sexually mature without attaining the adult body form.

palatoquadrate — the upper jaw element of primitive fishes and chondrichthyes, portions of which contribute to the palate, jaw articulation, and middle ear of other vertebrates.

patagium — a membranous flap of skin on the leading edge of a bird's wing, stretching between the upper arm and forearm.

pedicellate tooth — a form of tooth characteristic of lissamphibians which includes a narrow zone of uncalcified tissue between the upper and lower parts.

pharyngotremy — the condition in which the pharyngeal walls are perforated by slit-like openings; found in chordates and hemichordates.

philopatry — tendency of sexually mature animals to return to the locale of their birth or hatching.

photophore — a light-emitting organ.

phylogenetic — pertaining to the development of an evolutionary lineage (cf. ontogenetic).

physoclistic — (of fishes) lacking a connection from the gut to the swim bladder as adults (cf. physotomous).

physostomous — (of fishes) having a connection between the swim bladder and gut in adults (cf. physoclistic).

piloerection — contraction of muscles attached to hair follicles resulting in the erection of the hair shafts.

piscivorous — fish-eating.

placoid scales — a primitive type of scale found in elasmobranchs and homologous with vertebrate teeth.

plastron — a ventral shell, as of a turtle.

pleurotremate — having the main gill openings on sides of the body anterior to the pectoral fins as in sharks (cf. hypotremate).

polymorphism — the simultaneous occurrence of two or more distinct phenotypes in a population.

portal system — a portion of the venous system specialized for the transport of substances from the site of production to the site of action. A portal system begins and ends in capillary beds.

postzygapophysis — an articulating surface on the posterior face of a neural arch (cf. prezygapophysis).

prezygapophysis — an articulating surface on the anterior face of a neural arch (cf. postzygapophysis).

proteroglyph — a type of snake dentition characterized by hollow fangs near the anterior end of a maxilla that is usually relatively immobile (cf. solenoglyph).

Proterozoic — the later part of the Precambrian, from about 1.5 billion years ago until the beginning of the Cambrian 500 million years ago.

protostomy — the condition in which the embryonic blastopore forms the mouth of the adult animal (cf. deuterostomy).

prototherian mammals — the subclass Prototheria that includes the modern egg-laying mammals and various extinct forms.

protraction — movement away from the center of the body (cf. retraction).

protrusible — capable of being moved away (protruded) from the body.

pygostyle — terminal bone of the avian tail, formed by fusion of several vertebrae.

ratite — any of several flightless birds with a flat, keelless sternum.

refugium — an isolated area of habitat fragmented from a formerly more extensive biome.

rete mirabile — "marvelous net," a complex mass of intertwined capillaries specialized for countercurrent exchange.

retraction — movement toward the center of the body (cf. protraction).

rhachitomous — a vertebral structure characterized by the presence of both intercentrum and pleurocentrum.

rod — a photoreceptor cell in the vertebrate retina specialized to function effectively under conditions of dim light.

rostrum — the snout; especially an extension anterior to the mouth.

scapulocoracoid cartilage — in elasmobranchs and certain primitive gnathostomes, the single solid element of the pectoral girdle.

scutes — scales, especially broad or inflexible ones.

semicircular canals — the portion of the vertebrate inner ear associated with sensing acceleration in three-dimensional space.

sinus — an open space in a duct or tubular system.

solenoglyph — a type of snake dentition characterized by long, hollow fangs that are the only teeth on highly mobile maxillae (cf. Proteroglyph).

somite — one member of a series of paired segments of the embryonic dorsal mesoderm of vertebrates.

squamation — the scaly covering of the body.

stenohaline — capable of living only within a narrow range of salinity of surrounding water; not capable of surviving a great change in salinity.

stenophagous — eating a narrow range of food items; a food specialist.

stenothermal — capable of living or of being active in only a narrow range of temperatures (cf. eurythermal).

stereospondylous — a vertebral structure characterized by the evolutionary loss of pleurocentrum and retention of the intercentrum.

steroids — a group of cholesterollike compounds that includes vertebrate sex hormones.

stratigraphy — the classification, correlation, and interpretation of stratified rocks.

stratum (plural, strata) — a layer of material.

suspensorium — the part of the skull by which the lower jaw articulates with the cranium.

sympatry — the occurrence of two or more species in the same area.

symphysis — the fibrous union between two bones in a joint that allows only a small degree of movement.

talonid — a basinlike heel on a lower molar tooth, found in certain mammals.

tarsometatarsus — in birds and some dinosaurs, the bone formed by fusion of the distal tarsal elements with the metatarsals (cf. tibiotarsus).

tetraploidy — the condition of having a tetraploid chromosome number (4N) as a result of duplication of the diploid number (2N).

thecodont teeth — teeth set in bony sockets in the jaw.

therian mammals — mammals of the subclass Theria, including the living marsupials and placental mammals plus the extinct trituberculates.

tibiotarsus — in birds and some dinosaurs, the bone formed by fusion of the tibia and proximal tarsal elements (cf. tarsometatarsus).

troglodyte — an organism that lives in caves.

ureotelic —excreting nitrogenous wastes primarily as urea.

uricotelic — excreting nitrogenous wastes primarily as uric acid and its salts.

urodele — a salamander.

urogenital — pertaining to the organs, ducts, and structures of the excretory and reproductive systems.

vacuoles — membrane-bound spaces within cells containing secretions, storage products, etc.

visceral skeleton — the skeleton primitively associated with the gill arches, uniquely derived from the neural crest cells and forming in mesoderm immediately adjacent to the endoderm lining the gut.

visual releaser — a structure, color, pattern, or action that is especially effective in eliciting a specific kind of behavior.

viviparous — reproducing by retaining the young within the mother's uterus. Nourishment for the young is obtained through the placenta.

Weberian apparatus — the complex of ossicles, ligaments, gas bladder, and inner ear characteristic of ostariophysan teleosts and responsible for their sensitive hearing.

zygantrum — a posteriorly directed accessory articular recess on the neural arch of some squamate reptiles, especially snakes (cf. zygosphene).

zygapophysis — an articular process of the neural arch of a vertebra (cf. postzygapophysis and prezygapophysis).

zygodactylus — a type of foot, specialized for grasping branches, in which the toes are arranged in two opposable groups.

zygosphene — anteriorly projecting accessory articular surface on the neural arch of some squamate reptiles, especially snakes (cf. zygantrum).

Index

Page numbers in italics refer to illustrations.

LATIN AND GREEK GLOSSARY

Many biological names and terms are derived from Latin (L) or Greek (G). Learning even a few dozen of these roots is a great aid to a biologist. The following terms are often encountered in vertebrate biology. The words are presented in the spelling and form in which they are most often encountered; this is not necessarily the original form of the word in its etymologically pure state. Further information can be found in a reference such as Jaeger, E. C. [1955] *A Source-Book of Biological Names and Terms* 3rd ed., Charles C Thomas, Springfield, Illinois.

a, ab (L) away from
a, an (G) not, without
acanth (G) thorn
actin (G) a ray
ad (L) toward, at, near
al, alula (L) a wing
alveol (L) a pit
ampho (G) both, double
ana (G) up, upon, through
anthrac (G) coal
apsid (G) an arch, loop
ap, apo (G) away from
aqu (L) water
arch (G) beginning, first in time
arthr (G) a joint
av (L) a bird
bas (G) base, bottom
bi (G) life
blast (G) bud, sprout
brachi (G) arm
brachy (G) short
branchi (G) a fin
bucc (L) the cheek
cal (G) beautiful
calic (L) a cup
capit (L) head
carn (L) flesh
caud (L) tail
cephal (G) head
cer, cerae (G) a horn
cerc (G) tail
chir, cheir (G) hand
choan (G) funnel, tube
chord (G) guts, a string
chorio (G) skin, membrane
chrom (G) color
cloac (L) a sewer
coel (G) hollow
cornu (L) a horn
cortic, cortex (L) bark, rind

costa (L) a rib
cran (G) the skull
cten (G) a comb
cut, cutis (L) the skin
cyn (G) a dog
dactyl (G) a finger
de (L) down, away from
dent (L) a tooth
derm (G) skin
di (G) two, double
di, dia (G) through, across
din (G) terrible, powerful
duct (L) a leading
dur (L) hard
e, ex (L) out, without
ect (G) outside
eid (G) form, appearance
end (G) within
enter (G) bowel, intestine
ep (G) on, upon
erythr (G) red
eu, ev (G) good, true
eury (G) broad
fer (L) carrier of
fil (L) a thread
gaster (G) the belly
glob (L) a ball
glom, glomer (L) a ball of yarn
gloss (G) the tongue
gnath (G) the jaw
gracil (L) slender
gul (L) the throat
gymn (G) naked
gyr (G) round, a circle
haem (G) blood
hal (G) the sea
hemi (G) half
hepat (G) the liver
herp (G) to creep
hetero (G) other, different

hipp (G) a horse
hist (G) web, tissue
hol (G) whole, entire
homo (G) alike
hyp (G) under, beneath
hyper (G) above, beyond
hyps (G) high, height
ichthy (G) a fish
in (L) in, into, not, without
infra (L) below
inter (L) between, among
intr (L) inside
is, iso (G) similar, equal
kin (G) movement
lecith (G) yolk
lingu (L) tongue
liss (G) smooth
loph (G) crest, ridge
lut (L) yellow
macr (G) long, large
magn (L) great, large
mamm (L) breast
mast (G) a breast
medull (L) marrow, pith
mela (G) black
mer (G) a part
mes (G) middle
meta, met (G) next to
micr (G) small
mon (G) single
morph (G) shape
nect (G) swimming
neo (G) new, recent
not (G) the back
odont, don (G) tooth
omni (L) all
opercul (L) a cover, lid
opisth (G) behind
ops (G) appearance
ornith (G) a bird
orth (G) straight
oste (G) bone
palae, paleo (G) ancient
par, para (G) beside
peri (G) around, near
phag (G) to eat
phil (G) loving, friend
phor (G) to bear
phot (G) light
phyl (G) a tribe, race

platy (G) broad
pleur (G) a rib, the side
pod (G) a foot
poly (G) many
post (L) after, behind
prim (L) first
pro (G) before, in front of
prot (G) first, primary
pseud (G) false
pter, pteron (G) wing, feather
pyg (G) the rump
ram (L) a branch
rept (L) to crawl
retro (L) backward
rhin (G) a nose
rhynch (G) a beak, snout
sarc (G) flesh
saur (G) a lizard
scler (G) hard
som (G) the body
sphen (G) a wedge
splanchn (G) viscera
spondyl (G) vertebra
squam (L) a scale
steg (G) a roof
sten (G) narrow, straight
stom (G) mouth
styl (G) a pillar
sub (L) under, below
super (L) above, over
syn, sym (G) together
tele (G) perfect, entire
tethy (G) a sea goddess
tetr (G) four
thec (G) a case
theri (G) a wild animal
therm (G) heat
tom (G) a cut, slice
top (G) a place
trem (G) a hole
tri (G) three
trich (G) a hair
trop (G) a turn, change
troph (G) one who feeds
ultim (L) the farthest, last
vas (L) a vessel
ventr (L) the belly
vor (L) to devour
zyg (G) a yolk

MAY